Lecture Notes in Computer Science 656

Edited by G. Goos and J. Hartmanis

M. Rusinowitch J. L. Rémy (Eds.)

Conditional Term Rewriting Systems

Third International Workshop, CTRS-92
Pont-à-Mousson, France, July 8-10, 1992
Proceedings

Springer-Verlag
Berlin Heidelberg New York
London Paris Tokyo
Hong Kong Barcelona
Budapest

Series Editors

Gerhard Goos
Universität Karlsruhe
Postfach 69 80
Vincenz-Priessnitz-Straße 1
W-7500 Karlsruhe, FRG

Juris Hartmanis
Cornell University
Department of Computer Science
4130 Upson Hall
Ithaca, NY 14853, USA

Volume Editors

Michaël Rusinowitch
Jean-Luc Rémy
CRIN & INRIA-Lorraine
B. P. 239, F-54506 Vandoeuvre-Les-Nancy Cedex, France

CR Subject Classification (1991): F.4.1-2, D.3.1, I.2.3

ISBN 3-540-56393-8 Springer-Verlag Berlin Heidelberg New York
ISBN 0-387-56393-8 Springer-Verlag New York Berlin Heidelberg

Printed in Germany

Typesetting: Camera ready by author
45/3140-543210 - Printed on acid-free paper

To the memory of Stéphane Kaplan

Preface

This volume contains the papers presented at the Third International Workshop on Conditional Term Rewriting Systems held on 8-10 July 1992 at Pont-à-Mousson, France.

The first CTRS workshop was held in 1987 at the Université of Paris XI in Orsay and the second one took place in 1990 at Concordia University in Montreal. The proceedings have been published in the Lecture Notes in Computer Science, Springer Verlag, volume 308 and volume 516 respectively. Topics of these workshops include conditional rewriting and its applications to programming languages, specification languages, automated deduction, constrained rewriting, typed rewriting, higher-order rewriting, graph-rewriting.

We would like to thank the *Région Lorraine*, the *Centre de Recherche en Informatique de Nancy* (which is the joint computer science laboratory of the Universities of Nancy and the *Centre National de la Recherche Scientifique*) and the Lorraine research center of the *Institut National de la Recherche en Informatique et Automatique* for their support.

We are also grateful to Christiane Guyot and Véronique Poirel for their help with many tasks during the preparation of the workshop.

September 1992 The Organizers,

Michaël Rusinowitch
Jean-Luc Rémy

Table of Contents

July 8

Typed Systems and Graph Rewriting

Modularity and Termination

July 9

Proof Techniques and Extensions of Conditional Rewriting

Theorem-Proving and Normal Form Languages

Applications of Conditional Rewriting and New Formalisms

July 10

Contextual Rewriting and Constrained Rewriting

Applications to Logic Programming, Normalization Strategies and Unification

ALGEBRAIC SEMANTICS OF REWRITING TERMS AND TYPES

Karl Meinke
Department of Mathematics and Computer Science,
University College Swansea,
Singleton Park,
Swansea SA2 8PP,
Great Britain.
June 1992

Abstract We present a universal algebraic framework for rewriting terms and types over an arbitrary equational specification of types and typed combinators. Equational type specifications and their initial algebra semantics were introduced in Meinke [1991b]. For an arbitrary equational type specification (ε, E) we prove that the corresponding rewriting relation $\overset{R(\varepsilon,\ E)^*}{\longrightarrow}$ coincides with the provability relation $(\varepsilon, E) \vdash$ for the equational calculus of terms and types. Using completeness results for this calculus we deduce that rewriting for ground terms and ground types coincides with calculation in the initial model $I(\varepsilon, E)$ of the equational type specification.

0. INTRODUCTION

The basic theory of universal algebra in higher types was introduced in Poigné [1986], Möller [1987], Möller et al [1988] and Meinke [1990], [1992]. A higher type universal algebra is a many–sorted universal algebra in which the carrier sets have some type theoretic structure. Applications of the theory of higher type universal algebra to programming language definition and hardware verification are discussed in Meinke [1991a] and Meinke and Steggles [1992]. The basic theory of higher type universal algebra concerns only the propositional type constructions $\times$ and $\rightarrow$ for product and function spaces. In Meinke [1991b] a more general algebraic theory of abstract type constructions and typed combinator systems was introduced, and an equational specification theory for algebras of types and combinators was presented, including results on initial models and completeness of equation and type assignment calculi. Within this general framework, complex type constructions such as predicative and impredicative Π and Σ types and recursive types may be modelled algebraically. The universality of the theory allows for user defined type disciplines and the specification of abstract data types with a rich type structure.

An equational type specification is a pair $Spec = ((\Gamma, \varepsilon), (\Sigma, E))$ consisting of a type signature Γ, and equational theory ε of types, a combinator signature Σ and an equational theory E of combinators. The model class $Mod(Spec)$ forms a category which admits an initial algebra $I(Spec)$ which we may take as the initial algebra semantics of $Spec$. From the viewpoint of computation it is natural to consider term rewriting as a mechanism for computation in the initial model $I(Spec)$. In this paper we introduce a theory of term rewriting for equational type specifications. An important new feature of this theory is that both rewriting of types and rewriting of terms are allowed. Given a combinator term t over a specification $Spec = ((\Gamma,\ \varepsilon),\ (\Sigma,\ E))$ the oriented equations of ε allow us to rewrite type terms in t while the oriented equations of E allow us to rewrite combinator terms in t. Our main result is a Correspondence Theorem for rewriting which establishes that the rewrite relation $\overset{R(\varepsilon,\ E)^*}{\longrightarrow}$ on combinator terms using the oriented equations of ε and E coincides with the provability relation $(\varepsilon,\ E) \vdash$ in the equational calculus of terms and types. By a completeness theorem for this calculus it follows that rewriting coincides with calculation in the initial model $I(Spec)$.

The structure of this paper is as follows. In section 1 we present essential prerequisites from universal algebra and category theory. In section 2 we review basic definitions and results on the algebraic theory of types and combinators. In section 3 we introduce the fundamental definitions and concepts of rewriting for types and combinators. In section 4 we review the equational calculus of terms and types. Finally in section 5 we prove a Correspondence Theorem for rewriting and equational deduction. We have attempted to make the paper self contained. However the reader may wish to consult Meinke [1991b] and Meinke and Wagner [1992] for more detailed proofs and examples.

1. ALGEBRAIC AND CATEGORY THEORETIC PRELIMINARIES

The notation for many–sorted universal algebra that we use is taken from Meinke and Tucker [1992] where all of the basic technical definitions and results on algebra that we require may be found. Let us make our notation explicit.

An *S–sorted signature* Σ consists of a non–empty set S, the elements of which are ***sorts***, and an $S^* \times S$–indexed family $\langle\, \Sigma_{w,\,s} \mid w \in S^*,\ s \in S \,\rangle$ of sets (where S^* denotes the set of all words over S). For the empty word $\lambda \in S^*$ and any sort $s \in S$, each element $c \in \Sigma_{\lambda,\,s}$ is a ***constant symbol*** of sort s; for each non–empty word $w = s(1) \ldots s(n) \in S^+$ and any sort $s \in S$, each element $\sigma \in \Sigma_{w,\,s}$ is a ***function symbol*** of ***domain type*** w, ***codomain type*** s and ***arity*** n. Thus we can define Σ to be the pair $(S,\ \langle\, \Sigma_{w,\,s} \mid w \in S^*,\ s \in S \,\rangle)$.

Let Σ be an S–sorted signature. An *S–sorted* Σ ***algebra*** is an ordered pair (A, Σ^A), consisting of an S–indexed family $A = \langle\, A_s \mid s \in S \,\rangle$ of carrier sets A_s and an $S^* \times S$ indexed family $\Sigma^A = \langle\, \Sigma^A_{w,s} \mid w \in S^*,\ s \in S \,\rangle$ of sets of constants and algebraic operations. For each sort $s \in S$, $\Sigma^A_{\lambda,s} = \{\ c_A \mid c \in \Sigma_{\lambda,s}\ \}$, where $c_A \in A_s$ is a constant that interprets c. For each $w = s(1) \ldots s(n) \in S^+$ and each sort $s \in S$, $\Sigma^A_{w,s} = \{\ f_A \mid f \in \Sigma_{w,s}\ \}$, where $f_A : A^w \to A_s$ is an operation with domain $A^w = A_{s(1)} \times \ldots \times A_{s(n)}$ and codomain A_s which interprets f. As usual, we let A denote both a Σ algebra and its S–indexed family of carrier sets.

We require one essential construction from category theory, the cofibration construction, which we use to structure the model class of an equational type specification.

1.1. Definition. Let C be a category and Cat be a category of categories.

Let $F : C^{op} \to Cat$ be any functor. We can form the category $Gr(F)$ termed the *(Grothendieck) cofibration of* F. The class of objects of $Gr(F)$ is the coproduct of the classes $|F(c)|$ for all $c \in |C|$, i.e.

$$|Gr(F)| = \coprod_{c \in |C|} |F(c)|$$

$$= \{ \ (c,\ a) \ | \ c \in |C| \text{ and } a \in |F(c)| \ \}.$$

The morphisms of $Gr(F)$ are all pairs $(f,\ g) : (c,\ a) \to (c',\ a')$ such that $f : c \to c'$ is a morphism in C and $g : a \to (Ff)a'$ is a morphism in $F(c)$. Composition of morphisms is defined as follows. Given

$$(f_0,\ g_0) : (c_0,\ a_0) \to (c_1,\ a_1), \quad (f_1,\ g_1) : (c_1,\ a_1) \to (c_2,\ a_2)$$

we define the composition

$$(f_2,\ g_2) = (f_1,\ g_1)(f_0,\ g_0) : (c_0,\ a_0) \to (c_2,\ a_2)$$

by $f_2 : c_0 \to c_2$ where $f_2 = f_1 f_0$, and $g_2 : a_0 \to (F f_2) a_2$ in $F(c_0)$ is given by

$$g_2 = (F f_0) g_1 g_0.$$

□

In constructing model classes for equational type specifications we also require the *model functor* which maps each S–sorted signature Σ in the category *Sig* of all many–sorted signatures to the category $Alg(S,\ \Sigma)$ of all S–sorted Σ algebras.

1.2. Definition. Define the *model functor* $\mu : Sig^{op} \to Cat$ as follows. For any signature $(S,\ \Sigma) \in Sig$ define

$$\mu(S,\ \Sigma) = Alg(S,\ \Sigma).$$

Given a signature morphism $\theta : (S,\ \Sigma) \to (S',\ \Sigma')$ in *Sig* where $\theta = (\theta_S,\ \theta_\Sigma)$, we define the functor $\mu(\theta) : Alg(S',\ \Sigma') \to Alg(S,\ \Sigma)$. For any algebra $A \in Alg(S',\ \Sigma')$ define $\mu(\theta)A \in Alg(S,\ \Sigma)$ as follows. For each sort $s \in S$, define the carrier set $(\mu(\theta)A)_s = A_{\theta_S(s)}$. For each sort $s \in S$ and any constant symbol $c \in \Sigma_{\lambda,\ s}$ define the constant

$$c_{\mu(\theta)A} = \theta_\Sigma^{\lambda,\ s}(c)_A.$$

For any $n \geq 1$, any sorts $s_0, \ldots, s_n \in S$ and any function symbol $f \in \Sigma_{s_1 \ldots s_n,\ s_0}$ define the function

$$f_{\mu(\theta)A} = \theta_\Sigma^{s_1 \ldots s_n,\ s_0}(f)_A.$$

For any Σ' homomorphism $\phi : A \to B$ in $Alg(S',\ \Sigma')$ define $\mu(\theta)\phi : \mu(\theta)A \to \mu(\theta)B$ in $Alg(S,\ \Sigma)$ by

$$(\mu(\theta)\phi)_s = \phi_{\theta_S(s)}.$$

□

2. EQUATIONAL TYPE SPECIFICATIONS

An equational type specification consists of two components: (i) an *equational specification of types*; and, (ii) an *equational specification of typed combinators* over the type system of (i). Each component has its own formal syntax given by a signature, and semantics given by a model class of algebras. However the parameterisation of the combinator specification by the type specification requires us to structure within a single category of models, algebras of different, but related signature. For this we will use the cofibration construction of section 1.

To equationally specify a type system we formalise the syntax of types within a many–sorted signature.

2.1. Definition. By a *type signature* we mean a K–sorted signature Γ for some non–empty set K. In this context we term K a set of *kinds* and Γ a *K-kinded type signature.* □

We are only concerned with the semantics of types up to isomorphism. Thus, for a type signature Γ, we allow *any* Γ algebra G as a possible semantics for Γ.

2.2. Definition. Let Γ be a K–kinded type signature. By a *type algebra G* we mean a Γ algebra $G \in Alg(\Gamma)$. □

To equationally specify a typed combinator system we formalise the syntax of combinators within a many–sorted signature Σ. To impose the type notation of a K–kinded type signature Γ on Σ we take ground Γ terms as the sort set for Σ. Since not all the kinds in K are necessarily kinds of types we allow a restricted subset $K' \subseteq K$ of kinds to be the source of type names in Σ.

2.3. Definition. Let Γ be a K–kinded type signature, and $K' \subseteq K$ be any subset. By a *combinator signature* Σ over Γ and K' we mean a $\cup_{k \in K'} T(\Gamma)_k$–sorted signature Σ. □

In the sequel we consider a fixed, but arbitrarily chosen K–kinded type signature Γ, subset $K' \subseteq K$ and combinator signature Σ over Γ and K'.

Next we define the model class of a combinator signature Σ over a type signature Γ. For each type algebra $G \in Alg(\Gamma)$ we can construct a derived combinator signature $\Sigma(G)$ in which classes of combinators of semantically identical type in G are collected together. The construction is functorial.

2.4. Definition. Let Γ be a K–kinded type signature, $K' \subseteq K$ and Σ be a combinator signature over Γ and K'. Then Σ, Γ and K' uniquely determine a functor

$$\Sigma : Alg(\Gamma) \to Sig$$

where for any $G \in Alg(\Gamma)$ the signature $\Sigma(G)$ is the $\coprod_{k \in K'} G_k$–sorted signature defined as follows. For any $n \geq 0$, any kinds $k_0, \ldots, k_n \in K'$ and any $g_i \in G_{k_i}$ for $1 \leq i \leq n$,

$$\Sigma(G)_{(k_1, g_1)\ldots(k_n, g_n), (k_0, g_0)} =$$
$$\bigcup \{ \ \Sigma_{\tau_1 \ldots \tau_n, \tau_0} \ | \ \text{ for each } 1 \leq i \leq n, \ \ \tau_i \in T(\Gamma)_{k_i} \text{ and } \tau_{iG} = g_i \ \}.$$

For any algebras $G, G' \in Alg(\Gamma)$ and Γ homomorphism $\phi : G \to G'$, define the signature morphism $\Sigma(\phi) : \Sigma(G) \to \Sigma(G')$ where $\Sigma(\phi) = (\Sigma(\phi)_S, \Sigma(\phi)_\Sigma)$ and

$\Sigma(\phi)_S : \coprod_{k \in K'} G_k \to \coprod_{k \in K'} G'_k$ is given by

$$\Sigma(\phi)_S(k, g) = (k, \phi_k(g))$$

for any $k \in K'$ and $g \in G_k$. Also for any $n \geq 0$, any $k_0, \ldots, k_n \in K'$ and any $g_i \in G_{k_i}$ for $1 \leq i \leq n$ define $\Sigma(\phi)_\Sigma^{(k_1, g_1)\ldots(k_n, g_n), (k_0, g_0)}$ to be the inclusion mapping. □

We construct the model class of Σ over Γ as a cofibered category of algebras.

2.5. Definition. Let $F = \mu\Sigma : Alg(\Gamma) \to Cat$ be the composition of the functors $\mu : Sig^{op} \to Cat$ and $\Sigma : Alg(\Gamma) \to Sig$. The *model class* of Σ over Γ is the cofibered category

$$Mod(\Gamma, \Sigma) = Gr(F).$$

□

Let $Y = \langle Y_k \mid k \in K \rangle$ be a family of sets of variable symbols. We have the usual notions of *term algebra* $T(\Gamma, Y)$ over Γ and Y and *equation* over Γ and Y as a formula of the form $\tau = \tau'$ for $\tau, \tau' \in T(\Gamma, Y)_k$ and $k \in K$. Given a set ε of equations over Γ and Y we have the usual many–sorted equational calculus which is a sound and complete inference system for equational consequences of ε.

However, the notions of *term*, *equation* and *validity* for a combinator signature Σ over a type signature Γ are complicated by the parametric role of Γ. In the sequel we consider some fixed, but arbitrarily chosen family $X = \langle X_\tau \mid k \in K'$ and $\tau \in T(\Gamma)_k \rangle$ of disjoint sets of combinator variables. Since the standard definition of terms over Σ and X is too restrictive to deal with types as values, we use a more liberal notion of *preterm* over Σ and X.

2.6. Definition. Define the K' indexed family $T_{pre}(\Sigma, X) = \langle T_{pre}(\Sigma, X)_k \mid k \in K' \rangle$ of sets $T_{pre}(\Sigma, X)_k$ of *preterms of kind* k over Σ and X, and for each preterm $t \in T_{pre}(\Sigma, X)_k$ its *nominal type* $\tau_t \in T(\Gamma)_k$, inductively.

(i) For any kind $k \in K'$, any types $\sigma, \tau \in T(\Gamma)_k$ and any variable $x \in X_\sigma$,

$$x^\tau \in T_{pre}(\Sigma, X)_k,$$

and has nominal type τ.

(ii) For any kind $k \in K'$, any types $\sigma, \tau \in T(\Gamma)_k$ and any constant symbol $c \in \Sigma_{\lambda, \sigma}$,

$$c^\tau \in T_{pre}(\Sigma, X)_k,$$

and has nominal type τ.

(iii) For any $n \geq 1$, any kinds $k_0, \ldots, k_n \in K'$, any types $\tau_i \in T(\Gamma)_{k_i}$ for $1 \leq i \leq n$ and $\sigma, \tau \in T(\Gamma)_{k_0}$, any function symbol $f \in \Sigma_{\tau_1 \ldots \tau_n, \sigma}$ and any preterms $t_i \in T_{pre}(\Sigma, X)_{k_i}$ for $1 \leq i \leq n$,

$$f^\tau(t_1, \ldots, t_n) \in T_{pre}(\Sigma, X)_{k_0}$$

and has nominal type τ.

We let $T_{pre}(\Sigma)_k$ denote the set of all ground preterms over Σ of kind k. □

To check whether a preterm t is well typed in the context of particular type algebra $G \in Alg(\Gamma)$ and to calculate the value of the type of t, we make the following definition.

2.7. Definition. Consider any algebra $G \in Alg(\Gamma)$. Define the K'–indexed family $Type^G = \langle\, Type_k^G : T_{pre}(\Sigma, X)_k \rightsquigarrow G_k \mid k \in K' \,\rangle$ of partial functions $Type^G = Type_k^G$ inductively.

(i) For any kind $k \in K'$, any types $\sigma, \tau \in T(\Gamma)_k$ and any variable $x \in X_\sigma$,

$$Type^G(x^\tau) = \begin{cases} \tau_G, & \text{if } \tau_G = \sigma_G; \\ \uparrow, & \text{otherwise.} \end{cases}$$

(ii) For any kind $k \in K'$, any types $\sigma, \tau \in T(\Gamma)_k$ and any constant symbol $c \in \Sigma_{\lambda,\,\sigma}$,

$$Type^G(c^\tau) = \begin{cases} \tau_G, & \text{if } \tau_G = \sigma_G; \\ \uparrow, & \text{otherwise.} \end{cases}$$

(iii) For any $n \geq 1$, any kinds $k_0, \ldots, k_n \in K'$, any types $\tau_i \in T(\Gamma)_{k_i}$ for $1 \leq i \leq n$ and $\sigma, \tau \in T(\Gamma)_{k_0}$, any function symbol $f \in \Sigma_{\tau_1 \ldots \tau_n,\, \sigma}$ and any preterms $t_i \in T_{pre}(\Sigma, X)_{k_i}$ for $1 \leq i \leq n$,

$$Type^G(\, f^\tau(t_1, \ldots, t_n)\,) =$$

$$\begin{cases} \tau_G, & \text{if } \sigma_G = \tau_G \text{ and for each } 1 \leq i \leq n,\ Type^G(t_i) = \tau_{iG}; \\ \uparrow, & \text{otherwise.} \end{cases}$$

We say that a preterm $t \in T_{pre}(\Sigma, X)_k$ is *well typed* with respect to G if, and only if, $Type^G(t)$ is defined. □

Given a Γ algebra G the preterms which are well typed with respect to G are precisely the preterms which can be translated into well formed terms over the derived combinator signature $\Sigma(G)$ and a derived family $X(G)$ of sets of variables.

2.8. Definition. Let $G \in Alg(\Gamma)$ any type algebra. Define the derived family

$$X(G) = \langle\, X(G)_{(k,\, g)} \mid k \in K' \text{ and } g \in G_k \,\rangle$$

of sets of variables, where for any $k \in K'$ and any $g \in G_k$ we have

$$X(G)_{(k,\, g)} = \bigcup \{\, X_\tau \mid \tau \in T(\Gamma)_k \text{ and } \tau_G = g \,\}.$$

Define the K'–indexed family

$$Trans^G = \langle\, Trans_k^G : T_{pre}(\Sigma, X)_k \rightsquigarrow \cup_{g \in G_k} T(\Sigma(G), X(G))_{(k,\, g)} \mid k \in K' \,\rangle$$

of (partial) *translation mappings* $Trans^G = Trans_k^G$ inductively.

(i) For any kind $k \in K'$, any types $\sigma, \tau \in T(\Gamma)_k$ and any variable $x \in X_\sigma$, define

$$Trans^G(x^\tau) = \begin{cases} x^{(k,\, \tau_G)}, & \text{if } Type^G(x^\tau)\downarrow; \\ \uparrow, & \text{otherwise.} \end{cases}$$

(ii) For any kind $k \in K'$, any types $\sigma, \tau \in T(\Gamma)_k$ and any constant symbol $c \in \Sigma_{\lambda,\,\sigma}$, define

$$Trans^G(c^\tau) = \begin{cases} c^{(k,\, \tau_G)}, & \text{if } Type^G(c^\tau)\downarrow; \\ \uparrow, & \text{otherwise.} \end{cases}$$

(iii) For any $n \geq 1$, any kinds $k_0, \ldots, k_n \in K'$, any types $\tau_i \in T(\Gamma)_{k_i}$ for $1 \leq i \leq n$ and $\sigma, \tau \in T(\Gamma)_{k_0}$, any function symbol $f \in \Sigma_{\tau_1 \ldots \tau_n, \sigma}$ and any preterms $t_i \in T_{pre}(\Sigma, X)_{k_i}$ for $1 \leq i \leq n$, define

$$Trans^G(\, f^{\tau}(t_1, \ldots, t_n)\,) =$$

$$\begin{cases} f^{(k,\, \tau_G)}(\, Trans^G(t_1), \ldots, Trans^G(t_n)\,), & \text{if } Type^G(\, f^{\tau_0}(t_1, \ldots, t_n)\,) \downarrow; \\ \uparrow, & \text{otherwise.} \end{cases}$$

□

2.9. Proposition. *Let $G \in Alg(\Gamma)$, be any type algebra, $k \in K'$ be any kind and $t \in T_{pre}(\Sigma, X)_k$ be any preterm:*
(i) $Type^G(t) \downarrow$ if, and only if, $Trans^G(t) \downarrow$;
(ii) if $Type^G(t) \downarrow = g$ then $Trans^G(t) \in T(\, \Sigma(G), X(G)\,)_{(k,\, g)}$.

Proof. By induction on the complexity of t. □

We now use the semantical apparatus introduced for types to make precise the concepts of *well typedness* and *validity* for an *equation in preterms* with respect to a Γ type algebra G.

2.10. Definition.

(i) By an *equation in preterms* over Σ and X of kind $k \in K'$ we mean an expression of the form

$$t = t'$$

where $t, t' \in T_{pre}(\Sigma, X)_k$.

(ii) For any type algebra $G \in Alg(\Gamma)$ we say that $t = t'$ is *well typed with respect to G* if, and only if, $Type^G(t) \downarrow$ and $Type^G(t') \downarrow$ and $Type^G(t) = Type^G(t')$. If $C \subseteq Alg(\Gamma)$ is any class of algebras then we say that $t = t'$ is *well typed with respect to C* if, and only if, $t = t'$ is well typed with respect to every $G \in C$.

(iii) For any pair of algebras $(G, A) \in |Mod(\Gamma, \Sigma)|$ and any assignment $\alpha : X(G) \to A$ we say that $t = t'$ is *valid in* (G, A) under α and write

$$(G, A) \models_\alpha t = t'$$

if, and only if, $t = t'$ is well typed with respect to G and

$$Val^{A,\, \alpha}(\, Trans^G(t)) = Val^{A,\, \alpha}(\, Trans^G(t'))$$

(where $g = Type^G(t)$ and $Val^{A,\, \alpha} : T(\Sigma(G), X(G)) \to A$ is the term evaluation mapping in A under assignment α).

If $t = t'$ is valid in (G, A) under every assignment $\alpha : X(G) \to A$ we say that $t = t'$ is *valid in* (G, A) and write $(G, A) \models t = t'$. If $C \subseteq |Mod(\Gamma, \Sigma)|$ is any subclass then we say that $t = t'$ is *valid in* C and write $C \models t = t'$ if, and only if, $(G, A) \models t = t'$ for all $(G, A) \in C$.

(iv) If E is any set of equations over Σ and X then we say that E is valid in (G, A) (respectively C) if and only if $(G, A) \models e$ (respectively $C \models e$) for every equation $e \in E$. □

Combining a set of equations in types with a set of equations in combinator preterms we have the concept of an *equational type specification*.

2.11. Definition. By an *equational type specification* we mean a pair

$$Spec = (\,(\Gamma,\ \varepsilon),\ (\Sigma,\ E)\,)$$

where Γ is a K–kinded type signature, ε is an equational theory over Γ and some family Y of sets of variables, Σ is a combinator signature over Γ and $K' \subseteq K$, and E is a set of equations in preterms over Σ and some family X of sets of variables. □

To give an initial model semantics to a specification $Spec = (\,(\Gamma,\ \varepsilon),\ (\Sigma,\ E)\,)$ we define and structure the model class $Mod(\,(\Gamma,\ \varepsilon),\ (\Sigma,\ E)\,)$ as a cofibered category.

2.12. Definition. Let $Spec = (\,(\Gamma,\ \varepsilon),\ (\Sigma,\ E)\,)$ be an equational type specification. Define the functor $Spec : Alg(\Gamma,\ \varepsilon)^{op} \to Cat$ by

$$Spec(G) = Alg(\,\Sigma(G),\ E(G)\,)$$

for each $G \in Alg(\Gamma,\ \varepsilon)$, where

$$E(G) = \{\ Trans^{G}(t) = Trans^{G}(t')\ \mid\ t = t' \in E\ \}.$$

For any $G,\ G' \in Alg(\Gamma,\ \varepsilon)$ and Γ homomorphism $\phi : G \to G'$ the functor

$$Spec(\phi) : Alg(\,\Sigma(G'),\ E(G')\,) \to Alg(\,\Sigma(G),\ E(G)\,)$$

is the restriction of $\mu(\phi)$ (c.f. Definition 1.2) to $Alg(\,\Sigma(G'),\ E(G')\,)$. We define the *model class* of $Spec$ to be the cofibered category

$$Mod(Spec) = Gr(Spec).$$

□

2.13. Proposition. *Let $Spec = (\,(\Gamma,\ \varepsilon),\ (\Sigma,\ E)\,)$ be an equational type specification. Then the model class $Mod(Spec)$ admits an initial object $I(Spec)$.*

Proof. See Meinke [1991b]. □

Thus each equational type specification $Spec$ has an initial algebra semantics $I(Spec)$.

3. REWRITING TERMS AND TYPES

In this section we introduce the fundamental definitions and concepts for rewriting type terms and combinator preterms. In the sequel we consider a fixed, but arbitrarily chosen kind set K, K–kinded type signature Γ, K–indexed family Y of sets of type variables, kind subset $K' \subseteq K$, combinator signature Σ over Γ and K', and $T(\Gamma)_{K'}$–indexed family X of sets of variables.

Since type terms are simply expressions in the algebra $T(\Gamma,\ Y)$ of all many-sorted terms over Γ and Y, the usual definitions and concepts of term rewriting may be applied. In particular, for each kind $k \in K$ and each type term $\tau \in T(\Gamma,\ Y)_k$ we may define the set $Occ(\tau) \subseteq \mathbf{N}^*$ of *occurrences in* τ together with the map $Knd_\tau : Occ(\tau) \to K$, which assigns to each occurrence its kind, inductively as usual. Furthermore, for any occurrence $\bar{i} \in Occ(\tau)$ of kind $k' \in K$ and any type term

$\sigma \in T(\Gamma, Y)_{k'}$ we may inductively define the ***subterm of*** τ ***at*** $\bar{\imath}$, denoted by $\tau(\bar{\imath})$ and the ***substitution of*** σ ***into*** τ ***at*** $\bar{\imath}$, denoted by $\tau(\bar{\imath}/\sigma)$.

Recall that if ε is any set of equations over Γ and Y and $R(\varepsilon)$ is the rewrite rule set

$$R(\varepsilon) = \{ \ \tau \to \tau', \tau' \to \tau \ | \ \tau = \tau' \in \varepsilon \ \}$$

then we have a correspondence between many–sorted term rewriting with the rule set $R(\varepsilon)$ and deduction using ε and the many–sorted equational calculus.

3.1. Correspondence Theorem. *For any set* ε *of equations over* Γ *and* Y *and any terms* $\tau_1, \tau_2 \in T(\Gamma, Y)_k$,

$$\tau_1 \xrightarrow{R(\varepsilon)^*} \tau_2 \ \Leftrightarrow \ \varepsilon \vdash \tau_1 = \tau_2,$$

where $\xrightarrow{R(\varepsilon)^*}$ *is the reflexive, transitive closure of the rewrite relation* $\xrightarrow{R(\varepsilon)}$ *on terms induced by* $R(\varepsilon)$.

Proof. See for example Ehrig and Mahr [1985]. □

In this section we shall define a rewrite relation $\xrightarrow{R(\varepsilon, E)}$ on combinator preterms induced by an equational type specification $((\Gamma, \varepsilon), (\Sigma, E))$. In section 4 we review the equational calculus of types and terms introduced in Meinke [1991b]. In section 5 we prove a generalised Correspondence Theorem which establishes that the rewriting relation $\xrightarrow{R(\varepsilon, E)^*}$ coincides with deduction using (ε, E) and the equational calculus of types and terms. Completeness results for this calculus then provide a semantic characterisation of $\xrightarrow{R(\varepsilon, E)^*}$ in terms of the initial model $I((\Gamma, \varepsilon), (\Sigma, E))$.

We begin our analysis of rewriting for preterms by observing that in any combinator preterm $t \in T_{pre}(\Sigma, X)_k$ we have both combinator symbols from Σ and X and type symbols from Γ. We shall allow rewriting for both kinds of symbols using either rewrite rules for type terms or rewrite rules for combinator preterms. Thus we distinguish in a preterm t occurrences of type symbols in t and occurrences of combinator symbols in t.

3.2. Definition. For any kind $k \in K'$ and any combinator preterm $t \in T_{pre}(\Sigma, X)_k$ we define the sets $Occ^{com}(t) \subseteq \mathbf{N}^*$ and $Occ^{typ}(t) \subseteq \mathbf{N}^* \times \mathbf{N}^*$ of all ***combinator occurrences*** and ***type occurrences*** respectively in t. We also define the maps

$$Knd_t^{com} : Occ^{com}(t) \to K, \quad Knd_t^{typ} : Occ^{typ}(t) \to K$$

which assign kinds to combinator occurrences and type occurrences. These sets and mappings are defined inductively.

(i) For any kind $k \in K'$, any type term $\tau \in T(\Gamma)_k$ and any preterm $t \in T_{pre}(\Sigma, X)_k$ with nominal type τ,

$$\lambda \in Occ^{com}(t)$$

and

$$Knd_t^{com}(\lambda) = k.$$

Also for any $\bar{i} \in Occ(\tau)$,

$$(\lambda, \bar{i}) \in Occ^{typ}(t)$$

and

$$Knd_t^{typ}(\lambda, \bar{i}) = Knd_\tau(\bar{i}).$$

(ii) For any $n \geq 1$, any kinds $k_0, \ldots, k_n \in K'$, any types $\tau_i \in T(\Gamma)_{k_i}$ for $1 \leq i \leq n$ and $\sigma, \tau \in T(\Gamma)_{k_0}$, any function symbol $f \in \Sigma_{\tau_1 \ldots \tau_n, \sigma}$ and any preterms $t_i \in T_{pre}(\Sigma, X)_{k_i}$ for $1 \leq i \leq n$, let t denote the preterm $f^\tau(t_1, \ldots, t_n)$. For each $1 \leq m \leq n$, if $\bar{i} \in Occ^{com}(t_m)$ then

$$m.\bar{i} \in Occ^{com}(t)$$

and

$$Knd_t^{com}(m.\bar{i}) = Knd_{t_m}^{com}(\bar{i}).$$

Also for any $(\bar{i}, \bar{j}) \in Occ^{typ}(t_m)$,

$$(m.\bar{i}, j) \in Occ^{typ}(t)$$

and

$$Knd_t^{typ}(m.\bar{i}, \bar{j}) = Knd_{t_m}^{typ}(\bar{i}, \bar{j}).$$

□

To rewrite a combinator subexpression in a combinator preterm $t \in T_{pre}(\Sigma, X)_k$ we require substitution at a combinator occurrence $\bar{i}$ of t. To rewrite a type subexpression in t we require substitution at a type occurrence $(\bar{i}, \bar{j})$ of t. We define both substitutions by induction on the complexity of occurrences.

3.3. Definition. For any kind $k \in K'$, any preterm $t \in T_{pre}(\Sigma, X)_k$, any combinator occurrence $\bar{i} \in Occ^{com}(t)$ of kind $k' \in K'$ and any preterm $t' \in T_{pre}(\Sigma, X)_{k'}$ we define the *combinator subterm* of t at i, denoted by $t(\bar{i})$, and the *substitution of t' into t at $\bar{i}$* denoted by $t(\bar{i}/t')$ inductively as follows.

(i) For any kind $k \in K'$ and any preterm $t \in T_{pre}(\Sigma, X)_k$,

$$t(\lambda) = t,$$

and for any preterm $t' \in T_{pre}(\Sigma, X)_k$,

$$t(\lambda/t') = t'.$$

(ii) For any $n \geq 1$, any kinds $k_0, \ldots, k_n \in K'$, any types $\tau_i \in T(\Gamma)_{k_i}$ for $1 \leq i \leq n$, and $\sigma, \tau \in T(\Gamma)_{k_0}$, any function symbol $f \in \Sigma_{\tau_1 \ldots \tau_n, \sigma}$, any preterms $t_i \in T_{pre}(\Sigma, X)_{k_i}$ for $1 \leq i \leq n$, any $1 \leq m \leq n$ and any combinator occurrence $\bar{i} \in Occ^{com}(t_m)$,

$$f^\tau(t_1, \ldots, t_n)(m.\bar{i}) = t_m(\bar{i}),$$

and if $Knd_{t_m}^{com}(\bar{i}) = k'$ then for any preterm $t' \in T_{pre}(\Sigma, X)_{k'}$,

$$f^\tau(t_1, \ldots, t_n)(m.\bar{i}/t') = f^\tau(t_1, \ldots, t_{m-1}, t_m(\bar{i}/t'), t_{m+1}, \ldots, t_n).$$

For any type occurrence $(\bar{\imath}, \bar{\jmath}) \in Occ^{typ}(t)$ of kind $k' \in K'$ and any type term $\tau \in T(\Gamma)_{k'}$ we define the *type subterm* of t at $(\bar{\imath}, \bar{\jmath})$, denoted by $t(\bar{\imath}, \bar{\jmath})$ and the *substitution of τ into t at* $(\bar{\imath}, \bar{\jmath})$, denoted by $t(\bar{\imath}, \bar{\jmath}/\tau)$ inductively.

(iii) For any kind $k \in K'$, any type terms $\sigma, \tau \in T(\Gamma)_k$, any combinator constant symbol $c \in \Sigma_{\lambda, \sigma}$ and any occurrence $\bar{\jmath} \in Occ(\tau)$,

$$c^{\tau}(\lambda, \bar{\jmath}) = \tau(\bar{\jmath})$$

and if $Knd_{\tau}(\bar{\jmath}) = k'$ then for any type term $\tau' \in T(\Gamma)_{k'}$,

$$c^{\tau}(\lambda, \bar{\jmath}/\tau') = c^{\tau(\bar{\jmath}/\tau')}.$$

(iv) For any $n \geq 1$, any kinds $k_0, \ldots, k_n \in K'$, any type terms $\tau_i \in T(\Gamma)_{k_i}$ for $1 \leq i \leq n$ and $\sigma, \tau \in T(\Gamma)_{k_0}$, any combinator function symbol $f \in \Sigma_{\tau_1 \ldots \tau_n, \sigma}$, any preterms $t_i \in T_{pre}(\Sigma, X)_{k_i}$ for $1 \leq i \leq n$ and for any occurrence $\bar{\jmath} \in Occ(\tau)$,

$$f^{\tau}(t_1, \ldots, t_n)(\lambda, \bar{\jmath}) = \tau(\bar{\jmath})$$

and if $Knd_{\tau}(\bar{\jmath}) = k'$ then for any type term $\tau' \in T(\Gamma)_{k'}$

$$f^{\tau}(t_1, \ldots, t_n)(\lambda, \bar{\jmath}/\tau') = f^{\tau(\bar{\jmath}/\tau')}(t_1, \ldots, t_n).$$

For any $1 \leq m \leq n$ and any type occurrence $(\bar{\imath}, \bar{\jmath}) \in Occ^{typ}(t_m)$,

$$f^{\tau}(t_1, \ldots, t_n)(m.\bar{\imath}, \bar{\jmath}) = t_m(\bar{\imath}, \bar{\jmath})$$

and if $Knd^{typ}_{t_m}(\bar{\imath}, \bar{\jmath}) = k'$ then for any type term $\tau' \in T(\Gamma)_{k'}$,

$$f^{\tau}(t_1, \ldots, t_n)(m.\bar{\imath}, \bar{\jmath}/\tau') = f^{\tau}(t_1, \ldots, t_{m-1}, t_m(\bar{\imath}, \bar{\jmath}/\tau'), t_{m+1}, \ldots, t_n).$$

□

To rewrite type expressions over Γ we introduce *type rewrite rules* while to rewrite combinator expressions over Σ we introduce *combinator rewrite rules.*

3.4. Definition. A *type rewrite rule* ρ of kind $k \in K$ is an ordered pair of type terms (τ_1, τ_2) with $\tau_1, \tau_2 \in T(\Gamma, Y)_k$. We write $\tau_1 \to \tau_2$ to denote the rule (τ_1, τ_2).

A *combinator rewrite rule* r of kind $k \in K'$ is an ordered pair of combinator preterms (t_1, t_2) with $t_1, t_2 \in T_{pre}(\Sigma, X)_k$. We write $t_1 \to t_2$ to denote the rule (t_1, t_2). □

3.5. Definition. Let $k \in K'$ be any kind and $t, t' \in T_{pre}(\Sigma, X)_k$ be any combinator preterms of kind $k \in K'$.

(i) We say that t *rewrites* to t' *by a type rewrite rule* $\tau_1 \to \tau_2$ of kind $k' \in K$ at a type occurrence $(\bar{\imath}, \bar{\jmath}) \in Occ^{typ}(t)$ of kind k' if for some substitution $\theta : Y \to T(\Gamma)$,

$$t(\bar{\imath}, \bar{\jmath}) = \bar{\theta}(\tau_1)$$

and

$$t' = t(\bar{\imath}, \bar{\jmath}/\bar{\theta}(\tau_2)).$$

If t rewrites to t' by a type rewrite rule ρ at a type occurrence $(\bar{i}, \bar{j}) \in Occ^{typ}(t)$ then we write

$$t \xrightarrow{(\bar{i}, \bar{j}), \rho} t'$$

and when the occurence $(\bar{i}, \bar{j})$ is known or irrelevant we may simply write $t \xrightarrow{\rho} t'$.

(ii) We say that t *rewrites* to t' *by a combinator rewrite rule* $t_1 \rightarrow t_2$ of kind $k' \in K'$ at a combinator occurrence $\bar{i} \in Occ^{com}(t)$ of kind k' if for some substitution $\theta : X \rightarrow T_{pre}(\Sigma, X)$,

$$t(\bar{i}) = \bar{\theta}(t_1)$$

and

$$t' = t(\, \bar{i} / \bar{\theta}(t_2)\,).$$

If t rewrites to t' by a combinator rewrite rule r at a combinator occurrence $\bar{i} \in Occ^{com}(t)$ then we shall write

$$t \xrightarrow{\bar{i}, r} t'$$

and when the occurence $\bar{i}$ is known or irrelevant we may simply write $t \xrightarrow{r} t'$. □

Finally we can define the rewrite relation $\xrightarrow{R}$ on combinator preterms induced by a *type rewrite system* R on preterms.

3.6. Definition. By a *type rewrite system* R we mean an ordered pair $R = (R_{typ}, R_{com})$, where R_{typ} is a set of type rewrite rules and R_{com} is a set of combinator rewrite rules.

Define the K'-indexed family $\xrightarrow{R} = \langle \xrightarrow{R}_k \mid k \in K' \rangle$ of binary relations $\xrightarrow{R} = \xrightarrow{R}_k$ on $T_{pre}(\Sigma, X)$ by

$$t \xrightarrow{R} t' \;\Leftrightarrow\; t \xrightarrow{\rho} t' \text{ for some } \rho \in R_{typ} \text{ or } t \xrightarrow{r} t' \text{ for some } r \in R_{com},$$

for any $k \in K'$ and $t, t' \in T_{pre}(\Sigma, X)_k$. Let $\xrightarrow{R}{}^*$ denote the reflexive, transitive closure of $\xrightarrow{R}$.

For any equational type specification $Spec = ((\Gamma, \varepsilon), (\Sigma, E))$ we let $R(\varepsilon, E)$ denote the type rewrite system with

$$R(\varepsilon, E)_{typ} = \{\ \tau \rightarrow \tau',\ \tau' \rightarrow \tau \mid \tau = \tau' \in \varepsilon\ \}$$

and

$$R(\varepsilon, E)_{com} = \{\ t \rightarrow t',\ t' \rightarrow t \mid t = t' \in E\ \}.$$

□

4. THE EQUATIONAL CALCULUS OF TERMS AND TYPES

In this section we briefly review the type assignment calculus for preterms and the equational calculus of terms and types which were introduced in Meinke [1991b], and we recall the Soundness and Completeness Theorems for these calculi. We begin with the calculus of type assignment.

4.1. Definition. By a *type assignment* of kind $k \in K'$ we mean an expression of the form

$$t : \tau$$

where $t \in T_{pre}(\Sigma, X)_k$ is a combinator preterm of kind k and $\tau \in T(\Gamma)_k$. is a type term of kind k.

Given an algebra $G \in Alg(\Gamma)$ we say that the type assignment $t : \tau$ is *valid* in G and write $G \models t : \tau$ if, and only if,

$$Type^G(t) \downarrow \text{ and } Type^G(t) = \tau_G.$$

If $K \subseteq Alg(\Gamma)$ is any class of Γ algebras and $t : \tau$ is valid in G for every $G \in K$ we say that $t : \tau$ is valid in K and write $K \models t : \tau$. □

The type assignment calculus serves to identify those combinator preterms $t \in T_{pre}(\Sigma, X)_k$ which are well formed terms, in the context of a particular equational theory ε of types, as the *typable preterms*. The type assignment calculus makes use of the many–sorted equational calculus for type terms over Γ.

4.2. Definition. Let ε be an equational theory over Γ and Y. Define the inference rules of the *type assignment calculus*.

(i) For any kind $k \in K'$, any types $\sigma, \tau \in T(\Gamma)_k$ and any variable $x \in X_\sigma$,

$$\frac{\varepsilon \vdash \tau = \sigma}{\varepsilon \vdash x^\tau : \tau}$$

(ii) For any kind $k \in K'$, any types $\sigma, \tau \in T(\Gamma)_k$ and any constant symbol $c \in \Sigma_{\lambda, \sigma}$,

$$\frac{\varepsilon \vdash \tau = \sigma}{\varepsilon \vdash c^\tau : \tau}$$

(iii) For any $n \geq 1$, any kinds $k_0, \ldots, k_n \in K'$, any types $\tau_i \in T(\Gamma)_{k_i}$ for $1 \leq i \leq n$ and $\sigma, \tau \in T(\Gamma)_{k_0}$, any function symbol $f \in \Sigma_{\tau_1 \ldots \tau_n, \sigma}$ and any preterms $t_i \in T_{pre}(\Sigma, X)_{k_i}$ for $1 \leq i \leq n$,

$$\frac{\varepsilon \vdash \tau = \sigma, \quad \varepsilon \vdash t_1 : \tau_1 \ , \ldots, \ \varepsilon \vdash t_n : \tau_n}{\varepsilon \vdash f^\tau(t_1, \ldots, t_n) : \tau}$$

(iv) For any kind $k \in K'$, any types $\sigma, \tau \in T(\Gamma)_k$ and any preterm $t \in T_{pre}(\Sigma, X)_k$,

$$\frac{\varepsilon \vdash t : \tau, \quad \varepsilon \vdash \tau = \sigma}{\varepsilon \vdash t : \sigma}$$

□

4.3. Completeness Theorem. *For any equational theory ε over Γ and a family Y of infinite sets of variables, any kind $k \in K'$, any type $\tau \in T(\Gamma)_k$ and any preterm $t \in T_{pre}(\Sigma, X)_k$,*

$$\varepsilon \vdash t : \tau \Leftrightarrow Alg(\Gamma, \varepsilon) \models t : \tau.$$

Proof. See Meinke [1991b]. □

Next we review the equational calculus of terms and types in which formal derivations of equations in combinator preterms are carried out, starting from an equational type specification. This calculus augments the proof rules of the many-sorted equational calculus in two ways: (i) it makes use of the type assignment calculus of Definition 4.2 in order to avoid derivations of non-well typed equations from well typed equational type specifications; and, (ii) it includes additional proof rules to allow replacement of provably equal type expressions inside preterms. To define the substitution rule of the calculus we require a notion of substitution for preterms.

4.4. Definition. Let $\alpha = \langle\, \alpha_k : X_k \to T_{pre}(\Sigma,\, X)_k \mid k \in K' \,\rangle$ be any family of assignments (where $X_k = \cup_{\tau \in T(\Gamma)_k} X_\tau$). Define the family

$$\overline{\alpha} = \langle\, \overline{\alpha_k} : T_{pre}(\Sigma,\, X)_k \to T_{pre}(\Sigma,\, X)_k \mid k \in K' \,\rangle$$

of substitution mappings $\overline{\alpha} = \overline{\alpha_k}$ inductively.

(i) For each $k \in K'$ and $\sigma,\, \tau \in T(\Gamma)_k$ and constant symbol $c \in \Sigma_{\lambda,\, \sigma}$ define

$$\overline{\alpha}(c^\tau) = c^\tau.$$

(ii) For each $k \in K'$ and $\sigma,\, \tau \in T(\Gamma)_k$ and variable $x \in X_\sigma$ define

$$\overline{\alpha}(x^\tau) = \alpha_k(x).$$

(iii) For any $n \geq 1$, any $k_0, \ldots, k_n \in K'$, any $\tau_i \in T(\Gamma)_{k_i}$ for $1 \leq i \leq n$ and $\sigma,\, \tau \in T(\Gamma)_{k_0}$, and any $f \in \Sigma_{\tau_1,\ldots,\tau_n,\, \sigma}$ and any preterms $t_i \in T_{pre}(\Sigma,\, X)_{k_i}$ for $1 \leq i \leq n$ define

$$\overline{\alpha}(\, f^\tau(t_1, \ldots, t_n)\,) = f^\tau(\, \overline{\alpha}(t_1),\, \ldots,\, \overline{\alpha}(t_n)\,).$$

For any $k \in K'$, any type term $\tau \in T(\Gamma)_k$, any variable $x \in X_\tau$ and any preterm $t \in T_{pre}(\Sigma,\, X)_k$ we let $[x/t]$ denote the family of assignments such that for any $k' \in K'$, any $\sigma \in T(\Gamma)_{k'}$ and any variable $y \in X_\sigma$,

$$[x/t]_{k'}(y) = \begin{cases} t, & \text{if } k' = k \text{ and } \sigma = \tau \text{ and } y = x; \\ y^\sigma, & \text{otherwise.} \end{cases}$$

We write $t'[x/t]$ for $\overline{[x/t]}(t)$. □

The deduction rules of the equational calculus of terms and types are as follows.

4.5. Definition. Let ε be any set of equations over Γ and Y and E be any set of equations over Σ and X. Define the *deduction rules of the equational calculus of terms and types* as follows.

(i) For any equation $e \in E$,

$$\frac{}{(\varepsilon,\, E) \vdash e}$$

is an *axiom introduction rule.*

(ii) For any kind $k \in K'$ and any preterm $t \in T_{pre}(\Sigma,\, X)_k$,

$$\frac{\varepsilon \vdash t : \tau_t}{(\varepsilon,\, E) \vdash t = t}$$

is a *reflexivity rule*, where $\tau_t \in T(\Gamma)_k$ is the nominal type of t.

(iii) For any kind $k \in K'$ and any preterms $t_1, t_2 \in T_{pre}(\Sigma, X)_k$,

$$\frac{(\varepsilon, E) \vdash t_1 = t_2}{(\varepsilon, E) \vdash t_2 = t_1}$$

is a *symmetry rule.*

(iv) For any kind $k \in K'$ and any preterms $t_1, t_2, t_3 \in T_{pre}(\Sigma, X)_k$,

$$\frac{(\varepsilon, E) \vdash t_1 = t_2, \quad (\varepsilon, E) \vdash t_2 = t_3}{(\varepsilon, E) \vdash t_1 = t_3}$$

is a *transitivity rule.*

(v) For any kind $k \in K'$, any preterms $t, t' \in T_{pre}(\Sigma, X)_k$, any kind $k' \in K'$, any preterms $t_1, t_2 \in T_{pre}(\Sigma, X)_{k'}$, any type $\tau \in T(\Gamma)_{k'}$ and any variable $x \in X_\tau$,

$$\frac{(\varepsilon, E) \vdash t = t', \quad (\varepsilon, E) \vdash t_1 = t_2, \quad \varepsilon \vdash t_1 : \tau}{(\varepsilon, E) \vdash t[x/t_1] = t'[x/t_2]}$$

is a *substitution rule.*

(vi.a) For any $n \geq 0$ any kinds $k_0, \ldots, k_n \in K'$, any types $\tau_i \in T(\Gamma)_{k_i}$ for $1 \leq i \leq n$ and $\sigma, \tau \in T(\Gamma)_{k_0}$, any constant or operation symbol $f \in \Sigma_{\tau_1 \ldots \tau_n, \tau}$, and any preterms $t_i \in T_{pre}(\Sigma, X)_{k_i}$ for $1 \leq i \leq n$

$$\frac{\varepsilon \vdash \tau = \sigma}{(\varepsilon, E) \vdash f^\tau(t_1, \ldots, t_n) = f^\sigma(t_1, \ldots, t_n)}$$

is a *type replacement rule* (for constants and operations).

(vi.b) For any kind $k \in K'$, any types $\sigma, \tau \in T(\Gamma)_k$, any variable $x \in X_\tau$

$$\frac{\varepsilon \vdash \tau = \sigma}{(\varepsilon, E) \vdash x^\tau = x^\sigma}$$

is a *type replacement rule* (for variables). □

4.6. Completeness Theorem. *Let ε be any set of equations over Γ and a family Y of infinite sets of variables. Let E be any set of equations in preterms over Σ and a family X of infinite sets of variables. Suppose that each equation $e \in E$ is well typed with respect to $Alg(\Gamma, \varepsilon)$.*

For any $k \in K'$ and preterms $t, t' \in T_{pre}(\Sigma, X)_k$,

$$(\varepsilon, E) \vdash t = t' \quad \Leftrightarrow \quad Mod((\Gamma, \varepsilon), (\Sigma, E)) \models t = t'.$$

Proof. See Meinke [1991b]. □

A special case of the Completeness Theorem 4.6 for equations in ground preterms can be stated in terms of the initial model of an equational type specification.

4.7. Initiality Theorem. *Let ε be any set of equations over Γ and Y. Let E be any set of equations over Σ and X which are well typed with respect to $Alg(\Gamma, \varepsilon)$.*

For any $k \in K'$ and ground preterms $t,\ t' \in T_{pre}(\Sigma,\ X)_k$,

$$(\varepsilon,\ E) \vdash t = t' \ \Leftrightarrow\ I((\Gamma,\ \varepsilon),\ (\Sigma,\ E)) \models t = t'.$$

Proof. See Meinke [1991b]. □

5. CORRESPONDENCE OF REWRITING AND DEDUCTION

For any equational type specification $Spec = ((\Gamma,\ \varepsilon)(\Sigma,\ E))$ we will show that the rewrite relation $\overset{R(\varepsilon,\ E)^*}{\longrightarrow}$ coincides with the provability relation $(\varepsilon,\ E) \vdash$ of the equational calculus of terms and types. For ground preterms $t,\ t' \in T_{pre}(\Sigma)_k$ we show that provable equivalence coincides with semantical identity in the initial model $I(Spec)$. Thus rewriting for ground preterms using $R(\varepsilon,\ E)$ agrees with calculation in $I(Spec)$.

5.1. Correspondence Theorem. *Let $((\Gamma,\ \varepsilon),\ (\Sigma,\ E))$ be an equational type specification. Suppose that each equation $e \in E$ is well typed with respect to $Alg(\Gamma,\ \varepsilon)$. For any $k \in K'$ and any preterms $t,\ t' \in T_{pre}(\Sigma,\ X)_k$, if t and t' are well typed with respect to $Alg(\Gamma,\ \varepsilon)$ then*

$$(\varepsilon,\ E) \vdash t = t' \ \Leftrightarrow\ t \overset{R(\varepsilon,\ E)^*}{\longrightarrow} t'.$$

Proof. $\Leftarrow$ Suppose t and t' are identical preterms. Since t is well typed with respect to $Alg(\Gamma,\ \varepsilon)$ then by Theorem 4.3, $\varepsilon \vdash t : \tau_t$. So by the reflexivity rule $(\varepsilon,\ E) \vdash t = t'$.

Suppose t and t' are distinct preterms. We need only show that if $t \overset{R(\varepsilon,\ E)}{\longrightarrow} t'$ then $(\varepsilon,\ E) \vdash t = t'$.

(i) Suppose that for some $k \in K'$, type terms $\tau_1,\ \tau_2 \in T(\Gamma)_k$, type rewrite rule $\rho = \tau_1 \to \tau_2 \in R(\varepsilon,\ E)_{typ}$ and type occurrence $(\bar{\imath},\ \bar{\jmath}) \in Occ^{typ}(t)$,

$$t \overset{(\bar{\imath},\ \bar{\jmath}),\ \rho}{\longrightarrow} t'.$$

Then for some ground substitution $\theta : Y \to T(\Gamma)$,

$$t(\bar{\imath},\ \bar{\jmath}) = \overline{\theta}(\tau_1) \ \text{ and } \ t' = t(\bar{\imath},\ \bar{\jmath} / \overline{\theta}(\tau_2)).$$

Now $\tau_1 = \tau_2 \in \varepsilon$ and so $\varepsilon \vdash \tau_1 = \tau_2$. Hence $\varepsilon \vdash \overline{\theta}(\tau_1) = \overline{\theta}(\tau_2)$. We prove that $(\varepsilon,\ E) \vdash t = t'$ by induction on the complexity of $\bar{\imath}$.

Basis. Suppose $\bar{\imath} = \lambda$ and that the nominal type of t is τ for $\tau \in T(\Gamma)_k$. Then $\tau(\bar{\jmath}) = \overline{\theta}(\tau_1)$. So

$$t' = t(\lambda,\ \bar{\jmath} / \overline{\theta}(\tau_2)).$$

But $\varepsilon \vdash \tau = \tau(\bar{\jmath} / \overline{\theta}(\tau_2))$. So by the type replacement rule for constants and operations,

$$(\varepsilon,\ E) \vdash t = t'.$$

Induction Step. Suppose $\bar{i} = m.\bar{l}$ for some $1 \leq m \leq n$. Then for some $n \geq 1$ and $k_0, \ldots, k_n \in K'$ and type terms $\tau_i \in T(\Gamma)_{k_i}$ for $1 \leq i \leq n$ and $\sigma, \tau \in T(\Gamma)_{k_0}$ and combinator operation symbol $f \in \Sigma_{\tau_1 \ldots \tau_n, \sigma}$ and preterms $t_i \in T_{pre}(\Sigma, X)_{k_i}$ for $1 \leq i \leq n$ we must have $1 \leq m \leq n$ and t must be the preterm $f^\tau(t_1, \ldots, t_n)$ and $(\bar{l}, \bar{j}) \in Occ^{typ}(t_m)$. Then $t(\bar{i}, \bar{j}) = t_m(\bar{l}, \bar{j}) = \overline{\theta}(\tau_1)$ and

$$t' = t(\bar{i}, \bar{j} / \overline{\theta}(\tau_2))$$
$$= f^\tau(t_1, \ldots, t_{m-1}, t_m(\bar{l}, \bar{j} / \overline{\theta}(\tau_2)), t_{m+1}, \ldots, t_n).$$

By the induction hypothesis

$$(\varepsilon, E) \vdash t_m = t_m(\bar{l}, \bar{j} / \overline{\theta}(\tau_2)).$$

Hence by reflexivity and substitution

$$(\varepsilon, E) \vdash f^\tau(t_1, \ldots, t_n) = f^\tau(t_1, \ldots, t_{m-1}, t_m(\bar{l}, \bar{j} / \overline{\theta}(\tau_2)), t_{m+1}, \ldots, t_n).$$

i.e. $(\varepsilon, E) \vdash t = t'$.

(ii) Suppose that for some $k \in K'$, preterms $t_1, t_2 \in T_{pre}(\Sigma, X)_k$ and combinator rewrite rule $r = t_1 \to t_2 \in R(E)$ and for some combinator occurrence $\bar{i} \in Occ^{com}(t)$,

$$t \xrightarrow{\bar{i}, r} t'.$$

Then for some substitution $\theta : X \to T_{pre}(\Sigma, X)$,

$$t(\bar{i}) = \overline{\theta}(t_1) \text{ and } t' = t(\bar{i} / \overline{\theta}(t_2)).$$

Now $t_1 = t_2 \in E$ so $(\varepsilon, E) \vdash t_1 = t_2$. Since t and t' are well typed with respect to $Alg(\Gamma, \varepsilon)$ then so are $\overline{\theta}(t_1)$ and $\overline{\theta}(t_2)$. Thus by substitution,

$$(\varepsilon, E) \vdash \overline{\theta}(t_1) = \overline{\theta}(t_2).$$

Then since t is well typed with respect to $Alg(\Gamma, \varepsilon)$, by reflexivity and substitution,

$$(\varepsilon, E) \vdash t = t(\bar{i} / \overline{\theta}(t_2)),$$

i.e. $(\varepsilon, E) \vdash t = t'$.

$\Rightarrow$ We prove the result by induction on the complexity of derivations.

(i) Consider any derivation of the form

$$\frac{}{(\varepsilon, E) \vdash t = t'}$$

where $t = t' \in E$. Then $t \to t' \in R(E)$ so trivially $t \overset{R(\varepsilon, E)^*}{\longrightarrow} t'$.

(ii) Consider any kind $k \in K'$, any preterm $t \in T_{pre}(\Sigma, X)_k$ and a derivation of the form

$$\frac{\varepsilon \vdash t : \tau_t}{(\varepsilon, E) \vdash t = t.}$$

Then by the reflexive closure property of $\overset{R(\varepsilon, E)^*}{\longrightarrow}$ we have $t \overset{R(\varepsilon, E)^*}{\longrightarrow} t$.

(iii) Consider any kind $k \in K'$ and any preterms $t_1, t_2 \in T_{pre}(\Sigma, X)_k$ and a derivation ending in a deduction of the form

$$\frac{(\varepsilon, E) \vdash t_1 = t_2}{(\varepsilon, E) \vdash t_2 = t_1.}$$

By the induction hypothesis $t_1 \overset{R(\varepsilon, E)^*}{\longrightarrow} t_2$. So by definition of $R(\varepsilon, E)$, $t_2 \overset{R(\varepsilon, E)^*}{\longrightarrow} t_1$.

(iv) Consider any kind $k \in K'$, any preterms $t_1, t_2, t_3 \in T_{pre}(\Sigma, X)_k$ and a derivation ending in a deduction of the form

$$\frac{(\varepsilon, E) \vdash t_1 = t_2, \quad (\varepsilon, E) \vdash t_2 = t_3}{(\varepsilon, E) \vdash t_1 = t_3.}$$

By the induction hypothesis $t_1 \overset{R(\varepsilon, E)^*}{\longrightarrow} t_2$ and $t_2 \overset{R(\varepsilon, E)^*}{\longrightarrow} t_3$. So by the transitive closure of $\overset{R(\varepsilon, E)^*}{\longrightarrow}$ we have $t_1 \overset{R(\varepsilon, E)^*}{\longrightarrow} t_3$.

(v) Consider any kind $k \in K'$, any preterms $t, t' \in T_{pre}(\Sigma, X)_k$, any kind $k' \in K'$, any preterms $t_1, t_2 \in T_{pre}(\Sigma, X)_{k'}$ any type $\tau \in T(\Gamma)_{k'}$, any variable $x \in X_\tau$ and a derivation ending in a deduction of the form

$$\frac{(\varepsilon, E) \vdash t = t', \quad (\varepsilon, E) \vdash t_1 = t_2, \quad \varepsilon \vdash t_1 : \tau}{(\varepsilon, E) \vdash t[x/t_1] = t'[x/t_2].}$$

By the induction hypothesis $t \overset{R(\varepsilon, E)^*}{\longrightarrow} t'$ and $t_1 \overset{R(\varepsilon, E)^*}{\longrightarrow} t_2$. So $t[x/t_1] \overset{R(\varepsilon, E)^*}{\longrightarrow} t'[x/t_1]$ and $t'[x/t_1] \overset{R(\varepsilon, E)^*}{\longrightarrow} t'[x/t_2]$. Hence by the transitive closure of $\overset{R(\varepsilon, E)^*}{\longrightarrow}$ we have $t[x/t_1] \overset{R(\varepsilon, E)^*}{\longrightarrow} t'[x/t_2]$.

(vi) Consider any $n \geq 0$, any kinds $k_0, \ldots, k_n \in K'$, any type terms $\tau_i \in T(\Gamma)_{k_i}$ for $1 \leq i \leq n$ and $\sigma, \tau \in T(\Gamma)_{k_0}$, any combinator operation symbol $f \in \Sigma_{\tau_1 \ldots \tau_n, \tau}$, any preterms $t_i \in T_{pre}(\Sigma, X)_{k_i}$ for $1 \leq i \leq n$ and a derivation ending in a deduction of the form

$$\frac{\varepsilon \vdash \tau = \sigma}{(\varepsilon, E) \vdash f^\tau(t_1, \ldots, t_n) = f^\sigma(t_1, \ldots, t_n).}$$

Since $\varepsilon \vdash \tau = \sigma$ then by Theorem 3.1, $\tau \overset{R(\varepsilon)^*}{\longrightarrow} \sigma$. Suppose

$$\tau_1 \overset{\overline{j_1}, \rho_1}{\longrightarrow} \tau_2 \overset{\overline{j_2}, \rho_2}{\longrightarrow} \ldots \overset{\overline{j_{m-1}}, \rho_{m-1}}{\longrightarrow} \tau_m$$

is a rewrite sequence with $\tau_1 = \tau$ and $\tau_m = \sigma$ and $\rho_1, \ldots, \rho_{m-1} \in R(\varepsilon)$. Then

$$f^{\tau_1}(t_1, \ldots, t_n) \overset{(\lambda, \overline{j_1}), \rho_1}{\longrightarrow} f^{\tau_2}(t_1, \ldots, t_n) \overset{(\lambda, \overline{j_2}), \rho_2}{\longrightarrow} \ldots \overset{(\lambda, \overline{j_{m-1}}), \rho_{m-1}}{\longrightarrow} f^{\tau_m}(t_1, \ldots, t_n)$$

is a rewrite sequence. So $f^\tau(t_1, \ldots, t_n) \overset{R(\varepsilon, E)^*}{\longrightarrow} f^\sigma(t_1, \ldots, t_n)$.

(vii) Consider any $k \in K'$, any types $\sigma, \tau \in T(\Gamma)_k$, any variable $x \in X_\tau$ and a derivation ending in a deduction of the form

$$\frac{\varepsilon \vdash \tau = \sigma}{(\varepsilon, E) \vdash x^\tau = x^\sigma.}$$

The proof that $x^{\tau} \overset{R(\varepsilon, E)^*}{\longrightarrow} x^{\sigma}$ is similar to (vi). □

5.2. Corollary. *Let Spec* $= ((\Gamma, \varepsilon), (\Sigma, E))$ *be an equational type specification. Suppose that each equation* $e \in E$ *is well typed with respect to* $Alg(\Gamma, \varepsilon)$*. For any* $k \in K'$ *and any ground preterms* $t, t' \in T_{pre}(\Sigma)_k$*, if* t *and* t' *are well typed with respect to* $Alg(\Gamma, \varepsilon)$ *then the following are equivalent:*

(i) $(\varepsilon, E) \vdash t = t'$;
(ii) $I(Spec) \models t = t'$;
(iii) $t \overset{R(\varepsilon, E)^*}{\longrightarrow} t'$.

Proof. Immediate from the Correspondence Theorem 5.1 and the Initiality Theorem 4.7. □

We thank J.R. Hindley, J.V. Tucker and E.G. Wagner for helpful comments on this work. We also acknowledge the financial support of the Science and Engineering Research Council, the British Council and IBM T.J. Watson Research Center.

REFERENCES

H. Ehrig and B. Mahr, Fundamentals of Algebraic Specification 1, Equations and Initial Semantics, EATCS Monographs on Theoretical Computer Science 6, Springer Verlag, Berlin, 1985.

K. Meinke, Universal algebra in higher types, Theoretical Computer Science, 100, (1992) 385–417.

K. Meinke, Subdirect representation of higher type algebras, Report CSR 14–90, Dept. of Computer Science, University College Swansea, to appear in: K. Meinke and J.V. Tucker (eds), Many–Sorted Logic and its Applications, John Wiley, 1992.

K. Meinke, A recursive second order initial algebra specification of primitive recursion, Report CSR 8–91, Dept. of Computer Science, University College Swansea, 1991a.

K. Meinke, Equational specification of abstract types and combinators, Report CSR 11–91, Dept. of Computer Science, University College Swansea, to appear in: G. Jäger (ed), Proc. Computer Science Logic '91, Lecture Notes in Computer Science 626, Springer Verlag, Berlin, 1991b.

K. Meinke and L.J. Steggles, Specification and verification in higher order algebra: a case study of convolution, Report CSR 16–92, Dept. of Computer Science, University College Swansea, 1992.

K. Meinke and J.V. Tucker, Universal algebra, 189–411 in: S. Abramsky, D. Gabbay and T.S.E. Maibaum, (eds) Handbook of Logic in Computer Science, Oxford University Press, Oxford, 1992.

K. Meinke and E. Wagner, Algebraic specification of types and combinators, IBM research report, in preparation, 1992.

B. Möller, Higher–order algebraic specifications, Facultät für Mathematik und Informatik, Technische Universität München, Habilitationsschrift, 1987.

B. Möller, A. Tarlecki and M. Wirsing, Algebraic specifications of reachable higher–order algebras, in: D. Sannella and A. Tarlecki (eds), Recent Trends in Data Type Specification, Lecture Notes in Computer Science 332,(Springer Verlag, Berlin, 1988)

154–169.
A. Poigné, On specifications, theories and models with higher types, Information and Control 68, (1986) 1–46.

Context Rewriting

Stefan Kahrs*

University of Edinburgh
Laboratory for Foundations of Computer Science
King's Buildings, EH9 3JZ
email: smk@dcs.ed.ac.uk

Abstract. The usual definition of context as "a term with a hole" does not properly address the problems of applying rewrite rules to a context and is in particular inadequate in several respects when a context contains variable bindings. We claim that viewing TRSs as (free) preorder-enriched categories provides a smoother concept of context: on the one hand it allows to decompose a reduction step into redex (+ reduction) and context, on the other it allows to rewrite a context.
By generalising Klop's combinatory reduction systems, we can approach the problem of contexts with variable bindings in a similar way. The rewrite relation of a combinatory reduction system can be defined as the preorder of a certain free preorder-enriched category.

1 Introduction

A context is a term with a "hole" at a distinguished position (see [2]). There is one operation working on contexts, it is to fill the hole with a term, which gives a new term. This is the intuitive, informal meaning of context. Looking at the literature about term rewriting systems and λ-calculus, it also seems to be its established formal meaning.

This rather naïve definition of context has several drawbacks — in particular for λ-calculus (the $\Box$ denotes the hole):

- Contexts are not compatible with α-congruence: Let $C[\,] = \lambda x.\Box$ and $D[\,] = \lambda y.\Box$. We have $C[\,] \equiv D[\,]$, but $C[x] \not\equiv D[x]$.
- Contexts are not compatible with β-reduction: Let $C[\,] = (\lambda x.\Box)y$ and $D[\,] = \Box$. We have $C[\,] \to_\beta D[\,]$, but $C[x] \not\to_\beta D[x]$.

Even for ordinary term rewriting, the notion of context with a single hole at a particular position is sometimes too restrictive. Toyama defined for his proof of the modularity of the Church-Rosser property (see [18]) contexts with many holes; but this is still somewhat unsatisfactory, since rewriting can copy, delete or permute the holes of a context. To overcome this problem, one needs nameful holes — variables. This reduces the problem of filling holes with terms to ordinary substitutions. One can extend the rewrite relation to substitutions (if they rewrite pointwise) and call a relation $>$ *compatible*, if $\sigma > \tau$ implies $t\sigma > t\tau$. The main reason why such

* The research reported here was partially supported by SERC grant GR/E 78463.

a method has not been used to define term decomposition for term rewriting is probably historical: λ-calculus is the older theory with similar definitions and for λ-calculus it does not quite work this way.

In the following, we shall generalise the notion of context, such that (i) contexts allow variable binding, (ii) contexts are closed against rewriting, where the rewriting system is a combinatory reduction system, and (iii) contexts are compatible with rewriting, roughly speaking: if $C \twoheadrightarrow D$ and $t \twoheadrightarrow u$, then $C[t] \twoheadrightarrow D[u]$.

2 Preliminaries

The motivation for this work comes from modelling reductions with enriched categories, in particular preorder-enriched categories and 2-categories, see [5, 14, 15, 16]. Since we cannot assume full familiarity with the concepts of category theory, we adapt the basic definition from [13]:

Definition 1. A *category* $\mathcal{C}$ comprises:

1. a collection of *objects*;
2. for any two objects A and B a collection of *morphisms* $\mathcal{C}(A, B)$; (we write $f : A \rightarrow B$ for $f \in \mathcal{C}(A, B)$)
3. a composition operator assigning to each pair of morphisms $f : A \rightarrow B$ and $g : B \rightarrow C$ a *composite* morphism $g \circ f : A \rightarrow C$, satisfying the following *associative law*:

$$\forall A, B, C, D. \forall f : A \rightarrow B,\ g : B \rightarrow C,\ h : C \rightarrow D.$$
$$h \circ (g \circ f) = (h \circ g) \circ f$$

4. for each object A, an *identity* morphism $id_A : A \rightarrow A$ satisfying the following *identity law*:

$$\forall A, B. \forall f : A \rightarrow B.$$
$$id_B \circ f = f \ \wedge\ f \circ id_A = f$$

The two "collection" are usually replaced by "class" (for the collection of objects) and "set" (for the morphisms). In enriched category theory (see [7]) one considers collections with a richer structure than just sets, e.g. preordered sets, or even categories. Typically one is interested in categories in which this additional structure is preserved by the composition of the category. We shall restrict our attention to the case of preorder-enriched categories (remember that a preorder is just a reflexive and transitive relation):

Definition 2. A *preorder-enriched* category $\mathcal{C}_{\twoheadrightarrow}$ comprises:

1. a category $\mathcal{C}$, called the *underlying* category;
2. for any two objects A and B in $\mathcal{C}$ a preorder $\twoheadrightarrow_{AB}$ on $\mathcal{C}(A, B)$, such that:

$$\forall A, B, C. \forall f, g : A \rightarrow B.\ f \twoheadrightarrow_{AB} g \Rightarrow$$
$$\forall h : B \rightarrow C.\ \ h \circ f \twoheadrightarrow_{AC} h \circ g \ \wedge$$
$$\forall k : C \rightarrow A.\ \ f \circ k \twoheadrightarrow_{CB} g \circ k$$

In words: a preorder-enriched category is a category in which the morphisms are preordered and composition of morphisms is monotonic in both arguments.

Notation: in the following we shall drop the index of the preorder, partly because it is clear from the context, partly because for the preorder-enriched categories we are going to investigate the following two properties hold: (i) if a pair of morphisms happens to belong to two different collections of morphisms, then it is either related by both preorders, or by none of them; (ii) if $f \twoheadrightarrow g$ and $g \twoheadrightarrow h$ hold in two different collections of morphisms, then there is a third collection including all three morphisms. Both conditions together guarantee that the union of all the preorders is a preorder.

3 Viewing TRSs as preorder-enriched categories

Preorder-enriched categories capture some part of the structure of TRSs, because they allow some kind of interaction between rewrite steps and term structure.

Definition 3. Given a first-order signature Σ, we define the category $T(\Sigma)$ of terms as follows:

1. objects are: sequences of distinct variables;
2. a morphism $f : A \rightarrow B$ is a sequence of terms, such that $\mathrm{Var}(f) \subseteq A$ and $\#f = \#B$;
3. given $f : A \rightarrow B$ and $g : B \rightarrow C$, the morphism $g \circ f$ is defined as: $g[f_i/B_i]$
4. $id_A = A$, the identity on A is the object A itself.

Notation: we write A_i for the i-th component of a sequence A; $\#B$ is the length of the sequence B; $g[f_i/B_i]$ denotes pointwise substitution, i.e. it is the sequence of terms obtained from g by substituting each occurrence of a variable from B by the corresponding term from f. One can easily check that this forms indeed a category. This means to check the following laws:

$$M[N_i/X_i][P_j/Y_j] = M[N_i[P_j/Y_j]/X_i]$$
$$M[X_i/X_i] = M$$
$$X[M_i/X_i] = M$$

The first equation is only true under a certain proviso, but this proviso is satisfied by the definition.

Rydeheard and Stell [15] define a similar category (called T_Ω) by using sets of variables as objects and substitutions as morphisms. This is basically the same, as there is an obvious forgetful functor $U : T(\Sigma) \rightarrow T_\Omega$ which "forgets" the order of the variable sequences.

Definition 4. Given a TRS (Σ, R), we define the preorder-enriched category $T(\Sigma)_{\twoheadrightarrow}$ as follows:

1. the underlying category is $T(\Sigma)$;
2. the preorder $\twoheadrightarrow$ is the pointwise extension of $\rightarrow_R^*$ to sequences: $f \twoheadrightarrow g \iff \forall i.\, 1 \leq i \leq \#f \Rightarrow f_i \rightarrow_R^* g_i$.

Again we can easily check that $T(\Sigma)_{\twoheadrightarrow}$ satisfies the required properties: $\rightarrow_R^*$ is a preorder by construction and the extension to sequences preserves this property. Monotonicity of composition translates to the following:

$$t \rightarrow_R^* u \Rightarrow t\sigma \rightarrow_R^* u\sigma$$
$$t_i \rightarrow_R^* u_i \Rightarrow p[t_i/x_i] \rightarrow_R^* p[u_i/x_i]$$

Both is true for arbitrary term rewriting systems — a relation satisfying the first property is usually called "substitutive" [1] or "stable" [4], the second property is a slight generalisation of "compatible" [4].

So $T(\Sigma)_{\twoheadrightarrow}$ is indeed a preorder-enriched category, but this is only half the story — the relation $\twoheadrightarrow$ is also the smallest substitutive and compatible preorder containing R, i.e. $T(\Sigma)_{\twoheadrightarrow}$ can be freely generated from R, or more precisely: freely generated from the extension of the reflexive closure of R to sequences[2]. In other words: $T(\Sigma)_{\twoheadrightarrow}$ is initial amongst all preorder-enriched categories $T(\Sigma)_{\preceq}$ where $\preceq$ contains R.

This means that we are not only able to compose rewrite steps, but also to decompose them. Each single rewrite step $t \rightarrow_R u$ can be decomposed as $C[l\sigma] \rightarrow_R C[r\sigma]$ such that $l \rightarrow r$ is a rewrite rule in R. In $T(\Sigma)$ we can express the context $C[\,]$ as well as the substitution σ as morphisms; instead of singletons l and r we may have to use sequences $\langle \_, x_1, \cdots, x_n \rangle$ where the x_i are variables and the $\_$ is either l or r — this is the case when the context $C[\,]$ contains variables. So the general picture of a rewrite step looks like this:

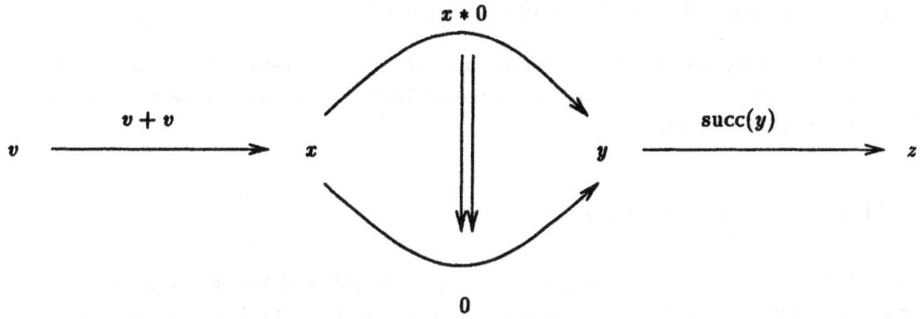

Fig. 1. Rewrite Step

The picture shows the decomposition of a rewrite step into substitution, rule and context: $\mathrm{succ}((v+v)*0) \rightarrow_R \mathrm{succ}(0)$ is decomposed into substitution $[v+v/x]$ (left), applied rewrite rule $x*0 \rightarrow 0$ (middle) and context $\mathrm{succ}(\Box)$ (right).

For first-order terms this is quite natural. But for λ-calculus, a corresponding definition, see [16], does not quite give the expected result. The problem is, that

[2] This observation is not new, see [15]. Rydeheard and Stell closed their relation also under coproducts, avoiding this "extension of ..." bit. Because we chose (roughly) the dual category, we would have to use *products* for a similar construction.

substitution in λ-calculus has to respect bound variables, in particular it implicitly renames bound variables to avoid name capture. If we replace in figure 1 the context morphism succ(y) for example by $\lambda x.y$ or $\lambda v.y$, then the entire rewrite step becomes $\lambda x'.((v+v)*0) \to \lambda x'.0$ for some fresh x'. The variables escape the dynamic binding.

As a consequence, it is not possible to decompose rewrite steps when bound variables are involved, e.g. $\lambda v.((v+v)*0) \to \lambda v.0$ is rather atomic. The corresponding preorder-enriched category (it is indeed one, see [16]) is not freely generated from the rules. The problem is that decomposing a term into context plus filling is not longer covered by decomposing it into term plus substitution.

To allow binding of variables by composition, it is necessary to provide more information in objects and/or morphisms. This can be achieved keeping ordinary (sequences of) terms as morphisms, as the author shows in [6]. But the construction there is rather tedious, since not using ordinary substitution for composition requires to re-invent (and re-prove) most substitution properties for this new composition.

Here we try a more straightforward approach: instead of having (sequences of) terms as morphisms, we use (sequences of) contexts. Since ordinary terms can be seen as nullary contexts, this is a proper generalisation. Composition of contexts is rather straightforward: if C and D are contexts, then $C \circ D$ is a context such that $(C \circ D)[t] = C[D[t]]$ for arbitrary terms t — this uniquely determines $C \circ D$. At least, this is the *intuitive* idea.

In the following, we shall *formalise* this idea. We need

- a formal notion of context, able to deal with bound variables,
- a notion of context composition,
- the extension of rewrite relations for such contexts,

such that rewriting and context composition fit nicely together (form a preorder-enriched category) and that terms and naïve contexts are in some sense special cases of the generalised contexts.

4 Higher Order Terms

To be able to treat bound variables properly, we adapt and generalise some definitions from Klop's Combinatory Reduction Systems (CRS), see [9]. Semantically the definition follows closely the Higher-Order Rewriting Systems (HRS) of Nipkow [12], but as we do not base it on λ-calculus but rather define HRSs directly, the style of the presentation is closer to CRSs. HRSs are — at least in the form defined here — a proper extension of TRSs, they are powerful enough to express β-reduction as a rewrite rule.

Definition 5. The set of types, *Typ*, is defined as the smallest set satisfying:

1. $O \in Typ$
2. $\alpha, \beta \in Typ \Rightarrow (\alpha \to \beta) \in Typ$

Typ comprises the types of simply typed λ-calculus. As usual, we drop many parentheses and take $\to$ to be right-associative.

Definition 6. *Ord*, $\#$: *Typ* $\to \mathcal{N}$ are the *order* and the *arity* of a type:

$$\begin{aligned} Ord\, O &= 1 \\ Ord(\alpha \to \beta) &= \max\{1 + Ord\,\alpha, Ord\,\beta\} \\ \#O &= 0 \\ \#(\alpha \to \beta) &= 1 + \#\beta \end{aligned}$$

Definition 7. A *signature* is a pair (*Sym*, *type*), where *Sym* is a set of symbols and *type* : *Sym* $\to$ *Typ* specifies the type of a symbol.

We assume the existence of a distinguished signature (*Var*, *type*), such that $type^{-1}(\tau)$ is an infinite set for every τ. We call the elements of *Var* *variables*.

Although those signatures assign a type to each symbol, we still think of them as *untyped* signatures, because there is only one elementary type O. An ordinary untyped TRS (or CRS) signature corresponds to the above one as follows: if the TRS signature assigns to F the arity n, then we have $type(F) = \tau$ with $\#\tau = n$ and $Ord\,\tau \leq 2$. This uniquely determines τ.

Definition 8. Given a signature $\Sigma = (Sym, type)$, the set of *terms* of a type τ and arity n, $Ter(\Sigma, \tau, n)$ is inductively defined over the alphabet $Var + Sym + \{[,]\}$ as follows:

1. $x \in Var \cup Sym \quad \Rightarrow \quad x \in Ter(\Sigma, type(x), \#type(x))$
2. $A \in Ter(\Sigma, \alpha \to \beta, n+1),\ B \in Ter(\Sigma, \alpha, 0) \quad \Rightarrow \quad AB \in Ter(\Sigma, \beta, n)$
3. $x \in Var,\ A \in Ter(\Sigma, \alpha, 0) \quad \Rightarrow \quad [x]A \in Ter(\Sigma, type(x) \to \alpha, 0)$

We call the terms of arity 0 *complete* and write $Ter(\Sigma, \tau)$ for $Ter(\Sigma, \tau, 0)$. We write $Ter(\Sigma)$ for the union of all $Ter(\Sigma, \tau, n)$.

Notation: in examples we shall add parentheses and commata to underline the structure of a complete term, i.e. we shall write $F(A_1, \cdots, A_n)$ for a complete term $FA_1 \cdots A_n$.

Complete terms as defined here can be seen as the η-expanded β-normal forms of simply typed λ-calculus, see [17]. They are β-normal forms, because in each term AB the subterm A has to be incomplete (of non-zero arity) and all abstractions are complete; they are η-expanded, because every subterm (of a complete term) of functional type is either complete and an abstraction, or incomplete but is applied to an argument, i.e. each further η-expansion would introduce a β-redex.

The following observations might help to understand the structure of terms and complete terms:

- For each term $t \in Ter(\Sigma, \alpha, n)$ we either have $\#\alpha = n$ or $n = 0$.
- For each (complete) term $t \in Ter(\Sigma, \alpha, 0)$ we either have $\alpha = O$ or that t is an abstraction $[x]t'$.
- Each complete term has the form $[x_1] \cdots [x_n]yt_1 \cdots t_k$ where $n \geq 0$ and y is either a variable or a function symbol, the type of which has arity k.
- Each incomplete term $t \in Ter(\Sigma, \alpha, n)$ has the form $yt_1 \cdots t_k$ where y is a variable or a function symbol, the type of which has arity $n + k$.

The metaterms of CRSs (as defined in [8]) are included in the above definition, i.e. they are expressible as complete terms, if we translate the signatures in the described way and add a distinguished symbol Λ of type $(O \rightarrow O) \rightarrow O$. The purpose of this symbol is to express CRS metaterms like $[x]t$ as $\Lambda[x]t$. Terms representing *CRS metaterms* are then restricted to use only bound variables of type O and free variables of at most order 2. We get the *CRS terms* by further restricting free variables to order 1 and *TRS terms* by further discarding the symbol Λ.

Instead of symbols with an arity, Klop has a built-in application for metaterms and symbols without arity. Functional CRSs have been introduced by Kennaway in [8]. We prefer this style, as it is more in the spirit of a generalised TRS definition. Functional CRSs relate to Klop's applicative CRSs, as ordinary TRSs do to applicative TRSs, see [10]. In this sense one could expect stronger modularity results for them, e.g. Church-Rosser might be a modular property of functional CRSs (given by disjoint union of signatures), but this goes beyond the scope of this paper.

Definition 9. FV : $Ter(\Sigma) \rightarrow \wp(Var)$ assigns each term its set of *free variables*:

$$\begin{aligned} \mathrm{FV}([v]t) &= \mathrm{FV}(t) \setminus \{v\} \\ \mathrm{FV}(AB) &= \mathrm{FV}(A) \cup \mathrm{FV}(B) \\ \mathrm{FV}(F) &= \emptyset, \quad F \in Sym \\ \mathrm{FV}(x) &= \{x\}, \quad x \in Var \end{aligned}$$

Definition 10. Given a term t and a variable x, we write $t =_\eta x$ if there is a sequence of distinct variables $y_1, \cdots, y_n$ $(n \geq 0)$, such that $t = [y_1] \cdots [y_n] x y_n \cdots y_1$.

We also define a notion of α-congruence for our terms. It is the usual one, but we shall use it in a slightly more general setting, based on proof rules.

Definition 11. *Sentences* are of the form $\Gamma \vdash t \equiv u$ or $\Gamma \vdash x = y$, where x and y are variables, t and u are terms of the same type and arity, and Γ is an *environment*[3]. An environment is a list $x_1 = y_1, \cdots, x_n = y_n$ of equations between variables. We write ϵ for the empty environment $(n = 0)$. A sentence *holds*, if it can be derived by the proof rules in figure 2.

The inequalities in the premise of the last rule of the second column (weakening) simply mean that v and x, and y and z are different variables. Notice that the sentence $\Gamma \vdash x = x$ does not necessarily hold, because weakening is restricted by these additional premises. For example $\epsilon \vdash [x]x \equiv [y]x$ does not hold, because for $x = y \vdash x = x$ weakening is not applicable.

Definition 12. Two complete terms t and u are *α-congruent* iff $\epsilon \vdash t \equiv u$.

Another difference to the definition of CRSs in [9, 8] is that we do not distinguish between variables (for abstraction, no arity) and metavariables (for rewrite rules). This makes the definition of substitution a bit compacter, because we do not need two notions of substitution.

[3] This is usually called a context in logic, but it would be confusing to overload this name here.

$$\frac{F \in Sym}{\Gamma \vdash F \equiv F} \qquad \frac{}{\epsilon \vdash x = x}$$

$$\frac{x, y \in Var \quad \Gamma \vdash x = y}{\Gamma \vdash x \equiv y} \qquad \frac{}{\Gamma, x = y \vdash x = y}$$

$$\frac{\Gamma, x = y \vdash t \equiv u}{\Gamma \vdash [x]t \equiv [y]u} \qquad \frac{v \neq x \quad y \neq z \quad \Gamma \vdash v = z}{\Gamma, x = y \vdash v = z}$$

$$\frac{\Gamma \vdash A \equiv C \quad \Gamma \vdash B \equiv D}{\Gamma \vdash AB \equiv CD}$$

Fig. 2. Proof rules for α-conguence

Definition 13. A *Σ-valuation* ρ is a partial mapping from variables to terms of the same type. We write Dom_ρ for the *domain* of ρ. Given a Σ-valuation ρ, its associated *substitution* $\widehat{\rho} : Ter(\Sigma) \rightarrow Ter(\Sigma)$ is defined as follows:

$$\begin{array}{lll}
\widehat{\rho}(x) & = \rho(x), & x \in Var,\ x \in Dom_\rho \\
\widehat{\rho}(x) & = x, & x \in Var,\ x \notin Dom_\rho \\
\widehat{\rho}(F) & = F, & F \in Sym \\
\widehat{\rho}([v]t) & = [z]\widehat{\rho'}(t), & type(v) = type(z),\ z \text{ fresh} \\
\widehat{\rho}(AB) & = \widehat{\rho}(A)\widehat{\rho}(B), & \widehat{\rho}(A) \in Ter(\Sigma, \tau, n),\ n > 0 \\
\widehat{\rho}(AB) & = \widehat{\sigma}(A'), & \widehat{\rho}(A) = [x]A',\ \sigma(x) = \widehat{\rho}(B),\ Dom_\sigma = \{x\}
\end{array}$$

In the fourth line, "z fresh" means:

$$\neg\exists y \in \mathrm{FV}([v]t).\ y \in Dom_\rho \wedge z \in \mathrm{FV}(\rho(y)) \ \vee\ y \notin Dom_\rho \wedge z = y$$

ρ' is defined as follows: $\rho'(v) = [x_1]\cdots[x_n]zx_1\cdots x_n$ where the x_i are distinct (and distinct from z) and $type(x_1) \rightarrow \cdots \rightarrow type(x_n) \rightarrow O = type(v)$; ρ' is as ρ on other variables.

Notation: we might omit the prefix Σ- if the signature is either clear from the context or irrelevant.

The fresh-ness of z avoids name capture by substitution. Different choices for z result in α-congruent substitution results — we could make choice and result unique by requiring a linear order on *Var* and then by choosing v if it is fresh and otherwise by taking the first fresh variable according to that order. This corresponds to Curry's definition of substitution for λ-calculus terms, see appendix C in [1].

Substitutions as defined above generalise the notion of substitution for CRS metaterms as we do not restrict the substituted variables to be at most second-order and as we also do not restrict the substitutes of variables not to contain higher-order variables. In particular the second restriction *need to* be dropped for our approach.

One could question whether this generalised notion of substitution is well-defined, as the application of a higher-order substitution to a variable may invoke further higher-order substitutions (see last equation of the definition). It is indeed well-defined, because if the substitution of an n-th order variable x invokes the substitution of another variable x', then x' has at most order $n-1$. One can think of the invoked substitutions as β-reductions, executed on the fly — β-reduction for simply typed λ-calculus is strongly normalising.

5 Higher-Order Reduction Systems

Based on these notions of term and substitution we can define a generalised notion of rewrite system.

Definition 14. A complete term t is called *simple* with respect to a set of variables U in the following cases: $[x]t$ is simple w.r.t. U, iff t is w.r.t. $U \cup \{x\}$; $FA_1 \cdots A_n$ (for $F \in Sym$) is simple w.r.t. U, if each A_i is simple w.r.t. U; $xA_1 \cdots A_n$ (for $x \in Var$) is simple w.r.t. U, if either $x \in U$ and each A_i is simple w.r.t. U, or if $x \notin U$ and for each A_i we have $A_i =_\eta y_i$ for some $y_i \in U$, such that for each $1 \leq j \leq n$ we have $y_i = y_j \iff i = j$.

Simple terms (w.r.t. $\emptyset$) have certain nice properties, e.g. unification is decidable and solutions to matching or unification problems are unique, see [11].

Definition 15. Given a signature $\Sigma = (Sym, \#)$, a *Σ-HRS-rule* $l \rightarrow r$ consists of a pair of complete terms $l, r \in Ter(\Sigma)$, such that:

1. $\mathrm{FV}(l) \supseteq \mathrm{FV}(r)$,
2. l is simple w.r.t. $\emptyset$,
3. $l \notin Var$,
4. $l, r \in Ter(\Sigma, O)$.

The first two conditions imply that $\widehat{\rho}(r)$ never contains more free variables than $\widehat{\rho}(l)$. The second and third condition exclude l to be of the form $xA_1 \cdots A_n$ for a variable x, $n \leq 0$. The last condition restricts l and r to be of type O, i.e. we do not allow to rewrite abstractions. Some arguments in favour of this restriction are given in [12].

The fact that only O is allowed as type of sides of a rule indicates that the system is still untyped — the types live only on meta-level. One could easily adapt terms and rules to a typed scenario by replacing O by type expressions of some type system, not necessarily simply typed λ-calculus. One could also use a more fancy type system for the meta-level, but in any case it is advisable not to confuse the type systems of object-level and meta-level.

Definition 16. A *Higher-Order Rewrite System* (short: HRS) is a pair (Σ, R) consisting of a signature Σ and a set of Σ-HRS-rules R. We define $\rightarrow_R$ as a relation on complete terms to be the compatible[4] closure of: $t \rightarrow_R u$, if $\epsilon \vdash t \equiv \widehat{\rho}(l)$ and $\epsilon \vdash u \equiv \widehat{\rho}(r)$ for some Σ-HRS-rule $l \rightarrow r \in R$ and some Σ-valuation ρ.

[4] As usual, a relation $>$ on complete terms is compatible, iff it is continued on complete subterms, e.g. $t_i > t_i' \Rightarrow Ft_1 \cdots t_{i-1}t_it_{i+1} \cdots t_n > Ft_1 \cdots t_{i-1}t_i't_{i+1} \cdots t_n$ etc.

Klop and Kennaway define such a relation $\rightarrow_R$ only on *CRS terms*, because their substitutions map CRS metaterms to CRS terms. For our purposes we shall need this more general definition.

Example: An HRS for $\beta\eta$-reduction of untyped λ-calculus is defined as follows: ($Sym = \{A, L\}$; $type(A) = O \rightarrow O \rightarrow O$, $type(L) = (O \rightarrow O) \rightarrow O$, $type(m) = O \rightarrow O$, $type(x) = type(y) = O$)

$$A(L([x]m(x)), y) \rightarrow m(y)$$
$$L([x]A(y, x)) \quad \rightarrow y$$

The first rule is β-, the second η-reduction. Note that the usual condition, that y in the η-redex $L([x]A(y, x))$ must not contain x freely, is here implicit — because substitution avoids name capture — and explicit as well, since (the type of) the variable y has arity 0 and cannot carry bound variables as m does in the β-rule.

The example is expressible as a CRS, because at most second-order variables are involved. Here is another (admittedly artificial) example which is not expressible as a CRS, because it uses a third-order variable and a fourth-order function symbol: $Sym = \{C, F, G\}$;
$type(C) = O$, $type(F) = O \rightarrow O \rightarrow O$, $type(G) = ((O \rightarrow O) \rightarrow O) \rightarrow O$,
$type(z) = O$, $type(x) = O \rightarrow O$, $type(y) = (O \rightarrow O) \rightarrow O$

The only rewrite rule is:

$$G([x]y([z]x(z))) \rightarrow y([z]F(z, z))$$

Using the HRS above, we can rewrite the term

$$G([x]F(x(C), x(x(z))))$$

in the empty context, using the valuation $\rho(y) = [x]F(x(C), x(x(z)))$, to

$$F(F(C, C), F(F(z, z), F(z, z)))$$

6 Contexts

We can use terms and substitutions to define contexts and their compositions. A context is simply a complete term that may contain certain free variables. In the following, we assume a fixed signature Σ.

We divide the set of variables *Var* into two disjoint subsets Var_0 and Var_1 that are isomorphic, i.e. there is a type-preserving bijection between them.

Definition 17. The category of contexts, **Cont**, has as objects pairs consisting of a variable $m \in Var_1$ and a sequence of distinct variables $x_1 \cdots x_k \in Var_0^*$, such that $type(m) = type(x_1) \rightarrow \cdots \rightarrow type(x_k) \rightarrow O$. In the following we abbreviate such a sequence as $\mathbf{x}$, such an object as $m\mathbf{x}$ and a nested abstraction $[x_1] \cdots [x_k]t$ as $[\mathbf{x}]t$. A morphism from $m\mathbf{x}$ to $n\mathbf{y}$ is a term t of type O, such that $\mathrm{FV}(t) \subseteq \{m, y_1, \cdots, y_{\#n}\}$.

Given two morphisms $t : m\mathbf{x} \rightarrow n\mathbf{y}$ and $u : n\mathbf{y} \rightarrow p\mathbf{z}$, the composite $u \circ t$ is the (complete) term $\widehat{\rho}(u)$, where the valuation ρ is defined as: $\rho(n) = [\mathbf{y}]t$.

For each object $A = mx_1 \cdots x_n$, the identity id_A is the term $mt_1 \cdots t_n$ such that $t_i =_\eta x_i$.

The definition of **Cont** contains an implicit theorem:

Theorem. **Cont** *forms a category, i.e. composition is associative and the identities are its neutral elements.*

Proof. (we only show associativity) $\Lambda\Sigma$ is the set of λ-calculus terms including symbols from Σ. We define an injective mapping $\overline{\cdot}: Ter(\Sigma) \to \Lambda\Sigma$ as follows:

$$\overline{[x]t} = (\lambda x \overline{t})$$
$$\overline{A\,B} = (\overline{A}\,\overline{B})$$
$$\overline{F} = F,\ F \in Sym$$
$$\overline{x} = x,\ x \in Var$$

We extend this mapping to valuations by applying the translation pointwise. By choosing ordinary substitution for valuation application in λ-calculus, we can state the following law: $\overline{\widehat{\rho}(t)} =_\beta \widehat{\overline{\rho}}(\overline{t})$, because the only difference between the two substitutions is that on HRS terms the β-redexes introduced by substitution are immediately contracted. Writing $f \bullet g$ for composing λ-calculus terms f and g by (ordinary) substitution and $f \circ g$ for composition in **Cont**, we can now derive:

$$\overline{x \circ (y \circ z)} =_\beta \overline{x} \bullet \overline{y \circ z} =_\beta \overline{x} \bullet (\overline{y} \bullet \overline{z}) \equiv (\overline{x} \bullet \overline{y}) \bullet \overline{z} =_\beta \overline{x \circ y} \bullet \overline{z} =_\beta \overline{(x \circ y) \circ z}$$

The above derivation exploited the mentioned law (four times), substitutivity and compatibility of $=_\beta$, and associativity of ordinary substitution composition. So we know $\overline{x \circ (y \circ z)} =_\beta \overline{(x \circ y) \circ z}$ — but in the image of $\overline{\cdot}$ are only β-normal-forms. As β-reduction is confluent, β-normal forms are unique, and as $\overline{\cdot}$ is injective, we finally have $x \circ (y \circ z) = (x \circ y) \circ z$. □

To model rewriting in **Cont**, we add a preorder corresponding to a given HRS:

Definition 18. Given an HRS (Σ, R), the preorder-enriched category Cont_R has **Cont** as its underlying category. For any two morphisms (terms) t and u, we have $t \twoheadrightarrow u$ iff $t \to_R^* u$.

Again we have an implicit theorem: Cont_R forms a preorder-enriched category for any HRS (Σ, R).

The monotonicity of composition means that $\widehat{\rho}(c) \to_R^* \widehat{\varrho}(d)$ holds, provided that $a \to_R^* b, c \to_R^* d$ and $\rho(m) = [x]a$ and $\varrho(m) = [x]b$. In words: if two substitutions rewrite pointwise and on their argument, then they also rewrite on their results.

7 Multi-Contexts

One interesting thing about modelling reduction with preorder-enriched categories is that they provide an alternative way of defining a rewrite relation. Instead of using the full $\to_R^*$ as preorder between morphisms, one constructs the free preorder-enriched category on top of the rules themselves.

This cannot work with our category **Cont**, because not all terms (of type O) are allowed as morphisms. In particular, any morphism in **Cont** can contain at most one variable from Var_1. To allow arbitrary terms as morphisms, we extend **Cont** as follows:

Definition 19. The category of multi-contexts, **MultiCont**, has as objects tuples of objects of **Cont**. Each object has the form $(m_1\mathbf{x}_1, \cdots, m_a\mathbf{x}_a)$ for some $a \geq 0$, where each $\mathbf{x}_i$ has the form $x_{i1} \cdots x_{ir_i}, r_i = \#m_i$. We additionally require $i \neq j \Rightarrow m_i \neq m_j$.

A morphism from $(m_1\mathbf{x}_1, \cdots, m_a\mathbf{x}_a)$ to $(n_1\mathbf{y}_1, \cdots, n_b\mathbf{y}_b)$ is a b-tuple of terms (of type O) $(t_1, \cdots, t_b)$, such that $\mathrm{FV}(t_i) \subseteq \{m_1, \cdots, m_a, y_{i1}, \cdots, y_{ir_i}\}$.

Given two morphisms $\mathbf{t} = (t_1, \cdots, t_a) : (m_1\mathbf{x}_1, \cdots, m_c\mathbf{x}_c) \to (n_1\mathbf{y}_1, \cdots, n_a\mathbf{y}_a)$ and $\mathbf{u} = (u_1, \cdots, u_b) : (n_1\mathbf{y}_1, \cdots, n_a\mathbf{y}_a) \to (p_1\mathbf{z}_1, \cdots, p_b\mathbf{z}_b)$ the composite $\mathbf{u} \circ \mathbf{t}$ is the b-tuple $(\widehat{\rho}(u_1), \cdots, \widehat{\rho}(u_b))$, where the valuation ρ is defined as $\rho(n_i) = [\mathbf{y}_i]t_i$.

For each object $A = (A_1, \cdots, A_a)$, the identity on A is the tuple of the corresponding identities in **Cont**, i.e. $(id_{A_1}, \cdots, id_{A_a})$.

We have again the implicit theorem that **MultiCont** forms indeed a category. Obviously, **Cont** is a full subcategory of **MultiCont**.

We can extend **MultiCont** to a preorder-enriched category in the same way as we did for **Cont**, but we have an alternative — we can define the free preorder-enriched category generated from **MultiCont** and the rewrite rules R of a HRS.

For ordinary TRSs, i.e. if we look at the subcategory of **MultiCont** with first-order terms and only 0-ary variables, this construction gives the same category and the same rewrite relation (provided the variables used in the rules are all in Var_1).

For full **MultiCont** this is not true. The problem is that a variable from Var_0 cannot be replaced (by substitution) for a variable from Var_1 within **MultiCont**. For example, if R is the rule $x * 0 \to 0$, where $x \in Var_1$, then we cannot deduce $z * 0 \to 0$ for $z \in Var_0$. For this reason, we close the relation also under renaming of *free* variables.

Definition 20. Given an HRS (Σ, R), MultiCont$_R$ has **MultiCont** as its underlying category. Its preorder is the smallest relation satisfying the inference rules in figure 3. (The objects are arbitrary, but we require that the compositions are well-formed.)

$$\frac{}{t \twoheadrightarrow t} \qquad \frac{t \twoheadrightarrow u \quad \Gamma \vdash t \equiv t' \quad \Gamma \vdash u \equiv u'}{t' \twoheadrightarrow u'}$$

$$\frac{s \twoheadrightarrow t \quad t \twoheadrightarrow u}{s \twoheadrightarrow u} \qquad \frac{r \twoheadrightarrow s \quad t \twoheadrightarrow u}{r \circ t \twoheadrightarrow s \circ u}$$

$$\frac{t \to u \in R}{t \twoheadrightarrow u} \qquad \frac{t_1 \twoheadrightarrow u_1 \quad \cdots \quad t_n \twoheadrightarrow u_n}{(t_1, \cdots, t_n) \twoheadrightarrow (u_1, \cdots, u_n)}$$

Fig. 3. Proof rules for $\twoheadrightarrow$ in MultiCont$_R$

The preorder-enriched category MultiCont$_R$ is now freely constructed. We do not have an implicit theorem again as for Cont$_R$ — MultiCont$_R$ is obviously preorder-enriched by construction, but we have an explicit one:

Theorem. *Let (Σ, R) be an HRS. For any two terms $t, u \in Ter(\Sigma, O)$ we have $t \rightarrow_R^* u$, if and only if $t \twoheadrightarrow u$ in* $\mathrm{MultiCont}_R$.

Proof. ($\Leftarrow$) We have to prove that $\rightarrow_R^*$ admits monotonicity and renaming. It is monotone, because it is substitutive (composition from the right) and compatible (special case of composition from the left). It admits renaming, because we can easily translate each environment Γ into a valuation γ, such that $\Gamma \vdash t \equiv t'$ (for complete terms t, t') if and only if $\widehat{\gamma}(t) = t'$.

($\Rightarrow$) The substitutivity of $\twoheadrightarrow$ seems to be obvious, but it is not quite. Suppose we have $t \twoheadrightarrow u$ and let ρ be an arbitrary valuation with a finite domain containing the free variables of t and u. Let ρ map the variables m_i, $1 \leq i \leq k$ to complete terms $s_i = [x_{i1}] \cdots [x_{ir_i}] b_i$. We can express ρ as a morphism $(b_1, \cdots, b_k)$ from some object containing the free variables of the s_i to $m_1 x_1 \cdots m_k x_k$, provided all free variables in s_i are in Var_1 and the x_{ij} are in Var_0. In that case, we can derive $\widehat{\rho}(t) \twoheadrightarrow \widehat{\rho}(u)$ by monotonicity. Otherwise, we choose another substitution ρ' with the same domain as ρ, $\rho'(m_i) = s_i'$, such that no s_i' contains a variable from Var_0 freely and for some environment Γ we have $\Gamma \vdash s_i \equiv s_i'$ for all i — in other words: we replace the variables from Var_0 one-to-one by fresh variables from Var_1. As we have $\Gamma \vdash \widehat{\rho'}(t) \equiv \widehat{\rho}(t)$ (similar for u), we can apply renaming and monotonicity and derive $\widehat{\rho}(t) \twoheadrightarrow \widehat{\rho}(u)$.

It remains to prove that $\twoheadrightarrow$ is compatible. Suppose we have $C[t] \rightarrow_R^* C[u]$ and $t \rightarrow_R^* u$, where $C[\,]$ is a context (in the usual sense), i.e. a term with a hole at a particular position. We can assume $t \twoheadrightarrow u$ and have to prove $C[t] \twoheadrightarrow C[u]$. For simplicity we also assume that $C[\,]$ does not contain any free variables.

Let $\{x_1, \cdots, x_n\}$ be the set of variables bound at the occurrence of $\square$ in $C[\,]$. Let $\{y_1, \cdots, y_n\} \subset Var_0$ be a set of variables not occurring in $C[\,]$, t, or u, such that $type(x_i) = type(y_i)$. Let $t' = t[y_i/x_i]$ and $u' = u[y_i/x_i]$. Let Γ be the environment $x_1 = y_1, \cdots, x_n = y_n$. We have $\Gamma \vdash t \equiv t'$ and $\Gamma \vdash u \equiv u'$, by assumption $t \twoheadrightarrow u$ and hence by renaming $t' \twoheadrightarrow u'$. Let $v \in Var_1$ be a variable of type $type(y_1) \rightarrow \cdots \rightarrow type(y_n) \rightarrow O$. We define a term c as $C[v x_1 \cdots x_n]$. Let Y be the object $v y_1 \cdots y_n$ in **MultiCont**. Let Z be an arbitrary singleton object in **MultiCont**. Clearly we have: $c : Y \rightarrow Z$. Let $X = (p_1 z_1 \cdots p_r z_r)$ be an object in **MultiCont** such that $\mathrm{FV}(t') \cup \mathrm{FV}(u') \subseteq \{p_1, \cdots, p_r, y_1, \cdots, y_n\}$. We have: $t', u' : X \rightarrow Y$.

Since the composition assigns to v the complete term $[y_1] \cdots [y_n] t'$ we get $t' \circ c = \widehat{\sigma}(c) = C[\widehat{\rho}(t')]$ where ρ maps each y_i to the corresponding x_i (η-expanded), hence $\widehat{\rho}(t') = t'[x'/y'] = t[y'/x'][x'/y'] = t$ and $t' \circ c = C[t]$. Notice that the application of σ to c does not change any bound variable: v is the only free variable in c, the bound variables at its occurrence are exactly $\{x_1, \cdots, x_n\}$ and none of them can occur freely in the substitute for v, $[y_1] \cdots [y_n] t'$ as they do not even occur freely in t'. Now we have $t' \circ c = C[t]$, by the same argument $u' \circ c = C[u]$, by the above argument $t' \twoheadrightarrow u'$ and can thus derive $C[t] \twoheadrightarrow C[u]$ by monotonicity. $\square$

Without the assumption that $C[\,]$ is closed (in the last case) the proof becomes less smooth: we have to add the k extra free variables of $C[\,]$ to the objects X and Y and the morphisms between X and Y are not longer t (and u) but $k+1$-tuples of the form $(t, o_1, \cdots, o_k)$, where the o_i are the identities on the corresponding object components.

8 Conclusion and Future Work

We showed how to model higher-order rewriting with enriched category theory. Destructuring a rewrite step into context, rule, and substitution has an analogous decomposition in the category **MultiCont**.

Therefore, contexts can be seen as morphisms in **MultiCont**. This notion of context allows many holes and many occurrences of a particular hole. It supports variable binding. As **MultiCont** is preorder-enriched, we also have the desired property that $C[\,] \twoheadrightarrow D[\,]$ and $t_i \twoheadrightarrow u_i$, $1 \leq i \leq n$ imply: $C[t_1, \cdots, t_n] \twoheadrightarrow D[u_1, \cdots, u_n]$.

Not everything in this approach is as elegantly presented as it ought to be, especially the rôle of renaming and the two disjoint variable sets. It should certainly be possible to replace this by a more elegant construction.

For several purposes, preorder-enriched categories provide not enough structure for modelling TRS or HRS. In particular, they ignore the redexes involved in a reduction, all reduction sequences between two terms are identified. For example, to express and prove in a categorical setting properties like CR^+ [1] for orthogonal TRSs, or the modularity of CR for arbitrary TRSs [18], one has to keep track of redexes. The appropriate structure for such a task seems to be a 2-category. There might be a connection to the explicit reductions introduced by Hindley to study the Church-Rosser property on an abstract level, see [3, 1].

References

1. Hendrik P. Barendregt. *The Lambda-Calculus, its Syntax and Semantics.* North-Holland, 1984.
2. Nachum Dershowitz and Jean-Pierre Jouannoud. Rewrite systems. In J. van Leeuwen, editor, *Handbook of Theoretical Computer Science*, chapter 6, pages 244–320. Elsevier Science Publishers, 1990.
3. J. R. Hindley. An Abstract Form of the Church-Rosser Theorem I. *Journal of Symbolic Logic*, 34(4):545–560, December 1969.
4. Gérard Huet. Confluent reductions: Abstract properties and applications to term rewriting systems. *Journal of the ACM*, 27:797–821, 1980.
5. C. Barry Jay. Modelling reduction in confluent categories. Technical Report ECS-LFCS-91-187, University of Edinburgh, LFCS, December 1991.
6. Stefan Kahrs. *λ-rewriting.* PhD thesis, Universität Bremen, 1991. (in German).
7. G. M. Kelly. *Basic Concepts of Enriched Category Theory*, volume 64 of *London Mathematical Society Lecture Notes Series.* Cambridge University Press, 1982.
8. J.R. Kennaway. Sequential evaluation strategies for parallel-or and related reduction systems. *Annals of Pure and Applied Logic*, 43:31–56, 1989.
9. Jan Willem Klop. *Combinatory Reduction Systems.* PhD thesis, Centrum voor Wiskunde en Informatica, 1980.
10. Jan Willem Klop. Term rewriting systems, a tutorial. *EATCS bulletin*, 32:143–183, 1987.
11. Dale Miller. A logic programming language with lambda-abstraction, function variables, and simple unification. In *LNCS 475*, pages 253–281, 1991.
12. Tobias Nipkow. Higher order critical pairs. In *Sixth Annual Symposium on Logic in Computer Science, Amsterdam*, 1991.
13. Benjamin C. Pierce. *Basic Category Theory for Computer Scientists.* MIT Press, 1991.

14. A. John Power. An abstract formulation for rewrite systems. In *Category Theory and Computer Science*, 1989. LNCS 389.
15. D. E. Rydeheard and J. G. Stell. Foundations of equational deduction: A categorical treatment of equational proofs and unification algorithms. In *Proceedings Category Theory and Computer Science*, pages 114–139. Springer, 1987. LNCS 283.
16. R. A. G. Seely. Modelling computations: A 2-categorical framework. In *Proceedings of the Second Annual Symposium on Logic in Computer Science*, pages 65–71, 1987.
17. W. Snyder and J. Gallier. Higher-order unification revisited: Complete sets of transformations. *Journal of Symbolic Computation*, 8:101–140, 1989.
18. Yoshihito Toyama. On the Chuch-Rosser property for the direct sum of term rewriting systems. *Journal of the ACM*, 34(1):128–143, 1987.

Explicit Cyclic Substitutions

Kristoffer Høgsbro Rose

DIKU, University of Copenhagen,
Universitetsparken 1, 2100 København Ø, Denmark
Internet: kris@diku.dk

Abstract. In this paper we consider rewrite systems that describe the λ-calculus enriched with recursive and non-recursive local definitions by generalizing the 'explicit substitutions' used by Abadi, Cardelli, Curien, and Lévy [1] to describe sharing in λ-terms. This leads to 'explicit cyclic substitutions' that can describe the mutual sharing of local recursive definitions. We demonstrate how this may be used to describe standard binding constructions (**let** and **letrec**)—directly using substitution and fixed point induction as well as using 'small-step' rewriting semantics where substitution is interleaved with the mechanics of the following β-reductions.

With this we hope to contribute to the synthesis of denotational and operational specifications of sharing and recursion.

1 Introduction

Most functional programming languages include some form of 'local binding' that makes sharing explicit, *e.g.*, the expression $(2^3 + 3^2) \times (2^3 - 3^2)$ can be computed by the following expression in a generic functional language with explicit sharing of the subexpressions 2^3 and 3^2:

$$\textbf{let } a = 2^3 \textbf{ and } b = 3^2 \textbf{ in } (a+b) \times (a-b)$$

It is clear that the two expressions have the same denotation since they compute the same value. But if we wish to give the two expressions a *rewrite semantics* in the form of a rewrite system describing how to obtain this value *operationally*, then the sharing matters because the two expressions rewrite in a different number of steps due to the duplication of computation in the first. The unshared version may rewrite in the following seven steps (where we underline the next

Supported by the Danish Research Council (SNF) under project DART.

subexpression to be reduced):

$$\begin{aligned}
(\underline{2^3} + 3^2) \times (2^3 - 3^2) &\Rightarrow (8 + \underline{3^2}) \times (2^3 - 3^2) \\
&\Rightarrow \underline{(8+9)} \times (2^3 - 3^2) \\
&\Rightarrow 17 \times (\underline{2^3} - 3^2) \\
&\Rightarrow 17 \times (8 - \underline{3^2}) \\
&\Rightarrow 17 \times \underline{(8-9)} \\
&\Rightarrow \underline{17 \times (-1)} \\
&\Rightarrow -17
\end{aligned}$$

But the shared version rewrites in five steps:

$$\begin{aligned}
&\textbf{let } a = \underline{2^3} \textbf{ and } b = 3^2 \textbf{ in } (a+b) \times (a-b) \\
&\quad \Rightarrow \textbf{let } a = 8 \textbf{ and } b = \underline{3^2} \textbf{ in } (a+b) \times (a-b) \\
&\quad \Rightarrow \textbf{let } a = 8 \textbf{ and } b = 9 \textbf{ in } \underline{(a+b)} \times (a-b) \\
&\quad \Rightarrow \textbf{let } a = 8 \textbf{ and } b = 9 \textbf{ in } 17 \times \underline{(a-b)} \\
&\quad \Rightarrow \textbf{let } a = 8 \textbf{ and } b = 9 \textbf{ in } \underline{17 \times (-1)} \\
&\quad \Rightarrow -17
\end{aligned}$$

The situation gets even more severe for *cyclic sharing*, *e.g.*, the following expression computes the parity of 37:

$$\begin{array}{lll}
\textbf{letrec} & even(n) & = \textbf{if } n = 0 \textbf{ then } 1 \textbf{ else } odd(n-1) \\
\textbf{and} & odd(n) & = \textbf{if } n = 0 \textbf{ then } 0 \textbf{ else } even(n-1) \\
\textbf{in } odd(37) & &
\end{array}$$

The only non-cyclic representation of this is the complete 'unfolding' of the problem into an infinite expression like the one shown in the box below.

$$\begin{array}{l}
\textbf{let}\quad even(n) = \textbf{if } n = 0 \textbf{ then } 1 \\
\qquad\qquad\qquad \textbf{else let}\quad even(n) = \textbf{if } n = 0 \textbf{ then } 1 \\
\qquad\qquad\qquad\qquad\qquad\qquad\qquad \textbf{else let}\quad even(n) = \cdots \\
\qquad\qquad\qquad\qquad\qquad\qquad\qquad\qquad \textbf{and } odd(n) = \cdots \\
\qquad\qquad\qquad\qquad\qquad\qquad\qquad\qquad \textbf{in } odd(n-1) \\
\qquad\qquad\qquad\qquad \textbf{and } odd(n) = \textbf{if } n = 0 \textbf{ then } 0 \\
\qquad\qquad\qquad\qquad\qquad\qquad\qquad \textbf{else let}\quad even(n) = \cdots \\
\qquad\qquad\qquad\qquad\qquad\qquad\qquad\qquad \textbf{and } odd(n) = \cdots \\
\qquad\qquad\qquad\qquad\qquad\qquad\qquad\qquad \textbf{in } even(n-1) \\
\qquad\qquad\qquad\qquad \textbf{in } odd(n-1) \\
\textbf{and}\quad odd(n) = \textbf{if } n = 0 \textbf{ then } 0 \\
\qquad\qquad\qquad \textbf{else let}\quad even(n) = \textbf{if } n = 0 \textbf{ then } 1 \\
\qquad\qquad\qquad\qquad\qquad\qquad\qquad \textbf{else let}\quad even(n) = \cdots \\
\qquad\qquad\qquad\qquad\qquad\qquad\qquad\qquad \textbf{and } odd(n) = \cdots \\
\qquad\qquad\qquad\qquad\qquad\qquad\qquad\qquad \textbf{in } odd(n-1) \\
\qquad\qquad\qquad\qquad \textbf{and } odd(n) = \textbf{if } n = 0 \textbf{ then } 0 \\
\qquad\qquad\qquad\qquad\qquad\qquad\qquad \textbf{else let}\quad even(n) = \cdots \\
\qquad\qquad\qquad\qquad\qquad\qquad\qquad\qquad \textbf{and } odd(n) = \cdots \\
\qquad\qquad\qquad\qquad\qquad\qquad\qquad\qquad \textbf{in } even(n-1) \\
\qquad\qquad\qquad\qquad \textbf{in } even(n-1) \\
\textbf{in } odd(37)
\end{array}$$

The traditional way to handle such infinite specifications is by least fixed point induction with which it is quite easy to see that the parity of 37 can be computed by unfolding 'just' 37 times—this is interesting because 37 is, after all, much less than infinity! But from a rewriting point of view this kind of argument is unsatisfactory because the unfolding clearly does not preserve sharing: we effectively *copy* the repeated piece of code an unlimited number of times—thus it is not possible to see what parts of the resulting term stems from the same piece of the original. Again—this may be a very useful property in denotational semantics where the goal is to identify algorithms with the same denotation but if we wish to study the rewrite system then we need insight into the mechanics of unfolding.

Thus we feel that there is a need for describing the relation between sharing in the forms discussed above and traditional ways of describing the meaning of such expressions. We will investigate this in the archetypical setting of 'enriched λ-calculi:' we will describe how the λ-calculus extended with explicit notation for the two kinds of sharing is related to the traditional λ-calculus where sharing is described implicitly using substitution and fixed point combinators.

Relation to other work

This work grew out of work in my thesis [22, part I] on generalizing the "Explicit Substitutions" rewrite system of Abadi, Cardelli, Curien, and Lévy [1] to handle cyclic structures. The chosen approach is similar to the "enriched λ-calculus" used to describe and implement functional programming languages by, *e.g.*, Peyton Jones [19], and to the "λ_B-calculus" graph reduction system of Ariola and Arvind [3] in that they intend to support mutual recursion directly. However, it differs in that there is no implied underlying 'execution model:' the description presented here is a pure, self-contained rewriting system. The work described here has been inspired by the 'substitution pushing' used by the "Call-by-Mix" strategy of Grue [10]. The current trend of 'optimal λ-reduction' also touches on sharing although only as a method to minimise the number of reductions [17, 16, 9, 18]

If fact a lot of the work done within **term graph rewriting** [5] seems to be aiming at achieving the same results as we are: to provide a convenient yet formal specification method that makes it possible to reason about rather than abstract away from sharing and recursion; *cf.* [12, 11, 14, 15, 25, 26, 27] and the work of this author [13, 23, 22, part II].

A different approach altogether is to consider **letrec** terms as *regular* terms in the sense of Courcelle [7] since such infinite terms have a finite number of distinct subterms—as we might expect from the fact that it is derived from a finite term.

Structure of the Paper

In §2 we summarize the λ-calculus as it is usually described using an implicit notion of substitution, we define acyclic and cyclic simultaneous substitutions in this tradition, and we extend standard results to these. In §3 we then discuss **let** and the relation between interpretations based on simulation through the λ-calculus and using explicit acyclic substitutions. In §4 we repeat the exercise for **letrec** interpretations where we need explicit cyclic substitutions. After concluding we discuss further work, and finally we briefly mention the relation to graph rewriting systems in an appendix.

2 Implicit Substitution and β-reduction

This section summarizes the λ-calculus and traditional λ-reduction based on "implicit" substitutions, *i.e.*, on substitution as metanotation. We first define the set Λ of λ-terms modulo renaming equipped with acyclic and cyclic simultaneous substitution. Then we define traditional β-reduction in terms of substitution and formulate usual theorems in the tradition of Barendregt [4].

2.1 Definition. Assume a set $\mathbb{V}$ of variables (ranged over by $x, y, z, \ldots$).

(i) The *λ-terms* (ranged over by $M, N, P, \ldots$) are defined inductively by

$$M \quad ::= \quad x \quad | \quad (\lambda x.M) \quad | \quad (MN)$$

We omit outermost ()s and ()s that may be placed using the following rules: (a) abstraction associates to the right, (b) application associates to the left, and (c) application takes precedence over abstraction. *E.g.*, $(\lambda xy.MNP) = (\lambda x.(\lambda y.((MN)P)))$.

(ii) The *free variables* of a term is the set of variables not bound by an enclosing abstraction; inductively:

$$\begin{aligned} \mathrm{FV}(x) &= \{x\} \\ \mathrm{FV}(\lambda x.M) &= \mathrm{FV}(M) \setminus \{x\} \\ \mathrm{FV}(MN) &= \mathrm{FV}(M) \cup \mathrm{FV}(N) \end{aligned}$$

(iii) A term with a *renaming* postfix $[x := y]$ denotes the term obtained by changing 'free occurrences' of x in it to y; inductively specified as $\Rrightarrow$ (assuming x, y, z are distinct variables) by:

$$\begin{aligned} x\,[x := y] &\Rrightarrow y \\ y\,[x := y] &\Rrightarrow y \\ z\,[x := y] &\Rrightarrow z \\ (\lambda x.M)\,[x := y] &\Rrightarrow \lambda x.M \\ (*) \qquad (\lambda y.M)\,[x := y] &\Rrightarrow \lambda y'.\,M[y := y'][x := y] \quad \text{if } y' \notin \mathrm{FV}(M) \\ (\lambda z.M)\,[x := y] &\Rrightarrow \lambda z.\,M[x := y] \\ (MN)\,[x := y] &\Rrightarrow M[x := y]\,N[x := y] \end{aligned}$$

Note: We use $\Rrightarrow$ to indicate the direction that should be used to interpret the renaming as a 'pure' λ-term.

(iv) *Equivalence* of terms means "syntactic equality modulo renaming" (*aka* "α-equivalence"), and is defined inductively by

$$\begin{aligned} x &\equiv x \\ \lambda x.M &\equiv \lambda y.N \quad \text{if } M[x := y] \equiv N \\ MN &\equiv PQ \quad\quad \text{if } M \equiv P \wedge N \equiv Q \end{aligned}$$

Clearly $M \Rrightarrow N$ implies $M \equiv N$.

(v) Λ is the set of *opaque λ-terms*, *i.e.*, the λ-terms modulo equivalence: $\Lambda = \lambda\text{-}terms/\equiv$.

The definition of renaming above carefully avoids 'variable capture,' *i.e.*, changing a free variable into a bound variable, at the cost of introducing an extra renaming (equation ($*$) of 2.1(iii) adds $[y := y']$ to rename the bound variables before it happens). In particular this means that every opaque λ-term has representative λ-terms where all bound variables are different. We will make use of this below.

2.2 Convention. All considerations in the following are over Λ, *i.e.*, we will consider λ-terms as representatives for the corresponding opaque λ-term. In particular sideconditions on freeness of variables introduced by "with" will always be assumed satisfied by picking an appropriate representative (this is essentially the *variable convention* [4, conventions 2.1.12–13]).

With this convention we are ready for a definition of 'proper' *simultaneous substitution* based on the 'variable substitution' (renaming) defined above.

2.3 Definitions.

(i) A *binder* is a pair $y := Q$ of a variable and a term. A *binder set* is an unordered collection of the form $y_1 := Q_1, y_2 := Q_2, \ldots, y_k := Q_k$ for some integer $k \geq 1$. We abbreviate binder sets as $y_1 := Q, \ldots_k$.

(ii) A term with an *acyclic simultaneous substitution* postfix $[y_1 := Q_1, \ldots_k]$ denotes the term obtained by replacing free occurrences of the variables $y_1, \ldots, y_k$ by the corresponding 'substituend' terms $Q_1, \ldots, Q_k$:

$$x\,[y_1 := Q_1, \ldots_k] \Rrightarrow x \tag{1}$$

$$y_i\,[y_1 := Q_1, \ldots_k] \Rrightarrow Q_i \tag{2}$$

$$(\lambda x.M)\,[y_1 := Q_1, \ldots_k] \Rrightarrow \lambda x'.\,M[x := x'][y_1 := Q_1, \ldots_k] \quad \text{with } x' \notin \mathrm{FV}(M\,Q_1 \ldots Q_k) \tag{3}$$

$$(MN)\,[y_1 := Q_1, \ldots_k] \Rrightarrow M[y_1 := Q_1, \ldots_k]\,N[y_1 := Q_1, \ldots_k] \tag{4}$$

(iii) A term with a *cyclic simultaneous substitution* postfix $[\![y_1 := Q_1, \ldots_k]\!]$, where the Q_i should be *proper terms*, *i.e.*, may not be any of the single variable terms $y_1, \ldots, y_k$, denotes the term obtained by replacing free occurrences of $y_1, \ldots, y_k$ in it by the corresponding 'substituend' terms

$Q_1, \ldots, Q_k$ infinitely:

$$(1) \qquad x\,[\![y_1 := Q_1, \ldots_k]\!] \Rrightarrow x$$

$$(2) \qquad y_i\,[\![y_1 := Q_1, \ldots_k]\!] \Rrightarrow Q_i[\![y_1 := Q_1, \ldots_k]\!]$$

$$(3) \qquad (\lambda x.M)\,[\![y_1 := Q_1, \ldots_k]\!] \Rrightarrow \lambda x'.\,M[x := x'][\![y_1 := Q_1, \ldots_k]\!] \quad \text{with } x' \notin \mathrm{FV}(M\,Q_1 \ldots Q_k)$$

$$(4) \qquad (MN)\,[\![y_1 := Q_1, \ldots_k]\!] \Rrightarrow M[\![y_1 := Q_1, \ldots_k]\!]\,N[\![y_1 := Q_1, \ldots_k]\!]$$

(iv) Given a term by a cyclic simultaneous substitution $M[\![y_1 := Q_1, \ldots_k]\!]$. y_{i_0} is a *cyclic variable* in a cycle of length n if there is a sequence $y_{i_0} \in \mathrm{FV}(Q_{i_1})$, $y_{i_1} \in \mathrm{FV}(Q_{i_2})$, ..., $y_{i_n} \in \mathrm{FV}(Q_{i_0})$ and $\{y_{i_0}, \ldots, y_{i_n}\} \cap \mathrm{FV}(M) \neq \emptyset$.

Again we avoid variable capture carefully by the somewhat 'brute force' approach of renaming the bound variable of *any* abstraction under which we substitute—it is clear that this does not change the opaque λ-term in question. This problem would dissappear altogether if we used de Bruijn indices [6], but we prefer proper variables to keep the connection to **let** and **letrec** obvious.

We have used $\Rrightarrow$ again to indicate the *direction* of the substitution unfolding here as well. The following properties of the defined renaming and substitution concepts are easy to verify and will prove useful later. They are generalisations of standard λ-calculus results to simultaneous acyclic and cyclic substitutions; the only complication is the technical restriction on the substituends of cyclic substitutions that is needed to ensure that there are no 'un-unfoldable' substitutions, *e.g.*, $x[\![x := x]\!]$ is not a term.

2.4 Propositions.

(i) *The unfolding sequence* $M[y_1 := Q_1, \ldots_k] \Rrightarrow \ldots \Rrightarrow N$ *is always finite.*

(ii) *The unfolding sequence* $M[\![y_1 := Q_1, \ldots_k]\!] \Rrightarrow \ldots \Rrightarrow N$ *is finite iff there are no cyclic variables in the substitution.*

(iii) *Acyclic substitution lemma, no* $x \in \mathrm{FV}(Q_1 \ldots Q_k)$*:*

$$M[x_1 := P_1, \ldots_k][y_1 := Q_1, \ldots_n] \equiv M[y_1 := Q_1, \ldots_n][x_1 := P_1[y_1 := Q_1, \ldots_n], \ldots_k]$$

(iv) *Cycle unfolding lemma:*

$$M[\![x_1 := P_1, \ldots_k]\!] \equiv M[x_1 := P_1, \ldots_k][\![x_1 := P_1, \ldots_k]\!]$$

(v) *Cyclic substitution lemma:*

$$M[\![x_1 := P_1, \ldots_k]\!][\![y_1 := Q_1, \ldots_n]\!] \equiv M[\![y_1 := Q_1, \ldots_n]\!][\![x_1 := P_1[\![y_1 := Q_1, \ldots_n]\!], \ldots_k]\!]$$

The traditional definition of and Church-Rosser theorem for β-reduction looks as follows in our notation.

2.5 Definition. The semantics of the λ-calculus is given by the *β-reduction step* relation

$$(\beta) \qquad (\lambda x.M)N \Rightarrow M[x := N]$$

We will also use the following derived relations: ***reduction*** $\twoheadrightarrow$ is $\Rightarrow$s transitive, reflexive closure, and ***conversion*** $\Leftrightarrow$ is $\Rightarrow$s transitive, reflexive, and symmetric closure. M is a ***redex*** iff $\exists N\colon M \Rightarrow N$, otherwise it is a ***normal form***: nf(M). Finally, ***completion*** $\twoheadrightarrow\!\!|$ is defined by $M \twoheadrightarrow\!\!| \, N$ iff $M \twoheadrightarrow N$ and nf(N).

Remark. We silently impose the usual restriction that a "notion of reduction" relation must always be transitive, reflexive, and ***compatible*** with (the operations of) the set it is defined over—the last simply means that the reduction step can be applied anywhere in terms. Other notions are possible, *cf.* Abramsky [2].

2.6 Theorem (Church-Rosser). *β-reduction is confluent, i.e., completes the 'diamond diagram'*

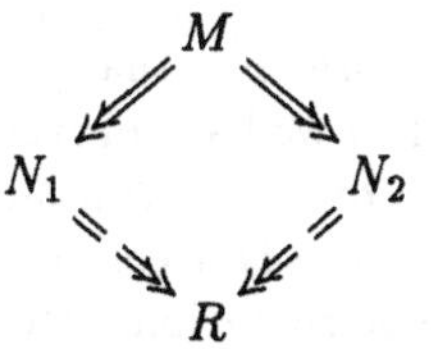

It is trivial to show that $Q_1 \Rightarrow Q'_1$ implies $M[y_1 := Q_1, \ldots_k] \twoheadrightarrow M[y_1 := Q'_1, \ldots_k]$, and the following are easy to verify:

2.7 Propositions. *β-reduction is substitutive with respect to both acyclic and cyclic substitution:*

(i) $M \twoheadrightarrow N$ *implies* $M[y_1 := Q_1, \ldots_k] \twoheadrightarrow N[y_1 := Q_1, \ldots_k]$.
(ii) $M \twoheadrightarrow N$ *implies* $M[\![y_1 := Q_1, \ldots_k]\!] \twoheadrightarrow N[\![y_1 := Q_1, \ldots_k]\!]$.

However, $Q_1 \twoheadrightarrow Q'_1$ does not in general imply $M[\![y_1 := Q_1, \ldots_k]\!] \twoheadrightarrow M[\![y_1 := Q'_1, \ldots_k]\!]$ because Q'_1 may be a variable making the second substitution undefined. *E.g.*, even though $Ix \twoheadrightarrow x$ we do not have that $x[\![x := Ix]\!] \equiv I(I(I \ldots))$ (where $I \equiv \lambda y.y$) reduces to $x[\![x := x]\!]$ since the latter is not a term.

3 Explicit Acyclic Substitution and "let"

We will now concentrate on the interpretation of **let** and acyclic substitution. First we show how we may simulate simultaneous substitution using β-reduction. Then we show how the mechanics of substitution may be build in by presenting an alternate formulation of the $\lambda\sigma$-calculus of Abadi, Cardelli, Curien, and Lévy [1].

Intuitively the denotation of **let** should be a standard λ-term with the same meaning as an acyclic substitution, *i.e.*,

$$\textbf{let } x_1 = N_1 \textbf{ and} \ldots \textbf{and } x_k = N_k \textbf{ in } M \;\equiv\; M[x_1 := N_1, \ldots_k]$$

It is easy to prove that such a term exist using the standard 'tupling trick.'

3.1 Definition. *k-tupling* $\langle M_1, \ldots, M_k \rangle$ of $M_1, \ldots, M_k$ with *projections* π_n^k, $1 \leq n \leq k$, are terms that satisfy $\pi_n^k \langle M_1, \ldots, M_k \rangle \twoheadrightarrow M_n$.

In fact simple versions of tupling exist, *e.g.*, $\langle M_1, \ldots, M_k \rangle \equiv \lambda z.z M_1 \ldots M_k$ with $\pi_n^k \equiv \lambda x.x(\lambda z_1 \ldots z_k.z_n)$.

3.2 Lemma. *k-tupling simulates acyclic simultaneous substitution:*

$$(\lambda y.P[y_1 := y\,\pi_1^k, \ldots, y_k := y\,\pi_k^k])(\lambda p.p\langle Q_1, \ldots, Q_k \rangle) \twoheadrightarrow P[y_1 := Q_1, \ldots_k]$$

Proof. Proposition 2.4(i) ensures that it is sufficient to complete

$$
\begin{array}{ccccccccc}
P[y_1 := Q_1, \ldots_k] & \Longrightarrow & \square & \Longrightarrow & \square & \Longrightarrow & \cdots & \Longrightarrow & P' \\
\big\uparrow & & \top & & \top & & & & \| \\
& & \vdots & & \vdots & & & & \| \\
R_0 & \Longrightarrow & R_1 & \Longrightarrow & R_2 & \Longrightarrow & \cdots & \Longrightarrow & M'
\end{array}
$$

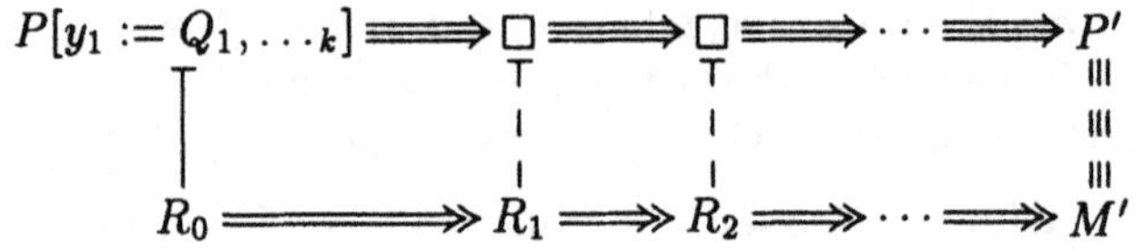

where $R_0 \equiv (\lambda y.P[y_1 := y\,\pi_1^k, \ldots, y_k := y\,\pi_k^k])\,(\lambda p.p\langle Q_1, \ldots, Q_k \rangle)$ and $P[y_1 := Q_1, \ldots_k] \vdash R$ means that R represents that particular 'unfoldedness' of the substitution. We are finished when we have proven the correctness of all possible such diagram 'cells' by induction over the structure of definition 2.3(ii): for each possible $\Rightarrow$-step we prove that there is a corresponding $\twoheadrightarrow$-reduction from the representation 'before' to the representation 'after.'

Case 1: $x[y_1 := Q_1, \ldots_k] \vdash (\lambda y.x[y_1 := y\,\pi_1^k, \ldots_k])(\lambda p.p\langle Q_1, \ldots, Q_k \rangle) \Rightarrow x$ ✓

Case 2:
$$
\begin{aligned}
y_i&[y_1 := Q_1, \ldots_k] \\
&\vdash (\lambda y.y_i[y_1 := y\,\pi_1^k, \ldots_k])(\lambda p.p\langle Q_1, \ldots, Q_k \rangle) \\
&\Rightarrow y_i[y_1 := y\,\pi_1^k, \ldots_k][y := \lambda p.p\langle Q_1, \ldots, Q_k \rangle] \\
&\Rightarrow (y\,\pi_i^k)[y := \lambda p.p\langle Q_1, \ldots, Q_k \rangle] \\
&\Rightarrow (\lambda p.p\langle Q_1, \ldots, Q_k \rangle)\,\pi_i^k \\
&\Rightarrow \pi_i^k \langle Q_1, \ldots, Q_k \rangle \twoheadrightarrow Q_i \quad \checkmark
\end{aligned}
$$

Case 3:
$$
\begin{aligned}
(\lambda x.M)&[y_1 := Q_1, \ldots_k] \\
&\vdash (\lambda y.(\lambda x.M))[y_1 := y\,\pi_1^k, \ldots_k])(\lambda p.p\langle Q_1, \ldots, Q_k \rangle) \\
&\Rightarrow (\lambda y.\lambda x'.M[x := x'][y_1 := y\,\pi_1^k, \ldots_k])(\lambda p.p\langle Q_1, \ldots, Q_k \rangle) \\
&\Rightarrow \lambda x'.M[x := x'][y_1 := (\lambda p.p\langle Q_1, \ldots, Q_k \rangle)\,\pi_1^k, \ldots_k] \\
&\Rightarrow \lambda x'.M[x := x'][y_1 := \pi_1^k \langle Q_1, \ldots, Q_k \rangle, \ldots_k] \quad \text{by 2.4(iii)} \\
&\twoheadrightarrow \lambda x'.M[x := x'][y_1 := Q_1, \ldots_k] \qquad \text{by 2.4(iii)} \quad \checkmark
\end{aligned}
$$

Case 4:
$$
\begin{aligned}
(MN)&[y_1 := Q_1, \ldots_k] \\
&\vdash (\lambda y.(MN)[y_1 := y\,\pi_1^k, \ldots_k])(\lambda p.p\langle Q_1, \ldots, Q_k \rangle) \\
&\Rightarrow (MN)[y_1 := y\,\pi_1^k, \ldots_k][y := (\lambda p.p\langle Q_1, \ldots, Q_k \rangle)] \\
&\Rightarrow (MN)[y_1 := (\lambda p.p\langle Q_1, \ldots, Q_k \rangle)\,\pi_1^k, \ldots_k] \\
&\twoheadrightarrow (MN)[y_1 := Q_1, \ldots_k] \qquad \text{by 2.4(iii)} \\
&\Rightarrow M[y_1 := Q_1, \ldots_k]\,N[y_1 := Q_1, \ldots_k] \qquad \blacksquare
\end{aligned}
$$

This shows that the above encoding of **let** is correct and that the 'interpretation overhead' using tupling is constant (actually linear in the number k of simultaneous binders). It also shows that the expressive power of the standard λ-calculus will not be extended when we add substitution explicitly.

3.3 Definition. Assume definition 2.1. The $\lambda\sigma$-calculus is derived by modifying the individual points as follows:

(i) The *$\lambda\sigma$-terms*: add acyclic binding as syntax, *i.e.*,

$$M \ ::= \ x \ | \ (\lambda x.M) \ | \ (MN) \ | \ M[x_1 := N_1, \ldots_k]$$

(ii) The *free variables*: add

$$\mathrm{FV}(M[x_1 := N_1, \ldots_k]) = \mathrm{FV}(M\,N_1 \ldots N_k) \setminus \{x_1, \ldots, x_k\}$$

(iii) The *renaming* postfix: add

$$\begin{aligned} &M[x_1 := N_1, \ldots_k]\,[x_i := y] \ \Rrightarrow M[x_1 := N_1, \ldots_k] \\ &M[y_1 := N_1, \ldots_k]\,[x := y_i] \\ &\qquad \Rrightarrow M[y_1 := y_1'] \ldots [y_k := y_k'][x := y_i][y_1' := N_1, \ldots_k] \\ &M[z_1 := N_1, \ldots_k]\,[x := y] \ \Rrightarrow M[x := y][z_1 := N_1, \ldots_k] \end{aligned}$$

(iv) *Equivalence*: add

$$M[x_1 := N_1, \ldots_k] \equiv P[y_1 := Q_1, \ldots_k]$$
$$\text{if } M[x_1 := y_1] \ldots [x_k := y_k] \equiv P \wedge \forall i\colon N_i \equiv Q_i$$

The *$\lambda\sigma$-reduction steps* are β-reduction (of definition 2.5) and syntactic versions of acyclic $\Rrightarrow$-unfolding (of definition 2.3(ii)):

$$(\beta\sigma) \qquad (\lambda x.M)N \Rightarrow M[x := N]$$

$$(\sigma_0) \qquad x[y_1 := Q_1, \ldots_k] \Rightarrow x$$

$$(\sigma_1) \qquad y_i[y_1 := Q_1, \ldots_k] \Rightarrow Q_i$$

$$(\sigma_2) \qquad (\lambda x.M)[y_1 := Q_1, \ldots_k] \Rightarrow \lambda x'.\, M[x := x'][y_1 := Q_1, \ldots_k]$$
$$\text{with } x' \notin \mathrm{FV}(M\,Q_1 \ldots Q_k)$$

$$(\sigma_3) \qquad (MN)[y_1 := Q_1, \ldots_k] \Rightarrow M[y_1 := Q_1, \ldots_k]\,N[y_1 := Q_1, \ldots_k]$$

The theory $\boldsymbol{\sigma}$ contains the σ-rules and $\boldsymbol{\lambda\sigma}$ contains all the above rules.

The remainder of this section is devoted to prove the confluence of $\boldsymbol{\lambda\sigma}$: first we prove that $\boldsymbol{\sigma}$ and $(\beta\sigma)$ seperately are confluent and strongly normalizing. Then we prove that they commute and consequently that $\boldsymbol{\lambda\sigma}$ is confluent. It is interesting to compare this proof with the confluence of the untyped $\lambda\sigma$ of [1] because their proof combines the *full* β-reduction with σ-reduction whereas our proof combines the *syntactic* $(\beta\sigma)$ with the (also syntactic) substitution rules $\boldsymbol{\sigma}$.

The key lemma making this possible is Hindley–Rosen [4, proposition 3.3.5] presented immediately hereafter. Notice how our presentation below is rather terse because most of the properties we prove are easily seen to be equivalent to standard substitution results above! This is the obvious advantage of the fact that $\lambda\sigma$ in a way is just a syntactic version of the standard calculus.

3.4 Lemma (Hindley–Rosen). *Given two reduction relations* $\twoheadrightarrow_1$ *and* $\twoheadrightarrow_2$ *and let* $\twoheadrightarrow_{12}$ *be the transitive reflexive closure of their union. Suppose* (1) *that each of the two relations are confluent and* (2) *that they commute, i.e.,*

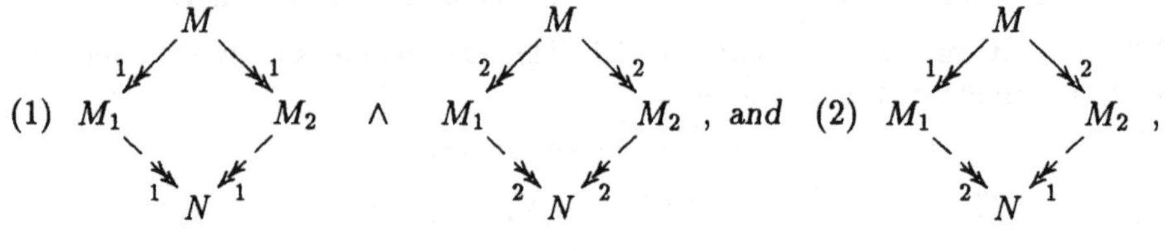

then their union is confluent:

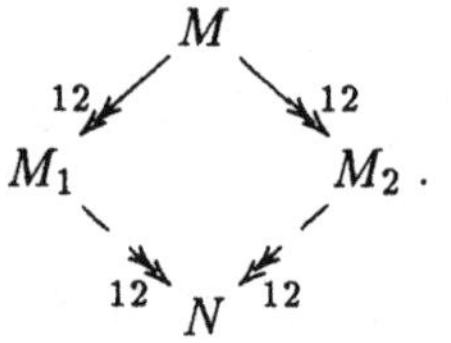

3.5 Theorem. $\lambda\sigma$ *is confluent.*

Proof. Use Hindley–Rosen 3.4 on $(\beta\sigma)$ and σ: We have

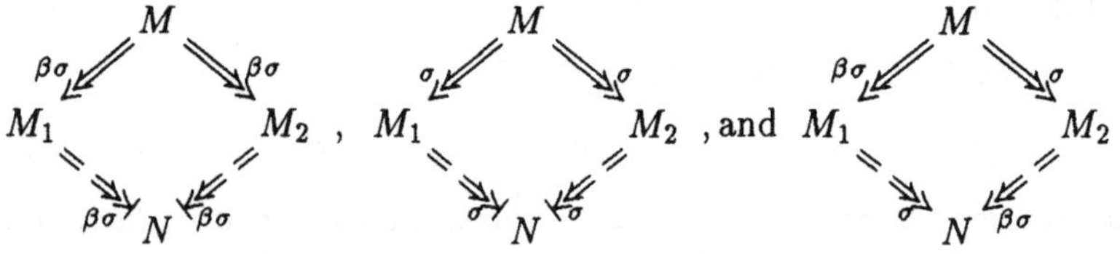

The left follows from the observation that $(\beta\sigma)$-reduction will replace some β-redex $(\lambda x.M)N$ with σ-redex $M[x := N]$; whether M and N contain β-redices themself is of no consequence. Since there are no more than a finite number of redexes in any term then this is also strongly normalizing. That σ is strongly normalizing follows immediately from proposition 2.4(i); the confluence then follows from a simple induction over the definition (it also follows from 2.6 and 3.2). Finally the leftmost diagram is just a rephrasing of 2.4(iii). ■

Thus it seems that standard concepts of the λ-calculus extend nicely to handle acyclic substitutions allowing the modeling of sharing through —let—.

4 Explicit Cyclic Substitution and "letrec"

In this section we will discuss the **letrec** interpretation, prove its correctness as far as it goes, and discuss the problems with it. Finally we present the $\lambda\sigma^*$ calculus, a cyclic version of $\lambda\sigma$ defined in the previous section, and prove that it is confluent.

Intuitively we should encode **letrec** like **let** but with a simultaneous cyclic substitution, *i.e.*,

$$\textbf{letrec } x_1 = N_1 \textbf{ and} \dots \textbf{and } x_k = N_k \textbf{ in } M \;\equiv\; M[\![x_1 := N_1, \dots_k]\!]$$

The problem with this is, of course that *any* term created by a (non-trivial) cyclic substitution is infinite (where 'non-trivial' means 'with cycles' such that the substitution is never 'exhausted') as the *parity* example of the introduction showed. In fact it is clear that the definition of **letrec** terms may yield infinite terms in much the same way as fixed point induction. The following example shows how this may be exploited to encode an 'immediate fixed point' operator, assuming $x \notin \mathrm{FV}(M)$:

$$\begin{aligned}\textbf{letrec } x = Mx \textbf{ in } x &\equiv x[\![x := Mx]\!]\\ &\Rrightarrow Mx[\![x := Mx]\!]\\ &\Rrightarrow M(Mx)[\![x := Mx]\!]\\ &\vdots\\ &\Rrightarrow M(M(M(M\ldots)))[\![x := Mx]\!]\\ &\Rrightarrow M(M(M(M\ldots)))\end{aligned}$$

The above involves an infinite number of $\Rrightarrow$ unfoldings to build the infinite term, of course; the last step is due to the fact that there are no free xs in the infinite sequence of Ms. Thus our simulation result is weaker since we can not hope to simulate infinite unfolding.

4.1 Lemma. *β-reduction using fixed point iteration and k-tupling may simulate the termination or nontermination of cyclic simultaneous substitution as follows depending on presense of cyclic variables in $P[\![y_1 := Q_1, \ldots_k]\!]$:*

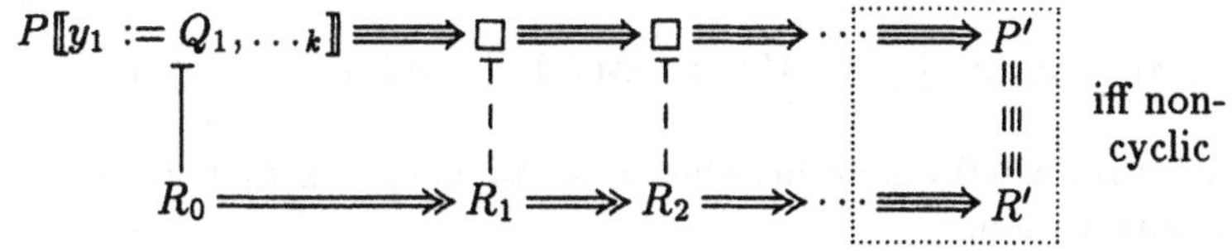

where $R_0 = \overline{P}[y := Y(\lambda yp.p\langle\overline{Q_1}, \ldots, \overline{Q_k}\rangle)]$, $\forall M: \overline{M} \equiv M[y_1 := y\,\pi^k_1, \ldots, y_k := y\,\pi^k_k]$.

Proof. Proceeds along the same lines as the proof for 3.2 except that we use proposition 2.4(v) which means that the second case is the only one that is substantially different—this is as should be expected since it corresponds to the only equation of 2.3(iii) that differs from 2.3(ii):

Case 2:

$$\begin{aligned}&y_i[\![y_1 := Q_1, \ldots_k]\!]\\ &\quad\vdash \overline{y_i}[y := Y(\lambda yp.p\langle\overline{Q_1}, \ldots, \overline{Q_k}\rangle)]\\ &\quad\Rightarrow (y\,\pi^k_i)[y := Y(\lambda yp.p\langle\overline{Q_1}, \ldots, \overline{Q_k}\rangle)]\\ &\quad\Rightarrow Y(\lambda yp.p\langle\overline{Q_1}, \ldots, \overline{Q_k}\rangle)\,\pi^k_i\\ &\quad\Rrightarrow (\lambda yp.p\langle\overline{Q_1}, \ldots, \overline{Q_k}\rangle)(Y(\lambda yp.p\langle\overline{Q_1}, \ldots, \overline{Q_k}\rangle))\,\pi^k_i\\ &\quad\Rrightarrow (\lambda p.p\langle\overline{Q_1}, \ldots, \overline{Q_k}\rangle)[y := Y(\lambda yp.p\langle\overline{Q_1}, \ldots, \overline{Q_k}\rangle)]\,\pi^k_i\\ &\quad\Rrightarrow (\lambda p.(p\langle\overline{Q_1}, \ldots, \overline{Q_k}\rangle)[y := Y(\lambda yp.p\langle\overline{Q_1}, \ldots, \overline{Q_k}\rangle)])\,\pi^k_i\end{aligned}$$

$$\Rightarrow (p\langle\overline{Q_1},\ldots,\overline{Q_k}\rangle)[y := Y(\lambda yp.p\langle\overline{Q_1},\ldots,\overline{Q_k}\rangle)][p := \pi_i^k]$$
$$\Rightarrow (p[y := Y(\lambda yp.p\langle\overline{Q_1},\ldots,\overline{Q_k}\rangle)]\,\langle\overline{Q_1},\ldots,\overline{Q_k}\rangle$$
$$[y := Y(\lambda yp.p\langle\overline{Q_1},\ldots,\overline{Q_k}\rangle)])[p := \pi_i^k]$$
$$\Rightarrow (p\,\langle\overline{Q_1},\ldots,\overline{Q_k}\rangle[y := Y(\lambda yp.p\langle\overline{Q_1},\ldots,\overline{Q_k}\rangle)])[p := \pi_i^k]$$
$$\Rightarrow (p[p := \pi_i^k]\,\langle\overline{Q_1},\ldots,\overline{Q_k}\rangle[y := Y(\lambda yp.p\langle\overline{Q_1},\ldots,\overline{Q_k}\rangle)][p := \pi_i^k])$$
$$\Rrightarrow (p[p := \pi_i^k]\,\langle\overline{Q_1},\ldots,\overline{Q_k}\rangle[y := Y(\lambda yp.p\langle\overline{Q_1},\ldots,\overline{Q_k}\rangle)])$$
$$\Rightarrow (\pi_i^k\,\langle\overline{Q_1},\ldots,\overline{Q_k}\rangle[y := Y(\lambda yp.p\langle\overline{Q_1},\ldots,\overline{Q_k}\rangle)])$$
$$\Rrightarrow \pi_i^k\,\langle\overline{Q_1}[y := Y(\lambda yp.p\langle\overline{Q_1},\ldots,\overline{Q_k}\rangle)],\ldots,$$
$$\overline{Q_k}[y := Y(\lambda yp.p\langle\overline{Q_1},\ldots,\overline{Q_k}\rangle)]\rangle$$
$$\Rrightarrow \overline{Q_i}[y := Y(\lambda yp.p\langle\overline{Q_1},\ldots,\overline{Q_k}\rangle)] \quad \checkmark$$

This reduction sequence also shows how all cyclic variables remain in the term after unfolding. ■

Now let us see what happens if we add cyclic substitution as *syntax* to the calculus.

4.2 Definition. Assume definition 2.1. The $\lambda\sigma^*$*-calculus* is derived by modifying the individual points as follows:

(i) The *$\lambda\sigma$-terms*: add acyclic binding as syntax, *i.e.*,

$$M \;::=\; x \;\mid\; (\lambda x.M) \;\mid\; (MN) \;\mid\; M[\![x_1 := N_1,\ldots_k]\!]$$

we say that a term is cyclic when it contains syntax that would be cyclic as a substitution.

(ii) The *free variables*: add

$$\mathrm{FV}(M[\![x_1 := N_1,\ldots_k]\!]) = \mathrm{FV}(M\,N_1\ldots N_k)\setminus\{x_1,\ldots,x_k\}$$

(iii) The *renaming* postfix: add

$$M[\![x_1 := N_1,\ldots_k]\!]\,[x_i := y] \;\Rrightarrow M[\![x_1 := N_1,\ldots_k]\!]$$
$$M[\![y_1 := N_1,\ldots_k]\!]\,[x := y_i]$$
$$\Rrightarrow M[y_1 := y_1']\ldots[y_k := y_k'][x := y_i][\![y_1' := N_1,\ldots_k]\!]$$
$$M[\![z_1 := N_1,\ldots_k]\!]\,[x := y] \;\;\Rrightarrow M[x := y][\![z_1 := N_1,\ldots_k]\!]$$

(iv) *Equivalence*: add

$$M[\![x_1 := N_1,\ldots_k]\!] \equiv P[\![y_1 := Q_1,\ldots_k]\!]$$
$$\text{if } M[x_1 := y_1]\ldots[x_k := y_k] \equiv P$$
$$\wedge\,\forall i\colon N_i[x_i := y_i]\ldots[x_k := y_k] \equiv Q_i$$

The $\lambda\sigma^*$*-reduction steps* are β-reduction (of definition 2.5) and syntactic versions of cyclic $\Rrightarrow$-unfolding (of definition 2.3(iii)):

$$(\beta\sigma^*) \qquad (\lambda x.M)N \Rightarrow M[x := x'][\![x' := N]\!] \quad \text{with } x' \notin \mathrm{FV}(MN)$$

$$(\sigma_0^*) \qquad x[\![y_1 := Q_1, \ldots_k]\!] \Rightarrow x$$

$$(\sigma_1^*) \qquad y_i[\![y_1 := Q_1, \ldots_k]\!] \Rightarrow Q_i[\![y_1 := Q_1, \ldots_k]\!]$$

$$(\sigma_2^*) \qquad (\lambda x.M)[\![y_1 := Q_1, \ldots_k]\!] \Rightarrow \lambda x'.M[x := x'][\![y_1 := Q_1, \ldots_k]\!] \quad \text{with } x' \notin \mathrm{FV}(M\,Q_1 \ldots Q_k)$$

$$(\sigma_3^*) \qquad (MN)[\![y_1 := Q_1, \ldots_k]\!] \Rightarrow M[\![y_1 := Q_1, \ldots_k]\!]N[\![y_1 := Q_1, \ldots_k]\!]$$

The theory $\boldsymbol{\sigma}^*$ contains the σ^*-rules and $\boldsymbol{\lambda\sigma}^*$ contains all the above rules.

Finally we will just sketch the Church-Rosser theorem for $\boldsymbol{\lambda\sigma}^*$ since it follows the same scheme of theorem 3.5.

4.3 Theorem. $\boldsymbol{\lambda\sigma}^*$ *is confluent.*

Proof. We invoke the Hindley–Rosen lemma 3.4 again, this time with $(\beta\sigma^*)$ and $\boldsymbol{\sigma}^*$. Again $(\beta\sigma^*)$ is confluent and strongly normalizing, and from 2.4(iv), 3.5, and 4.1 we conclude that $\boldsymbol{\sigma}^*$ is confluent even though it is strongly normalizing exactly for non-cyclic terms and divergent for cyclic terms. □

So it also seems that standard concepts of the λ-calculus extend nicely to cyclic substitutions allowing the modeling of recursive definitions through —letrec—.

5 Conclusion and Future Work

We have discussed different ways of describing sharing and recursion in the λ-calculus, and have shown rather brief and elegant proofs of confluency. We would like to improve the understanding as well as the presentation; in particular it appears promising to investigate the relation between fixed point iteration and the infinite unfolding we have presented.

The knowledge gained is that *explicitly represented sharing and recursion* can be maintained and exploited in a computational model—this is not new insight, but still there seems to be a need for better links from the operational insight of implementors and the denotational oversight of formalists.

The natural next step is thus to try to 'derive' an abstract reduction machine from our very simple $\lambda\sigma$ and $\lambda\sigma^*$-rewrite systems in a way similar to the way the Krivine machine is derived by Curien [8] from his $\lambda\rho$-calculus. It will be interesting to see how the cyclicity shows up in such a machine. It would also be interesting to try to combine current work on optimal reductions in acyclic explicit substitution calculi with the cyclicity notions discussed here. Finally the relation between this presentation, regular trees, and in particular graph reduction needs to be better understood—it unfortunately seems that there is a lot of duplication of work due to the lack of a current reference frame.

Appendix Graph Reduction

We mentioned in the introduction that our techniques are similar to operational techniques used within "graph reduction machine" descriptions of functional programming languages. A different approach is to describe sharing by using a *term graph* representation. The author's thesis [22, part II] shows how the substitution concepts above may be 'lifted' to term graph rewriting [5] such that we may represent the β-reduction rule as

(β)

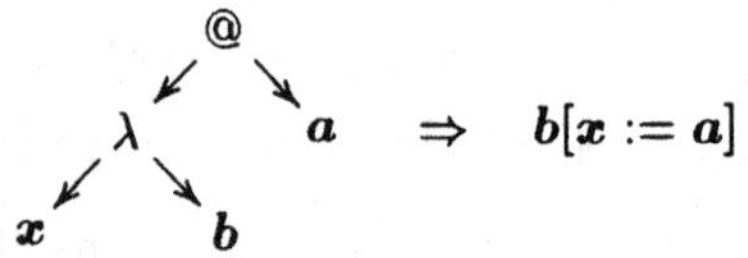

in a suitable representation (where λ denotes abstraction, @ denotes application, and bold variables denote subgraphs). The difficult bit is of course to define the substitution such that it faithfully generalizes standard substitution to the much richer domain of graphs. And the interesting but perhaps not suprising thing about generalizing term to graph rewriting is that the encountered difficulties resemble the restriction of definition 2.3(iii) that substituends of a simultaneous cyclic substitution postfix can not be the substitution variables, *cf.* [14].

References

1. M. Abadi, L. Cardelli, P.-L. Curien, and J.-J. Lévy, *Explicit Substitutions*, in [21], 31–46.
2. S. Abramsky, *The Lazy Lambda Calculus*, ch. 4, in D. A. Turner (ed.), *Research Topics in Functional Programming*, Addison-Wesley, 1990, pp. 65–116.
3. Z. M. Ariola and Arvind, *A Syntactic Approach to Program Transformations*, in *PEPM '91—Symposium on Partial Evaluation and Semantics-based Program Manipulation* (Yale University, New Haven, Connecticut, USA), 17–19 June 1991, pp. 116–129.
4. H. Barendregt, *The Lambda Calculus: Its Syntax and Semantics*, revised edition, North-Holland, 1984.
5. H. P. Barendregt, M. C. D. J. van Eekelen, J. R. W. Glauert, J. R. Kennaway, M. J. Plasmeijer, and M. R. Sleep, *Term Graph Rewriting*, in J. W. de Bakker, A. J. Nijman, and P. C. Treleaven (eds.), *PARLE '87—Parallel Architectures and Languages Europe* (Eindhoven, The Netherlands), vol. II, LNCS no. 256, Springer-Verlag, June 1987, pp. 141–158.
6. N. G. de Bruijn, *Lambda calculus notation with nameless dummies, a tool for automatic formula manipulation with application to the Church-Rosser theorem*, Koninkijke Nederlandse Akademie van Wetenschappen, Series A, Mathematical Sciences **75** (1972), 381–392.
7. B. Courcelle, *Fundamental Properties of Infinite Trees*, Theoretical Computer Science **25** (1983), 95–169.
8. P.-L. Curien, *An Abstract Framework for Environment Machines*, Unpublished note from LIENS/CNRS, July 1990.

9. G. Gonthier, M. Abadi, and J.-J. Lévy, *The Geometry of Optimal Lambda Reduction*, in *POPL '92—Nineteenth Annual ACM Symposium on Principles of Programming Languages* (Albuquerque, New Mexico), January 1992, pp. 1–14.
10. K. Grue, *Call-by-Mix: A Reduction Strategy for Pure λ-calculus*, Unpublished note from DIKU (University of Copenhagen), 1987.
11. C. Hankin, *Static Analysis of Term Graph Rewriting Systems*, ESPRIT "Semantique" working paper, 1990.
12. C. A. R. Hoare, *Recursive Data Structures*, Journal of Computer and Information Sciences **4** (1975), no. 2, 105–132.
13. K. H. Holm, *Graph Matching in Operational Semantics and Typing*, in A. Arnold (ed.), *CAAP '90—15th Colloqvium on Trees and Algebra in Programming* (Copenhagen, Denmark), LNCS no. 431, Springer-Verlag, March 1990, pp. 191–205.
14. J. R. Kennaway, J. W. Klop, M. R. Sleep, and F. J. de Vries, *On the Sdequacy of Graph Rewritng for Simulating Term Rewriting*, in [24] ch. 8 (to appear).
15. P. W. M. Koopman, S. E. W. Smetsers, M. C. D. J. van Eekelen, and M. J. Plasmeijer, *Efficient Graph Rewriting using the Annotated Functional Strategy*, in [20], 225–250, (available as nijmegen tech. report 91-25).
16. J. Lamping, *An Algorithm for Optimal Lambda Calculus Reduction*, in [21], 16–30.
17. J.-J. Lévy, *Optimal Reductions in the Lambda Calculus*, in J. P. Seldin and J. R. Hindley (eds.), *To H. B. Curry: Essays in Combinatory Logic, Lambda Calculus, and Formalism*, Academic Press, 1980, pp. 159–191.
18. L. Maranget, *Optimal Derivations in Weak Lambda-Calculi and in Orthogonal Terms Rewriting Systems*, in *POPL '91—Eightteenth Annual ACM Symposium on Principles of Programming Languages* (Orlando, Florida), January 1991, pp. 255–269.
19. S. L. Peyton Jones, *The Implementation of Functional Programming Languages*, Prentice-Hall, 1987.
20. M. J. Plasmeijer and M. R. Sleep (eds.), *SemaGraph '91 Symposium on the Semantics and Pragmatics of Generalized Graph Rewriting* (Nijmegen, Holland), December 1991, (available as nijmegen tech. report 91-25).
21. POPL, *POPL '90—Seventeenth Annual ACM Symposium on Principles of Programming Languages* (San Francisco, California), January 1990.
22. K. H. Rose, *GOS—Graph Operational Semantics*, Speciale 92-1-9, DIKU (University of Copenhagen), Universitetsparken 1, DK–2100 København Ø, Denmark, March 1992, (56pp).
23. ———, *Graph-based Operational Semantics for Lazy Functional Languages*, in [24] ch. 19 (to appear).
24. M. R. Sleep, M. J. Plasmeijer, and M. C. D. J. van Eekelen (eds.), *Term Graph Rewriting: Theory and Practice*, John Wiley & Sons, 1992 (to appear).
25. J. Staples, *A Graph-like Lambda Calculus for which Leftmost Outermost Reduction is Optimal*, in V. Claus, H. Ehrig, and G. Rozenberg (eds.), *1978 International Workshop in Graph Grammars and their Application to Computer Science and Biology* (Bad Honnef, F. R. Germany), LNCS no. 73, Springer-Verlag, 1978, pp. 440–454.
26. Y. Toyama, S. Smetsers, M. van Eekelen, and M. J. Plasmeijer, *The Functional Strategy and Transitive Term Rewriting Systems*, in [20], 99–114, (available as nijmegen tech. report 91-25).
27. C. P. Wadsworth, *Semantics and Pragmatics of the Lambda Calculus*, Ph.D. Thesis, Programming Research Group, Oxford University, 1971.

SIMPLE TYPE INFERENCE FOR TERM GRAPH REWRITING SYSTEMS

R. Banach[1]

Computer Science Department, Manchester University,

Manchester, M13 9PL, U.K.

Abstract

A methodology for polymorphic type inference for general term graph rewriting systems is presented. This requires modified notions of type and of type inference due to the absence of structural induction over graphs. Induction over terms is replaced by dataflow analysis.

1 Introduction

Term graphs are objects that locally look like terms, but globally have a general directed graph structure. Since their introduction in Barendregt *et al.* (1987), they have served the purpose of defining a rigorous framework for graph reduction implementations of functional languages (Peyton-Jones (1987)). This was the original intention. However the rewriting of term graphs defined in the operational semantics of the model, makes term graph rewriting systems (TGRSs) interesting models of computation in their own right. One can thus study all sorts of issues in the specific TGRS context. Typically one might be interested in how close TGRSs are to TRSs and this problem is examined in Barendregt et al. (1987), Farmer et al. (1990), or Kennaway et al. (1991).

In this paper we examine a related issue, that of type inference. There are two ways that we could approach this question. The first is to look for conditions on TGRSs that ensure we recover the results that exist for TRSs of various kinds. This has the virtue of finding the best generalizations of known TRS results in the graph world but leaves the question of what type inference for general TGRSs might look like, unanswered. The second approach addresses the latter question head on, and attempts to construct type theories for TGRSs directly, without close reference to the TRS results. This is the approach we will follow.

Two main obstacles have to be overcome when we consider the graph world as opposed to the term world. These are the collapse of structural induction, and the triviality of term matching compared with graph matching. For terms, structural induction is the mainstay of most proofs of significant results, and it is hard to see what to replace it with. A byproduct of structural induction is the fact that many results are context free in the sense that if Π is some property that holds for a set of terms T, then if $C[\ ]$ is some context, $C[t]$ satisfies Π for all $t \in T$ and contexts $C[\ ]$. Such results frequently hold in TRSs related to the lambda calculus (see Barendregt (1984), Girard et al. (1989), Huet (1990)).

For terms, we also have that if a rule matches some subterm, then apart from variable instantiations, the subterm is an exact copy of the rule LHS. This leads to a number of subtle properties that terms have compared with the more general term graphs, and is founded on the fact that terms are simply trees. In the more general graph world, we have to be more careful about the matching problem as matchings are not just isomorphisms with variable instantiation, and we have to confront the fact that important properties of a given rule are by no means required to hold for some arbitrary execution graph that its LHS may match.

1. Email: `rbanach@cs.man.ac.uk`

Given these shortcomings of the graph world, what weapons can we use to overcome the lack of structural induction and the complexity of matchings? The crude answer is that we must resort to induction over the structure of executions instead of induction over the structure of graphs (since the latter doesn't exist in any sense useful to us), and having that, to dataflow analysis of the rule system (the validity of which is itself established by the former technique). In fact, the former technique is not so different from what happens in the term world (where it leads to the subject reduction property), but in the term world, the triviality of several of the steps is such that they are passed over without mention, and the only issues of interest that remain can be dealt with by induction over terms. This is also related to the context freedom mentioned previously. A further issue related to this is rulewise modularity. Many term based systems have the property that certain semantic attributes of a rule can be considered in isolation from those of the rest of the rules in the system, again reducing induction over executions to induction over terms. This is a pleasing and desirable feature. Unfortunately context freedom and rulewise modularity are the main casualties of the passage to the graph world. There must be *some* casualties of course, we cannot expect significant generalization without paying some price. The lesson for the graph world though, boils down to the fact that we can no longer consider subsystems in isolation from one another as easily as we can in the term world. The entire system must in principle be taken into account when considering even some seemingly local properties of some small part of it.

The structure of the rest of the paper is as follows. In Section 2 we define our TGRSs as a generalisation of the model of Barendregt et al. (1987). The terminology is a little non-standard for convenience. Section 3 deals with simple dataflow analysis, showing how the structure of a rule system permits information about dynamic occurrences of so-called germ instances in execution graphs to be infered statically. Section 4 considers the question of how types are to be defined for TGRSs, and comes up with a simple scheme that is refined into two specific type disciplines in subsections 4.1 and 4.2. These are called the crude type discipline and the simple inperative type discipline respectively. Section 5 shows the results of section 3 can be used to show the soundness of a unification-based type inference algorithm for the type disciplines introduced in section 4. Section 6 contains a discussion, suggests extensions, and concludes.

2 Term Graph Rewriting

We assume we are given an alphabet $\mathbf{S} = \{S, T \ldots\}$ of **node symbols**. We write $\{1 \ldots n\}$ for the set of naturals between 1 and n inclusive. We let $\{1 \ldots 0\} = \varnothing$ and let $\mathbf{SeqN}$ be the set of all such subsets of $\mathbf{N}$, including $\varnothing$.

Definition 2.1 A **term graph** (or just **graph**) G is a triple (N, σ, α) where

(1) N is a set of nodes,

(2) σ is a map with signature $N \to \mathbf{S}$,

(3) α is a map with signature $N \to N^*$.

Thus σ maps a node to the node symbol that labels it, and α maps each node to its sequence of successors. We write $A(x)$, the **arity** of a node, for the domain of $\alpha(x)$, i.e. the member of $\mathbf{SeqN}$ consisting of indices of $\alpha(x)$. We also write $x \in G$ instead of $x \in N(G)$ etc. Lastly we subscript N, σ, α with the name of the graph in question whenever more than one graph is being discussed. A successor of a node determines an arc of the graph, and we will write such an arc as (p_k, c), to indicate that the child c is the k'th child of the parent p, i.e. that $c = \alpha(p)[k]$ for some $k \in A(p)$. The names are intended to be somewhat alliterative: N for nodes, σ for symbols, α for arcs.

We will assume for the remainder of this paper that symbols have fixed arities, i.e. that there is a map $A : \mathbf{S} \to \mathbf{SeqN}$ such that if $\sigma(x) = S$, then $A(x) = A(S)$ (the different A's should not cause confusion).

We will also assume there is a special node symbol Any, not normally considered to be in $\mathbf{S}$, and if any nodes of a graph are labeled with Any, we call such a graph a **pattern**. We further assume that

$$\sigma(x) = \mathsf{Any} \Rightarrow \alpha(x) = \varepsilon \qquad (\varepsilon = \text{the empty sequence}).$$

Nodes x with $\sigma(x)$ = Any are called **implicit**, whereas others are **explicit**. More generally we allow ourselves to regard a graph as a pattern for which the number of implicit nodes is zero when convenient, but we never normally regard a pattern as a graph. Patterns and graphs are not normally deemed to have roots unless we specifically say so or it is clear from context.

Definition 2.2 A **rule** D is a triple $(P, root, Red)$ where

(1) P is a pattern known as the pattern of the rule.

(2) $root$ is an explicit node of P called the **root** and all implicit nodes are accessible from $root$. Also if $\sigma(root) = S$, then D is a **rule for** S.

(3) Red is a set of pairs (called **redirections**) of nodes of P such that if $(x, y) \in Red$, then x is explicit, and accessible from $root$. Also if $(x, y), (u, v) \in Red$, then if $x = u$ then $y = v$, and if $x \neq u$, then $\sigma(x) \neq \sigma(u)$. For $(x, y) \in Red$, x is called the **LHS** of the redirection while y is the **RHS**.

The subpattern of P accessible from (and including) the root is called the **left subpattern** of the rule and is usually denoted by L, while nodes of P not in the left subpattern are called **contractum** nodes. Thus another, slightly redundant way of writing a rule is $D = (incl : L \to P, root, Red)$ where $incl$ is the inclusion of the left subpattern into P.

Definition 2.3 A **matching** or homomorphism of a pattern P with root $root$ say, to a graph (or pattern) G at a node $t \in G$ is a map $h : P \to G$ such that

(1) $h(root) = t$

(2) If $x \in P$ is explicit, then $\sigma(x) = \sigma(h(x))$, $A(x) = A(h(x))$, and for all $k \in A(x)$, $h(\alpha(x)[k]) = \alpha(h(x))[k]$.

Thus a matching is a type of graph homomphism in which implicit nodes may match anything but explicit nodes have to behave well. By dropping the condition (1) involving roots, this definition will also suffice for matching general patterns to graphs or other patterns, or for matching graphs to other graphs.

We now turn to the rewriting model itself. An informal summary and diagramatic example follow the formal definitions below.

Definition 2.4 (Contractum Building) Let $D = (incl : L \to P, root, Red)$ be a rule with left subpattern L and let G be a graph. Let $g : L \to G$ be a matching of L to G at some node of G. Then $g(L)$ is called the **redex**. We build first the graph G' given by

(1) $N_{G'} = (N_G \uplus N_P)/\approx$ which is the disjoint union of N_G and N_P factored by the equivalence relation $\approx$, where $\approx$ is the smallest equivalence relation such that[1] $(1, x) \approx (2, n)$ whenever $g(n) = x$.

(2) $\sigma_{G'}(\{(1, x)\}) = \sigma_G(x)$,
$\sigma_{G'}(\{(2, n)\}) = \sigma_P(n)$,
$\sigma_{G'}(\{(1, x), (2, n_1) \ldots (2, n_m)\}) = \sigma_G(x)$,
which is consistent because g is a matching.

(3) $\alpha_{G'}(\{(1, x)\})[k] = \{(1, \alpha_G(x)[k])\ldots\}$ for $k \in A(x)$,
$\alpha_{G'}(\{(2, n)\})[k] = \{(2, \alpha_P(n)[k])\ldots\}$ for $k \in A(n)$,
$\alpha_{G'}(\{(1, x), (2, n_1) \ldots (2, n_m)\})[k] = \{(1, \alpha_G(x)[k]) \ldots\}$ for $k \in A(x)$,
which is again consistent since g is a matching. The ... on the RHS of these cases indicates that the k'th child of say a singleton equivalence class node of G', eg. $\{(1, x)\}$, may not itself be a singleton.

Lemma 2.5 There is a matching $g' : P \to G'$ that extends $g : L \to G$ (allowing for obvious identification of nodes modulo disjoint union etc.).

Proof. Define $g'(2, n) = \{(2, n)\ldots\}$. It is clear that g' has the appropriate properties. ☺

1. The notation $(1, x)$ or $(2, n)$ tags each element of the binary disjoint union with the tag 1 or 2 to unambiguously indicate its origin. Likewise $\{(1, x) \ldots\}$ denotes the equivalence class containing $(1, x)$.

Definition 2.6 (Redirection) Let $D = (incl : L \to P, root, Red)$ be a rule with left subpattern L, G a graph, g a matching of L to G and G' the graph resulting from the construction of 2.4. We construct the graph H given by

(1) $N_H = N_{G'}$,

(2) $\sigma_H = \sigma_{G'}$,

(3) $$\alpha_H(\{(1,x)\})[k] = \begin{cases} \{(2,y)\ldots\} & \text{if } (u,y) \in Red \text{ for some } y \in P \text{ and } u \in g'^{-1}(\alpha_{G'}(\{(1,x)\})[k]) \\ \alpha_{G'}(\{(1,x)\})[k] & \text{otherwise} \end{cases}$$

$$\alpha_H(\{(2,n)\})[k] = \begin{cases} \{(2,y)\ldots\} & \text{if } (u,y) \in Red \text{ for some } y \in P \text{ and } u \in g'^{-1}(\alpha_{G'}(\{(2,n)\})[k]) \\ \alpha_{G'}(\{(2,n)\})[k] & \text{otherwise} \end{cases}$$

$$\alpha_H(\{(1,x),(2,n_1)\ldots(2,n_m)\})[k] =$$
$$\begin{cases} \{(2,y)\ldots\} & \text{if } (u,y) \in Red \text{ for some } y \in P \text{ and } u \in g'^{-1}(\alpha_{G'}(\{(1,x)\ldots\})[k]) \\ \alpha_{G'}(\{(1,x),(2,n_1)\ldots(2,n_m)\})[k] & \text{otherwise} \end{cases}$$

Definition 2.7 If D, G and g are as in 2.4, 2.5, then a **rewrite** of G at x according to D is defined to yield the graph H. G is said to be the **pre-graph**, and H the **post-graph** of the rewrite.

In plain language, the graph G' glues copies of the contractum nodes of P onto G, ensuring that arcs are introduced in such a way that the extended matching $g' : P \to G'$ exists. Another way of visualising the same thing is to take disjoint copies of G and P, and for all redex nodes x, to "pinch together" x and its primage under g. This is similar to the formal construction. Likewise, the redirection phase locates the images in G' of the redirection pairs of D, and swings all arcs incident on the image of the LHS to point to the image of the RHS. There is no ambiguity about redirection as the LHS's of distinct redirections are labelled with different symbols so their images under matching cannot coincide. The diagrams below provide a small example of rewriting. The objects occuring there are named after their role in the above theory. Note that the faint dotted arrows represent the redirections of the rule.

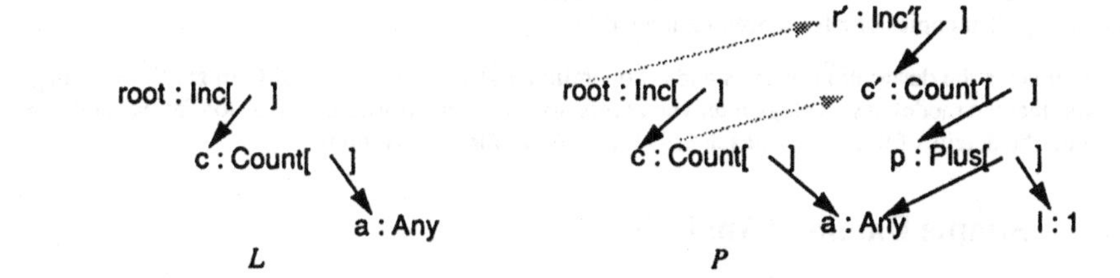

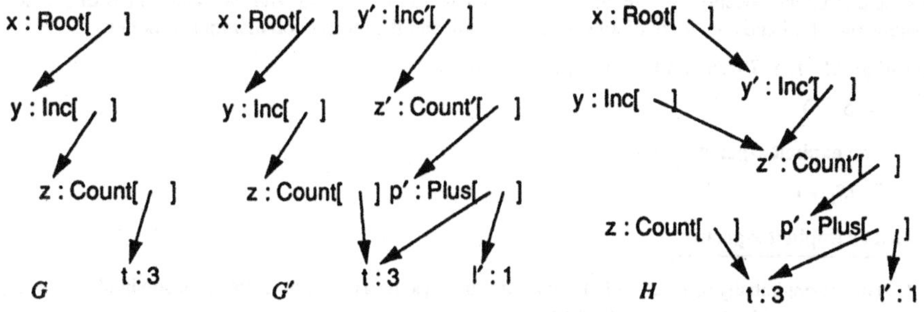

Definition 2.8 An **initial graph** is one consisting of an isolated node with empty arity, labelled by the symbol Initial.

Definition 2.9 A system is a set of rules **R**. An execution of **R** is a sequence of graphs $[G_0, G_1 \ldots]$ of maximal length such that

(1) G_0 is initial,

(2) For all $i \geq 0$ such that i+1 is an index, there is a rule $D \in$ **R** and $x \in G_i$ such that G_i is the pre-graph and G_{i+1} the post-graph of a rewrite of G_i at x according to D.

Note that the above definition does not address garbage collection or reduction strategy, let alone fairness or other issues of concern in concurrency theory[1]. TGRSs may be enhanced with notions that would shed light on these things (see eg. Glauert et al. (1988a, b), Banach (1991c, 1992)) but to do so here would clutter the exposition needlessly.

Definition 2.10 Any graph that occurs in an execution of some system is an **execution graph** of that system.

Definition 2.11 Let G_i be an execution graph and G_{i+1} be its successor in an execution. Let G'_i be the corresponding graph after contractum building. The functions $i_{G_i,G'_i}, r_{G_i,G'_i}, i_{G'_i,G_{i+1}}, r_{G'_i,G_{i+1}}$ are defined as follows:

(1) i_{G_i,G'_i} is the natural injection of nodes of G_i to nodes of G'_i.

(2) $r_{G_i,G'_i} = i_{G_i,G'_i}$.

(3) $i_{G'_i,G_{i+1}}$ is the natural bijection of nodes of G'_i to nodes of G_{i+1}.

(4) $r_{G'_i,G_{i+1}}(x) = i_{G'_i,G_{i+1}}(x)$ unless x has been redirected, i.e. it has had its incident arcs swung over to point to some other node $y \neq i_{G'_i,G_{i+1}}(x) \in G_{i+1}$ during the redirection phase, in which case $r_{G'_i,G_{i+1}}(x) = y$.

We define $i_{G_i,G_{i+1}}$ as the composition $i_{G'_i,G_{i+1}} \circ i_{G_i,G'_i}$, and similarly for $r_{G_i,G_{i+1}}, i_{G_i,G_{i+k}}, r_{G_i,G'_{i+k}}$ etc. Thus $i_{G_i,G_{i+n}}(x)$ is the copy of x in G_{i+n}, while $r_{G_i,G_{i+k}}(x)$ follows the redirection history of x and is the node of G_{i+n} that copies of x have been redirected to.

This completes the description of the standard operational semantics of general term graph rewriting. It turns out that the model has an elegant universal reformulation, but it would take us too far out of our way to go into the details of this. See Banach (1991a, b) for a fuller description.

3 Simple Dataflow Analysis

We shift our attention from execution graphs as such, to relations between symbols that estimate the possible local structures occurring in them. In particular we are concerned with the occurrences of symbols at parent and child ends of an arc of an execution graph and the redirection histories of nodes.

Definition 3.1 If $S, T \in$ **S** and $k \in A(S)$, then the object

$$\Gamma = (S_k, T)$$

is called an **explicit** k**-germ**. An object

$$\Gamma' = (S_k, -)$$

is called an **implicit** k**-germ**.

1. TGRSs were initially invented with the objective of providing a model for graph reduction implementations of functional languages, parallel as well as serial.

An arc (p_l, c) of a graph G is an instance of a germ (S_k, T) iff $\sigma(p) = S$, $\sigma(c) = T$ and $l = k$. It is an instance of $(S_k, -)$ iff $\sigma(p) = S$ and $l = k$. These concepts may also be applied to arcs in patterns, provided we accept that arcs in which the child node is implicit may be instances only of implicit germs. For instance:

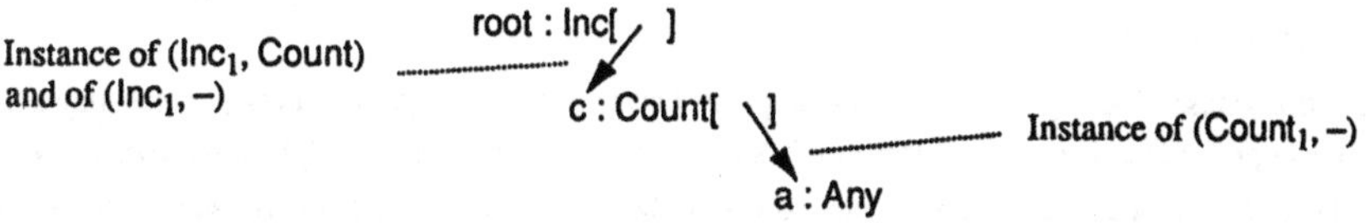

Let a fixed system be given. Our aim is to be able to estimate what germ instances actually occur during the rewriting of a system. Speaking loosely, we do this by estimating what symbols may occur at given child positions of any symbol and also what symbols a given symbol may get redirected to. This is done by an induction over the structure of executions.

Theorem 3.2 Assume a fixed system given. Suppose for all $S \in \mathbf{S}$, $k \in A(S)$ there are sets of symbols $R(S)$, $O_k(S)$ such that the following hold.

(R) $S \in R(S)$ and for all $T \in R(S)$, $R(T) \subseteq R(S)$.

(O_k) For all $T \in O_k(S)$, $R(T) \subseteq O_k(S)$.

Suppose also that for all rules $(incl : L \to P, root, Red)$ in the system we have (1) – (4) below.

(1) Let (p_k, c) be an instance of (S_k, T) in P where p and c are both explicit. Then $T \in O_k(S)$.

(2) Let (a, b) be in Red such that a and b are both explicit with $\sigma(a) = S$ and $\sigma(b) = T$. Then $T \in R(S)$.

(3) Let (p_k, c) be an instance of $(S_k, -)$ in P, and (q_l, c) be an instance of $(T_l, -)$ in L, where c is implicit and p is a contractum node. Then $O_l(T) \subseteq O_k(S)$.

(4) Let (a, b) be in Red such that b is implicit. Let (v_l, b) be an instance of $(T_l, -)$ in L and let $\sigma(a) = S$. Then $O_l(T) \subseteq R(S)$.

Then for every execution graph G_i, if $x \in G_i$ with $\sigma(x) = S$

(a) If (x_k, y) is an arc of G_i with $\sigma(y) = T$, then $T \in O_k(S)$,

(b) If G_j is some execution graph occurring later than G_i in the execution and $r_{G_i,G_j}(x) = z$ with $\sigma(z) = T$, then $T \in R(S)$.

Proof. The proof is by induction over the structure of executions. We will show that the contractum building and redirection phases of a rewrite, both preserve properties (a) and the one-step version of (b), (b1), i.e. where G_j is the successor of G_i in the execution; (b) then follows by induction and (R).

Base Case. The initial graph has no arcs so (a) holds trivially; and there is nothing to prove for (b1) since no rewrite created the initial graph.

Inductive Step. Suppose the hypotheses hold for $G_0 \ldots G_i$.

Contractum building Phase. Let G'_i be the graph after contractum building. Clearly (a) holds by hypothesis for $i_{G_i,G'_i}(G_i)$ so we need only consider the new arcs added during the building. Let (u_k, v) be such an arc and let it be the image under the extended matching $g' : P \to G'_i$ (where P is the pattern of the rule governing the rewrite) of an arc (p_k, c) in P. Now c is either explicit or implicit. If explicit, then (p_k, c) is an instance of some (S_k, T) and by (1), $T \in O_k(S)$. Since g' is an extended matching, $\sigma(u) = S$ and $\sigma(v) = T$ so $\sigma(v) \in O_k(\sigma(u))$. Alternatively, c is implicit. Since there must be a path of length at least one from the root of the left subpattern to c, c has an explicit parent in the left subpattern of the rule. Suppose (q_l, c) is the relevant arc. Since g' is an extended matching, $g'((q_l, c)) = (w_l, v)$ where $w \in i_{G_i,G'_i}(g(L))$. Let

$T = \sigma(w)$. Let $S = \sigma(u)$ and $g'((p_k, c)) = (u_k, v)$ as before. Now $\sigma(v) \in O_l(\sigma(w)) = O_l(T)$ by hypothesis, and $\sigma(v) = \sigma(c)$ because g' is an extended matching. By (3), $O_l(T) = O_l(\sigma(w)) = O_l(\sigma(q)) \subseteq O_k(\sigma(p)) = O_k(\sigma(u)) = O_k(S)$, the penultimate step because g' is an extended matching. Thus $\sigma(v) = \sigma(c) \in O_k(\sigma(u))$ as required for (a). Since there are no redirections in the contractum building phase, $r_{G_i,G'_i} = i_{G_i,G'_i}$ so (b1) holds trivially.

Redirection Phase. Clearly there is nothing to prove for the non-redirected arcs in G_{i+1} since they are just injective copies under $i_{G'_i,G_{i+1}}$, of arcs of G'_i so (a) holds for them. Let $x \in G'_i$ be a redirected node, i.e. one whose incident arcs are to be redirected to $y \in G'_i$ because $(x, y) = (g'(a), g'(b))$ where (a, b) is a redirection of the rule governing the rewrite. By 2.2.(3) a is explicit. Let $S = \sigma(a)$ so $S = \sigma(x)$. Suppose b is explicit and $\sigma(b) = T$. Then by (2), $T \in R(S)$ and so $\sigma(y) \in R(\sigma(x))$. Hence $\sigma(r_{G'_i,G_{i+1}}(x)) \in R(\sigma(x))$ as required for (b1). Suppose alternatively b is implicit. Then there is an arc (q_l, b) in the left subpattern L of the rule such that (q_l, b) is an instance of $(T_l, -)$ say. Since g' is an extended matching, there is an arc $(t_l, y) \in G'_i$ with $g'((q_l, b)) = (t_l, y)$ so $\sigma(t) = T$ and thus by hypothesis $\sigma(y) \in O_l(T)$ and $\sigma(b) \in O_l(T)$. By (4), $O_l(T) \subseteq R(S)$ which gives $\sigma(y) \in R(\sigma(x))$ and $\sigma(r_{G'_i,G_{i+1}}(x)) \in R(\sigma(x))$ as before. Thus (b1) holds for an arbitrary redirection.

We need to show that (a) holds for the redirected arcs. Consider an arc $(w_m, x) \in G'_i$ where x is redirected. We know that $\sigma(x) \in O_m(\sigma(w))$. But also $\sigma(r_{G'_i,G_{i+1}}(x)) \in R(\sigma(x))$. Now $R(\sigma(x)) \subseteq O_m(\sigma(w))$ by (O_k). So $\sigma(r_{G'_i,G_{i+1}}(x)) \in O_m(\sigma(w))$ and (a) holds for the redirected arcs of G_{i+1}. ☺

Example 3.3 We have not presented a complete rule system in the pictures that illustrated rewriting and germ instances above, so strictly speaking we can't give an example of dataflow analysis. Nevertheless we can draw certain conclusions about the $O_k(-)$ and $R(-)$ sets of any system that includes the one rule that we did give. For instance, refering to the pattern P illustrated above, we must have Count′ $\in O_1$(Inc′), Plus $\in O_1$(Count′), O_1(Count) $\subseteq O_1$(Plus), 1 $\in O_1$(Plus), and also Inc′ $\in R$(Inc), Count′ $\in R$(Count). In addition, if rather than (root, r′) $\in$ *Red* we had had (root, a) $\in$ *Red*, then we would have had O_1(Count) $\subseteq R$(Inc) instead of Inc′ $\in R$(Inc). What other relationships hold between symbols, $O_k(-)$ sets and $R(-)$ sets depends of course on what other rules are present in the system.

The above theorem can serve as a specification for an algorithm to determine suitable sets $R(-)$, $O_k(-)$, if the system is finite. The basic strategy is iteration until a least fixed point is reached using the conditions (R), (O_k), (1) – (4) as consistency conditions that drive various symbols into membership of the various $R(-)$ or $O_k(-)$ sets. See Banach (1992), Hankin (1991) for details. The analysis evidently provides a safe estimate of the sets of germ instances that actually occur during executions. In section 5, we will use the properties of the $R(-)$ and $O_k(-)$ sets as the basic driving engine for type inference once we have determined what types are to be in the present context.

4 Simple Type Theories For TGRSs

What should we mean by types for TGRSs? This is not a trivial question since type systems for conventional rewriting systems i.e. TRSs or the lambda calculus, are connected with issues such as strong normalisation. Since not even all TRSs (let alone TGRSs) have the strong normalisation property and since for those that do, the connection with type theories is built on precisely the cornerstone of structural induction over terms that we are forced to abandon, we must look for some weaker framework.

Since TGRSs are so general, it is not possible to demand the property that the type of an execution graph (whatever *that* might be) is an invariant of rewriting. This is the crucial (and suprisingly, seldom stated) subject reduction property of most type systems for TRSs that enables the strong normalisation results to go through. With its abandonment, the way is open to construct a more local concept of type, and for the particular notions of type that we will develop below, the invariant we will end up with, is that the types of graph nodes are invariants of rewriting (particularly of redirection). The proof of this though, is no longer something that can be swept under the carpet as happens for TRSs. Given that we must irrevocably lose context freedom in the passage to the graph world, this is about the best that we could hope for.

The type theories we will construct are about the simplest that one could imagine under these circumstances. They resemble to some degree the phenomena found in conventional imperative languages like Pascal or Ada. The locality of the meaning of types is certainly reminiscent, but the type inference and parametric polymorphism aspects of our schemes are more general. The rest of this subsection sets up the general framework within which both of the type disciplines that we will develop fit. The following two subsections specialise this to the specific disciplines in question.

N.B. Due to pressure of space, and also for reasons of technical convenience, little motivatory material will occur among the definitions and theorems of this section and the next. On a first reading, readers may find it more convenient to briefly skim the definitions in the rest of this section and then skip to the discussion in section 6, where the salient properties of our type system are highlighted and compared to those of conventional systems. The informal impression gained thereby should help to make the structure and detailed content of the intervening technical material more accessible and digestible.

We assume we have an alphabet $\mathbf{Z}^c$ of **type constants** and a disjoint alphabet $\mathbf{Z}^v$ of **type variables.** $\mathbf{Z} = \mathbf{Z}^c \cup \mathbf{Z}^v$. We will use letters from the middle of the Greek alphabet as meta-variables standing for members of either alphabet. If we want to emphasise membership of either $\mathbf{Z}^c$ or $\mathbf{Z}^v$, we will superscript in the appropriate way, eg. ξ^c is in $\mathbf{Z}^c$.

Definition 4.1 A **rule basis** Σ is a partial function $\Omega \times \mathbf{Z}^c \to \mathbf{Z}^*$ where Ω is a set to be specified later. If $\Sigma(\omega, \xi^c)$ is defined (for $\omega \in \Omega$), then $((\omega, \xi^c), \Sigma(\omega, \xi^c))$ is called a **type rule** for ω and ξ^c. We write such a rule using the notation $\omega^{\xi^c \leftarrow \xi_1 \ldots \xi_n}$, or $(\omega, \xi^c) \leftarrow \xi_1 \ldots \xi_n$ if we wish to be less cluttered, where $\xi_1 \ldots \xi_n$ is the value of $\Sigma(\omega, \xi^c)$. A type rule for ω and ξ^c is **linear** iff $\Sigma(\omega, \xi^c)$ contains no more than one occurence of any $\xi^v \in \mathbf{Z}^v$. A rule basis is **linear** iff all its type rules are linear. The **arity** of a type rule, written $A(\omega, \xi^c)$ is the domain of $\Sigma(\omega, \xi^c)$, i.e. the set of indices of $\Sigma(\omega, \xi^c)$. No confusion will arise from this yet other arity concept.

Definition 4.2 An **axiom basis** B, is a partial function $\mathbf{S} \to \mathbf{Z}^c$. When $B(S)$ is defined, if $\xi^c = B(S)$, we say that S^{ξ^c} is a **typed symbol (or t-symbol)**. More generally, we will also consider t-symbols S^θ for any $\theta \in \mathbf{Z}$.

Definition 4.3 A **type structure** (Σ, B) for a given set Ω fixed for the type discipline being considered, consists of a rule basis and an axiom basis. The structure is **linear** iff the rule basis is.

Definition 4.4 A **substitution** s is a map $\mathbf{Z} \to \mathbf{Z}$ which is the identity on $\mathbf{Z}^c$, i.e. $s \in [\mathbf{Z}^v \to \mathbf{Z}] \cup \{Id_{\mathbf{Z}^c}\}$.

4.1 A Crude Type Discipline

Let Ω be the one-point set $\{\bullet\}$. Then since $\Omega \times \mathbf{Z}^c \cong \mathbf{Z}^c$, it is preferable to drop all mention of Ω. We can therefore write a type rule as $(\xi^c, \Sigma(\xi^c))$ or as $\xi^c \leftarrow \xi_1 \ldots \xi_n$ instead of using the more elaborate forms.

Now let some TGRS **R**, and some type structure (Σ, B) be considered fixed, and let **S** and **Z** be the appropriate alphabets.

Definition 4.1.1 Let G be a graph and let $\tau : G \to \mathbf{Z}$ be a map from the nodes of G to type symbols. Then τ is a **typing** of G iff for all $x \in G$,

(1) If $B(\sigma(x))$ is defined, then $\tau(x) = B(\sigma(x))$.

(2) If $\Sigma(\tau(x))$ is defined, then $A(\tau(x)) = A(\sigma(x)) = A(x)$ and there is a(n x-dependent) substitution s_x such that for all $k \in A(x)$,

$$\tau(\alpha(x)[k]) = s_x(\Sigma(\tau(x))[k]).$$

We say that τ is a **concrete typing** of G iff $\tau(G) \subseteq \mathbf{Z}^c$.

Definition 4.1.2 Let typings be given for all execution graphs and let the following hold for all executions of $\mathbf{R}$. If G_i, G_j are in some execution with $i < j$, and τ_i, τ_j are the given typings, then for all $x \in G_i$

$$\tau_i(x) = \tau_j(i_{G_i,G_j}(x)) = \tau_j(r_{G_i,G_j}(x))$$

Then we say that $\mathbf{R}$ is **correctly typed** by (Σ, B). If all such typings are concrete, we say that $\mathbf{R}$ is **concretely correctly typed** by (Σ, B).

4.2 A Simple Imperative Type Discipline

Let Ω be $\mathbf{S}$ the node symbol alphabet, and let some TGRS $\mathbf{R}$, and some type theory (Σ, B) be considered fixed.

Definition 4.2.1 Let G be a graph and let $\tau : G \to \mathbf{Z}$ be a map from the nodes of G to type symbols. Then τ is a **typing** of G iff for all $x \in G$,

(1) If $B(\sigma(x))$ is defined, then $\tau(x) = B(\sigma(x))$.

(2) If $\Sigma(\sigma(x), \tau(x))$ is defined, then $A(\sigma(x), \tau(x)) = A(\sigma(x)) = A(x)$ and there is a(n x-dependent) substitution s_x such that for all $k \in A(x)$,

$$\tau(\alpha(x)[k]) = s_x(\Sigma(\sigma(x), \tau(x))[k]).$$

We say that τ is a **concrete typing** of G iff $\tau(G) \subseteq \mathbf{Z}^c$.

Definition 4.2.2 Let typings be given for all execution graphs and let the following hold for all executions of $\mathbf{R}$. If G_i, G_j are in some execution with $i < j$, and τ_i, τ_j are the given typings, then for all $x \in G_i$

$$\tau_i(x) = \tau_j(i_{G_i,G_j}(x)) = \tau_j(r_{G_i,G_j}(x))$$

Then we say that $\mathbf{R}$ is **correctly typed** by (Σ, B). If all such typings are concrete, we say that $\mathbf{R}$ is **concretely correctly typed** by (Σ, B).

5 The SDLPTI Algorithm

SDLPTI stands for Simple Dataflow Linear Polymorphic Type Inference. Simple and linear because those are the only types of dataflow analysis and polymorphism respectively that we are concerned with. Actually we will see a little later that we can extend the applicability of the algorithm to certain non-linear cases as well.

From now on, we will restrict our attention to the simple imperative type discipline. Results for the crude type discipline are easily recovered by simply forcing the value of the partial function $\Sigma(S, \xi^c)$ to be independent of S, and then discarding S. In fact for the algorithm below all one need do is to delete the short underlined passages in step [10i] and the algorithm becomes correct for the crude type discipline. We assume that $\mathbf{S}$, $\mathbf{Z}$, $\mathbf{R}$ and (Σ, B) are fixed as before. We also insist that $\mathbf{S}$ and $\mathbf{R}$ are finite and that there are at least recursive algorithms for deciding membership of $\mathbf{Z}$ and (Σ, B). Finiteness of $\mathbf{S}$ and $\mathbf{R}$ is sufficient if theorem 3.2 is to serve as the basis for an iterative algorithm for determining $O_k(-)$ and $R(-)$ sets, and a finite number of $O_k(-)$ and $R(-)$ sets is sufficient to ensure termination of algorithm 5.3 below. We assume henceforth that a suitable collection of $O_k(-)$ and $R(-)$ sets have been given for the system $\mathbf{R}$ but we don't care whether they were obtained using the algorithm suggested by 3.2 or by some other magic.

Definition 5.1 Let Q be a set of t-symbols. Then $Q^\tau = \{\theta \mid S^\theta \in Q\}$. We say that Q and Q^τ are **acceptable** iff Q^τ contains at most one type constant. We say Q and Q^τ are **type singletons** iff Q^τ is a singleton.

Theorem 5.2 Suppose (Σ, B) is linear. Suppose all symbols in **S** are decorated with a type, yielding a set of t-symbols t-**S**. For each t-symbol, let $R(-)$ and $O_k(-)$ sets of t-symbols be given that satisfy the hypotheses of theorem 3.2 (where we ignore the type decorations for the purposes of 3.2). Suppose also for each $S^{\xi} \in$ t-**S**,

(1) If $B(S)$ is defined, then $B(S) = \xi$,

(2) If $\Sigma(S, \xi)$ is defined, then $A(S) = A(S, \xi)$, and for each $k \in A(S, \xi)$ such that $\Sigma(S, \xi)[k] = \theta^c \in \mathbf{Z}^c$, for all $T^{\zeta} \in O_k(S^{\xi})$, $\zeta = \theta^c$,

(3) $R(S^{\xi})$ is a type singleton.

Then (Σ, B) correctly types **R** when each execution graph node x is typed by θ where $\sigma(x)^{\theta}$ is in t-**S**.

Proof. Let G_i be an execution graph and $x \in G_i$. Map all nodes to type symbols as suggested. Then (1) implies (1) of 4.2.1. By theorem 3.2, if y is the k'th child of x, then $\sigma(y) \in O_k(\sigma(x))$, and thus (2) trivially guarantees (2) of 4.2.1 because (Σ, B) is linear. So the map is indeed a typing of G_i. Theorem 3.2 also guarantees that for all $j > i$, $\sigma(r_{G_i,G_j}(x)) \in R(\sigma(x))$, hence (3) ensures that we have a correct typing of the system. ☺

We have suggested that for dataflow analysis, suitable $R(-)$ and $O_k(-)$ sets may be obtained by iteration. We now present an algorithm for determining suitable sets of t-symbols t-**S** for which the conditions of theorem 5.2 hold.

Algorithm 5.3 (The SDLPTI algorithm)

[1] Decorate each $S \in \mathbf{S}$ with a fresh type variable from $\mathbf{Z}^v$ which does not occur in Σ. Call the resulting set of t-symbols t-**S**.

[2] Let t-$R(-)$ and t-$O_k(-)$ sets be given by

$$T^{\zeta} \in \text{t-}R(S^{\xi}) \Leftrightarrow S^{\xi}, T^{\zeta} \in \text{t-}\mathbf{S} \text{ and } T \in R(S)$$
$$T^{\zeta} \in \text{t-}O_k(S^{\xi}) \Leftrightarrow S^{\xi}, T^{\zeta} \in \text{t-}\mathbf{S} \text{ and } T \in O_k(S)$$

[3] Define ρ on t-**S** by

$$T^{\zeta} \rho S^{\xi} \Leftrightarrow T^{\zeta} \in \text{t-}R(S^{\xi})$$

and let ρ_s be the symmetric closure of ρ. Then ρ_s is an equivalence relation. Write $[S^{\xi}]$ for the component of ρ_s containing S^{ξ}.

[4] For each $[S^{\xi}]$ let

$$[S^{\xi}]_B = [S^{\xi}] \cup \{T^{\zeta^c} \mid B(T) = \zeta^c \text{ and } T^{\theta^v} \in [S^{\xi}] \text{ for some } \theta^v \in \mathbf{Z}^v\}$$

[5] Unify each $[S^{\xi}]^{\xi}_B$. That is

If any $[S^{\xi}]_B$ is not acceptable **Then** FAIL and exit

Else Define a substitution I^0 by

For all $[S^{\xi}]_B$ **Do**

If $[S^{\xi}]^{\xi}_B$ contains a concrete type symbol θ^c

Then For all $\zeta \in \mathbf{Z}^v \cap [S^{\xi}]^{\xi}_B$, let $I^0(\zeta) = \theta^c$

Else Choose some $\theta^v \in [S^{\xi}]^{\xi}_B$ and for all $\zeta \in \mathbf{Z}^v \cap [S^{\xi}]^{\xi}_B$, let $I^0(\zeta) = \theta^v$

For all remaining $\theta \in \mathbf{Z}^v$, $I^0(\theta) = \theta$

[6] Let t-$\mathbf{S}^0$ be given by $S^\xi \in$ t-$\mathbf{S} \Leftrightarrow S^{I^0(\xi)} \in$ t-$\mathbf{S}^0$. Let ρ^0 be given by $T^\zeta \, \rho \, S^\xi \Leftrightarrow T^{I^0(\zeta)} \, \rho^0 \, S^{I^0(\xi)}$.
Let ρ^0_s be the symmetric closure of ρ^0 and write $[S^\xi]^0$ for a typical component. Let t-$O^0_k(-)$ sets be given by $T^\zeta \in$ t-$O_k(S^\xi) \Leftrightarrow T^{I^0(\zeta)} \in$ t-$O^0_k\,(S^{I^0(\xi)})$.

[7] Let $i = 0$.

[8i] **Repeat**

[9i] $i := i + 1$

[10i] **If** for any $S^\xi, T^\zeta \in$ t-$\mathbf{S}^{i-1}$ we have $\xi \in \mathbf{Z}^c$ and Σ contains a rule for $\underline{S}$ and ξ such that $A(\underline{S}, \xi) \neq A(S)$, or there is a $k \in A(S, \xi)$ such that $\Sigma(S, \xi)[k] = \theta^c \in \mathbf{Z}^c$ and $T^\zeta \in$ t-$O^{i-1}_k(S^\xi)$ with $\theta^c \neq \zeta \in \mathbf{Z}^c$
Then FAIL and exit
Else Define a substitution I^i by
  For all $S^\xi, T^\zeta \in$ t-$\mathbf{S}^{i-1}$ **Do**
    If $\zeta \in \mathbf{Z}^v, \xi \in \mathbf{Z}^c$ and Σ contains a rule for $\underline{S}$ and ξ such that there is a $k \in A(\underline{S}, \xi)$ such that $T^\zeta \in$ t-$O^{i-1}_k(S^\xi)$
    Then $I^i(\zeta) = \theta^c$ where θ^c is any such ξ
  For all remaining $\theta \in \mathbf{Z}^v$, $I^i(\theta) = \theta$

[11i] Let t-$\mathbf{S}^i$ be given by $S^\xi \in$ t-$\mathbf{S}^{i-1} \Leftrightarrow S^{I^i(\xi)} \in$ t-$\mathbf{S}^i$. Let ρ^i be given by $T^\zeta \, \rho^{i-1} \, S^\xi \Leftrightarrow T^{I^i(\zeta)} \, \rho^i \, S^{I^i(\xi)}$.
Let ρ^i_s be the symmetric closure of ρ^i and write $[S^\xi]^i$ for a typical component. Let t-$O^i_k(-)$ sets be given by $T^\zeta \in$ t-$O^{i-1}_k(S^\xi) \Leftrightarrow T^{I^i(\zeta)} \in$ t-$O^i_k\,(S^{I^i(\xi)})$.

[12i] **Until** $I^i = \mathrm{Id}_{\mathbf{Z}}$

[13] Let f be the final value for i. Output t-$\mathbf{S}^f$ and SUCCEED.

As a matter of teminology, if the SDLPTI algorithm SUCCEEDs on some system, then for every $S \in \mathbf{S}$, if $S^\xi \in$ t-$\mathbf{S}^f$ where t-$\mathbf{S}^f$ is the set output by the algorithm, we call ξ the **SDLPTI type** of S.

Theorem 5.4 Let $\mathbf{S}$, $\mathbf{Z}$, $\mathbf{R}$, (Σ, B) be given as before with $\mathbf{S}$ and $\mathbf{R}$ finite and membership of $\mathbf{Z}$ and (Σ, B) recursively decidable. Suppose (Σ, B) is linear.

(1) The SDLPTI algorithm terminates and either SUCCEEDs or FAILs.

(2) If it SUCCEEDs, then $\mathbf{R}$ is correctly typed by (Σ, B) and a suitable typing of any execution graph may be given by typing each node x of the graph by the SDLPTI type of $\sigma(x)$.

Proof Because $\mathbf{S}$ and $\mathbf{R}$ are finite and all the substitutions mentioned have finite support, all the individual steps of the algorithm are finitely computable. To show that there are a finite number of iterations of steps [9i] – [12i] we note that each iteration except the last retypes a finite non-zero number of symbols decorated with variables, to symbols decorated with constants. Since no symbol decorated with a constant is ever retyped, and there are only a finite number of symbols involved to start with, the loop must terminate. The only exit points of the algorithm are when it SUCCEEDs or FAILs, so we have (1).

We note that by 3.2.(R), the ρ^i relations are reflexively and transitively closed, so their symmetric closures are indeed equivalence relations.

As to the output when the algorithm SUCCEEDs, we need only show that the three conditions of theorem 5.2 hold with respect to the set of t-symbols t-$\mathbf{S}^f$, created by the algorithm.

To satisfy 5.2.(3), we must ensure that all members of an $R(-)$ set have the same type. Similarly, to satisfy 5.2.(1), we must ensure any symbol typed by the axiom basis B, has the same type in t-$\mathbf{S}^f$. Both tasks are accomplished by steps [1] – [5], which build the equivalence classes of ρ_s and perform the unification with the axiom basis.

It remains to unify the resulting type labelling of symbols with the rule basis Σ. This is accomplished by the loop [8i] – [12i]. To enforce conformance to Σ we have to check relationships between type symbols at parent and child ends of all potential arcs of execution graphs. All such arcs are instances of germs that can be constructed from the $O_k(-)$ sets. So let $(S^{\xi}{}_k, T^{\zeta})$ be such a germ where the symbols bear the type symbols appropriate to the i'th iteration of the algorithm. If $\xi \in \mathbf{Z}^v$ there is nothing to check. Otherwise if $\xi \in \mathbf{Z}^c$ and $\Sigma(S, \xi)$ is defined, then arities must match, and if $\Sigma(S, \xi)[k] \in \mathbf{Z}^c$, then if $\zeta \in \mathbf{Z}^c$ too, then ζ must $= \Sigma(S, \xi)[k]$ (else FAILure). If $\Sigma(S, \xi)[k] \in \mathbf{Z}^c$ but $\zeta \in \mathbf{Z}^v$, then ζ must be instantiated to $\Sigma(S, \xi)[k]$. There may be several such competing instantiations for ζ arising from other germs $(U^{\chi}{}_j, T^{\zeta})$. Any one such instantiation will do, since if they all agree (i.e. all wish to map ζ to the same $\theta^c \in \mathbf{Z}^c$) then there is no conflict, while if they disagree, FAILure is invevitable at the i+1'th iteration. This explains the form of the else branch of [10i].

So the loop progressively instantiates type variables, only when needed, until termination or FAILure. It clearly generates a most general unification of Σ with t-$\mathbf{S}^0$, hence a most general unification of t-$\mathbf{S}$ with (Σ, B). Thus t-$\mathbf{S}^f$ is a correct typing of **R**. ☺

It goes without saying that the above algorithm, is not one that would be implemented as given, being constructed primarily for readability. Many aspects can be optimised to produce a practical algorithm, but these issues are outside the scope of this paper.

As promised, we will strengthen theorem 5.4 to enable the SDLPTI algorithm to effectively generate correct typings in certain non-linear cases. The problem with non-linearity is that if $k_1 \neq k_2$ and $\Sigma(S, \xi^c)[k_1] = \zeta^v = \Sigma(S, \xi^c)[k_2]$ then if a graph node x is of type ξ^c then its k_1 and k_2'th children must be of the same type, θ_1 say, even if that type is different to the type, θ_2 say, of the k_1 and k_2'th children of y, where y is also of type ξ^c. Since dataflow analysis and hence the SDLPTI algorithm works on each germ (and perforce each germ instance) independently, they are unable to distinguish this case from the case where the k_1'th child of x is of type θ_1, the k_2'th child of x is of type θ_2, the k_1'th child of y is of type θ_1, and the k_2'th child of y is of type θ_2. The latter is a non-typing of the graph containing x and y according to 4.5.

However, if we can be sure that every node of type ξ^c created during a rewrite has its k_1 and k_2'th children of equal type at the moment of creation, then we can use the SDLPTI algorithm to check that the system is correctly typed. This is because the SDLPTI algorithm ensures that the functions $i_{G,H}(-)$, $r_{G,H}(-)$ preserve type, so if the children are of equal type at the point the parent is created, then they remain of equal type subsequently, despite redirection. Clearly this is an observation outside the remit of the standard SDLPTI algorithm.

Theorem 5.5 Let **S**, **Z**, **R** and (Σ, B) be as before, but suppose (Σ, B) is non-linear. For $\xi^c \in \mathbf{Z}^c$ and $\zeta^v \in \mathbf{Z}^v$, let $W_{\xi^c}(\zeta^v)$ be the set of indices $W_{\xi^c}(\zeta^v) = \{k \mid \Sigma(S, \xi^c)[k] = \zeta^v\}$. Suppose for all (S, ξ^c) and ζ^v such that $\Sigma(S, \xi^c)$ is defined and non-linear and $W_{\xi^c}(\zeta^v)$ is a non-singleton, and for all x a contractum node of the pattern P of a rule D of **R**, with $\sigma(x) = S$ and ξ^c the SDLPTI type of S, there is a $\theta \in \mathbf{Z}$ (depending on both x and $W_{\xi^c}(\zeta^v)$), such that for all $k \in W_{\xi^c}(\zeta^v)$, either

(1) The k'th child of x is an explicit node y_k, and θ is the SDLPTI type of $\sigma(y_k)$, or

(2) The k'th child of x is an implicit node, and $O_k(\sigma(x)^{\xi^c})^{\tau} = \{\theta\}$ (i.e. a singleton SDLPTI type).

Then the output of the SDLPTI algorithm is a correct typing of the system.

Proof. If the SDLPTI algorithm SUCCEEDs, then we know we have a correct typing provided we can be sure that the extra condition implicit in 4.5.(2) for non-linear rules holds, namely that if $\Sigma(\xi)[k_1] = \zeta^{\mathrm{v}} = \Sigma(\xi)[k_2]$ for $k_1 \neq k_2$, then any node of type ξ has children of identical type at its k_1 and k_2'th positions.

From the properties of dataflow analysis and the SDLPTI algorithm we know, as remarked previously, that if (p_k, c) is some arc of an execution graph G_i, then for all G_j with $j > i$, $i_{G_i,G_j}(p)$ is of the same type as p, and $r_{G_i,G_j}(c)$ is of the same type as c. So to check the property required, we need only check it at the points at which contractum nodes x are instantiated as nodes of execution graphs. But at all such points we have the structure of the rule governing the rewrite to help us, because it gives direct information about the children of x and hence about the children of the instantiation of x. If such a child is explicit, its symbol can be read off from the rule and its type determined (being the SDLPTI type), thus giving the type of the child of the instantiation. This corresponds to case (1) above. If the child is implicit however, we must refer to the set $O_k(\sigma(x)^{\xi^c})$. This may contain many symbols, but if it happens to be a type singleton, we can likewise determine the type of the child of the instantiation. This corresponds to case (2) above. Thus under the given hypotheses, the uniformity of the types of the children across the relevant set of positions of x gives us the equality we need for the extra condition. ☺

The above theorem has slightly different implications depending on whether the θ for a given $W_{\xi^c}(\zeta^{\mathrm{v}})$ is concrete or not. If it is concrete, then it clearly tells us the type of the relevant children of (instantiations of) x explicitly. On the other hand, if it is a type variable, then it merely tells us that it is *possible* to type the system consistently, without offering us an explicit type for the children in question. In such a case (as in others in which the SDLPTI type of some symbols are type variables) the type structure we started with is too weak to fully type the system. Only a less general unifier would concretely type the system, and for that we need some more axioms and/or some more rules. In any event, the success of the SDLPTI algorithm assures us that such a consistent extension of the type structure exists.

Note that the power of the above result for dealing with genuine non-linear polymorphism comes from the dependence of θ on x. This in turn only has any real force when condition (2) is never needed, i.e. when all relevant children of contractum nodes labelled by symbols of type ξ^c are explicit, freeing the dependence of θ upon x from $O_k(\sigma(x))$ which for a fixed symbol $\sigma(x)$, does not depend on the node x that happens to bear that symbol. Contrast this with the linear case where $O_k(\sigma(x))$ does not have to be a type singleton and can thus contain several differently typed $R(-)$ sets, this being the chief means by which polymorphism is achieved within our TGR framework.

6 Discussion and Conclusions

We have defined TGRSs, dataflow analysis, type structures, and we have shown how the properties of dataflow analysis are suited to doing type inference of the kind required by our particular brand of type theory. Some points are worthy of further discussion, so let's look at an example and compare the present framework to conventional ones.

Example 6.1 We'll continue our slightly half-hearted example introduced previously, constrained as we inevitably are by not having given a complete system. Again we'll refer to the P shown in section 2. Now our definitions for correctness made no mention of the typing of rules. This was deliberate, since the semantics of the variable Any nodes in a pattern (by which we mean the attributes of the execution graph nodes that they will match) are entirely dependent on the rest of the system, and trying to assign types to them cannot be contemplated until the existence of a correct typing for the whole system has been established. In this sense, talking of the typing of P is contingent on such a global typing. Assuming such, the dataflow analysis of example 3.3 and the properties of the unification algorithm 5.3 allow us to infer certain things about symbols and germ instances occuring in P.

Let us start with the crude discipline and let (Σ, B) be given by $(\Sigma_1 \cup \{\mathit{Int} \leftarrow \mathit{Int}, \mathit{Int}\}, B_1 \cup \{\mathsf{Inc}^{\mathit{Up}}, \mathsf{Plus}^{\mathit{Int}}, \mathsf{Count}^{C}\})$ where (Σ_1, B_1) refer to the other parts of the system. Since $\mathsf{Inc}' \in R(\mathsf{Inc})$, $\mathsf{Inc}' \rho^0_s \mathsf{Inc}$ whence $\mathsf{Inc}'^{\mathit{Up}}$ is an SDLPTI typing. Similarly Count'^{C}. We see also that if there were a monomorphic type rule for C, it would have to be $C \leftarrow \mathit{Int}$ because of the instance of the germ $(\mathsf{Count}'_1, \mathsf{Plus})$ in the contractum of P. From $\mathsf{Plus}^{\mathit{Int}}$ and $\mathit{Int} \leftarrow \mathit{Int}, \mathit{Int}$ we deduce that $\mathsf{a} : \mathsf{Any}^{\mathit{Int}}$ and $\mathsf{I} : 1^{\mathit{Int}}$. The former is consistent with the $C \leftarrow \mathit{Int}$ we discovered above and tells us that any node matched to the particular implicit node a in a redex for our rule must be of type Int if the system is to be SDLPTI-typeable. The latter spells trouble. From $\mathit{Int} \leftarrow \mathit{Int}, \mathit{Int}$ and 1^{Int} we deduce that $A(1) = \{1...2\}$. But this is nonsense since $\mathsf{I} : 1$ has no children in P and I is an explicit node. This means that any system including our rule is not SDLPTI-typeable using the crude discipline and given (Σ, B). We therefore see that the crude discipline is very severe in restricting the arities of similarly typed symbols to be the same.

Now let us examine the same example using the simple imperative discipline and let (Σ, B) be $(\Sigma_1 \cup \{(\mathsf{Plus}, \mathit{Int}) \leftarrow \mathit{Int}, \mathit{Int}, (1, \mathit{Int}) \leftarrow\ \}, B_1 \cup \{\mathsf{Inc}^{\mathit{Int}}, \mathsf{Count}^{C}, \mathsf{Plus}^{\mathit{Int}}, 1^{\mathit{Int}}\})$. Note that since the types of child nodes now depend on the parent type and parent symbol rather than just the parent type alone, the previous problem does not arise and we can even safely have $\mathsf{Inc}^{\mathit{Int}}$ in B. On the down side, we have to put more into the basis because of the weaker constraints enforced. Much as before, we can conclude $\mathsf{Inc}'^{\mathit{Int}}$, Count'^{C} and $\mathsf{a} : \mathsf{Any}^{\mathit{Int}}$, and can consistently postulate the type rules $(\mathsf{Inc}, \mathit{Int}) \leftarrow C$, $(\mathsf{Inc}', \mathit{Int}) \leftarrow C$, $(\mathsf{Count}, C) \leftarrow \mathit{Int}$, $(\mathsf{Count}', C) \leftarrow \mathit{Int}$.

So the imperative discipline allows much more natural looking rules to be well typed. One can see the similarity with imperative languages, where operators are endowed with a result type and sequence of operand types, and type checking proceeds by matching the type at a point in the program against the type demanded for a child at the relevant position of its parent (as in Pascal); or more generally some inference is done (as in Ada).

Note also that the presentation of type rules as $(S, \xi^c) \leftarrow \zeta_1 ... \zeta_n \in \mathbf{S} \times \mathbf{Z}^c \rightarrow \mathbf{Z}^*$ apparently allows overloading by permitting

$$(S, \xi^c_1) \leftarrow \zeta_1 ... \zeta_n \text{ and } (S, \xi^c_2) \leftarrow \zeta'_1 ... \zeta'_n \quad (\text{with } \xi^c_1 \neq \xi^c_2)$$

to coexist in Σ. This is true, but useless in the context of this paper since the first thing that algorithm 5.3 does is to attempt to unify the types of all occurrences of S in step [5] so at most one of the above rules would ever be used. To exploit potential overloading we need to re-engineer the dataflow analysis with a go-faster supercharged version, capable of discriminating different occurrences of the same node symbol in some way. A more finegrained version of dataflow analysis can do this but is beyond the scope of this paper. (In actual fact it would inevitably go slower rather than faster.)

The main reason we used the more relational presentation for rules, i.e. $\mathbf{S} \times \mathbf{Z}^c \rightarrow \mathbf{Z}^*$ rather than the functional $\mathbf{S} \rightarrow \mathbf{Z}^c \rightarrow \mathbf{Z}^*$, is the uniformity of presentation of the crude and imperative disciplines it allows. (We note that a fully relational $\mathbf{S} \times \mathbf{Z}^c \times \mathbf{Z}^*$ for the rule basis, and $\mathbf{S} \times \mathbf{Z}^c$ for the axiom basis could have been contemplated.) Presenting crude rules in the form $\{\bullet\} \rightarrow \mathbf{Z}^c \rightarrow \mathbf{Z}^*$ would clearly not have worked. As it is, (very nearly) the same theory will do for both disciplines, useful for comparison and for saving space. This also neatly illustrates the nice separation of concerns achieved by making the inference engine of the type discipline (dataflow analysis), independent of the correctness checker (unification). Either or both components may be traded up for a more powerful model relatively independently. (We ruefully observe that the first thing that the design of an efficient implementation of either discipline would do, would be to jam the loops implicit in dataflow analysis and unification, wrecking this independence.)

Now for some more general remarks. As with most useful type systems, typability of systems in the dynamic sense of 4.1.2 or 4.2.2 is undecidable (it is easy to reduce the halting problem to it); thus the SDLPTI algorithm provides an intentionally based approximation to typability. The algorithm reminds us of some features of other type systems. Prominent among these is the Hindley-Milner (H-M) system (Hindley

(1969), Milner (1978)), whose use of unification inspired our use of it here. The main difference between ours and the H-M system is of course the absence of structural induction and its replacement by dataflow analysis. This global analysis feature is somewhat reminiscent of the Milner-Tofte (M-T) type system for ML with assignments[1] (Milner and Tofte (1991), Tofte (1990)), though we have not used co-induction in the explicit way that they do. Moreover for us, even the patterns occurring in rules, which are the templates for the objects we want to type i.e. the execution graphs, can be cyclic, and fail to be freely generated by recursion. This is of course fundamental and is the main impetus for bringing in the dataflow analysis.

Turning now to the types themselves, the types in our system fail to have anything but a trivial structure, unlike the H-M case where there are typically formation rules such as

$$\frac{\alpha \text{ is a type} \quad \beta \text{ is a type}}{\alpha \times \beta \text{ is a type}} \qquad\qquad \frac{\alpha \text{ is a type} \quad \beta \text{ is a type}}{\alpha \rightarrow \beta \text{ is a type}}$$

(it might indeed have been more honest to call our system sort inference rather than type inference). However this is something we can emulate easily enough. Let us take an H-M style language of types **TL** given by the syntax

$$\alpha \in \mathbf{TL} = b \mid v \mid \alpha_1 \rightarrow \alpha_2 \mid \alpha_1 \times \alpha_2$$

where b represents base types (*Int*, *Bool* ...) and v represents variables. Let $\lceil\ \rceil$ be a Gödelisation of **TL** that maps each $\alpha \in \mathbf{TL}$ to its "Gödel number" $\lceil \alpha \rceil$ in $\mathbf{Z}^c$. With a suitable reinterpretation of the unification steps in algorithm 5.3 involving the Gödel number coding of unification in **TL**, algorithm 5.3 will do duty as a correctness checker/type inferer for the new scheme; the occurs check maintaining the termination properties of 5.3. We have thus another separation of concerns, the separation of unification in **TL** from other aspects of our system. Let us briefly look at how this emulation works. There are two basic styles we can use, the functional and the applicative. The functional style is similar to what we are used to already. Say we wish to apply an operator $F^{\alpha \times \beta \rightarrow \gamma}$ to operands A^{α} and B^{β}, giving $(F^{\alpha \times \beta \rightarrow \gamma}(A^{\alpha}, B^{\beta}))^{\gamma}$. We can emulate this H-M situation by having $\{F^{\lceil \gamma \rceil}, A^{\lceil \alpha \rceil}, B^{\lceil \beta \rceil}\} \subseteq B$, and a rule $(F, \lceil \gamma \rceil) \leftarrow \lceil \alpha \rceil, \lceil \beta \rceil \in \Sigma$. On the other hand, the applicative style introduces special purpose application symbols **App***n* (of arity $\{1 \ldots n\}$) to represent the explicit application of an operator to its arguments. With the previous $F^{\alpha \times \beta \rightarrow \gamma}$, A^{α}, B^{β}, our little example becomes $(\mathbf{App3}(F^{\alpha \times \beta \rightarrow \gamma}, A^{\alpha}, B^{\beta}))^{\gamma}$. Our emulation now requires $\{F^{\lceil \alpha \times \beta \rightarrow \gamma \rceil}, A^{\lceil \alpha \rceil}, B^{\lceil \beta \rceil}, \mathbf{App3}^{\lceil \gamma \rceil}\} \subseteq B$, and $(\mathbf{App3}, \lceil \gamma \rceil) \leftarrow \lceil \alpha \times \beta \rightarrow \gamma \rceil, \lceil \alpha \rceil, \lceil \beta \rceil \in \Sigma$. Note that in this case we really would need to overload the **App3** symbol, since we would have to assume that there were other operators such as $G^{\alpha' \times \beta' \rightarrow \gamma}$ of type different from F, which would also use the **App3** combinator. This in turn would require us to at least use the finegrained version of dataflow analysis to have a non-trivial system. In fact all the rules of the system would be instantiations of

$$(\mathbf{App}n, \lceil v_n \rceil) \leftarrow \lceil v_1 \times v_2 \times \ldots \times v_{n-1} \rightarrow v_n \rceil, \lceil v_1 \rceil, \lceil v_2 \rceil \ldots \lceil v_{n-1} \rceil$$

where n is a meta-variable and the v_i's are variables of **TL** (note that any non-ground instantiation of the above is non-linear modulo Gödel numbering). In such a case a strategy for for typechecking that just enforced the consistency condition on the children of an **App***n* node and infered the type of the parent, rather than concentrating on the parent-child type relationship, would work more simply. Modifying algorithm 5.3 to do this is not too hard.

One aspect of H-M and M-T systems definitely missing from ours is the necessity to deal with bound variables and non-trivial environments created by static scoping constructs such as (**let** $x = e$ **in** e') and lambda abstractions. The absence of these features, attributable to the "flatness" of TGRSs affords us some considerable simplification.

In case the reader by now thinks that the crude discipline has no merits not amply exceeded by the imperative discipline, we point out one *final* property not shared by its imperative brother or his fancier cousins.

1. The author is indebted to Kris Rose for bringing the Milner-Tofte work to his attention, and he intends to explore the connection between the techniques used here and those of the M-T system elsewhere.

Consider a crude style rule basis Σ that is concrete (no polymorphism) and complete (each type has a rule). Build the final type-graph Δ, whose nodes are the concrete type symbols, and whose arcs are given by: for each ξ, if $\xi \leftarrow \zeta_1 \ldots \zeta_n$ is the rule for ξ, then the k'th child of ξ is ζ_k. If a system **R** is well typed acording to Σ and a suitable B, then for every execution graph G of the system, there is a node-symbol-forgetting homomorphism $h : G \rightarrow \Delta$. Readers should convince themselves that this property fails for the imperative discipline.

To conclude then, we have discussed type inference for TGRSs and shown that credible type systems can be built using dataflow analysis and unification, even in the most general case where no particular nice structural properties are assumed for the system, fulfilling the promise in the conclusion of Banach (1989). We have also suggested that more powerful extensions of these systems can be contemplated. These extensions will be described elsewhere.

References

Banach R. (1989), Dataflow Analysis of Term Graph Rewriting Systems, *in* proc. PARLE-89, Odijk E., Rem M., Syre J-C. *eds.*, Springer, LNCS **366** 55-72.

Banach R. (1991a), DACTL Rewriting is Categorical, *in* proc. SemaGraph-91, University of Nijmegen Dept. of Informatics Technical Report **91-25** part II 339-357. Also see *in*: Term Graph Rewriting: Theory and Practice, John Wiley, 1992, *to appear.*

Banach R. (1991b), Term Graph Rewriting and Garbage Collection à la Grothendieck. *Submitted to TCS.*

Banach R. (1991c), MONSTR: Term Graph Rewriting for Parallel Machines, *in* proc. SemaGraph-91, University of Nijmegen Dept. of Informatics Technical Report **91-25** part II 251-260. Also see *in*: Term Graph Rewriting: Theory and Practice, John Wiley, 1992, *to appear.*

Banach R. (1992), MONSTR, *in preparation.*

Barendregt H.P. (1984), The Lambda Calculus. Its Syntax and Semantics, North-Holland.

Barendregt H.P., van Eekelen M.C.J.D., Glauert J.R.W., Kennaway J.R., Plasmeijer M.J., Sleep M.R. (1987), Term Graph Rewriting, *in* proc. PARLE-87, de Bakker J.W., Nijman A.J., Treleaven P.C. *eds.*, Springer, LNCS **259** 141-158.

Farmer W.M., Watro R.J. (1990), Redex Capturing in Term Graph Rewriting, Int. Jour. Found. Comp. Sci. **1** 369-386, and *in* proc. RTA-91, R.V. Book *ed.*, Springer, LNCS **488** 13-24.

Girard J-Y., Taylor P., Lafont Y. (1989), Proofs and Types, Cambridge Tracts in Theoretical Computer Science 7, CUP.

Glauert J.R.W., Kennaway J.R., Sleep M.R., Somner G.W. (1988a), Final Specification of DACTL, Internal Report SYS-C88-11, School of Information Systems, University of East Anglia, Norwich, U.K.

Glauert J.R.W., Hammond K., Kennaway J.R., Papdopoulos G.A., Sleep M.R. (1988b), DACTL: Some Introductory Papers, School of Information Systems, University of East Anglia, Norwich, U.K.

Hankin C. (1991), Static Analysis of Term Graph Rewriting Systems, *in* proc. PARLE-91, Aarts E.H.L., van Leeuwen J., Rem M. *eds.*, Springer, LNCS **506** 367-384.

Hindley R. (1969), The Principal Type-Scheme of an Object in Combinatory Logic, Trans. Amer. Math. Soc. **146** 29-60.

Huet G. (1990), Logical Foundations of Functional Programming, Addison-Wesley.

Kennaway J.R., Klop J-W., Sleep M.R., de Vries F-J. (1991), Transfinite Reductions in Orthogonal Term Rewrite Systems, *in* proc. RTA-91, R.V. Book *ed.*, Springer, LNCS **488** 1-12, and Report CS-R9041, CWI Amsterdam.

Milner R. (1978), A Theory of Type Polymorphism in Programming, Jour. Comp. Sys. Sci. **17** 348-375.

Milner R., Tofte M. (1991), Co-induction in Relational Semantics, Theor. Comp. Sci. **87** 209-220.

Peyton-Jones S.L. (1987), The Implementation of Functional Programming Languages, Prentice-Hall.

Tofte M. (1990), Type Inference for Polymorphic References, Inf. and Comp. **89** 1-34.

Consistency and Semantics of Equational Definitions over Predefined Algebras

Valentin Antimirov[1] * and Anatoli Degtyarev[2]

[1] Computer Science Department, Aarhus University, Aarhus DK-8000, Denmark
email : anti@daimi.aau.dk
[2] Department of Cybernetics, Kiev University, 252127, Kiev, Ukraine
email : caphedra%d105.icyb.kiev.ua

Abstract. We introduce and study the notion of an equational definition over a predefined algebra (EDPA) which is a modification of the notion of an algebraic specification enrichment. We argue that the latter is not quite appropriate when dealing with partial functions (in particular, with those defined by non-terminating functional programs), and suggest EDPA as a more adequate tool for specification and verification purposes. Several results concerning consistency of enrichments and correctness of EDPA are presented. The relations between EDPA and some other approaches to algebraic specification of partial functions are discussed.

1 Introduction

1.1 Motivation

Algebraic specification and term-rewriting methods seem very convenient to use in the following wide-spread situation: given a set A of data with several predefined functions $g_1, \ldots, g_k$ on it, one needs to define (sometimes constructively) a set of new functions on A (a "specification" or "programming" stage) and to analize their logical properties (a "verification" stage).

A standard algebraic specification approach to this task would consist of two steps:

i) to consider the set A with $g_1, \ldots, g_k$ as an algebra $\mathcal{A}$, and to specify it as an abstract data type – an *initial model* of some basic specification $SP_{\mathcal{A}}$;

ii) to construct an *enrichment* $SP' = SP_{\mathcal{A}} + (F, R)$ of the basic specification where a set of new axioms R (together with those of $SP_{\mathcal{A}}$) are supposed to define the meaning of the new function symbols $f \in F$.

This construction was introduced long ago (cf., e.g. [GTW78]) and is known to work quite well when all the functions to be defined in this way are total. It is possible in this case to find a so-called *conservative* (i.e. *sufficiently complete* and *consistent*) enrichment SP' which has "practically the same" initial model as the predefined algebra $\mathcal{A}$, and therefore the standard interpretation of F, defined by the initial model of SP', unambiguously defines also a corresponding interpretation $f^{\mathcal{A}}$ of each $f \in F$ on $\mathcal{A}$. This allows to make use of equational logic (with induction), as well as

* On leave from the V.M.Glushkov Institute of Cybernetics, Kiev, Ukraine

of various automated deduction procedures based on term-rewriting technique for proving logical properties of functions $f^{\mathcal{A}}$.

In this paper we intend to generalize this approach to the case of *partial* functions over a predefined algebra – the case which is known to be of big importance for practical applications of algebraic specifications. To illustrate that the task is not trivial and to point out some subtle problems one should expect on this way, let us consider an example.

1.2 An Instructive Example: Is Induction a Safe Tool for Functional Program Verification?

Consider the following functional program for integer division on natural numbers:

$$\begin{gathered}\texttt{fun div}: \texttt{Nat},\texttt{Nat}->\texttt{Nat};\\ \texttt{div}(\texttt{x},\texttt{y}) = \texttt{if } \texttt{x} < \texttt{y} \texttt{ then 0 else } 1 + \texttt{div}(\texttt{x}-\texttt{y},\texttt{y}),\end{gathered} \tag{1}$$

(here "–" denotes a natural minus, i.e. $m - n = 0$ for all $m < n$).

The program is not always terminating, still it can be used safely in a context where it is supposed to be called with positive second argument. Let's try to define *algebraic* (not denotational!) semantics of (1) within the "initial algebra approach" sketched above. For this purpose first we need to specify the predefined algebra (of natural numbers) with all operations involved in the program; let's take the following specification: [3]

```
spec NAT is
sorts Bool Nat
 ops   true false : -> Bool .
 ops   0  1 : -> Nat .
 op    suc_ : Nat -> Nat .
 ops (_+_), (_-_) : Nat Nat -> Nat .
 op  _<_ : Nat Nat -> Bool .
 op  if_then_else_ : Bool Nat Nat -> Nat .
vars x, y : Nat
eqs
 [e1] 1 = suc 0 .                    [e7]  x - 0 = x .
 [e2] x + 0 = x .                    [e8]  0 - x = 0 .
 [e3] x + suc y = suc(x + y).        [e9]  (suc x)-(suc y) = x - y .
 [e4] x < 0 = false .                [e10] x - x = 0 .
 [e5] 0 < suc x = true .             [e11] if true  then x else y = x .
 [e6] (suc x)<(suc y) = x < y .      [e12] if false then x else y = y .
end
```

It is not difficult to check that NAT is indeed a correct specification of natural numbers.

Now the enrichment DIV=NAT+({div}, (1)) is expected to define semantics of div. Obviously, the enrichment is not sufficiently complete (e.g., div(1,0) isn't

[3] We use OBJ-like syntax (cf. [GW88])

equal to any natural number), still it is consistent, [4] and for all pairs of natural numbers m, n (represented as canonical NAT-terms 0, suc 0, suc(suc 0),...), where n is not equal to 0, the ground term div(m, n) can be reduced in DIV to some (uniquely defined) natural number. This defines div(m,n) as a partial function on Nat (i.e., on the carrier of this sort in the initial algebra of NAT).

So far so good, but a problem arises if we try to use the enrichment for verification purposes. Suppose we have to prove the following correctness condition for div:

$$0 < y \;\Rightarrow\; \mathtt{div}(x * y,\; y) \;=\; x. \tag{2}$$

for all x,y:Nat, where _*_ denotes the multiplication operation which should also be specified. Let's add to NAT the following usual axioms for multiplication:

```
op (_*_) : Nat Nat -> Nat .
eqs [e13] 0 * y = 0 .          [e14] (suc x) * y = x * y + y .
```

(note, the enriched NAT is still correct) and try to prove (2) as a theorem of DIV = NAT+({div}, (1)) using induction on the variable x. The basic case, when $x = 0$, does not cause any difficulty, for

$$\mathtt{div}(0 * y, y) \;=\; \mathtt{div}(0, y) \;=\; \mathtt{if}\ 0 < y\ \mathtt{then}\ 0\ \mathtt{else}\ \ldots \;=\; 0$$

whenever the premise of (2) holds. Then, assuming (2) for x = x0, we need to prove

$$0 < y \;\Rightarrow\; \mathtt{div}((\mathtt{suc}\ x0) * y,\; y) \;=\; \mathtt{suc}\ x0.$$

Simplifying the term div((suc x0)*y,y) in DIV we obtain the expression div((x0*y+y)-y,y)+1; this could be simplified further to div(x0*y,y)+1 = x0+1 = suc x0 (that would complete the proof) if we had the following lemma:

$$(\forall\ x, y : \mathtt{Nat})\ (x + y) - y = x \tag{3}$$

Well, the lemma can be easily proved in NAT by induction on the variable y, so we are done. Or are we?

Once the lemma (3) has been obtained, one can use it for proving the following "powerful" theorem:

$$(\forall\ x, y : \mathtt{Nat})\ x = y \tag{4}$$

First, using the axioms e10, (1) and the lemma one derives

$$0 = \mathtt{div}(1,0) - \mathtt{div}(1,0) = (1 + \mathtt{div}(1,0)) - \mathtt{div}(1,0) = 1 = \mathtt{suc}(0),$$

then, using this with the basic axioms e4, e5 , one gets true=false , and then proves (4) using e11, e12 .

Now it is not a problem to prove (2) – as well as *any other* conditional equation in DIV – but who would accept this as a *verification* of the program (1)?

Of course, the point is that the enrichment of DIV by the lemma (3) is *inconsistent*, in spite of the fact that (3) is an inductive theorem of NAT. So one probably should prohibit to use it for proving theorems in DIV. But then we may not use it for proving (2) too; so our first proof was not correct?

[4] The fact which is not so easy to prove! We shall return to it in Sect. 3

Putting off the answer to the end of the paper, let us here just note that the example illustrates one of the main problem we are going to deal with: how to formalize the algebraic semantics of equational definitions of (possibly partial) functions over a predefined algebra within (some natural extension of) the "initial algebra" approach so that one could use safely inductive theorems of the predefined model for verification purposes.

1.3 Overview of the Paper

The example considered above gives rise to some general questions:

i) How to check consistency of (incomplete) enrichments?

ii) How to define an appropriate algebraic semantics of incomplete equational definitions like (1) ?

In our paper [AD92] we have already addressed the first question (in a slightly more general framework of Horn-equational logic), and suggested a model-theoretic technique for proving consistency of enrichments, as well as some sufficient conditions of consistency.

In the present paper we are going to extend and improve those results, as well as to suggest a very general approach to the problem formulated in the second question above. Specifically, after brief overview of basic notions and notations in Sect. 2, we shall introduce the notion of *equational definition over a predefined algebra* (EDPA), and study its semantics in Sect. 3. Using some model-theoretic technique, we shall demonstrate that a standard "free-extension" construction does not give appropriate semantics of EDPA. Our Instructive Example will be used again to illustrate this point: we shall prove that DIV is indeed a consistent enrichment of NAT, but the equation (3) is not valid in its initial algebra. In Sect. 4 we shall develop an approach that will allow to define a kind of "safe" semantics of EDPA. The approach is based on the idea of "restrictions on substitutions" which has also been exploited in several papers devoted to to algebraic specifications with partial functions [GDLE84, SNGM89] and term-rewriting systems over built-in algebras [AB92]; in Sect. 5 we consider the relations of our results with these papers.

2 Basic Notions and Notations

In this section we briefly recall some standard notions and notations of algebraic specification and term-rewriting theory [EM85], [Wir90], [DJ90].

Given a set of sorts S, a *many-sorted*, or *S-sorted signature* Σ is a disjoint union of sets $\Sigma_{w,s}$ of function symbols (or f-symbols, for short) of type $w \to s$ where $w \in S^*, s \in S$; constants are nullary functions of type $\to s$.

The notions of Σ-algebra A, Σ-subalgebra $B \subset A$, Σ-congruence on A, Σ-homomorphisms are supposed to be defined as usual [EM85, Wir90]; we shall use S-indexed notations for denoting carriers A_s of sort s and Cartesian products $A_w = A_{s_1} \times \ldots \times A_{s_n}$ (where $w = s_1 \ldots s_n$).

Given an S-sorted set of variables $X = \dot{\cup}_{s \in S} X_s$ (a disjoint union), $T_\Sigma(X)$ denotes the *absolutely free* Σ*-algebra over* X whose elements are Σ-terms; then T_Σ denotes the set (and the algebra) of *ground* Σ-terms (the latter is an initial object in the category $\mathbf{Alg}_\Sigma$ of all Σ-algebras). To avoid the "empty sorts" problem, we consider in this paper only those signatures Σ for which $(T_\Sigma)_s$ is non-empty for each $s \in S$.

An *(algebraic) specification* SP over a signature Σ is a pair (Σ, E) where E is a set of *axioms* of SP which are universal quasi-equations (often called conditional equations); sometimes we shall restrict E to be a set of atomic equations, then SP will be called as *purely equational* specification.

Given a specification $SP = (\Sigma, E)$, $\mathbf{Alg}(SP)$ denotes the category of all *SP-algebras* (also called *models of* SP), i.e. Σ-algebras satisfying all the axioms of E. This category has the initial object $I(SP)$ which can be represented (up to isomorphism) as the quotient $T_{SP} = T_\Sigma/_{\equiv_E}$ of the ground-term algebra T_Σ by the least Σ-congruence $\equiv_E$ generated by E. The uniquely defined homomorphism from $I(SP)$ to a given SP-algebra is called *initial*.

Given two S-sorted signatures Σ, Σ', a specification $SP' = (\Sigma', E')$ is called an *enrichment* of $SP = (\Sigma, E)$ if $\Sigma' \subseteq \Sigma$ (we also say that Σ' is an *enrichment* of Σ) and $E \subseteq E'$; SP' can be presented in the form $SP + (F, R)$ or $(\Sigma + F, E + R)$ where $F = \Sigma' \setminus \Sigma$ is the set of *new* f-symbols, and $R = E' \setminus E$ is a set of *new* axioms (then f-symbols and axioms of SP will be referred to as "old"). This enrichment is called

- *consistent* (wrt. SP) if it satisfies the "no-confusion" condition, i.e. if the restriction of the congruence $\equiv_{E+R}$ to the set T_Σ coincides with $\equiv_E$;
- *complete* (wrt. SP) if it satisfies the "no-junk" condition, i.e., if each E'-equivalence class $[t']_{E'} \in T_{SP'}$ contains some ground Σ-term;
- *conservative* (wrt. SP) if it is both consistent and complete (wrt. SP).

We say that $SP + (F, R)$ is a *functional* enrichment (or f-enrichment, for short), if for some F-indexed family of $(\Sigma + F)$-terms r_f the set R consists of (oriented) equations of the form $f(\mathbf{x}) = r_f$ where $\mathbf{x}$ is a list of distinct variables including all those occurring in r_f; then R itself will be called a *functional definition*.

A *forgetful functor* from $\mathbf{Alg}(\Sigma + F)$ to $\mathbf{Alg}(\Sigma)$ maps each $(\Sigma + F)$-algebra A' to its *Σ-reduct* $A = A'|_\Sigma$ which gets its carriers A_s and interpretations of function symbols f^A (for all $f \in \Sigma$) from A'; then A' is called an *enrichment of A by $F^{A'}$*. The forgetful functor is known to map any SP'-algebra to some SP-algebra for any enrichment $SP' = SP + (F, R)$; moreover, it has a *left adjoint* functor (also called a *free functor*) which maps each SP-algebra A to its *free SP'-enrichment*.

A Σ-reduct $I(SP')|_\Sigma$ of the initial algebra $I(SP')$ will be also denoted as $I_\Sigma(SP')$

An algebraic specification $SP = (\Sigma, E)$ can be considered as a *term-rewriting system* (t.r.s. for short) [DJ90] through orienting of equations in E from left to right. This t.r.s. defines the *rewrite relation* $\rightarrow_E$ on Σ-terms ; its symmetric (reflexive, transitive, reflexive transitive, symmetric reflexive transitive) closure is denoted by $\leftrightarrow_E$ (correspondingly by $\rightarrow_E^{\equiv}$, $\rightarrow_E^{+}$, $\rightarrow_E^{*}$, $\leftrightarrow_E^{*}$; the latter is known to coincide with $\equiv_E$).

The t.r.s. is called *normalizing* on some set of terms T if each term $t \in T$ has at least one *normal form*; it is called *confluent* if the composition $\leftarrow_E^{*} \circ \rightarrow_E^{*}$ is included into $\rightarrow_E^{*} \circ \leftarrow_E^{*}$ (where $\leftarrow_E^{*}$ denotes the inverse to $\rightarrow_E^{*}$).

We shall also use standard notations for a subterm $t|_\pi$ and a context $t[\ ]_\pi$ of a term t where π is some *position* in t (cf. [DJ90]); $\mathcal{V}(\varepsilon)$ denotes the set of variables occurring in a syntactic object (a term, a formula, a set of those, etc.) ε.

3 Equational Definitions over Predefined Algebras as Enrichments

Given an S_0-sorted signature Σ_0, a Σ_0-algebra $\mathcal{A}$, and a set of f-symbols F such that $\Sigma_0 + F$ is an enrichment of Σ_0, a triple $(\mathcal{A}, F, R)$ (denoted also $(F, R)_{\mathcal{A}}$), where R is a set of oriented $\Sigma_0 + F$-equations, is called a ***equational definition over*** $\mathcal{A}$ if the main f-symbol of the left-hand side of each $e \in R$ belong to F.

This gives only syntax; to define semantics of EDPA means to set a correspondence between triples $(\mathcal{A}, F, R)$ and sets of partial functions

$$F^{\mathcal{A}} = \{f^{\mathcal{A}} : \mathcal{A}_w \stackrel{\sim}{\rightarrow} \mathcal{A}_s \mid f \in F_{w,s}\}$$

Suppose $SP = (\Sigma, E)$ is an algebraic specification of $\mathcal{A}$ in the sense that Σ is a finite extension of Σ_0 and the Σ_0-reduct of $I(SP)$ is isomorphic to $\mathcal{A}$. Then we say that the enrichment $SP + (F, R)$ is an ***algebraic presentation*** of EDPA $(F, R)_{\mathcal{A}}$.

We are going to define semantics of EDPA using their algebraic presentations. For the sake of simplicity, we shall identify a predefined algebra $\mathcal{A}$ with the initial algebra of its basic specification SP (forgetting about the difference of their signatures).

3.1 Algebraic Semantics of Enrichments

Consider initial algebras $I(SP) \cong T_\Sigma/_{\equiv_E}$ and $I(SP') \cong T_{\Sigma+F}/_{\equiv_{E+R}}$ of a given specification $SP = (\Sigma, E)$ and its enrichment $SP' = SP + (F, R)$. They are known to relate in the following way: there is a (unique) homomorphism h from $I(SP)$ to $I_\Sigma(SP')$ which maps an equivalence class $[t]_E \in T_\Sigma/_{\equiv_E}$ to a corresponding equivalence class $[t]_{E+R} \in T_{\Sigma+F}/_{\equiv_{E+R}}$ for each Σ-term t, i.e., $h([t]_E) = [t]_{E+R}$. This homomorphism is known to be injective (surjective) iff the enrichment SP' is consistent (complete) wrt. SP.

The interpretation $f^{I(SP')}$ of a new f-symbol $f \in F$ on $I(SP')$ satisfies the following equation:

$$f^{I(SP')}(\,[t'_1]_{E+R}, \ldots [t'_n]_{E+R}) = [\,f(t'_1, \ldots t'_n)\,]_{E+R} \tag{5}$$

for all tuples of $(\Sigma + F)$-terms t'_i of appropriate sorts; the same equation defines the interpretation of f on $I_\Sigma(SP')$ that gives a free SP'-enrichment of $I(SP)$.

Now we need to define some ***basic interpretation*** $F^{\mathcal{A}}$ of F on $\mathcal{A}$, i.e. on $I(SP)$, such that the corresponding enrichment $\mathcal{A}' = \mathcal{A} + F^{\mathcal{A}}$ would be in "good relations" with the set of axioms $E + R$ (since we are going to use them for reasoning about $F^{\mathcal{A}}$).

Whenever SP' is a conservative enrichment, the basic interpretation is uniquely defined by (5) and gives a set of total functions $F^{\mathcal{A}}$. So, in order to capture the case of partially defined functions over $\mathcal{A}$, we should, at least, drop the "no-junk" condition and consider incomplete enrichments.

Still it seems quite reasonable to impose the "no-confusion" requirement on algebraic presentations, for in this case they would "preserve" the structure of $\mathcal{A}$ in the initial algebra $I(SP')$: the homomorphism $h : I(SP) \rightarrow I_\Sigma(SP')$ would be injective and its image would be a Σ-subalgebra of $I_\Sigma(SP')$ isomorphic to $\mathcal{A}$. The following proposition shows how to obtain from this a ***partial*** $(\Sigma + F)$-subalgebra of $I(SP')$.

Proposition 1. *Given a consistent enrichment* $SP' = SP + (F, R)$ *of* $SP = (\Sigma, E)$, *there exists a set of partial functions*

$$F^{I(SP)} = \{ f^{I(SP)} : I(SP)_w \xrightarrow{\sim} I(SP)_s \mid f \in F_{w,s} \}$$

defined as follows:

$$f^{I(SP)}(\,[t_1]_E, \ldots [t_m]_E) = [\, f(t_1, \ldots, t_m)\,]_{E+R} \cap T_\Sigma \tag{6}$$

for all tuples $t_1, \ldots t_m$ *of* Σ*-terms of appropriate sorts provided the right-hand side is not the empty set, otherwise* $f^{I(SP)}(\,[t_1]_E, \ldots [t_m]_E)$ *is undefined.*

Moreover, the enrichment of $I(SP)$ *with* $F^{I(SP)}$ *will be a partial* $(\Sigma + F)$*-subalgebra of* $I(SP')$.

Proof. The correctness of (6), as well as the statement in whole follow from the consistency condition: the initial homomorphism h is injective in this case, so $I(SP)$ is (isomorphic to) a Σ-subalgebra of $I_\Sigma(SP')$, and the equation (6) just defines the restriction of $f^{I(SP')}$ to $I(SP)$ considered as a subset of $I(SP')$.

□

As a matter of fact, this construction is similar to that in [Kre87] which was intended to provide an approach to formalize partial functions within "the simpler framework of total algebras and conventional specifications". It does seem natural to take (6) as the definition of semantics of $(F, R)_\mathcal{A}$. To explain why this would not be quite satisfactory, we need first to address the problem of how to check the conditions when this definition can be used, i.e. how to prove consistency of (possibly incomplete) enrichments.

3.2 Consistency of Enrichments

A general model-theoretic method for proving consistency of enrichments, which does not impose any requirements on specifications involved, was introduced in [GTW78] (cf. also [EM85]) and is based on the following sufficient condition.

Fact 1. (a sufficient condition of consistency of enrichments)
An enrichment $SP' = SP + (F, R)$ *is consistent wrt. an algebraic specification* $SP = (\Sigma, E)$ *if there exists an algebra* $A \in \mathbf{Alg}(SP')$ *such that its* Σ*-reduct* $A|_\Sigma$ *is isomorphic to* $I(SP)$.

Thus, to prove consistency of $SP' = SP + (F, R)$ it suffices to find some interpretation of new function symbols $f \in F$ on $I(SP)$ satisfying (together with the known interpretation of old symbols from SP) all axioms in R.

This technique can always be applied to complete enrichments, for in this case the condition gets necessary. Occasionally it can also be applied to some incomplete enrichments; still not to all of them – e.g., this does not work for the enrichment DIV considered in Sect. 1.2.

To overcome this disadvantage we have obtained the following *criterion* of consistency (the result was announced in our paper [AD92]; here we give a complete proof).

Theorem 2. *Given an algebraic specification $SP = (\Sigma, E)$, an enrichment $SP' = SP + (F, R)$ is consistent iff there exists an algebra $A \in \mathbf{Alg}(SP')$ such that its Σ-reduct $A|_\Sigma$ contains a subalgebra isomorphic to $I(SP)$.*

Proof. If the enrichment is consistent, then the initial homomorphism $h : I(SP) \rightarrow I_\Sigma(SP')$ is an injection and its image gives a subalgebra isomorphic to $I(SP)$. So we can take $A = I(SP')$ in this case.

For the converse, let the enrichment be inconsistent (so that h isn't injective). Then for any $A \in \mathbf{Alg}(SP')$ the initial homomorphism $k : I(SP) \rightarrow A|_\Sigma$ is not injective since it can be (uniquely) factored into the composition $h' \circ h$ where h' is the Σ-reduct of the initial homomorphism from $I(SP')$ to A. Thus $A|_\Sigma$ doesn't contain a subalgebra isomorphic to $I(SP)$ (since k is the only homomorphism from $I(SP)$ to A).

□

The corresponding technique for proving consistency of an enrichment $SP' = SP + (F, R)$ consists of the following steps:

1) To construct an extension $T^C_{SP} \in \mathbf{Alg}(SP)$ of the initial algebra $T_{SP} \cong I(SP)$ by a set C of new "non-standard" elements (i.e., to extend the interpretation of all basic operations $g \in \Sigma$ to the carrier $T_{SP} \cup C$);
2) To construct some interpretation of the new function symbols $f \in F$ on $T_{SP} \cup C$ such that the enrichment of T^C_{SP} with this interpretation would satisfy all axioms in R.

Theorem 2 guarantees that these steps can always be fulfilled whenever the enrichment $SP + (F, R)$ is consistent.

Let us apply this technique to confirm consistency of the enrichment DIV from Sect. 1.2.

3.3 Proving Consistency of the Instructive Example

The initial algebra of the basic specification NAT from Sect. 1.2 is isomorphic to a two-sorted algebra with carriers $\mathbf{N}$ of `Nat` (the set of natural numbers), $\mathbf{B} = \{true, false\}$ of `Bool` and usual interpretation of all the operations. To prove consistency of DIV=NAT+({`div`}, (1)), let's construct the following extension A of I(NAT) by one "non-standard" natural number $\mathbf{c}$: `Nat`, i.e. $A_{\mathtt{Bool}} = \mathbf{B}$, $A_{\mathtt{Nat}} = \mathbf{N} \cup \{\mathbf{c}\}$, and the extensions of all operations are defined by the following equations:

$$\mathbf{suc}^A(\mathbf{c}) = \mathbf{c};$$
$$+^A(n, \mathbf{c}) = +^A(\mathbf{c}, n) = +^A(\mathbf{c}, \mathbf{c}) = \mathbf{c};$$
$$^A(n', \mathbf{c}) = *^A(\mathbf{c}, n') = *^A(\mathbf{c}, \mathbf{c}) = \mathbf{c}; \quad *^A(0, \mathbf{c}) = *^A(\mathbf{c}, 0) = 0;$$
$$-^A(n, \mathbf{c}) = -^A(\mathbf{c}, \mathbf{c}) = 0; \quad -^A(\mathbf{c}, n) = \mathbf{c}$$
$$<^A (n, \mathbf{c}) = true; \quad <^A (\mathbf{c}, n) = <^A (\mathbf{c}, \mathbf{c}) = false;$$
$$if^A(true, \mathbf{c}, n) = \mathbf{c}; \quad if^A(true, n, \mathbf{c}) = n;$$
$$if^A(false, \mathbf{c}, n) = n; \quad if^A(false, n, \mathbf{c}) = \mathbf{c}$$

for all $n, n' \in \mathrm{N}$, $n' > 0$. (It is easy to check by direct calculations that the extensions of operations defined by these equations satisfy all the axioms of NAT, i.e. $A \in \mathrm{Alg(NAT)}$. In order to check the fact that $I(\mathrm{NAT}) \subset A$, one can observe that the enrichment of NAT by a constant $\mathtt{c} : \mathtt{Nat}$ these equations forms a *terminating* rewrite system – this just makes possible to prove its consistency wrt. NAT by methods suggested in [JK89, Kir92].)

To complete the proof we suggest the following interpretation of div on A that satisfies the equation (1):

$$\mathtt{div}^A(n, 0) = \mathtt{div}^A(\mathtt{c}, n) = \mathtt{c}; \quad \mathtt{div}^A(n, \mathtt{c}) = 0; \quad \mathtt{div}^A(\mathtt{c}, \mathtt{c}) = 1; \quad \mathtt{div}^A(n, n') = k$$

for all $n, n' \in \mathrm{N}$, $n' > 0$ where $k \in \mathrm{N}$ is the quotient of integer division n on n'.

Notice that (3) is not valid in A (consider $x = y = c$), therefore it is also not valid in $I(\mathrm{DIV})$, since A is its surjective image. Thus we have proved that a (rather ordinary) functional program can be consistent wrt. some basic specification of a predefined model and inconsistent wrt. some of its inductive consequences. That is why one couldn't use induction (over predefined model) for verification of functional programs (considered as EDPA) if (6) was taken as the definition of semantics of EDPA. Let's consider an approach to overcome this problem.

4 "Safe" Semantics of EDPA

One can guess that the basic reason of inconsistency of some f-enrichments is the opportunity to substitute terms with new functional symbols into old axioms: eventually such a term can denote a "junk" (a non-reachable value of $I(SP')$) that extends the range of interpretation of variables in the axioms and/or inductive theorems of the basic specification.

A natural idea, then, is to prohibit those substitutions. However, the restriction would be too strong, because in this case all old operations would get *strict* with respect to new terms: for instance, one couldn't simplify the term $0 + f(1)$ to $f(1)$ using an old axiom $0 + x = x$. The situation with conditionals (if-then-else) would be even worse: e.g., the equation (1) defines div as an empty function, provided if-then-else is strict.

In [AD92] we have already shown how to solve the problem with conditionals. Here we are going to suggest a more general solution that will allow most of old functions (axiomatized in some "safe" way) to be non-strict. The benefit of this approach is that it provides a wider class of possible operational semantics of predefined operations in EDPA (not only call-by-value).

4.1 Restricted Rewriting and Equality

In this section we consider purely equational specifications SP (of a predefined Σ-algebra $\mathcal{A}$) equipped with the following additional information: the set X of all variables (used in axioms of SP) contains a distinguished subset X^+ of *safe* variables (then variables in $X \setminus X^+$ will be called *unsafe*). When we need to reflect this information in the terminology and definitions, we shall be using the notation $\Sigma(X^+)$

for a signature Σ. (This construction, as well as the terminology, is inspired by the approach to partial functions suggested in [GDLE84].)

A $\Sigma(X^+)$-equation $e \in E$ will be called *safe* if it contains only safe variables (i.e., $\mathcal{V}(e) \subset X^+$). A substitution θ on $T_{\Sigma+F}(X)$ will be called *safe* if it maps safe variables into Σ-terms (i.e., $\theta(X^+) \subset T_\Sigma(X^+)$).

Now we introduce the following relations that will serve for term-rewriting and equational derivations with some restrictions on substitutions.

Definition 3. Given a basic specification $SP = (\Sigma,\ E)$ and its enrichment $SP' = SP + (F, R)$, let $\rightarrow_{E:}$ denote the following relation on $T_{\Sigma+F}(Y)$: $t \rightarrow_{E:} t'$ holds if there exists an equation $l = r \in E$, a safe substitution $\theta : X \rightarrow T_{\Sigma+F}(Y)$ and a position π such that $t|_\pi \doteq \theta(l)$ and $t' \doteq t[\theta(r)]_\pi$. Then a *restricted rewrite relation* $\rightarrow_{E:R}$ and a *restricted equality* $=_{E:R}$ (both on $T_{\Sigma+F}(Y)$) are defined as follows:

$$\rightarrow_{E:R} \;\rightleftharpoons\; \rightarrow_{E:} \cup \rightarrow_R; \quad t =_{E:R} t' \;\rightleftharpoons\; t \leftrightarrow^*_{E:R} t'$$

It follows from the definition that the restricted equality $=_{E:R}$ is a congruence on $T_{\Sigma+F}(Y)$ (included into that $=_{E+R}$). This makes possible to define a quotient $T_{\Sigma+F}/_{=_{E:R}}$ and to use its Σ-reduct (rather than $I_\Sigma(SP')$) in the definition of semantics of EDPA. To provide correctness of this new construction, an enrichment has to satisfy the "no-confusion" condition wrt. $=_{E:R}$. This is the matter of the next definition and proposition.

Definition 4. Given a specification $SP = (\Sigma(X^+), E)$, its enrichment $SP + (F, R)$ is said to be *safe-consistent* (wrt. SP) if $t_1 =_{E:R} t_2$ implies $t_1 =_E t_2$ for any pair of ground Σ-terms t_1, t_2.

Proposition 5. *Given a safe-consistent enrichment* $SP' = SP + (F, R)$ *of* $SP = (\Sigma(X^+), E)$, *there exists a set of partial functions*

$$F^{I(SP)} = \{f^{I(SP)} : I(SP)_w \xrightarrow{\sim} I(SP)_s \mid f \in F,\ f : w \rightarrow s\}$$

defined as follows:

$$f^{I(SP)}(\ [t_1]_E, \dots [t_m]_E) = [\ f(t_1, \dots, t_m)\ ]_{E:R} \cap T_\Sigma \tag{7}$$

for all tuples $t_1, \dots t_m$ *of ground* Σ*-terms of appropriate sorts provided the right-hand side is not the empty set, otherwise* $f^{I(SP)}(\ [t_1]_E, \dots [t_m]_E)$ *is undefined. Moreover, the enrichment of* $I(SP)$ *with* $F^{I(SP)}$ *will be a partial* $(\Sigma{+}F)$*-subalgebra of* $T_\Sigma/_{=_{E:R}}$.

□

We take (7) as the "generic" definition of semantics of an EDPA $(F, R)_A$ presented by a safe-consistent enrichment $(\Sigma(X^+), E) + (F, R)$; the set X^+ of safe variables is a parameter of this definition. As far as safe-consistency is in general weaker than consistency, this gives an opportunity to get a wider class of correct EDPA which will include *all* functional definitions. The following technical details are just steps toward this goal.

In what follows, we consider a specification $SP = (\Sigma(X^+), E)$ and its enrichment $SP' = SP + (F, R)$, where R is a rewrite system such that all its left-hand sides contain some $f \in F$ (this is a bit more general class of enrichments than algebraic presentations of EDPA).

Definition 6. We say that the t.r.s. R *respects* the set of equations E if the following holds:

$$\leftrightarrow_{E:} \circ \rightarrow_R \subseteq \rightarrow_R^= \circ \leftrightarrow_{E:}^*;$$

Given a set of terms $T \subset T_{\Sigma+F}(X)$, we say that R *respects* E *on* T if the same inclusion holds for restrictions of the relations involved on T.

Proposition 7. *If the system R respects the set of equations E on T, then for all $t_1, t_2, t_3 \in T$ there exists $t_4 \in T$ such that*

i) if $t_1 \leftrightarrow_{E:}^* t_2 \rightarrow_R t_3$ then $t_1 \rightarrow_R^= t_4 \leftrightarrow_{E:}^* t_3$;
ii) if $t_1 \leftrightarrow_{E:}^* t_2 \rightarrow_R^* t_3$ then $t_1 \rightarrow_R^* t_4 \leftrightarrow_{E:}^* t_3$.

Proof. (sketch).
i) By straightforward induction on the length of the derivation $t_1 \leftrightarrow_{E:}^* t_2$.
ii) By straightforward induction on the length of the derivation $t_2 \rightarrow_R^* t_3$ using (i).

□

Now we can formulate and prove the following fundamental property of the restricted rewriting relation $\rightarrow_R$ and the congruence $=_{E:R}$.

Lemma 8. *Suppose the system R is confluent and respects the set of equations E. Then the following holds:*

$$=_{E:R} \subseteq \rightarrow_R^* \circ =_{E:} \circ \leftarrow_R^*$$

(i.e., the relation $\rightarrow_R$ is Church-Rosser modulo $=_{E:}$.)

Proof. Let $t =_{E:R} t'$ hold for some $(\Sigma + F)$-terms t, t'. The derivation $t \leftrightarrow_{E:R}^* t'$ can be represented as a chain

$$t \doteq t_0 \sim t_1 \sim \ldots \sim t_n \doteq t', \tag{8}$$

where each occurrence of $\sim$ denotes either $\rightarrow_R^+$, $\leftarrow_R^+$, or $\leftrightarrow_{E:}^+$, and adjacent occurrences are different. Due to Prop. 7 and the confluence of R, the following transformation rules α_i, $i = 1, 2, 3$ can be applied to the chain:

$$\begin{aligned}
\alpha_1 :\ & t_1 \leftarrow_R^+ t_2 \rightarrow_R^+ t_3 \ \Rightarrow\ t_1 \rightarrow_R^* t_4 \leftarrow_R^* t_3; \\
\alpha_2 :\ & t_1 \leftrightarrow_{E:}^+ t_2 \rightarrow_R^+ t_3 \ \Rightarrow\ t_1 \rightarrow_R^* t_4 \leftrightarrow_{E:}^* t_3; \\
\alpha_3 :\ & t_1 \leftarrow_R^+ t_2 \leftrightarrow_{E:}^+ t_3 \ \Rightarrow\ t_1 \leftrightarrow_{E:}^* t_4 \leftarrow_R^* t_3.
\end{aligned}$$

This system of rules is normalizing on the set of chains of the form (8), because each application of α_1, α_2 (of α_1, α_3) to the rightmost occurrence of $\rightarrow_R^+$ (the leftmost occurrence of $\leftarrow_R^+$) reduces either the distance from that occurrence to the left (right) end of the chain, or the number of those in the chain. Therefore after a finite number of steps the chain (8) will get the form

$$t \rightarrow_R^* t_1 \leftrightarrow_{E:}^* t_2 \leftarrow_R^* t' \tag{9}$$

for some $(\Sigma + F)$-terms t_1, t_2.

□

Corollary 9. *The enrichment $SP + (F, R)$ of $SP = (\Sigma, E)$ is safe-consistent if R is confluent on $T_{\Sigma+F}$ and respects E on $T_{\Sigma+F}$.*

Proof. It suffices to observe that each derivation $t_1 \leftrightarrow^*_{E:R} t_2$, where the outermost terms t_1, t_2 belong to T_Σ, after transforming by α_i to (9) will get the form $t_1 \leftrightarrow^*_{E:} t_2$, because no one rule from R can be applied to t_1, t_2.

□

This lemma (with the corollary) can be used for inventing various sufficient conditions of safe-consistency (i.e., correctness of EDPA). We present the corresponding results in the next subsection.

4.2 Sufficient Conditions of Safe-Consistency

First, we show how to obtain the following result (announced in [AD92]) about correctness of a wide class of functional definitions with non-strict conditionals. To define this class, we suppose that the predefined algebra $\mathcal{A}$ and its specification SP satisfy the following requirements:

1) They contain the sort `Bool` of boolean values with constants `true` and `false` which are interpreted by two distinct values in $\mathcal{A}_{\texttt{Bool}}$. (Other total boolean operations may occur in $\mathcal{A}$ and SP too).

2) They contain the conditional functions `if` : `Bool`, $s, s \to s$ for each sort s with the usual axioms in SP:

$$\texttt{if}(\texttt{true}, \texttt{x}, \texttt{y}) = \texttt{x}; \quad \texttt{if}(\texttt{false}, \texttt{x}, \texttt{y}) = \texttt{y} \tag{10}$$

Let IF denote the set of equations (10) for all sorts; then the set of axioms of SP will be represented as $IF \cup E$ (where E is a set of other equations).

Theorem 10. *Any functional enrichment $SP + (F, R)$ of an algebraic specification $SP = (\Sigma(X^+), IF \cup E)$ is safe-consistent provided all axioms in E are safe.*

Proof. (sketch) The system R is obviously confluent; thus, due to the lemma, it suffices to check that R respects both E and IF on the set of ground $(\Sigma + F)$-terms.

The first fact is easy to prove by case analysis of possible overlappings of applications of $\to_R$ at a position π_1, and $\leftarrow_{E:}$ or $\to_{E:}$ at a position π_2 of some $(\Sigma + F)$-term t in the derivation $t_1 \leftrightarrow_{E:} t \to_R t_2$ (since each $e \in E$ is safe, the only possible non-trivial case is when $t|_{\pi_2}$ is a subterm of $t|_{\pi_1}$). In each case there exists a term t' such that $t_1 \to_R t' \leftrightarrow^*_{E:} t_2$.

To prove that R respects IF, we need to add to the above one additional case (since variables in IF are not supposed to be safe): a derivation $t_1 \leftrightarrow_{IF} t \to_R t_2$ where the arrow $\to_R$ is applied at a position π_1 of t, and the arrow $\leftarrow_{IF}$ (or $\to_{IF}$) is applied to a subterm $t|_{\pi_2}$ which contains the $(\Sigma + F)$-subterm $t|_{\pi_1}$. Again, one can show that in this case there exists a term t' such that $t_1 \to^{=}_R t' \leftrightarrow^*_{IF} t_2$

Thus the inclusion $\leftrightarrow_{E:\cup IF} \circ \to_R \subseteq \to^{=}_R \circ \leftrightarrow^*_{E:\cup IF}$; holds, so the enrichment $SP + (F, R)$ is safe-consistent.

□

We also announce here the following theorem, which offers even more general sufficient conditions of safe-consistency. Recall that an equation is called *left-linear* (*right-linear*) if its left-hand (right-hand) side is linear; it is called *linear* if it is both left- and right-linear.

Theorem 11. *Any f-enrichment $SP + (F, R)$ of an algebraic specification $SP = (\Sigma(X^+), E)$ is safe-consistent provided all non-linear axioms in E are safe.* □

This theorem gives the corresponding specialization of (7) which provides correctness of any functional EDPA wrt. a wide class of basic specifications. Using this, we can suggest the following solution of the puzzle with verification given in Sect. 1.2: to make NAT "safe" for (1), as well as for any functional definition, it suffices to mark the variable x in the non-linear axiom e10 as safe. The same should be done with all non-linear inductive theorems of NAT, in particular – with the lemma (3). This makes impossible to deduce the contradiction (4), but allows to use the lemma for proving the correctness condition (2).

However, we don't know at the moment the most general (syntactical) conditions that would provide *the widest* class of basic specifications "safe" for an arbitrary functional enrichment; this is one of interesting questions for further research.

5 Relations with Other Approaches

As we have already mentioned, the idea to impose restrictions on substitutions of new terms into old axioms in order to treat partial functions in algebraic specifications properly is not new. For instance, it was used in the fifth chapter of [SNGM89] within the framework of order-sorted equational logic [GM89], as well as in [AB92]. Let's recall the corresponding construction of [SNGM89] called there *stratification*.

Suppose a basic specification $SP = (\Sigma, E)$ with an S-sorted signature Σ should be enriched by a set F of some (possibly partial) functions. Then one should proceed as follows.

First, Σ should be extended by a set of new sorts and declarations: for each basic sort $s \in S$ its *error supersort* $s^?$ should be introduced (i.e., s is a subsort of $s^?$), and each old function symbol $g \in \Sigma$ of type $s_1, \ldots, s_n \to s$ gets an additional declaration

$$g : s_1^?, \ldots, s_n^? \to s^?.$$

Then, if one was going to specify a new (partial) function $f \in F$ of type $s_1, \ldots, s_n \to s$, one actually should introduce the following declaration:

$$f : s_1^?, \ldots, s_n^? \to s^? \tag{11}$$

and a set of corresponding axioms R.

As a consequence, any term of the form $f(t_1, \ldots, t_n)$ will have the sort $s^?$ even in the case when each t_i has the sort s_i. Since none of the axioms of SP contain variables of the error supersorts, it gets impossible to substitute terms with new functional symbols into them; so the corresponding congruence, specified by $SP + (F, R)$ in this way, is just our "restricted equality" $=_{E:R}$ constructed when *all* basic axioms are safe. [5]

However, as we have pointed out in Sect. 4, this approach is very restrictive since it makes all the old functions strict wrt. new terms; this, for instance, makes impossible to use non-strict conditionals and functional definitions like the program (1); e.g., if-then-else was modeled by a strict function in examples of stratified specifications in [SNGM89]. Our theorems 10 and 11 show that actually this is not necessary for consistency. To reformulate our results for order-sorted specifications, let's introduce the following construction: given an old axiom $e \in E$, let $e^?$ denote its "sort-lifted" version – the result of substitution $x : s^?$ instead of $x : s$ for each $x \in \mathcal{V}(e)$; let $E^?$ denote the set $\{e^? \mid e \in E\}$.

Proposition 12. *Let an f-enrichment $SP + (F, R)$ of $SP = (\Sigma, E)$ be obtained by stratification (where all new functional symbols $f \in F$ are declared as in (11)), and let $E_1 \subset E$ be a subset of linear equations. Then the enrichment $SP + (F, R + E_1^?)$ is consistent.* □

In particular, one can get non-strict conditionals by adding $IF^?$ – the set of sort-lifted versions of usual (linear!) axioms IF.

Thus, following [GM87], we can add one more problem (let us call it the "functional enrichment consistency" problem) to the long list of those solved by order-sorted algebra.

Still the stratification construction is not the only possible way to represent our "safe" semantics of EDPA. Another possibility (which seems simpler and more convenient for this specific task) is to make use of *algebras with Okay predicates* – the specification framework introduced in [GDLE84] and developed further in [ANK90]. Algebraic specifications in this approach make explicit syntactical distinction between safe/unsafe variables, functions and terms; this gives a direct way to implement the restrictions on substitutions and to represent EDPA (cf. more details in [AD92]).

Acknowledgements:

We are grateful to Natalya Soboleva for very helpful technical assistance, and to Michael Rusinowitch for careful reading this paper and useful remarks.

The first author is much obliged to Peter Mosses for valuable remarks on the text of the paper, as well as for permanent help and encouragement. He would like to thank Helene Kirchner for discussing some interesting examples concerning this work. He also acknowledges the financial support of the Danish Natural Sciences Research Council, grant No 11–9479.

References

[AD92] Antimirov V., Degtyarev A.: Consistency of equational enrichments. LPAR'92,

[5] to be very precise, we should also add here that all variables in new axioms $e \in R$ should be of "questioned" sorts, for otherwise the congruence will be even weaker then $=_{E:R}$; still the latter would be even "better" for consistency.

Proc. Int. Conf. on Logic Programming and Automated Reasoning, St. Petersburg, LNCS **624**, Springer-Verlag (1992) 393–402.

[ANK90] Antimirov V., Naidich D., Koval V.: Partial functions in simulation: formal models and calculi. Proc. IMACS European Simulation Meeting, Esztergom, Hungary, 1990, pp.143–148.

[AB92] Avenhaus J., Becker K.: Conditional rewriting modulo a built-in algebra. Technical report (SEKI Report SR–92–11), 1992, 23p.

[DJ90] Dershowitz N., Jouannaud J.-P.: Rewrite systems. J.van Leeuwen, A.Meyer, M.Nivat, M.Paterson, and D.Perrin editors, Handbook of Theoretical Computer Science, Vol B, chapter 6, Elsevier Sci. Pub. 1990.

[EM85] Ehrig H., Mahr B.: Fundamentals of algebraic specification: Vol. 1, Springer-Verlag, 1985.

[GDLE84] Gogolla M., Drosten K., Lipeck U., Ehrich H.-D.: Algebraic and operational semantics of specifications allowing exceptions and errors. TCS **34**(1984) 289–313.

[GM87] Goguen J., Meseguer J.: Order-sorted algebra solves the constructor selector, multiple representation and coercion problems. Proc. Second Symposium on Logic in Comp. Sci. IEEE Comp. Society Press, 1987, 18–29.

[GM89] Goguen J., Meseguer J.: Order-sorted algebra 1. SRI International, Technical Report SRI–CLS–89, July 1989.

[GTW78] Goguen J., Thatcher J., Wagner E.: An initial algebra approach to the specification, correctness and implementation of abstract data types. Current trends in programming methodology, Vol.4, Prentice-Hall, 1978, pp.80–149.

[GW88] Goguen J., Winkler T.: Introducing OBJ3. Technical report SRI-CSL-89-10, Comp. Sci. Lab., SRI International, 1988.

[JK89] Jouannaud J.-P., Kounalis E.: Automatic proofs by induction in theories without constructors. Information and Computation, **82**, 1 (1989) 1–33.

[Kir92] Kirchner H.: Proofs in parameterized specifications. Technical report (extended version) CRIN 91-R-045.

[Kre87] Kreowski H.-J.: Partial algebras flow from algebraic specifications. ICALP'87, Proc. Int. Coll. on Automata, Languages, and Programming, LNCS **267**, Springer-Verlag (1987) 521–530.

[SNGM89] Smolka J., Nutt W., Goguen J., Meseguer J.: Order-sorted equational computation. H.Aït-Kaci and M.Nivat, editors, Resolution of Equations in Algebraic Structures, Academic Press, New-York, 1989, 297–367.

[Wir90] Wirsing M.: Algebraic specification. J.van Leeuwen, A.Meyer, M.Nivat, M.Paterson, and D.Perrin editors, Handbook of Theoretical Computer Science, Vol B, chapter 13. Elsevier Sci. Pub. B.V.,1990,

Completeness of Combinations of Conditional Constructor Systems

Aart Middeldorp

Advanced Research Laboratory, Hitachi Ltd.
Hatoyama, Saitama 350-03, Japan
ami@harl.hitachi.co.jp

In this paper we extend the recent divide and conquer technique of Middeldorp and Toyama for establishing (semi-)completeness of constructor systems to conditional constructor systems. We show that both completeness (i.e. the combination of confluence and strong normalization) and semi-completeness (confluence plus weak normalization) are decomposable properties of conditional constructor systems without extra variables in the conditions of the rewrite rules.

1. Introduction

A property of term rewriting systems is *modular* if it is preserved under disjoint union. In the past few years the modularity of properties of term rewriting systems has been extensively studied. The first results in this direction were obtained by Toyama. In [15] he showed that confluence is a modular property (see Klop *et al.* [5] for a simplified proof) and in [16] he refuted the modularity of strong normalization. His counterexample inspired many researchers to look for conditions which are sufficient to recover the modularity of strong normalization (e.g. Rusinowitch [14], Middeldorp [8], Kurihara and Ohuchi [6], and Toyama *et al.* [17]). Recently Gramlich [3] proved an interesting theorem which generalizes the results of [6, 8, 14]. Another recent contribution to the topic of modularity is the work of Caron [1] who investigates the decidability problem of reachability from a modularity perspective.

The disjointness requirement limits the practical applicability of the results mentioned above. The results of [6, 8, 14] were generalized to combinations of term rewriting systems that possibly share constructors—function symbols which do not occur at the leftmost position in left-hand sides of rewrite rules—by Kurihara and Ohuchi [7] and Gramlich [3]. Middeldorp and Toyama [12] obtained a useful divide and conquer technique for establishing (semi-)completeness of term rewriting systems that adhere to the so-called constructor discipline. In such *constructor systems* all function symbols occurring at non-leftmost positions in left-hand sides of rewrite rules are constructors.

Several modularity results have been extended to conditional term rewriting systems by Middeldorp [9, 10], Ohlebusch [13] and Gramlich [4]. In the present paper we extend the results of [12] to conditional constructor systems. The paper is organized as follows. In the next section we briefly recapitulate the basic notions of (conditional) term rewriting. Section 3 presents the results that are proved in detail in the subsequent two sections. We conclude in Section 6 with suggestions for further research.

2. Preliminaries

Let $\mathcal{T}(\mathcal{F},\mathcal{V})$ be the set of terms built from a set of function symbols $\mathcal{F}$ and a countably infinite set of variables $\mathcal{V}$. The symbol $\equiv$ will be used to denote identity of terms. The *root symbol* of a term t is defined as follows: $root(t) = f$ if $t \equiv f(t_1,\ldots,t_n)$ and $root(t) = t$ if $t \in \mathcal{V}$.

Let $\Box$ be a fresh constant, named *hole*. A *context* C is a term in $\mathcal{T}(\mathcal{F}\cup\{\Box\},\mathcal{V})$. The designation *term* is restricted to members of $\mathcal{T}(\mathcal{F},\mathcal{V})$. A context may contain zero, one or more holes. If C is a context with n holes and $t_1,\ldots,t_n$ are terms then $C[t_1,\ldots,t_n]$ denotes the result of replacing from left to right the holes in C by $t_1,\ldots,t_n$. A term s is a *subterm* of a term t if there exists a context C such that $t \equiv C[s]$.

A *substitution* σ is a mapping from $\mathcal{V}$ to $\mathcal{T}(\mathcal{F},\mathcal{V})$ such that its *domain* $\mathcal{D}\sigma = \{x \in \mathcal{V} \mid \sigma(x) \not\equiv x\}$ is finite. Substitutions are extended to morphisms from $\mathcal{T}(\mathcal{F},\mathcal{V})$ to $\mathcal{T}(\mathcal{F},\mathcal{V})$. We call $\sigma(t)$ an *instance* of t. In the following we frequently write t^σ instead of $\sigma(t)$. Let $\mathcal{T}' \subseteq \mathcal{T}(\mathcal{F},\mathcal{V})$. The set of all substitutions σ with $\sigma(x) \in \mathcal{T}'$ for every $x \in \mathcal{D}\sigma$ is denoted by $\Sigma(\mathcal{T}')$.

A *term rewriting system* (TRS for short) is a pair $(\mathcal{F},\mathcal{R})$. Here $\mathcal{F}$ is a signature and $\mathcal{R}$ consists of pairs (l,r) with $l,r \in \mathcal{T}(\mathcal{F},\mathcal{V})$ such that the left-hand side l is not a variable and variables which occur in the right-hand side r also occur in l. Pairs (l,r) are called *rewrite rules* and will henceforth be written as $l \to r$. A rewrite rule $l \to r$ is *left-linear* if l does not contain multiple occurrences of the same variable. A *left-linear* TRS contains only left-linear rewrite rules.

The *rewrite relation* $\to_{\mathcal{R}}$ is defined as follows: $s \to_{\mathcal{R}} t$ if there exists a rewrite rule $l \to r \in \mathcal{R}$, a context C and a substitution σ such that $s \equiv C[l^\sigma]$ and $t \equiv C[r^\sigma]$. The subterm l^σ of s is called a *redex* and we say that s rewrites to t by *contracting* redex l^σ. We call $s \to_{\mathcal{R}} t$ a *rewrite step*. The transitive-reflexive closure of $\to_{\mathcal{R}}$ is denoted by $\twoheadrightarrow_{\mathcal{R}}$. If $s \twoheadrightarrow_{\mathcal{R}} t$ we say that s *reduces* to t. The transitive closure of $\to_{\mathcal{R}}$ is denoted by $\to^+_{\mathcal{R}}$. We write $s \leftarrow_{\mathcal{R}} t$ if $t \to_{\mathcal{R}} s$; likewise for $s \twoheadleftarrow_{\mathcal{R}} t$. The transitive-reflexive-symmetric closure of $\to_{\mathcal{R}}$ is called *conversion* and denoted by $=_{\mathcal{R}}$. If $s =_{\mathcal{R}} t$ then s and t are *convertible*. Two terms t_1, t_2 are *joinable*, denoted by $t_1 \downarrow_{\mathcal{R}} t_2$, if there exists a term t_3 such that $t_1 \twoheadrightarrow_{\mathcal{R}} t_3 \twoheadleftarrow_{\mathcal{R}} t_2$. Such a term t_3 is called a *common reduct* of t_1 and t_2. When no confusion can arise, we omit the subscript $\mathcal{R}$.

A term s is a *normal form* if there is no term t with $s \to t$. A TRS is *weakly normalizing* if every term reduces to a normal form. We write $s \to^! t$ if $s \twoheadrightarrow t$ and t is a normal form. A TRS is *strongly normalizing* if there are no infinite reduction sequences $t_1 \to t_2 \to t_3 \to \cdots$. A TRS is *locally confluent* if for all terms s, t_1, t_2 with $t_1 \leftarrow s \to t_2$ we have $t_1 \downarrow t_2$. A TRS is *confluent* or has the *Church-Rosser* property if for all terms s, t_1, t_2 with $t_1 \twoheadleftarrow s \twoheadrightarrow t_2$ we have $t_1 \downarrow t_2$. A well-known equivalent formulation of confluence is that every pair of convertible terms is joinable ($t_1 = t_2 \Rightarrow t_1 \downarrow t_2$). The renowned Newman's Lemma states that every locally confluent and strongly normalizing TRS is confluent. A *complete* TRS is confluent and strongly normalizing. A *semi-complete* TRS is confluent and weakly normalizing. Each term in a (semi-)complete TRS has a unique normal form. The above properties of TRSs specialize to terms in the obvious way.

A *constructor system* (CS for short) is a TRS $(\mathcal{F}, \mathcal{R})$ with the property that $\mathcal{F}$ can be partitioned into disjoint sets $\mathcal{D}$ and $\mathcal{C}$ such that every left-hand side $F(t_1, \ldots, t_n)$ of a rewrite rule of $\mathcal{R}$ satisfies $F \in \mathcal{D}$ and $t_1, \ldots, t_n \in \mathcal{T}(\mathcal{C}, \mathcal{V})$. Function symbols in $\mathcal{D}$ are called *defined symbols* and those in $\mathcal{C}$ *constructors*. To emphasize the partition of $\mathcal{F}$ into $\mathcal{D}$ and $\mathcal{C}$ we write $(\mathcal{D}, \mathcal{C}, \mathcal{R})$ instead of $(\mathcal{F}, \mathcal{R})$ and $\mathcal{T}(\mathcal{F}, \mathcal{V})$ is denoted by $\mathcal{T}(\mathcal{D}, \mathcal{C}, \mathcal{V})$.

The rules of a *conditional term rewriting system* (CTRS for short) have the form $l \to r \Leftarrow c$. Here the conditional part c is a (possibly empty) sequence $s_1 = t_1, \ldots, s_n = t_n$ of equations. We assume that l is not a variable and that variables occurring in r and c also occur in l. In other words, we do not allow *extra* variables in the conditions. The reasons for excluding conditional rewrite rules with extra variables will be explained later. A rewrite rule without conditions will be written as $l \to r$. The rewrite relation associated with a CTRS $\mathcal{R}$ is obtained by interpreting the equality signs in the conditional part of a rewrite rule as joinability. Formally: $s \to_{\mathcal{R}} t$ if there exist a rewrite rule $l \to r \Leftarrow c$ in $\mathcal{R}$, a context C and a substitution σ such that $s \equiv C[l^\sigma]$, $t \equiv C[r^\sigma]$ and $c_1^\sigma \downarrow_{\mathcal{R}} c_2^\sigma$ for every equation $c_1 = c_2$ in c. All notions defined for TRSs extend to CTRSs. For every CTRS $\mathcal{R}$ we inductively define TRSs $\mathcal{R}_n$ ($n \geqslant 0$) as follows: $\mathcal{R}_0 = \varnothing$ and $\mathcal{R}_{n+1} = \{ l^\sigma \to r^\sigma \mid l \to r \Leftarrow c \in \mathcal{R} \text{ and } c_1^\sigma \downarrow_{\mathcal{R}_n} c_2^\sigma \text{ for all } c_1 = c_2 \text{ in } c \}$. Observe that $\mathcal{R}_n \subseteq \mathcal{R}_{n+1}$ for all $n \geqslant 0$. We have $s \to_{\mathcal{R}} t$ if and only if $s \to_{\mathcal{R}_n} t$ for some $n \geqslant 0$. The minimum such n is called the *depth* of $s \to t$. Depths of conversions $s =_{\mathcal{R}} t$ and 'valleys' $s \downarrow_{\mathcal{R}} t$ are similarly defined.

A CTRS $(\mathcal{F}, \mathcal{R})$ is a *conditional constructor system* (CCS for short) if its underlying TRS (i.e. the TRS obtained from $\mathcal{R}$ by omitting all conditions) is a CS. Observe that we put no (further) limitations on the conditions of the rules in a CCS.

3. Decomposability

The following definition originates from Middeldorp and Toyama [12]. It expresses a natural way to divide a large (conditional) constructor system into smaller, not necessarily disjoint, parts.

DEFINITION 3.1.

- Let $(\mathcal{D}, \mathcal{C}, \mathcal{R})$ be a CCS and $\mathcal{D}'$ a set of function symbols. The subset of $\mathcal{R}$ consisting of all rewrite rules $l \to r \Leftarrow c$ that satisfy $root(l) \in \mathcal{D}'$ is denoted by $\mathcal{R} \mid \mathcal{D}'$.
- Two CCSs $(\mathcal{D}_1, \mathcal{C}_1, \mathcal{R}_1)$ and $(\mathcal{D}_2, \mathcal{C}_2, \mathcal{R}_2)$ are *composable* if $\mathcal{D}_1 \cap \mathcal{C}_2 = \mathcal{D}_2 \cap \mathcal{C}_1 = \varnothing$ and $\mathcal{R}_1 \mid \mathcal{D}_2 = \mathcal{R}_2 \mid \mathcal{D}_1$. The second requirement states that both CCSs should contain all rewrite rules which 'define' a defined symbol whenever that symbol is shared.
- A property $\mathcal{P}$ of CCSs is *decomposable* if for all composable CCSs CCS_1, CCS_2 with the property $\mathcal{P}$ we have that $CCS_1 \cup CCS_2$ has the property $\mathcal{P}$.

Notice that CCSs without common function symbols are trivially composable. Thus every decomposable property of CCSs is also a modular property of CCSs.

In this paper we extend the main results of Middeldorp and Toyama [12] from CSs to CCSs. That is, we will show that both completeness and semi-completeness are decomposable properties of CCSs.

EXAMPLE 3.2. Consider the CCS $(\mathcal{D}, \mathcal{C}, \mathcal{R})$ with $\mathcal{D} = \{+, \times, even, odd\}$, $\mathcal{C} = \{0, S, true, false\}$ and

$$\mathcal{R} = \left\{ \begin{array}{llllll} 0 + x & \to & x, & S(x) + y & \to & S(x+y) \\ 0 \times x & \to & 0, & S(x) \times y & \to & x \times y + y \\ even(0) & \to & true, & even(S(x)) & \to & odd(x) \\ odd(x) & \to & true & \Leftarrow & \multicolumn{2}{l}{even(x) = false} \\ odd(x) & \to & false & \Leftarrow & \multicolumn{2}{l}{even(x) = true} \end{array} \right\}.$$

Let $\mathcal{D}_1 = \{+, \times\}$, $\mathcal{D}_2 = \{even, odd\}$, $\mathcal{R}_1 = \mathcal{R} \mid \mathcal{D}_1$, $\mathcal{R}_2 = \mathcal{R} \mid \mathcal{D}_2$, $\mathcal{C}_1 = \{0, S\}$ and $\mathcal{C}_2 = \{0, S, true, false\}$. The CCSs $(\mathcal{D}_1, \mathcal{C}_1, \mathcal{R}_1)$, $(\mathcal{D}_2, \mathcal{C}_2, \mathcal{R}_2)$ constitute a decomposition of $(\mathcal{D}, \mathcal{C}, \mathcal{R})$. Standard rewriting techniques show the completeness of $(\mathcal{D}_1, \mathcal{C}_1, \mathcal{R}_1)$ and the strong normalization of $(\mathcal{D}_2, \mathcal{C}_2, \mathcal{R}_2)$. For (local) confluence of $(\mathcal{D}_2, \mathcal{C}_2, \mathcal{R}_2)$ we have to show that there exists no term t that satisfies both $even(t) \downarrow_{\mathcal{R}_2} false$ and $even(t) \downarrow_{\mathcal{R}_2} true$, which follows by an easy induction argument. Our main result now yields the completeness of $(\mathcal{D}, \mathcal{C}, \mathcal{R})$. We would like to stress that the problem of showing the unsatisfiability of $even(x) \downarrow_{\mathcal{R}} false \wedge even(x) \downarrow_{\mathcal{R}} true$ is considerably more complicated than showing the unsatisfiability of $even(x) \downarrow_{\mathcal{R}_2} false \wedge even(x) \downarrow_{\mathcal{R}_2} true$.

Unlike the extension of modularity results from TRSs to CTRSs in Middeldorp [9, 10], we do not make use of the decomposability of (semi-)completeness for CSs in our proofs. Rather, we extend the proof ideas in [12] to conditional systems. This is not entirely a routine matter. For instance, the proof of the decomposability of completeness for CSs in [12] employs the decomposability of local-confluence. That result, however, does not hold for CCSs as an easy adjustment of the counterexample in [10] against the modularity of local confluence for CTRSs shows. In [12] it is observed that neither strong normalization nor confluence are decomposable properties of CSs.

4. Marked Reduction

The proofs in [12] heavily depend on the notion of *marked reduction.* In this section we extend this notion to conditional systems and we show that the key properties of marked reduction as established in [12] are still valid in the conditional case. Throughout this section we will be dealing with the union $(\mathcal{D},\mathcal{C},\mathcal{R})$ of two composable CCSs $(\mathcal{D}_1,\mathcal{C}_1,\mathcal{R}_1)$ and $(\mathcal{D}_2,\mathcal{C}_2,\mathcal{R}_2)$.

DEFINITION 4.1.

- The set $\mathcal{D}^* = \{F^* \mid F \in \mathcal{D}_1\}$ consists of *marked* defined symbols. Terms in $\mathcal{T}(\mathcal{D}^* \cup \mathcal{D},\mathcal{C},\mathcal{V})$ are called *marked terms.* An *unmarked term* belongs to $\mathcal{T}(\mathcal{D},\mathcal{C},\mathcal{V})$. Observe that we do not mark symbols in $\mathcal{D}_2 - \mathcal{D}_1$. In the following we abbreviate $\mathcal{T}(\mathcal{D},\mathcal{C},\mathcal{V})$ to $\mathcal{T}$.
- If t is a marked term then $e(t) \in \mathcal{T}$ denotes the term obtained from t by erasing all marks and t^* denotes the term obtained from t by marking every unmarked defined symbol in t that belongs to $\mathcal{D}_1$. These notions are extended to other syntactic objects like substitutions and sequences of equations in the obvious way. The set $\{l^* \to r^* \Leftarrow c^* \mid l \to r \Leftarrow c \in \mathcal{R}_1\}$ of *marked rewrite rules* is denoted by $\mathcal{M}$.
- Two marked terms s and t are *similar*, denoted by $s \approx t$, if $e(s) \equiv e(t)$. If s and t are similar then their *intersection* is the unique term $s \wedge t$ such that $s \wedge t \approx s \approx t$ and a defined symbol occurrence in $s \wedge t$ is marked if and only if the corresponding occurrences in s and t are marked. Since $\wedge$ is easily shown to be associative and commutative, we can extend it to sets of pairwise similar terms in the obvious way, i.e. if $S = \{s_1,\ldots,s_n\}$ is a set of pairwise similar terms, then $\wedge S$ denotes $s_1 \wedge \ldots \wedge s_n$.

DEFINITION 4.2. If $t \equiv C[t_1,\ldots,t_n]$ such that all defined symbols in C are marked and every t_i $(i = 1,\ldots,n)$ is unmarked then we call t a *capped* term. Furthermore, if $root(t_i) \in \mathcal{D}$ for $i = 1,\ldots,n$ then we write $t \equiv C[\![t_1,\ldots,t_n]\!]$. Notice that every capped term can be written in this way. Adopting the terminology from Kurihara and Ohuchi [6, 7], the subterms $t_1,\ldots,t_n$ of $t \equiv C[\![t_1,\ldots,t_n]\!]$ will be called *aliens* of t. The set of all capped terms is denoted by $\hat{\mathcal{T}}$ and the set of all aliens in a capped term t is denoted by $aliens(t)$.

DEFINITION 4.3. A set $T' \subseteq \hat{\mathcal{T}}$ is said to be *alien defined* by a property P if a term t belongs to T' if and only if every alien of t satisfies P.

It is not difficult to see that every alien defined set T' satisfies the following closure properties:

- if $s,t \in T'$ are similar then $s \wedge t \in T'$,
- if $t \in T'$ and $s \subseteq t$ then $s \in T'$,
- if $t^* \equiv t$ and $\sigma \in \Sigma(T')$ then $t^\sigma \in T'$.

DEFINITION 4.4. Let $s \in \hat{\mathcal{T}}$.

- We write $s \to_m t$ if there exists a context C, a rewrite rule $C_1[x_1, \ldots, x_n] \to r \Leftarrow c$ in $\mathcal{M}$ (with all variables occurring in its left-hand side displayed) and terms $s_1, \ldots, s_n$ such that the following four conditions are satisfied:
  - $s \equiv C[C_1[s_1, \ldots, s_n]]$,
  - $s_i \approx s_j$ whenever $x_i \equiv x_j$ for $1 \leqslant i < j \leqslant n$,
  - $t \equiv C[r^\sigma]$,
  - $c_1^\sigma \downarrow_m^\approx c_2^\sigma$ for every equation $c_1 = c_2$ in c.

  Here σ is the substitution induced by l and s, i.e. $\sigma(x) \equiv \wedge\{s_i \mid x_i \equiv x\}$ for $x \in \{x_1, \ldots, x_n\}$ and $\sigma(x) \equiv x$ if $x \notin \{x_1, \ldots, x_n\}$, and $\downarrow_m^\approx$ denotes joinability with respect to $\to_m$ modulo $\approx$, i.e. the relation $\twoheadrightarrow_m \cdot \approx \cdot \twoheadleftarrow_m$. The relation $\to_m$ is called *marked reduction.*
- We write $s \to_u t$ if $s \to_{\mathcal{R}} t$. The relation $\to_u$ is called *unmarked reduction.* Clearly $s \to_u t$ if and only if one of the aliens in s is rewritten. In the following we restrict the use of $\to_{\mathcal{R}}$ to unmarked terms.

Specializing the above definition to CSs yields the relations $\to_m^o$ (for $\to_m$) and $\to_m^i$ (for $\to_u$) of [12]. The well-definedness of $\to_m$ and the closure of $\hat{\mathcal{T}}$ under $\to_m$ easily follow from the alien definedness of $\hat{\mathcal{T}}$. In particular, we have the following fact.

PROPOSITION 4.5. *Let $s \in \hat{\mathcal{T}}$. If $s \to_m t$ then aliens$(t) \subseteq$ aliens(s).*

PROOF. Routine. □

As a matter of fact, the above proposition shows that any alien defined subset of $\hat{\mathcal{T}}$ is closed under marked reduction. Closure of $\hat{\mathcal{T}}$ under $\to_u$ is obvious. The crucial point in the definition of marked reduction is that we allow terms in an instantiated condition to be joinable with respect to marked reduction *up to similarity.* In [12] the concept of similarity was introduced to cope with non-left-linear rewrite rules. However, in a conditional system non-left-linearity can be hidden in the conditions. A conditional rewrite rule $f(x, y) \to c \Leftarrow x = y$ resembles in many respects the unconditional but non-left-linear rule $f(x, x) \to c$.

PROPOSITION 4.6. *Let $s \in \hat{\mathcal{T}}$.*
(1) If $s \to_m t$ then $e(s) \to_{\mathcal{R}_1} e(t)$.
(2) If $s \to_u t$ then $e(s) \to_{\mathcal{R}} e(t)$.

PROOF. The first part is obtained by a straightforward induction on the depth of $s \to_m t$. The second part is trivial. □

Consider a capped term s and a reduction step $e(s) \to_{\mathcal{R}} t$. In the unconditional case (i.e. if $\mathcal{R}$ is a CS) it is easy to see that we can lift $e(s) \to_{\mathcal{R}} t$ to $s \to_m t'$ or $s \to_u t'$ for some term $t' \in \hat{\mathcal{T}}$ with $e(t') \equiv t$ $(*)$. Together with Proposition 4.6 this constitutes a very clear relationship between $\to_{\mathcal{R}}$-reduction on $\mathcal{T}$ and $\to_m$ and $\to_u$-reduction on $\hat{\mathcal{T}}$. In the conditional case this nice correspondence is lost, as shown in the next example.

EXAMPLE 4.7. Consider the CCS

$$\mathcal{R}_1 = \begin{cases} f(x) & \rightarrow & x & \Leftarrow & x = b \\ a & \rightarrow & b \end{cases}$$

with $\mathcal{D}_1 = \{f, a, b\}$ and $\mathcal{C}_1 = \varnothing$. We have $f(a) \rightarrow_{\mathcal{R}_1} a$, but $f^*(a)$ is a normal form with respect to marked reduction since $a \downarrow_m^{\approx} b^*$ does not hold.

Fortunately, we do not really need this one-to-one correspondence between ordinary reduction and reduction ($\rightarrow_m$ and $\rightarrow_u$) on capped terms. The relationship expressed in Lemma 4.17 below is sufficient for our purposes. The next few propositions pave the way for Lemma 4.17. First we show that $(*)$ does hold for "inside normalized" capped terms.

DEFINITION 4.8. A term in $\hat{\mathcal{T}}$ is called *inside normalized* if it is a normal form with respect to $\rightarrow_u$. In other words, a term in $\hat{\mathcal{T}}$ is inside normalized if and only if its aliens are normal forms. The subset of $\hat{\mathcal{T}}$ consisting of all inside normalized terms is denoted by $\hat{\mathcal{T}}_{in}$.

Observe that $\hat{\mathcal{T}}_{in}$ is alien defined by the property of being in normal form. Hence $\hat{\mathcal{T}}_{in}$ is closed under marked reduction.

EXAMPLE 4.9. Consider the CCS of Example 4.7. We have $f(b) \rightarrow_{\mathcal{R}_1} b$. The term $f^*(b)$ is inside normalized, and since $b \downarrow_m^{\approx} b^*$ we can indeed lift the step $f(b) \rightarrow_{\mathcal{R}_1} b$ to $f^*(b) \rightarrow_m b$.

PROPOSITION 4.10. *Let $s \in \hat{\mathcal{T}}_{in}$. If $e(s) \rightarrow_{\mathcal{R}} t$ then there exists a term $t' \in \hat{\mathcal{T}}_{in}$ such that $s \rightarrow_m t'$ and $e(t') \equiv t$.*

PROOF. We use induction on the depth of $e(s) \rightarrow_{\mathcal{R}} t$. In case of zero depth there is nothing to prove. Suppose the depth of $e(s) \rightarrow_{\mathcal{R}} t$ equals $n+1$ $(n \geqslant 0)$. By definition there exists a context C, a rewrite rule $l \rightarrow r \Leftarrow c \in \mathcal{R}$ and a substitution σ such that $e(s) \equiv C[l^\sigma]$, $t \equiv C[r^\sigma]$ and, for every equation $c_1 = c_2$ in c, $c_1^\sigma \downarrow_{\mathcal{R}} c_2^\sigma$ with depth at most n. Write $l \equiv C_1[x_1, \ldots, x_n]$ such that all variables in l are displayed. Since there are no extra variables in the rule $l \rightarrow r \Leftarrow c$, we may assume that $\mathcal{D}\sigma \subseteq \{x_1, \ldots, x_n\}$. We have $s \equiv C'[C_1^*[s_1, \ldots, s_n]]$ for some context C' and terms $s_1, \ldots, s_n \in \hat{\mathcal{T}}_{in}$ such that $e(C') \equiv C$ and $e(s_i) \equiv x_i^\sigma$ for $i = 1, \ldots, n$. Let τ be the substitution induced by l and s. Notice that $\tau \in \Sigma(\hat{\mathcal{T}}_{in})$ and $e(\tau) = \sigma$. Let $c_1 = c_2$ be an equation in c. We will show that $\tau(c_1^*) \downarrow_m^{\approx} \tau(c_2^*)$. We know that there exists a valley $c_1^\sigma \twoheadrightarrow_{\mathcal{R}} u \twoheadleftarrow_{\mathcal{R}} c_2^\sigma$ in which the depth of every step is at most n. Since $e(\tau(c_i^*)) \equiv c_i^\sigma$ and $\tau(c_i^*) \in \hat{\mathcal{T}}_{in}$, a straightforward induction argument shows the existence of a term u_i such that $\tau(c_i^*) \twoheadrightarrow_m u_i$ and $e(u_i) \equiv u$, for $i = 1, 2$. Clearly $u_1 \approx u_2$. Hence $\tau(c_1^*) \downarrow_m^{\approx} \tau(c_2^*)$. From the inside normalization of s we infer that $l \rightarrow r \Leftarrow c \in \mathcal{R}_1$ and hence $l^* \rightarrow r^* \Leftarrow c^* \in \mathcal{M}$. Define $t' \equiv C'[\tau(r^*)]$. We have $s \rightarrow_m t'$. Clearly $e(t') \equiv C[r^\sigma] \equiv t$. □

PROPOSITION 4.11. *Let $s \in \hat{\mathcal{T}}_{in}$. If $e(s) \to^!_{\mathcal{R}} t$ then there exists a term $t' \in \hat{\mathcal{T}}_{in}$ such that $s \to^!_m t'$ and $e(t') \equiv t$.*

PROOF. Repeated application of Proposition 4.10 yields a term $t' \in \hat{\mathcal{T}}_{in}$ such that $s \twoheadrightarrow_m t'$ and $e(t') \equiv t$. It remains to show that t' is a normal form with respect to $\to_m$. This follows from the assumption that t is a normal form, by means of Proposition 4.6(1). □

In the sequel we are mainly interested in capped terms whose aliens are semi-complete. The set of all such terms is denoted by $\hat{\mathcal{T}}_{sc}$. Clearly $\hat{\mathcal{T}}_{in} \subseteq \hat{\mathcal{T}}_{sc} \subseteq \hat{\mathcal{T}}$. Notice that $\hat{\mathcal{T}}_{sc}$ is alien defined by the property of being semi-complete. Hence $\hat{\mathcal{T}}_{sc}$ is closed under marked reduction. Closure under $\to_u$ follows from Proposition 4.16 below.

PROPOSITION 4.12. *Unmarked reduction is semi-complete on $\hat{\mathcal{T}}_{sc}$.*

PROOF. Obvious. □

So every term $t \in \hat{\mathcal{T}}_{sc}$ has a unique normal form in $\hat{\mathcal{T}}_{in}$ with respect to $\to_u$, which will be denoted by $\psi(t)$. Consider similar terms $s, t \in \hat{\mathcal{T}}_{sc}$. In general $\psi(s)$ and $\psi(t)$ are not similar. The next proposition states that they can be made similar by applying certain 'balancing' marked reduction steps.

PROPOSITION 4.13. *Let $s, t \in \hat{\mathcal{T}}_{sc}$. If $s \approx t$ then there exist similar terms $s', t' \in \hat{\mathcal{T}}_{sc}$ such that $\psi(s) \twoheadrightarrow_m s'$, $\psi(t) \twoheadrightarrow_m t'$ and $\psi(s \wedge t) \equiv s' \wedge t'$.*

PROOF. We may write $s \wedge t \equiv C⟦s_1 \wedge t_1, \ldots, s_n \wedge t_n⟧$, $s \equiv C[s_1, \ldots, s_n]$ and $t \equiv C[t_1, \ldots, t_n]$. We will define similar terms s'_i and t'_i such that $\psi(s_i) \twoheadrightarrow_m s'_i$, $\psi(t_i) \twoheadrightarrow_m t'_i$ and $\psi(s_i \wedge t_i) \equiv s'_i \wedge t'_i$, for every $i \in \{1, \ldots, n\}$. Fix i. We have $e(s_i) \equiv t_i$ or $s_i \equiv e(t_i)$. Assume without loss of generality the former. By definition $s_i \to^!_u \psi(s_i)$. Proposition 4.6 yields $t_i \twoheadrightarrow_u e(\psi(s_i))$ and since t_i is semi-complete we obtain $e(\psi(s_i)) \to^!_u \psi(t_i)$. According to Proposition 4.11 $\psi(s_i)$ has a normal from s'_i with respect to $\to_m$ such that $e(s'_i) \equiv \psi(t_i)$. Define $t'_i \equiv \psi(t_i)$. We clearly have $\psi(s_i \wedge t_i) \equiv \psi(t_i) \equiv s'_i \wedge \psi(t_i) \equiv s'_i \wedge t'_i$. Now that we have defined $s'_1, t'_1 \ldots, s'_n, t'_n$, let $s' \equiv C[s'_1, \ldots, s'_n]$ and $t' \equiv C[t'_1, \ldots, t'_n]$. Clearly $s' \approx t'$, $\psi(s) \equiv C[\psi(s_1), \ldots, \psi(s_n)] \twoheadrightarrow_m s'$, $\psi(t) \equiv C[\psi(t_1), \ldots, \psi(t_n)] \twoheadrightarrow_m t'$ and $\psi(s \wedge t) \equiv C[\psi(s_1 \wedge t_1), \ldots, \psi(s_n \wedge t_n)] \equiv C[s'_1 \wedge t'_1, \ldots, s'_n \wedge t'_n] \equiv s' \wedge t'$. □

The above proposition easily generalizes to the case of $n \geqslant 2$ pairwise similar terms.

PROPOSITION 4.14. *Let $s_1, \ldots, s_n \in \hat{\mathcal{T}}_{sc}$ be pairwise similar terms. There exist pairwise similar terms $t_1, \ldots, t_n \in \hat{\mathcal{T}}_{sc}$ such that $\psi(s_i) \twoheadrightarrow_m t_i$ for every $i \in \{1, \ldots, n\}$ and $\psi(\wedge\{s_i \mid 1 \leqslant i \leqslant n\}) \equiv \wedge\{t_i \mid 1 \leqslant i \leqslant n\}$.* □

DEFINITION 4.15. We define a relation $\rhd$ on terms in $\hat{\mathcal{T}}$ as follows: $s \rhd t$ if for every alien $t' \subseteq t$ there exists an alien $s' \subseteq s$ and a context C not containing defined symbols such that $s' \twoheadrightarrow_{\mathcal{R}} C[t']$. The relation $\rhd$ is clearly transitive. In words, $s \rhd t$ if every alien of t can be 'traced back' to some alien in s.

PROPOSITION 4.16. *Let* $s \in \hat{T}$. *If* $s \to_m t$ *or* $s \to_u t$ *then* $s \triangleright t$.

PROOF. If $s \to_m t$ then the result follows from Proposition 4.5. Suppose $s \to_u t$. Let $s \equiv C\llbracket s_1, \ldots, s_n \rrbracket$. We have $t \equiv C[s_1, \ldots, t_i, \ldots, s_n]$ for some term t_i with $s_i \to_{\mathcal{R}} t_i$. Let a be an alien in t. Either $a \equiv s_j$ for some $j \neq i$ or a is an alien in t_i. In the former case a is an alien in s. In the latter case we have $s_i \to_{\mathcal{R}} t_i \equiv C'[a]$ for some context C'. Observe that C' is unmarked and hence it does not contain defined symbols (otherwise a would not be an alien in t_i). □

LEMMA 4.17. *Let* $s \in \hat{T}_{sc}$. *If* $e(s) \to_{\mathcal{R}} t$ *then* $s \to_u t'$ *or* $s \to_u^! \cdot \to_m^+ \cdot \leftarrow_u^! t'$ *for some term* $t' \in \hat{T}_{sc}$ *such that* $e(t') \equiv t$. *Moreover, we may assume that* $s \triangleright t'$.

PROOF. We use induction on the depth of $e(s) \to_{\mathcal{R}} t$. The case of zero depth is trivial. Suppose the depth of $e(s) \to_{\mathcal{R}} t$ equals $n+1$ ($n \geqslant 0$). By definition there exists a context C, a rewrite rule $l \to r \Leftarrow c \in \mathcal{R}$ and a substitution σ such that $e(s) \equiv C[l^\sigma]$, $t \equiv C[r^\sigma]$ and, for every equation $c_1 = c_2$ in c, $c_1^\sigma \downarrow_{\mathcal{R}} c_2^\sigma$ with depth at most n. We distinguish two cases. If the 'redex occurrence' l^σ is not marked in s then we may write $s \equiv C'[l^\sigma]$ for some context C' with $e(C') \equiv C$. In this case we clearly have $s \to_u C'[r^\sigma]$ and $e(C'[r^\sigma]) \equiv t$. Moreover, $s \triangleright C'[r^\sigma]$ by the preceding proposition. Suppose l^σ is marked in s. Write $l \equiv C_1[x_1, \ldots, x_n]$ such that C_1 is variable-free. Without loss of generality we assume that $\mathcal{D}\sigma \subseteq \{x_1, \ldots, x_n\}$. We have $s \equiv C'[C_1^*[s_1, \ldots, s_n]]$ for some context C' and terms $s_1, \ldots, s_n \in \hat{T}_{sc}$ such that $e(C') \equiv C$ and $e(s_i) \equiv x_i^\sigma$ for $i = 1, \ldots, n$. Let τ be the substitution induced by l and s, and define a substitution v by $v(x) \equiv \psi(x^\tau)$ for every $x \in \mathcal{V}$. Notice that v is well-defined since $\tau \in \Sigma(\hat{T}_{sc})$. Let $c_1 = c_2$ be an equation in c. We will show that $v(c_1^*) \downarrow_m^{\approx} v(c_2^*)$, see Figure 1. We know the existence of a valley $c_1^\sigma \twoheadrightarrow_{\mathcal{R}} u \twoheadleftarrow_{\mathcal{R}} c_2^\sigma$ in which the depth of every step is at most n. Since $e(\tau(c_i^*)) \equiv c_i^\sigma$ and $\tau(c_i^*) \in \hat{T}_{sc}$, a straightforward induction argument shows the existence of a term u_i such that $\psi(\tau(c_i^*)) \twoheadrightarrow_m \psi(u_i)$ and $e(u_i) \equiv u$, for $i = 1,2$. Since $u_1 \approx u_2$, Proposition 4.13 yields $\psi(u_1) \downarrow_m^{\approx} \psi(u_2)$. Hence $v(c_1^*) \equiv \psi(\tau(c_1^*)) \downarrow_m^{\approx} \psi(\tau(c_2^*)) \equiv v(c_2^*)$. We have $\psi(s) \equiv \psi(C')[C_1^*[\psi(s_1), \ldots, \psi(s_n)]]$. With help of Proposition 4.14 we can find terms $t_1, \ldots, t_n$ such that $\psi(s_i) \twoheadrightarrow_m t_i$ for $i \in \{1, \ldots, n\}$ and $\psi(\wedge\{s_i \mid x_i \equiv x\}) \equiv \wedge\{t_i \mid x_i \equiv x\}$ for every $x \in \{x_1, \ldots, x_n\}$. Clearly $\psi(s) \twoheadrightarrow_m \psi(C')[C_1^*[t_1, \ldots, t_n]]$. Notice that $\wedge\{t_i \mid x_i \equiv x\} \equiv v(x)$ for every $x \in \{x_1, \ldots, x_n\}$. Hence $\psi(C')[C_1^*[t_1, \ldots, t_n]] \to_m \psi(C')[v(r^*)]$. Define $t' \equiv C'[\tau(r^*)]$. Clearly $\psi(t') \equiv \psi(C')[\psi(\tau(r^*))] \equiv \psi(C')[v(r^*)]$. One easily shows that $s \triangleright t'$. □

COROLLARY 4.18. *Let* $s \in \hat{T}_{sc}$. *If* $e(s) \twoheadrightarrow_{\mathcal{R}} t$ *then there exists a term* $t' \in \hat{T}_{sc}$ *such that* $e(t') \equiv t$ *and* $\psi(s) \twoheadrightarrow_m \psi(t')$. □

In the proof of our main results we transform reduction sequences in the union $(\mathcal{D}, \mathcal{C}, \mathcal{R})$ of two composable CCSs $(\mathcal{D}_1, \mathcal{C}_1, \mathcal{R}_1)$ and $(\mathcal{D}_2, \mathcal{C}_2, \mathcal{R}_2)$ having a certain property $\mathcal{P}$ by means of the previous results into $\mathcal{R}_1$-sequences. In order to employ the fact that $(\mathcal{D}_1, \mathcal{C}_1, \mathcal{R}_1)$ satisfies $\mathcal{P}$, we have to get rid of subterms that do not belong to $\mathcal{T}(\mathcal{D}_1, \mathcal{C}_1, \mathcal{V})$. In [12] the notion of *$\mathcal{D}'$-replacement* was introduced for this purpose. Since in the present setting we can only mark defined symbols in $\mathcal{D}_1$, this notion is amenable to a slight simplification.

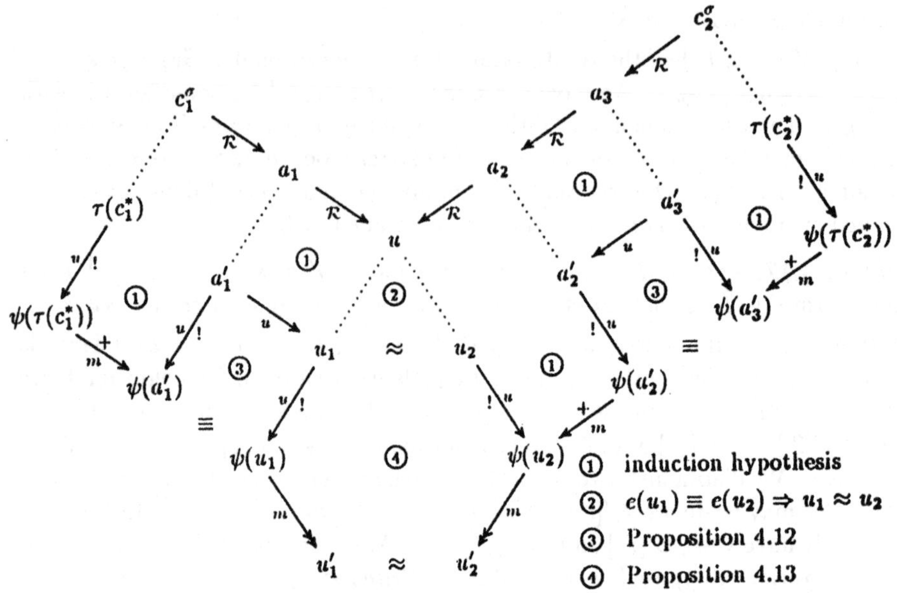

FIGURE 1.

DEFINITION 4.19.

- A set of pairs $\phi = \{\langle s_1, x_1\rangle, \ldots, \langle s_n, x_n\rangle\}$ is a *garbage collector* if $x_1, \ldots, x_n$ are mutually distinct variables and $s_1, \ldots, s_n$ are mutually distinct unmarked terms without occurrences of $x_1, \ldots, x_n$ such that $root(s_i) \in \mathcal{D}_2 - \mathcal{D}_1$ for $i = 1, \ldots, n$. Let $t \equiv C[t_1, \ldots, t_m]$ such that all maximal occurrences of unmarked subterms in t with root symbol in $\mathcal{D}_2 - \mathcal{D}_1$ are displayed. We say that ϕ is *applicable* to t, denoted by $\phi \diamond t$, if $x_1, \ldots, x_n$ do not occur in t and $\{t_1, \ldots, t_m\} \subseteq \{s_1, \ldots, s_n\}$. In this case we may write $t \equiv C[s_{i_1}, \ldots, s_{i_m}]$ with all occurrences of $s_1, \ldots, s_n$ in t displayed and we define $\phi(t) \equiv C[x_{i_1}, \ldots, x_{i_m}]$.
- Let $\phi = \{\langle s_1, x_1\rangle, \ldots, \langle s_n, x_n\rangle\}$ be a garbage collector. If $t \equiv C[x_{i_1}, \ldots, x_{i_m}]$ such that all occurrences of the variables $x_1, \ldots, x_n$ in t are displayed, then the term $C[s_{i_1}, \ldots, s_{i_m}]$ is denoted by $\phi^{-1}(t)$.

PROPOSITION 4.20. *Every capped term t has an applicable garbage collector ϕ.*

PROOF. Let $\{t_1, \ldots, t_n\}$ be the set of all maximal subterms of t with root symbol in $\mathcal{D}_2 - \mathcal{D}_1$. Choose fresh variables $x_1, \ldots, x_n$ and define $\phi = \{\langle t_1, x_1\rangle, \ldots, \langle t_n, x_n\rangle\}$. By construction ϕ is a garbage collector with $\phi \diamond t$. □

PROPOSITION 4.21. *The relations $\rightarrow_m$, $\rightarrow_{\mathcal{R}_1}$ and $\rightarrow_{\mathcal{R}_2}$ are closed under ϕ^{-1}.*

PROOF. Since the mapping ϕ^{-1} can be considered as a substitution in $\Sigma(\mathcal{T})$, the result follows from the closure of $\rightarrow_m$, $\rightarrow_{\mathcal{R}_1}$ and $\rightarrow_{\mathcal{R}_2}$ under such substitutions, which is easily shown by an induction on the depth of steps $s \to_m t$, $s \to_{\mathcal{R}_1} t$ and $s \to_{\mathcal{R}_2} t$. □

PROPOSITION 4.22. *Let t be a capped term and suppose that ϕ is a garbage collector.*
(1) *If $\phi \diamond t$ then $\phi^{-1}(\phi(t)) \equiv t$.*
(2) *We have $\phi \diamond t$ if and only if $\phi \diamond e(t)$. Moreover, if $\phi \diamond t$ then $e(\phi(t)) \equiv \phi(e(t))$.*
(3) *If t does not contain symbols in $\mathcal{D}_2 - \mathcal{D}_1$ then $\phi \diamond \phi^{-1}(t)$ and $\phi(\phi^{-1}(t)) \equiv t$.*

PROOF. Straightforward. □

PROPOSITION 4.23. *Let s and t be similar capped terms. If $\phi \diamond s$ then $\phi \diamond t$ and $\phi(s) \approx \phi(t)$.*

PROOF. We obtain $\phi \diamond t$ by two applications of Proposition 4.22(2). The similarity of $\phi(s)$ and $\phi(t)$ also follows from Proposition 4.22(2). □

PROPOSITION 4.24. *Let ϕ be a garbage collector that is applicable to a capped term s. If $s \to_m t$ then $\phi \diamond t$ and $\phi(s) \to_m \phi(t)$.*

PROOF. Since every unmarked subterm of t with root symbol in $\mathcal{D}_2 - \mathcal{D}_1$ occurs in some alien in t, we obtain $\phi \diamond t$ from Proposition 4.5. We prove that $\phi(s) \to_m \phi(t)$ by induction on the depth of $s \to_m t$. In case of zero depth we have nothing to prove. Suppose the depth of $s \to_m t$ equals $n + 1$ ($n \geqslant 0$). By definition there exists a context C, a rewrite rule $C_1[x_1, \ldots, x_n] \to r \Leftarrow c$ in $\mathcal{M}$ with C_1 variable-free and terms $s_1, \ldots, s_n$ such that $s \equiv C[C_1[s_1, \ldots, s_n]]$, $s_i \approx s_j$ whenever $x_i \equiv x_j$, $t \equiv C[r^\sigma]$ and, for every equation $c_1 = c_2$ in c, $c_1^\sigma \downarrow_m^\approx c_2^\sigma$ with depth at most n. Here σ is the substitution induced by l and s. $x \notin \{x_1, \ldots, x_n\}$. We have $\phi(s) \equiv \phi(C)[C_1[\phi(s_1), \ldots, \phi(s_n)]]$. Let τ be the substitution induced by l and $\phi(s)$. Proposition 4.23 yields $x^\tau \equiv \phi(x^\sigma)$, hence $\phi(t) \equiv \phi(C)[r^\tau]$. Let $c_1 = c_2$ be an equation in c. From $c_1^\sigma \downarrow_m^\approx c_2^\sigma$ we obtain $c_1^\tau \equiv \phi(c_1^\sigma) \downarrow_m^\approx \phi(c_2^\sigma) \equiv c_2^\tau$ by a routine induction argument. Therefore $\phi(s) \to_m \phi(t)$. □

5. Main Results

In this section we present our main results. We prove the decomposability of weak normalization, semi-completeness and completeness, in that order. The proof of the decomposability of completeness for CSs in [12] does not depend on the decomposability of semi-completeness. This is possible since the decomposability of local confluence—which holds in the unconditional case—enables one to circumvent a direct proof of confluence. However, local confluence is not a decomposable property of CCSs. Hence we show the decomposability of semi-completeness before we tackle completeness.

PROPOSITION 5.1. *Suppose $(\mathcal{D}_1, \mathcal{C}_1, \mathcal{R}_1)$ and $(\mathcal{D}_2, \mathcal{C}_2, \mathcal{R}_2)$ are composable CCSs and let $s \in \mathcal{T}(\mathcal{D}_i, \mathcal{C}_i, \mathcal{V})$ for some $i \in \{1,2\}$. If $s \to_{\mathcal{R}_1 \cup \mathcal{R}_2} t$ then $s \to_{\mathcal{R}_i} t$.*

PROOF. The proposition is easily proved by induction on the depth of the step $s \to_{\mathcal{R}_1 \cup \mathcal{R}_2} t$, using the equality $\mathcal{R}_1 \mid \mathcal{D}_2 = \mathcal{R}_2 \mid \mathcal{D}_1$. □

Lemma 5.2. *Weak normalization is a decomposable property of CCSs.*

Proof. Let $(\mathcal{D},\mathcal{C},\mathcal{R})$ be the union of weakly normalizing and composable CCSs $(\mathcal{D}_1,\mathcal{C}_1,\mathcal{R}_1)$ and $(\mathcal{D}_2,\mathcal{C}_2,\mathcal{R}_2)$. By induction on the structure of t we will show that every term $t \in \mathcal{T}$ has a normal form with respect to $\mathcal{R}$. The case $t \in \mathcal{C} \cup \mathcal{V}$ is trivial. If t is a defined constant then t belongs to some $\mathcal{D}_i$ and the result follows from the weak normalization of $(\mathcal{D}_i,\mathcal{C}_i,\mathcal{R}_i)$ and Proposition 5.1. For the induction step, suppose that $t \equiv F(t_1,\ldots,t_n)$ with $t_1,\ldots,t_n$ weakly normalizing. Let s_i be an $\mathcal{R}$-normal form of t_i for $i = 1,\ldots,n$. Clearly $t \twoheadrightarrow_{\mathcal{R}} F(s_1,\ldots,s_n)$. If $F \in \mathcal{C}$ then $F(s_1,\ldots,s_n)$ is an $\mathcal{R}$-normal form of t. Suppose $F \in \mathcal{D}$. Without loss of generality we may assume that $F \in \mathcal{D}_1$. Let $t' \equiv F^*(s_1,\ldots,s_n)$. From Proposition 4.20 we obtain a garbage collector ϕ which is applicable to t'. Since $e(\phi(t')) \in \mathcal{T}(\mathcal{D}_1,\mathcal{C}_1,\mathcal{V})$ and $(\mathcal{D}_1,\mathcal{C}_1,\mathcal{R}_1)$ is weakly normalizing, there exists a $\mathcal{R}_1$-normal form n such that $e(\phi(t')) \rightarrow^!_{\mathcal{R}_1} n$. According to Proposition 5.1 n is also normal form with respect to $\mathcal{R}_2$. Using Proposition 4.21, we easily infer $\phi(t') \in \hat{\mathcal{T}}_{in}$ from $t' \in \hat{\mathcal{T}}_{in}$. From Proposition 4.11 we obtain a term $n' \in \hat{\mathcal{T}}_{in}$ with $e(n') \equiv n$ and $\phi(t') \rightarrow^!_m n'$. Propositions 4.21 and 4.22(1) yield $t' \twoheadrightarrow_m \phi^{-1}(n')$. The inside normalization of $\phi^{-1}(n')$ follows from the inside normalization of t' by means of Proposition 4.5. From Proposition 4.6(1) we obtain $F(s_1,\ldots,s_n) \equiv e(t') \twoheadrightarrow_{\mathcal{R}_1} e(\phi^{-1}(n'))$. Clearly $e(\phi^{-1}(n')) \equiv \phi^{-1}(n)$. We conclude the proof by showing that $\phi^{-1}(n)$ is an $\mathcal{R}$-normal form. Suppose to the contrary that $\phi^{-1}(n) \rightarrow_{\mathcal{R}} u$ for some term u. Since $\phi^{-1}(n') \in \hat{\mathcal{T}}_{in}$, we can apply Proposition 4.10, which yields a term $u' \in \hat{\mathcal{T}}_{in}$ such that $e(u') \equiv u$ and $\phi^{-1}(n') \rightarrow_m u'$. Since n' does not contain function symbols in $\mathcal{D}_2 - \mathcal{D}_1$, we obtain $\phi \circ \phi^{-1}(n')$ and $\phi(\phi^{-1}(n')) \equiv n'$ from Proposition 4.22(3). Proposition 4.24 now yields $n' \rightarrow_m \phi(u')$, contradicting the fact that n' is a normal form with respect to $\rightarrow_m$. □

Theorem 5.3. *Semi-completeness is a decomposable property of CCSs.*

Proof. Let $(\mathcal{D},\mathcal{C},\mathcal{R})$ be the union of semi-complete and composable CCSs $(\mathcal{D}_1,\mathcal{C}_1,\mathcal{R}_1)$ and $(\mathcal{D}_2,\mathcal{C}_2,\mathcal{R}_2)$. From Lemma 5.2 we infer the weak normalization of $(\mathcal{D},\mathcal{C},\mathcal{R})$. By induction on the structure of t we will show that every term $t \in \mathcal{T}$ is confluent. The case $t \in \mathcal{D} \cup \mathcal{C} \cup \mathcal{V}$ is easy. Suppose $t \equiv F(t_1,\ldots,t_n)$ such that $t_1,\ldots,t_n$ are confluent and thus semi-complete. If F is a constructor then t is easily shown to be confluent. Suppose $F \in \mathcal{D}$. Without loss of generality we assume that $F \in \mathcal{D}_1$. Let $t' \equiv F^*(t_1,\ldots,t_n)$. Clearly $t' \in \hat{\mathcal{T}}_{sc}$. Proposition 4.20 yields a garbage collector ϕ which is applicable to $\psi(t')$. After these preliminary definitions, we consider diverging reductions $u_1 \twoheadleftarrow_{\mathcal{R}} t \twoheadrightarrow_{\mathcal{R}} u_2$. Figure 2 shows how to obtain a common reduct of u_1 and u_2. Therefore t is confluent. □

Theorem 5.4. *Completeness is a decomposable property of CCSs.*

Proof. Suppose $(\mathcal{D}_1,\mathcal{C}_1,\mathcal{R}_1)$ and $(\mathcal{D}_2,\mathcal{C}_2,\mathcal{R}_2)$ are complete and composable CCSs. From Theorem 5.3 we obtain the semi-completeness of their union $(\mathcal{D},\mathcal{C},\mathcal{R})$. Hence it remains to show that $(\mathcal{D},\mathcal{C},\mathcal{R})$ is strongly normalizing. This will be established by induction on the structure of terms $t \in \mathcal{T}$. The base case is easy. Let $t \equiv F(t_1,\ldots,t_n)$ such that $t_1,\ldots,t_n$ are strongly normalizing and thus complete. If F is a constructor then t is clearly strongly normalizing. So assume

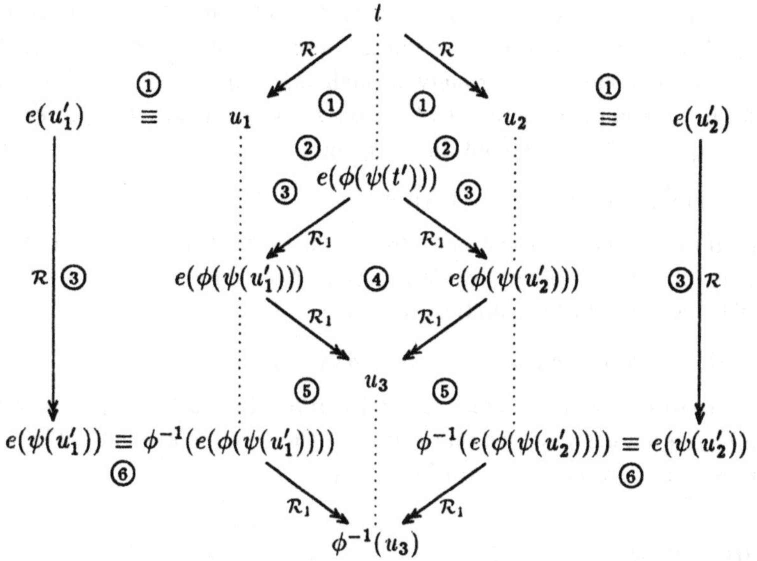

① Corollary 4.18 ③ Proposition 4.6 ⑤ Proposition 4.21
② Proposition 4.24 ④ confluence of $(\mathcal{D}_1, \mathcal{C}_1, \mathcal{R}_1)$ ⑥ Proposition 4.22(1,2)

FIGURE 2.

without loss of generality that $F \in \mathcal{D}_1$. If t is not strongly normalizing then there exists an infinite reduction sequence

$$t \equiv s_1 \to_{\mathcal{R}} s_2 \to_{\mathcal{R}} s_3 \to_{\mathcal{R}} \cdots .$$

Let $t' \equiv s_1' \equiv F^*(t_1, \ldots, t_n)$. Clearly $t' \in \hat{\mathcal{T}}_{sc}$. Repeated application of Lemma 4.17 yields terms $s_i' \in \hat{\mathcal{T}}_{sc}$ $(i > 1)$ with $e(s_i') \equiv s_i$ and, for all $i \geqslant 1$, $s_i' \triangleright s_{i+1}'$ and $s_i' \to_u s_{i+1}'$ or $\psi(s_i') \to_m^+ \psi(s_{i+1}')$, see Figure 3. We now show that the second

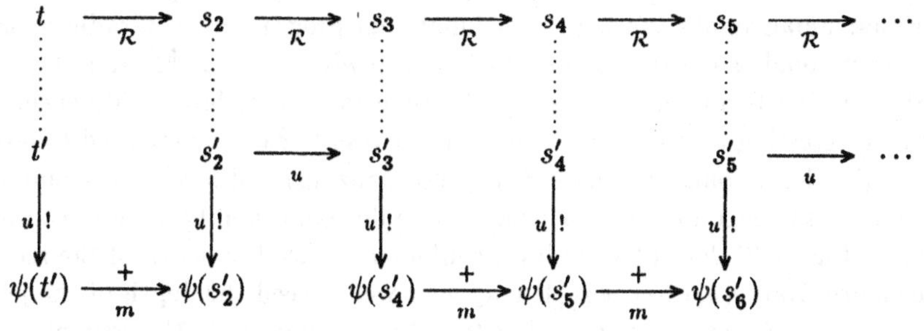

FIGURE 3.

possibility occurs infinitely often. Suppose to the contrary that there exists an index N such that $s_i' \to_u s_{i+1}'$ for all $i \geqslant N$. By the pigeon-hole principle there exists an alien a in s_N' with an infinite $\mathcal{R}$-reduction. Since $t' \triangleright s_N'$ there exists an alien a' in t' such that $a' \twoheadrightarrow_{\mathcal{R}} C[a]$ for some context C. Hence a' is not strongly normalizing. However, since a' is an unmarked subterm of $t' = F^*(t_1, \ldots, t_n)$, it must be a subterm of one of the strongly normalizing terms $t_1, \ldots, t_n$. This is impossible. From the semi-completeness of $\rightarrow_u$ on $\hat{\mathcal{T}}_{sc}$ we infer that $\psi(s'_i) \equiv \psi(s'_{i+1})$ whenever $s'_i \rightarrow_u s'_{i+1}$. Hence we obtain a sequence

$$\psi(t') \equiv \psi(s'_1) \twoheadrightarrow_m \psi(s'_2) \twoheadrightarrow_m \psi(s'_3) \twoheadrightarrow_m \cdots$$

containing infinitely many reduction steps. From Proposition 4.20 we obtain a garbage collector ϕ which is applicable to $\psi(t')$. According to Lemma 4.24 the term $\phi(\psi(t'))$ has an infinite reduction sequence

$$\phi(\psi(t')) \equiv \phi(\psi(s'_1)) \twoheadrightarrow_m \phi(\psi(s'_2)) \twoheadrightarrow_m \phi(\psi(s'_3)) \twoheadrightarrow_m \cdots .$$

Employing Proposition 4.6, the erasure of all markers in the last sequence yields an infinite $\mathcal{R}_1$-reduction sequence starting from the term $e(\phi(\psi(t')))$. This contradicts the strong normalization of $(\mathcal{D}_1, \mathcal{C}_1, \mathcal{R}_1)$. □

6. Conclusion

The results presented in this paper should be extended to CCSs with extra variables in the conditions of the rewrite rules. This will not be an easy matter. First of all, the results on marked reduction in Section 4 do not extend to such CCSs (after an obvious adjustment of the substitution σ in Definition 4.4). However, all counterexamples (against Proposition 4.10 and Lemma 4.17) that we could think of involved a non-confluent CCS. Another problem is that Proposition 5.1 and Lemma 5.2 do not allow this relaxation in the distribution of variables. For instance, in [9, 10] it is observed that weak normalization is not a modular property of CTRSs with extra variables in the conditions, but semi-completeness is modular. We strongly believe that both semi-completeness and completeness are decomposable properties of CCSs with extra variables in the conditions.

A CTRS $\mathcal{R}$ is called *level-confluent* if every $\mathcal{R}_n$ is confluent. Level-confluence (Giovannetti and Moiso [2]) is a key property for ensuring completeness of languages that amalgamate the logic and functional programming paradigms and whose operational semantics is conditional narrowing, see e.g. [2] and [11]. At present very few techniques are available for establishing level-confluence. It would be interesting to investigate whether our results can be extended to level-complete (i.e. level-confluent and strongly normalizing) and level-semi-complete CCSs. The confluence part of our proof of the decomposability of semi-completeness (see Figure 2) doesn't yield level-confluence, since the depth of the reduction sequence from $e(u'_i)$ to $e(\psi(u'_i))$ may very well exceed the depth of the peak $u_1 \twoheadleftarrow_{\mathcal{R}} t \twoheadrightarrow_{\mathcal{R}} u_2$. A possible approach is to reduce u'_i only to its $\mathcal{R}_n$-normal form, where n is the depth of $u_1 \twoheadleftarrow_{\mathcal{R}} t \twoheadrightarrow_{\mathcal{R}} u_2$, but so far we haven't been able to put this idea into a proof.

Acknowledgements. The presentation of the paper benefitted from the comments of Vincent van Oostrom.

References

1. A.C. Caron, *Decidability of Reachability and Disjoint Union of Term Rewriting Systems*, Proc. CAAP'92, LNCS **581**, pp. 86–101, 1992.
2. E. Giovannetti and C. Moiso, *A Completeness Result for E-Unification Algorithms based on Conditional Narrowing*, Proc. Workshop on Foundations of Logic and Functional Programming, Trento, LNCS **306**, pp. 157–167, 1986.
3. B. Gramlich, *Generalized Sufficient Conditions for Modular Termination of Rewriting*, Proc. ALP'92, LNCS, 1992, to appear. Full version: *A Structural Analysis of Modular Termination of Term Rewriting Systems*, SEKI Report SR-91-15, Universität Kaiserslautern, 1991.
4. B. Gramlich, *Sufficient Conditions for Modular Termination of Conditional Term Rewriting Systems*, this volume.
5. J.W. Klop, A. Middeldorp, Y. Toyama and R.C. de Vrijer, *A Simplified Proof of Toyama's Theorem*, report CS-R9156, CWI, Amsterdam, 1991.
6. M. Kurihara and A. Ohuchi, *Modularity of Simple Termination of Term Rewriting Systems*, Journal of IPS Japan **31**(5), pp. 633–642, 1990.
7. M. Kurihara and A. Ohuchi, *Modularity of Simple Termination of Term Rewriting Systems with Shared Constructors*, Report SF-36, Hokkaido University, Sapporo, 1990. To appear in Theoretical Computer Science, 1992.
8. A. Middeldorp, *A Sufficient Condition for the Termination of the Direct Sum of Term Rewriting Systems*, Proc. LICS'89, Pacific Grove, pp. 396–401, 1989.
9. A. Middeldorp, *Modular Properties of Term Rewriting Systems*, Ph.D. thesis, Vrije Universiteit, Amsterdam, 1990.
10. A. Middeldorp, *Modular Properties of Conditional Term Rewriting Systems*, report CS-R9105, CWI, Amsterdam, 1991. To appear in Information and Computation.
11. A. Middeldorp and E. Hamoen, *Counterexamples to Completeness Results for Basic Narrowing*, report CS-R9154, CWI, Amsterdam, 1991. Extended abstract to appear in Proc. ALP'92, LNCS.
12. A. Middeldorp and Y. Toyama, *Completeness of Combinations of Constructor Systems*, Proc. RTA'91, Como, LNCS **488**, pp. 188–199, 1991.
13. E. Ohlebusch, *Combinations of Simplifying Conditional Term Rewriting Systems*, this volume.
14. M. Rusinowitch, *On Termination of the Direct Sum of Term Rewriting Systems*, Information Processing Letters **26**, pp. 65–70, 1987.
15. Y. Toyama, *On the Church-Rosser Property for the Direct Sum of Term Rewriting Systems*, Journal of the ACM **34**(1), pp. 128–143, 1987.
16. Y. Toyama, *Counterexamples to Termination for the Direct Sum of Term Rewriting Systems*, Information Processing Letters **25**, pp. 141–143, 1987.
17. Y. Toyama, J.W. Klop and H.P. Barendregt, *Termination for the Direct Sum of Left-Linear Term Rewriting Systems*, Proc. RTA'89, Chapel Hill, LNCS **355**, pp. 477–491, 1989.

Collapsed Tree Rewriting: Completeness, Confluence, and Modularity

Detlef Plump*
Universität Bremen

Abstract

Collapsed trees are (hyper)graphs which represent functional expressions such that common subexpressions can be shared. Rewrite steps with collapsed trees include applications of term rewrite rules as well as "folding steps" which identify common subexpressions. Different aspects of this model of computation are considered: (1) It is shown that collapsed tree rewriting is complete with respect to equational validity in the same sense as term rewriting is. (2) Collapsed tree rewriting and term rewriting are compared with respect to confluence. (3) Termination and convergence of collapsed tree rewriting turn out to be modular properties, in contrast to the situation for term rewriting.

1 Introduction

The representation of functional expressions by *collapsed trees* (also known as *directed acyclic graphs*) is well-known from implementations of term rewriting and functional programming languages. The advantage over trees is that common subexpressions can be shared, so that evaluation costs are cut down both in time and space. However, while for term rewriting there is a comprehensive theory which addresses various operational and semantical aspects, comparatively little is known about collapsed tree rewriting. This appears unsatisfactory as the two models of computation do behave different in several respects. For instance, there are non-terminating term rewriting systems that become terminating under collapsed tree rewriting, and confluence may get lost by passing from terms to collapsed trees.

This paper makes three contributions to the theory of collapsed tree rewriting. First of all, it is shown that collapsed tree rewriting is complete with respect to equational validity in the same sense as term rewriting is. This result requires the "collapsing" of collapsed trees during evaluation, and hence does not hold

*Author's address: Fachbereich Mathematik und Informatik, Universität Bremen, Postfach 33 04 40, 2800 Bremen 33, Germany. e-mail: `det@informatik.uni-bremen.de`. Research supported by Deutsche Forschungsgemeinschaft and by ESPRIT Basic Research Working Group 3299, COMPUGRAPH.

in most of the existing approaches to function evaluation by graph rewriting ([BvEG+87, FW91, CR92] to mention a few).

Then the relationship between term rewriting and collapsed tree rewriting with respect to confluence is clarified. It turns out that term rewriting is confluent whenever collapsed tree rewriting is, but the converse does not hold. Sufficient conditions for confluence are presented.

Finally, collapsed tree rewriting is shown to behave modular when terminating or convergent (i.e. terminating and confluent) systems are combined; in both cases the combined systems are allowed to share certain function symbols. In contrast, it is known that even the disjoint union of term rewriting systems may destroy termination and convergence.

2 Collapsed Tree Rewriting

Collapsed tree rewriting is similar to *jungle evaluation* as it is described in [HP91]. Jungles are acyclic hypergraphs which represent collections of many-sorted terms, and collapsed trees are single-rooted jungles over unsorted signatures. While the restriction to the unsorted case is immaterial (and could be easily dropped at the costs of more complex definitions), a significant difference is that the rewrite steps used here include "garbage collection". The latter is natural from the perspective of term rewriting and simplifies the formulation of confluence results, as otherwise in most cases the garbage in jungles must be ignored explicitly.

Let X be a fixed set the elements of which are called *variables*. A *signature* Σ is a set of function symbols disjoint from X. Each function symbol f comes with an integer $arity(f) \geq 0$; f is called a *constant* if $arity(f) = 0$. For each variable x, define $arity(x) = 0$. From now on Σ stands for an arbitrary signature.

Definition 2.1 (hypergraph) A *hypergraph* G over Σ is a system $\langle V, E, s, t, l\rangle$, where V, E are finite sets of *nodes* and *hyperedges* (or *edges* for short) , $s: E \to V$, $t: E \to V^*$ are mappings that assign a source node and a string of target nodes to each hyperedge, and $l: E \to \Sigma \cup X$ is a mapping that labels each hyperedge e such that $arity(l(e))$ is the length of $t(e)$.

The components of a hypergraph G are referred to as V_G, E_G, s_G, t_G, and l_G. A node v is a *predecessor* of a node v' if there is an edge e with source v such that v' occurs in $t_G(e)$. The relations $>_G$ and $\geq_G$ are the transitive respectively reflexive-transitive closure of the predecessor relation. G is *acyclic* if there is no node v with $v >_G v$. For each node v, G/v is the subhypergraph[1] of G consisting of all nodes v' with $v \geq_G v'$ and of all edges outgoing from these nodes.

Definition 2.2 (collapsed tree) A hypergraph C over Σ is a *collapsed tree* if

(1) there is a node $root_C$ such that $root_C \geq_C v$ for each node v,

[1] Given hypergraphs U, G, U is a *subhypergraph* of G if $V_U \subseteq V_G$, $E_U \subseteq E_G$, and if s_U, t_U, l_U are restrictions of the corresponding mappings of G.

(2) C is acyclic, and

(3) each node has a unique outgoing edge, that is, s_C is bijective.

Example 2.3 Figure 1 shows a collapsed tree with function symbols $+$, $\times$, 0, and a variable x, where $arity(+) = arity(\times) = 2$ and $arity(0) = 0$. Hyperedges are depicted as boxes with inscribed labels, and circles represent nodes. A line connects each edge with its source node, while arrows point to target nodes. The order in a target string is given by the left-to-right order of the arrows leaving a box, unless it is indicated by numbered arrows.

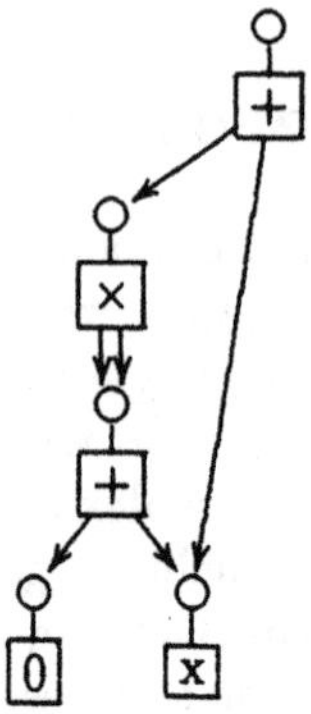

Figure 1: Collapsed tree

As usual, a *term* over Σ and X is a string that consists of a single variable or constant, or has the form $f(t_1, \ldots, t_n)$ for a function symbol f with arity n and terms $t_1, \ldots, t_n$. $T(\Sigma, X)$ is the set of all terms over Σ and X. In examples, terms with binary function symbols are written in infix notation if convenient.

Each node v in a collapsed tree represents a term which can be obtained by traversing all edges "below" v, analogously to a preorder tree traversal.

Definition 2.4 (from nodes to terms) Let C be a collapsed tree. Then the mapping $term_C\colon V_C \to T(\Sigma, X)$ is defined by

$$term_C(v) = \begin{cases} l_C(e) & \text{if } t_C(e) = \lambda,^2 \\ l_C(e)\,(term_C(v_1), \ldots, term_C(v_n)) & \text{if } t_C(e) = v_1 \ldots v_n, \end{cases}$$

where e is the unique edge with source v.

In the following $term(C)$ stands for $term_C(root_C)$.

As an example, $term(\mathrm{C}) = ((0 + \mathrm{x}) \times (0 + \mathrm{x})) + \mathrm{x}$ for the collapsed tree C in Figure 1.

[2] λ is the empty string.

A *rewrite rule* $l \to r$ over Σ consists of two terms $l, r \in T(\Sigma, X)$ such that l is not a variable and all variables in r occur also in l. Given a set R of rewrite rules, $\mathcal{R} = \langle \Sigma, R \rangle$ is a *term rewriting system*. $\to_{\mathcal{R}}$ is the rewrite relation on $T(\Sigma, X)$ associated with $\mathcal{R}$, and $\to^{+}_{\mathcal{R}}, \to^{*}_{\mathcal{R}}$ are the transitive respectively reflexive-transitive closure of $\to_{\mathcal{R}}$. (The reader is assumed to be familiar with basic concepts of term rewriting; for an introduction, see for example [DJ90, HO80, Klo90].) Until the end of section 4, $\mathcal{R}$ denotes an arbitrary term rewriting system over Σ.

Given some rewrite rule $l \to r$ and a collapsed tree C, the matching of l with a subterm of $term(C)$ is realized by a hypergraph morphism $\underline{\lozenge l} \to C$, where $\underline{\lozenge l}$ is a particular hypergraph representing l.

Definition 2.5 (hypergraph morphism) Let G,H be hypergraphs. A *hypergraph morphism* $g: G \to H$ is a pair of mappings $\langle g_V: V_G \to V_H, g_E: E_G \to E_H \rangle$ that preserve sources, targets, and labels; that is, $s_H \circ g_E = g_V \circ s_G$, $t_H \circ g_E = g_V^* \circ t_G$, and $l_H \circ g_E = l_G$.[3]

Definition 2.6 (tree with shared variables) A collapsed tree C is a *tree with shared variables* if (1) for each node v, $indegree_C(v) > 1$ implies $term_C(v) \in X$, and (2) for all nodes v,v', $term_C(v) = term_C(v') \in X$ implies $v = v'$.[4]

For every term t, $\lozenge t$ is a tree with shared variables such that $term(\lozenge t) = t$.

To model the matching of l in C, $\lozenge l$ is not yet the right representation of l because hypergraph morphisms preserve labels. The solution is simple: cut off all edges labelled with variables.

For every collapsed tree C, let $\underline{C}$ be the hypergraph that is obtained from C by removing all edges labelled with variables.

Definition 2.7 (evaluation step) Let C, D be collapsed trees. Then there is an *evaluation step* from C to D, denoted by $C \Rrightarrow_{\mathcal{E}} D$, if there is a rewrite rule $l \to r$ in $\mathcal{R}$ and a hypergraph morphism $g: \underline{\lozenge l} \to C$ such that D is isomorphic[5] to the collapsed tree constructed as follows:

(1) Remove the hyperedge outgoing from $g(root_{\lozenge l})$, yielding a hypergraph C'.

(2) Build the disjoint union $C' + \underline{\lozenge r}$ and

- identify $g(root_{\lozenge l})$ with $root_{\lozenge r}$,
- for each pair $\langle v, v' \rangle \in V_{\lozenge l} \times V_{\lozenge r}$ with $term_{\lozenge l}(v) = term_{\lozenge r}(v') \in X$, identify $g(v)$ with v'.

Let C'' be the resulting hypergraph.

(3) Remove garbage, that is, the resulting collapsed tree is $C''/root_C$.

---

[3] Given a mapping $f: A \to B$, $f^*: A^* \to B^*$ sends λ to λ and $a_1 \dots a_n$ to $f(a_1) \dots f(a_n)$.

[4] $indegree_C(v)$ is defined as $\sum_{e \in E_C} \#(v, t_C(e))$, where $\#(v, t_C(e))$ is the number of occurrences of v in $t_C(e)$.

[5] Two hypergraphs G, H are *isomorphic*, denoted by $G \cong H$, if there is a hypergraph morphism $g: G \to H$ with g_V and g_E bijective.

Example 2.8 Figure 2 shows an evaluation step based on the rewrite rule $x \times (y + z) \to (x \times y) + (x \times z)$. Note that the morphism locating $\underline{\Diamond x \times (y + z)}$ identifies the nodes representing x and $(y + z)$.

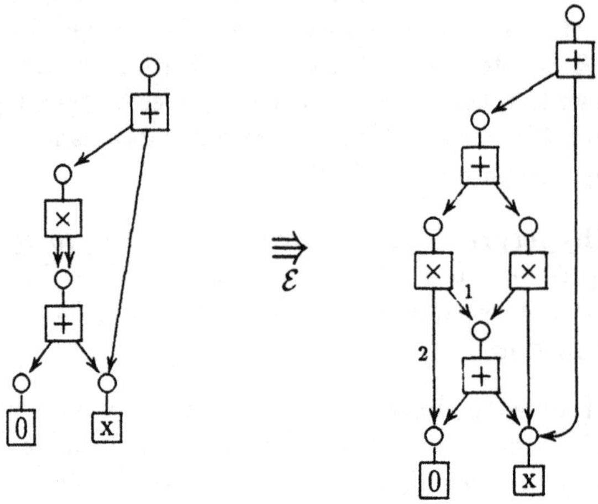

Figure 2: Evaluation step

Evaluation steps as defined above work well for *left-linear* rewrite rules: if no variable occurs more than once in a term l, then there is a morphism $\underline{\Diamond l} \to C$ into a collapsed tree C whenever l can be matched with a subterm of $term(C)$ (see Lemma 3.8 in the next section). But if l is non-linear, then no such morphism needs to exist. (For example, there is no morphism into the tree representing l.) This problem is overcome by "folding" collapsed trees prior to evaluation steps.

Definition 2.9 (folding step) Let C, D be collapsed trees. Then there is a *folding step* $C \Rrightarrow_{\mathcal{F}} D$ if there are distinct edges e, e' in C with $l_C(e) = l_C(e')$ and $t_C(e) = t_C(e')$, and if D is isomorphic to the collapsed tree that is obtained from C by identifying e with e' and $s_C(e)$ with $s_C(e')$.

As an example, Figure 3 shows a folding step.

It is easy to check that folding steps preserve represented terms.

Lemma 2.10 *For every folding step $C \Rrightarrow_{\mathcal{F}} D$, $term(C) = term(D)$.*

3 Soundness and Completeness

In this section it is shown that collapsed tree rewriting is sound and complete for function evaluation in the same sense as term rewriting is. Soundness is obtained by showing that collapsed tree rewriting implements a special kind of parallel

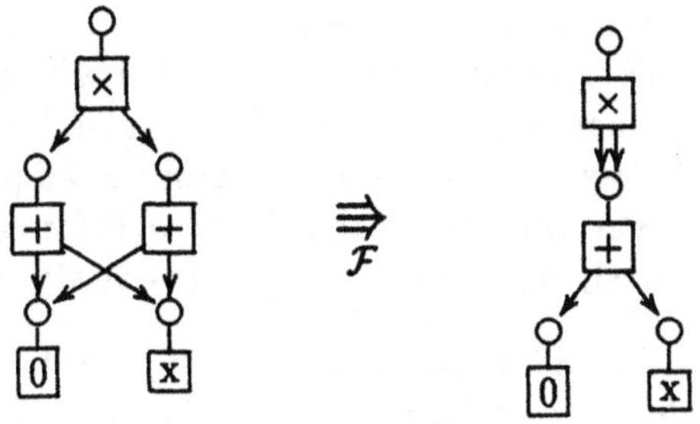

Figure 3: Folding step

term rewriting. Completeness (with respect to equational validity) follows from the fact that for every term rewrite step $t \rightarrow_{\mathcal{R}} u$, the trees representing t and u are convertible by collapsed tree rewriting.

In order to state precisely the effect of an evaluation step $C \Rrightarrow_{\mathcal{E}} D$ on *term*(C), some notions are needed for relating positions in terms with nodes in collapsed trees.

Given a term t, the set $Pos(t)$ of *positions* in t is the subset of $\mathbb{N}^*$ defined by (1) $Pos(t) = \{\lambda\}$ if t is a variable or a constant, and (2) $Pos(t) = \{\lambda\} \cup \{i.\pi \mid 1 \leq i \leq n \text{ and } \pi \in Pos(t_i)\}$[6] if t is a composite term $f(t_1, \ldots, t_n)$. For each $\pi \in Pos(t)$, the subterm t/π is given by (1) $t/\pi = t$ if $\pi = \lambda$ and (2) $t/\pi = t_i/\rho$ if $\pi = i.\rho$ and $t = f(t_1, \ldots, t_n)$.

Definition 3.1 (from positions to nodes) Let C be a collapsed tree. Then the mapping $node_C\colon Pos(term(C)) \rightarrow V_C$ is defined by $node_C(\lambda) = root_C$ and $node_C(i.\pi) = node_{C/v_i}(\pi)$, where $v_1 \ldots v_n$ is the target string of the edge with source v.

For a node v, the number n of positions in $node_C^{-1}(v)$ is a measure for the degree of sharing at v: C/v represents n occurrences of $term_C(v)$ in *term*(C). n is the number of paths from $root_C$ to v.

Definition 3.2 (path) Given two nodes v, v' in a collapsed tree C, a *path* from v to v' is a (possibly empty) sequence of pairs $\langle e_1, i_1\rangle, \ldots, \langle e_n, i_n\rangle$ such that $e_1, \ldots, e_n \in E_C$, $v = s_C(e_1)$, $v' = t_C(e_n)|_{i_n}$, and $t_C(e_j)|_{i_j} = s_C(e_{j+1})$ for $j = 1, \ldots, n-1$.[7]

Lemma 3.3 *Let C be a collapsed tree. Then*

(1) *for each position π in term(C), term(C)$/\pi = term_C(node_C(\pi))$, and*

(2) *for each node v in C, $node_C^{-1}(v)$ contains as many positions as there are paths from $root_C$ to v.*

[6] $i.\pi$ denotes the left addition of i to the string π.

[7] For a string w, $w|_i$ is the i-th element in w.

Now the application of a rewrite rule $l \to r$ to a collapsed tree C can be explained as the parallel application of $l \to r$ at all positions in $term(C)$ that are represented by the image of $root_{\Diamond l}$ in C. To this end, let $t \rightrightarrows_\Delta u$ denote the parallel application of a rewrite rule $l \to r$ to a term t at all positions in Δ; more precisely, there is a substitution σ such that $t/\pi = \sigma(l)$ for each $\pi \in \Delta$, and u is obtained by simultaneously replacing $\sigma(l)$ with $\sigma(r)$ at all positions in Δ.

Theorem 3.4 (Soundness Theorem) *Let $C \Rrightarrow_{\mathcal{E}} D$ be an evaluation step based on a rewrite rule $l \to r$ and a morphism $g: \underline{\Diamond l} \to C$. Then*

$$term(C) \underset{\Delta}{\rightrightarrows} term(D)$$

through $l \to r$, where $\Delta = node_C^{-1}(g(root_{\Diamond l}))$.

Proof By a straightforward adaptation of the corresponding proof in [HP91]. □

As argued in the previous section, evaluation steps have to be combined with folding steps to cope with arbitrary term rewriting systems. A priori, there should be no restriction for mixing evaluation and folding steps; therefore the following deals with properties of the union of both relations.

Definition 3.5 $\Rrightarrow_{\mathcal{R}} = \Rrightarrow_{\mathcal{E}} \cup \Rrightarrow_{\mathcal{F}}$.

Given collapsed trees C, D, $C \Rrightarrow_{\mathcal{R}}^* D$ expresses that there is a sequence $C \Rrightarrow_{\mathcal{R}} \ldots \Rrightarrow_{\mathcal{R}} D$ or that $C \cong D$.

Corollary 3.6 *For all collapsed trees C, D,*

$$C \Rrightarrow_{\mathcal{R}}^* D \text{ implies } term(C) \to_{\mathcal{R}}^* term(D).$$

Proof By the Soundness Theorem and Lemma 2.10, $term(C) \to_{\mathcal{R}}^* term(D)$ whenever $C \Rrightarrow_{\mathcal{R}}^+ D$. Moreover, $C \cong D$ implies $term(C) = term(D)$. □

Corollary 3.6 establishes the soundness of collapsed tree rewriting in the sense of first-order logic. For it is well-known that whenever $t \to_{\mathcal{R}}^* u$ for two terms t, u, then t and u are equal in every model of $\mathcal{R}$, where the rules in $\mathcal{R}$ are considered as universally quantified equations. By the completeness theorem of Birkhoff [Bir35], term rewriting is also complete in the sense that if two terms t, u are equal in all models, then t and u are equivalent with respect to the equivalence generated by $\to_{\mathcal{R}}$. In the rest of this section it is shown that collapsed tree rewriting is complete in the same sense: two collapsed trees are equivalent under the equivalence generated by $\Rrightarrow_{\mathcal{R}}$ whenever their represented terms are equivalent.

Definition 3.7 Let $=_{\mathcal{R}}$ and $\equiv_{\mathcal{R}}$ be the reflexive-symmetric-transitive closure of $\to_{\mathcal{R}}$ and $\Rrightarrow_{\mathcal{R}}$, respectively.

While soundness of collapsed tree rewriting follows from the fact that every sequence of $\Rrightarrow_{\mathcal{R}}$-steps corresponds to a sequence of $\rightarrow_{\mathcal{R}}$-steps, there is no such direct correspondence from which completeness could be obtained. Indeed, the converse of Corollary 3.6 does not hold. As a simple counterexample, consider the rewrite rules $\mathrm{f(x)} \rightarrow \mathrm{g(x,x)}$ and $\mathrm{a} \rightarrow \mathrm{b}$, where x is a variable and a, b are constants. Then $\mathrm{f(a)} \rightarrow_{\mathcal{R}} \mathrm{g(a,a)} \rightarrow_{\mathcal{R}} \mathrm{g(a,b)}$, but there is no corresponding sequence of $\Rrightarrow_{\mathcal{R}}$-steps because of the sharing introduced by an evaluation step with the first rule.

Nevertheless, collapsed tree rewriting is complete with respect to equational validity. The point is that $\equiv_{\mathcal{R}}$ comprises "reversed folding steps" which allow to unfold collapsed trees between evaluation steps. As a consequence, every rewrite step $t \rightarrow_{\mathcal{R}} u$ can be lifted to a $\equiv_{\mathcal{R}}$-conversion of the trees representing t and u, leading straightforward to the completeness result.

Before the "Lifting Lemma" can be stated, the different kinds of matching for collapsed trees and terms have to be related. The following lemma is proved similar to Theorem 3.12 in [HP91].

Lemma 3.8 (morphisms and substitutions) *Let C be a collapsed tree and l be a term. Then there is a morphism $g: \underline{\Diamond l} \rightarrow C$ with $g(root_{\Diamond l}) = root_C$ if and only if there is a substitution σ such that*

(1) *$term(C) = \sigma(l)$, and*

(2) *for all positions π, ρ in l, $l/\pi = l/\rho \in X$ implies $node_C(\pi) = node_C(\rho)$.*

In the simulation of a single term rewrite step by collapsed tree rewriting, two distinguished representations of terms are particularly useful.

Definition 3.9 (tree and fully collapsed tree) A collapsed tree C is a *tree* if no node has an indegree greater than one. C is a *fully collapsed tree* if each two different nodes represent different subterms, that is, if $term_C$ is injective.

For every term t, Δt and $\blacktriangle t$ denote a tree respectively a fully collapsed tree representing t.

It can be shown that Δt and $\blacktriangle t$ are unique up to isomorphism. Moreover, Δt can be transformed by folding steps into each collapsed tree representing t, which in turn can be transformed into $\blacktriangle t$.

Lemma 3.10 (existence of foldings) *For every collapsed tree C,*

$$\Delta term(C) \overset{*}{\underset{\mathcal{F}}{\Rrightarrow}} C \overset{*}{\underset{\mathcal{F}}{\Rrightarrow}} \blacktriangle term(C).$$

Lemma 3.11 (Lifting Lemma) *For every term rewrite step $t \rightarrow_{\mathcal{R}} u$ there are collapsed trees C, D such that*

$$\Delta t \overset{*}{\underset{\mathcal{F}}{\Rrightarrow}} C \underset{\mathcal{E}}{\Rrightarrow} D \overset{*}{\underset{\mathcal{F}}{\Lleftarrow}} \Delta u.$$

Proof Let $l \to r$ be the rule applied in $t \to_{\mathcal{R}} u$, σ be the associated substitution, and π be the position in t where $\sigma(l)$ is replaced by $\sigma(r)$. Then $\triangle t/node_{\triangle t}(\pi) \cong \triangle\sigma(l)$ and the folding $\triangle\sigma(l) \Rrightarrow^*_{\mathcal{F}} \blacktriangle\sigma(l)$ can be embedded in $\triangle t$, yielding a folding $\triangle t \Rrightarrow^*_{\mathcal{F}} C$. Then $C/node_C(\pi) \cong \blacktriangle\sigma(l)$ and each variable in l corresponds to a unique node in $\blacktriangle\sigma(l)$. Hence, by Lemma 3.8, there is a morphism from $\underline{\Diamond l}$ to $C/node_C(\pi)$ that maps $root_{\Diamond l}$ to $node_C(\pi)$; let $g\colon \underline{\Diamond l} \dashrightarrow C$ be the extension of this morphism to C. Then there is an evaluation step $C \Rrightarrow_{\mathcal{E}} D$ based on $l \to r$ and g. As there is only one path from $root_C$ to $node_C(\pi)$, the Soundness Theorem 3.4 yields $term(C) \rightrightarrows_{\{\pi\}} term(D)$; thus $term(D) = u$. Finally, $\triangle u \Rrightarrow^*_{\mathcal{F}} D$ holds by Lemma 3.10. □

For simplicity, the folding $\triangle t \Rrightarrow^*_{\mathcal{F}} C$ in the proof of the Lifting Lemma makes $\triangle t/node_{\triangle t}(\pi)$ fully collapsed. But it is often possible to do with a non-complete folding, as Lemma 3.8 requires only that multiple occurrences of a variable in l correspond to the same node in C. In particular, this folding can be omitted if l is a linear term.

Theorem 3.12 (Completeness Theorem) *For all collapsed trees C, D,*

$$C \equiv_{\mathcal{R}} D \text{ if and only if } term(C) =_{\mathcal{R}} term(D).$$

Proof If $C \equiv_{\mathcal{R}} D$, then $term(C) =_{\mathcal{R}} term(D)$ follows from Corollary 3.6 by induction on the number of $\Rrightarrow_{\mathcal{R}}$-steps constituting the equivalence.

Conversely, assume that $term(C) =_{\mathcal{R}} term(D)$. Then, by induction on the number of $\to_{\mathcal{R}}$-steps, the Lifting Lemma 3.11 yields $\triangle term(C) \equiv_{\mathcal{R}} \triangle term(D)$. With Lemma 3.10 follows $C \equiv_{\mathcal{R}} \triangle term(C) \equiv_{\mathcal{R}} \triangle term(D) \equiv_{\mathcal{R}} D$. □

It is important to notice that completeness fails when folding steps are discarded from $\Rrightarrow_{\mathcal{R}}$. This is demonstrated by the following example.

Example 3.13 Let $\mathcal{R}$ contain the following rules:

$$\begin{aligned} &\mathrm{a} \to \mathrm{f(a)} \\ &\mathrm{g(x)} \to \mathrm{p(x, f(x))} \\ &\mathrm{h(x)} \to \mathrm{p(f(x), x)} \end{aligned}$$

Then $\mathrm{g(a)} =_{\mathcal{R}} \mathrm{h(a)}$ since both terms are rewritable to $\mathrm{p(f(a), f(a))}$. But $\triangle\mathrm{g(a)}$ and $\triangle\mathrm{h(a)}$ are not convertible by evaluation steps alone, as can be easily checked.

4 Confluence

By the Completeness Theorem 3.12 it is possible to prove equality of terms by collapsed tree rewriting, but in general such proofs are impractical since they involve "reversed" rewrite and folding steps. As for term rewriting, the remedy is to restrict to confluent systems; then, as a consequence of completeness, equality can be proved by searching for common reducts of collapsed trees.

It turns out, though, that $\Rrightarrow_{\mathcal{R}}$ needs not to be confluent when $\rightarrow_{\mathcal{R}}$ is. The main result in this section shows that this phenomenon is ruled out when $\Rrightarrow_{\mathcal{R}}$ is weakly terminating. As a result, confluence of collapsed tree rewriting can be established for two important classes of confluent term rewriting systems.

Recall that $\rightarrow_{\mathcal{R}}$ is *confluent* if for all terms t, t_1, t_2 with $t_1 \;{}_{\mathcal{R}}^{*}\!\!\leftarrow t \rightarrow_{\mathcal{R}}^{*} t_2$, there is a term t_3 such that $t_1 \rightarrow_{\mathcal{R}}^{*} t_3 \;{}_{\mathcal{R}}^{*}\!\!\leftarrow t_2$. Confluence of $\Rrightarrow_{\mathcal{R}}$ is defined analogously.

Corollary 4.1 *If $\Rrightarrow_{\mathcal{R}}$ is confluent, then the following holds:*

(1) *For all collapsed trees C, D, $term(C) =_{\mathcal{R}} term(D)$ if and only if there is a collapsed tree E such that $C \Rrightarrow_{\mathcal{R}}^{*} E \;{}_{\mathcal{R}}^{*}\!\!\Lleftarrow D$.*

(2) *$\rightarrow_{\mathcal{R}}$ is confluent.*

Proof (1) It is well-known that confluence is equivalent to the *Church-Rosser property*, the latter meaning here that $C \equiv_{\mathcal{R}} D$ if and only if $C \Rrightarrow_{\mathcal{R}}^{*} E \;{}_{\mathcal{R}}^{*}\!\!\Lleftarrow D$ for some collapsed tree E. So the proposition follows from the Completeness Theorem.

(2) Let $t_1 \;{}_{\mathcal{R}}^{*}\!\!\leftarrow t \rightarrow_{\mathcal{R}}^{*} t_2$ for some terms t, t_1, t_2, and consider collapsed trees T_1, T_2 with $term(T_i) = t_i$, for $i = 1, 2$. Then $t_1 =_{\mathcal{R}} t_2$ and hence, by (1), there is a collapsed tree T_3 such that $T_1 \Rrightarrow_{\mathcal{R}}^{*} T_3 \;{}_{\mathcal{R}}^{*}\!\!\Lleftarrow T_2$. With Corollary 3.6 follows $t_1 \rightarrow_{\mathcal{R}}^{*} term(T_3) \;{}_{\mathcal{R}}^{*}\!\!\leftarrow t_2$. Thus $\rightarrow_{\mathcal{R}}$ is confluent. □

The following counterexample reveals that the converse of Corollary 4.1(2) does not hold.

Example 4.2 Suppose that $\mathcal{R}$ contain the following rules:

$$\begin{aligned} &\mathrm{f(x)} \rightarrow \mathrm{g(x,x)} \\ &\mathrm{a} \rightarrow \mathrm{b} \\ &\mathrm{g(a,b)} \rightarrow \mathrm{c} \\ &\mathrm{g(b,b)} \rightarrow \mathrm{f(a)} \end{aligned}$$

Based on the following rewrite steps, structural induction shows that every term has a unique normal form:

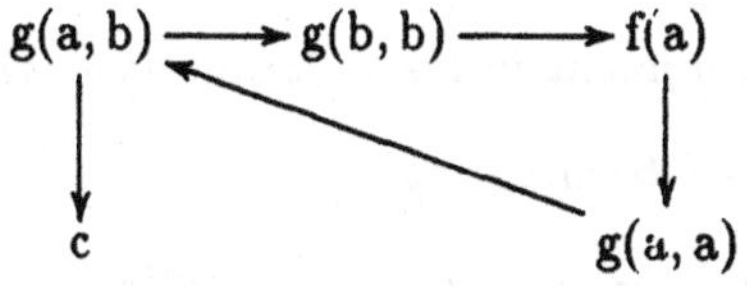

So $\rightarrow_{\mathcal{R}}$ is confluent, but $\Rrightarrow_{\mathcal{R}}$ is not. To see this, consider the rewrite steps $\vartriangle\mathrm{c} \Lleftarrow \vartriangle\mathrm{g(a,b)} \Rrightarrow \vartriangle\mathrm{g(b,b)} \Rrightarrow \vartriangle\mathrm{f(a)} \Rrightarrow \blacktriangle\mathrm{g(a,a)}$. There is no step from $\blacktriangle\mathrm{g(a,a)}$ back to $\vartriangle\mathrm{g(a,b)}$, and hence the only collapsed trees derivable from $\blacktriangle\mathrm{g(a,a)}$ are $\blacktriangle\mathrm{g(b,b)}$, $\vartriangle\mathrm{f(a)}$, and $\blacktriangle\mathrm{g(a,a)}$ itself. Thus $\Rrightarrow_{\mathcal{R}}$ is non-confluent.

Call a term t a *normal form* if there is no rewrite step $t \rightarrow_{\mathcal{R}} t'$. Analogously, a collapsed tree C is a normal form if there is no step $C \Rrightarrow_{\mathcal{R}} C'$. $\rightarrow_{\mathcal{R}}$ respectively

$\Rrightarrow_{\mathcal{R}}$ is *weakly terminating* if every term respectively collapsed tree is rewritable to a normal form.

It turns out that a counterexample as above is only possible if some collapsed trees are not reducible to a normal form. The following lemma is needed for the proof of this result.

Lemma 4.3 (1) *A collapsed tree C is fully collapsed if and only if there is no folding step $C \Rrightarrow_{\mathcal{F}} D$.*

(2) *A term t is a normal form if and only if $\blacktriangle t$ is a normal form.*

Proof See Theorems 4.7 and 6.3 in [HP91]. □

Theorem 4.4 *If $\Rrightarrow_{\mathcal{R}}$ is weakly terminating, then $\Rrightarrow_{\mathcal{R}}$ is confluent if and only if $\rightarrow_{\mathcal{R}}$ is confluent.*

Proof One direction of the proposition follows directly from Corollary 4.1. For the other direction, assume that $\rightarrow_{\mathcal{R}}$ is confluent and $\Rrightarrow_{\mathcal{R}}$ weakly terminating. Let C, D, E be collapsed trees such that $D\ {}_{\mathcal{R}}^{*}\!\Lleftarrow C \Rrightarrow_{\mathcal{R}}^{*} E$. Then there are normal forms D', E' with $D \Rrightarrow_{\mathcal{R}}^{*} D'$ and $E \Rrightarrow_{\mathcal{R}}^{*} E'$. Corollary 3.6 yields $term(D')\ {}_{\mathcal{R}}^{*}\!\leftarrow term(C) \rightarrow_{\mathcal{R}}^{*} term(E')$. By Lemma 4.3(1), $D' \cong \blacktriangle term(D')$ and $E' \cong \blacktriangle term(E')$. Hence, according to Lemma 4.3(2), $term(D')$ and $term(E')$ are normal forms. So $term(D') = term(E')$ since $\rightarrow_{\mathcal{R}}$ is confluent. It follows $D' \cong E'$ by the uniqueness of fully collapsed trees. Thus $\Rrightarrow_{\mathcal{R}}$ is confluent. □

It is interesting that the above proof fits into Curien and Ghelli's schema for showing confluence of weakly terminating systems [CG91] (although the proof was obtained prior to the publication of [CG91]). The interpretation function used in their method is here the *term*-function which assigns to every collapsed tree the represented term.

With Theorem 4.4 collapsed tree rewriting can be shown to be confluent for two important classes of confluent term rewriting systems. Call $\mathcal{R}$ *terminating* if there is no infinite sequence $t_1 \rightarrow_{\mathcal{R}} t_2 \rightarrow_{\mathcal{R}} \ldots$ Furthermore, $\mathcal{R}$ is *orthogonal* if all rules are left-linear and if left-hand sides of distinct rules do not overlap. (Two terms l and l' *overlap* if there are substitutions σ, σ' such that $\sigma(l/\pi) = \sigma'(l')$ for some position π in l with $l/\pi \notin X$.)

Corollary 4.5 *$\Rrightarrow_{\mathcal{R}}$ is confluent in the following two cases:*

(1) *$\rightarrow_{\mathcal{R}}$ is confluent and terminating.*

(2) *$\mathcal{R}$ is orthogonal and $\rightarrow_{\mathcal{R}}$ is weakly terminating.*

Proof (1) Let $\rightarrow_{\mathcal{R}}$ be confluent and terminating and suppose there were an infinite sequence $C_1 \Rrightarrow_{\mathcal{R}} C_2 \Rrightarrow_{\mathcal{R}} \ldots$ Then this sequence contains infinitely many evaluation steps because $\Rrightarrow_{\mathcal{F}}$ is terminating. Since folding preserves terms, the Soundness Theorem implies the existence of an infinite sequence of $\rightarrow_{\mathcal{R}}$-steps, contradicting the assumption. So $\Rrightarrow_{\mathcal{R}}$ is terminating, and hence confluent by Theorem 4.4.

(2) Let C be some collapsed tree and $\Omega = \{\omega_1, \ldots, \omega_n\}$ be a set of outermost redexes in $term(C)$ such that (i) for each outermost redex ω in $term(C)$ there is $\omega_i \in \Omega$ with $term(C)/\omega = term(C)/\omega_i$, and (ii) $term(C)/\omega_i \neq term(C)/\omega_j$ for $i \neq j$. Then there is a sequence $term(C) = t_1 \rightrightarrows_{\Delta_1} t_2 \rightrightarrows_{\Delta_2} \ldots \rightrightarrows_{\Delta_n} t_{n+1}$ with $\Delta_i = \{\pi \mid t_i/\pi = t_i/\omega_i\}$ (this sequence is well-defined because $\mathcal{R}$ is orthogonal). By the Soundness Theorem and Lemma 3.10, there is a sequence

$$C = C_1 \Rrightarrow^*_{\mathcal{F}} \overline{C_1} \Rrightarrow_{\mathcal{E}} C_2 \Rrightarrow^*_{\mathcal{F}} \overline{C_2} \Rrightarrow_{\mathcal{E}} \ldots \Rrightarrow_{\mathcal{E}} C_{n+1} \Rrightarrow^*_{\mathcal{F}} \overline{C_{n+1}}$$

such that for $i = 1, \ldots, n+1$, $\overline{C_i}$ is fully collapsed and $term(\overline{C_i}) = t_i$. Define C' as $\overline{C_{n+1}}$. Then the sequence $C \Rrightarrow^*_{\mathcal{R}} C' \Rrightarrow^*_{\mathcal{R}} C'' \Rrightarrow^*_{\mathcal{R}} \ldots$ terminates in a normal form whenever $term(C)$ has a normal form; this is because $t_1 \rightrightarrows_{\Delta_1} t_2 \rightrightarrows_{\Delta_2} \ldots$ is *eventually outermost* in the sense of O'Donnell, see Theorem 10 in [O'D77]. Hence, if $\rightarrow_{\mathcal{R}}$ is weakly terminating, then so is $\Rrightarrow_{\mathcal{R}}$. Now the proposition follows by Theorem 4.4 and the well-known fact that orthogonal term rewriting systems are confluent. □

In point (1) of Corollary 4.5, termination cannot be relaxed to weak termination since $\rightarrow_{\mathcal{R}}$ is weakly terminating in Example 4.2. In (2), weak termination cannot be dropped as otherwise folding steps may cause non-confluence. This is demonstrated by the following example.

Example 4.6 Let $\mathcal{R}$ contain the single rule $\mathrm{a} \rightarrow \mathrm{f(a)}$ and assume that there is a binary function symbol g. Then

$$\blacktriangle\mathrm{g(f(a), f(a))} \underset{\mathcal{E}}{\Lleftarrow} \blacktriangle\mathrm{g(a, a)} \underset{\mathcal{F}}{\Lleftarrow} \Delta\mathrm{g(a, a)} \underset{\mathcal{E}}{\Rrightarrow} \Delta\mathrm{g(a, f(a))} \underset{\mathcal{F}}{\Rrightarrow} \blacktriangle\mathrm{g(a, f(a))},$$

but the outer collapsed trees do not have a common reduct.

5 Modularity

This section is concerned with modular properties of collapsed tree rewriting, that is, properties that are preserved when different term rewriting systems are combined (see [Mid90] for an excellent survey on modular properties of term rewriting). The first result establishes the modularity of termination, and can be proved by adapting the proof of the corresponding result for jungle evaluation in [Plu91].

Definition 5.1 Two term rewriting systems $\langle \Sigma_0, R_0 \rangle$ and $\langle \Sigma_1, R_1 \rangle$ are *crosswise disjoint* if the function symbols in the left-hand sides of R_i do not occur in the right-hand sides of R_{1-i}, for $i = 0, 1$.

Theorem 5.2 (modularity of termination) *Let $R = \langle \Sigma_0 \cup \Sigma_1, R_0 \cup R_1 \rangle$ and $\mathcal{R}_i = \langle \Sigma_i, R_i \rangle$, $i = 0, 1$, be term rewriting systems such that $\mathcal{R}_0$ and $\mathcal{R}_1$ are crosswise disjoint. Then $\Rrightarrow_{\mathcal{R}}$ is terminating if and only if $\Rrightarrow_{\mathcal{R}_0}$ and $\Rrightarrow_{\mathcal{R}_1}$ are terminating.*

This result does not hold for term rewriting, even if Σ_0 and Σ_1 are disjoint [Toy87]. Even worse, if both systems are terminating and confluent, their disjoint union still needs not to terminate [Toy87, Dro89].

In the rest of this section it is shown that confluence together with termination is a modular property of collapsed tree rewriting, provided the given systems are crosswise disjoint and their left-hand sides do not mutually overlap. This result is not obvious as confluence alone may get lost already by signature extensions (see the example below Lemma 5.8).

At first an excursion to jungle evaluation is made in order to show that evaluation steps with non-overlapping rules commute. Call a hypergraph G a *jungle* if conditions (2) and (3) of Definition 2.2 are satisfied, and write $G \Rightarrow H$ if H is isomorphic to the jungle constructed in steps (1) and (2) of Definition 2.7 (with C standing for G).

The following lemma is an immediate consequence of the definition of $\Rrightarrow_{\mathcal{E}}$.

Lemma 5.3 *For every evaluation step $C \Rrightarrow_{\mathcal{E}} D$ there is a jungle $\overline{D}$ such that $C \Rightarrow \overline{D}$ and $D = \overline{D}/root_C$.*

For collapsed trees C, D, write $C \Rrightarrow_{\mathcal{E}}^{\lambda} D$ if $C \Rrightarrow_{\mathcal{E}} D$ or $C \cong D$.

Lemma 5.4 *Let G, H be jungles such that $G \Rightarrow H$. Then $G/v \Rrightarrow_{\mathcal{E}}^{\lambda} H/v$ for each node v in G.*

Proof Let $l \to r$ and $g: \underline{\Diamond l} \to G$ be the rewrite rule and the morphism associated with $G \Rightarrow H$.

Case 1: $g(root_{\Diamond l}) \in V_{G/v}$. Then g can be restricted to a morphism $\underline{\Diamond l} \to G/v$, so $G/v \Rightarrow X$ for some jungle X. By the construction of $G \Rightarrow H$, the removal of $G - G/v$ from H yields a well-defined jungle X'. Moreover, this removal commutes with the two construction steps that transform G into H. Hence $X = X'$. Then $X/v = (H - (G - G/v))/v = H/v$ because the nodes and edges in $G - G/v$ are unreachable from v in H. Thus $G/v \Rrightarrow_{\mathcal{E}} X/v = H/v$.

Case 2: $g(root_{\Diamond l}) \notin V_{G/v}$. Then the edge removed by $G \Rightarrow H$ is in $G - G/v$. Also, none of the inserted nodes and edges is reachable from v. Hence $G/v \cong H/v$. □

Theorem 5.5 (commutation of evaluation steps) *Let $C \Rrightarrow_{\mathcal{E}} D_i$ be an evaluation step based on a rewrite rule $l_i \to r_i$ and a morphism $g_i: \underline{\Diamond l_i} \to C$, and let e_i be the edge removed in C ($i = 0, 1$). If $e_i \notin g_{1-i}(\underline{\Diamond l_{1-i}})$, then there is a collapsed tree E such that $D_0 \Rrightarrow_{\mathcal{E}}^{\lambda} E \; {}_{\mathcal{E}}^{\lambda}\!\Lleftarrow D_1$.*

Proof By Lemma 5.3 there are jungles $\overline{D}_0, \overline{D}_1$ such that $C \Rightarrow \overline{D}_i$ and $D_i = \overline{D}_i/root_C$, for $i = 0, 1$. By choosing suitable hypergraph rules, these steps become direct derivations in the "Berlin approach" to (hyper)graph rewriting, see [HP91]. In particular, the above condition guarantees that the two derivations are "parallel independent" in the sense of [EK76]; then the Church-Rosser Theorem I in that paper[8] shows that there is a jungle F such that $\overline{D}_0 \Rightarrow E \Leftarrow \overline{D}_1$. With Lemma 5.4 follows $D_0 \Rrightarrow_{\mathcal{E}}^{\lambda} E/root_C \; {}_{\mathcal{E}}^{\lambda}\!\Lleftarrow D_1$. □

[8] The result is formulated for graphs, but the category-theoretic proof allows a straightforward adaptation to the hypergraph case.

Definition 5.6 Two term rewriting systems $\langle\Sigma_0, R_0\rangle$ and $\langle\Sigma_1, R_1\rangle$ are *non-interfering* if no left-hand side of R_i overlaps a left-hand side of R_{1-i}, for $i = 0, 1$.

Given a term rewriting system $\mathcal{R}$, $\Rrightarrow_{\mathcal{R}}$ is *locally confluent* if for all collapsed trees C, D_0, D_1 with $D_0 \;{}_{\mathcal{R}}\!\Lleftarrow C \Rrightarrow_{\mathcal{R}} D_1$, there is a collapsed tree E such that $D_0 \Rrightarrow^*_{\mathcal{R}} E \;{}^*_{\mathcal{R}}\!\Lleftarrow D_1$. $\Rrightarrow_{\mathcal{R}}$ is *convergent* if it is terminating and confluent.

Theorem 5.5 implies that local confluence of collapsed tree rewriting is preserved by the union of non-interfering term rewriting systems over the same signature.

Corollary 5.7 *Let $\mathcal{R} = \langle\Sigma, R_0 \cup R_1\rangle$ be a term rewriting system such that $\mathcal{R}_0 = \langle\Sigma, R_0\rangle$ and $\mathcal{R}_1 = \langle\Sigma, R_1\rangle$ are non-interfering. If $\Rrightarrow_{\mathcal{R}_0}$ and $\Rrightarrow_{\mathcal{R}_1}$ are locally confluent, then so is $\Rrightarrow_{\mathcal{R}}$.*

Proof Let $\Rrightarrow_{\mathcal{R}_0}, \Rrightarrow_{\mathcal{R}_1}$ be locally confluent and C, D_0, D_1 be collapsed trees such that $D_0 \;{}_{\mathcal{R}}\!\Lleftarrow C \Rrightarrow_{\mathcal{R}} D_1$. By assumption, D_0 and D_1 have a common reduct if one of the two steps is a folding step, or if both steps are applications of rewrite rules from either $\mathcal{R}_0$ or $\mathcal{R}_1$. Assume therefore $C \Rrightarrow_{\mathcal{E}} D_i$, based on a rule $l_i \to r_i$ from $\mathcal{R}_i$ and a morphism $g_i\colon \underline{\Diamond l_i} \to C$, for $i = 0, 1$. Then, by Theorem 5.5, $D_0 \Rrightarrow^{\lambda}_{\mathcal{E}} E \;{}^{\lambda}_{\mathcal{E}}\!\Lleftarrow D_1$ for some collapsed tree E, provided that the edge removed by $C \Rrightarrow_{\mathcal{E}} D_i$ is not in $g_{1-i}(\underline{\Diamond l_{1-i}})$, for $i = 0, 1$. The latter holds as otherwise l_0 and l_1 would overlap by the substitutions induced by g_0 and g_1. □

Lemma 5.8 *Let $\mathcal{R} = \langle\Sigma, R\rangle$ and $\mathcal{R}' = \langle\Sigma', R\rangle$ be term rewriting systems such that $\Sigma \subseteq \Sigma'$. If $\Rrightarrow_{\mathcal{R}}$ is convergent, then so is $\Rrightarrow_{\mathcal{R}'}$.*

Proof Let $\Rrightarrow_{\mathcal{R}}$ be convergent. As $\mathcal{R}'$ is the union of $\mathcal{R}$ and $\langle\Sigma', \emptyset\rangle$, Theorem 5.2 shows that $\Rrightarrow_{\mathcal{R}'}$ is terminating (since collapsed tree rewriting for $\langle\Sigma', \emptyset\rangle$ consists only of folding steps). Furthermore, $\to_{\mathcal{R}}$ is confluent by Corollary 4.1. Then $\to_{\mathcal{R}'}$ is also confluent, see Proposition 3.1.2 in [Mi 190], so Theorem 4.4 yields the confluence of $\Rrightarrow_{\mathcal{R}'}$. □

In contrast to the situation for term rewriting, Lemma 5.8 does not hold for confluence alone. This can be seen with the one rule system a $\to$ f(a) over the signature {a, f}, for which collapsed tree rewriting is clearly confluent. But Example 4.6 demonstrates that an additional binary function symbol causes non-confluence.

Theorem 5.9 (modularity of convergence) *Let $\mathcal{R} = \langle\Sigma_0 \cup \Sigma_1, R_0 \cup R_1\rangle$ and $\mathcal{R}_i = \langle\Sigma_i, R_i\rangle$, $i = 0, 1$, be term rewriting systems such that $\mathcal{R}_0$ and $\mathcal{R}_1$ are crosswise disjoint and non-interfering. If $\Rrightarrow_{\mathcal{R}_0}$ and $\Rrightarrow_{\mathcal{R}_1}$ are convergent, then so is $\Rrightarrow_{\mathcal{R}}$.*

Proof Let $\Rrightarrow_{\mathcal{R}_0}$ and $\Rrightarrow_{\mathcal{R}_1}$ be convergent. By Theorem 5.2, $\Rrightarrow_{\mathcal{R}}$ is terminating. Hence, by Newman's Lemma (see for example [Hue80]), it remains to show that $\Rrightarrow_{\mathcal{R}}$ is locally confluent. Let $\mathcal{R}'_i = \langle\Sigma_0 \cup \Sigma_1, R_i\rangle$, for $i = 0, 1$; by Lemma 5.8, $\Rrightarrow_{\mathcal{R}'_0}$ and $\Rrightarrow_{\mathcal{R}'_1}$ are convergent. Then Corollary 5.7 shows that $\Rrightarrow_{\mathcal{R}}$ is locally confluent. □

6 Conclusion

Collapsed tree rewriting is an efficient model of function evaluation which is sound and complete in the same sense as term rewriting is. Folding steps that identify common subexpressions are necessary to achieve completeness and to cope with non-left-linear rewrite rules. Rewriting collapsed trees rather than terms or trees makes a difference for operational properties like confluence and termination. The class of confluent term rewriting systems properly includes those systems which are confluent under collapsed tree rewriting; but to the latter belong all convergent and all weakly terminating orthogonal systems. In contrast to confluence, collapsed tree rewriting terminates for more rule systems than term rewriting does; in particular, termination is preserved by combinations of systems under suitable conditions. This allows also to establish modularity of convergence. Both modularity results do not hold for term rewriting.

References

[Bir35] Garrett Birkhoff. On the structure of abstract algebras. *Proceedings of the Cambridge Philosophical Society*, 31:433–454, 1935.

[BvEG+87] H.P. Barendregt, M.C.J.D. van Eekelen, J.R.W. Glauert, J.R. Kennaway, M.J. Plasmeijer, and M.R. Sleep. Term graph rewriting. In *Proc. Parallel Architectures and Languages Europe*, pages 141–158. Springer Lecture Notes in Computer Science 259, 1987.

[CG91] Pierre-Louis Curien and Giorgio Ghelli. On confluence for weakly normalizing systems. In *Proc. Rewriting Techniques and Applications*, pages 215–225. Springer Lecture Notes in Computer Science 488, 1991.

[CR92] Andrea Corradini and Francesca Rossi. Hyperedge replacement jungle rewriting for term rewriting systems and logic programming. *Theoretical Computer Science*, 1992. To appear.

[DJ90] Nachum Dershowitz and Jean-Pierre Jouannaud. Rewrite systems. In Jan van Leeuwen, editor, *Handbook of Theoretical Computer Science*, vol. B, chapter 6. Elsevier, 1990.

[Dro89] Klaus Drosten. *Termersetzungssysteme*. Informatik-Fachberichte 210. Springer-Verlag, 1989.

[EK76] Hartmut Ehrig and Hans-Jörg Kreowski. Parallelism of manipulations in multidimensional information structures. In *Proc. Mathematical Foundations of Computer Science*, pages 284–293. Springer Lecture Notes in Computer Science 45, 1976.

[FW91] William M. Farmer and Ronald J. Watro. Redex capturing in term graph rewriting. In *Proc. Rewriting Techniques and Applications*, pages 13–24. Springer Lecture Notes in Computer Science 488, 1991.

[HO80] Gérard Huet and Derek C. Oppen. Equations and rewrite rules, a survey. In Ronald V. Book, editor, *Formal Language Theory: Perspectives and Open Problems*, pages 349–405. Academic Press, 1980.

[HP91] Berthold Hoffmann and Detlef Plump. Implementing term rewriting by jungle evaluation. *RAIRO Theoretical Informatics and Applications*, 25(5):445–472, 1991.

[Hue80] Gérard Huet. Confluent reductions: Abstract properties and applications to term rewriting systems. *Journal of the ACM*, 27(4):797–821, 1980.

[Klo90] Jan Willem Klop. Term rewriting systems — from Church-Rosser to Knuth-Bendix and beyond. In *Proc. Automata, Languages, and Programming*, pages 350–369. Springer Lecture Notes in Computer Science 443, 1990.

[Mid90] Aart Middeldorp. Modular properties of term rewriting systems. Dissertation, Vrije Universiteit, Amsterdam, 1990.

[O'D77] Michael J. O'Donnell. *Computing in Systems Described by Equations*. Lecture Notes in Computer Science 58. Springer-Verlag, 1977.

[Plu91] Detlef Plump. Implementing term rewriting by graph reduction: Termination of combined systems. In *Proc. Conditional and Typed Rewriting Systems*, pages 307–317. Springer Lecture Notes in Computer Science 516, 1991.

[Toy87] Yoshihito Toyama. Counterexamples to termination for the direct sum of term rewriting systems. *Information Processing Letters*, 25:141–143, 1987.

Combinations of Simplifying Conditional Term Rewriting Systems

Enno Ohlebusch

Universität Bielefeld, 4800 Bielefeld 1, Germany,
e-mail: enno@techfak.uni-bielefeld.de

Abstract. A conditional term rewriting system (CTRS) is called simplifying if there exists a simplification ordering $>$ on terms such that the left-hand side of any rewrite rule is greater than the right-hand side and the terms occurring in the conditions of that rule. If a simplifying join CTRS consists of finitely many rules, it is terminating and the applicability of a rewrite rule is decidable by recursively reducing the terms in the conditions. Consider two finite CTRSs $\mathcal{R}_1$ and $\mathcal{R}_2$ which may share constructors (symbols that do not occur at the root position of the left-hand side of any rewrite rule) but no other function symbols. It will be shown that the combined CTRS $\mathcal{R} = \mathcal{R}_1 \cup \mathcal{R}_2$ is simplifying if and only if $\mathcal{R}_1$ and $\mathcal{R}_2$ are simplifying. Moreover, confluence is a modular property of finite simplifying join CTRSs.

1 Introduction

During the past decade, term rewriting has gained an enormous importance in fields of computer science concerned with symbolic manipulation. Among others, it may be viewed as a way of executing prototypes of entities that are specified via an algebraic specification. In order to handle large algebraic specifications, modularization concepts were introduced. This entails the interest in modular properties of term rewriting systems. A property $\mathcal{P}$ of CTRSs is called modular if for all disjoint CTRSs $\mathcal{R}_1$ and $\mathcal{R}_2$ the union $\mathcal{R}_1 \cup \mathcal{R}_2$ has the property $\mathcal{P}$ if and only if both $\mathcal{R}_1$ and $\mathcal{R}_2$ have the property $\mathcal{P}$. While confluence is a modular property (cf. [Toy87b, Mid90a]), termination – the other important property of CTRSs – is not modular (cf. [Toy87a]). A comprehensive survey of the results obtained so far can be found in [Mid90b]. Most of the results deal with unconditional TRSs, whereas fewer is known about modular properties of CTRSs which arise naturally in the context of algebraic specifications. In this paper, we extend a theorem achieved by Kurihara and Ohuchi [KO90a] on the modularity of simplifying unconditional TRSs which may share constructors. Precisely, we show that two finite CTRSs $\mathcal{R}_1$ and $\mathcal{R}_2$ (possibly sharing constructors) are simplifying if and only if the combined system $\mathcal{R} = \mathcal{R}_1 \cup \mathcal{R}_2$ is simplifying. In order to prove our generalization, we can reuse the proof of the aforementioned theorem. Additionally, it is shown that confluence is preserved under combinations of finite simplifying join CTRSs with shared constructors. All in all, the combined system of two finite simplifying and confluent join CTRSs with shared constructors is again simplifying (hence terminating) and confluent, and thus, in particular convergent. Some counterexamples show that a weakening of these premises is impossible.

Most of the orderings used for semi-mechanical termination proofs are simplification orderings. Among those are the recursive path ordering, the lexicographic path ordering, and the recursive decomposition ordering. For an overview and a comparison of the orderings we refer to [Ste89].

We give a small example to show the practical relevance of our theorem. Suppose we have implemented the recursive path ordering to show termination of CTRSs. Consider the systems $\mathcal{R}_1 = \{F(F(C,x),x) \to F(x,F(D,x)) \Leftarrow x{\downarrow}C\}$ and $\mathcal{R}_2 = \{g(D) \to g(C)\}$ which share the constructors C and D[1]. Using the recursive path ordering, it is easy to prove that both systems are simplifying, whereas the combined system $\mathcal{R} = \mathcal{R}_1 \cup \mathcal{R}_2$ cannot be shown to be simplifying by the recursive path ordering (the first rule requires the precedence $C > D$ and the second one needs $D > C$). However, the obtained result tells us that there is a simplification ordering proving that $\mathcal{R}$ is simplifying (in the example for instance a lexicographic path ordering), but we do not have to explicitly construct it.

Moreover, it should be pointed out that we define simplification orderings directly on sets of terms with variables, not on sets of ground terms as is usually done in the literature. Thus, we do not need the existence of at least one constant to show the termination of a term rewriting system by a simplification ordering.

2 Preliminaries

In this section, we briefly recall the basic notions of unconditional term rewriting as surveyed in e.g. Dershowitz and Jouannaud [DJ90] and Klop [Klo90]. We use the notations suggested in [DJ91], but there are various other notations used by different authors.

A *signature* is a countable set $\mathcal{F}$ of *function symbols* or *operators*, where every $f \in \mathcal{F}$ is associated with a natural number denoting its arity. $\mathcal{F}_n$ denotes the set of all function symbols of arity n, hence $\mathcal{F} = \bigcup_{n \geq 0} \mathcal{F}_n$. Elements of $\mathcal{F}_0$ are called *constants*. We sometimes assume $\mathcal{F}$ to be the disjoint union of two sets $\mathcal{D}$ and $\mathcal{C}$. The set $\mathcal{T}(\mathcal{F},\mathcal{V})$ of *terms* built from a signature $\mathcal{F}$ and a countable set of *variables* $\mathcal{V}$ with $\mathcal{F} \cap \mathcal{V} = \emptyset$ is the smallest set such that $\mathcal{V} \subseteq \mathcal{T}(\mathcal{F},\mathcal{V})$ and if $f \in \mathcal{F}$ has arity n and $t_1,\ldots,t_n \in \mathcal{T}(\mathcal{F},\mathcal{V})$ then $f(t_1,\ldots,t_n) \in \mathcal{T}(\mathcal{F},\mathcal{V})$. We write f instead of $f(\,)$ whenever f is a constant. The set of variables occurring in a term $t \in \mathcal{T}(\mathcal{F},\mathcal{V})$ is denoted by $\mathcal{V}ar(t)$. Terms without variables are called *ground* terms. The set of all ground terms is denoted by $\mathcal{T}(\mathcal{F})$. Identity of terms and function symbols is denoted by $\equiv$. To emphasize that $\mathcal{F} = \mathcal{D} \cup \mathcal{C}$, we write $\mathcal{T}(\mathcal{D},\mathcal{C},\mathcal{V})$ instead of $\mathcal{T}(\mathcal{F},\mathcal{V})$ at the appropriate places; likewise $\mathcal{T}(\mathcal{D},\mathcal{C})$ for $\mathcal{T}(\mathcal{F})$. Let $t \in \mathcal{T}(\mathcal{F},\mathcal{V})$. Then $root(t)$ is defined as: $root(t) \equiv t$ if $t \in \mathcal{V}$, and $root(t) \equiv f$ if $t \equiv f(t_1,\ldots,t_n)$.

[1] This example is a slight modification of the one in [KO90a]. By replacing $\mathcal{R}_1$ with $\mathcal{R}_1' = \{F(C) \to F(D)\}$, we obtain an example which shows that in general total orderings are not sufficient to show the termination of the combined system.

A *substitution* σ is a mapping from $\mathcal{V}$ to $\mathcal{T}(\mathcal{F},\mathcal{V})$ such that $\{x \in \mathcal{V} \mid \sigma(x) \not\equiv x\}$ is finite. This set is called the *domain* of σ and will be denoted by $\mathcal{D}om(\sigma)$. Occasionally we present a substitution σ as $\{x \mapsto \sigma(x) \mid x \in \mathcal{D}om(\sigma)\}$. The substitution with empty domain will be denoted by ε. Substitutions are extended to morphisms from $\mathcal{T}(\mathcal{F},\mathcal{V})$ to $\mathcal{T}(\mathcal{F},\mathcal{V})$, i.e. $\sigma(f(t_1,\ldots,t_n)) \equiv f(\sigma(t_1),\ldots,\sigma(t_n))$ for every n-ary function symbol f and terms $t_1,\ldots,t_n$. We call $\sigma(t)$ an *instance* of t. We frequently write $t\sigma$ instead of $\sigma(t)$.

In order to describe subterm occurrences of a term, we introduce the notationally convenient notion "context" instead of the more precise notion "position" (cf. [DJ90]). Let $\Box$ be a special constant symbol. A *context* $C[,\ldots,]$ is a term in $\mathcal{T}(\mathcal{F}\cup\{\Box\},\mathcal{V})$. If $C[,\ldots,]$ is a context with n occurrences of $\Box$ and $t_1,\ldots,t_n$ are terms, then $C[t_1,\ldots,t_n]$ is the result of replacing from left to right the occurrences of $\Box$ with $t_1,\ldots,t_n$. A context containing precisely one occurrence of $\Box$ is denoted by $C[\,]$. A term s is a *subterm* of a term t if there exists a context $C[\,]$ such that $t \equiv C[s]$. By abuse of notation we write $\mathcal{T}(\mathcal{F},\mathcal{V})$ for $\mathcal{T}(\mathcal{F}\cup\{\Box\},\mathcal{V})$, interpreting $\Box$ as a special constant which is always available but used only for the aforementioned purpose.

Let $\to_{\mathcal{R}}$ be a relation on terms, i.e., $\to_{\mathcal{R}} \subseteq \mathcal{T}(\mathcal{F},\mathcal{V})\times\mathcal{T}(\mathcal{F},\mathcal{V})$. The transitive-reflexive closure of $\to_{\mathcal{R}}$ is denoted by $\to_{\mathcal{R}}^*$. If $s \to_{\mathcal{R}}^* t$, we say that s *reduces* to t and we call t a *reduct* of s. We write $s \leftarrow_{\mathcal{R}} t$ if $t \to_{\mathcal{R}} s$; likewise for $s\,{}_{\mathcal{R}}^*\!\!\leftarrow t$. The transitive closure of $\to_{\mathcal{R}}$ is denoted by $\to_{\mathcal{R}}^+$, and $\leftrightarrow_{\mathcal{R}}$ denotes the symmetric closure of $\to_{\mathcal{R}}$ (i.e., $\leftrightarrow_{\mathcal{R}} = \to_{\mathcal{R}} \cup \leftarrow_{\mathcal{R}}$). The transitive-reflexive closure of $\leftrightarrow_{\mathcal{R}}$ is called *conversion* and denoted by $=_{\mathcal{R}}$. If $s =_{\mathcal{R}} t$, then s and t are *convertible*. Two terms t_1, t_2 are *joinable*, denoted by $t_1 \downarrow_{\mathcal{R}} t_2$, if there exists a term t_3 such that $t_1 \to_{\mathcal{R}}^* t_3 \,{}_{\mathcal{R}}^*\!\!\leftarrow t_2$. Such a term t_3 is called a *common reduct* of t_1 and t_2. The relation $\downarrow_{\mathcal{R}}$ is called *joinability*. A term s is a *normal form* if there is no term t with $s \to_{\mathcal{R}} t$. A term s has a normal form if $s \to_{\mathcal{R}}^* t$ for some normal form t. The set of all normal forms of $\to_{\mathcal{R}}$ is denoted by $NF(\to_{\mathcal{R}})$. $\to_{\mathcal{R}}$ is *terminating*, if there is no infinite reduction sequence $t_1 \to_{\mathcal{R}} t_2 \to_{\mathcal{R}} t_3 \to_{\mathcal{R}} \ldots$. The relation $\to_{\mathcal{R}}$ is *confluent* if for all terms s, t_1, t_2 with $t_1 \,{}_{\mathcal{R}}^*\!\!\leftarrow s \to_{\mathcal{R}}^* t_2$ we have $t_1 \downarrow_{\mathcal{R}} t_2$. It is well-known that $\to_{\mathcal{R}}$ is confluent if and only if every pair of convertible terms is joinable. $\to_{\mathcal{R}}$ is *locally confluent* if for all terms s, t_1, t_2 with $t_1 \leftarrow_{\mathcal{R}} s \to_{\mathcal{R}} t_2$ we have $t_1 \downarrow_{\mathcal{R}} t_2$. If $\to_{\mathcal{R}}$ is confluent and terminating, it is called *convergent*.

A *term rewriting system* (TRS for short) is a pair $(\mathcal{F},\mathcal{R})$ consisting of a signature $\mathcal{F}$ and a set $\mathcal{R} \subset \mathcal{T}(\mathcal{F},\mathcal{V})\times\mathcal{T}(\mathcal{F},\mathcal{V})$ of *rewrite rules* or *reduction rules*. Every rewrite rule (l,r) must satisfy the following two constraints: (i) the left-hand side l is not a variable, and (ii) variables occurring in the right-hand side r also occur in l. Rewrite rules (l,r) will be denoted by $l \to r$. The rewrite rules of a TRS $(\mathcal{F},\mathcal{R})$ define a *rewrite relation* $\to_{\mathcal{R}}$ on $\mathcal{T}(\mathcal{F},\mathcal{V})$ as follows: $s \to_{\mathcal{R}} t$ if there exists a rewrite rule $l \to r$ in $\mathcal{R}$, a substitution σ and a context $C[\,]$ such that $s \equiv C[l\sigma]$ and $t \equiv C[r\sigma]$. We call $s \to_{\mathcal{R}} t$ a *rewrite step* or *reduction step*. A TRS $(\mathcal{F},\mathcal{R})$ has one of the above properties if its rewrite relation has the respective property.

A *partial ordering* $>$ is a transitive and irreflexive relation on $\mathcal{T}(\mathcal{F},\mathcal{V})$. If a partial ordering is terminating, it is also called *well-founded*.

3 Simplifying Conditional Term Rewriting Systems

Definition 1. A *conditional term rewriting system* is a pair $(\mathcal{F}, \mathcal{R})$ consisting of a signature $\mathcal{F}$ and a set $\mathcal{R}$ of *conditional rewrite rules*. Each of these rules is of the form $l \to r \Leftarrow s_1 \simeq t_1, \ldots, s_n \simeq t_n$ with $l, r, s_1, \ldots, s_n, t_1, \ldots, t_n \in \mathcal{T}(\mathcal{F}, \mathcal{V})$, where $\simeq$ denotes either $\downarrow$ (joinability) or $=$ (conversion)[2]. $s_1 \simeq t_1, \ldots, s_n \simeq t_n$ are the *conditions* of the rewrite rule . If a rewrite rule has no conditions, that is to say, $n = 0$, we write $l \to r$. As for unconditional TRSs, we impose two restrictions on each rewrite rule in $\mathcal{R}$: the left-hand side l is not a variable, and every variable which occurs in the right-hand side r also occurs in l. If $\simeq$ denotes $\downarrow$, then $(\mathcal{F}, \mathcal{R})$ is said to be a *join* CTRS. The rewrite relation associated with $(\mathcal{F}, \mathcal{R})$ is defined by: $s \to_{\mathcal{R}} t$ if there exists a rewrite rule $l \to r \Leftarrow s_1 \downarrow t_1, \ldots, s_n \downarrow t_n$ in $\mathcal{R}$, a substitution σ and a context $C[\,]$ such that $s \equiv C[l\sigma], t \equiv C[r\sigma]$ and $s_j\sigma \downarrow_{\mathcal{R}} t_j\sigma$ for all $j \in \{1, \ldots, n\}$. If $\simeq$ denotes $=$, then $(\mathcal{F}, \mathcal{R})$ is called *semi-equational* CTRS, and $\to_{\mathcal{R}}$ is defined analogously.

We often simply write $\mathcal{R}$ if the underlying signature $\mathcal{F}$ is clear from the context. Note that if every rewrite rule in a CTRS $\mathcal{R}$ has no conditions, then $\mathcal{R}$ is an unconditional TRS in the usual sense. A least fixpoint characterization of the associated rewrite relation can be found in [Mid90a]. With regard to Proposition 5 we confine the considerations in this paper to finite CTRSs. Thus, in the sequel we will tacitly assume all systems considered to consist of finitely many rules. Exceptions from this assumption will be explicitly indicated. Some considerations concerning infinite TRSs in [Ohl92a] show that the restriction to finite systems may be weakened at some places.

Definition 2. A *rewrite ordering* is a partial ordering $>$ on $\mathcal{T}(\mathcal{F}, \mathcal{V})$ which is *closed under contexts* (i.e., if $s > t$ then $C[s] > C[t]$ for all contexts $C[\,]$) and *closed under substitutions* (i.e., if $s > t$ then $s\sigma > t\sigma$ for all substitutions σ). A *simplification* ordering $>$ is a rewrite ordering possessing the *subterm property*, i.e., $C[t] > t$ for all contexts $C[\,] \not\equiv \Box$.

Lemma 3. *Let $>$ be a simplification ordering on $\mathcal{T}(\mathcal{F}, \mathcal{V})$.*

1. *Distinct variables $x, y \in \mathcal{V}$ are incomparable w.r.t. $>$.*
2. *If $s > t$, then $Var(t) \subseteq Var(s)$.*

Proof. 1. Suppose $x > y$. Then it follows that $x\sigma > y\sigma$ for any substitution σ. With $\sigma = \{x \mapsto y; y \mapsto x\}$ this implies $y > x$, a contradiction.
2. Let $s > t$. Assume that there is a variable $x \in Var(t) \setminus Var(s)$. Then $t \equiv C[x]$ for some context $C[\,]$. With $\sigma = \{x \mapsto s\}$ it follows that $s \equiv s\sigma > t\sigma \equiv C[s]$, contradicting the subterm property or irreflexivity of $>$. □

Definition 4 (Kaplan). A CTRS $(\mathcal{F}, \mathcal{R})$ is *simplifying*[3] if there exists a simplification ordering $>$ on $\mathcal{T}(\mathcal{F}, \mathcal{V})$ with $l > r, l > s_j$, and $l > t_j$, for each rewrite rule $l \to r \Leftarrow s_1 \simeq t_1, \ldots, s_n \simeq t_n$ in $\mathcal{R}$ and each $j \in \{1, \ldots, n\}$.

[2] We restrict ourselves to these two most common interpretations, cf. also [DOS88].
[3] For unconditional TRSs Kurihara and Ohuchi (see [KO90b]) introduced the notion *simply terminating*.

Note that according to the above lemma, a simplifying CTRS does not have extra-variables in its conditions, i.e., the variables appearing in the conditions of a rule must also appear in the left-hand side of that rule.

The next proposition shows that simplifying join CTRSs possess two interesting properties. Unfortunately, simplifying semi-equational CTRSs lack the second property as is demonstrated in the example below.

Proposition 5. *Let the finite CTRS $\mathcal{R}$ be simplifying. Then $\mathcal{R}$ is terminating, and if additionally $\mathcal{R}$ is a join system, then the applicability of a conditional rewrite rule $l \to r \Leftarrow s_1 \downarrow t_1, \ldots, s_n \downarrow t_n$ to a term $s \equiv C[l\sigma]$ is decidable by recursively reducing the terms $s_j\sigma$ and $t_j\sigma$ for all $j \in \{1, \ldots, n\}$.*

Proof. $\mathcal{R}$ is certainly terminating if the unconditional TRS obtained from $\mathcal{R}$ by removing the conditions from the rules is terminating. But finite simplifying unconditional TRSs are terminating as was first proved by Dershowitz in [Der82]. For the second statement we refer to [Kap87]. □

Example 6. Consider $\mathcal{R} = \{a \to b \Leftarrow g = h, f \to g \Leftarrow a = b, f \to h \Leftarrow a = b\}$. $\mathcal{R}$ is simplifying as is shown by the simplification ordering $f > a > b > g > h$. Does a reduce to b?

Next we show that a CTRS $\mathcal{R}$ is simplifying if and only if the transitive closure of a certain rewrite relation is irreflexive. This characterization of simplifying CTRSs is the key to the proof of Theorem 23 and does not depend on the interpretation of $\simeq$.

Definition 7. Let $\mathcal{R}$ be a CTRS. With $\mathcal{R}$ we associate the following TRS

$$\begin{aligned} \mathcal{S} := \; & \{l \to r \mid l \to r \Leftarrow s_1 \simeq t_1, \ldots, s_n \simeq t_n \in \mathcal{R}\} \\ & \cup \{l \to s_i \mid l \to r \Leftarrow s_1 \simeq t_1, \ldots, s_n \simeq t_n \in \mathcal{R};\ i \in \{1, \ldots, n\}\} \\ & \cup \{l \to t_i \mid l \to r \Leftarrow s_1 \simeq t_1, \ldots, s_n \simeq t_n \in \mathcal{R};\ i \in \{1, \ldots, n\}\}. \end{aligned}$$

We are also interested in another unconditional TRS related to $(\mathcal{F}, \mathcal{R})$ which depends only on the underlying signature (and thus may consist of infinitely many rules if the signature is infinite):

$$\mathcal{F}^e := \{f(x_1, \ldots, x_n) \to x_i \mid f \in \mathcal{F}_n;\ n \in \mathbb{N};\ i \in \{1, \ldots, n\}\}.$$

Lemma 8. *Let $(\mathcal{F}, \mathcal{S})$ be an unconditional TRS, and let $(\mathcal{F}, \mathcal{S}e) := (\mathcal{F}, \mathcal{S} \cup \mathcal{F}^e)$. $\to_{\mathcal{S}e}^+$ is a simplification ordering if it is irreflexive.*

Proof. $\to_{\mathcal{S}e}^+$ is transitive by definition. Any rewrite relation is closed under contexts and substitutions. Hence the same holds for $\to_{\mathcal{S}e}^+$ (being the transitive closure of a rewrite relation). Finally, $\to_{\mathcal{S}e}^+$ has the subterm property because $\to_{\mathcal{F}^e}$ is contained in $\to_{\mathcal{S}e}^+$. Thus, the assertion follows. □

Lemma 9. *Let the unconditional TRS $(\mathcal{F}, \mathcal{S})$ be simplifying w.r.t. the simplification ordering $>$. Then (i) $\to_{\mathcal{S}e}^+ \subseteq >$ and (ii) $\to_{\mathcal{S}e}^+$ is a simplification ordering. In other words, $\to_{\mathcal{S}e}^+$ is the smallest simplification ordering showing that $(\mathcal{F}, \mathcal{S})$ is simplifying.*

Proof. (i) Since $>$ is transitive, it suffices to show that $s \rightarrow_{\mathcal{S}e} t$ implies $s > t$. So let $s \rightarrow_{\mathcal{S}e} t$. If $s \rightarrow_{\mathcal{S}} t$, then clearly $s > t$. If otherwise $s \rightarrow_{\mathcal{F}^e} t$, that is, $s \equiv C[f(s_1, \ldots, s_n)]$ and $t \equiv C[s_i]$ where $f \in \mathcal{F}_n$, then the assertion follows from the fact that $>$ has the subterm property and is closed under contexts.
(ii) Follows from Lemma 8 because (i) implies that $\rightarrow^+_{\mathcal{S}e}$ is irreflexive. □

Proposition 10. *Let $\mathcal{R}$ be a CTRS and $\mathcal{S}$ the associated TRS. Then the following statements are equivalent:*

1. *$\mathcal{R}$ is simplifying.*
2. *$\mathcal{S}$ is simplifying.*
3. *$\mathcal{S}e$ is terminating.*
4. *$\rightarrow^+_{\mathcal{S}e}$ is irreflexive.*

Proof. 1. ⇔ 2. : By definition of $\mathcal{S}$.
2. ⇒ 3. : Let $>$ be a simplification ordering showing that $\mathcal{S}$ is simplifying. Assume that $\mathcal{S}e$ is not terminating, i.e., there is an infinite reduction sequence $s_1 \rightarrow_{\mathcal{S}e} s_2 \rightarrow_{\mathcal{S}e} s_3 \rightarrow_{\mathcal{S}e} \ldots$, where $s_i \in \mathcal{T}(\mathcal{F}, \mathcal{V})$. According to Lemma 9 this implies $s_1 > s_2 > s_3 > \ldots$. Note that Lemma 3 states that $\mathcal{V}ar(s_i) \subseteq \mathcal{V}ar(s_1)$ for all $i \in \mathbb{N}$. Let $\mathcal{F}'$ be the set of function symbols occurring in $\mathcal{S}$, and let $\mathcal{F}''$ be the set of function symbols occurring in s_1. Then each s_i ($i \in \mathbb{N}$) is an element of $\mathcal{T}(\mathcal{F}' \cup \mathcal{F}'', \mathcal{V}ar(s_1))$. This, however, contradicts the fact that the restriction of $>$ to $\mathcal{T}(\mathcal{F}' \cup \mathcal{F}'', \mathcal{V}ar(s_1))$ is a well-founded simplification ordering (since $\mathcal{F}', \mathcal{F}''$, and $\mathcal{V}ar(s_1)$ are finite sets, cf. also [Der82, Ohl92a]).
3. ⇒ 4. : Evidently, if $\rightarrow^+_{\mathcal{S}e}$ is not irreflexive, then $\mathcal{S}e$ is not terminating.
4. ⇒ 2. : By Lemma 8, $\rightarrow^+_{\mathcal{S}e}$ is a simplification ordering. Hence the assertion follows from $\rightarrow_{\mathcal{S}} \subseteq \rightarrow^+_{\mathcal{S}e}$. □

The following example was first given by Toyama in [Toy87a] for the purpose of showing that termination is not modular.

Example 11. $\mathcal{R}_1 = \{F(0,1,x) \rightarrow F(x,x,x)\}$ is evidently terminating but not simplifying, since $\rightarrow^+_{\mathcal{S}_1 e_1}$ is not irreflexive: If $t \equiv F(0,1,1)$ and $s \equiv F(t,t,t)$, then there is the cyclic reduction sequence $s \rightarrow_{\mathcal{F}_1{}^e} F(0,t,t) \rightarrow_{\mathcal{F}_1{}^e} F(0,1,t) \rightarrow_{\mathcal{S}_1} s$. The TRS $(\mathcal{F}_2, \mathcal{R}_2) = (\{g\}, \{g(x,y) \rightarrow x, g(x,y) \rightarrow y\})$ is simplifying because $\mathcal{R}_2$ is terminating and $\mathcal{R}_2 = \mathcal{S}_2 = \mathcal{F}_2{}^e = \mathcal{S}_2 e_2$.
The union of the TRSs $\mathcal{R}_1$ and $\mathcal{R}_2$ is not terminating, for there is a cyclic reduction sequence starting with $F(g(0,1), g(0,1), g(0,1))$.

For the sake of simplicity we will later introduce a new binary constructor Cons and a constant Nil. That this does no harm is shown in the next lemma.

Lemma 12. *Let $(\mathcal{F}, \mathcal{S})$ be an unconditional TRS. Set $\mathcal{F}' := \mathcal{F} \cup \{\mathit{Cons}, \mathit{Nil}\}$. Then $\mathcal{S}e$ is terminating if and only if $\mathcal{S}e'$ is terminating, where $\mathcal{S}e' = \mathcal{S} \cup \mathcal{F}'^e$.*

Proof. Case 1: $\mathcal{F}$ contains at most constants and function symbols of arity 1. Then $\mathcal{S} \cup \mathcal{F}^e$ has no duplicating rule (i.e., no right-hand side of a rule contains more occurrences of some variable than the left-hand side of that rule). The

same holds for $\{\textsc{Cons}(x,y) \to x, \textsc{Cons}(x,y) \to y\}$. The union (direct sum) of these terminating TRSs without duplicating rules is again terminating due to a theorem of Rusinowitch [Rus87].
Case 2: $\mathcal{F}$ contains at least one function symbol f of arity $n \geq 2$. Suppose that $\mathcal{S}e'$ is not terminating. By Proposition 10, there is a cyclic reduction sequence $s \equiv s_1 \to_{\mathcal{S}e'} s_2 \to_{\mathcal{S}e'} \dots \to_{\mathcal{S}e'} s_n \equiv s$ of terms $s_i \in \mathcal{T}(\mathcal{F}',\mathcal{V})$. In every s_i replace each $\textsc{Cons}(t_1,t_2)$ with $f(t_1,t_2,z,\dots,z)$ and each $\textsc{Nil}$ with z, where z is a fresh variable, and denote this term by s_i'. Note that $\textsc{Cons}$ and $\textsc{Nil}$ do not occur in any rule of $\mathcal{S}$. Then $s' \equiv s_1' \to_{\mathcal{S}e} s_2' \to_{\mathcal{S}e} \dots \to_{\mathcal{S}e} s_n' \equiv s'$ is a cyclic sequence of terms $s_i' \in \mathcal{T}(\mathcal{F},\mathcal{V})$, where s_i' is rewritten to s_{i+1}' by the rule $f(x,y,z,\dots,z) \to x$ (resp. $f(x,y,z,\dots,z) \to y$) if s_i is reduced to s_{i+1} using the rule $\textsc{Cons}(x,y) \to x$ (resp. $\textsc{Cons}(x,y) \to y$). This contradicts the termination of $\mathcal{S}e$. □

4 Combined Conditional Term Rewriting Systems

In this section, we prove that the property to be simplifying is a modular property (Theorem 23). The proof depends on Proposition 10 and is thus essentially the same as in [KO90a]. We point out that Theorem 23 holds even for infinite term rewriting systems (cf. [Ohl92a]). As a second main result (Theorem 35), we obtain that confluence is a modular property of simplifying join CTRSs with shared constructors. But first of all, we need some prerequisites.

4.1 Technical Prerequisites

Now we consider the union $(\mathcal{F},\mathcal{R}) := (\mathcal{F}_1 \cup \mathcal{F}_2, \mathcal{R}_1 \cup \mathcal{R}_2)$ of two CTRSs $(\mathcal{F}_1,\mathcal{R}_1)$ and $(\mathcal{F}_2,\mathcal{R}_2)$, where $\mathcal{F}_1 = \mathcal{D}_1 \cup \mathcal{C}$, $\mathcal{F}_2 = \mathcal{D}_2 \cup \mathcal{C}$, and moreover

- $\mathcal{D}_1$, $\mathcal{D}_2$, and $\mathcal{C}$ are pairwise disjoint, and
- $\mathcal{C} \subseteq \mathcal{F} \setminus \{root(l) \mid l \to r \Leftarrow s_1 \simeq t_1, \dots, s_n \simeq t_n \in \mathcal{R}\}$.

Elements from $\mathcal{C}$ do not occur at the root position of the left-hand side of any rewrite rule and are called (shared) *constructors*. $\mathcal{R}$ is called the *combined CTRS of* $\mathcal{R}_1$ *and* $\mathcal{R}_2$ *with shared constructors* $\mathcal{C}$. To be able to distinguish between symbols from different sets, we use capitals $F, G, \dots$ for function symbols from $\mathcal{D}_1$, small letters $f, g, \dots$ for those from $\mathcal{D}_2$, and small capitals $\textsc{c}, \textsc{d}, \dots$ for constructors. As conventional $x, y, z, x_1, y_1, z_1, \dots$ will denote variables.

Definition 13. Let $t \in \mathcal{T}(\mathcal{D}_1 \cup \mathcal{D}_2, \mathcal{C}, \mathcal{V})$. We color each symbol in t. Function symbols from $\mathcal{D}_1$ are colored black, those from $\mathcal{D}_2$ white, and constructors as well as variables are transparent. If t does not contain white (black) function symbols, we speak of a *black (white) term*. t is said to be *transparent* if it only contains constructors and variables. Consequently, a transparent term may be regarded as black or white, this is convenient for later purposes. t is called *top black (top white, top transparent)* if $root(t)$ is black (white, transparent).

Several definitions and considerations are symmetrical in the colors black and white. Therefore, we state the respective definitions and considerations only for the color black (the same applies cum grano coloris for the color white).

Definition 14. Let s be a top black term. There is a uniquely determined black context $C[,\ldots,]$ such that $s \equiv C[s_1,\ldots,s_n]$ and $root(s_j) \in \mathcal{D}_2$ for $j \in \{1,\ldots n\}$. We denote this by $s \equiv C[\![s_1,\ldots,s_n]\!]$. Moreover, we define the set $S(s)$ of all *principal* subterms of s to be $S(s) := \{s_1,\ldots,s_n\}$. Now let s be an arbitrary term. There is a (possibly empty) transparent context $C[,\ldots,]$ such that $s \equiv C[s_1,\ldots,s_n]$, where $root(s_j) \in \mathcal{D}_1 \cup \mathcal{D}_2$ for $j \in \{1,\ldots n\}$; from now on such a transparent context will be denoted by $\overline{C}[,\ldots,]$. A subterm t of s is called an *inner* subterm of s if it is a subterm of some principal subterm of an s_j, $j \in \{1,\ldots n\}$. Otherwise t is said to be an *outer* subterm of s. We denote the set of all outer subterms of s by $O(s)$.

Definition 15. Let $s \equiv C[\![s_1,\ldots,s_n]\!]$ and $s \rightarrow_{\mathcal{R}} t$ by an application of a rewrite rule of $\mathcal{R}$. We write $s \rightarrow^{i}_{\mathcal{R}} t$ if the rule is applied in one of the s_j and we write $s \rightarrow^{o}_{\mathcal{R}} t$ otherwise. The relation $\rightarrow^{i}_{\mathcal{R}}$ is called *inner* reduction and $\rightarrow^{o}_{\mathcal{R}}$ is called *outer* reduction. Now let $s \equiv \overline{C}[s_1,\ldots,s_n]$ with $\overline{C}[,\ldots,] \not\equiv \Box$ and $s \rightarrow_{\mathcal{R}} t$, i.e., $t \equiv \overline{C}[s_1,\ldots,s_{j-1},t_j,s_{j+1},\ldots,s_n]$ for some $j \in \{1,\ldots n\}$. Then we write $s \rightarrow^{i}_{\mathcal{R}} t$ if $s_j \rightarrow^{i}_{\mathcal{R}} t_j$ and $s \rightarrow^{o}_{\mathcal{R}} t$ if $s_j \rightarrow^{o}_{\mathcal{R}} t_j$.

Definition 16. Let s be a top black or top white term. We define

$$rank(s) = \begin{cases} 1 & \text{, if } s \in \mathcal{T}(\mathcal{D}_1,\mathcal{C},\mathcal{V}) \cup \mathcal{T}(\mathcal{D}_2,\mathcal{C},\mathcal{V}) \\ 1 + max\{rank(s_j) \mid 1 \le j \le n\}, & \text{if } s \equiv C[\![s_1,\ldots,s_n]\!] \end{cases}$$

Now let t be a top transparent term. Then we define

$$rank(t) = \begin{cases} 0 & \text{, if } t \in \mathcal{T}(\mathcal{C},\mathcal{V}) \\ max\{rank(t_j) \mid 1 \le j \le m\}, & \text{if } t \equiv \overline{C}[t_1,\ldots,t_m] \end{cases}$$

The proofs of the next lemmata are straightforward. $\mathcal{R}$ denotes as usual a combined CTRS and s,t are terms from $\mathcal{T}(\mathcal{F},\mathcal{V})$.

Lemma 17. $s \rightarrow_{\mathcal{R}} t \Rightarrow rank(s) \ge rank(t)$.

Lemma 18. *Let $root(s) \in \mathcal{D}_1$ and let $s \equiv s_1 \rightarrow_{\mathcal{R}} s_2 \rightarrow_{\mathcal{R}} \ldots \rightarrow_{\mathcal{R}} s_n \equiv s$ be a cyclic reduction sequence. Then $rank(s_j) = rank(s)$ and $root(s_j) \in \mathcal{D}_1$ for every index $j \in \{1,\ldots n\}$.*

Example 19. Consider the CTRSs $\mathcal{R}_1 = \{F(c(x,y)) \rightarrow c(F(x),y) \Leftarrow x\downarrow y\}$ and $\mathcal{R}_2 = \{g(x) \rightarrow c(x,x)\}$ as well as the reduction sequence

$$s_1 \equiv F(c(g(A),g(A))) \rightarrow^{o}_{\mathcal{R}} s_2 \equiv c(F(g(A)),g(A)) \rightarrow^{i}_{\mathcal{R}} s_3 \equiv c(F(c(A,A)),g(A))$$

We have $rank(s_i) = 3$ for $i \in \{1,2\}$ and $rank(s_3) = 2$. $S(s_1) = \{g(A)\}$ whereas $S(s_i)$ is not defined for $i \in \{2,3\}$. Furthermore, e.g. $O(s_1) = \{s_1, c(g(A),g(A))\}$ and $O(s_3) = \{s_3, F(c(A,A)), c(A,A), A, g(A)\}$.

Definition 20. Let σ and τ be substitutions. We write $\sigma \propto \tau$ if $x\sigma \equiv y\sigma$ implies $x\tau \equiv y\tau$ for all $x,y \in \mathcal{V}$. The notation $\sigma \rightarrow^{*}_{\mathcal{R}} \tau$ is used if $x\sigma \rightarrow^{*}_{\mathcal{R}} x\tau$ for all $x \in \mathcal{V}$. Obviously, $\sigma \rightarrow^{*}_{\mathcal{R}} \tau$ implies $t\sigma \rightarrow^{*}_{\mathcal{R}} t\tau$ for all $t \in \mathcal{T}(\mathcal{F},\mathcal{V})$.

Definition 21. A substitution σ is said to be *black* if $x\sigma$ is black for every $x \in \mathcal{D}om(\sigma)$ and it is said to be *top black* if $x\sigma$ is top black for all $x \in \mathcal{D}om(\sigma)$.

Proposition 22. *Every substitution σ can be decomposed into $\sigma_2 \circ \sigma_1$ such that σ_1 is black and σ_2 is top white and $\sigma_2 \propto \epsilon$ (recall that ϵ has empty domain).*

Proof. The proof can be found in [Mid90a]. □

4.2 To be Simplifying is a Modular Property

Theorem 23. *Let $\mathcal{R}_1$ and $\mathcal{R}_2$ be two CTRSs and let $\mathcal{R}$ be the combined CTRS of $\mathcal{R}_1$ and $\mathcal{R}_2$ with shared constructors. Then $\mathcal{R}$ is simplifying if and only if both $\mathcal{R}_1$ and $\mathcal{R}_2$ are simplifying.*

Proof. Let $\mathcal{S}$,$\mathcal{S}_1$, and $\mathcal{S}_2$ be the unconditional TRSs associated with $\mathcal{R}$, $\mathcal{R}_1$, and $\mathcal{R}_2$, respectively. We have

$$\rightarrow_{\mathcal{S}e} = \rightarrow_{(\mathcal{S}_1e_1)\cup(\mathcal{S}_2e_2)} = \rightarrow_{\mathcal{S}_1e_1} \cup \rightarrow_{\mathcal{S}_2e_2} = \rightarrow_{\mathcal{S}_1} \cup \rightarrow_{\mathcal{F}_1^e} \cup \rightarrow_{\mathcal{S}_2} \cup \rightarrow_{\mathcal{F}_2^e}$$

The only-if part is trivial so let us consider the if direction. According to Proposition 10, it must be shown that $\rightarrow^+_{\mathcal{S}e}$ is irreflexive. Without loss of generality (Lemma 12) we may assume that the set $\mathcal{C}$ of constructors contains two symbols not used previously, namely a constructor $Cons$ of arity 2 and a constant Nil. Assuming that $\rightarrow^+_{\mathcal{S}e}$ is not irreflexive, we will derive a contradiction. We assume that there is a cyclic reduction sequence

$$s \equiv s_1 \rightarrow_{\mathcal{S}e} \dots \rightarrow_{\mathcal{S}e} s_n \equiv s,$$

$n > 1$, with $s_j \in \mathcal{T}(\mathcal{D}_1 \cup \mathcal{D}_2, \mathcal{C}, \mathcal{V})$ for all $j \in \{1, \dots, n\}$. Without loss of generality we may further assume that s is top black, and that rank$(s) =: k$ is minimal, i.e., there is no cyclic sequence $t \equiv t_1 \rightarrow_{\mathcal{S}e} \dots \rightarrow_{\mathcal{S}e} t_m \equiv t, m > 1$, with rank$(t) < k$. Consequently, $\rightarrow^+_{\mathcal{S}e}$ is irreflexive on $T = \{t \mid rank(t) < rank(s)\}$ (or equivalently $\rightarrow^+_{\mathcal{S}e}$ is terminating on T). According to Lemma 18, every $s_j, 0 \le j \le n$, is top black and has rank k. The proof idea is to construct a cyclic reduction sequence of black terms

$$\rho(s) \equiv \rho(s_1) \rightarrow^+_{\mathcal{S}_1e_1} \dots \rightarrow^+_{\mathcal{S}_1e_1} \rho(s_n) \equiv \rho(s)$$

yielding a contradiction to the irreflexivity of $\rightarrow^+_{\mathcal{S}_1e_1}$. First we transform each term s_j into a black term $\rho(s_j)$ and then show that $s_j \rightarrow_{\mathcal{S}e} s_{j+1}$ implies $\rho(s_j) \rightarrow^+_{\mathcal{S}_1e_1} \rho(s_{j+1})$. In order to define the transformation, we consider the sets $S = \bigcup_{j=1}^n S(s_j)$ and $O = \bigcup_{j=1}^n O(s_j)$ of all principal and outer subterms occurring in the sequence. Notice that $root(t) \in \mathcal{D}_2$ for all $t \in S$ and $root(t) \in \mathcal{D}_1 \cup \mathcal{C} \cup \mathcal{V}$ for all $t \in O$ as well as $rank(t) \le rank(s)$ for each $t \in S \cup O$. Furthermore, for every $t \in S$ let

$$\Delta(t) = \{t' \mid \exists j \in \{1, \dots, n-1\} \text{ with } s_j \equiv C[\![\dots, t, \dots]\!] \rightarrow^i_{\mathcal{S}e} C[\dots, t', \dots] \equiv s_{j+1} \text{ and } t \rightarrow_{\mathcal{S}e} t'\}$$

be the set of successors of t. Note that $\Delta(t)$ is finite for any $t \in S$. Then we define $\rho : S \cup O \rightarrow \mathcal{T}(\mathcal{D}_1, \mathcal{C}, \mathcal{V})$ recursively by:

1. $\rho(t) = [\rho(t'_1), \ldots, \rho(t'_m)]$ [4] if $t \in S$ and $\Delta(t) = \{t'_1, \ldots, t'_m\}$.
   Note that $t'_1, \ldots, t'_m \in S \cup O$.
2. $\rho(t) = F(\rho(t_1), \ldots, \rho(t_m))$ if $t \in O$ and $t \equiv F(t_1, \ldots, t_m)$ [5] .
   Again observe that $t_1, \ldots, t_m \in S \cup O$.

Here $[t_1, \ldots, t_m]$ is an abbreviation for $\mathit{Cons}(t_1, \mathit{Cons}(\ldots, \mathit{Cons}(t_m, \mathit{Nil}) \ldots))$, where $[\,]$ stands for Nil. This definition is illustrated in Example 24.

We have to show that the recursion eventually stops, i.e., ρ is well-defined for any $t \in S \cup O$. First we prove that (i) ρ is well-defined for any element t of $T' = \{t \mid rank(t) < rank(s), t \in S \cup O\}$ and then that (ii) ρ is well-defined for any $t \in O' = \{t \mid rank(t) = rank(s), t \in S \cup O\} = \{t \mid rank(t) = rank(s), t \in O\} \subseteq O$.
(i) $\rightarrow^{+}_{\mathcal{S}e}$ is terminating on T' since $T' \subseteq T$. Let $\rightarrow_{arg}$ be the smallest relation on T' with

- $t \rightarrow_{arg} t'_j$ if $t \in S$ and $t'_j \in \Delta(t)$,
- $t \rightarrow_{arg} t_j$ if $t \in O$ and $t \equiv F(t_1, \ldots, t_m)$.

Notice that $t \rightarrow_{arg} u$ implies $t \rightarrow^{+}_{\mathcal{S}e} u$. Thus $\rightarrow_{arg}$ is terminating on T'. Consequently, ρ is well-defined on T'.
(ii) Assume that ρ is not well-defined on O', that is to say, $\rightarrow_{arg}$ (now defined on O') is not terminating on O'. Hence there exists an infinite reduction sequence $t_1 \rightarrow_{arg} t_2 \rightarrow_{arg} \ldots$ of terms $t_j \in O'$. Since $O' \subseteq O$, it follows that t_1 must contain infinitely many function symbols (since the second clause of the definition of ρ applies infinitely many times), a contradiction.

Next we show that $s_j \rightarrow_{\mathcal{S}e} s_{j+1}$ implies $\rho(s_j) \rightarrow^{+}_{\mathcal{S}_1 e_1} \rho(s_{j+1})$. Again, we consider two cases, namely (i) $s_j \rightarrow^{o}_{\mathcal{S}e} s_{j+1}$ and (ii) $s_j \rightarrow^{i}_{\mathcal{S}e} s_{j+1}$:
(i) $s_j \rightarrow^{o}_{\mathcal{S}e} s_{j+1}$
If $s_j \rightarrow^{o}_{\mathcal{S}_1} s_{j+1}$, then $\rho(s_j) \rightarrow_{\mathcal{S}_1} \rho(s_{j+1})$. For if $s_j \equiv C[\![t_1, \ldots, t_m]\!]$ then we have $s_{j+1} \equiv C'[t_{i_1}, \ldots, t_{i_l}]$ for some black context $C'[, \ldots,]$ and some indices $i_1, \ldots, i_l \in \{1, \ldots, m\}$. Clearly, it follows from $\rho(s_j) \equiv C[\rho(t_1), \ldots, \rho(t_m)]$ and $\rho(s_{j+1}) \equiv C'[\rho(t_{i_1}), \ldots, \rho(t_{i_l})]$ that the rule that reduced s_j to s_{j+1} also reduces $\rho(s_j)$ to $\rho(s_{j+1})$. If otherwise $s_j \rightarrow^{o}_{\mathcal{F}^e} s_{j+1}$, then $\rho(s_j) \rightarrow_{\mathcal{F}^e_1} \rho(s_{j+1})$, since only symbols from $\mathcal{F}_1$ are involved in the outer reduction step.
(ii) $s_j \rightarrow^{i}_{\mathcal{S}e} s_{j+1}$
Then $s_j \equiv C[\![t_1, \ldots, t_m]\!]$ and $s_{j+1} \equiv C[t_1, \ldots, t_{l-1}, t'_l, t_{l+1}, \ldots, t_m]$ for some index $l \in \{1, \ldots, m\}$ and some term t'_l, where $t_l \rightarrow_{\mathcal{S}e} t'_l$. Furthermore, $\rho(s_j) \equiv C[\rho(t_1), \ldots, \rho(t_m)]$ and $\rho(s_{j+1}) \equiv C[\rho(t_1), \ldots, \rho(t_{l-1}), \rho(t'_l), \rho(t_{l+1}), \ldots, \rho(t_m)]$. Since $t'_l \in \Delta(t_l)$, we have by definition of ρ that $\rho(t_l) = [\ldots, \rho(t'_l), \ldots]$, i.e., $\rho(t'_l)$ is a subterm of $\rho(t_l)$. Hence $\rho(t_l) = [\ldots, \rho(t'_l), \ldots] \rightarrow^{+}_{\mathcal{F}^e_1} \rho(t'_l)$, where only the rules $\mathit{Cons}(x, y) \rightarrow x$ and $\mathit{Cons}(x, y) \rightarrow y$ are used. Therefore, $\rho(s_j) \rightarrow^{+}_{\mathcal{F}^e_1} \rho(s_{j+1})$.
All in all, we have shown that $\rho(s) \rightarrow^{+}_{\mathcal{S}_1 e_1} \rho(s)$, a contradiction to the fact that $\mathcal{R}_1$ is simplifying (Proposition 10). □

[4] The order is arbitrary but fixed.
[5] $F \in \mathcal{D}_1 \cup \mathcal{C} \cup \mathcal{V}$.

Example 24. Consider $\mathcal{S}_1 = \{F(\text{C}(x), F(y, \text{D}(z))) \to F(y, x), A \to B\}$ as well as $\mathcal{S}_2 = \{g(x) \to \text{C}(x),\ g(x) \to \text{D}(x)\}$. In the reduction sequence

$$\begin{aligned} s_1 \equiv F(g(A), F(b, g(A))) &\to^i_{\mathcal{S}} s_2 \equiv F(g(B), F(b, g(A))) \\ &\to^i_{\mathcal{S}} s_3 \equiv F(\text{C}(B), F(b, g(A))) \\ &\to^i_{\mathcal{S}} s_4 \equiv F(\text{C}(B), F(b, \text{D}(A))) \\ &\to^o_{\mathcal{S}} s_5 \equiv F(b, B), \end{aligned}$$

the sets of all outer and principal subterms occurring in the reduction sequence are given by $O = \{s_1, s_2, s_3, s_4, s_5, F(b, g(A)), \text{C}(B), B, F(b, (\text{D}(A))), \text{D}(A), A\}$ and $S = \{g(A), b, g(B)\}$, respectively. Moreover, we have $\Delta(b) = \{\}$ as well as $\Delta(g(A)) = \{g(B), \text{D}(A)\}$, and $\Delta(g(B)) = \{\text{C}(B)\}$.
Therefore, ρ is completely defined by $\rho(b) = [\,]$, $\rho(g(B)) = [\rho(\text{C}(B))] = [\text{C}(B)]$, and $\rho(g(A)) = [\rho(g(B)), \rho(\text{D}(A))] = [[\text{C}(B)], \text{D}(A)]$.
As in the above proof, we get a reduction sequence of black terms

$$\begin{aligned} \rho(s_1) &\equiv F([[\text{C}(B)], \text{D}(A)], F([\,], [[\text{C}(B)], \text{D}(A)])) \\ \to^+_{\mathcal{F}_1^e} \rho(s_2) &\equiv F([\text{C}(B)], F([\,], [[\text{C}(B)], \text{D}(A)])) \\ \to^+_{\mathcal{F}_1^e} \rho(s_3) &\equiv F(\text{C}(B), F([\,], [[\text{C}(B)], \text{D}(A)])) \\ \to^+_{\mathcal{F}_1^e} \rho(s_4) &\equiv F(\text{C}(B), F([\,], \text{D}(A))) \\ \to_{\mathcal{S}_1} \rho(s_5) &\equiv F([\,], B) \end{aligned}$$

4.3 Confluence is a Modular Property of Simplifying Join CTRSs with Shared Constructors

Confluence is a modular property of disjoint TRSs as was shown by Toyama (Toyama's Theorem, see [Toy87b]). Middeldorp ([Mid90a]) extended this result to disjoint CTRSs. However, confluence is not modular if the TRSs share constructors as was pointed out by Kurihara and Ohuchi [KO90a] using the next example.

Example 25. Let $\mathcal{R}_1 = \{F(x, x) \to A, F(x, \text{C}(x)) \to B\}$ and let $\mathcal{R}_2 = \{a \to \text{C}(a)\}$. $\mathcal{R}_1$ and $\mathcal{R}_2$ are confluent, but $\mathcal{R}_1 \cup \mathcal{R}_2$ is not confluent since $F(a, a)$ has two normal forms A and B.

Fortunately, confluence is a modular property of *simplifying unconditional* TRSs with shared constructors.

Corollary 26. *Let $\mathcal{R}_1$ and $\mathcal{R}_2$ be two TRSs and let $\mathcal{R}$ be the combined TRS of $\mathcal{R}_1$ and $\mathcal{R}_2$ with shared constructors. Then $\mathcal{R}$ is simplifying and confluent (hence convergent) if and only if both $\mathcal{R}_1$ and $\mathcal{R}_2$ are simplifying and confluent.*

Proof. The only-if case is trivial. We show the if direction. By Theorem 23, $\mathcal{R}$ is simplifying and thus terminating. Since rewrite rules of $\mathcal{R}_1$ do not overlap with rewrite rules of $\mathcal{R}_2$ and vice versa, the set of critical pairs of $\mathcal{R}$ consists of the critical pairs of $\mathcal{R}_1$ and those of $\mathcal{R}_2$. Hence the assertion follows from Newman's

Lemma [6] in conjunction with the Critical Pair Lemma [7] because $\mathcal{R}_1$ and $\mathcal{R}_2$ are confluent. □

In the context of simplifying join CTRSs with shared constructors, the modularity of confluence is not so easily obtained; although a Critical Pair Lemma holds for simplifying CTRSs (cf. [Kap87]) and the set of conditional critical pairs of $\mathcal{R}$ consists of the conditional critical pairs of $\mathcal{R}_1$ and those of $\mathcal{R}_2$, we cannot argue as in the unconditional case. This is due to the fact that we have to deal with *contextual* critical pairs (see [Kap87]) and that the fundamental property of TRSs, that is, $s \rightarrow_{\mathcal{R}} t$ either implies $s \rightarrow_{\mathcal{R}_1} t$ or $s \rightarrow_{\mathcal{R}_2} t$, does not hold for join CTRSs (cf. [Mid90a]). To overcome this obstacle, we use the structure of the proof of modularity of confluence for disjoint CTRSs ([Mid90a]). It goes without saying that we have to modify parts of the proof. The proofs of some lemmata hold cum grano salis, whereas some intricate technical details simplify due to the termination property of the combined system. The proof idea is to construct two rewrite relations $\rightarrow_1$ and $\rightarrow_2$ such that their union is confluent, and reduction in the combined system $\mathcal{R}$ corresponds to joinability with respect to $\rightarrow_1 \cup \rightarrow_2$. From these two properties the modularity of confluence for simplifying join CTRSs with shared constructors is easily inferred.

Definition 27. The rewrite relation $\rightarrow_1$ is defined as follows: $s \rightarrow_1 t$ if there is a rewrite rule $l \rightarrow r \Leftarrow s_1 \downarrow t_1, \ldots, s_n \downarrow t_n$ in $\mathcal{R}_1$ such that $s \equiv C[l\sigma], t \equiv C[r\sigma]$ and $s_i\sigma \downarrow_1^o t_i\sigma$ for $i = 1, \ldots, n$. Here the superscript o in $s_i\sigma \downarrow_1^o t_i\sigma$ means that $s_i\sigma$ and $t_i\sigma$ are joinable using only *outer* $\rightarrow_1$ -reduction steps. The relation $\rightarrow_2$ is defined analogously. The union of $\rightarrow_1$ and $\rightarrow_2$ is denoted by $\rightarrow_{1,2}$.

Example 28. Let $\mathcal{R}_1 = \{F(x, c) \rightarrow G(x) \Leftarrow x \downarrow c\}$ and $\mathcal{R}_2 = \{a \rightarrow c\}$. We have $F(a, c) \rightarrow_{\mathcal{R}} G(a)$ but neither $F(a, c) \rightarrow_1 G(a)$ nor $F(a, c) \rightarrow_2 G(a)$. However, the terms are joinable with respect to $\rightarrow_{1,2}$: $F(a, c) \rightarrow_2 F(c, c) \rightarrow_1 G(c) \leftarrow_2 G(a)$.

Proposition 29. *If $s \rightarrow_{1,2} t$ then $s \rightarrow_{\mathcal{R}} t$.*

Proof. Trivial. □

Proposition 30. *Let s, t be black terms and let σ be a top white substitution with $s\sigma \rightarrow_1^o t\sigma$. If τ is a substitution with $\sigma \propto \tau$ then $s\tau \rightarrow_1^o t\tau$.*

Proof. Essentially the same as in [Mid90a]. □

Lemma 31. *The relation $\rightarrow_{1,2}$ is convergent.*

Proof. We define two unconditional TRSs $\mathcal{S}_1$ and $\mathcal{S}_2$ by

$$\mathcal{S}_i = \{u \rightarrow v \mid u, v \in \mathcal{T}(\mathcal{D}_i, \mathcal{C}, \mathcal{V}), root(u) \notin \mathcal{C} \text{ and } u \rightarrow_i v\}.$$

First of all note that the $\mathcal{S}_i$ are infinite in general. It is easy to show that the restrictions of the relations $\rightarrow_{\mathcal{S}_i}$, $\rightarrow_i$, and $\rightarrow_{\mathcal{R}_i}$ to $\mathcal{T}(\mathcal{D}_i, \mathcal{C}, \mathcal{V}) \times \mathcal{T}(\mathcal{D}_i, \mathcal{C}, \mathcal{V})$

[6] A terminating TRS is confluent if and only if it is locally confluent.

[7] A TRS is locally confluent if and only if all its critical pairs are convergent.

coincide (note that $\mathcal{R}_i$ is convergent). In particular, the TRS $\mathcal{S}_i$ is convergent on $\mathcal{T}(\mathcal{D}_i, \mathcal{C}, \mathcal{V}) \times \mathcal{T}(\mathcal{D}_i, \mathcal{C}, \mathcal{V})$.

Furthermore, it can be shown as in [Mid90a] that the rewrite relations $\rightarrow_{\mathcal{S}_i}$ and $\rightarrow_i$ are also the same on $\mathcal{T}(\mathcal{D}_1 \cup \mathcal{D}_2, \mathcal{C}, \mathcal{V}) \times \mathcal{T}(\mathcal{D}_1 \cup \mathcal{D}_2, \mathcal{C}, \mathcal{V})$. If we can show that $\mathcal{S}_1 \cup \mathcal{S}_2$ is convergent, then it follows with

$$\rightarrow_{\mathcal{S}_1 \cup \mathcal{S}_2} = \rightarrow_{\mathcal{S}_1} \cup \rightarrow_{\mathcal{S}_2} = \rightarrow_1 \cup \rightarrow_2 = \rightarrow_{1,2}$$

that $\rightarrow_{1,2}$ is also convergent on $\mathcal{T}(\mathcal{D}_1 \cup \mathcal{D}_2, \mathcal{C}, \mathcal{V}) \times \mathcal{T}(\mathcal{D}_1 \cup \mathcal{D}_2, \mathcal{C}, \mathcal{V})$. Since $\mathcal{R}$ is simplifying and hence terminating, it follows from $\rightarrow_{\mathcal{S}_1 \cup \mathcal{S}_2} = \rightarrow_{1,2}$ in conjunction with Proposition 29 that $\mathcal{S}_1 \cup \mathcal{S}_2$ is terminating. Since $\mathcal{S}_1$ and $\mathcal{S}_2$ are confluent, and since rewrite rules of $\mathcal{S}_1$ do not overlap with rewrite rules of $\mathcal{S}_2$ (and vice versa), it follows as in the proof of Corollary 26 that $\mathcal{S}_1 \cup \mathcal{S}_2$ is also confluent. □

Lemma 32. *If $s \rightarrow_{\mathcal{R}} t$ then $s \downarrow_{1,2} t$.*

Proof. The same as in [Mid90a]. The proof uses the next proposition (the proof of which is technically intricate without the termination property of $\mathcal{R}$). □

Proposition 33. *Let $s_1, \ldots, s_n$, $t_1, \ldots, t_n$ be black terms. For every substitution σ with $s_i\sigma \downarrow_{1,2} t_i\sigma$ for every $i \in \{1, \ldots n\}$ there is a substitution τ such that $\sigma \rightarrow^*_{1,2} \tau$ and $s_i\tau \downarrow^o_1 t_i\tau$ for every $i \in \{1, \ldots n\}$.*

Proof. Let $x \in Dom(\sigma)$. Since $\rightarrow_{1,2}$ is convergent, $x\sigma$ has a unique normal form u with respect to $\rightarrow_{1,2}$. We define τ by

$$\tau = \{x \mapsto u \mid x \in Dom(\sigma), x\sigma \rightarrow^*_{1,2} u, u \in NF(\rightarrow_{1,2})\}$$

Clearly, $\sigma \rightarrow^*_{1,2} \tau$. Moreover, $s_i\tau \, {}^*_{1,2}\!\leftarrow s_i\sigma \downarrow_{1,2} t_i\sigma \rightarrow^*_{1,2} t_i\tau$. The confluence of $\rightarrow_{1,2}$ guarantees $s_i\tau \downarrow_{1,2} t_i\tau$ for every $i \in \{1, \ldots n\}$. We show that this implies $s_i\tau \downarrow^o_1 t_i\tau$ for every $i \in \{1, \ldots n\}$. Proposition 22 yields a decomposition of τ into $\tau_2 \circ \tau_1$ such that τ_1 is black and τ_2 is top white. Evidently, $x\tau_2 \in NF(\rightarrow_{1,2})$ for all $x \in Dom(\tau_2)$. Let v be any term with $s_i\tau \equiv (s_i\tau_1)\tau_2 \rightarrow_{1,2} v$. Since $x\tau_2 \in NF(\rightarrow_{1,2})$ for all $x \in Dom(\tau_2)$, we have $(s_i\tau_1)\tau_2 \rightarrow^o_1 v$ and further $v \equiv v'\tau_2$ for some black term v'. Now the claim follows by induction on the length of the conversion $s_i\tau \downarrow_{1,2} t_i\tau$. □

Proposition 34. *The relations $=_{\mathcal{R}}$ and $\downarrow_{1,2}$ coincide.*

Proof. This is a consequence of Proposition 29 and Lemmata 31 and 32. □

Theorem 35. *Confluence is a modular property of simplifying join CTRSs with shared constructors.*

Proof. Let $(\mathcal{D}_1 \cup \mathcal{C}, \mathcal{R}_1)$ and $(\mathcal{D}_2 \cup \mathcal{C}, \mathcal{R}_2)$ be two simplifying join CTRSs with shared constructors. We have to show that the combined system $\mathcal{R}$ (which is simplifying) is confluent if and only if both $\mathcal{R}_1$ and $\mathcal{R}_2$ are confluent. The only-if case is straightforward. In order to show the other implication we consider a conversion $t_1 \, {}^*_{\mathcal{R}}\!\leftarrow s \rightarrow^*_{\mathcal{R}} t_2$. According to Proposition 34 we have $t_1 \downarrow_{1,2} t_2$ and thus Proposition 29 implies $t_1 \downarrow_{\mathcal{R}} t_2$. □

4.4 Attempts to weaken the Premises

Naturally the question arises whether Theorem 23 also holds for reductive or decreasing CTRSs (see [DOS88]). The answer is negative as is shown by the "classical" counterexample of Toyama (cf. Example 11).

Additionally, one may ask whether $\mathcal{R}_1$ and $\mathcal{R}_2$ may share defined function symbols (symbols that do occur at the root position of the left-hand side of a rewrite rule). Especially in the context of parameterized algebraic specifications it would be nice if at least one of the CTRSs could contain defined function symbols of the other in its conditions (but not vice versa). Unfortunately, there is no affirmative answer to that question, not even in the restricted case that the systems do not share constructors in the left- and right-hand sides of rules (i.e., shared symbols occur only in the conditions).

Example 36. Let $\mathcal{R}_1 = \{F(x,x) \to x \Leftarrow F(a,b) \downarrow x\}$ and $\mathcal{R}_2 = \{a \to b\}$. $\mathcal{R}_1$ and $\mathcal{R}_2$ are simplifying but the union $\mathcal{R} = \mathcal{R}_1 \cup \mathcal{R}_2$ is not (see [Ohl92b]).

In the above example there is a "mixed" term $F(a,b)$ in the condition of the only rule of $\mathcal{R}_1$. It needs some further investigations if the union $(\mathcal{F},\mathcal{R})$ of two simplifying CTRSs $(\mathcal{F}_1,\mathcal{R}_1)$ and $(\mathcal{F}_2,\mathcal{R}_2)$ is again simplifying provided that

- $\mathcal{F}_1$ is the disjoint union of $\mathcal{F}_1'$ and $\mathcal{F}_2$ (in particular, $\mathcal{R}_2$ does not contain any symbol from $\mathcal{F}_1'$), and
- if $l \to r \Leftarrow s_1 \downarrow t_1, \ldots, s_n \downarrow t_n$ is a rule of $\mathcal{R}_1$ then $l, r \in \mathcal{T}(\mathcal{F}_1', \mathcal{V})$ and we have $s_i, t_i \in \mathcal{T}(\mathcal{F}_1', \mathcal{V})$ or $s_i, t_i \in \mathcal{T}(\mathcal{F}_2, \mathcal{V})$.

However, we note that even if the union of such two simplifying CTRSs is again simplifying, confluence is not preserved in general. Consider for instance the term rewriting systems $\mathcal{R}_1 = \{F(x) \to C \Leftarrow x \downarrow a, F(x) \to D \Leftarrow x \downarrow b\}$ and $\mathcal{R}_2 = \{a \to b\}$.

5 Conclusions

We have shown that the combined CTRS $\mathcal{R} = \mathcal{R}_1 \cup \mathcal{R}_2$ with shared constructors is simplifying if and only if $\mathcal{R}_1$ and $\mathcal{R}_2$ are simplifying, and that confluence is a modular property of simplifying join CTRSs. By counterexamples it was also shown that a weakening of the premises leads to the loss of the modularity of at least one of these two properties. Because of Proposition 5, we have focused on finite simplifying *join* systems. However, confluence seems to be a modular property of simplifying semi-equational CTRSs, too (this must be checked in detail along the lines of the proof of modularity of confluence for disjoint semi-equational CTRSs given in [Mid90b]). Moreover, we point out that the results obtained for join CTRSs also hold for *normal* CTRSs (since every normal system can be viewed as a join system, cf. [Mid90a]).
Recently Gramlich (cf. [Gra91]) obtained several sufficient conditions for the modularity of termination of (finite and finitely branching, respectively) unconditional term rewriting systems (possibly sharing constructors) and of disjoint conditional term rewriting systems. The careful reader will observe that his results and those given in the present paper complement one another.

Acknowledgements: The author is grateful to Robert Giegerich, Stefan Kurtz, Jörg Süggel, and Aart Middeldorp for valuable comments on previous versions of the paper. Thanks also go to Anke Bodzin for typesetting the manuscript.

References

[Der82] N. Dershowitz. Orderings for Term-Rewriting Systems. *Theoretical Computer Science* **17***(3)*, pages 279–301, 1982.

[DJ90] N. Dershowitz and J.P. Jouannaud. Rewrite Systems. In L. van Leeuwen, editor, *Handbook of Theoretical Computer Science, Vol. B*, chapter 6. North-Holland, 1990.

[DJ91] N. Dershowitz and J.P. Jouannaud. Notations for Rewriting. *Bulletin of the EATCS* **43**, pages 162–172, February 1991.

[DOS88] N. Dershowitz, M. Okada, and G. Sivakumar. Canonical Conditional Rewrite Systems. In *Proceedings of the 9th Conference on Automated Deduction*, pages 538–549. Lecture Notes in Computer Science **310**, Springer Verlag, 1988.

[Gra91] B. Gramlich. A Structural Analysis of Modular Termination of Term Rewriting Systems. SEKI Report SR-91-15, Universität Kaiserslautern, 1991.

[Kap87] S. Kaplan. Simplifying Conditional Term Rewriting Systems: Unification, Termination and Confluence. *Journal of Symbolic Computation* **4***(3)*, pages 295–334, 1987.

[Klo90] J.W. Klop. Term Rewriting Systems. Report CS-R9073, Centre for Mathematics and Computer Science, 1990.

[KO90a] M. Kurihara and A. Ohuchi. Modularity of Simple Termination of Term Rewriting Systems with Shared Constructors. Report SF-36, Hokkaido University, Sapporo, 1990.

[KO90b] M. Kurihara and A. Ohuchi. Modularity of Simple Termination of Term Rewriting Systems. *Journal of IPS Japan* **31***(5)*, pages 633–642, 1990.

[Mid90a] A. Middeldorp. Confluence of the Disjoint Union of Conditional Term Rewriting Systems. In *Proceedings of the 2nd International Workshop on Conditional and Typed Rewriting Systems*, pages 295–306. Lecture Notes in Computer Science **516**, Springer Verlag, 1990.

[Mid90b] A. Middeldorp. *Modular Properties of Term Rewriting Systems.* PhD thesis, Free University Amsterdam, 1990.

[Ohl92a] E. Ohlebusch. A Note on Simple Termination of Infinite Term Rewriting Systems. Report Nr. 7, Technische Fakultät, Universität Bielefeld, 1992.

[Ohl92b] E. Ohlebusch. Combinations of Simplifying Conditional Term Rewriting Systems. Report Nr. 6, Technische Fakultät, Universität Bielefeld, 1992.

[Rus87] M. Rusinowitch. On Termination of the Direct Sum of Term Rewriting Systems. *Information Processing Letters* **26**, pages 65–70, 1987.

[Ste89] J. Steinbach. Extensions and Comparison of Simplification Orderings. In *Proceedings of the 3rd International Conference on Rewriting Techniques and Applications*, pages 434–448. Lecture Notes in Computer Science **355**, Springer Verlag, 1989.

[Toy87a] Y. Toyama. Counterexamples to Termination for the Direct Sum of Term Rewriting Systems. *Information Processing Letters* **25**, pages 141–143, 1987.

[Toy87b] Y. Toyama. On the Church-Rosser Property for the Direct Sum of Term Rewriting Systems. *Journal of the ACM* **34***(1)*, pages 128–143, 1987.

Sufficient Conditions for Modular Termination of Conditional Term Rewriting Systems

Bernhard Gramlich*

Fachbereich Informatik, Universität Kaiserslautern
Erwin Schrödinger Straße, W-6750 Kaiserslautern, Germany
gramlich@informatik.uni-kl.de

Abstract. Recently we have shown the following abstract result for unconditional term rewriting systems (TRSs). Whenever the disjoint union $\mathcal{R}_1 \oplus \mathcal{R}_2$ of two (finite) terminating TRSs $\mathcal{R}_1$, $\mathcal{R}_2$ is non-terminating, then one of the systems, say $\mathcal{R}_1$, enjoys an interesting property, namely it is not termination preserving under non-deterministic collapses, i.e. $\mathcal{R}_1 \oplus \{G(x,y) \to x, G(x,y) \to y\}$ is non-terminating, and the other system $\mathcal{R}_2$ is collapsing, i.e. contains a rule with a variable right hand side. This result generalizes known sufficient syntactical criteria for modular termination of rewriting. Here we extend this result and derived sufficient criteria for modularity of termination to the case of conditional term rewriting systems (CTRSs). Moreover we relate various definitions of notions related to termination of CTRSs to each other and discuss some subtleties and problems concerning extra variables in the rules.

1 Introduction

From a theoretical point of view and also for efficiency reasons it is very useful to know whether a combined TRS has some property whenever this property already holds for the single 'modules'. A simple and natural way of such 'modular' constructions is given by the concept of 'direct sum' ([17]) or 'disjoint union'. Two TRSs $\mathcal{R}_1$ and $\mathcal{R}_2$ over signatures $\mathcal{F}_1$ and $\mathcal{F}_2$, respectively, are said to be *disjoint* if $\mathcal{F}_1$ and $\mathcal{F}_2$ are disjoint, i.e. $\mathcal{F}_1 \cap \mathcal{F}_2 = \emptyset$ (in that case the rule sets of $\mathcal{R}_1$ and $\mathcal{R}_2$ are necessarily disjoint, too). The (disjoint) union of two disjoint TRSs $\mathcal{R}_1$, $\mathcal{R}_2$ is denoted by $R_1 \oplus R_2$. We shall also speak of the disjoint union of $\mathcal{R}_1$ and $\mathcal{R}_2$ using the implicit convention that $\mathcal{R}_1$ and $\mathcal{R}_2$ are assumed to be disjoint TRSs. A property P of TRSs is said to be *modular* if the following holds for all disjoint TRSs $\mathcal{R}_1$, $\mathcal{R}_2$: $\mathcal{R}_1 \oplus \mathcal{R}_2$ has property P iff both $\mathcal{R}_1$ and $\mathcal{R}_2$ have property P. Toyama [17] has shown that confluence is modular. The termination property, however, is in general not modular as witnessed by the following counterexample of [17]:

* This research was supported by the 'Deutsche Forschungsgemeinschaft, SFB 314 (D4-Projekt)'.

Example 1. $\mathcal{R}_1 : \quad f(a,b,x) \to f(x,x,x) \qquad \mathcal{R}_2 : \quad G(x,y) \to x$
$\qquad G(x,y) \to y$

Clearly, both R_1 and R_2 are terminating, but $\mathcal{R}_1 \oplus \mathcal{R}_2$ admits e.g. the following infinite derivation:

$$\begin{aligned} f(a,b,G(a,b)) &\to_{\mathcal{R}_1} f(G(a,b),G(a,b),G(a,b)) \\ &\to_{\mathcal{R}_2} f(a,G(a,b),G(a,b)) \\ &\to_{\mathcal{R}_2} f(a,b,G(a,b)) \\ &\to_{\mathcal{R}_1} \cdots \end{aligned}$$

In the last years a couple of sufficient conditions for modularity of termination of (unconditional) TRSs have been found. Some of these criteria are formulated syntactically (in terms of collapsing and duplicating rules, see Rusinowith [16], Middeldorp [11]). Other criteria involve more semantical requirements. For instance, it is shown by Toyama, Klop & Barendregt [18] that for left-linear and confluent TRSs termination (and hence completeness, i.e. confluence plus termination) is modular. In Gramlich [6] we have shown that termination (and hence completeness) is modular for locally confluent overlay systems (a TRS is an overlay system if all its critical pairs are obtained by 'root overlaps'). Kurihara & Ohuchi [10] proved that simple termination, i.e. termination by means of some simplification ordering, is a modular property of (finite) TRSs. A structural analysis of potential counterexamples and a couple of derived sufficient conditions generalizing results from [16], [11] and [10] have been presented in Gramlich [4], [5]. Below we shall show how the main ideas and results of this approach can be extended to CTRSs.

For CTRSs results from [16] and [11] concerning collapse-free and non-duplicating systems for the unconditional case have been adapted and extended in Middeldorp [12], [13]. Recently Middeldorp has shown in [14] that the main result of Middeldorp & Toyama [15] stating that completeness is invariant under the (non-disjoint) combination of constructor systems (which may share common constructors) does also hold for CTRSs without extra variables in the conditions of the rewrite rules.

Before going into details now, let us motivate and sketch our approach for analyzing modularity of termination. Having again a closer look on example 1 above and the non-terminating derivation indicated there, it is obvious that the collapsing $\mathcal{R}_2$-steps using the rules $G(x,y) \to x, G(x,y) \to y$ play an essential role for enabling the derivation to be infinite. In fact, this observation can be generalized to arbitrary situations where we have (finite) terminating (C)TRSs $\mathcal{R}_1, \mathcal{R}_2$ over disjoint signatures $\mathcal{F}_1$ and $\mathcal{F}_2$, respectively, such that the disjoint union $\mathcal{R}_1 \oplus \mathcal{R}_2$ is non-terminating. In any infinite $(\mathcal{R}_1 \oplus \mathcal{R}_2)$-derivation $s_0 \to s_1 \to s_2 \to s_3 \to \ldots$, all the s_i's must be 'mixed' terms, i.e. involve function symbols from both signatures $\mathcal{F}_1$ and $\mathcal{F}_2$. We shall show that for any counterexample satisfying a certain minimality property concerning the 'layer structure' of its terms, one can construct from this counterexample an infinite derivation in $R_i \oplus \{G(x,y) \to x, G(x,y) \to y\}$ for $i = 1$ or $i = 2$, let's say for $i = 1$. This is achieved by an appropiate transformation from terms over $\mathcal{F}_1 \uplus \mathcal{F}_2$ into terms over $\mathcal{F}_1 \uplus \{A, G\}$ (here A is a new constant symbol and G is a new binary function symbol) which abstracts from the concrete form of $\mathcal{F}_2$-layers but retains enough relevant information for the translation of the reduction steps. This characteristic property of minimal counterexamples provides the basis for a couple of modularity results derived subsequently. It also corresponds nicely to the intuition that the existence of counterexamples crucially depends on 'non-deterministic collapsing' reduction steps. Hence, example 1 above is in a sense the simplest conceivable counterexample.

In the next section we briefly recall some basic notions, definitions and facts for TRSs, CTRSs and for disjoint unions needed later on. In section 3 the main results and their applications for CTRSs will be presented. Finally some open problems and perspectives will be discussed. For a more detailed presentation and discussion of the whole approach see [4]).

2 Preliminaries

2.1 Basic Notations and Definitions for TRSs

We briefly recall the basic terminology needed for dealing with TRSs (e.g. [9]). Let $\mathcal{V}$ be a countably infinite set of *variables* and $\mathcal{F}$ be a set of *function symbols* with $\mathcal{V} \cap \mathcal{F} = \emptyset$. Associated to every $f \in \mathcal{F}$ is a natural number denoting its arity. Function symbols of arity 0 are called *constants*. The set $\mathcal{T}(\mathcal{F}, \mathcal{V})$ of *terms* over $\mathcal{F}$ and $\mathcal{V}$ is the smallest set with (1) $\mathcal{V} \subseteq \mathcal{T}(\mathcal{F}, \mathcal{V})$ and (2) if $f \in \mathcal{F}$ has arity n and $t_1, \ldots, t_n \in \mathcal{T}(\mathcal{F}, \mathcal{V})$ then $f(t_1, \ldots, t_n) \in \mathcal{T}(\mathcal{F}, \mathcal{V})$. If some function symbols are allowed to be varyadic then the definition of $\mathcal{T}(\mathcal{F}, \mathcal{V})$ is generalized in an obvious way. The set of all *ground terms* (over $\mathcal{F}$), i.e. terms with no variables, is denoted by $\mathcal{T}(\mathcal{F})$. In the following we shall always assume that $\mathcal{T}(\mathcal{F})$ is non-empty, i.e. there is at least one constant in $\mathcal{F}$. Identity of terms is denoted by $\equiv$. The set of variables occurring in a term t is denoted by $V(t)$.

A *context* $C[, \ldots,]$ is a term with 'holes', i.e. a term in $\mathcal{T}(\mathcal{F} \uplus \{\square\}, \mathcal{V})$ where $\square$ is a new special constant symbol. If $C[, \ldots,]$ is a context with n occurrences of $\square$ and $t_1, \ldots, t_n$ are terms then $C[t_1, \ldots, t_n]$ is the term obtained from $C[, \ldots,]$ by replacing from left to right the occurrences of $\square$ by $t_1, \ldots, t_n$. A context containing precisely one occurrence of $\square$ is denoted by $C[]$. For the set $\mathcal{T}(\mathcal{F} \uplus \{\square\}, \mathcal{V})$ we also write $\mathcal{CON}(\mathcal{F}, \mathcal{V})$. A *non-empty* context is a term from $\mathcal{CON}(\mathcal{F}, \mathcal{V}) \backslash \mathcal{T}(\mathcal{F}, \mathcal{V})$ which is different from $\square$. A term s is a *subterm* of a term t if there exists a context $C[]$ with $t \equiv C[s]$. If in addition $C[] \not\equiv \square$ then s is a *proper* subterm of t. A substitution σ is a mapping from $\mathcal{V}$ to $\mathcal{T}(\mathcal{F}, \mathcal{V})$ such that its domain $dom(\sigma)$ $\{x \in \mathcal{V} | \sigma x \not\equiv x\}$ is finite. Its homomorphic extension to a mapping from $\mathcal{T}(\mathcal{F}, \mathcal{V})$ to $\mathcal{T}(\mathcal{F}, \mathcal{V})$ is also denoted by σ.

A *term rewriting system (TRS)* is a pair $(\mathcal{R}, \mathcal{F})$ consisting of a signature $\mathcal{F}$ and a set $\mathcal{R} \subseteq \mathcal{T}(\mathcal{F}, \mathcal{V}) \times \mathcal{T}(\mathcal{F}, \mathcal{V})$ of (rewrite) rules (l, r) denoted by $l \rightarrow r$ with $l \notin \mathcal{V}$ and $V(r) \subseteq V(l)$. [2]

[2] This restriction of excluding variable left-hand sides and right-hand side extra-variables is not a severe one. In particular, concerning termination of rewriting it only excludes trivial cases.

Instead of $(\mathcal{R}, \mathcal{F})$ we also write $\mathcal{R}^{\mathcal{F}}$ or simply $\mathcal{R}$ when $\mathcal{F}$ is clear from the context or irrelevant. Given a TRS $\mathcal{R}^{\mathcal{F}}$ the rewrite relation $\rightarrow_{\mathcal{R}^{\mathcal{F}}}$ for terms $s, t \in \mathcal{T}(\mathcal{F}, \mathcal{V})$ is defined as follows: $s \rightarrow_{\mathcal{R}^{\mathcal{F}}} t$ if there exists a rule $l \rightarrow r \in \mathcal{R}$, a substitution σ and a context $C[]$ such that $s \equiv C[\sigma l]$ and $t \equiv C[\sigma r]$. We also write $\rightarrow_{\mathcal{R}}$ or simply $\rightarrow$ when $\mathcal{F}$ or $\mathcal{R}^{\mathcal{F}}$ is clear from the context, respectively. The symmetric, transitive and transitive-reflexive closures of $\rightarrow$ are denoted by $\leftrightarrow$, $\rightarrow^+$ and $\rightarrow^*$, respectively. Two terms s, t are *joinable in* $\mathcal{R}$, denoted by $s \downarrow_{\mathcal{R}} t$, if there exists a term u with $s \,{}_{\mathcal{R}}^{*}\!\leftarrow u \rightarrow_{\mathcal{R}}^{*} t$. A term s is *irreducible* or a *normal form* if there is no term t with $s \rightarrow t$. A TRS $\mathcal{R}$ is *terminating* or *strongly normalizing* if $\rightarrow$ is noetherian, i.e. if there is no infinite reduction sequence $s_1 \rightarrow s_2 \rightarrow s_3 \rightarrow \cdots$.

A *partial ordering* $>$ on a set D is a transitive and irreflexive binary relation on D. A partial ordering $>$ on $\mathcal{T}(\mathcal{F}, \mathcal{V})$ is said to be *monotonic (w.r.t. the term structure)* if it possesses the *replacement property*

$$s > t \implies C[s] > C[t]$$

for all $s, t, C[]$. It is *stable (w.r.t. substitutions)* if

$$s > t \implies \sigma s > \sigma t$$

for all s, t, σ. A *term ordering on* $\mathcal{T}(\mathcal{F}, \mathcal{V})$ is a monotonic and stable partial ordering on $\mathcal{T}(\mathcal{F}, \mathcal{V})$. A *reduction ordering* is a well-founded term ordering. A term ordering $>$ is said to be a simplification ordering if it additionally enjoys the subterm property

$$C[s] > s$$

for any s and any non-empty context $C[]$.[3]

A TRS is confluent if ${}^{*}\!\leftarrow \circ \rightarrow^{*} \subseteq \rightarrow^{*} \circ {}^{*}\!\leftarrow$ and locally confluent if $\leftarrow \circ \rightarrow \subseteq \rightarrow^{*} \circ {}^{*}\!\leftarrow$.[4] A confluent and terminating TRS is said to be *convergent* or *complete*.

2.2 Conditional Term Rewriting Systems

Moreover, we need some basic terminology about CTRSs.

Definition 1. A *CTRS* is a pair $(\mathcal{R}, \mathcal{F})$ consisting of a signature $\mathcal{F}$ and a set of *conditional rewrite rules* of the form

$$s_1 = t_1 \wedge \ldots \wedge s_n = t_n \implies l \rightarrow r$$

with $s_1, \ldots, s_n, t_1, \ldots, t_n, l, r \in \mathcal{T}(\mathcal{F}, \mathcal{V})$. Moreover, we require $l \notin \mathcal{V}$ and $V(r) \subseteq V(l)$ as for unconditional TRSs, i.e. no variable left hand sides and no extra variables on the right hand side. For our purposes it will be useful to exclude extra variables in the conditions, too. This means to require additionally

[3] For the case that varyadic function symbols are allowed one additionally requires here the so-called 'deletion'-property (cf. Dershowitz [1]).

[4] Here, '$\circ$' denotes relation composition.

$\bigcup_{i=1}^{n}\{V(s_i), V(t_i)\} \subseteq V(l)$.[5] If the condition is empty, i.e. $n = 0$, we simply write $l \to r$. Instead of $(\mathcal{R}, \mathcal{F})$ we also write $\mathcal{R}^{\mathcal{F}}$ or simply $\mathcal{R}$ when $\mathcal{F}$ is clear from the context or irrelevant.

Depending on the interpretation of the equality sign in the conditions of rewrite rules, different reduction relations may be associated with a given CTRS.

Definition 2.

(1) In a *join CTRS* $\mathcal{R}$ the equality sign in the conditions of rewrite rules is interpreted as joinability. Formally this means: $s \to_{\mathcal{R}} t$ if there exists a rewrite rule $s_1 = t_1 \wedge \ldots \wedge s_n = t_n \implies l \to r \in \mathcal{R}$, a substitution σ and a context $C[]$ such that $s \equiv C[\sigma l]$, $t \equiv C[\sigma r]$ and $\sigma s_i \downarrow_{\mathcal{R}} \sigma t_i$ for all $i \in \{1, \ldots, n\}$. For rewrite rules of a join CTRS we shall use the notation $s_1 \downarrow t_1 \wedge \ldots \wedge s_n \downarrow t_n \implies l \to r$.

(2) Semi-equational CTRSs are obtained by interpreting the equality sign in the conditions as convertibility, i.e. as $\stackrel{*}{\leftrightarrow}$.

Definition 3. The reduction relation corresponding to a given CTRS $\mathcal{R}$ is inductively defined as follows ($\Box$ denotes $\downarrow$ or $\stackrel{*}{\leftrightarrow}$, respectively):

$$\begin{array}{rcl} \mathcal{R}_0 & = & \{l \to r \mid l \to r \in \mathcal{R}\}\,, \\ \mathcal{R}_{i+1} & = & \{\sigma l \to \sigma r \mid s_1 \Box t_1 \wedge \ldots \wedge s_n \Box t_n \implies l \to r \in \mathcal{R}, \\ & & \sigma s_j \Box_{\mathcal{R}_i} \sigma t_j \text{ for } j = 1, \ldots, n\}\,, \\ s \to_{\mathcal{R}} t :\iff & & s \to_{\mathcal{R}_i} t \text{ for some } i \geq 0, \text{i.e.} \to_{\mathcal{R}} = \bigcup_{i \geq 0} \to_{\mathcal{R}_i}\,. \end{array}$$

The *depth* of a rewrite step $s \to_{\mathcal{R}} t$ is defined to be the minimal i with $s \to_{\mathcal{R}_i} t$.

The other basic notions for unconditional TRSs generalize in a straightforward manner to CTRSs.

In general, conditional rewriting is much more complicated than unconditional rewriting. For instance, the rewrite relation may be undecidable even for complete CTRSs without extra variables in the conditions (cf. [8]).

Definition 4. ([3]) A CTRS $\mathcal{R}$ is *decreasing* if there exists an extension $>$ of the reduction relation induced by $\mathcal{R}$ which satisfies the following properties:

(1) $>$ is noetherian.
(2) $>$ has the subterm property, i.e. $C[s] > s$ for every term s and every non-empty context $C[]$.
(3) If $s_1 = t_1 \wedge \ldots \wedge s_n = t_n \implies l \to r$ is a rule in $\mathcal{R}$ and σ is a substitution then $\sigma l > \sigma s_i$ and $\sigma l > \sigma t_i$ for $i = 1, \ldots, n$.

---

[5] Extra variables in the conditions may be quite natural in many situations, in particular from a specification or programming point of view. Later on we will discuss the reason for excluding them here.

A CTRS $\mathcal{R}$ is *reductive* (cf. [7]) if there exists a well-founded monotonic extension $>$ of the reduction relation induced by $\mathcal{R}$ satisfying (3).

Clearly, every reductive system is decreasing and any decreasing system is terminating. Decreasingness exactly captures the finiteness of recursive evaluation of terms (cf. [2]). For decreasing (join) CTRSs all the basic notions are decidable, e.g. reducibility and joinability. Moreover, fundamental results like the critical pair lemma hold for decreasing (join) CTRSs which is not the case in general for arbitrary (terminating join) CTRSs.

In the following we shall tacitly assume that all CTRSs considered are join CTRSs (which is the most important case in practice), except for cases where another kind of CTRSs is explicitly mentioned.

2.3 Basic Notations for Disjoint Unions

The following notations and definitions for dealing with disjoint unions of (C)TRSs mainly follow [17] and [13]. Let $\mathcal{R}_1^{\mathcal{F}_1}$, $\mathcal{R}_2^{\mathcal{F}_2}$ be (C)TRSs with disjoint signatures $\mathcal{F}_1$, $\mathcal{F}_2$. Their *disjoint union* $\mathcal{R}_1 \oplus \mathcal{R}_2$ is the (C)TRS $(\mathcal{R}_1 \uplus \mathcal{R}_2, \mathcal{F}_1 \uplus \mathcal{F}_2)$.[6]

Let $t \equiv C[t_1, \ldots, t_n]$ with $C[, \ldots,] \not\equiv \Box$. We write $t \equiv C[\![t_1, \ldots, t_n]\!]$ if $C[, \ldots,] \in CON(\mathcal{F}_a, \mathcal{V})$ and $root(t_1), \ldots, root(t_n) \in \mathcal{F}_b$ for some $a, b \in \{1, 2\}$ with $a \neq b$. In this case the t_i's are the *principal subterms* or *principal aliens* of t. Note that every $t \in \mathcal{T}(\mathcal{F}_1 \uplus \mathcal{F}_2, \mathcal{V}) \setminus (\mathcal{T}(\mathcal{F}_1, \mathcal{V}) \cup \mathcal{T}(\mathcal{F}_2, \mathcal{V}))$ has a unique representation of the form $t \equiv C[\![t_1, \ldots, t_n]\!]$. The set of all *special subterms* or *aliens* of t is recursively defined in a straightforward manner.

The *rank* of a term $t \in \mathcal{T}(\mathcal{F}_1 \uplus \mathcal{F}_2, \mathcal{V})$ is defined by

$$rank(t) = \begin{cases} 1 & \text{if} \quad t \in \mathcal{T}(\mathcal{F}_1, \mathcal{V}) \cup \mathcal{T}(\mathcal{F}_2, \mathcal{V}) \\ 1 + max\{rank(t_i) \mid 1 \leq i \leq n\} & \text{if} \quad t \equiv C[\![t_1, \ldots, t_n]\!] \end{cases}$$

An important basic fact about the rank of terms occurring in a $(\mathcal{R}_1 \oplus \mathcal{R}_2)$-derivation is the following ([17]): If $s \rightarrow^* t$ then $rank(s) \geq rank(t)$. Moreover, if $s \in \mathcal{T}(\mathcal{F}_1 \uplus \mathcal{F}_2, \mathcal{V})$ with $rank(s) = n$ then there exists a ground instance σs of s with $rank(\sigma s) = n$, too. [7] A (finite or infinite) derivation $D: s_1 \rightarrow s_2 \rightarrow s_3 \ldots$ is said to have rank n ($rank(D) = n$) if n is the minimal rank of all the s_i's, i.e. $n = \min\{rank(s_i) \mid 1 \leq i\}$.

The topmost homogeneous part of a term $t \in \mathcal{T}(\mathcal{F}_1, \mathcal{V}) \cup \mathcal{T}(\mathcal{F}_2, \mathcal{V})$, denoted by $top(t)$, is obtained from t by replacing all principal subterms by $\Box$, i.e.

$$top(t) = \begin{cases} t & \text{if} \quad rank(t) = 1 \\ C[, \ldots,] & \text{if} \quad t \equiv C[\![t_1, \ldots, t_n]\!] \end{cases}$$

Furthermore we shall use the abbreviations $\mathcal{GT}_\oplus$ for $\mathcal{T}(\mathcal{F}_1 \uplus \mathcal{F}_2)$, $\mathcal{GT}_\oplus^n$ for $\{t \in \mathcal{GT}_\oplus \mid rank(t) = n\}$, $\mathcal{GT}_\oplus^{<n}$ for $\{t \in \mathcal{GT}_\oplus \mid rank(t) < n\}$ and $\mathcal{GT}_\oplus^{\leq n}$ for

[6] The symbol '$\uplus$' denotes union of disjoint sets.

[7] This is easily verified by substituting appropriately $\mathcal{F}_1$- or $\mathcal{F}_2$-ground terms for those variables which occur in the 'deepest layer' of s.

$\{t \in \mathcal{GT}_{\oplus} | rank(t) \leq n\}$. For the sake of better readability the function symbols from $\mathcal{F}_1$ are considered to be black and those of $\mathcal{F}_2$ to be white. Variables have no colour. A *top black (white)* term has a black (white) root symbol.

For $s, t \in \mathcal{GT}_{\oplus}$ the one-step reduction $s \to t$ is said to be *inner* – denoted by $s \xrightarrow{i} t$ – if the reduction takes place in one of the principal subterms of s. Otherwise, we speak of an *outer* reduction step and write $s \xrightarrow{o} t$. A rewrite step $s \to t$ is *destructive at level 1* if the root symbols of s and t have different colours. The step $s \to t$ is *destructive at level* $n+1$ (for $n \geq 1$) if $s \equiv C[\![s_1, \ldots, s_j, \ldots, s_n]\!] \xrightarrow{i} C[s_1, \ldots, t_j, \ldots, s_n] \equiv t$ with $s_j \to t_j$ destructive at level n. Clearly, if a rewrite step is destructive (at some level) then the applied rewrite rule is collapsing, i.e. has a variable right-hand side. This is a basic fact which should be kept in mind subsequently.

3 Modularity of Termination for Conditional Term Rewriting Systems

For generalizing results concerning modular properties of TRSs to the conditional case a careful analysis is necessary. As mentioned by Middeldorp (cf. [13]), the additional complications mainly arise from the fact that the fundamental property

$$s \to_{\mathcal{R}_1 \oplus \mathcal{R}_2} t \quad \Longrightarrow \quad s \to_{\mathcal{R}_1} t \quad \vee \quad s \to_{\mathcal{R}_2} t \quad (*)$$

only holds for unconditional TRSs but not for CTRSs in general. This is due to the fact that for verifying the applicability of an $\mathcal{R}_1$-rule, i.e. for proving the corresponding instantiated conditions, rules from $\mathcal{R}_2$ may be crucial. Consider for instance

Example 2.

$$\mathcal{R}_1 = \{\ x \downarrow b \wedge x \downarrow c \Longrightarrow a \to a\ \} \text{ over } \mathcal{F}_1 = \{a, b, c\},$$
$$\mathcal{R}_2 = \{G(x,y) \to x, G(x,y) \to y\} \text{ over } \mathcal{F}_2 = \{G, A\}.$$

Here, we have $a \to_{\mathcal{R}_1 \oplus \mathcal{R}_2} a$ by applying the $\mathcal{R}_1$-rule (x is substituted by $G(b,c)$), but neither $a \to_{\mathcal{R}_1} a$ nor $a \to_{\mathcal{R}_2} a$. Hence, this is also a very simple counterexample to modularity of termination for CTRSs, because both $\mathcal{R}_1$ and $\mathcal{R}_2$ are terminating (the reduction relation of $\mathcal{R}_1$ is empty). Moreover, the infinite $(\mathcal{R}_1 \oplus \mathcal{R}_2)$-derivation $a \to a \to a \to \ldots$ has rank one, a phenomenon which cannot occur in the unconditional case. Note that the system $\mathcal{R}_1$ above has the extra variable x in the condition part of its rule.[8] Therefore it cannot be decreasing. When we forbid extra variables in the conditions then the minimal rank of potential counterexamples is at least 3 as we shall see. But the fundamental property $(*)$ above may still be violated as shown by

Example 3. (see [13] for similar counterexamples)

[8] Hence, strictly spoken $\mathcal{R}_1$ is no CTRS in the sense of definition 1.

$$\mathcal{R}_1 = \{ \ x \downarrow a \ \wedge \ x \downarrow b \implies f(x) \to f(x) \ \} \text{ over } \mathcal{F}_1 = \{a, b, f\},$$
$$\mathcal{R}_2 = \{G(x, y) \to x, G(x, y) \to y\} \text{ over } \mathcal{F}_2 = \{G, A\}.$$

Here, both $\mathcal{R}_1$ and $\mathcal{R}_2$ are clearly decreasing (and even reductive), hence terminating, but $\mathcal{R}_1 \oplus \mathcal{R}_2$ is non-terminating since we have e.g. the infinite $(\mathcal{R}_1 \oplus \mathcal{R}_2)$-derivation

$$f(G(a, b)) \to f(G(a, b)) \to f(G(a, b)) \to \dots$$

where the $\mathcal{R}_1$-rule is enabled because the instantiated conditions can be verified using $\mathcal{R}_2$.

For unconditional TRSs it is known that each of the conditions

(a) neither $\mathcal{R}_1$ nor $\mathcal{R}_2$ contains a duplicating[9] rule ([16]),
(b) neither $\mathcal{R}_1$ nor $\mathcal{R}_2$ contains a collapsing rule, and
(c) one of the system $\mathcal{R}_1$, $\mathcal{R}_2$ contains neither collapsing nor duplicating rules ([11])

is sufficient for ensuring modularity of termination. The above example shows that (a) and (c) do not generalize to the conditional case. In [12] it is shown that (b) is still sufficient in the conditional case, and that (a) and (c) are sufficient under the additional assumption that both CTRSs are confluent. Moreover, confluence turns out to be a modular property of CTRSs as shown by Middeldorp in [13].

In the following we shall show that the essential features and results of our structural analysis of modular termination for the unconditional case presented in [5] can be generalized to CTRSs in a rather straightforward manner.

It is easy to see that conditional reduction steps are rank decreasing, i.e. $s \to_{\mathcal{R}_1 \oplus \mathcal{R}_2} t$ implies $rank(s) \geq rank(t)$.

Next some structural properties of minimal counterexamples are exhibited.

Lemma 5. *Let $\mathcal{R}_1$, $\mathcal{R}_2$ be two terminating disjoint CTRSs such that*

$$D: s_1 \to s_2 \to s_3 \to \dots$$

is an infinite derivation in $\mathcal{R}_1 \oplus \mathcal{R}_2$ (involving only ground terms) of minimal rank, i.e. any derivation in $\mathcal{R}_1 \oplus \mathcal{R}_2$ of smaller rank is finite. Then we have:

(a) $rank(D) \geq 3$.
(b) Infinitely many steps in D are outer steps.

Proof.

(b) Assume for a proof by contradiction that only finitely many steps in D are outer ones. W.l.o.g. we may further assume that no step in D is an outer one. Hence, for $s_1 \equiv C[\![t_1, \dots, t_n]\!]$ all reductions in D are inner ones and take place below one of the positions of the t_i's. Since D is infinite we conclude by the pigeon hole principle that at least one of the t_i's initiates an infinite derivation whose rank is smaller than $rank(D)$. But this is a contradiction to the minimality assumption concerning $rank(D)$.

[9] A rule is said to be duplicating if some variable occurs in the right side strictly more often than in the left side.

(a) For proving (a) we first remark that $rank(D) = 1$ is impossible.[10] For showing by contradiction that $rank(D) = 2$ is impossible, too, let us assume w.l.o.g. that all s_j's are top black and have $rank\, 2$. Then it is easy to see that for any step $s_j \xrightarrow{i}_{\mathcal{R}_1 \oplus \mathcal{R}_2} s_{j+1}$ we have $top(s_j) \equiv top(s_{j+1})$. Moreover, for $s_j \xrightarrow{o}_{\mathcal{R}_1 \oplus \mathcal{R}_2} s_{j+1}$ using an $\mathcal{R}_1$-rule we can show by induction over the depth of rewriting that $top(s_j) \xrightarrow{o}_{\mathcal{R}_1} top(s_{j+1})$ holds using the same $\mathcal{R}_1$-rule. Hence, the $(\mathcal{R}_1 \oplus \mathcal{R}_2)$-derivation D translates into the $\mathcal{R}_1$-derivation

$$D' : top(s_1) \rightarrow^*_{\mathcal{R}_1} top(s_2) \rightarrow^*_{\mathcal{R}_1} top(s_3) \rightarrow^*_{\mathcal{R}_1} \cdots .$$

which according to (b) contains infinitely many proper $\mathcal{R}_1$-steps. But this contradicts termination of $\mathcal{R}_1$. ∎

Note, that for unconditional TRSs in any minimal counterexample there are infinitely many inner reduction steps which are destructive at level 2. This property does not hold for CTRSs in general. To wit, consider the infinite $(\mathcal{R}_1 \oplus \mathcal{R}_2)$-derivation

$$f(G(a,b)) \rightarrow f(G(a,b)) \rightarrow f(G(a,b)) \rightarrow \ldots$$

in example 3 above where all reductions are outer steps.

In order to be able to state our main result we need the following definitions.

Definition 6. A CTRS $\mathcal{R}$ is said to be *termination preserving under non-deterministic collapses* if termination of $\mathcal{R}$ implies termination of $\mathcal{R} \oplus \{G(x,y) \rightarrow x, G(x,y) \rightarrow y\}$.

Definition 7. Let $\mathcal{R}_1$, $\mathcal{R}_2$ be two (finite) terminating disjoint CTRSs, $\mathcal{R} := \mathcal{R}_1 \oplus \mathcal{R}_2$ and $n \in N$ such that for every $s \in \mathcal{T}(\mathcal{F}_1 \uplus \mathcal{F}_2)$ with $rank(s) \leq n$ there is no infinite $\mathcal{R}$-derivation starting with s. Moreover, let $<_{\mathcal{T}(\mathcal{F}_1 \uplus \{A,G\})}$ be some arbitrary, but fixed total ordering on $\mathcal{T}(\mathcal{F}_1 \uplus \{A,G\})$. Then the $\mathcal{F}_2$- (or *white*) *abstraction* is defined to be the mapping

$$\Phi : \mathcal{GT}_{\oplus}^{\leq n} \uplus \{t \in \mathcal{GT}_{\oplus}^{n+1} | root(t) \in \mathcal{F}_1\} \longrightarrow \mathcal{T}(\mathcal{F}_1 \uplus \{A,G\})$$

given by

$$\Phi(t) := \begin{cases} t & \text{if } t \in \mathcal{T}(\mathcal{F}_1) \\ A & \text{if } t \in \mathcal{T}(\mathcal{F}_2) \\ C[\![\Phi(t_1), \ldots, \Phi(t_m)]\!] & \text{if } t \equiv C[\![t_1, \ldots, t_m]\!], root(t) \in \mathcal{F}_1 \\ CONS(SORT(\Phi^*(SUCC^{\mathcal{F}_1}(t)))) & \text{if } t \equiv C[\![t_1, \ldots, t_m]\!], root(t) \in \mathcal{F}_2 \end{cases}$$

with

$$\begin{aligned} SUCC^{\mathcal{F}_1}(t) &:= \{t' \in \mathcal{T}(\mathcal{F}_1 \uplus \mathcal{F}_2) | t \rightarrow^*_{\mathcal{R}} t', root(t') \in \mathcal{F}_1\}, \\ \Phi^*(M) &:= \{\Phi(t) | t \in M\} \quad \text{for} \quad M \subseteq dom(\Phi), \\ CONS(\langle\rangle) &:= A, \\ CONS(\langle s_1, \ldots, s_{k+1}\rangle) &:= G(s_1, CONS(\langle s_2, \ldots, s_{k+1}\rangle)) \quad \text{and} \\ SORT(\{s_1, \ldots, s_k\}) &:= \langle s_{\pi(1)}, \ldots, s_{\pi(k)}\rangle, \end{aligned}$$

[10] Here our general assumption that extra variables in the conditions of rules are forbidden is crucial! See also example 2.

such that $s_{\pi(j)} \leq_{\mathcal{T}(\mathcal{F}_1 \uplus \{A,G\})} s_{\pi(j+1)}$ for $1 \leq j < k$.

Intuitively, for computing $\Phi(t)$ one proceeds top-down in a recursive fashion. Top black layers are left invariant whereas (for the case of top black t) the principal top white subterms are transformed by computing for every such top white subterm the set of possible top black successors, abstracting the resulting terms recursively, sorting the resulting set of abstracted terms and finally constructing again an ordinary term by means of using the new constant symbol A (for empty arguments sets) and the new binary function symbol G (for non-empty argument sets). The sorting process and the total ordering involved here are due to some proof-technical subtleties which are explained and detailed in [4]. Note moreover, that the finiteness assumption in definition 7 concerning the involved CTRSs ensures that Φ is well-defined since the set of possible top black successors of top white principal aliens is finite. In fact, this finiteness assumption can be considerably weakened as shown in [4], [5].

Now we can formalize some properties of the above defined abstracting transformation which are technically crucial for proving the main structural result.

Lemma 8. *Let $\mathcal{R}_1$, $\mathcal{R}_2$, $\mathcal{R} := \mathcal{R}_1 \oplus \mathcal{R}_2$, n and Φ be given as in definition 7. Then, for any $s, t \in \mathcal{T}(\mathcal{F}_1 \uplus \mathcal{F}_2)$ with $rank(s) \leq n$ and $root(s) \in \mathcal{F}_2$ we have:*

$$s \rightarrow_{\mathcal{R}} t \quad \Longrightarrow \quad \Phi(s) \rightarrow^{*}_{\mathcal{R}'_2} \Phi(t),$$

where $\mathcal{R}'_2 := \mathcal{R}^{G}_{sub} := \{G(x,y) \rightarrow x, G(x,y) \rightarrow y\}$.

Proof. By an easy case analysis according to the definition of Φ. ■

Lemma 9. *Let $\mathcal{R}_1$, $\mathcal{R}_2$ and Φ be given as in definition 7. Then, Φ is rank decreasing, i.e. for any $s \in dom(\Phi) := \mathcal{GT}^{\leq n}_{\oplus} \uplus \{t \in \mathcal{GT}^{n+1}_{\oplus} | root(t) \in \mathcal{F}_1\}$ we have $rank(\Phi(s)) \leq rank(s)$.*

Proof. By an easy induction on $rank(s)$ using the definition of Φ. ■

Lemma 10. *Let $\mathcal{R}_1$, $\mathcal{R}_2$, $\mathcal{R} = \mathcal{R}_1 \oplus \mathcal{R}_2$, $\mathcal{R}'_2 = \mathcal{R}^{G}_{sub} = \{G(x,y) \rightarrow x,\ G(x,y) \rightarrow y\}$, $\mathcal{R}' = \mathcal{R}_1 \uplus \mathcal{R}'_2$ be given. Moreover, let n and the $\mathcal{F}_2$-abstraction Φ be given as in definition 7. Then, for any $s, t \in \mathcal{T}(\mathcal{F}_1 \uplus \mathcal{F}_2)$ with $rank(s) \leq n+1$, $root(s) \in \mathcal{F}_1$ and $s \rightarrow_{\mathcal{R}} t$ we have:*

(a) If $s \xrightarrow{o}_{\mathcal{R}} t$ using an $\mathcal{R}_1$-rule is not destructive at level 1 then $\Phi(s) \xrightarrow{o}_{\mathcal{R}'} \Phi(t)$ using the same $\mathcal{R}_1$-rule, and moreover this step is also not destructive at level 1.

(b) If $s \xrightarrow{o}_{\mathcal{R}} t$ using an $\mathcal{R}_1$-rule is destructive at level 1 then $\Phi(s) \xrightarrow{o}_{\mathcal{R}'} \Phi(t)$ using the same $\mathcal{R}_1$-rule, and moreover this step is also destructive at level 1.

(c) If $s \xrightarrow{i}_{\mathcal{R}} t$ is not destructive at level 2 then $\Phi(s) \xrightarrow{i}{}^{*}_{\mathcal{R}'_2} \Phi(t)$ with all steps not destructive at level 2.

(d) If $s \xrightarrow{i}_{\mathcal{R}} t$ is destructive at level 2 then $\Phi(s) \xrightarrow{i}{}^{+}_{\mathcal{R}'_2} \Phi(t)$ such that exactly one of these steps is destructive at level 2.

Proof. The general proof structure is as follows: We show by induction over the depth of rewriting steps the implication

$$s \rightarrow_{\mathcal{R}} t \quad \Longrightarrow \quad \Phi(s) \rightarrow^*_{\mathcal{R}'} \Phi(s) \, .$$

The case distinction for inner and outer as well as for destructive and non-destructive reduction steps proceeds according to the definition of Φ yielding a proof of (a)-(d).[11] ■

Finally we obtain the general structure theorem for CTRSs.

Theorem 11. *Let $\mathcal{R}_1, \mathcal{R}_2$ be two disjoint (finite) CTRSs which are both terminating such that their disjoint union $\mathcal{R}_1 \oplus \mathcal{R}_2$ is non-terminating. Then $\mathcal{R}_j$ is not termination preserving under non-deterministic collapses for some $j \in \{1,2\}$ and the other system $\mathcal{R}_k$, $k \in \{1,2\} \setminus \{j\}$, is collapsing. Moreover, the minimal rank of counterexamples in $\mathcal{R}_j \oplus \{G(x,y) \rightarrow x, G(x,y) \rightarrow y\}$ is less than or equal to the minimal rank of counterexamples in $\mathcal{R}_1 \oplus \mathcal{R}_2$.*

Proof. Let $\mathcal{R}_1, \mathcal{R}_2$ with $\mathcal{R} := \mathcal{R}_1 \oplus \mathcal{R}_2$ be given as stated above. We consider a minimal counterexample, i.e. an infinite $\mathcal{R}$-derivation

$$D: \quad s_1 \rightarrow s_2 \rightarrow s_3 \rightarrow \quad \ldots$$

of minimal rank, let's say $n+1$. W.l.o.g. we may assume that all the s_i's are top black, i.e. $\mathcal{F}_1$-rooted ground terms having *rank* $n+1$. Since the preconditions of definition 7 are satisfied we may apply the white ($\mathcal{F}_2$-) abstraction function Φ to the s_i's. Using lemma 10 we conclude that this yields an $\mathcal{R}'$-derivation

$$D': \quad \Phi(s_1) \rightarrow^* \Phi(s_2) \rightarrow^* \Phi(s_3) \rightarrow^* \quad \ldots$$

where $\mathcal{R}' := \mathcal{R}_1 \oplus \mathcal{R}'_2$ with $\mathcal{R}'_2 := \mathcal{R}^G_{sub} = \{G(x,y) \rightarrow x, G(x,y) \rightarrow y\}$. From lemma 10 we know that for any step $s_j \rightarrow s_{j+1}$ in D we have

$$s_j \xrightarrow{o}_{\mathcal{R}} s_{j+1} \Longrightarrow \Phi(s_j) \xrightarrow{o}_{\mathcal{R}'} \Phi(s_{j+1}) \, ,$$
$$s_j \xrightarrow{i}_{\mathcal{R}} s_{j+1} \Longrightarrow \Phi(s_j) \xrightarrow{i}{}^{*}_{\mathcal{R}'} \Phi(s_{j+1}).$$

Since according to lemma 5 (b) infinitely many steps in D are outer ones it follows that D' is an infinite $\mathcal{R}'$-derivation. But this means that $\mathcal{R}_1$ is not termination preserving under non-deterministic collapses. Moreover, under the assumption that $\mathcal{R}_2$ is non-collapsing the $\mathcal{F}_2$-abstraction of principal subterms of a minimal counterexample always yields the constant A which implies that the transformed infinite derivation is an $\mathcal{R}_1$-derivation contradicting termination of $\mathcal{R}_1$. Thus $\mathcal{R}_2$ must be collapsing. Lemma 9 finally implies $rank(D') \leq rank(D)$ which finishes the proof. ■

[11] Note that the assumption of having no extra variables in the conditions is important because this would cause problems with the *rank* of instantiated condition terms. In that case substitution of the extra variables in the condition part which are implicitly existentially quantified might yield terms of arbitrarily high *rank* which in turn might prevent Φ from being well-defined for these instantiated terms.

This general result can now be exploited for deriving a couple of sufficient criteria for modular termination of CTRSs. For instance, termination (and hence also completeness) is modular for the class of (finite) CTRSs which are termination preserving under non-deterministic collapses as well as for the class of collapse-free CTRSs[12] (cf. [12]). It is also modular for the class of (finite) CTRSs which are non-deterministically collapsing where a CTRS $\mathcal{R}^{\mathcal{F}}$ is said to be *non-deterministically collapsing* if some term can be reduced to two distinct variables.

Definition 12. Let $\mathcal{R}^{\mathcal{F}}$ be a CTRS and $f \in \mathcal{F}, \mathcal{F}' \subseteq \mathcal{F}$. Then, $\mathcal{R}^{\mathcal{F}}$ is said to be *f-simply terminating* if $\mathcal{R}^{\mathcal{F}} \cup \mathcal{R}^{f}_{sub}$ with $\mathcal{R}^{f}_{sub} := \{f(x_1, \ldots, x_j, \ldots, x_n) \rightarrow x_j \,|\, 1 \leq j \leq n\}$ is terminating. $\mathcal{R}^{\mathcal{F}}$ is *$\mathcal{F}'$-simply terminating* if $\mathcal{R}^{\mathcal{F}} \cup \bigcup_{f \in \mathcal{F}'} \mathcal{R}^{f}_{sub}$ is terminating. $\mathcal{R}^{\mathcal{F}}$ is *simply terminating* if $\mathcal{R}^{\mathcal{F}}$ is $\mathcal{F}$-simply terminating (see [10]), i.e. if $\mathcal{R}^{\mathcal{F}} \cup \bigcup_{f \in \mathcal{F}} \mathcal{R}^{f}_{sub}$ is terminating.

Clearly, simple termination implies termination for CTRSs. Moreover, we easily obtain the following results.

Lemma 13. *Let $\mathcal{R}^{\mathcal{F}}$ be a f-simply terminating CTRS for some $f \in \mathcal{F}$ of arity greater than 1. Then $\mathcal{R}^{\mathcal{F} \cup \mathcal{F}'}$ is $(\{f\} \cup \mathcal{F}')$-simply terminating for any $\mathcal{F}'$ with $\mathcal{F}' \cap \mathcal{F} = \emptyset$.*

Corollary 14. *Let $\mathcal{R}_1^{\mathcal{F}_1}$, $\mathcal{R}_2^{\mathcal{F}_2}$ be two (finite) disjoint CTRSs with $f_1 \in \mathcal{F}_1, f_2 \in \mathcal{F}_2$ of arity greater than 1 such that $\mathcal{R}_i$ is f_i-simply terminating for $i = 1, 2$. Then the disjoint union $\mathcal{R}_1 \oplus \mathcal{R}_2$ is (f_1- and f_2-simply) terminating, too.*

By specializing corollary 14 we finally obtain

Theorem 15. *Simple termination is modular for the class of (finite) CTRSs $\mathcal{R}^{\mathcal{F}}$ such that there exists at least one function symbol $f \in \mathcal{F}$ of arity greater than 1.*

Kurihara & Ohuchi [10] have shown that for finite unconditional TRSs simple termination can be characterized by means of simplification orderings. This yields in slightly generalized form:

Lemma 16. *A (possibly infinite) TRS $\mathcal{R}^{\mathcal{F}}$ over some finite signature $\mathcal{F}$ is ($\mathcal{F}-$) simply terminating if and only if there exists a simplification ordering $\succ$ with $l \succ r$ for every rule $l \rightarrow r \in \mathcal{R}^{\mathcal{F}}$.*

An analogous characterization of simple termination for CTRSs does not seem to be possible in a straightforward manner. Obviously, any (finite) simply terminating CTRS can be shown to be terminating by some simplification ordering, but not vice-versa in general. To see this, let us have again a look on example 3.

[12] Note that for collapse-free CTRSs the finiteness assumption in definition 7 is not necessary for ensuring well-definedness of the abstraction function Φ.

Example 4. (example 3 continued)
Consider $\mathcal{R}_1 = \{\ x \downarrow a \ \wedge \ x \downarrow b \implies f(x) \to f(x)\ \}$ over the extended signature $\mathcal{F}_1 = \{a, b, f, G\}$, with G binary. Here the reduction relation induced by $\mathcal{R}_1$ is empty, hence any simplification ordering trivially suffices for ensuring termination of $\mathcal{R}_1$. But, due to the non-termination of $\mathcal{R}_1 \cup \{f(x) \to x, G(x,y) \to x, G(x,y) \to y\}$, $\mathcal{R}_1$ is not simply terminating .

In order to provide a better intuition about simple termination of CTRSs we relate this notion to other ones. For that purpose we need the following definitions and abbreviations:

A (finite) CTRS $\mathcal{R}$ is *simplifying (SIMP)* (cf. [8]) if there exists a simplification ordering $>$ satisfying $l > r, s_i, t_i, 1 \leq i \leq n$ for all rules $s_1 \downarrow t_1 \wedge \ldots \wedge s_n \downarrow t_n \implies l \to r \in \mathcal{R}$. $\mathcal{R}$ is *terminating (or strongly normalizing) with a simplification ordering (SOSN)* if there exists a simplification ordering $>$ with $\to_{\mathcal{R}} \subseteq >$. It is *simply reducing (SRED)* if there exists a simplification ordering $>$ with $l > r$ for every rule $s_1 \downarrow t_1 \wedge \ldots \wedge s_n \downarrow t_n \implies l \to r \in \mathcal{R}$, i.e. if the corresponding unconditional TRS can be oriented by some simplification ordering. For terminating (or strongly normalizing), decreasing and simply terminating CTRSs we use the abbreviations *SN*, *DEC* and *SISN*, respectively. Then we obtain the following easily verifiable result.

Lemma 17. *For any (finite) CTRS $\mathcal{R}$ we have:*
(a) $SIMP(\mathcal{R}) \implies SRED(\mathcal{R}) \implies SISN(\mathcal{R}) \implies SOSN(\mathcal{R}) \implies SN(\mathcal{R})$.
(b) All implications in (a) are proper.
(c) $SIMP(\mathcal{R}) \implies DEC(\mathcal{R}) \implies SN(\mathcal{R})$.
(d) The properties $SRED(\mathcal{R}), SISN(\mathcal{R})$ and $SOSN(\mathcal{R})$ are incomparable with $DEC(\mathcal{R})$.

Note in particular, that every simplifying CTRS is clearly simply terminating but not vice-versa in general. Simple termination even does not imply decreasingness as shown e.g. by the CTRS consisting of the single rule $a \downarrow b \implies a \to a$.

Finally us mention some aspects not yet handled. Firstly our modularity results have only been proved for join CTRS. But it should (at least be partially) possible to extend them to the semi-equational case. Note that again new subtle effects may occur in semi-equational CTRSs. For instance, lemma 5 (a) does not hold any more. To wit, consider

Example 5. $\mathcal{R}_1 = \{\ b \stackrel{*}{\leftrightarrow} c \implies a \to a\ \}, \mathcal{R}_2 = \{G(x,y) \to x, G(x,y) \to y\}$.

Both $\mathcal{R}_1$ and $\mathcal{R}_2$ are terminating (and decreasing) but $\mathcal{R}_1 \oplus \mathcal{R}_2$ is non-terminating. We even have a counterexample of *rank* 1, namely $a \to_{\mathcal{R}_1 \oplus \mathcal{R}_2} a \to_{\mathcal{R}_1 \oplus \mathcal{R}_2} a \ldots$ because we have $b \leftarrow G(b,c) \to c$ in $\mathcal{R}_2$, hence $b \stackrel{*}{\leftrightarrow}_{\mathcal{R}_2} c$. Considered as join CTRS, $\mathcal{R}_1 \oplus \mathcal{R}_2$ is terminating, however!

Secondly, the variable conditions required for CTRSs should be investigated in more detail. It should be possible to allow extra variables in the conditions. Extra variables in right hand sides seem to be much more difficult to handle than extra variables only in the conditions (see the discussion in [13]).

Moreover, the relationship between the results in [12] and ours remains to be clarified.

Acknowledgements: I would like to thank Aart Middeldorp for fruitful discussions on modularity topics and on some subtleties of CTRSs.

References

1. N. Dershowitz. Orderings for term-rewriting systems. *Theoretical Computer Science*, pages 279–301, 1982.
2. N. Dershowitz and M. Okada. A rationale for conditional equational programming. *Theoretical Computer Science*, 75:111–138, 1990.
3. N. Dershowitz, M. Okada, and G. Sivakumar. Canonical conditional rewrite systems. In E. Lusk and R. Overbeek, editors, *Proc. of the 9th Int. Conf. on Automated Deduction*, volume 310 of *Lecture Notes in Computer Science*, pages 538–549. Springer, 1988.
4. B. Gramlich. A structural analysis of modular termination of term rewriting systems. SEKI Report SR-91-15, Dept. of Comp. Science, Univ. of Kaiserslautern, 1991.
5. B. Gramlich. Generalized sufficient conditions for modular termination of rewriting. In *Proc. of 3rd Int. Conf. on Algebraic and Logic Programming, Pisa, Italy*, Lecture Notes in Computer Science. Springer-Verlag, 1992. to appear.
6. B. Gramlich. Relating innermost, weak, uniform and modular termination of term rewriting systems. In A. Voronkov, editor, *Conference on Logic Programming and Automated Reasoning, St. Petersburg*, volume 624 of *Lecture Notes in Artificial Intelligence*, pages 285–296. Springer-Verlag, 1992.
7. J.-P. Jouannaud and B. Waldmann. Reductive conditional term rewriting systems. In *Proceedings of the 3rd IFIP Working Conference on Formal Description of Programming Concepts*, pages 223–244. North-Holland, 1986.
8. S. Kaplan. Simplifying conditional term rewriting systems: Unification, termination and confluence. *Journal of Symbolic Computation*, 4:295–334, 1987.
9. J.W. Klop. Term rewriting systems. In S. Abramsky, D. Gabbay, and T. Maibaum, editors, *Handbook of Logic in Computer Science*, volume I. Oxford University Press, 1990.
10. M. Kurihara and A. Ohuchi. Modularity of simple termination of term rewriting systems. *Journal of IPS, Japan*, 34:632–642, 1990.
11. A. Middeldorp. A sufficient condition for the termination of the direct sum of term rewriting systems. In *Proceedings of the 4th IEEE Symposium on Logic in Computer Science*, pages 396–401, Pacific Grove, 1989.
12. A. Middeldorp. Termination of disjoint unions of conditional term rewriting systems. Technical Report CS-R8959, Centre for Mathematics and Computer Science, Amsterdam, 1989.
13. A. Middeldorp. *Modular Properties of Term Rewriting Systems*. PhD thesis, Centre for Mathematics and Computer Science, Amsterdam, 1990.
14. A. Middeldorp. Completeness of combinations of conditional constructor systems. Advanced Research Laboratory, Hitachi Ltd., Hatoyama, Japan, draft version, see also this volume, February 1992.

15. A. Middeldorp and Y. Toyama. Completeness of combinations of constructor systems. In R.V. Book, editor, *Proc. of the 4th Int. Conf. on Rewriting Techniques and Applications*, volume 488 of *Lecture Notes in Computer Science*, pages 174–187. Springer, 1991.
16. M. Rusinowitch. On termination of the direct sum of term rewriting systems. *Information Processing Letters*, 26:65–70, 1987.
17. Y. Toyama. On the Church-Rosser property for the direct sum of term rewriting systems. *Journal of the ACM*, 34(1):128–143, 1987.
18. Y. Toyama, J.W. Klop, and H.P. Barendregt. Termination for the direct sum of left-linear term rewriting systems. In N. Dershowitz, editor, *Proc. of the 3rd Int. Conf. on Rewriting Techniques and Applications*, volume 355 of *Lecture Notes in Computer Science*, pages 477–491. Springer, 1989.

Termination of Combined (Rewrite and λ-Calculus) Systems

Carlos Loria-Saenz* and Joachim Steinbach**

Universität Kaiserslautern
Fachbereich Informatik, Postfach 3049
6750 Kaiserslautern (Germany)

Abstract. We consider the termination problem of combinations of term-rewriting systems and the λ-calculus employing standard methods of the theory of term rewriting systems in contrast to (extensions of) Tait-Girard's technique. In particular for some class of higher-order rule systems, we explicitly construct a well-founded ordering over λ-terms whose combination with the β-reduction is terminating. Then, by embedding the higher-order rewriting relation into this ordering, we can prove termination for combinations of such a class of higher-order term rewriting systems and the λ-calculus.

1 Introduction

Term rewriting systems (TRS) and the λ-calculus are important abstract models for computation and logic and their combination can be used to obtain a general and rich representation of functional programming languages. As an illustration, using first-order rule systems, we can specify usual arithmetic operations such as the addition on natural numbers. With the help of higher-order rules, we can additionally specify the generating function *iter*[1] which allows to define many other functions over lists by only instantiating the variables F and x (like *length*):

$$x + 0 \rightarrow x \quad (1) \qquad iter(F, x, []) \rightarrow x \quad (3)$$

$$s(x) + y \rightarrow s(x + y) \quad (2) \qquad iter(F, x, y :: l) \rightarrow iter(F, (Fxy), l) \quad (4)$$

$$length(l) \rightarrow iter(\lambda v.\lambda w.s(v), 0, l) \quad (5)$$

In order to use such systems we need to combine the corresponding relations. One way to achieve that, is to take the union of the R-reduction (the application of rules to algebraic or higher-order terms) and the standard λ-reduction (the application to λ-terms). Thus, the *termination problem* in this case is identical

* Supported by DAAD-ITCR, e-mail: loria@informatik.uni-kl.de
** Supported by SFB 314 (D4-Projekt), e-mail: steinba@informatik.uni-kl.de

[1] (F is a higher-order variable, x, y and l are first-order ones, $[]$ represents nil and $::$ the cons operation)

to the termination proof of the union between a terminating R-reduction and the typed λ-reduction. The results of this paper refer to this approach. Note that the termination of the combination has been proved by means of an extension of the Tait-Girard technique based on computability predicates (see [BG91] and [JO91, DO90]).

In the following, we assume familiarity with all the notations and definitions used in connection with TRSs and the λ-calculus. For simplicity, we consider the denominated ***simply-typed lambda-calculus***. The relation $\Rightarrow_\beta$ denotes the β-reduction. Note that for the typed λ-calculus, a well-known termination result of $\Rightarrow_\beta$ holds (see [HS86] for a proof using the mentioned Tait-Girard technique). We also assume that the β-normal forms are given in ***long-β-normal form***. The combination of a rule system R and the (typed) λ-calculus is represented by the union of the corresponding reduction relations and denoted by $\Rightarrow_{\beta,R}$. $FV(M)$ denotes the set of free variables occurring in the well-typed term M. Usually, we employ the following conventions: x, y, z are first-order variables, s, t, l, r are first-order terms. X, Y, Z are higher-order variables. M, N, P are well-typed terms, σ is a well-typed substitution.

2 An Ordering for Higher-Order Rewriting and the λ-Calculus

The problem of proving termination of higher-order combinations is more difficult than that of the first-order case: arbitrary combinations of terminating higher-order systems are not terminating in general. What we do in the higher-order case is to explicitly construct a well-founded ordering over λ-terms such that the combination of this ordering with the β-reduction is terminating. Then, to achieve termination, the rewriting relation must be included into this ordering.

2.1 Sketch of an Ordering

To construct an ordering over λ-terms, we extend the usual idea for first-order systems using orderings based on precedences (especially the recursive path ordering RPO of [Der82]). We extend these orderings to higher-order terms. We illustrate the basic ideas of the construction of the ordering by an example. Suppose, we want to compare the terms $map(X, y :: L)$ and $X(y) :: map(X, L)$. To do that, we proceed as in the first-order case using an RPO, say $>_1$. Assume further that we have a quasi-ordering $\gtrsim_\mathcal{F}$ over the operators of $\mathcal{F}$. In order to compare higher-order terms, we apply some **term transformations** (see [Der82]). The main feature of our technique is the following one: An operator is closely connected with the higher-order operators of its subterms. This way, operators are interpreted as special terms[2]. This leads to an extended set of operators denoted by $\mathcal{F}_0$. Then, we construct a quasi-ordering $\gtrsim_0$ over $\mathcal{F}_0$. Thus, $>_0$ (the strict part of $\gtrsim_0$) is intended to operate as a precedence ordering for our RPO $>_1$. For instance, by requiring that $map >_\mathcal{F} ::$ holds w.r.t. the given precedence

[2] For example, the extended operator for $map(X, y :: L)$ is of the form $map_{\{X\}}$.

$>_{\mathcal{F}}$, we must prove $map(X, y :: L) >_1 X(y)$ and $map(X, y :: L) >_1 map(X, L)$ when proving $map(X, y :: L) >_1 X(y) :: map(X, L)$. Now, in order to compare $map(X, y :: L)$ and $X(y)$, we transform the terms by considering $map_{\{X\}}$ as the (extended) operator of $map(X, y :: L)$ and X as the operator of $X(y)$[3]. Now, we compare these subscripted operators as terms by using the extended precedence ordering $>_0$. This way, $map_{\{X\}} >_0 X$ holds (because of the subterm property of $>_0$). Furthermore, $map(X, y :: L) >_1 y$ holds. Thus, we can prove $map(X, y :: L) >_1 X(y)$. Similarly, we prove $map(X, y :: L) >_1 map(X, L)$ since the corresponding extended operators are the same. Then, we compare the arguments as multisets using the multiset extension of $>_1$.

Now, we describe the extended operators for λ-abstractions and how to compare them. For example, we want to compare $length(l)$ and $iter(\lambda v\, w.s(v), 0, l)$. Suppose $length >_{\mathcal{F}} iter, s$. As a subgoal of this proof, we need to compare $length(l)$ and $\lambda v\, w.s(v)$. The extended operator for $length(l)$ is $length_{\{\}}$. For the λ-abstraction, we interpret the λ-term as a unary operator $\lambda v\, w$ evaluating to $s(v)$. Considering this interpretation, the extended operator for $\lambda v\, w.s(v)$ is $\lambda v\, w.s_{\{\}}$. Again, we compare these operators as terms using $>_0$ together with $length >_{\mathcal{F}} iter$. To do that, we admit $length_{\{\}} >_0 \lambda v\, w$ whenever the type of the bounded variables v and w is *simpler* than the type of $length$ ($length : list(nat) \rightarrow nat$ and $v, w : nat$). Note that bounded variables are interpreted as constants. Now, we can prove $length_{\{\}} >_0 s_{\{\}}$ since $length >_{\mathcal{F}} s$. The remaining part of the proof can be realized similarly to the above technique[4].

Thus, our ordering $>_1$ is defined over terms by extending the operators[5]. The set of such terms is denoted by $\mathcal{T}(\mathcal{F}_0, \mathcal{X})$ (or $\mathcal{T}(\mathcal{F}_0)$ for closed terms). As we can see from this description, the ordering $>_1$ is defined over β-normalized terms, only. Therefore, we extend $>_1$ to $>_2$ which is defined over non-normalized terms by comparing the unique β-normal forms. Hence, $M >_2 N$ if, and only if, $\beta\text{-}nf(M) >_1 \beta\text{-}nf(N)$. We can show that $>_2 \cup \Rightarrow_\beta$ is terminating over $\mathcal{T}(\mathcal{F}, \mathcal{X})$ ($>_2$ is a well-founded partial ordering over $\mathcal{T}(\mathcal{F}, \mathcal{X})$). The hierarchical construction of $>_2$ is sketched in figure 1.

2.2 Embedding the Rewriting Relation into $>_2$

Unfortunately, the ordering $>_2$ cannot directly be used as a reduction ordering. In general, $>_2$ is neither compatible with the term structure nor with substitutions. However, we are able to define an interesting class of terms for which these properties can be satisfied. We say that a β-normalized term $M \in \mathcal{T}(\mathcal{F}, \mathcal{X})$ is *simple* if no subterm of M is of the form $X(M_1, \ldots, M_n)$ where X is a higher-order variable and $n > 0$. A rule $M \rightarrow N$ is called simple whenever M is a non-variable simple term and N is a β-normalized term. A TRS is simple if each rule is

[3] The index $\{X\}$ of *map* characterizes the dependence of the operator *map* upon the higher-order subterm X.

[4] In order to prove the termination of *iter* (contained in the example of the introduction), an extension of $>_1$ by integrating a status function (and requiring right-to-left status for *iter*) can be used.

[5] A detailed formal definition of $>_1$ is contained in the appendix.

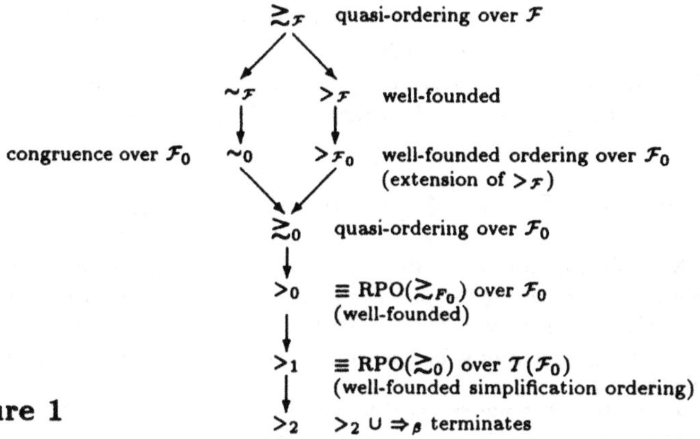

Figure 1

simple. Note that each first-order term is simple. The rules showed by the example of section 1 are also simple. In contrast, the rule $f(X(y), X(y)) \rightarrow f(X(y), y)$ is not simple[6]. Now, we can prove the following result.

Theorem 2.1 *Let M be a simple term and N be a β-normalized term. Suppose that M and N have the same type α. Then, the following conditions hold whenever $M >_2 N$:*

- *$M\sigma >_2 N\sigma$ for σ over $FV(M) \cup FV(N)$.*
- *$M'[M]_p >_2 M'[N]_p$ for each term M' and for each non-vanishing position $p \in \mathcal{P}os(M')$ such that $M'|_p : \alpha$.*

Using the above result and requiring the higher-order relation $\Rightarrow_R$ to be applied only at the so-called non-vanishing positions (positions belonging to the β-normal form of the term), we can guarantee termination of the combined relation $\Rightarrow_{\beta,R}$.

Corollary 2.1 *Let R be a simple TRS such that $M >_2 N$ holds for each rule $M \rightarrow N \in R$. Further, suppose that the $\Rightarrow_R$-reduction is applied only at non-vanishing positions. Then, the relation $\Rightarrow_{\beta,R}$ is terminating.*

References

[BG91] V. Breazu-Tannen and J. Gallier. Polymorphic rewriting conserves algebraic strong normalization. *TCS*, 83:3–28, 1991.

[Der82] N. Dershowitz. Orderings for term-rewriting systems. *TCS*, 17(3):279–301, 1982.

[6] Note that the rule is terminating since $f(X(y), X(y)) >_2 f(X(y), y)$ holds. However, we cannot prove the termination of the combined relation $\Rightarrow_{\beta,R}$.

[DO90] N. Dershowitz and M. Okada. A rationale for conditional equational programming. *TCS*, 75:111–138, 1990.

[HS86] J. R. Hindley and J. P. Seldin. *Introduction to Combinators and λ-Calculus.* London Mathematical Soc., 1986.

[JO91] J.-P. Jouannaud and M. Okada. A computation model for executable higher-order algebraic specification languages. *LICS*, pages 350–361, 1991.

A Definitions of Orderings and Congruences

Definition A.1 (Congruence $\sim_0$ over $\mathcal{T}(\mathcal{F}_0)$) *We define the congruence $\sim_0$ as the smallest congruence containing the following rules (we use subscripted symbols such as G_L to denote extended operators. L represents a list of extended operators).*

- $\lambda\mathbf{x}.G_L \sim_0 \lambda\mathbf{y}.H_M$ *if* $\mathbf{x}$ *and* $\mathbf{y}$ *are permutatively identical and* $G_L \sim_0 H_M$
- $\lambda\mathbf{x}.G_L \sim_0 H_M$ *if*
  - $x_i : \alpha_i$ *for each component* $x_i \in \mathbf{x}$, $i = 1, \ldots, n$ *and*
  - $H : \alpha_1, \ldots, \alpha_n \to \beta$ *and*
  - $G_L \sim_0 H_M$
- $G_L \sim_0 H_M$ *if* $G \sim_{\mathcal{F}} H$ *and* $L \sim_0 M$

where G and H are operators in $\mathcal{F}$ or higher-order variables.

Definition A.2 (Ordering $>_0$ over $\mathcal{F}_0$) *To prove $G_1 >_0 G_2$ match the terms G_1 and G_2 according to the following cases:*

- $\lambda\mathbf{x}.G_L >_0 \lambda\mathbf{y}.H_M$ *if* $\mathbf{x}$ *and* $\mathbf{y}$ *are permutatively identical and* $G_L >_0 H_M$
- $\lambda\mathbf{x}.G_L >_0 H_M$ *if* $G_L \gtrsim_0 H_M$
- $G_L >_0 \lambda\mathbf{x}.H_M$ *if*
  - $x_i : \alpha_i$ *for each component* $x_i \in \mathbf{x}$, $i = 1, \ldots, n$ *and* $H : \alpha_1, \ldots, \alpha_n \to \beta$ *and*
  - $G_L >_0 H_M$
- $G_L >_0 N$ *if* $\exists N' \in L$ *s.t.* $N' \gtrsim_0 N$
- $G_L >_0 H_M$ *if* $G >_{\mathcal{F}} H$ *and* $G_L >_0 N', \forall N' \in M$ *or* $G \sim_{\mathcal{F}} H$ *and* $L >_0^{mul} M$

where $\gtrsim_{\mathcal{F}}$ represents the given quasi-ordering over $\mathcal{F}$ and $>_0^{mul}$ the multiset extension of $>_0$.

Definition A.3 (Ordering $>_1$ over $\mathcal{T}(\mathcal{F}_0)$) *To prove $M >_1 N$ match the terms M and N according to the following cases:*

- $\lambda\mathbf{x}.G_L(M_1, \ldots, M_m) >_1 \lambda\mathbf{y}.H_M(N_1, \ldots, N_n)$ *if*
  - $G_L(M_1, \ldots, M_m) \gtrsim_1 \lambda\mathbf{y}.H_M(N_1, \ldots, N_n)$ *or*
  - $\lambda\mathbf{x}.G_L >_0 \lambda\mathbf{y}.H_M$ *and* $\lambda\mathbf{x}.G_L(M_1, \ldots, M_m) >_1 N_k, \forall k$ *or*
  - $\lambda\mathbf{x}.G_L \sim_0 \lambda\mathbf{y}.H_M$ *and* $\{M_1, \ldots, M_m\} >_1^{mul} \{N_1, \ldots, N_n\}$
- $\lambda\mathbf{x}.G_L(M_1, \ldots, M_m) >_1 H_M(N_1, \ldots, N_n)$ *if*
  - $G_L(M_1, \ldots, M_m) \gtrsim_1 H_M(N_1, \ldots, N_n)$ *or*
  - $\lambda\mathbf{x}.G_L >_0 H_M$ *and* $\lambda\mathbf{x}.G_L(M_1, \ldots, M_m) >_1 N_k, \forall k$ *or*
  - $\lambda\mathbf{x}.G_L \sim_0 H_M$ *and* $\{M_1, \ldots, M_m\} >_1^{mul} \{N_1, \ldots, N_n\}$
- $G_L(M_1, \ldots, M_m) >_1 \lambda\mathbf{y}.H_M(N_1, \ldots, N_n)$ *if*
  - $M_k \gtrsim_1 \lambda\mathbf{y}.H_M(N_1, \ldots, N_n)$ *for some* k *or*
  - $G_L >_0 \lambda\mathbf{y}.H_M$ *and* $G_L(M_1, \ldots, M_m) >_1 N_k, \forall k$ *or*
  - $G_L \sim_0 \lambda\mathbf{y}.H_M$ *and* $\{M_1, \ldots, M_m\} >_1^{mul} \{N_1, \ldots, N_n\}$
- $G_L(M_1, \ldots, M_m) >_1 H_M(N_1, \ldots, N_n)$ *if*
  - $M_k \gtrsim_1 H_M(N_1, \ldots, N_n)$ *for some* k *or*
  - $G_L >_0 H_M$ *and* $G_L(M_1, \ldots, M_m) >_1 N_k, \forall k$ *or*
  - $G_L \sim_0 H_M$ *and* $\{M_1, \ldots, M_m\} >_1^{mul} \{N_1, \ldots, N_n\}$

where $\sim_1$ is a natural extension of $\sim_0$ to terms over $\mathcal{T}(\mathcal{F}_0)$.

Type removal in term rewriting

H. Zantema
Utrecht University, Department of Computer Science
P.O. box 80.089, 3508 TB Utrecht, The Netherlands
e-mail: hansz@cs.ruu.nl

Abstract. A property of many-sorted term rewriting systems is called persistent if it is not affected by removing the corresponding typing restriction. Persistency turns out to be a generalization of direct sum modularity. It is a more powerful tool for proving confluence and normalization properties. Strong normalization is persistent for the class of term rewriting systems for which not both duplicating rules and collapsing rules occur, generalizing a similar result of Rusinowitch for modularity. This result can be used for simplifying proofs on undecidability.

Introduction

Usually term rewriting systems are one-sorted: all terms and subterms are of the same type. The notion of term rewriting systems extends in a natural way to many-sorted terms. In this case a set of sorts, a set of operation symbols and a set of variable symbols is given. Each operation symbol has a sort (one of the sorts) and an arity. This arity is not simply a number, but a sequence of sorts. Each variable symbol has a sort. For every sort the set of terms of that sort is defined inductively in a straightforward way. This definition of terms is standard in the theory of algebraic specifications ([2]).

A many-sorted term rewriting system (TRS) is a set of pairs $l \to r$, where l and r are terms of the same sort. As usual l is not allowed to be a single variable and all variables in r also occur in l. Contexts, substitutions and the reduction relation are defined as expected, inducing definitions of normal forms, weak and strong normalization and confluence.

This is a very natural definition. One important application of TRS theory is in algebraic specifications, and the nature of algebraic specifications is many-sorted. The above notion of a many-sorted TRS is exactly what is needed for automatic implementation of many-sorted algebraic specifications and for applying Knuth-Bendix completion for the many-sorted case. Further, many variations of the λ-calculus are closely related to many-sorted TRS's.

Any many-sorted TRS is trivially mapped to a one-sorted TRS by removing all sort information and keeping the same reduction rules. The many-sorted terms can be considered as a subset of all (one-sorted) terms: the terms that satisfy the typing restriction. We call a property of many-sorted TRS's *persistent* if a many-sorted TRS has the property if and only if its adjoined one-sorted TRS has the property.

In this paper we show that every reasonable persistent property of term rewriting systems is also a modular property, i.e., invariant under taking disjoint unions of TRS's. In our view persistency is more basic than modularity; modularity can be considered as a particular case of persistency, while persistency turns out to be a more powerful tool for proving confluence or strong normalization than modularity.

Since strong normalization is not modular we conclude that it is neither persistent. Restricting to TRS's in which not both duplicating rules and collapsing rules occur, strong normalization is persistent, generalizing the similar result ([7]) for modularity.

Many-sorted term rewriting

First we introduce some standard terminology. Let S be a finite set representing the set of types or sorts. An *S-sorted set* $\mathcal{X}$ is defined to be a family of sets $(\mathcal{X}_s)_{s \in S}$. For S-sorted sets $\mathcal{X}$ and $\mathcal{Y}$ an *S-sorted map* $\phi : \mathcal{X} \to \mathcal{Y}$ is defined to be a family of maps $(\phi_s : \mathcal{X}_s \to \mathcal{Y}_s)_{s \in S}$.

By S^* we denote the set of finite sequences of elements of S, including the empty sequence. Let $\mathcal{F}$ be a set of symbols, called *operation symbols*. For every operation symbol an *arity* and a *sort* is given, described by functions

$$ar : \mathcal{F} \to S^* \quad \text{and} \quad st : \mathcal{F} \to S.$$

Let $\mathcal{X}$ be an S-sorted set of symbols, called *variables*. We define the S-sorted set $\mathcal{T}(\mathcal{F}, \mathcal{X})$ of *terms* inductively by

- $\mathcal{X}_s \subseteq \mathcal{T}(\mathcal{F}, \mathcal{X})_s$ for $s \in S$;
- $f(t_1, \ldots, t_n) \in \mathcal{T}(\mathcal{F}, \mathcal{X})_s$ for $f \in \mathcal{F}$ with $ar(f) = (s_1, \ldots, s_n)$ and $st(f) = s$, and $t_i \in \mathcal{T}(\mathcal{F}, \mathcal{X})_{s_i}$ for $i = 1, \ldots, n$.

An S-sorted *term rewriting system* (TRS) is defined to be an S-sorted set of rules: a rule $l \to r$ of sort s consists of two terms l and r of sort s for which l is no variable and r contains no variables that do not occur in l. The corresponding reduction relation is defined as expected.

Persistency

By removing all sort information every many-sorted term can be mapped to a one-sorted term as follows. Let $\mathcal{F}'$ be the set of symbols obtained by adding a prime (′) to every symbol of $\mathcal{F}$. For $f \in \mathcal{F}$ with $ar(f) = (s_1, \ldots, s_n)$ we define the arity of $f' \in \mathcal{F}'$ to be n. In this way $\mathcal{F}'$ defines a one-sorted signature. Since there is only one sort there is no need for an explicit notation for the sort. We choose $\mathcal{X}' = \bigcup_{s \in S} \mathcal{X}_s$ to be the set of one-sorted variable symbols; recall that the sets $\mathcal{X}_s$ are assumed to be disjoint.

Every term over $\mathcal{F}$ of any sort can be mapped to a term over $\mathcal{F}'$ by adding prime symbols to all operation symbols. This map

$$\Theta : \bigcup_{s \in S} \mathcal{T}(\mathcal{F}, \mathcal{X})_s \to \mathcal{T}(\mathcal{F}', \mathcal{X}'),$$

is inductively defined by

- $\Theta(x) = x$ for every $x \in \mathcal{X}_s$ for every $s \in S$;
- $\Theta(f(t_1,\ldots,t_n)) = f'(\Theta(t_1),\ldots,\Theta(t_n))$ for all $f \in \mathcal{F}$ and terms $t_1,\ldots,t_n$ of the right sort.

The set $\bigcup_{s\in S} \mathcal{T}(\mathcal{F},\mathcal{X})_s$ can be considered as a subset of $\mathcal{T}(\mathcal{F}',\mathcal{X}')$: the set of well-typed terms.

The map Θ is defined on TRS's in an obvious way: for any many-sorted TRS R the one-sorted TRS $\Theta(R)$ is defined to consist of the rules $\Theta(l) \rightarrow \Theta(r)$ for the rules $l \rightarrow r$ from R. One easily observes that for $t_1, t_2 \in \mathcal{T}(\mathcal{F},\mathcal{X})_s$:

$$t_1 \rightarrow_{R,s} t_2 \iff \Theta(t_1) \rightarrow_{\Theta(R)} \Theta(t_2).$$

A property of binary relations is called *component closed* if it holds for $(\mathcal{T},\rightarrow)$ if and only if it holds for all of its components. Here a component is defined to be an equivalence class of the equivalence relation generated by $\rightarrow$. For example, confluence, and weak and strong normalization are component closed properties.

A property of binary relations is called *persistent* if for every many-sorted TRS R the property holds for R if and only if it holds for $\Theta(R)$.

The notion of persistency is closely related to the notion of *modularity* as it is discussed in [5, 6]. Modularity has been extensively studied in [7, 9, 8, 6]. A (component closed) property of one-sorted TRS's is called *modular* if for every pair of (one-sorted) TRS's R_1 and R_2 with disjoint sets of operation symbols the property holds for both R_1 and R_2 if and only if it holds for $R_1 \oplus R_2$. Here $R_1 \oplus R_2$ denotes the union of both TRS's; it is a one-sorted TRS over the disjoint union of the sets of operation symbols.

Theorem 1 *Every component closed persistent property is modular.*

Proof: Let $\mathcal{P}$ be any component closed persistent property. Let R_1 and R_2 be one-sorted TRS's with disjoint sets of operation symbols. We define a two-sorted TRS R as follows. The sorts are denoted by s_1 and s_2, the operation symbols are the operation symbols from both R_1 and R_2. The arity of an operation symbol from R_1 of arity n is defined to be $(s_1, s_1, \ldots, s_1)$, and its sort is defined to be s_1. Similarly the arities and sorts of the operation symbols from R_2 are defined to consist solely from s_2. Now the terms of sort s_1 of R correspond one-to-one to the terms of R_1. Further the reduction relation at sort s_1 of R corresponds one-to-one to the reduction relation of R_1. The same holds for '1' replaced by '2'; the components corresponding to R are the disjoint union of the components of R_1 and the components of R_2. We conclude that

$$\mathcal{P}(R) \iff \mathcal{P}(R_1) \wedge \mathcal{P}(R_2).$$

On the other hand the terms of $\Theta(R)$ correspond one-to-one to the terms of $R_1 \oplus R_2$ and the reduction relation of $\Theta(R)$ corresponds one-to-one to the reduction relation of $R_1 \oplus R_2$. So

$$\mathcal{P}(\Theta(R)) \iff \mathcal{P}(R_1 \oplus R_2).$$

Since $\mathcal{P}$ is persistent $\mathcal{P}(R)$ and $\mathcal{P}(\Theta(R))$ are equivalent; combining this with the above results gives

$$\mathcal{P}(R_1 \oplus R_2) \iff \mathcal{P}(R_1) \wedge \mathcal{P}(R_2),$$

which we had to prove. □

Strong normalization

Since strong normalization is not a modular property we conclude from theorem 1 that strong normalization is neither a persistent property. The basic counterexample in [8] and the proof of theorem 1 lead to the following counterexample. Let $S = \{s_1, s_2\}$; the following variables and operation symbols are defined:

- x is a variable of sort s_1;
- y, z are variables of sort s_2;
- $0, 1$ are constants of sort s_1;
- f is an operation symbol of sort s_1 and arity (s_1, s_1, s_1);
- g is an operation symbol of sort s_2 and arity (s_2, s_2).

Let the S-sorted TRS R consist of the following rules:

$$\begin{array}{ll} f(0,1,x) & \to f(x,x,x) \\ g(y,z) & \to y \\ g(y,z) & \to z. \end{array}$$

One easily shows that the S-sorted TRS R is strongly normalizing, while

$$f(g(0,1),g(0,1),g(0,1)) \to f(0,g(0,1),g(0,1)) \to f(0,1,g(0,1)) \to$$

$$f(g(0,1),g(0,1),g(0,1)) \to \cdots$$

is an infinite reduction in $\Theta(R)$. This implies that strong normalization is not a persistent property.

However, we can define a particular class of many-sorted TRS's such that strong normalization is persistent for that class. First we need some definitions. A reduction rule is called a *collapsing* rule if its right hand side is a single variable. A reduction rule is called a *duplicating* rule if for some variable the number of occurrences in the right side is greater than the number of occurrences in the left hand side. For example, in the above example the first rule is duplicating and the second and the third are collapsing rules.

In [7] it is shown that strong normalization is modular in the class of one-sorted TRS's without collapsing rules and also in the class of one-sorted TRS's without duplicating rules. We generalize this result:

Theorem 2 *Strong normalization is persistent for the class of many-sorted TRS's not containing both collapsing rules and duplicating rules.*

Any infinite reduction of R is trivially translated to an infinite reduction of $\Theta(R)$. As a consequence, strong normalization of $\Theta(R)$ implies strong normalization of R. The difficult part is the converse: assume strong normalization of R and derive strong normalization of $\Theta(R)$. The full proof can be found in [10]; a brief sketch is the following.

Each one-sorted term can be considered as a tree. This tree can uniquely be split up into maximal many-sorted parts, i.e., a parent and a child are in the same many-sorted part if and only if the parent expects the same sort as the child yields. These many-sorted building blocks are organized in a tree structure. Since all left hand sides of the rules are well-typed, every redex pattern is contained in one such a building block. If there are no collapsing rules each building block of a term in a reduction is a descendant of exactly one building block of the starting term of the reduction. Since reductions inside a single building block will always terminate, the same holds for a tree of such building blocks, using some multiset argument.

If there are no duplicating rules, the total number of building blocks will never increase. If a building block collapses, the total number of building blocks strictly decreases. So in any reduction the number of such collapses is finite, and behind that finite part of the reduction the above argument can be applied.

The original proof in [7] for modularity, which is presented in more detail in [6], can be modified to cover our result. However, our proof was found independently, and is proof-theoretically more elementary.

Applications

Theorem 2 can be used as a tool for proving strong normalization of TRS as follows. Given a (one-sorted) TRS not containing both collapsing and duplicating rules. Then try to find a typing in such a way that all rules are well-typed; this can be done systematically in a way similar to unification. In this way we get a many-sorted TRS; theorem 2 states that for proving strong normalization of the original one-sorted system it suffices to prove strong normalization of the many-sorted system.

For example, strong normalization of the (one-sorted) system

$$f(0,1,x) \rightarrow f(x,x,x)$$

from Toyama's example can be proven as follows. Consider the typing in which 0 and 1 have sort s_1, and f has sort s_2 and arity (s_1, s_1, s_1). Then this many-sorted system is trivially strongly normalizing since at most one reduction step can be done. From theorem 2 we conclude that also the original one-sorted system is strongly normalizing.

A next example is

$$s(x) + (y + z) \rightarrow x + (s(s(y)) + z)$$
$$s(x) + (y + (z + w)) \rightarrow x + (z + (y + w))$$

as studied in [3]. The strong normalization proof can be given via the typing

$$s : s_1 \rightarrow s_1, \quad + : s_1 \times s_2 \rightarrow s_2;$$

many-sorted terms of sort s_2 then represent lexicographically decreasing sequences of natural numbers.

From these examples we see that persistency is a far more powerful tool for proving strong normalization than the similar modularity result.

As another application we mention a simplification of some undecidability proofs. We sketch a short proof of the undecidability of strong normalization of arbitrary TRS. Let any arbitrary Turing machine be given; we will associate a TRS such that the TRS is strongly normalizing if and only if the Turing machine terminates on all arbitrary initial states. For every machine state a binary symbol q_i is introduced; for every alphabet symbol a unary symbol a_i. The end of the tape will be denoted by a constant 0. Every configuration of the Turing machine is uniquely coded by a term $q_i(S_l(0), S_r(0))$ where S_l is the string of a_i symbols denoting the tape left of the head and S_r is the string of a_i symbols denoting the tape right of the head. Now the transition function can be described by a finite number of rewrite rules, each of the shape

$$q_i(S_l(x), S_r(y)) \rightarrow q_j(S'_l(x), S'_r(y)),$$

where S_l, S_r, S'_l, S'_r consist of zero, one or two symbols a_i. We have to prove that the TRS constructed in this way is strongly normalizing if and only if the Turing machine terminates on arbitrary input. This is not trivial since not every term corresponds to a state of the Turing machine. However, from theorem 2 we know that the (one-sorted) TRS is strongly normalizing if and only if the many-sorted TRS obtained by the typing

$$0 : \rightarrow s_1, \quad a_i : s_1 \rightarrow s_1, \quad q_i : s_1 \times s_1 \rightarrow s_2$$

is strongly normalizing. The terms of this many-sorted system of sort s_2 correspond exactly to the states of the Turing machine, while on terms of sort s_1 no rules are applicable, proving the statement.

The first proof of undecidability of strong normalization was given in [4], using only unary symbols. The suggestion to represent machine states by binary symbols like we did above can also be found in [5].

Our theorem 2 can also be applied for simplifying the proof of undecidability of strong normalization of a single left-linear rule as given in [1]; the notion of M-component and all its technical details can simply be removed there.

Confluence

Although it has not yet been executed in full detail, it seems that Toyama's proof of modularity of confluence ([9]) can straightforwardly be modified to a proof of persistency of confluence.

Persistency of confluence can be used as a tool for proving confluence of (one-sorted) TRS. We illustrate this by an example, due to Y. Toyama. Consider the one-sorted TRS:

$$\begin{aligned} a(x,y) &\rightarrow a(f(x), f(x)) \\ f(x) &\rightarrow g(x) \\ b(f(x), x) &\rightarrow b(x, f(x)) \\ b(g(x), x) &\rightarrow b(x, g(x)). \end{aligned}$$

Proving confluence of this TRS is not easy, and modularity of confluence can not be applied. However, we can define a many-sorted version of this TRS with the same rules, in which f and g both have arity and sort s_1, a and b have arity (s_1, s_1), a has sort s_2 and b has sort s_3. Proving confluence of this many-sorted TRS can be done using standard techniques:

- terms of sort s_1 are confluent since only the second rule is applicable,

- terms of sort s_2 are confluent since the first two rules of the system form an orthogonal TRS,
- terms of sort s_3 are confluent since the TRS consisting of the last three rules is strongly normalizing and all critical pairs have a common reduct.

Now the confluence of the original TRS follows from persistency of confluence. From this example we see that persistency provides a more powerful tool for proving confluence of one-sorted TRS's than applying modularity.

Concluding remarks

We introduced the notion of persistency and proved that it is a generalization of direct sum modularity. Further strong normalization is a persistent property for a similar class of TRS's of which Rusinowitch ([7]) proved strong normalization to be modular; the full proof is in [10]. This result provides a new and powerful tool for proving strong normalization of (one-sorted) TRS.

Proving persistency of weak normalization and weak confluence is not difficult; proving persistency of confluence seems to be feasible. It is unknown whether there are modular properties that are not persistent.

As a conjecture we state that strong normalization is a persistent property for the class of TRS's in which all variables are of the same sort.

References

1. DAUCHET, M. Simulation of Turing machines by a left-linear rewrite rule. In *Proceedings of the 3rd Conference on Rewriting Techniques an Applications* (1989), N. Dershowitz, Ed., vol. 355 of *Lecture Notes in Computer Science*, Springer, pp. 109–120.
2. EHRIG, H., AND MAHR, B. *Fundamentals of Algebraic Specification*, vol. 1 of *EATCS monographs on Theoretical Computer Science*. Springer, 1985.
3. HOFBAUER, D., AND LAUTEMANN, C. Termination proofs and the length of derivations (preliminary version). In *Proceedings of the 3rd Conference on Rewriting Techniques an Applications* (1989), N. Dershowitz, Ed., vol. 355 of *Lecture Notes in Computer Science*, Springer, pp. 167–177.
4. HUET, G., AND LANKFORD, D. S. On the uniform halting problem for term rewriting systems. Rapport Laboria 283, INRIA, 1978.
5. KLOP, J. W. Term rewriting systems. In *Handbook of Logic in Computer Science*, D. G. S. Abramski and T. Maibaum, Eds., vol. 1. Oxford University Press, 1991.
6. MIDDELDORP, A. *Modular Properties of Term Rewriting Systems*. PhD thesis, Free University Amsterdam, 1990.
7. RUSINOWITCH, M. On termination of the direct sum of term rewriting systems. *Information Processing Letters 26* (1987), 65–70.
8. TOYAMA, Y. Counterexamples to termination for the direct sum of term rewriting systems. *Information Processing Letters 25* (1987), 141–143.
9. TOYAMA, Y. On the Church-Rosser property for the direct sum of term rewriting systems. *Journal of the ACM 34*, 1 (1987), 128–143.
10. ZANTEMA, H. Termination of term rewriting: from many-sorted to one-sorted. In *Computing Science in the Netherlands* (November 1991), J. van Leeuwen, Ed., vol. 2, pp. 617 – 629. Also appeared as report RUU-CS-91-18, Utrecht University.

Termination of term rewriting by interpretation

H. Zantema
Utrecht University, Department of Computer Science
P.O. box 80.089, 3508 TB Utrecht, The Netherlands
e-mail: hansz@cs.ruu.nl

Abstract. We investigate how to prove termination of term rewriting systems by interpretation of terms. This can be considered as a generalization of polynomial interpretations. A classification of types of termination is proposed built on properties in the semantic level. A transformation on term rewriting systems eliminating distributive rules is introduced. Using this distribution elimination a new termination proof of SUBST from [10] is given.

1 Introduction

One of the main problems in the theory of term rewriting systems is the detection of termination: for a fixed system of rewrite rules, detect whether there exist infinite rewrite chains or not. In general this problem is undecidable ([12, 4]). However, there are several methods for deciding termination that are successful for many special cases. Roughly these methods can be divided into two main types: *syntactical* methods and *semantical* methods. In a syntactical method terms are ordered by a careful analysis of the term structure. A well-known representative of this type is the *recursive path order* ([5]). All of these orderings are simplification orderings, i.e., a term is always greater than its proper subterms. An overview and comparison of simplification orderings is given in [17].

In a semantical method terms are interpreted in some well-known well-founded ordered set in such a way that each rewrite chain will map to a descending chain, and hence will terminate. Until now most semantical methods have focussed on choosing the natural numbers as the well-founded ordered set. The method of *polynomial interpretations* ([15, 2]) can be seen as a particular case of a semantical method on natural numbers. In this paper we introduce the notion of a *monotone algebra* as the natural concept for semantical methods. Though we focus on 'pure' TRS, the ideas are easily extended to conditional TRS, typed TRS and TRS modulo equations. We propose a classification of types of termination based upon the types of orderings of the underlying monotone algebras.

We present a transformation of term rewriting systems eliminating a particular operation symbol. Using the framework of monotone algebras we prove that under some restrictions termination of the original system follows from termination of the eliminated system. Since in this construction distributive rules are removed it is called *distribution elimination*.

As an application of monotone algebras and distribution elimination we give a new termination proof for the systems studied in [10, 3, 1]. Our proof is simpler than the existing proofs and gives a stronger result: we prove that the system is simply terminating.

2 Monotone algebras

Let $\mathcal{F}$ be a set of operation symbols each having a fixed arity. We define a *well-founded monotone $\mathcal{F}$-algebra* $(A, >)$ to be an $\mathcal{F}$-algebra A for which the underlying set is provided with a well-founded order $>$ and each algebra operation is strictly monotone in all of its coordinates, more precisely: for each operation symbol $f \in \mathcal{F}$ and all $a_1, \ldots, a_n, b_1, \ldots, b_n \in A$ for which $a_i > b_i$ for some i and $a_j = b_j$ for all $j \neq i$ we have

$$f_A(a_1, \ldots, a_n) > f_A(b_1, \ldots, b_n).$$

Let $(A, >)$ be a well-founded monotone $\mathcal{F}$-algebra. Let $A^{\mathcal{X}} = \{\alpha : \mathcal{X} \to A\}$. We define

$$\phi : T(\mathcal{F}, \mathcal{X}) \times A^{\mathcal{X}} \to A$$

inductively by

$$\phi(x, \alpha) = x^{\alpha},$$
$$\phi(f(t_1, \ldots, t_n), \alpha) = f_A(\phi(t_1, \alpha), \ldots, \phi(t_n, \alpha))$$

for $x \in \mathcal{X}, \alpha : \mathcal{X} \to A, f \in \mathcal{F}, t_1, \ldots, t_n \in T(\mathcal{F}, \mathcal{X})$. This function induces a partial order $>_A$ on $T(\mathcal{F}, \mathcal{X})$ as follows:

$$t >_A t' \iff (\forall \alpha \in A^{\mathcal{X}} : \phi(t, \alpha) > \phi(t', \alpha)).$$

Intuitively: $t >_A t'$ means that for each interpretation of the variables in A the interpreted value of t is greater than that of t'.

We say that a non-empty well-founded monotone algebra $(A, >)$ *normalizes* a TRS if $l >_A r$ for every rule $l \to r$ of the TRS. This terminology is motivated by the following proposition.

Proposition 1 *A TRS is terminating if and only if it is normalized by a non-empty well-founded monotone algebra.*

In order to give the proof we need two lemmas.

Lemma 2 *Let $\sigma : \mathcal{X} \to T(\mathcal{F}, \mathcal{X})$ be any substitution and let $\alpha : \mathcal{X} \to A$. Define $\beta : \mathcal{X} \to A$ by $\beta(x) = \phi(x^{\sigma}, \alpha)$ for $x \in \mathcal{X}$. Then*

$$\phi(t^{\sigma}, \alpha) = \phi(t, \beta)$$

for all $t \in T(\mathcal{F}, \mathcal{X})$.

Proof: Induction on the structure of t. □

Lemma 3 *The partial order $>_A$ is closed under substitution and context.*

Proof: Let $t >_A t'$ for $t, t' \in \mathcal{T}(\mathcal{F}, \mathcal{X})$ and let $\sigma : \mathcal{X} \to \mathcal{T}(\mathcal{F}, \mathcal{X})$ be any substitution. Let $\alpha : \mathcal{X} \to A$. From lemma 2 we obtain

$$\phi(t^\sigma, \alpha) = \phi(t, \beta) > \phi(t', \beta) = \phi(t'^\sigma, \alpha).$$

The key point here is that β does not depend on t. This holds for all $\alpha : \mathcal{X} \to A$, so $t^\sigma >_A t'^\sigma$. Hence $>_A$ is closed under substitution.

For proving closedness under context let $t >_A t'$ for $t, t' \in \mathcal{T}(\mathcal{F}, \mathcal{X})$, and let $f \in \mathcal{F}$. Since $t >_A t'$ we have $\phi(t, \alpha) > \phi(t', \alpha)$ for all $\alpha : \mathcal{X} \to A$. Applying the monotonicity condition of f_A we obtain

$$\phi(f(\ldots, t, \ldots), \alpha) = f_A(\ldots, \phi(t, \alpha), \ldots) > f_A(\ldots, \phi(t', \alpha), \ldots) = \phi(f(\ldots, t', \ldots), \alpha).$$

This holds for all $\alpha : \mathcal{X} \to A$, so

$$f(\ldots, t, \ldots) >_A f(\ldots, t', \ldots),$$

which we had to prove. □

Now we give the proof of proposition 1.

Proof: Assume the TRS is normalized by a non-empty well-founded monotone algebra, and allows an infinite reduction. Then according to lemma 3 we have $t >_A t'$ for every reduction step $t \to t'$. Now applying any substitution in A (here we use non-emptiness) on the infinite reduction gives rise to an infinite decreasing chain in A, contradiction.

On the other hand, assume the system is terminating. Define $A = \mathcal{T}(\mathcal{F}, \mathcal{X})$, and define $>$ to be the transitive closure of the rewrite relation. One easily verifies that $(A, >)$ is a non-empty well-founded monotone algebra. We still have to prove that $l >_A r$ for each rewrite rule $l \to r$. Let $\alpha : \mathcal{X} \to A$. Since $A = \mathcal{T}(\mathcal{F}, \mathcal{X})$ we see that α is a substitution. Then

$$\phi(t, \alpha) = t^\alpha$$

for each term t, which is easily proved by induction on the structure of t. Since $l \to r$ is a rewrite rule, the term l^α can be reduced in one step to r^α. So

$$\phi(l, \alpha) = l^\alpha > r^\alpha = \phi(r, \alpha).$$

This holds for every $\alpha : \mathcal{X} \to A$, so $l >_A r$, which we had to prove. □

The way of proving termination of a TRS is now as follows: choose a well-founded poset A, define for each operation symbol a corresponding operation that is strictly monotone in all of its coordinates, and for which $\phi(l, \alpha) >_A \phi(r, \alpha)$ for all rewrite rules $l \to r$ and all $\alpha : \mathcal{X} \to A$. Then according to the above proposition the TRS is terminating. Often we choose A to be $\mathbb{N}$, which is defined to be the set of strictly positive integers. For example, for the system

$$f(f(x, y), z) \to f(x, f(y, z))$$

choose $(A, >) = (\mathbb{N}, >)$ and $f_A(x, y) = 2x + y$. Clearly f_A is strictly monotone in both coordinates, and

$$f_A(f_A(x, y), z) = 4x + 2y + z > 2x + 2y + z = f_A(x, f_A(y, z))$$

for all $x, y, z \in A$. Hence $f(f(x,y),z) >_A f(x, f(y,z))$, proving termination.

Next we give three examples in which we choose A to be two copies of the naturals; define for all three cases:

$$A = \{0,1\} \times \mathbb{N} \quad \text{and} \quad (a,n) > (b,m) \iff a = b \wedge n > m.$$

- $f(f(x)) \rightarrow f(g(f(x)))$. This system is proved to be terminating by choosing

$$f_A(0,n) = (0, n+1), \quad f_A(1,n) = (0,n), \quad g_A(0,n) = g_A(1,n) = (1,n).$$

- $f(g(x)) \rightarrow f(f(x)), g(f(x)) \rightarrow g(g(x))$. Choose

$$f_A(0,n) = (1, 2n), \quad f_A(1,n) = (1, n+1),$$

$$g_A(0,n) = (0, n+1), \quad g_A(1,n) = (0, 2n).$$

- $f(0,1,x) \rightarrow f(x,x,x)$. Choose $0_A = (0,1)$, $1_A = (1,1)$, and

$$f_A((a,n),(b,m),(c,k)) = \begin{cases} (0, n+m+k) & \text{if } a = b \\ (0, n+m+3k) & \text{if } a \neq b \end{cases}$$

3 Simple termination

If $\mathcal{F}$ is finite it is sometimes convenient to replace the well-foundedness condition in the definition of a well-founded monotone algebra by a simplicity condition as follows. A *simple monotone $\mathcal{F}$-algebra* $(A, >)$ is defined to be an $\mathcal{F}$-algebra A for which the underlying set is provided with a partial order $>$ such that each algebra operation is strictly monotone in all of its coordinates, and

$$f_A(a_1, \ldots, a_n) \geq a_i$$

for each $f \in \mathcal{F}$, $a_1, \ldots, a_n \in A$, and $i \in \{1, \ldots, n\}$. The corresponding reduction order $>_A$ is called a *simplification ordering*. This definition coincides with that in [6]. These definitions are motivated by the following two propositions.

Proposition 4 *Let $\mathcal{F}$ be finite and let $(A, >)$ be a simple monotone $\mathcal{F}$-algebra. Let A' be the smallest subalgebra of A, i.e., A' is the homomorphic image of the ground terms. Then $(A', >)$ is a well-founded monotone $\mathcal{F}$-algebra.*

Proof: The only property to prove is well-foundedness. Assume the restriction of $>$ to A' is not well-founded. Then there is an infinite chain

$$h(t_0) > h(t_1) > h(t_2) > h(t_3) > \cdots,$$

where h is the homomorphism from ground terms to A. The key argument is Kruskal's tree theorem ([13]); the relevance for termination of term rewriting systems is explained in [6]. This theorem states that there is some $i < j$ such that t_i can be homeomorphically embedded in t_j. Since $(A, >)$ is a simple monotone algebra and h is a homomorphism, we conclude that $h(t_j) \geq h(t_i)$, contradicting irreflexivity and transitivity of $>$. $\Box$

Proposition 5 *Let $\mathcal{F}$ be finite and let $(A, >)$ be a non-empty simple monotone $\mathcal{F}$-algebra. Let R be a TRS such that $l >_A r$ for all rewrite rules $l \to r$ of R. Then R is terminating.*

Proof: Apply proposition 4: A' is a well-founded monotone algebra normalizing R. In the case that $\mathcal{F}$ does not contain constants, add one dummy constant symbol forcing $A' \neq \emptyset$. □

For a set $\mathcal{F}$ of operation symbols we define $Emb(\mathcal{F})$ to be the TRS consisting of all the rules

$$f(x_1, \ldots, x_n) \quad \to \quad x_i$$

with $f \in \mathcal{F}$ and $i \in \{1, \ldots, n\}$.

Proposition 6 *Let R be a TRS over a set $\mathcal{F}$ of operation symbols. Then the following assertions are equivalent:*

(1) *R is simply terminating;*
(2) *$R \cup Emb(\mathcal{F})$ is simply terminating;*
(3) *$R \cup Emb(\mathcal{F})$ is terminating.*

Proof: The implication (2) ⇒ (1) is trivial. For proving (1) ⇒ (2) let $(A, >)$ be a simple monotone $\mathcal{F}$-algebra normalizing R. Since we allow equality in the definition of simplicity, we have to modify A in order to normalize $R \cup Emb(\mathcal{F})$. Choose

$$B = A \times \mathbb{N}$$

having the lexicographic order

$$(a, k) > (a', k') \iff a > a' \vee (a = a' \wedge k > k').$$

Define

$$f_B((a_1, k_1), \ldots, (a_n, k_n)) = (f_A(a_1, \ldots, a_n), 1 + \sum_{i=1}^{n} k_i).$$

Now $(B, >)$ is a simple monotone algebra normalizing both R and $Emb(\mathcal{F})$, proving (2).

The implication (2) ⇒ (3) is trivial. Finally, assume that (3) holds. Then according to proposition 1 there is a non-empty well-founded monotone $\mathcal{F}$-algebra $(A, >)$ normalizing $R \cup Emb(\mathcal{F})$. Since it normalizes $Emb(\mathcal{F})$ it is also a simple monotone $\mathcal{F}$-algebra. This implies (2). □

4 The hierarchy

Let $(A,>)$ be a monotone algebra. Depending on its properties we propose a hierarchy of types of termination. If $A = \mathbb{N}$ and $>$ is the ordinary order on $\mathbb{N}$ and f_A is a polynomial for all $f \in \mathcal{F}$, we speak about *polynomial termination*. If $A = \mathbb{N}$ and $>$ is the ordinary order on $\mathbb{N}$, we speak about *ω-termination*. In these cases we may have $\{n \in \mathbb{N} \mid n > N\}$ instead of $\mathbb{N}$, which gives equivalent definitions due to linear transformation. An implementation based on polynomial termination is described in [2]; a recent extension to elementary functions in which also exponents may occur is given in [16].

If the order $>$ on A is total and well-founded, we speak about *total termination*. If $(A,>)$ is a simple monotone algebra, we speak about *simple termination*.

The following implications hold:

polynomial termination
$\Longrightarrow$ ω-termination
$\Longrightarrow$ total termination
$\Longrightarrow$ simple termination
$\Longrightarrow$ termination.

The only non-trivial implication is the implication of simple termination from total termination. This follows immediately from the following proposition.

Proposition 7 *Let $(A,>)$ be a well-founded monotone $\mathcal{F}$-algebra for which the order $>$ is total on A. Then $(A,>)$ is a simple monotone $\mathcal{F}$-algebra.*

Proof: Assume it is not simple. Then there exist $f \in \mathcal{F}, a_1, \ldots, a_n \in A$ and $i \in \{1, \ldots, n\}$ such that $f_A(a_1, \ldots, a_n) \geq a_i$ does not hold. From totality we conclude:

$$a_i > f_A(a_1, \ldots, a_n).$$

Define $g : A \rightarrow A$ by $g(x) = f_A(a_1, \ldots, a_{i-1}, x, a_{i+1}, \ldots, a_n)$, then g is strictly monotone. We obtain an infinite chain

$$a_i > g(a_i) > g(g(a_i)) > g(g(g(a_i))) > \cdots,$$

contradicting the well-foundedness of $(A,>)$. □

A useful consequence of which the proof is immediate from the proof of proposition 6 is the following:

Proposition 8 *Let R be a TRS over a set $\mathcal{F}$ of operation symbols. Then R is totally terminating if and only if $R \cup Emb(\mathcal{F})$ is totally terminating.*

None of the implications in the hierarchy holds in the reverse direction:

$a(f(x), y) \rightarrow f(a(x, a(x, y)))$	is ω-terminating but not polynomially terminating;
$f(g(x)) \rightarrow g(f(f(x)))$	is totally terminating but not ω-terminating;
$f(a) \rightarrow f(b), g(b) \rightarrow g(a)$	is simply terminating but not totally terminating;
$f(f(x)) \rightarrow f(g(f(x)))$	is terminating but not simply terminating.

For more details on these examples we refer to [19].

Note that in our notion of ω-termination there is no restriction on the kind of monotone functions allowed. Hofbauer ([11]) proved that any TRS which can be proven terminating by a recursive path order, is ω-terminating. By giving restrictions on the kind of functions allowed, a more detailed hierarchy between polynomial termination and ω-termination can be given. An extensive study of the hierarchy between omega-termination and total termination will be given in [8].

5 Basic constructions

In order to be able find termination proofs for term rewriting systems by monotone algebras, it is useful to investigate some basic constructions of how to build a monotone algebra. For choosing the well-founded set $(A, >)$ we have:

- The natural numbers.
- Lexicographic order: if $(A, >)$ and $(B, >)$ are well-founded, then so is $A \times B$ with the lexicographic order.
- Multiset order: if $(A, >)$ is well-founded, then so is

  $$M(A) = \{X : A \to \mathbb{N} | \#\{a \in A | X(a) \neq 0\} < \infty\},$$

  with the order

  $$X > Y \iff X \neq Y \wedge (\forall x \in A : X(x) \geq Y(x) \vee (\exists a \in A : a > x \wedge X(a) > Y(a));$$

  for the proof we refer to [7]. This construction corresponds to exponentiation in ordinal arithmetic as discussed in [14, 8]: if $(A, >)$ corresponds to the ordinal α, then $(M(A), >)$ corresponds to the ordinal ω^α.

One of the results of [8] states that any useful $(A, >)$ in which the order is total and the order type is below ϵ_0, can be constructed by only the above constructions in a finite number of steps.

Until now we investigated only the underlying well-founded set; the next step is how to choose appropriate monotone functions. A lot of possibilities are easily found; now we introduce a a useful kind of lifting of monotone functions to multisets.

Write $[]$ for the empty multiset, defined by $[](a) = 0$ for all $a \in A$. For every $a \in A$ the singleton $[a]$ is defined by $[a](a) = 1$ and $[a](x) = 0$ for $x \neq a$. Multiset union $\cup$ is defined by $(X \cup Y)(a) = X(a) + Y(a)$; multiset union is associative and commutative. Every non-empty finite multiset can be obtained as a finite multiset union of singletons.

Write $M'(A) = M(A) \setminus \{[]\}$, the set of finite non-empty multisets over A. For every $f : A^k \to A$ the function $\bar{f} : (M'(A))^k \to M'(A)$ is defined as follows:

$$\bar{f}([a_1], [a_2], \ldots [a_k]) = [f(a_1, a_2, \ldots, a_k)],$$

$$\bar{f}(X_1, \ldots X_{i-1}, Y \cup Z, X_{i+1}, \ldots X_k) =$$

$$\bar{f}(X_1, \ldots X_{i-1}, Y, X_{i+1}, \ldots X_k) \cup \bar{f}(X_1, \ldots X_{i-1}, Z, X_{i+1}, \ldots X_k).$$

Intuitively: to compute $\bar{f}(X_1,\dots X_k)$, apply f to all possible choices of elements of $X_1,\dots X_k$, and collect all results in one multiset. One easily shows that if f is monotone in every coordinate then also $\bar{f}$ is monotone in every coordinate; here for $k > 1$ it is essential to restrict to non-empty multisets.

In the next section this lifting of f to $\bar{f}$ plays an essential role.

6 Distribution elimination

In this section we introduce a transformation of TRS's in which a particular operation symbol is eliminated, and prove that if the eliminated TRS is terminating then the original TRS is also terminating. Since the eliminated TRS is simpler then the original one, this provides a method for proving termination of complicated TRS's by proving termination of simple ones.

Let a be any fixed operation symbol of arity $n \geq 1$. A rewrite rule is called a *distribution rule* for a if it can be written as

$$C[a(x_1,\dots,x_n)] \rightarrow a(C[x_1],\dots,C[x_n])$$

for some non-trivial context $C[]$ in which the symbol a does not occur. For example,

$$b(z, f(a(x,y)) \rightarrow a(b(z, f(x)), b(z, f(y)))$$

is a distribution rule for a. Problems with distribution rules have been recognized before; for example in [9] a particular ordering for proving termination of AC rewriting systems is introduced for systems containing distribution rules.

Write $\mathcal{P}$ for powerset. The function $E_a : \mathcal{T}(\mathcal{F},\mathcal{X}) \rightarrow \mathcal{P}(\mathcal{T}(\mathcal{F},\mathcal{X}))$ is defined inductively as follows:

$$\begin{array}{lll} E_a(x) & = \{x\} & \text{for all } x \in \mathcal{X}, \\ E_a(f(t_1,\dots,t_k)) & = \{f(u_1,\dots,u_k) \mid \forall i : u_i \in E_a(t_i)\} & \text{for all } f \in \mathcal{F}, f \neq a \\ E_a(a(t_1,\dots,t_n)) & = \bigcup_{i=1}^{n} E_a(t_i). & \end{array}$$

Let R be a TRS for which each rule is either a distribution rule for a or a rule in which a does not occur in the left hand side. Then the TRS $E_a(R)$ is defined by

$$E_a(R) = \{l \rightarrow u \mid l \rightarrow r \text{ is a non-distribution rule of } R \text{ for } a \text{ and } u \in E_a(r)\}.$$

For example, if R is defined by

$$\begin{array}{ll} f(a(x,y)) & \rightarrow a(f(x), f(y)) \\ g(a(x,y)) & \rightarrow a(g(x), g(y)) \\ f(f(x)) & \rightarrow f(a(g(f(x)), g(f(x)))) \end{array}$$

then $E_a(R)$ consists only of the rule $f(f(x)) \rightarrow f(g(f(x)))$. This system is known to be terminating; the next proposition states that we can conclude that also R is terminating. As usual a term is defined to be linear if no variable occurs more than once, and a TRS is defined to be right-linear if for every rule the right hand side is linear.

Theorem 9 *Let R be a TRS for which each rule is either a distribution rule for a or a rule in which a does not occur in the left hand side. Then*

- *if $E_a(R)$ is totally terminating then R is totally terminating;*
- *if $E_a(R)$ is simply terminating and right-linear then R is simply terminating;*
- *if $E_a(R)$ is terminating and right-linear then R is terminating.*

Before giving the proof we give examples showing that the converse does not hold and the right-linearity requirement is essential. First choose R to be $f(f(x)) \to f(a(f(x)))$. Then R is terminating and satisfies the conditions (there are no distribution rules), while $E_a(R)$ consists of $f(f(x)) \to f(f(x))$ which is not terminating.

Next choose R to be:

$$f(0,1,x,x) \to f(x,x,a(0,1),a(0,1))$$
$$f(a(x,y),z,v,w) \to a(f(x,z,v,w),f(y,z,v,w))$$
$$f(x,a(y,z),v,w) \to a(f(x,y,v,w),f(x,z,v,w))$$

The second and third rule are distribution rules for a; the system $E_a(R)$ is

$$f(0,1,x,x) \to f(x,x,0,0)$$
$$f(0,1,x,x) \to f(x,x,0,1)$$
$$f(0,1,x,x) \to f(x,x,1,0)$$
$$f(0,1,x,x) \to f(x,x,1,1),$$

and is terminating, even simply terminating since the size of a term is not changed by reduction. However, in R we have the reduction

$$\begin{aligned} f(0,1,a(0,1),a(0,1)) &\to f(a(0,1),a(0,1),a(0,1),a(0,1)) \\ &\to a(f(0,a(0,1),a(0,1),a(0,1)),f(1,\ldots)) \\ &\to a(a(f(0,0,a(0,1),a(0,1)),\underbrace{f(0,1,a(0,1),a(0,1))}),f(1,\ldots)) \end{aligned}$$

in which the starting term occurs as a subterm. This can be expanded to an infinite reduction. This example was inspired by Toyama's example showing that termination is not modular ([18]). We conclude that the right-linearity requirement is essential in both the second and the third assertion of theorem 9. A simpler example is possible; we chose an example in which $E_a(R)$ is non-duplicating, also showing that the right-linearity restriction in the theorem may not be weakened to non-duplication.

Now we prove theorem 9.

Proof: Let $(A,>)$ be a well-founded monotone algebra for $E_a(R)$. Again write $M'(A)$ for the set of finite non-empty multisets over A.

We define the well-founded monotone algebra for R to be

$$B = M'(A) \times \mathbb{N}$$

where $\mathbb{N}$ consists of the strictly positive integers, with the lexicographic order

$$(X,k) > (Y,m) \iff X > Y \vee (X = Y \wedge k > m).$$

As operations we define

$$a_B((X_1,m_1),\ldots,(X_n,m_n)) = (\bigcup_{i=1}^{n} X_i, 1+\sum_{i=1}^{n} m_i)$$

and

$$f_B((X_1,m_1),\ldots,(X_k,m_k)) = (\bar{f}_A(X_1,\ldots,X_k), \prod_{i=1}^{k} m_i^2)$$

for all $f \in \mathcal{F}, f \neq a$. Note that these operations are strictly monotone in all coordinates. Further if $(A,>)$ is a simple monotone algebra then $(B,>)$ is also simple; if $(A,>)$ is total then $(B,>)$ is also total. We have to prove that for every rule of R the left hand side is greater than the right hand side, interpreted in B.

Let $\beta : \mathcal{X} \to B$ be arbitrary, and write $\phi_B(t,\beta) = (\phi_1(t), \phi_2(t)) \in M'(A) \times \mathbb{N}$ for $t \in \mathcal{T}(\mathcal{F},\mathcal{X})$, where ϕ_B is from the definition of well-founded monotone algebra. By definition we have

$$\phi_1(x) = \pi(x^\beta) \quad \text{for } x \in \mathcal{X}, \text{ where } \pi : B \to M'(A) \text{ is the projection on the first coordinate}$$

$$\phi_1(a(t_1,\ldots,t_n)) = \bigcup_{i=1}^{n} \phi_1(t_i)$$

$$\phi_1(f(t_1,\ldots,t_k)) = \bar{f}_A(\phi_1(t_1),\ldots,\phi_1(t_k)) \quad \text{for } f \in \mathcal{F}, f \neq a.$$

Let $C[]$ be any non-trivial context in which a does not occur. Since

$$\bar{f}_A(X_1,\ldots X_{i-1}, Y \cup Z, X_{i+1},\ldots X_k) =$$

$$\bar{f}_A(X_1,\ldots X_{i-1}, Y, X_{i+1},\ldots X_k) \cup \bar{f}_A(X_1,\ldots X_{i-1}, Z, X_{i+1},\ldots X_k)$$

for all operation symbols f occurring in $C[]$, we see that

$$\phi_1(C[a(x_1,\ldots,x_n)] = \bigcup_{i=1}^{n} \phi_1(C[x_i]) = \phi_1(a(C[x_1],\ldots,C[x_n])).$$

Since a does not occur in $C[]$, for every term t we have $\phi_2(C[t]) = c * \phi_2(t)^p$ for some $c \geq 1$ and $p > 1$. As a consequence we have

$$\phi_2(C[a(x_1,\ldots,x_n)] = c * (1 + \sum_{i=1}^{n} \phi_2(x_i))^p$$

$$> 1 + \sum_{i=1}^{n} c * \phi_2(x_i)^p \;=\; \phi_2(a(C[x_1],\ldots,C[x_n])).$$

As a consequence, for any distribution rule

$$C[a(x_1,\ldots,x_n)] \to a(C[x_1],\ldots,C[x_n])$$

in R we have

$$C[a(x_1,\ldots,x_n)] >_B a(C[x_1],\ldots,C[x_n]).$$

Every other rule in R is of the shape $l \to r$ in which a does not occur in l. Choose $s \in \phi_1(r)$ arbitrarily. We shall prove that in any case $\phi_1(l)$ contains an element strictly greater than s.

From the construction of E_a we obtain

$$\phi_1(r) = \bigcup_{u \in E_a(r)} \phi_1(u),$$

which is proved by induction on the structure of r. As a consequence, there is some $u \in E_a(r)$ such that $s \in \phi_1(u)$. We shall construct $\alpha : \mathcal{X} \to A$ such that $s \leq \phi_A(u, \alpha)$ and $x^\alpha \in \phi_1(x)$; here we need to distinguish between the cases of right-linearity and totality. For the first case we need to prove the following fact:

Fact. Let t be a linear term in which the symbol a does not occur and let $e \in \phi_1(t)$. Define $\mathcal{X}_t$ to be the set of variables occurring in t. Then there exists $\alpha : \mathcal{X}_t \to A$ such that $x^\alpha \in \phi_1(x)$ for all $x \in \mathcal{X}_t$ and $e = \phi_A(t, \alpha)$.

We prove this fact by induction on the structure of t. For the basis of the induction the term t is a variable x and we define $x^\alpha = e$. For the induction step we have $t = f(t_1, \ldots, t_k)$; then we have

$$e \in \phi_1(t) = \phi_1(f(t_1, \ldots, t_k)) = \bar{f}_A(\phi_1(t_1), \ldots, \phi_1(t_k)).$$

From the definition of $\bar{f}_A$ follows that there are $e_i \in \phi_1(t_i)$ for $i = 1, \ldots, k$ such that $e = f_A(e_1, \ldots, e_k)$. From the induction hypothesis we obtain $\alpha_i : \mathcal{X}_{t_i} \to A$ with $e_i = \phi_A(t_i, \alpha_i)$ for $i = 1, \ldots, k$. Since the term t is linear all $\mathcal{X}_{t_i}$ are disjoint. So there exists $\alpha : \mathcal{X}_t \to A$ such that $x^\alpha = x^{\alpha_i}$ if $x \in \mathcal{X}_{t_i}$. This gives $x^\alpha \in \phi_1(x)$; further we obtain

$$e = f_A(e_1, \ldots, e_k) = f_A(\phi_A(t_1, \alpha), \ldots, \phi_A(t_k, \alpha)) = \phi_A(t, \alpha),$$

proving the fact.

If $E_a(R)$ is right-linear, the term u is linear and we can apply the fact giving a particular $\alpha : \mathcal{X}_u \to A$. By choosing x^α arbitrarily in $\phi_1(x)$ for $x \in \mathcal{X} \setminus \mathcal{X}_u$ we obtain $\alpha : \mathcal{X} \to A$ for which $x^\alpha \in \phi_1(x)$ for all $x \in \mathcal{X}$ and $s = \phi_A(u, \alpha)$.

In the other case we assumed that $(A, >)$ is total. Then for every $x \in \mathcal{X}$ the finite non-empty multiset $\phi_1(x)$ has a unique maximum. Define $\alpha : \mathcal{X} \to A$ by choosing x^α to be this maximum for every $x \in \mathcal{X}$. Then one easily shows by induction on t that for this α we have $e \leq \phi_A(t, \alpha)$ for all terms t not containing a and all $e \in \phi_1(t)$; here we use again that every $e \in \phi_1(f(t_1, \ldots, t_k))$ can be written as $f_A(e_1, \ldots, e_k)$ for some $e_i \in \phi_1(t_i)$.

In all cases we have constructed some $\alpha : \mathcal{X} \to A$ for which $x^\alpha \in \phi_1(x)$ for all $x \in \mathcal{X}$ and $s \leq \phi_A(u, \alpha)$. Since $(A, >)$ is a well-founded monotone algebra for $E_a(R)$ and $l \to u$ is a rule of $E_a(R)$, we obtain

$$\phi_A(l, \alpha) > \phi_A(u, \alpha) \geq s.$$

One easily shows by induction on t that $\phi_A(t, \alpha) \in \phi_1(t)$ for all terms t not containing a. Since l is a term not containing a, we conclude that we have found an element $\phi_A(l, \alpha)$ in $\phi_1(l)$ which is strictly greater than s. Since this construction can be done for every $s \in \phi_1(r)$, we conclude that the multiset $\phi_1(l)$ is strictly greater than the multiset $\phi_1(r)$. Hence

$$\phi_B(l,\beta) > \phi_B(r,\beta).$$

This holds for all $\beta : \mathcal{X} \to B$, so $l >_B r$, which concludes the proof of theorem 9. □

7 Termination of SUBST

Let $\circ$ and $\cdot$ be binary symbols, λ a unary symbol, and 1, *id* and $\uparrow$ constants. Consider the TRS

$$\begin{array}{lll|ll}
\lambda(x) \circ y & \to \lambda(x \circ (1 \cdot (y \circ \uparrow))) & \text{(Abs)} & 1 \circ id & \to 1 \\
(x \cdot y) \circ z & \to (x \circ z) \cdot (y \circ z) & \text{(Map)} & \uparrow \circ id & \to \uparrow \\
(x \circ y) \circ z & \to x \circ (y \circ z) & \text{(Ass)} & 1 \circ (x \cdot y) & \to x \\
id \circ x & \to x & & \uparrow \circ (x \cdot y) & \to y,
\end{array}$$

called σ_0 in [3], which is essentially the same as the system SUBST in [10]. This system describes the process of substitution in combinatory categorical logic. Here 'λ' corresponds to Currying, '$\circ$' to composition, '*id*' to the identity, '$\cdot$' to pairing and '1' and '$\uparrow$' to projections. The original termination proof of SUBST in [10] is very complicated; the same holds for a newer proof in [3]. This result also implies termination of the process of explicit substitution in untyped λ-calculus; a recent overview of explicit substitutions can be found in [1].

In this section we prove that this system is totally terminating. According to proposition 8 it suffices to prove total termination of the system consisting only of the three rules (Abs), (Map) and (Ass), since for all other rules the right hand side can be embedded in the left hand side.

We see that the rule (Map) is a distribution rule for the operation '$\cdot$', and that '$\cdot$' does not occur in the left hand sides of (Abs) and (Ass). Hence according to theorem 9 it suffices to prove total termination of the eliminated system:

$$\begin{array}{lll}
\lambda(x) \circ y & \to \lambda(x \circ 1) & \text{(Abs1)} \\
\lambda(x) \circ y & \to \lambda(x \circ (y \circ \uparrow)) & \text{(Abs2)} \\
(x \circ y) \circ z & \to x \circ (y \circ z) & \text{(Ass)}.
\end{array}$$

As a total well-founded monotone algebra we choose $\mathbb{N} \times \mathbb{N} \times \mathbb{N}$, where $\mathbb{N}$ consists of the integers ≥ 0, with the lexicographic order

$$(x_1,x_2,x_3) > (y_1,y_2,y_3) \iff x_1 > y_1 \vee (x_1 = y_1 \wedge (x_2 > y_2 \vee (x_2 = y_2 \wedge x_3 > y_3))).$$

We define the operations in this algebra as follows:

$$1 = \uparrow = (0,0,0), \quad \lambda(x_1,x_2,x_3) = (x_1+1, x_2, x_3),$$

$$(x_1,x_2,x_3) \circ (y_1,y_2,y_3) = (x_1 + y_1, x_1 * (y_1+1) + x_2 + y_2, 2 * x_3 + y_3 + 1).$$

One easily shows that these operations are strictly monotone in all coordinates and that every left hand side of a rule in the eliminated system is strictly greater than the corresponding right hand side. This proves that the system is totally terminating, and hence σ_0 and SUBST are totally terminating.

References

1. ABADI, M., CARDELLI, L., CURIEN, P.-L., AND LÉVY, J.-J. Explicit substitutions. *Journal of Functional Programming 1*, 4 (1991), 375–416.
2. BEN-CHERIFA, A., AND LESCANNE, P. Termination of rewriting systems by polynomial interpretations and its implementation. *Science of Computing Programming 9*, 2 (1987), 137–159.
3. CURIEN, P.-L., HARDIN, T., AND RÍOS, A. Normalisation forte du calcul des substitutions. Tech. rep., LIENS Ecole Normale Supérieure, 1991.
4. DAUCHET, M. Simulation of Turing machines by a left-linear rewrite rule. In *Proceedings of the 3rd Conference on Rewriting Techniques an Applications* (1989), N. Dershowitz, Ed., vol. 355 of *Lecture Notes in Computer Science*, Springer, pp. 109–120.
5. DERSHOWITZ, N. Termination of rewriting. *Journal of Symbolic Computation 3*, 1 and 2 (1987), 69–116.
6. DERSHOWITZ, N., AND JOUANNAUD, J.-P. Rewrite systems. In *Handbook of Theoretical Computer Science*, J. van Leeuwen, Ed., vol. B. Elsevier, 1990, ch. 6, pp. 243–320.
7. DERSHOWITZ, N., AND MANNA, Z. Proving termination with multiset orderings. *Communications ACM 22*, 8 (1979), 465–476.
8. FERREIRA, M. C. F., AND ZANTEMA, H. Total termination of term rewriting. In preparation.
9. GNAEDIG, I., AND LESCANNE, P. Proving termination of associative commutative rewriting systems by rewriting. In *Proceedings of the 8th Conference on Automated Deduction* (1986), J. H. Siekmann, Ed., vol. 230 of *Lecture Notes in Computer Science*, Springer, pp. 52–61.
10. HARDIN, T., AND LAVILLE, A. Proof of termination of the rewriting system SUBST on CCL. *Theoretical Computer Science 46* (1986), 305–312.
11. HOFBAUER, D. Termination proofs by multiset path orderings imply primitive recursive derivation lengths. In *Algebraic and Logic Programming* (1990), H. Kirchner and W. Wechler, Eds., vol. 463 of *Lecture Notes in Computer Science*, Springer, pp. 347 – 358.
12. HUET, G., AND LANKFORD, D. S. On the uniform halting problem for term rewriting systems. Rapport Laboria 283, INRIA, 1978.
13. KRUSKAL, J. Well-quasi-ordering, the tree theorem, and Vazsonyi's conjecture. *Trans. American Mathematical Society 95* (1960), 210–225.
14. KURATOWSKI, K., AND MOSTOWSKI, A. *Set Theory*. North-Holland Publishing Company, 1968.
15. LANKFORD, D. S. On proving term rewriting systems ar noetherian. Tech. Rep. MTP-3, Louisiana Technical University, Ruston, 1979.
16. LESCANNE, P. Termination of rewrite systems by elementary interpretations. Tech. Rep. 91-R-168, CRIN, 1991.
17. STEINBACH, J. Extensions and comparison of simplification orderings. In *Proceedings of the 3rd Conference on Rewriting Techniques an Applications* (1989), N. Dershowitz, Ed., vol. 355 of *Lecture Notes in Computer Science*, Springer, pp. 434–448.
18. TOYAMA, Y. Counterexamples to termination for the direct sum of term rewriting systems. *Information Processing Letters 25* (1987), 141–143.
19. ZANTEMA, H. Termination of term rewriting by interpretation. Tech. Rep. RUU-CS-92-14, Utrecht University, April 1992.

Path Orderings for Termination of Associative-Commutative Rewriting*

Nachum Dershowitz, Subrata Mitra
Department of Computer Science
University of Illinois at Urbana-Champaign
1304 West Springfield Avenue
Urbana, Illinois 61801, USA.
{nachum, mitra}@cs.uiuc.edu

Abstract

We show that a simple, and easily implementable, restriction on the recursive path ordering, which we call the "binary path condition," suffices for establishing termination of extended rewriting modulo associativity and commutativity.

1 Introduction

Rewrite systems find application to various aspects of theorem proving and programming language semantics. The essential idea in rewriting is to use an asymmetric directed equality ($\rightarrow$), rather than the usual symmetric equality relation ($\approx$).

Termination of a system consisting of such directed equations means that no infinite sequences of left-to-right replacements are possible for any term. Termination is important for using rewriting as a computational tool, and for simplification in theorem provers. One popular way of proving termination of a rewrite system is to use *path orderings*, based on a precedence relation on the function symbols of the system. Another common approach interprets function symbols as multivariate polynomials. For a survey of these techniques, see [Der87].

A binary function f is said to be associative and commutative (AC for short) if f obeys the equations

$$\begin{aligned} f(x, f(y,z)) &= f(f(x,y),z) \\ f(x,y) &= f(y,x). \end{aligned}$$

Since it is not possible to orient the second equation without losing termination, rewriting modulo such a congruence has to be handled in a special way. In essence, we rewrite AC equivalence classes, rather than terms.

*This research was supported in part by the U. S. National Science Foundation under Grants CCR-90-07195 and CCR-90-24271.

Polynomials can be used to prove termination of rewriting modulo AC when AC-equivalent terms have the same interpretation. But this severely restricts the degree of polynomial that can be used. (See [Lan79] and [BL87].) Path orderings have been commonly used in theorem provers, even for AC-rewriting (see the discussion in [Bjo82, page 350]), despite the fact that they do not establish termination in the AC case (see the counterexamples in [DHJP83]). Extensions of path orderings have been proposed that do handle associative and commutative functions properly ([BP85], for example), most recently in [KSZ90]. However, the ordering of [KSZ90] is difficult to implement, because it requires many nondeterministic operations (like *pseudocopying*; see Section 2).

In this paper, we show that if a rewrite system can be proved terminating using the recursive path ordering (RPO), then it is also AC-terminating—provided that when comparing two terms with the same (or equivalent) AC symbol at their roots, we compare subterms componentwise, rather than as multisets. This criterion can be easily implemented.

We write $s \sim_{ac} t$ to denote that s and t are rearrangements using the AC axioms. AC-rewriting ($\rightarrow_{R/AC}$) can be defined as follows: $u[s]_\pi \rightarrow_{R/AC} u[t]_\pi$, for terms s, t, context u and position π, if $s \sim_{ac} s'$, $s' \rightarrow_R t'$ and $t' \sim_{ac} t$. When dealing with AC systems, it is often convenient to treat AC symbols as functions with variable arity by considering only *flattened* terms. We use $\bar{t}$ to denote the flattened version of t. An ordering $\succ$ is *AC-compatible* if, for all terms s, s', t, t', $s \sim_{ac} s' \succ t' \sim_{ac} t$ implies $s \succ t$, in which case, we can also say that $\bar{s} \succ \bar{t}$. A rewrite system is *AC-terminating* if and only if the relation $\rightarrow_{R/AC}$ is contained in an AC-compatible reduction ordering. In general, we use standard terminology and notation for rewrite systems. For a survey of the field, refer to [DJ90].

2 Binary Path Condition

In this section we develop a restricted version of RPO—called "binary path condition"—which can be extended to an AC-compatible reduction ordering.

We first show that RPO, in general, is not AC-compatible. Consider the rule

$$f(a, f(a, b)) \quad \rightarrow \quad f(b, f(a, a))$$

If we consider $b \succ_f a$, then we can show that $f(a, f(a, b)) \succ_{rpo} f(b, f(a, a))$, assuming multiset status for f. However, we also have that $f(a, f(a, b)) \sim_{ac} f(b, f(a, a))$. Clearly, RPO with lexicographic status is not compatible with the commutativity axiom:

$$f(a, b) \quad \rightarrow \quad f(b, a)$$

If we now have $a \succ_f b$, then using left-to-right status for f, we have $f(a, b) \succ_{rpo} f(b, a) \sim_{ac} f(a, b)$, which violates irreflexivity. Finally, we show that RPO on flattened terms is not AC-compatible:

$$\begin{array}{rcl} f(a, b) & \rightarrow & g(a, b) \\ f(a, g(a, b)) & \rightarrow & f(a, a, b) \end{array}$$

Here $f \succ_f g$, and f is AC. Now, we have $f(a, a, b) = \overline{f(a, f(a, b))} \succ_{rpo} f(a, g(a, b)) \succ_{rpo} f(a, a, b)$, which violates irreflexivity.

These counterexamples show that RPO with status cannot be extended to an AC-compatible ordering. We therefore define a restricted version of it ($\succ_{bpc}$), which uses RPO with status for the non-AC symbols, but uses RPO without status to compare terms which have equivalent top-level AC operators. Here we use $t \succsim_{bpc} s$ to mean $t \sim_{ac} s$ or $t \succ_{bpc} s$.

Definition 2.1 (Binary Path Condition). Let $\succ_f$ be a well-founded precedence ordering on the function symbols. We have $t = f(t_1, \ldots, t_n) \succ_{bpc} g(s_1, \ldots, s_m) = s$ iff one of the following holds:

1. $t_i \succsim_{bpc} s$ for some i, $1 \leq i \leq n$.
2. $f \succ_f g$, and $t \succ_{bpc} s_j$ for all j, $1 \leq j \leq m$.
3. $f \sim_f g$, f and g are non-AC, and have the same status, and either
   - f has multiset status, and $\{t_1, \ldots, t_n\} \succ_{mul} \{s_1, \ldots, s_m\}$, or,
   - f has lexicographic status, and
     - $(t_1, \ldots, t_n) \succ_{lex} (s_1, \ldots, s_m)$, and
     - $t \succ_{bpc} s_j$ for all j, $1 \leq j \leq m$.
4. $f \sim_f g$, f, g are AC, $t = f(t_1, t_2)$ and $s = g(s_1, s_2)$, and either $(t_1, t_2) \succ_{comp} (s_1, s_2)$ or $(t_1, t_2) \succ_{comp} (s_2, s_1)$, where $(t_1, t_2) \succ_{comp} (s_1, s_2)$ iff either $t_1 \succ_{bpc} s_1$ and $t_2 \succsim_{bpc} s_2$, or, $t_1 \succ_{bpc} s_2$ and $t_2 \succsim_{bpc} s_1$.

To compare terms with variables, we can use the fact that a ground term $t\sigma$ is greater under $\succ_{bpc}$ than $x\sigma$ (x is a variable), for any substitution σ, whenever x occurs in t.

Theorem 2.2. *Let R be a rewrite system. If for each rule $l \rightarrow r \in R$ we have $l \succ_{bpc} r$, then R is AC-terminating.*

We first recall the definition of AC-RPO ($\succ_{ac}$), on ground terms, due to [KSZ90]. This ordering compares flattened terms.

Definition 2.3. Let $\succ_f$ be a well-founded precedence ordering on the function symbols. We have $t = f(t_1, \ldots, t_n) \succ_{ac} g(s_1, \ldots, s_m) = s$ iff one of the following holds:

1. $t_i \succsim_{ac} s$ for some i, $1 \leq i \leq n$, where $t_i \succsim_{ac} s$ iff $t_i \sim_{ac} s$ or $t_i \succ_{ac} s$.
2. $f \succ_f g$, and $t \succ_{ac} s_j$ for all j, $1 \leq j \leq m$.
3. $f \sim_f g$, f and g are non-AC and have the same status, and either
   - f has multiset status, and $\{t_1, \ldots, t_n\} \succ_{mul} \{s_1, \ldots, s_m\}$, or,
   - f has lexicographic status, and
     - $(t_1, \ldots, t_n) \succ_{lex} (s_1, \ldots, s_m)$, and
     - $t \succ_{ac} s_j$ for all j, $1 \leq j \leq m$.
4. $f \sim_f g$, f, g are AC, $t = f(T)$, $s = g(S)$, $S' = S - T = \{s'_1, ..., s'_k\}$ (where "$-$" denotes the multiset difference performed using $\sim_{ac}$, i.e., terms equivalent with respect to $\sim_{ac}$ can be dropped from both T and S), and either

- $k = 0$ and $n > m$ (i.e., $S - T = \emptyset$ and $T - S \neq \emptyset$), or
- $f(T - S) \Rightarrow^* f(T')$, and $T' = T_1 \cup \ldots \cup T_k$ and for all i $(1 \leq i \leq k)$ either
  - $T_i = \{u\}$ and $u \succsim_{ac} s'_i$, or
  - $T_i = \{u_1, \ldots, u_l\}$ and $f(u_1, \ldots, u_l) \succsim_{ac} s'_i$.

  Also, in this case, either $t \Rightarrow^+ f(T')$, or, for at least one i, we have instead a strict decrease in $\succ_{ac}$.

Case 4 of this definition uses the operation $\Rightarrow$, which may be one of the following: *pseudo-copying*, *elevation* or *flattening*. Here we briefly explain these notions; for details refer to [KSZ90]. Pseudo-copying is used to allow a single *big* (that is, with a top-level function which is higher than f in the precedence relation $\succ_f$) subterm on the left-hand side to handle multiple subterms on the right. For example, while comparing the terms $t_1 = f(g(x))$ and $t_2 = f(h(x), h(x))$, where f is AC, and $g \succ_f f \succ_f h$, we can say that $t_1 \succ_{ac} t_2$, since $t_1 \Rightarrow f(gg(x), gg(x)) \succ_{ac} t_2$, where $gg(x)$ is a pseudo-copy of $g(x)$. Note that pseudo-copying is allowed only for big terms which are immediate subterms of the top-level AC operator of the left-hand side term. At times, a big term may be nested further down, in which case elevation is used to bring it up. For example, in comparing $f(c(g(x)))$ with $f(h(x), h(x))$, where $g \succ_f f \succ_f h \succ_f c$, we can use the following steps: $f(c(g(x))) \Rightarrow f(g(x)) \succ_{ac} f(h(x), h(x))$. Finally, flattening can be used to remove immediate nesting of different AC functions which have the same precedence. For example, we could say $f(g(x), y) \Rightarrow f(x, y)$, if $f \sim_f g$, and f and g are AC. The essential idea in this ordering is to partition the subterms of the AC functions and compare the components, using $\Rightarrow$ to make the relation transitive. It is shown in [KSZ90] that this ordering is well-founded and AC-compatible.

We are ready for a proof of the theorem:

Proof. We show that for any two terms s and t, if $t \succ_{bpc} s$ then $\bar{t} \succ_{ac} \bar{s}$, by induction on the sizes of s and t. There are several cases to be considered, depending on the possibile reasons why $t \succ_{bpc} s$. We assume that t is of the form $f(t_1, \ldots, t_n)$ and s is $g(s_1, \ldots, s_m)$.

1. If $t_i \succsim_{bpc} s$, then, by the inductive hypothesis we have $t_i \succsim_{ac} s$, and hence $\bar{t} \succ_{ac} \bar{s}$, by Case 1 of Definition 2.3.

2. If $f \succ_f g$ and $t \succ_{bpc} s_j$, then by the inductive hypothesis, we have $\bar{t} \succ_{ac} \overline{s_j}, 1 \leq j \leq m$. There are two further possibilities:

   - g is not AC. In this case, $\bar{s} = g(\overline{s_1}, \ldots, \overline{s_m})$, and therefore we have $\bar{t} \succ_{ac} \bar{s}$, by Case 2 of Definition 2.3.
   - Suppose g is AC (thus $m = 2$). In this case, there are various possibilities for s, for example we could have: $s = g(g(s_{1.1}, s_{1.2}), s_2)$, or $s = g(s_1, g(s_{2.1}, s_{2.2}))$, or $s = g(g(s_{1.1}, s_{1.2}), g(s_{2.1}, s_{2.2}))$, and so forth. However, in each case, we have $\bar{s} = g(s'_1, \ldots, s'_k)$, where each $s'_j, 1 \leq j \leq k$, is either a subterm of s_1 or of s_2. Therefore, we have $\bar{t} \succ_{ac} \bar{s}$.

3. If f and g are both non-AC, and $f \sim_f g$, then we can use the inductive hypothesis on the flatten subterms of s, and then the proposition follows.

4. If f and g are both AC, and $f \sim_f g$ then $n = m = 2$. Furthermore, without loss of generality, we can assume that $t_1 \succ_{bpc} s_1$, *and* $t_2 \succsim_{bpc} s_2$. (The other cases admit similar proofs.) There are two further possibilities:

    - If $t_2 \succ_{bpc} s_2$, then by the inductive hypothesis we have: $\overline{t_1} \succ_{ac} \overline{s_1}$ and $\overline{t_2} \succ_{ac} \overline{s_2}$. Thus, we could use this partitioning of t to show that $\overline{t} \succ_{ac} \overline{s}$.
    - If $t_2 \sim_{ac} s_2$, then again the proposition holds, because we could ignore $\overline{t_2}$ and $\overline{s_2}$ when comparing $\overline{t}$ with $\overline{s}$.

□

Since $\succ_{ac}$ is AC-compatible, we have $\overline{t} \succ_{ac} \overline{s}$, not only when $t \succ_{bpc} s$, but also if $t \sim_{ac} t' \succ_{bpc} s$. In order to prove termination of a system using $\succ_{bpc}$, it is therefore sufficient to use any rearrangement of the left- and right-hand side terms. We also have, for any AC function symbol f and terms s and t, that if $t \succ_{bpc} s$, then $f(t, X) \succ_{bpc} f(s, X)$ (and therefore, $\overline{f(t, X)} \succ_{ac} \overline{f(s, X)}$).

We have shown that the relation $\succ_{bpc}$ is embedded in the AC-compatible reduction ordering $\succ_{ac}$. Therefore, the binary path condition is sufficient for proving AC-termination. It is important to note that the relation ($\succ_{bpc}$) defined here is not really an ordering, because it is not transitive. For example, if we have the precedence relation $g \succ_f f \succ_f h$, then we can show that (here f is AC, while g and h are non-AC)

$$f(g(x), g(y)) \succ_{bpc} f(f(x,x), f(y,y)) \sim_{ac} f(f(x,y), f(x,y)) \succ_{bpc} f(h(x,y), h(x,y)).$$

However, it is the case that $f(g(x), g(y)) \not\succ_{bpc} f(h(x,y), h(x,y))$. The interesting point about $\succ_{bpc}$ is that it is easy to implement; much easier than the ordering of [KSZ90].

3 Examples

The binary path condition developed in the previous section, like $\succ_{ac}$, and unlike [BP85], has no restriction on the precedence relation $\succ_f$, and can therefore be used to prove termination of a large class of rewrite systems.

Two examples follow:

Example 3.1 (Arithmetic over natural numbers). *Here* $*$ *and* $+$ *are AC, and* $* \succ_f + \succ_f s \succ_f 0$.

$$\begin{aligned} 0 + x &\to x \\ s(x) + y &\to s(x + y) \\ 0 * x &\to 0 \\ s(x) * y &\to y + (x * y) \\ (x + y) * z &\to (x * z) + (y * z) \end{aligned}$$

Example 3.2 (Free commutative ring). *Here $*$ and $+$ are AC, and $* \succ_f - \succ_f + \succ_f 0$.*

$$\begin{aligned} 0+x &\rightarrow x \\ -x+x &\rightarrow 0 \\ -0 &\rightarrow 0 \\ --x &\rightarrow x \\ -(x+y) &\rightarrow -x+-y \\ 0*x &\rightarrow 0 \\ -x*y &\rightarrow -(x*y) \\ x*(y+z) &\rightarrow (x*y)+(x*z) \end{aligned}$$

4 Discussion

We have considered a simple restriction on RPO that can be extended to an AC-compatible ordering. The restriction disallows comparison of two terms with equivalent top-level AC functions when both subterms on the right-hand side are dominated by only one subterm on the left-hand side.

Independently, Bachmair [Bac92] has presented an AC-compatible rewrite-relation, also based on [KSZ90], and proved its termination using a minimal counterexample argument. Our termination condition is essentially the same as one rewrite step of [Bac92], with the possibility of multiset status added. It is believed that the transitive closure of this relation is identical to the ordering in [KSZ90], but this remains to be proved.

It will be interesting to be able to extend the relation defined here to cases where simple multiset comparisons may be allowed for subterms of the AC-terms.

References

[Bac92] Leo Bachmair. Associative-commutative reduction orderings. *Information Processing Letters.* To appear.

[BP85] Leo Bachmair and David A. Plaisted. Termination orderings for associative-commutative rewrite systems. *J. of Symbolic Computation*, vol. 1, pages 329–349 (1985).

[BL87] Ahlem Ben Cherifa and Pierre Lescanne. Termination of rewriting systems by polynomial interpretations and its implementation. *Science of Computer Programming*, vol. 9, pages 137–159 (1987).

[Bjo82] Dines Bjorner, editor. Proceedings of the IFIP Working Conference on Formal Description of Programming Concepts–II. Garmisch-Partenkirchen, West Germany, North-Holland 1982.

[Der87] Nachum Dershowitz. Termination of rewriting. *J. of Symbolic Computation*, vol. 3, pages 69–116 (1987).

[DHJP83] Nachum Dershowitz and Jieh Hsiang and N. Alan Josephson and David A. Plaisted. Associative-commutative rewriting. In *Proceedings of the Eighth International Joint Conference on Artificial Intelligence,* Karlsruhe, West Germany, pages 940–944, 1983.

[DJ90] Nachum Dershowitz and Jean-Pierre Jouannaud. Rewrite systems. In J. van Leeuwen, editor, *Handbook of Theoretical Computer Science*, chapter 6, pages 243–320, North-Holland, Amsterdam, 1990.

[KSZ90] Deepak Kapur, G. Sivakumar and Hantao Zhang. A new method for proving termination of AC-rewrite systems. In *Proceedings of the Tenth International Conference of Foundations of Software Technology and Theoretical Computer Science*, vol. 472 of *Lecture Notes in Computer Science*, pages 133–148, Springer-Verlag, Berlin, 1990.

[Lan79] Dallas S. Lankford. On proving term rewriting systems are Noetherian. Memo MTP-3, Mathematics Department, Louisiana Tech. University, Ruston, LA, 1979.

Generic Induction Proofs

Peter Padawitz
Fachbereich Informatik
Universität Dortmund
Germany

Abstract. We summarize a number of new results concerning inductive-theorem proving in the area of design specifications using Horn logic with equality. Induction is explicit here because induction orderings must be integrated into the specification. However, the proofs need less guidance if the specification is ground confluent and strongly terminating. Calculi for verifying these conditions are presented along with a list of useful applications.

Introduction

We present a calculus for proofs of Gentzen clauses with respect to the inductive theory of a Horn clause specification with equality. The calculus is generic insofar as it can be actualized by different ground theories (= Herbrand interpretations). Two actual parameters are discussed in detail: the set of all valid ground goals and the set of *strongly convergent* ground goals. Previous versions of the first actualization were presented in [Pad 91, 92]. The non-inductive part of the second one was introduced and applied in [Pad 92]. In contrast with inductive completion, the second actualization allows us to prove that a specification is ground confluent without mixing up axioms, which are restricted to Horn clauses, and theorems, which may be arbitrary Gentzen clauses. This distinction is strongly suggested by the practice of program design where the equations *axioms* = *programs* = *Horn clauses* and *theorems* = *requirements* = *Gentzen clauses* turn out to be a both simple and effective basis of formal reasoning.

The paper is organized as follows. Section 1 yields preliminaries and the general conditions on ground theories to actualize the generic calculus. Section 2 provides the non-inductive rules of the calculus, Section 3 the inductive ones. Section 4 deals with the first actualization, which allows us to prove inductive theorems w.r.t. a Horn axiom set AX. Here the generic calculus leads to *classical expansions,* i.e. explicit induction proofs. Section 5 reviews the notions and results of [Pad 92] concerning reductions upon AX. In Section 6, a powerful method for proving that AX is ground confluent is derived from the second actualization of the generic calculus. Since its soundness relies on the strong termination of AX, we present the *path calculus* for proving this condition in Section 7. It combines syntactic with semantic properties insofar as it generates a term ordering, which is based both on a signature ordering and on well-founded relations specified by AX.

Sections 2-4 also provide a theoretical basis of parts of EXPANDER, a prototyping system for testing and verifying declarative programs (cf. [Pad 92a]).

1. Preliminaries

We assume some familiarity with the basic notions of algebraic specification and Horn logic (cf., e.g., [EM 85], [Pad 88], [Wir 90]). Given sets A and B, *A⊎B* denotes the disjoint union of A and B, i.e., A⊎B is defined as $A \cup B$ iff $A \cap B = \emptyset$.

Let S be a set and *SIG* be an S-sorted signature with function as well as predicate symbols. ≡ denotes an equality predicate. Let X be an infinite set of variables. *Term* (*GTerm*) denotes the S-sorted set of (ground) terms over SIG and X. SIG is assumed to be *inhabited*, i.e. for all sorts s there is a ground term of sort s.

Sub (*GSub*) stands for the set of (ground) substitutions over SIG, i.e. S-sorted functions from X to Term (GTerm). As usual we write $x\sigma$ for $\sigma(x)$. For all $\sigma \in$ Sub, *dom(σ)* $= \{x \in X \mid x\sigma \neq x\}$. Given $t \in$ Term and $x \in X$, *[t/x]* denotes the substitution σ defined by $x\sigma = t$ and $dom(\sigma) = \{x\}$. Let $\sigma,\tau \in$ Sub such that for all $x \in dom(\sigma) \cap dom(\tau)$, $x\sigma = x\tau$. Then for all $x \in X$, $x(\sigma+\tau) = x\sigma$ if $x \in dom(\sigma)$ and $x(\sigma+\tau) = x\tau$ if $x \in dom(\tau)$.

Let $Y \subseteq X$. Then $Y\sigma = \{x\sigma \mid x \in Y\}$ and for all $x \in X$, $x(\sigma \upharpoonright Y) = x\sigma$ if $x \in Y$ and $x(\sigma \upharpoonright Y) = x$ if $x \in X-Y$. σ also denotes its unique SIG-homomorphic extension σ^* to Term. Hence for all $\sigma,\tau \in$ Sub and $x \in X$, $x\sigma\tau = \tau^*(x\sigma)$. σ *subsumes* τ, written as $\sigma \leq \tau$, if $\sigma\rho = \tau$ for some $\rho \in$ Sub. σ *unifies* the terms t and u if $t\sigma = u\sigma$. σ is a *most general unifier (mgu)* of t and u if σ unifies t and u and subsumes all unifiers of t and u.

Let $\sigma \in$ Sub. Then *EQ(σ)* $= \{x \equiv x\sigma \mid x \in dom(\sigma)\}$. Given $Y,Z \subseteq X$, σ is a *renaming of Y away from Z* if $Y\sigma \subseteq X$, $|Y\sigma| = |Y|$ and $Y\sigma \cap Z = \emptyset$. A set of atom(ic formula)s over SIG (and X) is called a *(ground) goal*. If g is a goal and σ is a renaming, then the goal $g\sigma = \{p\sigma \mid p \in g\}$ is a *variant* of g. A goal set gs stands for the disjunction of its elements. If $g \in gs$ is a variant of $g' \in gs$, then g is identified with g'.

A *Horn clause* is an expression of the form $g \Leftarrow d$ where g and d are goals. A *Gentzen clause* c has the form $\exists X_1 g_1 \vee ... \vee \exists X_n g_n \Leftarrow d$ where for all $1 \leq i \leq n$, $X_i \subseteq X$ is the set of existential(ly quantified) variables occurring in the goal g_i, but (w.l.o.g) not in the goal d.

Assumptions of the generic calculus $X_{in} \subseteq X$ is a given set of *input variables*. $X_{out} = X - X_{in}$ is called the set of *output variables*. For all $\sigma \in$ Sub, let $\sigma_{in} = \sigma\upharpoonright X_{in}$ and $\sigma_{out} = \sigma\upharpoonright X_{out}$. Let TH be a set of ground goals and GS $\subseteq$ GS' be sets of ground substitutions such that

- $h \subseteq g \in$ TH implies $h \in$ TH,
- for all $\sigma,\tau \in$ GS', $\sigma_{in}+\tau_{out} \in$ GS',
- for all $\sigma \in$ GS and $\tau \in$ GS', $\sigma_{in}+\tau_{out} \in$ GS. □

Definition *Gen(TH)* denotes the set of Gentzen clauses such that

$$\exists X_1 g_1 \vee ... \vee \exists X_n g_n \leftarrow g \quad \in \quad \text{Gen(TH)}$$

iff for all $\sigma \in$ GS (!), $g\sigma \in$ TH implies $g_i\tau \in$ TH for some $1 \le i \le n$ and $\tau \in$ GS' with $\tau\upharpoonright(X-X_i) = \sigma\upharpoonright(X-X_i)$. □

Note that TH is a subset of Gen(TH).

2. Deductive steps

Let **Pred** be a predicate on goals such that for all goals g, terms t,u and $x \in var(g)$,

(*) $\quad Pred(g[t/x]) \quad$ implies $\quad g[t/x] \leftarrow g[u/x] \cup \{t \equiv u\} \in$ Gen(TH).

In particular, for all $\sigma \in$ GSub,

(**) $\quad Pred(g) \quad$ implies $\quad g\sigma_{out} \leftarrow g\sigma \cup EQ(\sigma_{in}) \in$ Gen(TH).

Definition The *deductive expansion calculus upon (TH,Pred)* consists of the following two inference rules, which transform sets of goals.

Deductive resolution Let $\sigma \in$ Sub and $\exists X_1 h_1\sigma \vee ... \vee \exists X_n h_n\sigma \leftarrow h\sigma \in$ Gen(TH) such that for all $1 \le i \le n$, $X_i \cap var(g_i\sigma \cup X_{in}\sigma \cup X_{in}) = \emptyset$.

$$\frac{\{g_1 \uplus h_1, ..., g_n \uplus h_n\} \uplus rest}{\{(g_1 \cup ... \cup g_n \cup h)\sigma \cup EQ(\sigma_{in})\} \cup rest}$$

Deductive paramodulation Let $\sigma \in$ Sub, $x \in var(g_1 \cap ... \cap g_n)$ and $\exists X_1(t_1 \equiv x)\sigma \vee ... \vee \exists X_n(t_n \equiv x)\sigma \leftarrow h\sigma \in$ Gen(TH) such that for all $1 \le i,j \le n$, $X_i \cap var(g_j\sigma \cup h\sigma \cup X_{in}\sigma \cup X_{in}) = \emptyset$.

$$\frac{\{g_1[t_1/x], ..., g_n[t_n/x]\} \uplus rest}{\{(g_1 \cup ... \cup g_n \cup h)\sigma \cup EQ(\sigma_{in})\} \cup rest}$$

A sequence $gs_1,\ldots,gs_n$ of goal sets is called a *deductive expansion of* gs_1 *into* gs_n *upon (TH,Pred)* if

- for all $1 \le i < n$, gs_{i+1} is obtained from gs_i by a single deductive resolution or paramodulation step,
- for all $1 \le i \le n$ and $g \in gs_i$, Pred(g) holds true.

The corresponding inference relation is denoted by $\vdash_{TH}$. □

Theorem 1 (Soundness of deductive expansions) *If*

$$gs' = \{g_1',\ldots,g_m'\} \vdash_{TH} gs,$$

then for all $g \in gs$,

$$\exists X_{out} g_1' \vee \ldots \vee \exists X_{out} g_m' \Leftarrow g \in Gen(TH).$$ □[1]

3. Inductive steps

Suppose that there is a bijection * on a subset Y_{in} of X_{in} such that $X_{in} = Y_{in} \uplus Y_{in}^*$ and L_{in} is a list representation of Y_{in}, $s = sort(L_{in})$ and $>> : s \times s$ is a predicate of SIG such that the *induction ordering* $>>^{TH} \subseteq GTerm_s \times GTerm_s$ defined by:

$$t >>^{TH} t' \text{ iff } (t >> t') \in TH$$

is well-founded and for all $\sigma \in GS$ and $\tau \in GS'$,

$$(***) \qquad L_{in}\sigma >>^{TH} L_{in}\tau \text{ implies } \tau \in GS.$$

Moreover let *Sub'* be a set of substitutions including all renamings such that for all $\sigma \in$ Sub' and $\tau \in$ GS', $\sigma\tau \in$ GS'.

Definition Let CS be a set of Gentzen clauses. The *inductive expansion calculus upon (TH,Pred) for CS* consists of the deductive resolution and paramodulation rules and the following two inference rules, which also transform sets of goals.

Inductive resolution Let $c = \exists X_1 h_1 \vee \ldots \vee \exists X_n h_n \Leftarrow h \in CS$ and $\sigma \in$ Sub' be a renaming of $Z = X_1 \cup \ldots \cup X_n$ away from var(c) such that for all $1 \le i \le n$, $f_i\sigma = h_i{}^*\sigma$, $X_i \subseteq X_{out}$ and $Z\sigma \cap var(g_i\sigma \cup X_{in}\sigma \cup X_{in}) = \emptyset$.

---

[1]All proofs appear or is referred to in the full version of this paper.

$$\frac{\{g_1 \uplus f_1,..., g_n \uplus f_n\} \uplus rest}{\{(g_1 \cup...\cup g_n \cup h^* \cup \{L_{in}>>L_{in}^*\})\sigma \cup EQ(\sigma_{in})\} \cup rest}$$

Inductive paramodulation Let $c = \exists X_1u_1{\equiv}t\vee...\vee\exists X_nu_n{\equiv}t \leftarrow h \in CS$, $x \in var(g_1\cap...\cap g_n)$ and $\sigma \in$ Sub' be a renaming of $Z = X_1\cup...\cup X_n$ away from var(c) such that for all $1\le i\le n$, $t_i\sigma = u_i{}^*\sigma$, $X_i \subseteq X_{out}$ and $Z\sigma \cap var(g_i[t^*/x]\sigma\cup h^*\sigma\cup X_{in}\sigma\cup X_{in}) = \emptyset$.

$$\frac{\{g_1[t_1/x],..., g_n[t_n/x]\} \uplus rest}{\{(g_1[t^*/x] \cup...\cup g_n[t^*/x] \cup h^* \cup \{L_{in}>>L_{in}^*\})\sigma \cup EQ(\sigma_{in})\} \cup rest}$$

A sequence $gs_1,...,gs_n$ of goal sets is called an *inductive expansion of gs_1 into gs_n upon (TH,Pred) for CS* if

- for all $1\le i<n$, gs_{i+1} is obtained from gs_i by a single deductive or inductive resolution or paramodulation step,
- for all $1\le i\le n$ and $g \in gs_i$, Pred(g) holds true.

The corresponding inference relation is denoted by $\vdash_{TH,CS}$. □

Theorem 2 (Soundness of inductive expansions) *Given goals $g_1,...,g_m,g$ with all their variables in $Y_{in}\cup X_{out}$, let $CS = \{d_1\leftarrow h_1,...,d_m\leftarrow h_m\}$ be a set of Gentzen clauses such that*

- *for all $1\le i,j\le m$ and $\tau \in GSub$, $d_i\leftarrow g_j\tau_{out}\cup h_i$, $g\leftarrow h_i \in Gen(TH)$ and the existential variables of d_i are output variables,*
- *$var(CS) \subseteq Y_{in}\cup X_{out}$.*

$CS \subseteq Gen(TH)$ if $\{g_1,...,g_m\} \vdash_{TH,CS} \{g\}$. □

A special case of Theorem 2 is given by

Theorem 3 *Given goals $g_1,...,g_n,g$ with all their variables in $Y_{in}\cup X_{out}$, let*

$$CS = \{\exists X_1h_1\vee...\vee\exists X_nh_n \leftarrow g \mid \text{for all } 1\le i\le n,\ h_i \subseteq g_i\}.$$

$CS \subseteq Gen(TH)$ if $\{g_1,...,g_n\} \vdash_{TH,CS} \{g\}$. □

Moreover, m Horn clauses $d_1\leftarrow h_1,...,d_m\leftarrow h_m$ with different premises can be proved simultaneously if they can be related to goals $g_1,...,g_m,g$ as in Theorem 2. $h_1,...,h_m$ must have a common *guard* g such that the rest of

$h_1,...,h_m$ constitutes a *g-minimal* goal set (cf. [Pad 92], Chapter 2).

Definition A Horn clause $d \leftarrow g \uplus h$ is a *guarded clause* if $var(d) \subseteq X_{in} \cup var(h)$. Given a goal g, a set gs of goals is *g-minimal w.r.t. TH* if for all h,h' $\in$ gs and $\sigma,\tau \in GS$, $\sigma_{in} = \tau_{in}$ and $g\sigma \cup h\sigma \cup h'\tau \in TH$ imply $h = h'$ and $x\sigma \equiv x\tau \in TH$ for all $x \in var(h) \cap X_{out}$. □

Lemma 4 *Let* $CS = \{d_1 \leftarrow g \uplus h_1,...,d_n \leftarrow g \uplus h_n\}$ *be a set of guarded clauses such that for all* $1 \le i \le n$, $Pred(d_i)$ *holds true and* $\{h_1,...,h_n\}$ *is g-minimal w.r.t. TH. Let*

$$c = \exists X_{out}(d_1 \cup h_1) \vee ... \vee \exists X_{out}(d_n \cup h_n) \leftarrow g$$

and $c' = \exists X_{out}h_1 \vee ... \vee \exists X_{out}h_n \leftarrow g$.

(1) *For all* $1 \le i,j \le n$ *and* $\tau \in GSub$, $d_i \leftarrow (d_j \cup h_j)\tau_{out} \cup g \cup h_i \in Gen(TH)$.
(2) $CS \subseteq Gen(TH)$ *if* $c \in Gen(TH)$.
(3) $c \in Gen(TH)$ *if* $CS \cup \{c'\} \subseteq Gen(TH)$ *and* $var(g) \subseteq X_{in}$. □

Lemma 4(1) and Theorem 2 yield

Theorem 5 *Let* $CS = \{d_1 \leftarrow g \uplus h_1,...,d_n \leftarrow g \uplus h_n\}$ *be a set of guarded clauses such that for all* $1 \le i \le n$, $Pred(d_i)$ *holds true,* $\{h_1,...,h_n\}$ *is g-minimal w.r.t. TH and* $var(CS) \subseteq Y_{in} \cup X_{out}$. $CS \subseteq Gen(TH)$ *if* $\{d_1 \cup h_1,...,d_n \cup h_n\} \vdash_{TH,CS} \{g\}$. □

In each of the following two sections we consider an actual parameter (TH,Pred) and the respective inductive expansion calculus upon (TH,Pred).

4. Classical expansions

Let *AX* be a set of Horn clauses, usually constituting the axioms that interpret SIG as a set of functions and predicates on *Ini(SIG,AX)*, the *initial (SIG,AX)-structure* (cf. [EM 85], [Pad 88], [Wir 90]). Since we restrict AX to Horn clauses, Ini(SIG,AX) always exists and is represented as the quotient of GTerm by the *AX-equivalence* $\equiv_{AX}$:

$$t \equiv_{AX} t' \quad \text{iff} \quad AX \vdash_{cut} t \equiv t'$$

where $\vdash_{cut}$ denotes the inference relation of the *cut calculus for Horn clauses*, which consists of the congruence axioms for $\equiv$ and two inference rules for deriving Horn clauses:

Cut

$$\frac{g \leftarrow d \cup g',\ g' \leftarrow d'}{g \leftarrow d \cup d'}$$

Substitution Let $\sigma \in$ Sub.

$$\frac{g \leftarrow d}{g\sigma \leftarrow d\sigma}$$

In terms of our generic approach, let TH be the *ground deductive theory of AX*, i.e. the set of ground goals g such that AX $\vdash_{cut}$ g, and let GS = GS' = GSub and Sub' = Sub. Then Gen(TH) as defined in Section 1 is the set of Gentzen clauses that are valid in Ini(SIG,AX) (cf. [Pad 92], Section 3.1). In particular, AX $\subseteq$ Gen(TH). The clauses of Gen(TH) are called *inductively valid w.r.t. AX*. A goal g is called *solvable* if $g\sigma \in$ TH for some $\sigma \in$ GSub.

Define **Pred** as "constantly true". Since the cut calculus includes the congruence axioms for $\equiv$, (*) holds true (cf. Section 2). An inductive expansion upon (TH,Pred) is called a *classical expansion*.

Given a SIG-structure that A satisfies AX, the well-foundedness of an induction ordering (cf. Section 3) can be reduced to the well-foundedness of the relation $>>^A \subseteq A_s \times A_s$ defined by:

$$a >>^A b \text{ iff } A \models a >> b$$

(cf. [Pad 92], Proposition 5.3), i.e., one may refer to a given model of AX at this point.

By Theorems 3 and 5, the construction of classical expansions yields a proof method for inductive validity:

Theorem 6 (Soundness of classical expansions) *Let TH be the ground deductive theory of AX.*

(1) Given goals $g_1,\ldots,g_n,g$, $CS = \{\exists X_1 h_1 \vee \ldots \vee \exists X_n h_n \leftarrow g \mid$ *for all* $1 \le i \le n$, $h_i \subseteq g_i\}$ *is inductively valid w.r.t. AX if* $\{g_1,\ldots,g_n\} \vdash_{TH,CS} \{g\}$.

(2) Let $CS = \{d_1 \leftarrow g \uplus h_1,\ldots,d_n \leftarrow g \uplus h_n\}$ *be a set of guarded clauses such that* $\{h_1,\ldots,h_n\}$ *is g-minimal w.r.t. TH. CS is inductively valid w.r.t. AX if* $\{d_1 \cup h_1,\ldots,d_n \cup h_n\} \vdash_{TH,CS} \{g\}$. □

Since for all goals g,h, atoms p, terms t,u and variables x, the clauses $p \leftarrow p$, $x \equiv x$, $t \equiv u \leftarrow t \equiv u$ and $g \vee h \leftarrow g$ are inductively valid, the following *derived* rules are special cases of deductive resolution respectively paramodulation:

Factoring Let $\sigma \in$ Sub such that $p\sigma = q\sigma$.

$$\frac{\{g \uplus \{p,q\}\} \uplus rest}{\{g\sigma \cup \{p\sigma\}\} \cup rest}$$

Unification Let $\sigma \in$ Sub such that $t\sigma = u\sigma$.

$$\frac{\{g \uplus \{t\equiv u\}\} \uplus rest}{\{g\sigma\} \cup rest}$$

Term replacement Let $x \in var(g)$.

$$\frac{\{g[t/x]\} \uplus rest}{\{g[u/x] \cup \{t\equiv u\}\} \cup rest}$$

Goal elimination

$$\frac{\{g, h\} \uplus rest}{\{g\} \cup rest}$$

Example 7 Using the syntax of EXPANDER (cf. [Pad 92a]) we present a specification of the *greater* relation on naturals:

```
GREATER
   preds²          > 1 2 >> 3
   infixes         > >>
   vars            x y z z' ex
   axioms (1)      {s(x) > 0}
          (2)      {s(x) > s(y)} <== {x > y}
          (3)      {(s(x),s(y),z) >> (x,y,z')}
   theorems
          (1)      {x > y} <== {s(x) > s(y)}
          (2)      {x = s(ex!)} <== {x > y}
          (3)      {x = 0}V{x = s(ex!)}³
   conjects (1)    {x > z} <== {x > y, y > z}
```

Here is a classical expansion, which, by Theorem 6(1), implies that Conjecture 1 is inductively valid w.r.t. GREATER:

```
initial conclusion:
 (1) { x > z }
```

[2]The numbers following a function or predicate symbol P denote the axioms that specify P.
[3]The exclamation mark identifies *ex* as an existentially quantified variable.

```
initial premise:
 (1) { x > y, y > z }
atom 1 in conclusion goal 1 resolved with axiom 1
atom 1 in conclusion goal 1 resolved with axiom 2
 (1) { x1 > y1, x = s(x1), z = s(y1) }
 (2) { x = s(x1), z = 0 }
atom 1 in conclusion goal 1 resolved with conjecture 1
 (1) { x = s(x1), z = s(y1), x1 > y2, y2 > y1, (x, y, z) >> (x1, y2, y1) }
 (2) { x = s(x1), z = 0 }
atom 5 in conclusion goal 2 resolved with axiom 3
 (1) { z = s(y1), x1 > y2, y2 > y1, x = s(x1), y = s(y2) }
 (2) { x = s(x1), z = 0 }
atom 2 in conclusion goal 1 resolved with theorem 1
 (1) { z = s(y1), y2 > y1, x = s(x1), y = s(y2), s(x1) > s(y2) }
 (2) { x = s(x1), z = 0 }
atom 2 in conclusion goal 1 resolved with theorem 1
 (1) { z = s(y1), x = s(x1), y = s(y2), s(x1) > s(y2), s(y2) > s(y1) }
 (2) { x = s(x1), z = 0 }
term at position 4 1 replaced with equation 2 in conclusion goal 1
 (1) { z = s(y1), x = s(x1), y = s(y2), x > s(y2), s(y2) > s(y1) }
 (2) { x = s(x1), z = 0 }
terms at positions 4 2 , 5 1 replaced with equation 3 in conclusion goal 1
 (1) { z = s(y1), x = s(x1), y = s(y2), x > y, y > s(y1) }
 (2) { x = s(x1), z = 0 }
term at position 5 2 replaced with equation 1 in conclusion goal 1
 (1) { z = s(y1), x = s(x1), y = s(y2), x > y, y > z }
 (2) { x = s(x1), z = 0 }
atoms 2 1 in conclusion goals 2 1 resolved with theorem 3
 (1) { x = s(x1), y = s(y2), x > y, y > z }
atom 1 in conclusion goal 1 resolved with theorem 2
 (1) { y = s(y2), x > y, y > z, x > y3 }
atom 1 in conclusion goal 1 resolved with theorem 2
 (1) { x > y, y > z, x > y3, y > y4 }
conjecture 1 has been proved □
```

Example 8 Using the syntax of EXPANDER (cf. [Pad 92a]) we present a specification of the *division-and-remainder* function on naturals:

```
DIVISION
    functs        + 1 2 - * 3 4 div 5 6
    preds         < > >= >> 7
    infixes       + - * < > >= >>
    vars          x y z q r cq cr
    axioms (1)    { 0+x = x }
           (2)    { s(x)+y = s(x+y) }
           (3)    { 0*x = 0 }
           (4)    { s(x)*y = (x*y)+y }
           (5)    { div(x,y) = (0,x) } <== { x < y }
           (6)    { div(x,y) = (s(q),r) }<== { x >= y, y > 0, div(x-y,y) = (q,r) }
           (7)    { (x,y) >> (x,y-z) } <== { y >= z, z > 0 }
    theorems
           (1)    { (x+y)+z = (x+z)+y }
```

```
          (2)   { x = z+y } <== { x-y = z, x >= y }
          (3)   { div(x,y) = (cq!,cr!) } <== { y > 0 }
          (4)   { x < y }V{ x >= y }
conjects (1)    { x = (q*y)+r, r < y } <== { y > 0, div(x,y) = (q,r) }
```

Here is a classical expansion implying that Conjecture 1 is inductively valid w.r.t. DIVISION. In terms of Theorem 6(2) we choose $n = 1$, $d_1 = \{x = (q*y)+r,\ r < y\}$, $g = \{y > 0\}$ and $h_1 = \{div(x, y) = (q, r)\}$ because h_1 is g-minimal w.r.t. TH:

```
initial conclusion:
 (1) { x = (q*y)+r, r < y, div(x, y) = (q, r) }
initial premise:
 (1) { y > 0 }
term at position 3 1 in conclusion goal 1 paramodulated with axiom 5
term at position 3 1 in conclusion goal 1 paramodulated with axiom 6
current conclusion:
 (1) { x = (q*y)+r, r < y, (s(q1), r1) = (q, r), x >= y, y > 0,
       div(x-y, y) = (q1, r1) }
 (2) { x = (q*y)+r, r < y, (0, x) = (q, r), x < y }
equation 3 in conclusion goal 1 unified
 (1) { x = (s(q1)*y)+r1, r1 < y, x >= y, y > 0, div(x-y, y) = (q1, r1) }
 (2) { x = (q*y)+r, r < y, (0, x) = (q, r), x < y }
equation 3 in conclusion goal 2 unified
 (1) { x = (s(q1)*y)+r1, r1 < y, x >= y, y > 0, div(x-y, y) = (q1, r1) }
 (2) { x = (0*y)+x, x < y }
term at position 1 2 1 in conclusion goal 1 paramodulated with axiom 4
current conclusion:
 (1) { r1 < y, x >= y, y > 0, div(x-y, y) = (q1, r1), x = ((q1*y)+y)+r1 }
 (2) { x = (0*y)+x, x < y }
term at position 1 2 1 in conclusion goal 2 paramodulated with axiom 3
current conclusion:
 (1) { r1 < y, x >= y, y > 0, div(x-y, y) = (q1, r1), x = ((q1*y)+y)+r1 }
 (2) { x < y, x = 0+x }
term at position 2 2 in conclusion goal 2 paramodulated with axiom 1
current conclusion:
 (1) { r1 < y, x >= y, y > 0, div(x-y, y) = (q1, r1), x = ((q1*y)+y)+r1 }
 (2) { x < y }
term at position 5 2 in conclusion goal 1 paramodulated with theorem 1
 (1) { r1 < y, x >= y, y > 0, div(x-y, y) = (q1, r1), x = ((q1*y)+r1)+y }
 (2) { x < y }
atom 5 in conclusion goal 1 resolved with theorem 2
 (1) { r1 < y, x >= y, y > 0, div(x-y, y) = (q1, r1), x-y = (q1*y)+r1 }
 (2) { x < y }
atoms 1 5 in conclusion goal 1 resolved with conjecture 1
 (1) { x >= y, y > 0, div(x-y, y) = (q1, r1), (y, x) >> (y, x-y) }
 (2) { x < y }
atom 4 in conclusion goal 1 resolved with axiom 7
current conclusion:
 (1) { x >= y, y > 0, div(x-y, y) = (q1, r1) }
 (2) { x < y }
```

atom 3 in conclusion goal 1 resolved with theorem 3
(1) { x >= y, y > 0 }
(2) { x < y }
atoms 1 1 in conclusion goals 2 1 resolved with theorem 4
(1) { y > 0 } □

5. Reductions

An atomic formula p is called a *logical atom* if p is not an equation. A variable x is called a *fresh variable* of a Horn clause c = p⟵d if p is a logical atom and x ∈ var(d) - var(p) or p is an equation, say p = t≡u, and x ∈ var({u}∪d) - var(t). *fresh(c)* denotes the set of fresh variables of c.

General Assumption As in Section 4, let AX be a set of Horn clauses. Moreover, *NF* is a set of terms such that X ⊆ NF and NF is *closed under composition*, i.e. for all t ∈ NF, x ∈ var(t) and u ∈ NF, t[u/x] ∈ NF. □

NF is usually given by a set of *constructors* (cf. [Pad 92], Section 2.3).

Definition The *reduction calculus upon AX and NF* consists of the following inference rules, which transform goals.

Reduction Let c = t≡u⟵d ∈ AX, x ∈ var(t) and σ ∈ Sub such that fresh(c)σ ⊆ NF.

$$\frac{g[t\sigma/x]}{g[u\sigma/x] \cup d\sigma}$$

Resolution Let p be a logical atom, c = p⟵d ∈ AX and σ ∈ Sub such that fresh(c)σ ⊆ NF.

$$\frac{g \uplus \{p\sigma\}}{g \cup d\sigma}$$

Reflection (resolution with x≡x)

$$\frac{g \uplus \{t\equiv t\}}{g}$$

A sequence $g_1,\ldots,g_n$ of gs is called a *goal reduction of* g_1 *into* g_n *upon AX* if for all $1 \le i < n$, g_{i+1} is obtained from g_i by a single reduction, resolution or reflection step. The corresponding inference relation is denoted by $\vdash_{AX}$.

The *AX-reduction relation* $\longrightarrow_{AX}$ is a binary relation on terms *and* goals, which is defined as follows:

- $g \longrightarrow_{AX} g'$ iff $g = h[t\sigma/x]$ and $g' = h[u\sigma/x]$ for some goal h, $x \in var(t)$, $c = t \equiv u \Leftarrow d \in AX$ and $\sigma \in Sub$ with $d\sigma \vdash_{AX} \emptyset$ and $fresh(c)\sigma \subseteq NF$.

If $g \longrightarrow_{AX}^* g'$, then g' is called an *AX-reduct* of g. A substitution τ is an AX-reduct of a substitution σ if $x\sigma \longrightarrow_{AX}^* x\tau$ for all $x \in X$. Then we write $\sigma \longrightarrow_{AX}^* \tau$. A term or goal g is *AX-reduced* if g is the only AX-reduct of g. Otherwise g is *AX-reducible.* g is *AX-narrowed* if for all substitutions with $X\sigma \subseteq NF$, $g\sigma$ is AX-reduced. Otherwise g is *AX-narrowable.*

A goal g is *AX-convergent* if $g \vdash_{AX} \emptyset$. g is *strongly AX-convergent* if all AX-reducts of g are AX-convergent. AX is *ground confluent* if all AX-convergent ground goals are strongly AX-convergent. □

Since NF is closed under composition, $g \vdash_{AX} h$ implies $g\sigma \vdash_{AX} h\sigma$ for all $\sigma \in Sub$ with $X\sigma \subseteq NF$. However, $g \vdash_{AX} \emptyset$ implies $g\sigma \vdash_{AX} \emptyset$ for *all* $\sigma \in Sub$. By [Pad 92], Proposition 6.1, $\vdash_{AX}$ is sound w.r.t. the cut calculus (cf. Section 4), i.e. $g \vdash_{AX} \emptyset$ implies $AX \vdash_{cut} g$. Moreover, we have:

Theorem 9 (Church-Rosser Theorem for Horn clauses) *Suppose that each ground term has an AX-reduct in NF. AX is ground confluent iff for all ground goals g, $AX \vdash_{cut} g$ iff $g \vdash_{AX} \emptyset$.* □

If AX is ground confluent, then the following questions can be answered effectively:

- Given a *guard* g, is a goal set *g-minimal* (cf. Lemma 4)? *Constructor-based matrices* are g-minimal (cf. [Pad 92], Lemma 2.2). Constructor criteria, in turn, are based on ground confluence (cf. [Pad 92], Corollary 7.5).
- Is a given goal g *solvable* w.r.t. AX, i.e., does exist $\sigma \in Sub$ such that $AX \vdash_{cut} g\sigma$? A negative answer to this question implies that the *negation* of g holds true in the initial (SIG,AX)-structure (cf. Section 4). If AX is ground confluent and *strongly terminating* (see below), then the answer is negative if there is no successful *narrowing* expansion of g (cf. Section 7).
- Is AX *consistent w.r.t.* a set $BAX \subseteq AX$ of *base axioms* over a *base signature* $BSIG \subseteq SIG$, i.e., does $AX \vdash_{cut} g$ imply $BAX \vdash_{cut} g$ for all ground goals over BSIG? A "syntactic" consistency criterion also requires the ground confluence and strong termination of AX (cf. [Pad 92], Corollary 6.19). Consistency is crucial for proving program equivalence (cf. [Pad 92], Theorem 3.17) and *implementation correctness* (cf. [Pad 92], Section 7.6).
- Given a conjecture $c = gs_1 \Leftarrow gs_1'$, a proof of c using EXPANDER (cf.

[Pad 92a]) starts out from singleton lists *front* = $[gs_1]$ and *rear* = $[gs_1']$, extends them stepwise into, say *front* = $[gs_k,...,gs_1]$ and *rear* = $[gs_n',...,gs_1']$, consisting of successively inferred goal sets. *front* represents a backward proof, i.e. a classical expansion, of $c_1 = gs_1 \Leftarrow gs_k$, while *rear* represents a forward proof of $c_3 = gs_n' \Leftarrow gs_1'$. The proof of c is complete if certain syntactic conditions on gs_k and gs_n' ensure the validity of $c_2 = gs_k \Leftarrow gs_n'$. From the validity of c_1, c_2 and c_3 one concludes that c is also valid. Here the question is whether all the applied inference rules are sound insofar as they guarantee the inductive validity of c_1 and c_3. In fact, if AX is ground confluent and strongly terminating, all rules are sound.

The inductive expansion calculus handled in this section allows us to check whether axiom sets are ground confluent. For this purpose, let us figure out those Gentzen clauses whose validity is sufficient for concluding ground confluence.

For notational convenience, we identify a logical atom p with the equation $p \equiv true$ where *true* is a constant not occurring in SIG. Correspondingly, for all goals g and $x \in var(g)$ we have $g[true/x] = \emptyset$, and $g \longrightarrow_{AX}^* \emptyset$ means $g \vdash_{AX} \emptyset$.

Definition Let $t \equiv u \Leftarrow d$, $t' \equiv u' \Leftarrow e \in AX$, v be a term, $x \in var(v)$ and $\sigma,\tau \in$ Sub such that $t\sigma = v[t'\tau/x]$ and the occurrence of x in v is a function symbol occurrence in t. Then $c = v[u'\tau/x] \equiv u\sigma \Leftarrow d\sigma \cup e\tau$ is an *AX-critical clause* and $(t\sigma,c)$ is an *AX-reduction ambiguity*. A set RA of AX-reduction ambiguities is *complete* if each ground AX-reduction ambiguity is subsumed by an element of RA. □

The validity of AX-critical clauses is sufficient for ground confluence if AX satisfies the following condition:

Definition AX is *strongly terminating* if there is a transitive and well-founded relation $>_{AX}$ on ground terms and goals, called the *AX-reduction ordering*, such that *true* is minimal w.r.t. $>_{AX}$ and the following conditions hold true:

Reduction compatibility Let $c = t \equiv u \Leftarrow d \in AX$, v be a term or logical atom, $x \in var(v)$ and $\sigma \in$ GSub such that $d\sigma$ is AX-convergent and $fresh(c)\sigma \subseteq NF$. Then

$$v[t\sigma/x] >_{AX} v[u\sigma/x] \quad \text{and} \quad t\sigma >_{AX} d\sigma.$$

Subterm property For each ground term or logical atom t and all proper subterms u of t, $t >_{AX} u$. □

By reduction compatibility, $g \rightarrow_{AX}^{+} g'$ implies $g >_{AX} g'$.[4] Moreover, if $g \vdash_{AX} g' \neq g$ *and g' is AX-convergent*, then again $g >_{AX} g'$.

If an equation $t \equiv t'$ occurs as an argument of $>_{AX}$, it is identified with the multiset {t,t'}. Moreover, $>_{AX}$ extends to multisets of terms or logical atoms in the usual way: $M >_{AX} N$ iff $M \neq N$ and for all $u \in N\text{-}M$ there is $t \in M\text{-}N$ with $t >_{AX} u$.

In which sense must AX-critical clauses be valid so that we can conclude the ground confluence of AX?

Definition Let *t* be a term or a logical atom with $var(t) \subseteq Y_{in}$. The Gentzen clause $c = \exists X_1 g_1 \vee ... \vee \exists X_n g_n \Leftarrow g$ is *sub-t-reductively valid w.r.t. AX* if for all $\sigma \in GSub$ such that AX is *subconfluent at* $t\sigma$, i.e. each AX-convergent ground goal h with $t\sigma >_{AX} h$ is strongly AX-convergent, the following condition holds true:

- If $g\sigma$ is strongly AX-convergent, then $g_i\tau$ is AX-convergent for some $1 \le i \le n$ and $\tau \in GSub$ with $\tau|(X\text{-}X_i) = \sigma|(X\text{-}X_i)$.

c is *reductively valid w.r.t. AX* if c is sub-*true*-reductively valid w.r.t. AX. □

Theorem 10 (Superposition Theorem for Horn clauses) *Suppose that each ground term has an AX-reduct in NF, AX is strongly terminating and RA is a complete set of AX-reduction ambiguities. If for all (t,c) ∈ RA, c is sub-t-reductively valid w.r.t. AX, then AX is ground confluent.* □

6. Reductive expansions

In terms of the generic approach of Section 1, let

- TH be the set of strongly AX-convergent ground goals,
- Sub' and GS' be the sets of all $\sigma \in Sub$ (respectively GSub) with $X\sigma \subseteq NF$,
- for a term or logical atom t, let $GS = GS_t$ be the set of all $\sigma \in GS'$ such that AX is subconfluent at $t\sigma$.[5] (GS_{true} is GS' because *true* is minimal w.r.t. $>_{AX}$.)

[4]This implication seems to be sufficient for a transitive and well-founded relation $>_{AX}$ to "make AX terminating". However, $>_{AX}$ is introduced for a special purpose, namely for reducing ground confluence to a property of critical clauses. *To this end* the full reduction compatibility and the subterm property are essential as well (cf. [1], Section 9.2, and [9], Section 6.3).

[5]Note that for satisfying the closure condition on GS (cf. Section 1) we need the assumption that the variables of t are input variables.

Gen(TH) = $Gen_t(TH)$ - as defined in Section 1 - consists of Gentzen clauses that are sub-t-reductively valid w.r.t. AX:

Lemma 11 *Suppose that each ground term has an AX-reduct in NF. If c ∈ $Gen_t(TH)$, then c is sub-t-reductively valid w.r.t. AX.* □

Define the predicate Pred (cf. Section 2) as follows: $\mathbf{Pred}_t$(g) iff for all σ ∈ GS' such that gσ is AX-convergent, $t\sigma >_{AX} g\sigma$.

It is an immediate consequence of the definitions of AX-convergence and the AX-reduction relation that for all goals g, terms u,v and x ∈ var(g), g[u/x] is AX-convergent if u≡v is AX-convergent and g[v/x] is strongly AX-convergent. Hence, for σ ∈ GS_t with (g[v/x]∪{u≡v})σ ∈ TH, $Pred_t$(g) implies $t\sigma >_{AX} g[u/x]\sigma$ and thus g[u/x]σ ∈ TH. In other words, the clause g[u/x] ←g[v/x]∪{u≡v} belongs to Gen_t(TH), i.e. (*) holds true (cf. Section 1).

An inductive expansion upon (TH,$Pred_t$) is called a *sub-t-reductive expansion*. Since the corresponding calculus depends on t, the corresponding inference relation is denoted by $\vdash^t_{TH}$.

Since for each AX-convergent goal g, AX $\vdash_{cut}$ g, the well-foundedness of an induction ordering again reduces to the well-foundedness of the relation $>>^A$ in a given model A of AX (cf. Section 4). But since GS_t ≠ GS', we must explicitly provide for condition (***) of Section 3. It is an immediate consequence of the definitions of GS_t and GS' that (***) holds true if >> and the AX-reduction ordering are related to each other as follows:

(+) For all σ,τ ∈ GS', AX $\vdash_{cut}$ ($L_{in}\sigma >> L_{in}\tau$) implies $t\sigma >_{AX} t\tau$.

By Theorems 3 and 5 and Lemma 11, the construction of sub-t-reductive expansions yields a proof method for sub-t-reductive validity whenever (+) holds true:

Theorem 12 (Soundness of sub-t-reductive expansions) *Let TH and $Pred_t$ be defined as above. Suppose that (+) holds true.*

(1) Given goals $g_1,...,g_n,g$, let CS = {∃X_1h_1∨...∨∃X_nh_n←g | for all 1≤i≤n, $h_i \subseteq g_i$}. If {$g_1,...,g_n$} $\vdash^t_{TH,CS}$ {g}, then CS ⊆ Gen_t(TH) and CS is sub-t-reductively valid w.r.t. AX.

(2) Let CS = {d_1←g⊎h_1,...,d_n←g⊎h_n} be a set of guarded clauses such that for all 1≤i≤n, $Pred_t(d_i)$ holds true, {$h_1,...,h_n$} is g-minimal w.r.t. TH. If {d_1∪h_1,...,d_n∪h_n} $\vdash^t_{TH,CS}$ {g}, then CS ⊆ Gen_t(TH) and CS is sub-t-reductively valid w.r.t. AX. □

For starting a proof by building expansions one needs some basic lemmas, which are already known to be in Gen(TH). In case of a classical expansion the axioms of AX are inductively valid and can thus be used as lemmas. In case of a sub-t-reductive expansion we may apply an axiom $p \Leftarrow d$ if $Pred_t(g)$ holds true:

Lemma 13 *Let $p \Leftarrow d \in AX$ such that $Pred_t(p)$ holds true. Then $p \Leftarrow d \in Gen_t(TH)$.* □

From Theorems 10 and 12 and Lemma 11 we obtain a powerful method for proving ground confluence:

Theorem 14 (Criterion for ground confluence) *Suppose that each ground term has an AX-reduct in NF, (+) holds true, AX is strongly terminating and RA is a complete set of AX-reduction ambiguities. If for all $(t,g \Leftarrow d) \in RA$, $\{g\} \vdash_{TH,\{g \Leftarrow d\}} \{d\}$, then AX is ground confluent.* □

Since for all goals g,h, atoms p and terms u,v, the clauses $g \vee h \Leftarrow g$, $p \Leftarrow p$ and $u \equiv v \Leftarrow u \equiv v$ belong to $Gen_t(TH)$, the derived rules *goal elimination, factoring* and *term replacement* (cf. Section 4) can be used in a reductive expansion without violating its soundness. Applying the *unification* rule, however, is not always sound. Since for all $\sigma \in$ GSub, $x\sigma \equiv x\sigma$ is AX-convergent, $x \equiv x \in Gen(TH_t)$ means that $x\sigma \equiv x\sigma$ is strongly AX-convergent if $t\sigma >_{AX} x\sigma$. This holds true if $x \neq t$ and x occurs in t. Hence, when the unification rule is applied, the sides of the equation to be unified must be terms of the sort of some variable occurring in t.

7. Strong termination

Theorem 14 leaves us with two proof obligations, which require tractable sufficient conditions: the strong termination of AX and the validity of $Pred_t$ for all goals of the expansions demanded by Theorem 14. For meeting both proof obligations one may refer to the *path calculus for* $\geq_{SIG}$ *and* $\succ$ where $\geq_{SIG}$ is a reflexive and transitive relation on SIG-$\{\equiv\}$[6] and the *semantic ordering* $\succ$ is a (set of) binary predicate(s) of SIG such that the relation $\succ^{AX} \subseteq$ GTerm×GTerm defined by:

$$t \succ^{AX} t' \quad \text{iff} \quad AX \vdash_{cut} t \succ t'$$

is transitive and well-founded (cf. Section 4).

Definition Let t,u be terms, logical atoms or multisets of terms and logical atoms. Given $\geq_{SIG}$ and $\succ$ as above, the *path calculus for* $\geq_{SIG}$ *and* $\succ$

[6] $\equiv$ stands for each equality predicate of SIG.

consists of the following inference rules, which transform expressions of the form $t \gg u$ into goals. Let $>_{SIG} = \gtrsim_{SIG} - \lesssim_{SIG}$ and $\approx_{SIG} = \gtrsim_{SIG} \cap \lesssim_{SIG}$.

Multiset rules Let $t \neq \emptyset$.

$$\frac{(t \uplus u) \gg (t' \uplus u')}{t \gg t',\ (t \cup u) \gg t'} \qquad \frac{t \gg (t' \uplus \{true\})}{t \gg t'}$$

$$\frac{(t \uplus u) \gg u}{\emptyset}$$

Subterm rules Let $F(t_1,\dots,t_k)$ be a term or a logical atom.

$$\frac{F(t_1,\dots,t_k) \gg t}{t_i \gg t} \qquad 1 \le i \le k \qquad \frac{F(t_1,\dots,t_k) \gg t_i}{\emptyset}$$

Signature rules Let $F(t_1,\dots,t_k)$ and $G(u_1,\dots,u_n)$ be terms or logical atoms.

$$\frac{F(t_1,\dots,t_k) \gg G(u_1,\dots,u_n)}{F(t_1,\dots,t_k) \gg \{u_1,\dots,u_n\}} \qquad F >_{SIG} G$$

$$\frac{F(t_1,\dots,t_k) \gg G(u_1,\dots,u_n)}{(t_1,\dots,t_k) \succ (u_1,\dots,u_n),\ F(t_1,\dots,t_k) \gg \{u_1,\dots,u_n\}} \qquad F \approx_{SIG} G$$

$$\frac{F(t_1,\dots,t_k) \gg G(u_1,\dots,u_n)}{t_{i_1} \equiv u_1,\ \dots,\ t_{i_k} \equiv u_n,\ \{t_1,\dots,t_k\} \gg \{u_1,\dots,u_n\}} \qquad F \approx_{SIG} G,\ 1 \le i_1 < \dots < i_n \le k$$

Fresh variable rule

$$\frac{t \gg x}{\{t \gg x\sigma \mid \sigma \in GS'\}} \qquad x \in X\text{-var}(t)$$

The corresponding inference relation is denoted by $\vdash_{path}$. Remember that GS' is the set of all $\sigma \in$ GSub with $X\sigma \subseteq NF$ (cf. Section 6). Define the relation $>_{AX}$ on ground terms and goals as follows:

$g >_{AX} h$ iff there are terms or goals g',h',d and $\sigma \in$ GSub such that $g = g'\sigma$, $h = h'\sigma$, $(g' \succ h') \vdash_{path} d$, $AX \vdash_{cut} d\sigma$ and $\sigma(var(h')-var(g')) \in GS'$. □

Theorem 15 *Let* $>_{AX}$ *be defined as above.* $>_{AX}$ *is an AX-reduction ordering if for all* $c = t{\equiv}u \Leftarrow d \in AX$ *and* $\sigma \in GSub$ *such that* $d\sigma$ *is AX-convergent and* $\sigma fresh(c) \in GS'$, $t\sigma >_{AX} \{u\sigma\} \cup d\sigma$. □

The definition of $>_{AX}$, Theorem 15 and a special case of Theorem 6(1) immediately imply

Corollary 16 (Criterion for strong termination) *Let* $>_{AX}$ *be defined as above.* $>_{AX}$ *is an AX-reduction ordering and thus AX is strongly terminating if for all* $t{\equiv}u \Leftarrow d \in AX$ *there is a goal h such that*

$$(t \succ \{u\} \cup d) \vdash_{path} h \vdash_{TH} d. \qquad (++)$$

where TH is the set of ground goals g with $AX \vdash_{cut} g$ *(cf. Section 4).* □

By combining Theorem 14 with Corollary 16 we obtain

Theorem 17 (Criterion for ground confluence and strong termination) *Suppose that each ground term has an AX-reduct in NF, (+) holds true, RA is a complete set of AX-reduction ambiguities and for all* $t{\equiv}u \Leftarrow d \in AX$ *there is a goal h such that (++) holds true.*

(1) Let t be a term or logical atom with $var(t) \subseteq Y_{in}$ *and let g be a goal. If* $(t \succ g) \vdash_{path} h \vdash_{TH} g$ *for some goal h, then* $Pred_t(g)$ *holds true.*

(2) If for all $(t,g \Leftarrow d) \in RA$, $\{g\} \vdash_{TH,\{g \Leftarrow d\}} \{d\}$, *then AX is ground confluent.*

□

Since $var(t) = var(L_{in})$, (+) holds true if >> is a subrelation of ≻ and t is regarded as a function resp. predicate of SIG.

Example 18 Using the syntax of EXPANDER (cf. [Pad 92a]) we present a specification of stacks:

```
STACK
      consts   eps push
      functs   pop 1 top 2
      preds    Rep 3 4
      vars     x st
      axioms (1)  {pop(push(x,st)) = st}
             (2)  {top(push(x,st)) = x}}
             (3)  {Rep(eps)}
             (4)  {Rep(push(x,st))} <== {Rep(st)}
```

Stacks are to be implemented by arrays:

```
ARRAY
     consts   mt put
     functs   get 3 4
     preds    =/= >[7]
     infixes  =/= >
     vars     a m n x y
     axioms (1)  {put(put(a,n,x),n,y) = put(a,n,y)}
            (2)  {put(put(a,m,x),n,y) = put(put(a,n,y),m,x)} <== {m > n}
            (3)  {get(put(a,n,x),n) = x}
            (4)  {get(put(a,m,x),n) = get(a,n)} <== {m =/= n}
```

An abstract implementation of stacks by arrays is given by the following extension of STACK and ARRAY (cf. [Pad 92], Section 7.6):

```
ARRAY_TO_STACK
     base     ARRAY
     functs   eps 5 push 6 pop 7 top 8 abs 9
     preds    Rep >=
     infixes  >=
     vars     a m n x st ca cn
     axioms  (5)  {eps = abs(mt,0)}
             (6)  {push(x,abs(a,n)) = abs(put(a,n,x),s(n))}
             (7)  {pop(abs(a,s(n))) = abs(a,n)}
             (8)  {top(abs(a,s(n))) = get(a,n)}
             (9)  {abs(put(a,m,x),n) = abs(a,n)} <== {m >= n}
     theorems(1)  {n >= n}
             (2)  {st  = abs(ca!,cn!)} <== {Rep(st)}
     conjects (1) {pop(push(x,st)) = st} <== {Rep(st)}
             (2)  {top(push(x,st)) = x}} <== {Rep(st)}
```

The correctness of this implementation amounts to the following conditions (cf. [Pad 92], Example 7.20):

(1) Conjectures 1 and 2 are inductively valid w.r.t. ARRAY_TO_STACK.
(2) The set *AX* consisting of the axioms of STACK and ARRAY_TO_STACK except Axiom 9 is ground confluent and strongly terminating.
(3) Axiom 9 of ARRAY_TO_STACK is inductively valid w.r.t. AX.

First we show (1) by constructing classical expansions for Conjectures 1 and 2 (cf. Theorem 6(1)):

Conjecture 1:

```
initial conclusion:
 (1) { pop(push(x, st)) = st }
initial premise:
```

---

[7]STACK, ARRAY and ARRAY_TO_STACK are parameterized specifications. Hence the axioms for certain functions and predicates are left to extensions of these specifications. Conjectures are always proved with respect to *any* extension whose initial model statisfies the lemmas applied.

(1) { Rep(st) }
term at position 1 1 1 in conclusion goal 1 paramodulated with axiom 6
current conclusion:
(1) { pop(abs(put(a, n, x), s(n))) = abs(a, n), st = abs(a, n) }
term at position 1 1 in conclusion goal 1 paramodulated with axiom 7
current conclusion:
(1) { st = abs(a, n), abs(put(a, n, x), n) = abs(a, n) }
term at position 2 1 in conclusion goal 1 paramodulated with axiom 9
current conclusion:
(1) { st = abs(a, n), n >= n }
atom 2 in conclusion goal 1 resolved with theorem 1
(1) { st = abs(a, n) }
atom 1 in conclusion goal 1 resolved with theorem 2
(1) { Rep(st) }

Conjecture 2:
initial conclusion:
(1) { top(push(x, st)) = x }
initial premise:
(1) { Rep(st) }
term at position 1 1 1 in conclusion goal 1 paramodulated with axiom 6
current conclusion:
(1) { top(abs(put(a, n, x), s(n))) = x, st = abs(a, n) }
term at position 1 1 in conclusion goal 1 paramodulated with axiom 8
current conclusion:
(1) { st = abs(a, n), get(put(a, n, x), n) = x }
term at position 2 1 in conclusion goal 1 paramodulated with axiom 3
term at position 2 1 in conclusion goal 1 paramodulated with axiom 4
current conclusion:
(1) { st = abs(a, n), get(a, n) = x, n =/= n }
(2) { st = abs(a, n) }
conclusion goal 1 deleted
(1) { st = abs(a, n) }
atom 1 in conclusion goal 1 resolved with theorem 2
(1) { Rep(st) }

The reduction ambiguities of ARRAY and the pairs (t_i,c_i), $1 \le i \le 4$, where

$$\begin{aligned} t_1 &= pop(abs(put(a,n,x),s(n))) \\ t_2 &= top(abs(put(a,n,x),s(n))) \\ t_3 &= abs(put(put(a,n,x),n,y),s(n)) \\ t_4 &= abs(put(put(a,m,x),n,y),s(n)) \end{aligned}$$

and

$$\begin{aligned} c_1 &= \{pop(push(x,abs(a,n))) = abs(put(a,n,x),n)\} \\ c_2 &= \{top(push(x,abs(a,n))) = get(put(a,n,x),n)\} \\ c_3 &= \{abs(put(a,n,y),s(n)) = push(y,abs(put(a,n,x),n))\} \\ c_4 &= \{abs(put(put(a,n,y),m,x),s(n)) = push(y,abs(put(a,m,x),n))\} \Leftarrow\!\!= \{m > n\} \end{aligned}$$

yield a complete set of AX-reduction ambiguities. For proving the strong termination of AX one may use Corollary 16 based on suitable relations $\gtrsim_{SIG}$ and $\succ$.[8] For proving the ground confluence of AX we apply

Theorem 17(2) and present a subreductive expansion, which implies that Conjecture 5 is subreductively valid w.r.t. AX. Conjecture 5 is crucial for proving c_1-c_4. The underlying specification with axiom set AX reads as follows:

```
ARRAY_TO_STACK'
     base          STACK ARRAY
     functs        abs -
     preds         >> 5 >=
     infixes       - >> >=
     vars          a k m n n' x y st st'
     axioms (1)    {eps = abs(mt,0)}
            (2)    {push(x,abs(a,n)) = abs(put(a,n,x),s(n))}
            (3)    {pop(abs(a,s(n))) = abs(a,n)}
            (4)    {top(abs(a,s(n))) = get(a,n)}
            (5)    {(m,n,a,x) >> (m,n',a,x)} <== {m-n > m-n'}
     theorems
            (1)    {abs(put(a,n,x),s(n)) = push(x,abs(a,n))}
            (2)    {abs(a,n) = pop(abs(a,s(n)))}
            (3)    {n >= n}
            (4)    {m >= n, m >= s(n)} <== {m > n}
            (5)    {pop(st) = pop(st')} <== {st = st'}
            (6)    {m-n > m-s(n)} <== {m >= s(n)}
            (7)    {m >= s(n)}V{m = n} <== {m >= n}
     conjects (1)  {pop(push(x,abs(a,n))) = abs(put(a,n,x),n)}
            (2)    {top(push(x,abs(a,n))) = get(put(a,n,x),n)}
            (3)    {abs(put(a,n,y),s(n)) = push(y,abs(put(a,n,x),n))}
            (4)    {abs(put(put(a,n,y),m,x),s(n)) = push(y,abs(put(a,m,x),n))}
                                                    <== {m > n}
            (5)    {abs(put(a,m,x),n) = abs(a,n)} <== {m >= n}
```

Conjecture 5:

initial conclusion:
(1) { abs(put(a, m, x), n) = abs(a, n) }
initial premise:
(1) { m >= n }
term at position 1 1 in conclusion goal 1 paramodulated with theorem 2
(1) { pop(abs(put(a, m, x), s(n))) = abs(a, n) }
term at position 1 2 in conclusion goal 1 paramodulated with theorem 2
(1) { pop(abs(put(a, m, x), s(n))) = pop(abs(a, s(n))) }
term at position 1 1 1 in conclusion goal 1 paramodulated with theorem 1

(1) { pop(abs(put(a, m, x), s(n))) = pop(abs(a, s(n))) }
(2) { pop(push(x, abs(a, n))) = pop(abs(a, s(n))), m = n }
term at position 1 1 in conclusion goal 2 paramodulated with axiom 1
current conclusion:
(1) { pop(abs(put(a, m, x), s(n))) = pop(abs(a, s(n))) }
(2) { m = n, abs(a, n) = pop(abs(a, s(n))) }
atom 2 in conclusion goal 2 resolved with theorem 2

[8] For that purpose, Axioms 5 and 6 are inverted so that $>_{SIG}$ and ≻ may include the pairs (pop,abs), (top,abs), (top,get), (abs,eps) and (abs,push) respectively the predicate > of ARRAY.

(1) { pop(abs(put(a, m, x), s(n))) = pop(abs(a, s(n))) }
(2) { m = n }
atom 1 in conclusion goal 1 resolved with theorem 5
(1) { abs(put(a, m, x), s(n)) = abs(a, s(n)) }
(2) { m = n }
atom 1 in conclusion goal 1 resolved with conjecture 5
(1) { m >= s(n), (m, n, a, x) >> (m, s(n), a, x) }
(2) { m = n }
atom 2 in conclusion goal 1 resolved with axiom 13
current conclusion:
(1) { m >= s(n), m-n > m-s(n) }
(2) { m = n }
atom 2 in conclusion goal 1 resolved with theorem 6
(1) { m >= s(n) }
(2) { m = n }
atoms 1 1 in conclusion goals 1 2 resolved with theorem 7
(1) { m >= n }

Note that among all these expansions establishing the correctness of our implementation only the last one includes an induction step.

This finishes the proof of (2) and also of (3) because the preceding expansion of Theorem 8 does not only imply its *reductive* but also its *inductive* validity w.r.t. AX: The ground confluence of AX implies that inductive validity coincides with reductive validity (cf. Theorem 9). □

When proving the ground confluence of AX by *completion* (cf. [DJ 90], Section 8) one must actually show the ground confluence of AX∪L where L is the set of all lemmas used for proving that the critical clauses are reductively valid. When proving ground confluence by constructing subreductive expansions (cf. Theorem 17(2)) we only consider the original axiom set AX.

Not for proving ground confluence, but for proving inductive theorems, completion has been extended to *inductive completion* or *proof by consistency* (cf. [DJ 90], Section 8.5; [Duf 91], Section 7.3). In [Pad 92], Section 7.4, we have generalized this method to the inductive theory of Horn clause specifications.

Roughly said, a proof of a clause set CS by inductive completion (called a *reductive expansion* in [Pad 92]) turned out as a subreductive expansion of proper AX-reducts of CS where the underlying ground theory TH consists of all strongly (AX∪CS)-convergent ground goals. This contrasts Section 6 where TH is given by all *AX*-convergent ground goals. Hence in a reductive expansion, CS is applied via *deductive* resolution or paramodulation and not via *inductive* resolution or paramodulation as in a classical expansion of CS. However, the comparison of classical and reductive expansions of the same clauses reveals that both methods employ nearly the same steps, while reductive expansions enforce additional proof obligations such as the ground confluence of AX and the strong termination of AX∪CS. Inductive completion avoids the necessity of introducing *induction* orderings (cf. Section 3). But inductive completion relies on a *reduction* ordering, which must be taken into account in each step of the subreductive expansion that is part of a proof by inductive completion (cf. Section 6). At least in the range of applications we have in mind, inductive completion is not an improvement over classical induction proofs (cf. [Pad 92], Section 7.5).

The benefit from a ground confluent specification is *not* a proof procedure that is totally different from classical induction. Instead, it is the possibility of "speeding up" classical proofs by using additional rules whose soundness is only guaranteed if the axiom set is ground confluent (cf. Section 5).

References

[DJ 90] N. Dershowitz, J.-P. Jouannaud, *Rewrite Systems,* in: J. van Leeuwen, Handbook of Theoretical Computer Science, Elsevier (1990) 243-320

[Duf 91] D. Duffy, *Principles of Automated Theorem Proving,* John Wiley & Sons 1991

[EM 85] H. Ehrig, B. Mahr, *Fundamentals of Algebraic Specification 1,* Springer 1985

[Pad 88] P. Padawitz, *Computing in Horn Clause Theories,* Springer 1988

[Pad 91] P. Padawitz, *Inductive Expansion: A Calculus for Verifying and Synthesizing Functional and Logic programs,* Journal of Automated Reasoning 7 (1991) 27-103

[Pad 91a] P. Padawitz, *Reduction and Narrowing for Horn Clause Theories,* The Computer Journal 34 (1991) 42-51

[Pad 92] P. Padawitz, *Deduction and Declarative Programming,* Cambridge University Press 1992

[Pad 92a] P. Padawitz, *EXPANDER: User's Manual,* Universität Dortmund 1992

[Wir 90] M. Wirsing, *Algebraic Specification,* in: J. van Leeuwen, ed., Handbook of Theoretical Computer Science, Elsevier (1990) 675-788

A Constructor-Based Approach for Positive/Negative-Conditional Equational Specifications

Claus-Peter Wirth, Bernhard Gramlich

Fachbereich Informatik, Universität Kaiserslautern,
Postfach 3049, W-6750 Kaiserslautern, Germany
{wirth, gramlich}@informatik.uni-kl.de

Abstract: We present a constructor-based approach for assigning appropriate semantics to algebraic specifications given by finite sets of positive/negative-conditional equations. Under the assumption of confluence of the reduction relation we define, the factor algebra of the ground term algebra modulo the congruence of this reduction relation is a minimal model which is (beyond that) the minimum of all models that do not identify more constructor ground terms than necessary.

1 Introduction and Overview

We present a new uniform (i. e. not using "built-in"s or specification hierarchies) constructor-based approach for assigning appropriate semantics to algebraic specifications with finite sets of positive/negative-conditional ("p/n-conditional" for short) equations. A reduction relation is defined which allows us to generalize the fundamental results for positive-conditional reduction systems (cf. [5, 3]). Under the assumption of confluence, the factor algebra of the ground term algebra modulo the congruence of our reduction relation is an (up to isomorphism) uniquely determined minimal model, which (as developed in Wirth[8]) can be used to establish inductive validity and allows a straightforward abstraction for the treatment of non-ground equational clauses by contextual rewriting.

The class of models of a finite set R of p/n-conditional equations does not contain in general a minimum model (w. r. t. homomorphic mapping). The most promising attempt in literature to overcome this problem has been that in Kaplan[6]. There, one of the minimal ("quasi-initial") models is distinguished from the other minimal models by the use of control information extracted from the rules, which must be compatible with a noetherian ordering. In addition, Kaplan gives a straightforward ground term reduction relation. However, the approach of Kaplan[6] violates the paradigm of separation of logic and control and does not allow to express the distinction of the minimal model without the control part of the specification. Therefore, one loses the nice aspects of the pure logic view on the specification which becomes a mere program instead (cf. sect. 3). For that reason, we choose a new different approach. Instead of using control information we introduce two syntactic restrictions: Firstly, for

This research was supported by the Deutsche Forschungsgemeinschaft, SFB 314 (D4-Projekt)

a condition to be fulfilled we require the evaluation values of the terms of the negative equations to be in the set of evaluation values of constructor ground terms by adding condition literals expressing this property. Secondly, we restrict the constructor rules (which express equalities among the constructor terms) to have "Horn"-form and to be "constructor-preserving". We can then carry over the results of Kaplan[6] without using control information or noetherian orderings anymore (cf. sect. 5, 6). As a consequence, our reduction relation does not need to be noetherian or to be restricted to ground terms. Contrary to [6], we can show the monotony of this reduction relation w. r. t. the extension of the specification. As in [6], under the assumption of confluence of our reduction relation, the factor algebra of the ground term algebra modulo the congruence of our reduction relation is a minimal model for our specification. Unlike [6], however, it is also the minimum of all models that do not identify more constructor ground terms than necessary. To achieve decidability of reducibility we define several kinds of compatibility of R with a reduction ordering and present a complete critical-pair test à la Knuth-Bendix for the confluence of our reduction relation. The proofs, further concepts, and a deeper discussion can be found in Wirth&Gramlich[9].

2 Basic Notions and Notations

We will consider terms of fixed arity over a *many-sorted signature* $\mathrm{sig} = (\mathrm{F}, \mathrm{S}, \alpha)$ consisting of an enumerable set of function symbols F, a finite set of sorts S (disjoint from F), and a computable arity-function $\alpha : \mathrm{F} \to \mathrm{S}^+$. F has a decidable subset C of *constructor symbols*, which constitute their own signature $\mathrm{cons} = (\mathrm{C}, \mathrm{S}, \alpha|_{\mathrm{C}})$. We assume a fixed S-sorted variable-system V with disjoint infinite decidable ranges only. For the family of variables occurring in A we will use $\mathcal{V}(A)$. $\mathcal{T}(\mathrm{sig})$ (resp. $\mathcal{T}(\mathrm{cons})$) denotes the S-sorted family of all (well-sorted) *terms* over sig and V (resp. *constructor terms*), while $\mathcal{GT}(\mathrm{sig})$ (resp. $\mathcal{GT}(\mathrm{cons})$) denotes the S-sorted family of all *ground terms* over sig (resp. *constructor ground terms*). We assume $\forall s \in \mathrm{S} : \mathcal{GT}(\mathrm{cons})_s \neq \emptyset$. For a term t we denote by $\mathcal{O}(t)$ the set of its *occurrences*, by t/p the subterm of t at occurrence p, and by $t[p \leftarrow t']$ the result of replacing t/p with t' in t. The set of *substitutions* from a variable-system X to an S-sorted family of sets T is denoted by $\mathcal{SUB}(X, T)$. A relation R on $\mathcal{T}(\mathrm{sig})$ is called: *stable (w. r. t. substitution)* :iff $\forall (t, t') \in R : \forall \sigma \in \mathcal{SUB}(\mathrm{V}, \mathcal{T}(\mathrm{sig})) : (t\sigma, t'\sigma) \in R$; *monotonic (w. r. t. replacement)* :iff $\forall (t', t'') \in R : \forall t \in \mathcal{T}(\mathrm{sig}) : \forall p \in \mathcal{O}(t) : \forall s \in \mathrm{S} : (t/p, t', t'' \in \mathcal{T}(\mathrm{sig})_s \Rightarrow (t[p \leftarrow t'], t[p \leftarrow t'']) \in R)$. A relation R is called: *noetherian* :iff there is no $a : \mathbb{N} \to \mathrm{dom}(R)$ with $\forall i \in \mathbb{N} : (a_i, a_{i+1}) \in R$. An ordering $<$ or $>$ is called *well-founded* :iff $>$ is noetherian. A *congruence* is a monotonic equivalence. The symmetric closure of a relation $\Longrightarrow$ will be denoted by $\Longleftrightarrow$; its transitive closure by $\overset{\oplus}{\Longrightarrow}$; its reflexive transitive closure by $\overset{\circledast}{\Longrightarrow}$. Two terms v, w are called *joinable w. r. t.* $\Longrightarrow$:iff $v \downarrow w$:iff $v \overset{\circledast}{\Longrightarrow} \circ \overset{\circledast}{\Longleftarrow} w$. $\Longrightarrow$ is called *confluent below* u :iff $\forall v, w : ((v \overset{\oplus}{\Longleftarrow} u \overset{\oplus}{\Longrightarrow} w) \Rightarrow (v \downarrow w))$; it is called *confluent* :iff it is confluent below all u. When $\Longrightarrow$ is a relation on $\mathcal{T}(\mathrm{sig})$, we say $\Longrightarrow$ to be *ground* ... :iff $\Longrightarrow \cap (\mathcal{GT}(\mathrm{sig}) \times \mathcal{GT}(\mathrm{sig}))$ is ....

3 Motivation

Kaplan[6] defines a [negative] equation in the condition of a p/n-conditional equation to hold :iff its terms [do not] have a common reduct. If the resulting reduction relation is confluent and the rules are *decreasing* (cf. [3]), then its congruence closure is minimal (w. r. t. '$\subseteq$') among the congruence relations whose factor algebra (w. r. t. $\mathcal{GT}$(sig)) is a model of R. Confluence is essential here: Firstly, for the correctness of testing semantic equality of two condition terms by looking for a common reduct. Secondly, for the congruence to be minimal (cf. [9]). While confluence can be dropped for merely positive conditional equations by testing for congruence instead for a common reduct, it is worse for p/n-conditional equations: As any congruence yielding a model of the rules $c=d \longleftarrow d\neq e$ and $c=e$ must contain (d,e), the test of $d\neq e$ by such a congruence will always fail, so that we cannot establish $c=d$ by testing the condition of the first rule. But the rules have the minimum model '$c=d=e$', which cannot be obtained by condition-testing anymore, but only by paramodulation and factoring instead.

The major shortcoming of the reduction relation of [6], however, is that its congruence closure is not a minimum (smaller than anything else) but only minimal (there is nothing smaller) among the congruences yielding a model of R: There might be reductions $s\Longrightarrow t$ with $s=t$ not holding in all models logically specified by R. Kaplan[6] argues as follows: By writing $c=d \longleftarrow d\neq e$ instead of the logically equivalent $c=d \vee d=e$ the specifier adds some "operational" information to control the choice of the intended minimal congruence '$c=d$' of the congruences yielding a model of R ('$c=d$', '$d=e$', and '$c=d=e$'). This means that semantics is given by control and not expressible without; thereby violating the paradigm of separation of logic and control: It is not a *logical* specification (suitable for computation) anymore (as it was the case with positive conditional equations), but a program with none but operational semantics. Therefore, we lose the nice aspects of the pure logic view, e. g. the monotony of logic is lost:

Example 3.1 R: $x-0=x$; $\mathsf{s}(x)-\mathsf{s}(y)=x-y$;
$\mathsf{memberp}(x,\mathsf{nil})=\mathsf{false}$; $\mathsf{memberp}(x,\mathsf{cons}(y,l))=\mathsf{true} \longleftarrow x=y$;
$\mathsf{memberp}(x,\mathsf{cons}(y,l))=\mathsf{memberp}(x,l) \longleftarrow x\neq y$.
We have $\mathsf{memberp}(0,\mathsf{cons}(0-\mathsf{s}(0),\mathsf{nil})) \overset{\circledast}{\Longrightarrow} \mathsf{false}$ which no longer holds after adding the rule $0-\mathsf{s}(x)=0$.

Example 3.2 (Minimal Non-Minimum Models in Practice of Specification) (continuing Example 3.1)
We exclude the function symbols 0, s, and −, and enrich the signature with two constants a, b. Consider the following two congruence relations on ground terms, given by their congruence classes:
$\overset{\circledast}{\Longleftrightarrow}$: $\{\mathsf{a}\}$; $\{\mathsf{b}\}$; $\{\mathsf{false}\}\cup\{\mathsf{memberp}(x,l)\mid(x\in\{\mathsf{a},\mathsf{b}\}\wedge(x$ doesn't occur in $l))\}$;
$\{\mathsf{true}\}\cup\{\mathsf{memberp}(x,l)\mid(x\in\{\mathsf{a},\mathsf{b}\}\wedge(x$ does occur in $l))\}$;
$\{\mathsf{nil}\}$; $\{\mathsf{cons}(\mathsf{a},\mathsf{nil})\}$; $\{\mathsf{cons}(\mathsf{b},\mathsf{nil})\}$; $\{\mathsf{cons}(\mathsf{a},\mathsf{cons}(\mathsf{a},\mathsf{nil}))\}$; $\{\mathsf{cons}(\mathsf{a},\mathsf{cons}(\mathsf{b},\mathsf{nil}))\}$; ...
$\sim$: $\{\mathsf{a},\mathsf{b}\}$; $\{\mathsf{false},\mathsf{memberp}(\mathsf{a},\mathsf{nil}),\mathsf{memberp}(\mathsf{b},\mathsf{nil})\}$;
$\{\mathsf{true}\}\cup\{\mathsf{memberp}(x,l)\mid(x\in\{\mathsf{a},\mathsf{b}\}\wedge l\neq\mathsf{nil})\}$;
$\{\mathsf{nil}\}$; $\{\mathsf{cons}(\mathsf{a},\mathsf{nil}),\mathsf{cons}(\mathsf{b},\mathsf{nil})\}$; $\{\mathsf{cons}(x,\mathsf{cons}(y,\mathsf{nil}))\mid x,y\in\{\mathsf{a},\mathsf{b}\}\}$; ...

Now, both $\overset{*}{\Longleftrightarrow}$ and $\sim$ yield a model of R. By a$\sim$b; a$\overset{*}{\not\Longleftrightarrow}$b; we know that $\sim$ is not a minimum. Considering memberp(a, cons(b, nil)) and false, we know that $\overset{*}{\Longleftrightarrow}$ isn't a minimum either. But both are minimal among the congruences that yield a model of R. Hence their intersection does not yield a model of R. Thus, we have to choose between a $\neq$ b and memberp(a, cons(b, nil)) $\neq$ false. As $\overset{*}{\Longleftrightarrow}$ is somehow more appealing than $\sim$, one may argue that a $\neq$ b is somewhat more important than memberp a cons b nil $\neq$ false by stating a, b to be constructors and thinking freeness of constructors to be more important than that of non-constructors. But this treatment does not solve the problem in general: If

1. a or b is changed into a (composite) non-constructor term,

or

2. memberp is stated to be a constructor symbol too,

then the very same problem arises again.

Now, while the simple attempt above fails, the intended bias towards freeness of constructor terms can be achieved the following way:

A. Adding condition literals (cf. sect. 4) expressing *definedness* for all terms of negative equations in the condition. A term t is *defined* :iff 'Def t' holds :iff t has a congruent constructor ground term. For our example above this means that the last memberp-rule is not applicable if a or b is undefined, thereby avoiding the problem of (1) above.

B. Forcing rules whose left-hand sides are constructor terms to have no negative equations in their conditions and to be constructor-preserving[1]. For our example above this means that memberp cannot be a constructor symbol, thereby avoiding the problem of (2) above.

If a, b are defined terms, then $\overset{*}{\Longleftrightarrow}$ becomes the *minimum* congruence of those congruences that do not identify more defined terms than necessary. Contrariwise, if a or b is undefined, then the intersection of $\overset{*}{\Longleftrightarrow}$ and $\sim$ becomes a model because the last memberp-rule now reads

$$\text{'memberp}(x, \text{cons}(y, l)) = \text{memberp}(x, l) \longleftarrow x \neq y,\ \text{Def}\, x, \text{Def}\, y\text{'};$$

thus, memberp(a, cons(b, nil)) is undefined now. (B) is purely syntactic and not very restrictive in practice as it only limits congruences between constructor terms (and this even less restrictively than usual). (A) is not a usage of control information as before. It just means that '$\neq$' is syntactically restricted to defined terms. Undefined terms are due to partially specified functions or incomplete knowledge about the model world. They are often thought to be equal to some unknown defined terms (cf. Kapur&Musser[7]). Based on this tradition, our approach can be justified the following way: If two terms can be shown equal, they will stay equal even if an undefined term will be identified with a defined term later on. On the other hand might an undefined term become equal (w. r. t. $\overset{*}{\Longleftrightarrow}$) to a previously unequal term when identifying an undefined term with a defined term.

[1]i. e.: all terms in such a "constructor rule" are constructor terms and all variables of a constructor rule occur in its left-hand side

The changes of (A),(B) have the following advantages (compared to Kaplan[6]) (cf. sect. 5):

1. $\overset{\circledast}{\Longleftrightarrow}$ yields a minimal model that is (beyond that) the minimum of all models of R that do not identify more defined terms than necessary.
2. $\Longrightarrow$ is monotonic w. r. t. the addition of new rules that don't have left-hand sides that are old constructor terms.
3. $\Longrightarrow$ is stable when defined also on non-ground terms

If there aren't any undefined ground terms, then there is no difference from the reduction relation of Kaplan[6]: In this important case we offer semantics for [6] not using any control information.

Finally, we are not only able to specify our semantics without using control information, but also able to remove the control aspect of requiring the rules to be decreasing. As a consequence, our reduction relation does not need to be noetherian or restricted to ground terms.

4 Syntax and Semantics of the Specification Language

Definition 4.1 (Syntax of CRS) *A (p/n-)*conditional rule system *(*CRS*)* R *over* sig, cons *and* V *is a finite subset of the* set of rules $\mathcal{RUL}(\text{sig}, \text{cons})$ *over* sig, cons *and* V, *that will be defined in Definition 5.1. The only thing we have to know about it now is:* $\mathcal{RUL}(\text{sig}, \text{cons}) \subseteq \text{DEq}(\text{sig}) \times (\mathcal{LIT}(\text{sig}))^*$, *where* DEq(sig) *is the set of* directed equations *and* $\mathcal{LIT}(\text{sig})$ *is the set of* condition literals *over the meta-predicate*[2] *symbols on terms '=', '≠' (binary, commutative), and* 'Def' *(unary). The set of* condition terms *of a rule* $((l,r),C)$ *is denoted by* $\mathcal{TERMS}(C)$. $((l,r),C)$ *is said to be* extra-variable free <u>*:iff*</u> $\mathcal{V}(\mathcal{TERMS}(C),r) \subseteq \mathcal{V}(l)$. R *is* extra-variable free <u>*:iff*</u> *all its rules are extra-variable free.*

A rule $((l,r),C)$ from $\mathcal{RUL}(\text{sig}, \text{cons})$ expresses a universally quantified implication with the conjunction of the literals in C as the condition and with 'l=r' as the conclusion. The meaning of the meta-predicate symbols '=', '≠', and 'Def' is not open to interpretation: '=' is standard; '≠' is its negation; 'Def', however, is the "definedness" predicate which states that the evaluation of its argument belongs (with sort invariant) to the cons-term-generated cons-subalgebra of $\mathcal{A}$. We speak of our new kind of model just as a "sig/cons-model", because it is an upward-compatible extension (for 'Def') of the usual model concept of algebra.

[2] It is also possible to put 'Def' into the signature 'sig' as a predicate symbol whose semantics must then be given by each sig/cons-algebra $\mathcal{A}$ as a subfamily of $(\mathcal{A}[\mathcal{GT}(\text{cons})_s])_{s \in S}$. E. g. for finitely generated categories the composition of two morphisms should only be "defined" if the codomain of the first is equal to the domain of the second:

$$\text{Def}(f \circ g) \longleftarrow \text{Def} f,\ \text{Def} g,\ \text{Cod}(f)=\text{Dom}(g)$$

Since this more flexible approach requires a lot of additional terminology and basic theory, we choose the simpler approach due to lack of space here.

Definition 4.2 (Semantics of CRS) *Let* R *be a CRS over* sig, cons *and* V; *let* $\mathcal{A}$ *be a* sig*-algebra. Now* $\mathcal{A}$ *is a* sig/cons-model of R *:iff*
$\forall((l,r),C) \in \mathrm{R} : \forall b\ \mathcal{A}\text{-valuation of } \mathrm{V}:$

$$((C \textit{ is } \textbf{true w. r. t. } \mathcal{A}_b) \Rightarrow \mathcal{A}_b(l) = \mathcal{A}_b(r)) \ ,$$

where C *is* **true w. r. t.** $\mathcal{A}_b$ *:iff*

$$\forall s \in \mathrm{S}: \forall u,v \in \mathcal{T}(\mathrm{sig})_s : \begin{pmatrix} ((\ (u{=}v\ )\ in\ C) \Rightarrow \mathcal{A}_b(u) = \mathcal{A}_b(v)\ ) & \wedge \\ ((\ (u{\neq}v\ )\ in\ C) \Rightarrow \mathcal{A}_b(u) \neq \mathcal{A}_b(v)\ ) & \wedge \\ ((\ (\mathrm{Def}\, u)\ in\ C) \Rightarrow \mathcal{A}_b(u) \in \mathcal{A}[\mathcal{GT}(\mathrm{cons})_s]\ ) & \end{pmatrix}$$

As we have negative equations in our conditions, we cannot hope to get a minimum model because we can express things like '$a = b \vee b = c$', which has the incomparable minimal models '$a{=}b{\neq}c$' and '$a{\neq}b{=}c$'. What we will get instead is a model that is the minimum of all models that do not identify more defined terms than necessary. For formally expressing this minimality-property, we need the following definition.

Definition 4.3 *Define the quasi-orderings* $\lesssim_{\mathrm{H}}$ *and* $\lesssim_{\mathrm{cons}}$ *as (proper class) relations on* sig*-algebras by:* $\mathcal{A} \lesssim_{\mathrm{H}} \mathcal{B}$ **:iff** *there is a* sig*-homomorphism from* $\mathcal{A}$ *to* $\mathcal{B}$. $\mathcal{A} \lesssim_{\mathrm{cons}} \mathcal{B}$ **:iff** *there is a* cons*-homomorphism from the* cons*-subalgebra of* $\mathcal{A}$ *with the universes* $\mathcal{A}[\mathcal{GT}(\mathrm{cons})_s]$ *to the analogous subalgebra of* $\mathcal{B}$. *We trivially get* $\lesssim_{\mathrm{H}} \subseteq \lesssim_{\mathrm{cons}}$. *The corresponding equivalences, orderings, and reflexive orderings will be denoted by* $\approx$, $<$, $\leq$, *resp., with the corresponding subscript. A* sig*-algebra* $\mathcal{A}$ *will be called a* minimum model *(or else a* constructor-minimum model*) of a CRS* R *over* sig/cons/V **:iff** $\mathcal{A}$ *is a* $\lesssim_{\mathrm{H}}$*-minimum (or else* $\lesssim_{\mathrm{cons}}$*) of the class of all* sig/cons*-models of* R. *Similarly, a* sig*-algebra* $\mathcal{A}$ *will be called a* minimal model *(or else a* constructor-minimal model*) of a CRS* R *over* sig/cons/V **:iff** $\mathcal{A}$ *is a* sig/cons*-model of* R *and there is no* sig/cons*-model* $\mathcal{B}$ *of* R *with* $\mathcal{B} <_{\mathrm{H}} \mathcal{A}$ *(or else* $\mathcal{B} <_{\mathrm{cons}} \mathcal{A}$ *).*

The following lemma tells us that, considering minimum models, we can think in terms of congruences on $\mathcal{GT}(\mathrm{sig})$ instead of algebras:

Lemma 4.4 *Let* $\mathcal{B}$ *be a* sig/cons*-model of the CRS* R *over* sig/cons/V. *Define* $\sim$ *by* $(s,t \in \mathcal{GT}(\mathrm{sig}))$*:* $s \sim t$ *:iff* $\mathcal{B}(s) = \mathcal{B}(t)$. *Define the factor algebra* $\mathcal{A} := \mathcal{GT}(\mathrm{sig})/\!\sim$. *Now:*

1. $\mathcal{A}$ *is a* sig/cons*-model of* R.
2. $\mathcal{A} \lesssim_{\mathrm{H}} \mathcal{B}$. *Moreover, there is a unique (injective)* sig*-homomorphism from* $\mathcal{A}$ *to* $\mathcal{B}$.
3. $\mathcal{A} \approx_{\mathrm{cons}} \mathcal{B}$

Our sig/cons-model concept is algebraically well-behaved in that the trivial algebra is always a sig/cons-model and for each model there is a smaller one that is minimal (cf. [9]).

5 The Reduction Relation

Many authors impose rather strong restrictions on constructor equations, such as "no equations between constructors" ("free constructors") or "unconditional equations between constructors only". Compared to these, our restrictions are very weak. They serve to guarantee a constructor-minimum model for the constructor equations that is unique modulo $\approx_{\text{cons}}$, by requiring the constructor equations to be "constructor-preserving"[3] and to have "Horn"-form.

Definition 5.1 (Set of Rules)
(continuing Definition 4.1 by adding the restrictions on constructor equations)
The set of rules *over* sig, cons *and* V *is defined to be:* $\mathcal{RUL}(\text{sig}, \text{cons}) := \{((l,r),C) \in (\text{DEq}(\text{sig}) \times (\mathcal{LIT}(\text{sig}))^*) \mid (\, l \in \mathcal{T}(\text{cons}) \;\Rightarrow\; (\mathcal{V}(r, \mathcal{TERMS}(C)) \subseteq \mathcal{V}(l) \;\wedge\; r \in \mathcal{T}(\text{cons}) \;\wedge\; \mathcal{TERMS}(C) \subseteq \mathcal{T}(\text{cons}) \;\wedge\; \forall L \text{ in } C : \forall u,v:\; L \neq (u \neq v)))\}$

We are now going to define our reduction relation, having in mind to require it to be confluent in the sequel, whereas we do not require confluence for the definition because we cannot prove confluence criteria if the non-confluent case is undefined. Therefore, we have to be explicit how we test the condition literals — even if this testing is not straightforward when confluence is not provided. Our "operational" semantics for testing condition literals is the following: '$u = v$' is true if u, v have reducts $\hat{u}, \hat{v}$, resp., which are syntactically equal. 'Def u' is true if u has a constructor ground reduct. '$u \neq v$' is true if u, v have constructor ground reducts $\hat{u}, \hat{v}$, resp., which are not joinable. Thus, two terms in a condition literal are "operationally" equal if they are joinable, whereas they are unequal if they are not joinable after some reduction to constructor ground terms. The non-joinability alone of two terms is not sufficient for regarding them as unequal because we are never sure about the inequality of "undefined" terms. As it often occurs, our operational logic is four-valued (i. e. = and $\neq$ can independently be true or false), but in case of confluence: *tertium non datur*. In case of free or confluent constructors, the case of both '$u = v$' and '$u \neq v$' simultaneously being true means that we have something like an ambiguous function definition.

---

[3]The constructor-preservation is really necessary here for guaranteeing a constructor-minimum model as in Theorem 5.14: Let $0, 1, \text{true}, \text{false}$ be constructor constants, let weirdp be a defined constant, and take R:

$$\begin{array}{lcll} 1 & = 0 & \longleftarrow & \text{weirdp} = \text{true} \quad ; \\ \text{weirdp} & = \text{true} & \longleftarrow & \text{true} \neq \text{false} \quad . \end{array}$$

Now there are models of R with $0 \neq 1$ and models with $\text{true} \neq \text{false}$ but no models with $0 \neq 1$ and $\text{true} \neq \text{false}$. Also notice, that the constructor-preservation has some additional advantages, e. g.:

1. For $u \in \mathcal{GT}(\text{sig})$ and computable and unique normal form $\text{NF}(u)$ we can test " $\exists \hat{u} \in \mathcal{GT}(\text{cons}) : u \overset{*}{\Longrightarrow} \hat{u}$ " by " $\text{NF}(u) \in \mathcal{GT}(\text{cons})$ ".

2. Theorem 5.15 has no reasonable analogue for CRSs which are not constructor-preserving.

Another important aspect of our reduction relation is that it depends on the constructor subsignature 'cons' beyond the signature 'sig' — just as our notion of a "sig/cons-model" does.

Definition 5.2 (Our Reduction Relation $\Longrightarrow_{\mathrm{R}}$)
Let R *be a CRS over* sig/cons/V. *The reduction relation* $\Longrightarrow_{\mathrm{R}}$ *on* $\mathcal{T}(\mathrm{sig})$ *($\Longrightarrow$ for short) is defined to be the smallest relation satisfying the following requirement (:#):* $s \Longrightarrow t$ *<u>if</u>*

$$\exists((l,r),C)\in\mathrm{R}\colon \exists\sigma\in\mathcal{SUB}(\mathrm{V},\mathcal{T}(\mathrm{sig}))\colon \exists p\in\mathcal{O}(s)\colon \begin{pmatrix} s/p & = l\sigma & \wedge \\ t & = s[p\leftarrow r\sigma] & \wedge \\ C\sigma \text{ is } \textbf{fulfilled w. r. t. } \Longrightarrow \end{pmatrix},$$

where "C is **fulfilled w. r. t.** $\Longrightarrow$*" is a shorthand for* $\forall u,v\in\mathcal{T}(\mathrm{sig})$:

$$\begin{pmatrix} (\ (\ (u{=}v\ )\ in\ C) & \Rightarrow & & u\downarrow v\ ) & \wedge \\ (\ (\ (\mathrm{Def}\,u)\ in\ C) & \Rightarrow & \exists\hat{u}\in\mathcal{GT}(\mathrm{cons}): & u\overset{\circledast}{\Longrightarrow}\hat{u}\ ) & \wedge \\ (\ (\ (u{\neq}v\ )\ in\ C) & \Rightarrow & \exists\hat{u},\hat{v}\in\mathcal{GT}(\mathrm{cons}): & u\overset{\circledast}{\Longrightarrow}\hat{u}\not\downarrow\hat{v}\overset{\circledast}{\Longleftarrow}v\ )^{4} & \end{pmatrix}$$

Usually one tries to find a minimal reduction relation by taking the closure over a finitary generating relation. This is not possible here, because we have a negative condition ($\not\downarrow$). By the "Horn"-form of our constructor equations (and the constructor-preservation), however, we know that this negative condition does not influence the reduction of constructor terms; and (in the definition) '$\not\downarrow$' is applied to constructor (ground) terms only. Thus, we can get our minimal reduction relation by a double closure: first for constructor terms only; second for general terms knowing the constructor reduction to remain unchanged. This two step construction does not destroy the uniformity of the defining requirement(#), which allows us to write uniform normal form procedures, provided that the normal forms are computable.

Define $\Longrightarrow_{\mathrm{R},0} := \emptyset$ and $\Longrightarrow_{\mathrm{R},i+1}$ to be the left-hand side[5] of the requirement(#) of Definition 5.2 with $\Longrightarrow_{\mathrm{R},i}$ substituted for $\Longrightarrow$ on the right-hand side and the additional restriction of $s\in\mathcal{T}(\mathrm{cons})$; formally: $s\Longrightarrow_{\mathrm{R},i+1}t$ <u>:iff</u> $s\in\mathcal{T}(\mathrm{cons})\wedge\exists((l,r),C)\in\mathrm{R}\colon\exists\sigma\in\mathcal{SUB}(\mathrm{V},\mathcal{T}(\mathrm{sig}))\colon\exists p\in\mathcal{O}(s)\colon(s/p=l\sigma\ \wedge\ t=s[p\leftarrow r\sigma]\ \wedge\ C\sigma$ is **fulfilled w. r. t.** $\Longrightarrow_{\mathrm{R},i}$).

[4]This formulation requires confluence and constructor-preservation to make sense in two-valued logic. While other formulations (e. g. a universal instead of the existential quantification) might seem more satisfactory, this is the one required for a correct definition.

One might have expected $u\not\downarrow v$ instead of $\exists\hat{u},\hat{v}\in\mathcal{GT}(\mathrm{cons}):\ u\overset{\circledast}{\Longrightarrow}\hat{u}\not\downarrow\hat{v}\overset{\circledast}{\Longleftarrow}v$ for "$u{\neq}v$ in C" here, but this modification would not allow us to conclude that $\Longrightarrow_{\mathrm{R}}$ is uniquely defined as can be seen from the following example:

Example 5.3
Let a,b,c,d,e *be constants and take* R:

$$\begin{aligned} \mathsf{a} &= \mathsf{c} \longleftarrow \mathsf{b}\neq\mathsf{d} \ ; \\ \mathsf{b} &= \mathsf{d} \longleftarrow \mathsf{e}\neq\mathsf{c} \ ; \\ \mathsf{e} &= \mathsf{a} \ . \end{aligned}$$

The intersection of { (a,c), (e,a) } *and* { (b,d), (e,a) } *(which would be minimal relations satisfying the modified requirement(#) of Definition 5.2) would not satisfy the modified requirement of Definition 5.2 anymore.*

[5]the '<u>if</u>' replaced by '<u>:iff</u>'

Define $\Longrightarrow_{R,\omega} := \bigcup_{i\in\mathbb{N}} \Longrightarrow_{R,i}$ and $\Longrightarrow_{R,\omega+i+1}$ to be the union of $\Longrightarrow_{R,\omega}$ and the left-hand side[5] of the requirement(#) of Definition 5.2 with $\Longrightarrow_{R,\omega+i}$ substituted for $\Longrightarrow$ on the right-hand side. Finally, define $\Longrightarrow_{R} := \Longrightarrow_{R,\omega+\omega} := \bigcup_{i\in\mathbb{N}} \Longrightarrow_{R,\omega+i}$. Now $\Longrightarrow_{R}$ satisfies the requirement(#) of Definition 5.2, and every relation satisfying this requirement must contain $\Longrightarrow_{R}$. Hence $\Longrightarrow_{R}$ is the intended smallest relation.

By double induction over the above construction process it is trivial to verify the following corollaries.

Corollary 5.4 (Monotony of $\Longrightarrow_{R}$ w. r. t. Replacement)
$\Longrightarrow_{R,i}$, $\Longrightarrow_{R,\omega+i}$ *(for* $i\in\mathbb{N}$*),* $\Longrightarrow_{R}$ *are monotonic.*

Corollary 5.5 (Stability of $\Longrightarrow_{R}$) $\Longrightarrow_{R,i}$, $\Longrightarrow_{R,\omega+i}$ *(for* $i \in \mathbb{N}$*),* $\Longrightarrow_{R}$ *and their respective fulfilledness-predicates are stable.*

Lemma 5.6 (Monotony of $\Longrightarrow_{R,\beta}$ in β) *For two ordinal numbers* β, γ *smaller than* $\omega + \omega$*, if* β *is smaller than* γ *(i. e. if* $\beta \preceq \gamma$*), then* $\Longrightarrow_{R,\beta} \subseteq \Longrightarrow_{R,\gamma}$.

The following technical lemmas state constructor-preservation and that there is no need for a second closure for reduction of constructor terms.

Lemma 5.7
$\forall n \in \mathbb{N} : \forall s \in \mathcal{T}(\text{cons}) : \forall t : (\ s \overset{n}{\Longrightarrow} t \ \ \Rightarrow \ \ (s \overset{n}{\Longrightarrow}_{R,\omega} t \in \mathcal{T}(\text{cons}))\ )$

Lemma 5.8
$\forall n \in \mathbb{N} : \forall s \in \mathcal{GT}(\text{cons}) : \forall t : (\ s \overset{n}{\Longrightarrow} t \ \ \Rightarrow \ \ (s \overset{n}{\Longrightarrow}_{R,\omega} t \in \mathcal{GT}(\text{cons}))\ )$

Lemma 5.9 $\downarrow \cap (\mathcal{T}(\text{cons}) \times \mathcal{T}(\text{cons})) \ = \ \downarrow_{R,\omega}$

Lemma 5.10 (Fulfilledness Test may be Simple)
Let $C \in \mathcal{CIT}(\text{sig})^*$. *If for each element* $u \in \mathcal{TERMS}(C)$*:*

1. *u has a [ground] normal form* NF(*u*)
   (i. e. $u \overset{\circledast}{\Longrightarrow} \text{NF}(u) \notin \text{dom}(\Longrightarrow)$*[,* $\text{NF}(u) \in \mathcal{GT}(\text{sig})$*])*

and

2. $\Longrightarrow$ *is [ground] confluent below u [and* $\mathcal{V}(\mathcal{TERMS}(C)) = \emptyset$*],*

then C is fulfilled w. r. t. $\Longrightarrow$ *<u>iff</u>*
$\forall u, v \in \mathcal{T}(\text{sig})$:

$$\begin{pmatrix} (\ (\ (u{=}v\ )\ \mathit{in}\ C) \ \Rightarrow\ \text{NF}(u) = \text{NF}(v)\ ) & \wedge \\ (\ (\ (\text{Def}\, u)\ \mathit{in}\ C) \ \Rightarrow\ \text{NF}(u) \in \mathcal{GT}(\text{cons})\ ) & \wedge \\ (\ (\ (u{\neq}v\ )\ \mathit{in}\ C) \ \Rightarrow\ (\text{NF}(u), \text{NF}(v) {\in} \mathcal{GT}(\text{cons}) \wedge \text{NF}(u){\neq}\text{NF}(v))\ ) & \end{pmatrix}$$

Considering ground confluence, it is important to realize that for reduction on ground terms we can restrict ourselves to ground substitutions:

Lemma 5.11
(Ground Substitutions are Sufficient for Ground Reduction)
Let $\Longrightarrow_{\mathrm{R,ground}}$ *be defined just like* $\Longrightarrow_{\mathrm{R}}$ *with the exception of the additional restriction "*$\sigma \in \mathcal{SUB}(\mathrm{V}, \mathcal{GT}(\mathrm{sig})) \ \wedge\ s \in \mathcal{GT}(\mathrm{sig})$*" in the requirement(#) of Definition 5.2.*

Now $\Longrightarrow_{\mathrm{R,ground}}$ *is uniquely defined,* $\Longrightarrow_{\mathrm{R,ground}} = \Longrightarrow_{\mathrm{R}} \cap (\mathcal{GT}(\mathrm{sig}) \times \mathcal{GT}(\mathrm{sig}))$, $\overset{\oplus}{\Longrightarrow}_{\mathrm{R,ground}} = \overset{\oplus}{\Longrightarrow}_{\mathrm{R}} \cap (\mathcal{GT}(\mathrm{sig}) \times \mathcal{GT}(\mathrm{sig}))$, *and* $\mathrm{dom}(\Longrightarrow_{\mathrm{R,ground}}) = \mathrm{dom}(\Longrightarrow_{\mathrm{R}}) \cap \mathcal{GT}(\mathrm{sig})$. *Furthermore,* $\Longrightarrow_{\mathrm{R,ground}}$ *is confluent* <u>*iff*</u> $\Longrightarrow_{\mathrm{R}}$ *is ground confluent.*

By Example 5.3 we know that (for guaranteeing a constructor-minimum model) we have to restrict the terms of the negated equations to be "defined". This semantic restriction is made syntactically explicit in the following definition that specifies a "well-behaved" subclass of the class of CRSs, in which inequalities are founded on constructor ground terms.

Definition 5.12 (Moderate Conditional Rule Systems (MCRS))
A CRS R *is a* moderate conditional rule system *(*MCRS*)* <u>*:iff*</u>

$$\forall((l,r),C) \in \mathrm{R} : \forall(u{\neq}v) \textit{ in } C : (\ \mathrm{Def}\,u, \mathrm{Def}\,v \textit{ are in } C\ )\ .$$

For a motivation of MCRSs see item (A) in section 3, where we discussed the problems involved.

The following definition allows us to use a CRS as a shorthand for a MCRS:

Definition 5.13 (Moderated Version of a CRS)
The moderated version of a CRS R *is the MCRS* R′ *obtained by replacing each sequence of the form '*$u{\neq}v$*' in each condition of each rule in* R *by the*[6] *sequence '*$u{\neq}v$, $\mathrm{Def}\,u$, $\mathrm{Def}\,v$*'.*

Now we are able to state the fundamental theorem about $\Longrightarrow$. It says that for moderate CRSs R with (ground) confluent $\Longrightarrow_{\mathrm{R}}$, the factor algebra $\mathcal{GT}(\mathrm{sig})/\overset{\circledast}{\Longleftrightarrow}_{\mathrm{R}}$ is an (up to isomorphism) uniquely determined sig/cons-model of R.

[6] We think the symmetry in u and v to be removed in some deterministic manner

Theorem 5.14 (Minimal Model being the Minimum of the Constructor-Minimal Models)

Let R *be a MCRS over* sig/cons/V.
Furthermore, assume $\Longrightarrow_{\mathrm{R}}$ *to be ground confluent.*

1. $\mathcal{GT}(\mathrm{sig})/\overset{\circledast}{\Longleftrightarrow}$ *is a constructor-minimum*[7] *model of* R.

2. *Let* $\mathcal{B}$ *be a model of* R *with*

   (a) $\mathcal{B}$ *a constructor-minimal*[7] *model of* R
   or
   (b) $\mathcal{B} \lesssim_{\mathrm{H}} \mathcal{GT}(\mathrm{sig})/\overset{\circledast}{\Longleftrightarrow}$.

   Now $\mathcal{GT}(\mathrm{sig})/\overset{\circledast}{\Longleftrightarrow} \lesssim_{\mathrm{H}} \mathcal{B}$.

   In other words: $\mathcal{GT}(\mathrm{sig})/\overset{\circledast}{\Longleftrightarrow}$ *is (by 2b) a minimal model and (by 2a) the minimum of all models that do not identify more defined terms than necessary.*

Theorem 5.15
(Monotony of $\Longrightarrow_{\mathrm{R}}$ w. r. t. Consistent Extension of the Specification)

Let R *be a CRS over* sig, cons *and* V. *Let* R′ *be another CRS, but over* sig′, cons′ *and* V′; *with*

$$\left|\begin{array}{l} \mathrm{sig}' = (\mathrm{F}',\mathrm{S}',\alpha') \\ \mathrm{cons}' = (\mathrm{C}',\mathrm{S}',\alpha'|_{\mathrm{C}'}) \\ \mathrm{V}'|_{\mathrm{S}} = \mathrm{V} \end{array}\right| \begin{array}{l} \mathrm{F} \subseteq \mathrm{F}' \\ \mathrm{C} \subseteq \mathrm{C}' \subseteq \mathrm{F}' \\ \mathrm{S} \subseteq \mathrm{S}' \\ \alpha \subseteq \alpha' \end{array}\left| \begin{array}{l} \mathrm{R} \subseteq \mathrm{R}' \end{array}\right|$$

Thus, sig′/cons′/V′ *is an enrichment of* sig/cons/V *in the most general*[8] *sense we can think of.*

Moreover, assume[9]*:* $\forall((l,r),C) \in (\mathrm{R}' \setminus \mathrm{R}) : l \notin \mathcal{T}(\mathrm{cons})$ *(:\$)*

Now we have:

1. $\forall s \in \mathcal{T}(\mathrm{cons}) : \forall t : \left((s \overset{\circledast}{\Longrightarrow}_{\mathrm{R}} t) \Leftrightarrow (s \overset{\circledast}{\Longrightarrow}_{\mathrm{R}'} t)\right)$
   "no change on old constructor terms"

2. $\Longrightarrow_{\mathrm{R}} \subseteq \Longrightarrow_{\mathrm{R}'}$ *"monotony"*

---

[7] Cf. Definition 4.3

[8] One may even define new constructor function symbols for the old sorts and take them from the old defined symbols.

[9] This has to be required for keeping the negative conditions fulfilled: Having founded our inequalities on old constructor ground terms, all we have to take care of now is not to confuse these terms.

6 Confluence

We are now going to define critical peaks that consist of the conditional critical pair, its peak, and the overlap position. The other notions we will use are also standard, with the exception of the notion of "quasi overlay joinable" which is a slight weakening (allowing a constant non-overlay part) of "joinable overlays" in Dershowitz[3].

Definition 6.1 *The set of (non-trivial)* critical peaks *between two rules* $((l_k, r_k), C_k) \in \mathrm{R}$ *(whose common variables we assume to be renamed) is defined by:* $\mathrm{Cp}(\ ((l_0, r_0), C_0),\ ((l_1, r_1), C_1)\ ) := \{(((l_1[p \leftarrow r_0], r_1), C_0 C_1)\sigma,\ l_1\sigma,\ p) \mid p \in \mathcal{O}(l_1) \wedge l_1/p \notin \mathrm{V} \wedge (\sigma \text{ most general unifier for } l_0, l_1/p) \wedge l_1[p \leftarrow r_0]\sigma \neq r_1\sigma\}$. *The set of all critical peaks of* R *is* $\mathrm{CP}(\mathrm{R}) := \bigcup_{\mathrm{rule}_0 \in \mathrm{R}} \bigcup_{\mathrm{rule}_1 \in \mathrm{R}} \mathrm{Cp}(\mathrm{rule}_0, \mathrm{rule}_1)$. R *is said to be* overlapping <u>:iff</u> $\mathrm{CP}(\mathrm{R}) \neq \emptyset$. *A critical peak* $(((t_0, t_1), D), \hat{t}, p)$ *is* weakly [ground] joinable *(w. r. t.* $\Longrightarrow$*)* <u>:iff</u> $\forall \varphi \in \mathcal{SUB}(\mathrm{V}, \mathcal{T}(\mathrm{sig})[\cap \mathcal{GT}(\mathrm{sig})])$: $((D\varphi \text{ fulfilled} \wedge \forall u \in \mathcal{TERMS}(D\varphi) : \Longrightarrow \text{ is confluent below } u) \Rightarrow t_0\varphi \downarrow t_1\varphi)$. *A critical peak* $(((t_0, t_1), D), \hat{t}, p)$ *is* quasi overlay [ground] joinable *(w. r. t.* $\Longrightarrow$*)* <u>:iff</u> $\forall \varphi \in \mathcal{SUB}(\mathrm{V}, \mathcal{T}(\mathrm{sig})[\cap \mathcal{GT}(\mathrm{sig})])$:

$$\left(D\varphi \text{ fulfilled} \Rightarrow \left(\begin{array}{l} t_1\varphi = t_0\varphi[p \leftarrow t_1\varphi/p] \quad \wedge \\ (t_0/p)\varphi \downarrow t_1\varphi/p \overset{\oplus}{\Longleftarrow} (\hat{t}/p)\varphi \end{array} \right) \right).$$

The proof of the following theorem is similar to that of Theorem 4 in Dershowitz[3]:

Theorem 6.2 (Confluence Criterion) *Let* R *be a CRS over* sig, cons *and* V. *If* $\Longrightarrow_{\mathrm{R}}$ *is [ground]*[10] *noetherian and all critical peaks in* CP(R) *are quasi overlay [ground] joinable, then* $\Longrightarrow_{\mathrm{R}}$ *is [ground] confluent.*

In the sequel we assume two fixed stable orderings $>$ and $\rhd$ on terms, such that $>$ is monotonic and $\rhd$ is well-founded and contains $>$ and the proper subterm ordering. We call a rule $((l, r), C) \in \mathcal{RUL}(\mathrm{sig}, \mathrm{cons})$ *aligned* (w. r. t. $(>, \rhd)$) <u>:iff</u>

$$l > r \ \wedge\ \forall u \in \mathcal{TERMS}(C) :\ l \rhd u\ .$$

We have proper reason for this form (cf. [9]) but not the space to explain it here.

The following kind of compatibility is a generalization to negative conditions and a slight weakening of the notion of "decreasingness" in Dershowitz[3].

Definition 6.3 ([Ground] Compatibility of a CRS)
A CRS R *over* sig/cons/V *is* [ground] compatible <u>:iff</u>
$\forall ((l, r), C) \in \mathrm{R} : \forall \tau \in \mathcal{SUB}(\mathrm{V}, \mathcal{T}(\mathrm{sig})[\cap \mathcal{GT}(\mathrm{sig})])$:

$$(C\tau \text{ fulfilled} \Rightarrow (\ ((l, r), C)\tau \text{ is aligned}\ ))$$

---

[10] doesn't make a difference of course (cf. Corollary 5.5)

The [ground] compatibility of a CRS R guarantees the alignment of a [ground] instantiated rule of R when its condition is fulfilled. But, while this kind of compatibility is convenient for obtaining further theoretical properties of the reduction relation, we have a problem when using this kind of compatibility of R in practice of reduction: The terms in $C\tau$ must be smaller than $l\tau$ only if $C\tau$ is fulfilled; but for easily deciding whether $C\tau$ is fulfilled we need its terms to be smaller than $l\tau$ and the analogous property for the other rules. That this need not be a vicious circle is shown by the following definition, which allows us to test the literals in the condition from left to right, using the old programming trick of a sequential "short-circuiting" AND-operator[11]. Notice that the difference to Definition 6.3 is in the quantified variable i occurring as an index which allows us to step inductively from $(< i)$ to i.

Definition 6.4 ([Ground] Left-Right-Compatibility)
A CRS R *over* sig/cons/V *is* [ground] left-right-compatible *:iff*
$\forall((l,r), L_0 \ldots L_{n-1}) \in \mathrm{R} : \forall \tau \in \mathcal{SUB}(\mathrm{V}, \mathcal{T}(\mathrm{sig})[\cap \mathcal{GT}(\mathrm{sig})]) :$

$$\left(\begin{array}{lll} \forall i < n : (& (L_0 \ldots L_{i-1})\tau \textit{ fulfilled } \Rightarrow \forall u \in \mathcal{TERMS}(L_i) : l\tau \rhd u\tau &) \\ \wedge & ((L_0 \ldots L_{n-1})\tau \textit{ fulfilled } \Rightarrow l\tau > r\tau &) \end{array} \right)$$

Definition 6.5 ([Ground] Don't-Care-Compatibility)
A CRS R *over* sig/cons/V *is* [ground] don't-care-compatible *:iff*
$\forall((l,r), L_0 \ldots L_{n-1}) \in \mathrm{R} : \forall \tau \in \mathcal{SUB}(\mathrm{V}, \mathcal{T}(\mathrm{sig})[\cap \mathcal{GT}(\mathrm{sig})]) :$

$$\left(\begin{array}{lll} \forall i < n : & \forall u \in \mathcal{TERMS}(L_i) : l\tau \rhd u\tau & \\ \wedge & ((L_0 \ldots L_{n-1})\tau \textit{ fulfilled } \Rightarrow l\tau > r\tau &) \end{array} \right)$$

Having a left-right-compatible CRS, we don't have to test the instantiated rules for alignment anymore, provided that we test the literals of the instantiated conditions from left to right until one of them fails. Having a don't-care-compatible CRS, we can even test the literals of the instantiated conditions in parallel and don't have to care for the position of these equations in the condition list. The don't-care-compatibility is conceptionally the same as the "decreasingness" in Dershowitz[3].

The "compatibility" of Definition 6.3 (which seems to be the least restrictive one tractable in theory[12]) is intended to be an interface for the generation of logically stronger kinds of "compatibility" that are useful in practice (cf. definitions 6.4 and 6.5), where the don't-care-compatibility of Definition 6.5 seems to be the most important one. For restrictions of $\Longrightarrow_{\mathrm{R}}$, however, even weaker kinds of "compatibility" than the one of Definition 6.3 may be sufficient.

For compatible CRSs we can now give a complete critical peak test à la Knuth-Bendix:

Theorem 6.6 (Confluence Test)
Let R *be [ground] compatible a CRS over* sig/cons/V*. Now:*
$\Longrightarrow_{\mathrm{R}}$ *is [ground] confluent iff all critical peaks in* CP(R) *are weakly [ground] joinable.*

[11] '&&' (instead of '&') in C. 'AND' in LISP.

[12] allowing a complete confluence test and appropriate inference systems for inductive theorem proving using the same orderings

7 Decidability of Reducibility

While reducibility of ground terms is in general neither semi- nor co-semi-decidable for non-overlapping, extra-variable free MCRSs with noetherian and confluent reduction relation, we can generalize Theorem 3.4 in Kaplan[5] in various directions:

Lemma 7.1 *Let* R *be a CRS over* sig/cons/V.

1. $\Longrightarrow_{\mathrm{R}}$*[-ground]-reducibility*[13] *is co-semi-decidable if a [ground] normal form for each [ground] term is computable*
   (i. e. there is a computable (partial) function f *with* $\mathrm{dom}(f) = \{s \in \mathcal{T}(\mathrm{sig})[\cap\mathcal{GT}(\mathrm{sig})] \mid \exists t \in \mathcal{T}(\mathrm{sig})[\cap\mathcal{GT}(\mathrm{sig})] : (s \overset{\circledast}{\Longrightarrow}_{\mathrm{R}} t \wedge t \notin \mathrm{dom}(\Longrightarrow_{\mathrm{R}}))\}$
   such that $\forall s \in \mathrm{dom}(f) : s \overset{\circledast}{\Longrightarrow}_{\mathrm{R}} f(s) \in (\mathcal{T}(\mathrm{sig})[\cap\mathcal{GT}(\mathrm{sig})]) \setminus \mathrm{dom}(\Longrightarrow_{\mathrm{R}})$*).*
2. *A [ground] normal form for each [ground] term is computable (see above) if* $\Longrightarrow_{\mathrm{R}}$*[-ground]-reducibility is co-semi-decidable and*
   $\forall s \in \mathcal{GT}(\mathrm{cons}) : \exists t : s \overset{\circledast}{\Longrightarrow}_{\mathrm{R}} t \notin \mathrm{dom}(\Longrightarrow_{\mathrm{R}})$.

Corollary 7.2 *Let* R *be a CRS over* sig/cons/V. *Assume* $\Longrightarrow_{\mathrm{R}}$ *to be noetherian. Now, co-semi-decidability of* $\Longrightarrow_{\mathrm{R}}$*[-ground]-reducibility is logically equivalent to computability of a [ground] normal form for each [ground] term.*

The usefulness of our kinds of compatibility is shown (w. r. t. Corollary 7.2) by:

Lemma 7.3 *Let* R *be a [ground] compatible CRS over* sig/cons/V.
Now we have $(\Longrightarrow_{\mathrm{R}}[\cap(\mathcal{GT}(\mathrm{sig}) \times \mathcal{GT}(\mathrm{sig}))]) \subseteq >$, *and* $\Longrightarrow_{\mathrm{R}}$ *is noetherian.*
Furthermore, if R *is extra-variable free and if*
(a) $\rhd[\cap(\mathcal{GT}(\mathrm{sig}) \times \mathcal{GT}(\mathrm{sig}))]$ *is decidable*
or
(b) R *is [ground] left-right-compatible,*
then $\Longrightarrow_{\mathrm{R}}$*[-ground]-reducibility is decidable and* $\{\, t \mid s \overset{\oplus}{\Longrightarrow}_{\mathrm{R}} t \,\}$ *is a finite computable set for all* $s \in \mathcal{T}(\mathrm{sig})[\cap\mathcal{GT}(\mathrm{sig})]$.

8 Conclusion

The presented constructor-based approach for positive/negative-conditional equational specifications provides the theoretical background for inductive theorem proving in such specifications as described in Wirth[8]. Various conceivable notions of inductive validity are presented and discussed there. Moreover, an abstract inference system in the style of [1] for proving inductive (constructor) validity of first-order clauses is given. Several refined versions of this inference system incorporating powerful contextual rewriting and simplification rules as well as a marking scheme are developed. In case of a (ground confluent, noetherian and) sufficiently complete base specification the notions of (ordinary) inductive validity and (inductive) constructor validity coincide.

---

[13] Following our notational conventions, *ground-reducibility* is reducibility w. r. t. $\Longrightarrow_{\mathrm{R,ground}}$ or reducibility of ground terms (cf. Lemma 5.11), which is different from inductive reducibility.

The idea of choosing a constructor-based approach for assigning adequate semantics to p/n-conditional equational specifications was heavily inspired by previous work of Kapur&Musser[7] (for the case of unconditional equations only) and Zhang[10] where problems with the usual notion of inductive (initial) validity, in particular concerning its non-monotonic behaviour w. r. t. consistent extensions, and partially defined functions are treated. Under the appropriate syntactical restrictions (mentioned above) concerning the form of p/n-conditional rules it turns out that the combination of these ideas with the approach of Kaplan[6] becomes very fruitful and also relevant for practical purposes since many natural specifications involve both conditional equations (rules) with positive and negative conditions and partially defined functions.

The perfect model semantics approach of Bachmair&Ganzinger[2], which also includes a completion procedure, generalizes Kaplan's approach [6] by abstracting the control information hidden in the syntactic form of the rules into a reduction ordering which must be total on ground terms and which determines the construction process of the perfect model. In our analysis, however, reduction orderings used for orienting p/n-systems do not necessarily have to be total on ground terms. Very recently, another method for inductive theorem proving of first-order clauses has been developed in [4] along the lines of [2] by means of a transformational approach. Its relation to our approach in theory and practice remains to be investigated.

Acknowledgement: We would like to thank Ulrich Kühler for many valuable discussions and detailed criticisms on earlier versions of this paper.

References

[1] Leo Bachmair. *Proof by Consistency in Equational Theories.* 3rd LICS 1988, pp. 228-233.

[2] Leo Bachmair, Harald Ganzinger. *Perfect Model Semantics for Logic Programs with Equality.* Proc. of 8th Int. Conf. on Logic Programming, pp. 645-659, MIT Press, 1991.

[3] Nachum Dershowitz, Mitsuhiro Okada, G. Sivakumar. *Confluence of Conditional Rewrite Systems.* LNCS 308. Springer-Verlag, Berlin 1988.

[4] Harald Ganzinger, Jürgen Stuber. *Inductive Theorem Proving by Consistency for First-Order Clauses.* This volume.

[5] Stéphane Kaplan. *Conditional Rewrite Rules.* Theoretical Computer Science 33 (1984), pp. 175-193. North-Holland.

[6] Stéphane Kaplan. *Positive/Negative-Conditional Rewriting.* LNCS 308. Springer-Verlag, Berlin 1988.

[7] Deepak Kapur, David R. Musser. *Proof by Consistency.* Artificial Intelligence 31 (1987), pp. 125-157.

[8] Claus-Peter Wirth. *Inductive Theorem Proving in Theories specified by Positive/Negative Conditional Equations.* Diplomarbeit, 1991, Universität Kaiserslautern, Fachbereich Informatik, Postfach 3049, W-6750 Kaiserslautern.

[9] Claus-Peter Wirth, Bernhard Gramlich. *A Constructor-Based Approach for Positive/Negative-Conditional Equational Specifications.* SEKI-Report SR-92-10 (SFB), May 4, 1992, Universität Kaiserslautern, Fachbereich Informatik.

[10] Hantao Zhang. *Reduction, Superposition and Induction: Automated Reasoning in an Equational Logic.* Rensselaer Polytech. Inst., Dept. of Comp. Sci., Troy, NY, PhD thesis, 1988.

Semantics for Positive/Negative Conditional Rewrite Systems

Klaus Becker
Fachbereich Informatik
Universität Kaiserslautern
6750 Kaiserslautern
Germany

Abstract

We present two approaches to define the semantics for positive/negative conditional rewrite systems. The first — the recursive evaluation approach — is based on a well-founded ordering and leads to a perfect model semantics. The second — the built-in evaluation approach — is based on a set of predefined inequations and leads to an initial model semantics provided the inequations are interpreted in a non-standard way. Both approaches coincide if a suitable inequation covering property is satisfied.

1 Introduction

When programming with specifications, the specifier creates syntactic objects such as rewrite rules or clauses, keeping always in mind a certain meaning associated with these syntactic objects. To capture this meaning, a privileged model (usually a canonical term algebra) is distinguished from all the models of the specification in order to represent the semantics of the specification. The property of being privileged is reflected by a model theoretical characterization like initiality or perfectness representing a kind of minimality.

The specifications to be treated in this paper consist of positive/negative conditional equations resp. rewrite rules having positive as well as negative literals in their conditional parts. Whereas the semantics can be achieved in the mere positive conditional case by a fixpoint construction using only the logical (implicational) structure of the syntactic objects, some extra-logical informations due to the negative literals are required in the positive/negative case. The reason for that is that one must differentiate between the negative literals in the condition and the conclusion — which does not exist from the logical point of view — in order to get a unique minimal model (see [Ka88]).

The main idea of the construction process to yield the canonical term algebra is what we informally call "supportedness": The evaluation of a conditional equation in the construction process requires that the condition is supported in order to get a new equivalence. In other words, the condition has to be satisfied. The main problem here is how to support the inequivalences due to the negative literals in the condition. We will present two approaches to solve this problem corresponding to two different support strategies.

The support source of the first approach, the "recursive evaluation approach", is the equivalence relation to be defined, thus "negation as failure" is used as support strategy. In order to be sound, the construction process must be controlled by a well-founded ordering representing the fact that the literals in the condition are in a certain sense smaller than the conclusion. The basic ideas of this approach were made precise in the case of relational logic programming by Apt, Blair and Walker [ABW88] and Przymusinski [Pr88]. They were generalized by Bachmair and Ganzinger [BaGa91] for logic programs with equality and negation. Implicitely they were also used by Kaplan [Ka88] in the positive/negative conditional rewrite case. Our aim in this paper is to adopt the concept of Bachmair and Ganzinger [BaGa91] and clarify the influence that the ordering has on the semantics.

The support source for inequivalences in the second approach, called "built-in evaluation approach", are predefined inequations possibly induced by a predefined structure. This approach is related to the constructor-based approach for positive/negative conditional specifications of Wirth and Gramlich [WiGr92].

The paper presents the two approaches in the sections 3 and 4. Section 2 introduces some basic notions. Section 5 finally compares and relates the two approaches. Proofs are omitted where they are straightforward.

2 Basic Notions

We assume that the reader is familiar with the basic concepts of term rewriting (see [AvMa90, DeJo90]) and mathematical logic. Notions and notations not defined here are standard.

Throughout the paper let Σ be a given signature and let $TERM(\Sigma)$ denote the set of Σ-ground terms.

Definition 2.1 *A* positive/negative conditional rewrite system (over Σ) *is a set of formulas of the form*

$$\bigwedge_{i=1}^{m} u_i = v_i \bigwedge_{j=1}^{n} \overline{u}_j \neq \overline{v}_j \Rightarrow u = v$$

where $u, v, u_i, v_i, \overline{u}_j, \overline{v}_j$ are Σ-terms such that u is no variable and such that all variables occuring in $v, u_i, v_i, \overline{u}_j, \overline{v}_j$ also occur in u.

In the sequel $\mathcal{R}$ always denotes a given positive/negative conditional rewrite system. By $INST(\mathcal{R}) = \{\tau(R) \mid R \in \mathcal{R},\ \tau \text{ a ground substitution}\}$ we denote the set of ground instances of $\mathcal{R}$.

Next we review some relations to compare Σ-algebras. Let $\mathcal{A}$ and $\mathcal{B}$ denote Σ-algebras. We write (as in [Ka88]) $\mathcal{B} \geq \mathcal{A}$ iff there exists a unique homomorphism from $\mathcal{A}$ into $\mathcal{B}$.

Definition 2.2 *Let $\mathcal{K}$ be a class of Σ-algebras. An element $\mathcal{A}$ of $\mathcal{K}$ is said to be (a)* initial in *$\mathcal{K}$ iff $\mathcal{B} \geq \mathcal{A}$ for any $\mathcal{B} \in \mathcal{K}$,*
(b) quasi-initial in *$\mathcal{K}$ iff there exists no $\mathcal{B} \in \mathcal{K}$ such that $\mathcal{A} > \mathcal{B}$.*

Each Σ-algebra $\mathcal{A}$ induces a congruence relation on the Σ-ground terms by $\sim_{\mathcal{A}} = \{\{s,t\} \mid s,t \in TERM(\Sigma) \text{ and } s^{\mathcal{A}} = t^{\mathcal{A}}\}$, where $s^{\mathcal{A}}$ and $t^{\mathcal{A}}$ are the values of s and t in $\mathcal{A}$. If we are given an ordering $\succ$ on the Σ-ground terms, then we can use it to compare such congruence relations. For that purpose let $\succ^i = \succ^p \succ^p$ with $\succ^p = \prec\prec$ (see. [BaGa91]).

Definition 2.3 *Let $\mathcal{K}$ be a class of Σ-algebras, $\mathcal{A}$ an element of $\mathcal{K}$ and $\succ$ a (well-founded) ordering on $TERM(\Sigma)$. $\mathcal{A}$ is said to be* perfect in $\mathcal{K}$ wrt. $\succ$ *iff $\sim_{\mathcal{B}} \succeq^i \sim_{\mathcal{A}}$ for any $\mathcal{B} \in \mathcal{K}$.*

Notice that $\mathcal{A}$ is quasi-initial in $\mathcal{K}$ whenever $\mathcal{A}$ is perfect in $\mathcal{K}$ wrt. a suitable ordering.

The following example illustrates the last definition.

Example 2.1 *Let*

$$\begin{array}{lcl} \mathcal{R} & : & b \neq c \Rightarrow a = b \\ \succ & : & a \succ b, c \\ \mathcal{A} & : & a \sim_{\mathcal{A}} b \not\sim_{\mathcal{A}} c \\ \mathcal{B} & : & a \sim_{\mathcal{B}} b \sim_{\mathcal{B}} c \\ \mathcal{C} & : & a \not\sim_{\mathcal{C}} b \sim_{\mathcal{C}} c \end{array}$$

Obviously, $\mathcal{A}, \mathcal{B}, \mathcal{C}$ are models of $\mathcal{R}$. Let $\mathcal{K}$ be the class of all models of $\mathcal{R}$. There is no initial model of $\mathcal{R}$ in $\mathcal{K}$. Whereas $\mathcal{A}$ and $\mathcal{C}$ are quasi-initial in $\mathcal{K}$, only $\mathcal{A}$ is perfect in $\mathcal{K}$ wrt. $\succ$. One easily shows that $\sim_{\mathcal{B}} \succ^i \sim_{\mathcal{A}}$, $\sim_{\mathcal{C}} \succ^i \sim_{\mathcal{A}}$. Note that the ordering $\succ^i$ just reflects the supportedness requirement.

3 Recursive Evaluation Approach

3.1 Definition of the Semantics

Let $\mathcal{R}$ be a positive/negative conditional rewrite system. The definition of the semantics will be guided by a given well-founded ordering $\succ$ on $TERM(\Sigma)$. This ordering will not be given totally independently from the system $\mathcal{R}$. As we consider the elements of $\mathcal{R}$ as rewrite rules, we choose $\succ$ such that it reflects the ordering aspects intended to be inherent in $\mathcal{R}$. The main question that arises in this context is how to make this ordering aspect precise in order to achieve a satisfactory semantics. There has been an analogous question in the positive conditional rewrite case where some different notions (reductive [JoWa86], simplifying [Ka87], decreasing [DeOk90]) have been proposed in order to get a well-behaved rewrite relation.

The following notion (which is similar to that in [JoWa86]) establishes compatibility of $\succ$ with the structure of $\mathcal{R}$ and makes the rewrite relation to be introduced below terminating and decidable.

Definition 3.1 *$\mathcal{R}$ is said to be* reductive wrt. $\succ$ *iff*

$$s[u] \succ s[v], u_i, v_i, \overline{u}_j, \overline{v}_j$$

for any instance $\bigwedge_i u_i = v_i \bigwedge_j \overline{u}_j \neq \overline{v}_j \Rightarrow u = v \in INST(\mathcal{R})$ *and any context s.*

Next we adopt the concept from [BaGa91] to assign to the pair $(\mathcal{R}, \succ)$ an unconditional ground rewrite system $\mathcal{G}_{(\mathcal{R},\succ)}$ (or simply $\mathcal{G}$ if $(\mathcal{R}, \succ)$ is clear from the context). As $\mathcal{G}$ has a well known intended semantics — the equivalence relation $\sim$ induced by $\mathcal{G}$ on the Σ-ground terms resp. the canonical term algebra induced by $\sim$ on $TERM(\Sigma)$ — $(\mathcal{R}, \succ)$ is assigned a semantics too.

$\mathcal{G}$ will be constructed by Noetherian induction wrt. $\succ$ using the auxiliary ground rewrite systems $\mathcal{G}_{\prec u}$ and $\mathcal{G}_u$ (see below) where $u \in TERM(\Sigma)$.

Let $u \in TERM(\Sigma)$. Suppose that $\mathcal{G}_v$ is already defined for all $v \in TERM(\Sigma)$ with $v \prec u$. Then let

$$\mathcal{G}_{\prec u} = \bigcup_{v \prec u} \mathcal{G}_v$$

$$\mathcal{G}_u = \mathcal{G}_{\prec u} \cup \{u = v \mid \exists \ \bigwedge_i u_i = v_i \bigwedge_j \overline{u}_j \neq \overline{v}_j \Rightarrow u = v \ \in INST(\mathcal{R}) : \; u_i \sim_{\prec u} v_i \ for \ i = 1 \ldots m \ \text{and} \ \overline{u}_j \not\sim_{\prec u} \overline{v}_j \ for \ j = 1 \ldots n\}$$

$$\mathcal{G} = \bigcup_{u \in TERM(\Sigma)} \mathcal{G}_u .$$

Note that we write $\sim_{\prec u}$ for the congruence relation on $TERM(\Sigma)$ induced by $\mathcal{G}_{\prec u}$. $\mathcal{G}$ is called *the ground rewrite system induced by* $(\mathcal{R}, \succ)$. $\mathcal{G}$ induces a congruence relation $\sim = \overset{*}{\longleftrightarrow}_{\mathcal{G}}$ on $TERM(\Sigma)$ and henceforth via $\sim$ a canonical term algebra $\mathcal{T}_{(\mathcal{R},\succ)}$ (or $\mathcal{T}$ for short) with carrier $TERM(\Sigma)_{/\sim}$. $\mathcal{T}_{(\mathcal{R},\succ)}$ resp. $\mathcal{T}$ will be called *the canonical term algebra induced by* $(\mathcal{R}, \succ)$.

The definition process reflects the intuition that according to the ordering the conditions are evaluated before the conclusion is drawn. But in general this process is not sound, as the canonical term algebra need not be a model of the specification.

Example 3.1 *Let:*

$$\mathcal{R}: \quad c = d \Rightarrow b = c \qquad\qquad \succ: \quad b \succ c, d$$
$$a = c \qquad\qquad a \succ c$$
$$a = d \qquad\qquad a \succ d$$

In this example we get $\mathcal{G} = \{a = c, a = d\}$. Thus $\mathcal{T}_{(\mathcal{R},\succ)}$ is not a model of $\mathcal{R}$. Notice that this fact has nothing to do with negative literals in the condition; it is just a consequence of the construction process: terms that cannot be made equivalent at a certain stage in the construction process can become equivalent later on in the process. We exclude such unintended cases by introducing the notion of generic consistency.

Definition 3.2 *$\mathcal{R}$ is said to be* generically consistent wrt. $\succ$ *iff for all $s,t \in TERM(\Sigma)$ we have: $s \sim t$ iff $\forall u \succ s,t : s \sim_{\prec u} t$.*

Lemma 3.1 *Let $\mathcal{R}$ be reductive and generically consistent wrt. $\succ$. Then $\mathcal{T}$ is a model of $\mathcal{R}$.*

3.2 Model Theoretical Characterization

We first state an auxiliary lemma.

Lemma 3.2 *Let $\mathcal{R}$ be reductive and generically consistent wrt. $\succ$. Then $\mathcal{G}$ is confluent.*

Proof: Let $P(s)$ iff G_s is confluent (for any $s \in TERM(\Sigma)$). We prove $P(s)$ for all ground terms s by Noetherian induction wrt. $\succ$. Suppose that $P(s')$ for all $s' \prec s$. Note that $\mathcal{G}_{\prec s}$ then is confluent. It suffices to consider only the critical pairs induced by $\mathcal{G}_s$. Let $u = v, u' = v' \in \mathcal{G}_s$ such that u' is a subterm of u. Then $v = u[v']$ is a critical pair to be considered. As $v \sim u[v']$ and $s \succeq u \succ v, u[v']$ we get $v \sim_{\prec s} u[v']$ by generic consistency. Confluence of $\mathcal{G}_{\prec s}$ then provides $v \downarrow_{\mathcal{G}_{\prec s}} u[v']$ and hence $v \downarrow_{\mathcal{G}_s} u[v']$. □

Note that the converse of the statement made in the lemma is not true, as can be seen from example 3.2 below.

The following theorem states the fundamental property of $\mathcal{T}$ in being the privileged model of $\mathcal{R}$.

Theorem 3.1 *Let $\mathcal{R}$ be reductive and generically consistent wrt. $\succ$. Then $\mathcal{T}$ is perfect in the class of models of $\mathcal{R}$ wrt. $\succ$.*

Proof: Let $\mathcal{B} \models \mathcal{R}$. Assume that $\sim \neq \sim_{\mathcal{B}}$. Let $\{s,t\} \in \sim \setminus \sim_{\mathcal{B}}$. We construct $\{s',t'\} \in \sim_{\mathcal{B}} \setminus \sim$ with $\{s',t'\} \succ^p \{s,t\}$ resp. $\{s,t\} \succ\succ \{s',t'\}$. We only consider the case that $\{s,t\}$ is a (wrt. $\succ\succ$) minimal element of $\sim \setminus \sim_{\mathcal{B}}$. The other case then is simple.

By assumption we have $s \sim t$. As $\mathcal{G}$ is confluent by lemma 3.2 we get $s \downarrow_{\mathcal{G}} t$. As $s \not\sim_{\mathcal{B}} t$ there exists $w \in TERM(\Sigma)$ and $u = v \in \mathcal{G}$ such that w.l.o.g $s \xrightarrow{*}_{\mathcal{G}} w[u] \longrightarrow_{\mathcal{G}} w[v] \downarrow_{\mathcal{G}} t$ and $w[u] \not\sim_{\mathcal{B}} w[v]$. By the definition of $\mathcal{G}$ there exists an instance $\bigwedge_i u_i = v_i \bigwedge_j \overline{u}_j \neq \overline{v}_j \Rightarrow u = v \in INST(\mathcal{R})$ with $u_i \sim_{\prec u} v_i$ for all i and $\overline{u}_j \not\sim_{\prec u} \overline{v}_j$ for all j. Note that $\{s,t\} \succ\succ \{u_i, v_i\}, \{\overline{u}_j, \overline{v}_j\}$.

We get $u \not\sim_{\mathcal{B}} v$ for otherwise $w[u] \sim_{\mathcal{B}} w[v]$. Further $u_i \sim_{\mathcal{B}} v_i$ for all i for otherwise $\{s,t\}$ would not be a minial element of $\sim \setminus \sim_{\mathcal{B}}$. We have $\overline{u}_j \sim_B \overline{v}_j$

for some j for otherwise $u \sim_{\mathcal{B}} v$, as $\mathcal{B} \models \mathcal{R}$. Finally $\overline{u}_j \not\succ \overline{v}_j$ for otherwise $\overline{u}_j \sim_{\prec u} \overline{v}_j$. Thus we have $\{\overline{u}_j, \overline{v}_j\} \in \sim_{\mathcal{B}} \setminus \sim$ with $\{s,t\} \succ\succ \{\overline{u}_j, \overline{v}_j\}$. Now let $s' = \overline{u}_j$ and $t' = \overline{v}_j$. □

This result is strongly related to the main result in [BaGa91], which says that every consistent and saturated set of equational clauses has a perfect model wrt. the given ordering. This given ordering is assumed to be total on the set of ground terms and thus contains the proper subterm ordering on the ground terms. In our context, which starts with rewrite rules instead of clauses and does not necessiate a superposition calculus for pre-processing, we see that the ordering does not have to satisfy the totality requirement in order to induce a well-behaved semantics. However, if it does, then generic consistency in our context is equivalent to saturatedness in the context of [BaGa91] and our result becomes a special case of that in [BaGa91]. The differences due to the ordering assumptions are quite subtle and so will be investigated further in the sections to come.

3.3 Dependency on the Ordering

We start with two independency results.

Lemma 3.3 *Let $\mathcal{R}$ be reductive and generically consistent wrt. the two well-founded orderings $\succ^1$ and $\succ^2$. Then $\sim^1 = \sim^2$ (where $\sim^1$ and $\sim^2$ are defined via $\mathcal{G}^1 = \mathcal{G}_{(\mathcal{R},\succ^1)}$ resp. $\mathcal{G}^2 = \mathcal{G}_{(\mathcal{R},\succ^2)}$).*

Proof: First note that $\succ = \succ^1 \cap \succ^2$ is a well-founded ordering. Now suppose that $\sim^1 \neq \sim^2$. Then $D = (\sim^1 \setminus \sim^2) \cup (\sim^2 \setminus \sim^1)$ is not empty. Let (s,t) be an element of D which is minimal wrt. $\succ$. Assume w.l.o.g. that $(s,t) \in \sim^1 \setminus \sim^2$.

As $\mathcal{G}^1$ is confluent by lemma 3.2 we get $s \downarrow_{\mathcal{G}^1} t$. As $s \not\sim^2 t$ there exists $w \in TERM(\Sigma)$ and $u = v \in \mathcal{G}^1$ such that w.l.o.g $s \xrightarrow{*}_{\mathcal{G}^1} w[u] \longrightarrow_{\mathcal{G}^1} w[v] \downarrow_{\mathcal{G}^1} t$ and $w[u] \not\sim^2 w[v]$. In particular $u = v \notin \mathcal{G}^2$. By the definition of $\mathcal{G}^1$ there exists an instance $\bigwedge_i u_i = v_i \bigwedge_j \overline{u}_j \neq \overline{v}_j \Rightarrow u = v \in INST(\mathcal{R})$ with $u_i \sim^1_{\prec^1 u} v_i$ for all i and $\overline{u}_j \not\sim^1_{\prec^1 u} \overline{v}_j$ for all j. Note that $\{s,t\} \succ\succ \{u_i, v_i\}, \{\overline{u}_j, \overline{v}_j\}$. Thus $u_i \sim^1 v_i$ for all i and (by generic consistency wrt. $\succ^1$) $\overline{u}_j \not\sim^1 \overline{v}_j$ for all j.

As $u = v \notin \mathcal{G}^2$ and $\mathcal{R}$ is generically consistent wrt. $\succ^2$ there exists i with $u_i \not\sim^2 v_i$ or there exists j with $\overline{u}_j \sim^2 \overline{v}_j$. Thus there exists $(s',t') \in D$ with $(s,t) \succ (s',t')$ — which contradicts the assumption. □

Lemma 3.4 *Let $\mathcal{R}$ be reductive and generically consistent wrt. $\succ$. Let $\succ^*$ be a well-founded refinement of $\succ$. Then $\mathcal{G} = \mathcal{G}^*$ and hence $\sim = \sim^*$.*

Proof: We define two predicates P and Q on the Σ-ground terms by:
$P(u)$ iff $\mathcal{G}^*_u \subseteq \mathcal{G}$ for all $u \in TERM(\Sigma)$ and
$Q(u)$ iff $\mathcal{G}_u \subseteq \mathcal{G}^*_u$ for all $u \in TERM(\Sigma)$,
where $\mathcal{G}^*_u$ is defined wrt. $\succ^*$.

Then we prove $P(u)$ and $Q(u)$ for all $u \in TERM(\Sigma)$ simultaneously by Noetherian induction wrt. $\succ^*$. Let $P(u')$ and $Q(u')$ for all $u' \prec^* u$.

To prove $P(u)$ let $a = b \in \mathcal{G}_u^*$. The case $a = b \in \mathcal{G}_{\prec \cdot u}^*$ is easy, so let $a = u$ and $b = v$ for some $\bigwedge_i u_i = v_i \bigwedge_j \overline{u}_j \neq \overline{v}_j \Rightarrow u = v \in INST(\mathcal{R})$. In particular we have $u_i \sim_{\prec \cdot u}^* v_i$ for all i and $\overline{u}_j \not\sim_{\prec \cdot u}^* \overline{v}_j$ for all j. We want to show that $u = v \in \mathcal{G}$, that is that $u_i \sim_{\prec_u} v_i$ for all i and $\overline{u}_j \not\sim_{\prec_u} \overline{v}_j$ for all j. First, as $u_i \sim_{\prec \cdot u}^* v_i$ we get $u_i \sim v_i$ by the induction hypothesis $P(u')$ for some appropriate $u' \prec^* u$ and thus $u_i \sim_{\prec_u} v_i$ by generic consistency wrt. $\succ$. Second, if we had $\overline{u}_j \sim_{\prec_u} \overline{v}_j$ we would get $\overline{u}_j \sim_{u''} \overline{v}_j$ for some $u'' \prec u$. As $\succ \subseteq \succ^*$ we can use the induction hypothesis $Q(u'')$ to conclude $\overline{u}_j \sim_{\prec \cdot u}^* \overline{v}_j$. Hence $u = v \in \mathcal{G}$.

The proof of $Q(u)$ proceeds analogously. □

The first result can be interpretated as follows: any two "sound situations" lead to the same semantics. The second result tells us that if we start with a "sound situation", then refining the ordering does not effect the semantics.

Unfortunately, semantics can change in an essential way if we start with an "unsound situation".

Example 3.2 *Let:*

$$\mathcal{R} : \quad f(c) = f(d) \Rightarrow a = b$$
$$\qquad\qquad\qquad\qquad c = d$$

$$\succ \; : \; a \succ b, f(c), f(d) \quad ; \quad c \succ d$$
$$\succ^* \; : \; a \succ^* f(c) \succ^* c \succ^* d \quad ; \quad a \succ^* b, f(d)$$

Note that $\mathcal{R}$ is generically consistent wrt. $\succ^*$, but not wrt. $\succ$. The example shows that generic consistency and thus the semantics depends on the ordering that is chosen to make $\mathcal{R}$ reductive. In the next section we show that this unsatisfactory behaviour is no longer present if we restrict the class of orderings to be taken as induction guide in the semantics construction, that is if we impose stronger ordering requirements on the rewrite rules. Note that our notion of decreasingness does not coincide with the corresponding notion in [DeOk90].

Definition 3.3 *$\mathcal{R}$ is said to be* decreasing wrt. $\succ$ *iff $\mathcal{R}$ is reductive wrt. $\succ$ and $s[u] \succeq u$ for any context s and any u that is the left hand side of an instance of a rule from $\mathcal{R}$. $\mathcal{R}$ is said to be* decreasing *iff there exists a well-founded ordering $\succ$ such that $\mathcal{R}$ is decreasing wrt. $\succ$.*

3.4 Relationship to Kaplan's Approach

We first review the rewrite relation defined in [Ka88].

Definition 3.4 *For $s, t \in TERM(\Sigma)$ let $s \longrightarrow_{\mathcal{R}} t$ iff there exists an occurence $p \in O(s)$, a substitution σ and a rule $\bigwedge_i u_i = v_i \bigwedge_j \overline{u}_j \neq \overline{v}_j \Rightarrow u = v \;\in \mathcal{R}$ such that $s/p \equiv \sigma(u)$, $s[\sigma(v)] \equiv t$, $\sigma(u_i) \downarrow_{\mathcal{R}} \sigma(v_i)$ and not $\sigma(\overline{u}_j) \downarrow_{\mathcal{R}} \sigma(\overline{v}_j)$.*

The next lemma strengthens the result of lemma 3.2.

Lemma 3.5 *Let $\mathcal{R}$ be decreasing wrt. $\succ$. Then $\mathcal{R}$ is generically consistent wrt. $\succ$ iff $\mathcal{G}$ is confluent.*

The following lemma forms the basis for the main results to come.

Lemma 3.6 *Let $\mathcal{R}$ be decreasing wrt. $\succ$. Let $t \in TERM(\Sigma)$ and let $\mathcal{G}_{\preceq t}$ be confluent. Then for all $s \preceq t$ and all s' we have: $s \longrightarrow_{\mathcal{G}_*} s'$ iff $s \longrightarrow_{\mathcal{R}} s'$.*

Proof: Let $P(s)$ iff for all s': $s \longrightarrow_{\mathcal{G}_*} s'$ iff $s \longrightarrow_{\mathcal{R}} s'$. Then one proves $P(s)$ for all $s \preceq t$ by Noetherian induction wrt. $\succ$ in a straightforward way using lemma 3.5. □

As a consequence we now easily get the following results.

Theorem 3.2 *Let $\mathcal{R}$ be decreasing and generically consistent wrt. $\succ$. Then $\longrightarrow_{\mathcal{R}} = \longrightarrow_{\mathcal{G}}$ on the Σ-ground terms.*

Theorem 3.3 *Let $\mathcal{R}$ be decreasing wrt. $\succ$. Then $\mathcal{G}$ is confluent iff $\mathcal{R}$ is ground confluent wrt. $\longrightarrow_{\mathcal{R}}$.*

Theorem 3.2 together with lemma 3.5 show that the semantics does not depend on the ordering provided we consider only orderings that make $\mathcal{R}$ decreasing. This result can be interpretated in the following way: A stratification for $\mathcal{R}$ is a well-founded ordering that makes $\mathcal{R}$ decreasing. If $\mathcal{R}$ is generically consistent wrt. one of the possible stratifications, then the semantics does not depend on the specific stratification chosen. So we may view the semantics completely determined by the structure of the rules.

Theorem 3.3 together with lemma 3.5 show that generic consistency (wrt. $\succ$) is equivalent to ground confluence of $\longrightarrow_{\mathcal{R}}$ provided we have decreasingness. Now, ground confluence of $\longrightarrow_{\mathcal{R}}$ is equivalent to ground joinability of the critical pairs induced by $\mathcal{R}$ (see [Ka88]). So we get by these results a method to check generic consistency.

The last result in this section relates the semantics defined in section 3.1 to the normal-form semantics $\mathcal{T}_{NF(\mathcal{R})}$ (wrt. $\longrightarrow_{\mathcal{R}}$) introduced in [Ka88].

Corollary 3.1 *Let $\mathcal{R}$ be decreasing (wrt. $\succ$) and let $\longrightarrow_{\mathcal{R}}$ be ground confluent. Then $\mathcal{T}_{(\mathcal{R},\succ)} = \mathcal{T}_{NF(\mathcal{R})}$.*

Corollary 3.2 *Let $\mathcal{R}$ be decreasing wrt. $\succ$ and let $\longrightarrow_{\mathcal{R}}$ be ground confluent. Then $\mathcal{T}_{NF(\mathcal{R})}$ is perfect in $\{\mathcal{B} \mid \mathcal{B} \models \mathcal{R}\}$ wrt. $\succ$ and thus quasi-initial in $\{\mathcal{B} \mid \mathcal{B} \models \mathcal{R}\}$.*

4 Built-in Evaluation Approach

4.1 Motivation

First we describe the situation we have in mind when introducing this approach (see also [AvBe92]). We assume that Σ is a signature enrichment $\Sigma_0 + \Sigma_1$ with a base signature Σ_0. The symbols from Σ_0 are intended to be predefined: we are given a term-generated Σ_0-algebra $\mathcal{A}$ — the built-in algebra — inducing a set $\mathcal{R}_\mathcal{A}$ of rules that define the symbols from Σ_0. The "new symbols" from Σ_1 are intended to be defined by the rules from $\mathcal{R}$.

Example 4.1 *Let*

$$
\begin{array}{lll}
\Sigma_0 & : & (\{nat\}, \{+, 0, 1, 2, \ldots\}) \\
\mathcal{A} & : & \textit{natural number interpretation} \\
\mathcal{R}_\mathcal{A} & : & 0+0=0, 1+0=1, \ldots \\
\Sigma_1 & : & (\emptyset, \{div\}) \\
\mathcal{R} & : & y \neq 0 \Rightarrow div(x+y, y) = div(x, y) + 1 \\
 & & y \neq 0 \Rightarrow div(x, x+y) = x
\end{array}
$$

The Σ-algebras of interest should not destroy the built-in structure. They are the consistent extensions of the built-in algebra $\mathcal{A}$ that satisfy $\mathcal{R}$. In particular, different elements from $\mathcal{A}$ should not be identified in these algebras. We capture this fact by introducing a suitable set $\mathcal{N}$ of predefined inequations. First let $\sim_0$ be the equivalence relation induced by $\mathcal{A}$: let $u \sim_0 v$ iff $\mathcal{A} \models u = v$ where $u, v \in TERM(\Sigma_0)$. Then $u \neq v \in \mathcal{N}$ iff $u \not\sim_0 v$ where $u, v \in TERM(\Sigma_0)$. Now each Σ-algebra of interest must satisfy in addition to $\mathcal{R}$ (and $\mathcal{R}_\mathcal{A}$) the set $\mathcal{N}$ of ground inequations.

Within this just described context we can use $\mathcal{N}$ as source for supporting inequivalences. In order to apply a rule with an instatiation τ we may require that there exists a natural number constant n different from 0 such that $\tau(x) \sim n \neq 0$. So in addition to using the "positive" support $\tau(x) \sim n$, we use the "negative" support $n \not\sim_0 0$ given by the predefined algebra.

An interesting approach which fits in this framework is given in [WiGr92]. If $\mathcal{R}$ can be split into two parts $\mathcal{R}_0$ and $\mathcal{R}_1$ where $\mathcal{R}_0$ defines the relations among the so-called constructors by mere positive conditional equations, then $\mathcal{A}$ can be taken as the canonical constructor algebra induced by $\mathcal{R}_0$.

4.2 Definition of the Semantics

Let $\mathcal{R}$ be given as usual. Let in addition $\mathcal{N}$ be a set of ground inequations over Σ. The congruence relation capturing the semantics is defined with the aid of an inference system that acts on ground equations. Notice that the last rule depends on $\mathcal{R}$, $\mathcal{N}$ and — for technical reasons — on a set S of ground equations.

(Reflexivity)

$$\frac{}{u = u}$$

(Symmetry)

$$\frac{u = v}{v = u}$$

(Transitivity)

$$\frac{u = v, v = w}{u = w}$$

(Congruence)

$$\frac{u_1 = v_1, \ldots, u_n = v_n}{f(u_1, \ldots, u_n) = f(v_1, \ldots, v_n)}$$

$$if\ f(u_1, \ldots, u_n)\ and\ f(v_1, \ldots, v_n)\ are\ both\ well-formed\ \Sigma-terms.$$

(Substitutivity)

$$\frac{u_i = v_i (i = 1 \ldots m); \overline{u}_j = a_j, b_j = \overline{v}_j (j = 1 \ldots n)}{u = v}$$

$$if\ \bigwedge_i u_i = v_i \bigwedge_j \overline{u}_j \neq \overline{v}_j \Rightarrow u = v\ \in INST(\mathcal{R}),\ a_j \neq b_j \in \mathcal{N}\ (j = 1 \ldots n)$$

$$and\ u_i = v_i, \overline{u}_j = a_j, b_j = \overline{v}_j \in S\ (i = 1 \ldots m, j = 1 \ldots n).$$

We write $S \vdash_{\mathcal{R},\mathcal{N}} s = t$ if $s = t$ can be derived by the above inference rules. The inductive definition of the approximations of the intended congruence relation proceeds as usual. Let $\sim_0$ be the syntactic identity relation $\equiv$ on $TERM(\Sigma)$. Let $s \sim_{i+1} t$ iff $\sim_i \vdash_{\mathcal{R},\mathcal{N}} s = t$ for all $s, t \in TERM((\Sigma)$. The limit $\sim_{\mathcal{R},\mathcal{N}} = \bigcup_{i \geq 0} \sim_i$ is then a congruence relation on $TERM(\Sigma)$, the *congruence relation induced by $\mathcal{R}$ and $\mathcal{N}$*. It induces a canonical term algebra $\mathcal{T}_{(\mathcal{R},\mathcal{N})}$ which is intended to capture the semantics of $\mathcal{R}$ wrt. $\mathcal{N}$.

4.3 Model Theoretical Characterization

In general $\mathcal{T}_{(\mathcal{R},\mathcal{N})}$ is not a model of $\mathcal{R}$ as can be easily seen from the following example.

Example 4.2 *Let*

$$\begin{array}{lll} \mathcal{R} & : & x \neq 0 \Rightarrow f(x) = 0 \\ \mathcal{A} & : & natural\ number\ interpretation \\ \mathcal{N} & : & 0 \neq 1,\ 1 \neq 0,\ \ldots \end{array}$$

If x is instantiated by $f(0)$, then $\mathcal{T}_{(\mathcal{R},\mathcal{N})}$ satisfies the condition of the rule but not the conclusion. The reason for that is that the sign $\neq$ is semantically treated inversely to $=$, but not operationally in the definition above. To make apparent the non-standard interpretation which heavily depends on the set $\mathcal{N}$, we introduce a new predicate symbol $\neq_{\mathcal{N}}$ and write $\neq_{\mathcal{N}} (u, v)$ instead of $u \neq v$. Let $\mathcal{R}^{\neq_{\mathcal{N}}}$ be the rewrite system resulting from $\mathcal{R}$ by this syntactic change.

Next we dictate how the new predicate symbol has to be interpreted in the algebras of interest (so that the interpretation of the symbol still reflects a kind of negation). Any algebra $\mathcal{A}$ that is a model of $\mathcal{N}$ is enriched by an interpretation $\neq_{\mathcal{N}}{}^{\mathcal{A}}$ of the new predicate symbol, where $\neq_{\mathcal{N}}{}^{\mathcal{A}}(a,b)$ holds for $a,b \in A$ iff there exists an inequation $u_0 \neq v_0 \in \mathcal{N}$ such that $u_0{}^{\mathcal{A}} = a$ and $v_0{}^{\mathcal{A}} = b$. We write $\mathcal{A}^{\neq_{\mathcal{N}}}$ for the structure resulting from $\mathcal{A}$ by adding the interpretation $\neq_{\mathcal{N}}{}^{\mathcal{A}}$.

The next notion captures compatibility of $\mathcal{R}$ with $\mathcal{N}$.

Definition 4.1 *$\mathcal{R}$ is said to be* $\mathcal{N}$-consistent *iff $\mathcal{T}_{(\mathcal{R},\mathcal{N})}$ is a model of $\mathcal{N}$.*

The following results now provide a model theoretical characterization of $\mathcal{T}_{(\mathcal{R},\mathcal{N})}$. Note that initiality in this context results from the fact that we, because of the above changes, now have conditional equations (with a kind of built-in predicate) which are essentially positive conditional.

Lemma 4.1 *Let $\mathcal{R}$ be $\mathcal{N}$-consistent. Then $\mathcal{T}_{(\mathcal{R},\mathcal{N})}{}^{\neq_{\mathcal{N}}}$ is a model of $\mathcal{R}^{\neq_{\mathcal{N}}}$.*

Theorem 4.1 *Let $\mathcal{R}$ be $\mathcal{N}$-consistent. Then $\mathcal{T}_{(\mathcal{R},\mathcal{N})}$ is initial in the class of algebras $\{\mathcal{A} \mid \mathcal{A} \models \mathcal{N}$ and $\mathcal{A}^{\neq_{\mathcal{N}}} \models \mathcal{R}^{\neq_{\mathcal{N}}}\}$.*

5 Relationship Between the Approaches

We have developed two approaches to define the semantics of positive/negative conditional rewrite rules. The recursive evaluation approach, based on a well-founded ordering, leads to a perfect model semantics and a standard interpretation of $\neq$. The built-in evaluation approach, based on a set of ground inequations, leads to an initial model semantics and a non-standard interpretation of $\neq$. In general the two approaches lead to different semantics, as can be seen by their characterizing properties, but also directly from example 4.2: $f(0)$ is "undefined" wrt. $\mathcal{A}$. In the recursive evaluation approach we would use this fact to derive a "new" equivalence $f(f(0)) \sim 0$. But this derivation is not possible in the built-in evaluation approach due to the "undefinedness".

The difference between the two approaches results from the fact that "not equivalent" (negation as failure) is in general not the same as "provably inequivalent" (wrt. predefined inequivalences) if partial functions are present. Conversely, if "not equivalent" is always the same as "provably inequivalent", then of course the two approaches lead to the same semantics. This is made precise in the following definition which is formulated within the context of section 3.

Definition 5.1 *$\mathcal{R}$ is said to be* operationally covered by $\mathcal{N}$ (wrt. $\succ$) *iff for any instance $\bigwedge_i u_i = v_i \bigwedge_j \overline{u}_j \neq \overline{v}_j \Rightarrow u = v \in INST(\mathcal{R})$, if $u_i \sim_{\prec_u} v_i$ for $i = 1 \ldots m$ and $\overline{u}_j \not\sim_{\prec_u} \overline{v}_j$ for $j = 1 \ldots n$, then there exist $a_j \neq b_j \in \mathcal{N}$ $(j = 1, \ldots, n)$ such that $\overline{u}_j \sim_{\prec_u} a_j$ and $b_j \sim_{\prec_u} \overline{v}_j$ for $j = 1, \ldots, n$.*

The operational covering property can be interpreted as a kind of sufficient completeness property. Certain terms, that result from inequations occuring in the conditional part of the given rules, can be made equivalent to terms that are given by the set $\mathcal{N}$.

Theorem 5.1 *Let $\mathcal{R}$ be reductive and generically consistent wrt. $\succ$. Let $\mathcal{R}$ be operationally covered by $\mathcal{N}$ (wrt. $\succ$) and let $\mathcal{T}_{(\mathcal{R},\succ)} \models \mathcal{N}$. Then $\mathcal{T}_{(\mathcal{R},\succ)} = \mathcal{T}_{(\mathcal{R},\mathcal{N})}$.*

Consequently, if the mentioned conditions are satisfied, the recursive evaluation approach leads to an initial model semantics and the built-in evaluation approach interpretes $\neq$ in a standard way.

Acknowledgements: I would like to thank J. Avenhaus for many valuable discussions and helpful suggestions.

References

[ABW88] K. R. Apt, H. Blair and A. Walker, Towards a Theory of Declarative Knowledge, in: J. Minker, ed., *Foundations of Deductive Databases and Logic Programming* (Morgan Kaufmann, Los Altos, 1988) pp. 89-148.

[AvBe92] J. Avenhaus and K. Becker, Conditional rewriting modulo a built-in algebra, SEKI Report SR-92-11.

[AvMa90] J. Avenhaus and K. Madlener, Term rewriting and equational reasoning, in: R. B. Banerji, ed., *Formal Techniques in Artificial Intelligence* (North-Holland, Amsterdam, 1990) pp. 1-43.

[BaGa91] L. Bachmair and H. Ganzinger, Perfect model semantics for logic programs with equality, in: *Proc. of 8th Int. Conf. on Logic Programming* (MIT Press, 1991) pp. 645-659.

[DeJo90] N. Dershowitz and J. P. Jouannaud, Rewriting systems, in: J. van Leeuwen, ed., *Handbook of Theoretical Computer Science, Vol. B* (Elsevier, Amsterdam, 1990) pp. 241-320.

[DeOk90] N. Dershowitz and M. Okada, A rationale for conditional equational rewriting, *Theoret. Comput. Science* 75(1/2) (1990) pp. 111-137.

[JoWa86] J.-P. Jouannaud and B. Waldmann, Reductive conditional term rewriting systems, in: *Proc. Third IFIP Working Conference on Formal Description of Programming Concepts*, Ebberup, Denmark (1986).

[Ka87] S. Kaplan, Simplifying conditional term rewriting systems: unification, termination and confluence, *J. Symbolic Computation* 4(3) (1987) pp. 295-334.

[Ka88] S. Kaplan, Positive/negative conditional rewriting, in: *Conditional Term Rewriting Systems*, LNCS 308 (Springer, Berlin, 1987) pp. 129-143.

[Pr88] T. C. Przymusinski, On the declarative semantics of deductive databases and logic programming, in: J. Minker, ed., *Foundations of Deductive Databases and Logic Programming* (Morgan Kaufmann, Los Altos, 1988) pp. 193-216.

[WiGr92] C.-P. Wirth and B. Gramlich , A constructor-based approach for positive/negative conditional equational specifications, this volume.

Inductive Theorem Proving by Consistency for First-Order Clauses*†

Harald Ganzinger
Jürgen Stuber‡

Abstract. We show how the method of proof by consistency can be extended to proving properties of the perfect model of a set of first-order clauses with equality. Technically proofs by consistency will be similar to proofs by case analysis over the term structure. As our method also allows to prove sufficient-completeness of function definitions in parallel with proving an inductive theorem we need not distinguish between constructors and defined functions. Our method is linear and refutationally complete with respect to the perfect model, it supports lemmas in a natural way, and it provides for powerful simplification and elimination techniques.

1 Introduction

For proving inductive theorems of equational theories "proof by consistency" is a particularly powerful method. The method has been engineered during the last decade by gradually removing restrictions on the specification side, by reducing the search space for inferences, and by including methods from term rewriting for the simplification and elimination of conjectures. Musser [15] requires the specifications to contain a completely defined equality predicate. During completion inconsistency results in the equation true $\approx$ false. Huet and Hullot [12] assume a signature to be divided into constructors and defined functions. An equation between constructor terms signals an inconsistency. Jouannaud and Kounalis [13] admit arbitrary convergent rewrite system for presenting a theory. They introduce the notion of *inductive reducibility* to detect inconsistencies. Plaisted [18], among others, has shown that inductive reducibility is decidable for finite unconditional term rewriting systems. Fribourg [9] is the first to notice that not all critical pairs need to be computed for inductive completion. It suffices to consider only linear inferences for selected *complete positions*. Bachmair [1] refines this method to cope with unorientable equations; as a result his method is refutationally complete. His method of proof orderings admits powerful techniqes of simplification and removal of redundant equations without loosing refutation completeness. The latter is essential for verifying nontrivial inductive properties in finite time.

* This research was funded by the German Ministry for Research and Technology (BMFT) under grant ITS 9103. The responsibility for the contents of this publication lies with the authors.

† A full version of this paper appeared in: *Informatik—Festschrift zum 60. Geburtstag von Günter Hotz*, Teubner-Verlag, Stuttgart 1992

‡ Authors' address: Max-Planck-Institut für Informatik, Im Stadtwald, D-W-6600 Saarbrücken, Germany, {Harald.Ganzinger|Juergen.Stuber}@mpi-sb.mpg.de

More recently there have been some attempts to extend these techniques to Horn clauses. Orejas [16] places similar restrictions on specifications as Huet and Hullot. Bevers and Lewi [6] build on inductive reducibility, which is a severe restriction as inductive reducibility is in general undecidable for Horn clauses [14].

In this paper we extend the method described by Stuber [20] from Horn clauses to full first-order clauses with equality by adapting the method of Bachmair and Ganzinger [3, 5] for Knuth/Bendix-like completion for first-order clauses. Completion —*saturation up to redundancy*, as we prefer to call this process from now on—serves an important purpose. It produces a representation of a certain minimal model of the given (consistent) first-order theory and allows to prove the validity of ground equations in this model by conditional term rewriting with negation as failure. This distinguished minimal model is called the *perfect model*, and it depends on a given reduction ordering on terms. By inductive theorem proving for first-order theories we mean to prove *validity in the perfect model*, and the method consists in showing that enriching a given theory by a given set of conjectures does not change the perfect model, hence the name *proof by consistency.*

Unlike many other methods of inductive theorem proving [7, 11, 17], our method of proof by consistency does not require that constructors be given explicitly. Moreover we always generate a counterexample if the conjecture is false. In other words, our method is refutationally complete. It also is linear; neither inferences between axioms nor between conjectures have to be computed. The method is rather flexible as it is based on a very general notion of *fair inductive theorem proving derivations* and allows for powerful simplification and elimination techniques. The latter is provided by the notion of redundancy as developed in [3, 5]. In fact we will show that redundancy and inductive validity of clauses are equivalent concepts.

Technically the approach is based on the inference systems for first-order refutation theorem proving presented by Bachmair and Ganzinger [5] and briefly summarized in the appendix.

2 The Method

Clauses are implicitly universally quantified. We make quantifiers explicit and restrict them to generated values by adding a constraint $\mathsf{gnd}(x)$ for every variable x in the clause. We add clauses which define these *type predicates* such that $\models \mathsf{gnd}(t)$ if and only if t is (equivalent to) a ground term of the sort of x. More precisely, for each operator f of arity n a clause

$$\mathsf{gnd}(x_1), \ldots, \mathsf{gnd}(x_n) \rightarrow \mathsf{gnd}(f(x_1, \ldots, x_n))$$

is added, and a conjecture $\Gamma \rightarrow \Delta$ containing variables $x_1, \ldots, x_n$ becomes

$$\mathsf{gnd}(x_1), \ldots, \mathsf{gnd}(x_n), \Gamma \rightarrow \Delta.$$

A clause that is *closed* by explicit quantifiers in this way is valid if and only if it is valid in all generated models (Herbrand models over the given signature.) Validity in all generated models implies validity in the perfect model of a set of clauses, and is a key step towards second-order reasoning.

The perfect model of a set of clauses N is represented, in a sense that will become clear below, by a certain subset N' of N. These clauses define a canonical set R of ground rewrite rules such that the congruence generated by R is the perfect model of N. To prove the inductive validity of a conjecture H, we take its closed version H' and attempt to prove the validity of a set of instances of H' that covers all ground instances of H', assuming that H' is true for all smaller instances. (Here, "smaller" refers to some well-founded ordering on clauses.) The key points of our method are as follows:

(i) The covering set of instances of H' is generated by a narrowing-like process which enumerates the solutions to the antecedent of H' in the perfect model. By closing H we achieve that conjectures are either ground or else have a non-empty antecedent.

(ii) We eliminate an instance of H' if it follows from N and from smaller instances of H'. In this case we call the particular instance of H' ***composite***.

(iii) We assume that for a ground instance of H' it is decidable whether or not H' is true in the perfect model. In particular, we assume that N' and R are effectively given in a certain technical sense. This restricts our method, but makes it refutationally complete. If validity of ground clauses were undecidable for a theory N, the problem of inductive theorem proving for N would be hopeless anyway.

(iv) We saturate $N \cup H'$ by applying a positive superposition strategy. To enumerate the solutions of the antecedent of H' we allow to ***select*** an arbitrary atom A of the antecedent so as to guide the enumeration process to first concentrate on the solutions of A. If A is a type predicate $\mathsf{gnd}(x)$, then the effect is to enumerate all ground substitutions for the variable x in H. It may happen that some type clause

$$C_f = \mathsf{gnd}(x_1), \ldots, \mathsf{gnd}(x_n) \rightarrow \mathsf{gnd}(f(x_1, \ldots, x_n))$$

corresponding to some function f itself is an inductive consequence (with respect to N) of some subset B of other type clauses. This is the case if f is a function symbol that is ***sufficiently completely*** defined relative to the (in some sense more primitive) functions in B. In this case, the C_f needs ***not*** be superposed on A. In other words, only type clauses for constructors need to be considered for superposition. This optimization is implicitly built into our method as the inductive validity of C_f can be proved in parallel with H. No explicit distinction between constructor symbols and defined symbols is required. Moreover, equalities between constructor terms pose no problem in our framework.

For instance, consider the following specification for natural numbers.

```
natbase =
    sorts
        nat
    ops
        0 :   → nat
        s : nat → nat
```

The enrichment by type clauses yields

```
natbase^g = natbase +
        gnd : nat
```

axioms $\forall$ n : nat

$\mathsf{gnd}(0)$ (1)

$\mathsf{gnd}(n) \rightarrow \mathsf{gnd}(s(n))$ (2)

Consider the enrichment of the above specification by a definition of $\leq$.

$\mathbf{natleq}^g = \mathbf{natbase}^g +$

$\leq$: nat × nat

axioms $\forall$ m, n : nat

$0 \leq n$ (3)

$m \leq n \rightarrow s(m) \leq s(n)$ (4)

Suppose we would like to prove that $\leq$ is total, i.e. that

$$\rightarrow m \leq n, n \leq m,$$

which becomes

$$\mathsf{gnd}(n), \mathsf{gnd}(m) \rightarrow m \leq n, n \leq m$$

after closing, is inductively valid. In this particular case the theory is of Horn clause type so that the perfect model is the initial model.

Whenever a clause is added during a consistency proof, an equation in its antecedent for which solutions are to be enumerated is selected. (The selection will below be indicated by underlining.)

(5)	$\underline{\mathsf{gnd}(m)}, \mathsf{gnd}(n) \rightarrow m \leq n, n \leq m$	conjecture
	$\mathsf{gnd}(n) \rightarrow 0 \leq n, n \leq 0$	selective resolution (1) on (5) *composite* because of (3)
(6)	$\mathsf{gnd}(m), \underline{\mathsf{gnd}(n)} \rightarrow s(m) \leq n, n \leq s(m)$	selective resolution (2) on (5)
	$\mathsf{gnd}(m) \rightarrow s(m) \leq 0, 0 \leq s(m)$	selective resolution (1) on (6) *composite* because of (3)
	$\mathsf{gnd}(m), \mathsf{gnd}(n) \rightarrow s(m) \leq s(n), s(n) \leq s(m)$	selective resolution (2) on (6) *composite* because of (4) and (5)

We have seen that all clauses that can be enumerated by superposition on selected atoms are composite, i.e. follow from the theory and from smaller instances of the conjecture. For instance, for all ground terms N and M, the clause $C = \mathsf{gnd}(M), \mathsf{gnd}(N) \rightarrow s(M) \leq s(N), s(N) \leq s(M)$ follows from (4) and the instance $D = \mathsf{gnd}(M), \mathsf{gnd}(N) \rightarrow M \leq N, N \leq M$ of (5). D is emdedded in C, hence smaller than C.

The example demonstrates the strong analogy to classical methods of inductive theorem proving. Selecting one of the $\mathsf{gnd}(x)$ corresponds to the selection of an induction variable. Superposition with the type clauses results in a set of new instances representing the different cases to be proved. The elimination of a clause corresponds to an induction step for which the induction hypothesis may be used. Basis of the induction are well-founded orderings on terms which are extended to well-founded orderings on clauses.

For a second example, consider the usual definition of addition for natural numbers.

$$
\begin{array}{ll}
\textbf{natplus} = \textbf{natbase} + & \\
\quad \textbf{ops} & \\
\qquad + : \mathrm{nat} \times \mathrm{nat} \rightarrow \mathrm{nat} & \\
\quad \textbf{axioms} \quad \forall \mathrm{m}, \mathrm{n} : \mathrm{nat} & \\
\qquad 0 + \mathrm{n} \approx 0 & (1) \\
\qquad \mathrm{s(m)} + \mathrm{n} \approx \mathrm{s(m+n)} & (2)
\end{array}
$$

Extending the specification by type clauses yields

$$
\begin{array}{ll}
\textbf{natplus}^g = \textbf{natplus} + & \\
\quad \textbf{ops} & \\
\qquad \mathsf{gnd} : \mathrm{nat} & \\
\quad \textbf{axioms} \quad \forall \mathrm{m}, \mathrm{n} : \mathrm{nat} & \\
\qquad \mathsf{gnd}(0) & (3) \\
\qquad \mathsf{gnd}(\mathrm{m}) \rightarrow \mathsf{gnd}(\mathrm{s(m)}) & (4) \\
\qquad \underline{\mathsf{gnd}(\mathrm{m})}, \mathsf{gnd}(\mathrm{n}) \rightarrow \mathsf{gnd}(\mathrm{m+n}) & (5)
\end{array}
$$

We prove that + is a defined operator, that is that (5) is an inductive consequence of (1)–(4). We apply the same method and select the first literal in the antecedent of (5).

$\mathsf{gnd}(n) \rightarrow \mathsf{gnd}(0+n)$	selective resolution (3) on (5), *composite* because of (1) and (3)
$\mathsf{gnd}(m), \mathsf{gnd}(n) \rightarrow \mathsf{gnd}(s(m)+n)$	selective resolution (4) on (5), *composite* because of (2),(4) and (5)

Clauses which have been proved may be kept and used (as lemmas) for proving compositeness in a subsequent inductive proof. Moreover parallel induction is supported as we allow for arbitrary sets of conjectures to start with.

3 Preliminaries

3.1 Equational clauses

A *signature* Σ is a set of sorts together with a set of operator declarations $f : s_1, \ldots, s_n \rightarrow s$ over these sorts. $s_1, \ldots, s_n$ is called the *arity*, s the *coarity* of f. A Σ-*term* is a term built according to the operator declarations in Σ, possibly with variables. By a *ground* expression (i.e., a term, equation, formula, etc.) we mean an expression containing no variables. For simplicity we do not allow operator overloading and assume that all sorts are *inhabited*, i.e., admit ground terms. For the moment we will assume a fixed signature Σ. Where necessary, we will use the signature as a prefix, like in Σ-term.

We will define equations and clauses in terms of multisets. A *multiset* over X is an unordered collection with possible duplicate elements of X. Formally a multiset is given as a function M from X to the natural numbers. Intuitively, $M(x)$ specifies the number of occurrences of x in M. An *equation* is an expression $s \approx t$, which we identify with the multiset $\{s, t\}$. A *clause* is a pair of multisets of equations, written

$\Gamma \to \Delta$, where Γ is the *antecedent* and Δ the *succedent*. We usually write Γ_1, Γ_2 and A, Γ instead of $\Gamma_1 \cup \Gamma_2$ and $\Gamma \cup \{A\}$.

A clause represents an implication $A_1 \wedge \cdots \wedge A_m \supset B_1 \vee \cdots \vee B_m$; the empty clause, a contradiction. Clauses of the form $\Gamma, A \to A, \Delta$ or $\Gamma \to \Delta, t \approx t$ are called *tautologies*. A *specification* is a set of clauses together with the signature the clauses are defined over.

An inference π is a pair written as

$$\frac{C_1 \ldots C_n}{C}$$

where the *premises* $C_1, \ldots, C_n$ and the *conclusion* C are clauses. An *inference system* $\mathcal{I}$ is a set of inferences. An *instance* of an inference π in $\mathcal{I}$ is any inference in $\mathcal{I}$ with premises $C_1\sigma, \ldots, C_n\sigma$ and conclusion $C\sigma$.

3.2 Clause Orderings

Any ordering $\succ$ on a set S can be extended to an ordering $\succ_{mul}$ on finite multisets over S as follows: $M \succ_{mul} N$ if (i) $M \neq N$ and (ii) whenever $N(x) > M(x)$ then $M(y) > N(y)$, for some y such that $y \succ x$. If $\succ$ is a total [well-founded] ordering, so is $\succ_{mul}$. Given a set (or multiset) S and an ordering $\succ$ on S, we say that x is *maximal* relative to S if there is no y in S with $y \succ x$; and *strictly maximal* if there is no y in S with $y \succeq x$.

If $\succ$ is an ordering on terms, then the corresponding multiset ordering $\succ_{mul}$ is an ordering on equations, which we denote by $\succ^e$.

We have defined clauses as pairs of multisets of equations. Alternatively, clauses may also be thought of as multisets of *occurrences* of equations. We identify an occurrence of an equation $s \approx t$ in the antecedent of a clause with the multiset (of multisets) $\{\{s, \bot\}, \{t, \bot\}\}$, and an occurrence in the succedent with the multiset $\{\{s\}, \{t\}\}$, where $\bot$ is a new symbol.[4] We identify clauses with finite multisets of occurrences of equations. By $\succ^o$ we denote the twofold multiset ordering $(\succ_{mul})_{mul}$ of $\succ$, which is an ordering on occurrences of equations; by $\succ^c$ we denote the multiset ordering $\succ^o_{mul}$, which is an ordering on clauses. If $\succ$ is a well-founded [total] ordering, so are $\succ^e$, $\succ^o$, and $\succ^c$. From now on we will only consider orderings $\succ$ on terms which are reduction orderings and total on ground terms.

We say that a clause $C = \Gamma \to s \approx t, \Delta$ is *reductive* for $s \approx t$ if $t \not\succeq s$ and $s \approx t$ is a strictly maximal occurrence of an equation in C. For example, if $s \succ t \succ u$ and $s \succ v$ for every term v occurring in Γ, then $\Gamma \to s \approx t, s \approx u$ is reductive for $s \approx t$, but $\Gamma, s \approx u \to s \approx t$ is not. Since the ordering is total on ground terms, a ground clause is reductive if and only if $s \succ t$ and $s \approx t$ is greater than any other occurence of an equation. A nonreductive clause has no reductive ground instances.

3.3 Equality Herbrand Interpretations

We write $A[s]$ to indicate that A contains s as a subexpression and (ambiguously) denote by $A[t]$ the result of replacing a particular occurrence of s by t. By $A\sigma$ we

[4] The symbol $\bot$ is not part of the vocabulary of the given first-order language. It is assumed to be minimal with respect to any given ordering. Thus $t \succ \bot$, for all terms t.

denote the result of applying the substitution σ to A and call $A\sigma$ an *instance* of A. If $A\sigma$ is ground, we speak of a *ground instance*. Composition of substitutions is denoted by juxtaposition. Thus, if τ and ρ are substitutions, then $x\tau\rho = (x\tau)\rho$, for all variables x.

An *equivalence* is a reflexive, transitive, symmetric binary relation. An equivalence $\sim$ on terms is called a *congruence* if $s \sim t$ implies $u[s] \sim u[t]$, for all terms u, s, and t. If E is a set of ground equations, we denote by E^* the smallest congruence containing E.

By an (*equality Herbrand*) *interpretation* we mean a congruence on ground terms. An interpretation I is said to *satisfy* a ground clause $\Gamma \rightarrow \Delta$ if either $\Gamma \not\subseteq I$ or else $\Delta \cap I \neq \emptyset$. We also say that a ground clause C is *true in* I, if I satisfies C; and that C is *false in* I, otherwise. An interpretation I is said to satisfy a non-ground clause $\Gamma \rightarrow \Delta$ if it satisfies all ground instances $\Gamma\sigma \rightarrow \Delta\sigma$. An interpretation I is called a (*equality Herbrand*) *model* of N if it satisfies all clauses of N.

A set N of clauses is called *consistent* if it has a model; and *inconsistent* (or *unsatisfiable*), otherwise. We say that N *implies* C, and write $N \models C$, if every model of N satisfies C.

3.4 Convergent Rewrite Systems

A binary relation $\Rightarrow$ on terms is called a *rewrite relation* if $s \Rightarrow t$ implies $u[s\sigma] \Rightarrow u[t\sigma]$, for all terms s, t and u, and substitutions σ. A transitive, well-founded rewrite relation is called a *reduction ordering*. By $\Leftrightarrow$ we denote the symmetric closure of $\Rightarrow$; by $\overset{*}{\Rightarrow}$ the transitive, reflexive closure; and by $\overset{*}{\Leftrightarrow}$ the symmetric, transitive, reflexive closure. Furthermore, we write $s \Downarrow t$ to indicate that s and t can be rewritten to a common form: $s \overset{*}{\Rightarrow} v$ and $t \overset{*}{\Rightarrow} v$, for some term v. A rewrite relation $\Rightarrow$ is said to be *Church-Rosser* if the two relations $\overset{*}{\Leftrightarrow}$ and $\Downarrow$ are the same.

A set of equations E is called a *rewrite system* with respect to an ordering $\succ$ if we have $s \succ t$ or $t \succ s$, for all equations $s \approx t$ in E. If all equations in E are ground, we speak of a ground rewrite system. Equations in E are also called (*rewrite*) *rules*. When we speak of "the rule $s \approx t$" we implicitly assume that $s \succ t$. By $\Rightarrow_{E\succ}$ (or simply $\Rightarrow_E$) we denote the smallest rewrite relation for which $s \Rightarrow_E t$ whenever $s \approx t$ is in E and $s \succ t$. A term s is said to be in *normal form* (with respect to E) if it can not be rewritten by $\Rightarrow_E$, i.e., if there is no term t such that $s \Rightarrow_E t$. A term is also called *irreducible*, if it is in normal form, and *reducible*, otherwise. A rewrite system E is said to be *convergent* if the rewrite relation $\Rightarrow_E$ is well-founded and Church-Rosser. Convergent rewrite systems define unique normal forms.

3.5 Predicates

We allow that in addition to function symbols a signature may contain predicate symbols, which will be declared to have a special coarity pred. Thus we also consider expressions $P(t_1, \ldots, t_n)$, where P is some predicate symbol and $t_1, \ldots, t_n$ are terms built from function symbols and variables. We then have equations $s \approx t$ between (non-predicate) terms, called *function equations*, and equations $P(t_1, \ldots, t_n) \approx \mathrm{tt}$, called *predicate equations*, where tt is a distinguished unary predicate symbol that is taken to be minimal in the given reduction ordering $\succ$. For simplicity, we usually abbreviate $P(t_1, \ldots, t_n) \approx \mathrm{tt}$ by $P(t_1, \ldots, t_n)$.

4 The Perfect Model

The proof by consistency method proves properties of a standard model of a specification, which for unconditional equations and horn clauses is the initial model. The initial model can be characterized as the unique minimal (with respect to set inclusion) Herbrand interpretation satisfying N, and for Horn clause specifications it always exists if N is consistent. In the case of first-order clauses more than one minimal model may exist. For instance, if N consists of the single clause $\rightarrow p, q$ then both $\{p\}$ and $\{q\}$ are minimal models of N. We will use the ordering $\succ^e$ to single out one of the minimal models as the *perfect model*.

Let $\succ^p = (\succ^e)^{-1}$, then a model I is called *preferable* to J if $J \succ^p_{mul} I$. A *perfect* model (corresponding to $\succ$) is a minimal model with respect to $\succ^p_{mul}$. For instance, if we assume $q \succ p$ then $\{p\} \succ^p_{mul} \{q\}$, i.e., $\{q\}$ is the perfect model. It is important to see that different orderings may yield different perfect models. This is an essential difference to the case of Horn clauses. For general clauses the ordering $\succ$ must be explicitly given in order to uniquely identify the standard model one has in mind.

As $\succ^e$ is well-founded and total, $\succ^p_{mul}$ is a total ordering [19]. Hence there exists at most one perfect model for a set of clauses N. Since $\succ^p_{mul}$ contains $\subseteq$, a perfect model is also minimal.

In the remainder of this section we present methods and techniques for constructing, given a consistent set of clauses and an ordering, the corresponding perfect model. We also explain how to compute in this model. The proofs of the lemmas which justify our techniques may be found in [4] and [5].

4.1 Construction of the Perfect Model

Let N be a set of clauses and $\succ$ be a reduction ordering which is total on ground terms. We shall define an interpretation I for N by means of a convergent rewrite system R. For certain N, I will be the perfect model of N with respect to $\succ$.

First, we use induction on the clause ordering $\succ^c$ to define sets of equations E_C, R_C and I_C, for all ground clauses C over the given signature (not necessarily instances of N). Let C be such a ground clause and suppose that $E_{C'}$, $R_{C'}$ and I'_C have been defined for all ground clauses C' for which $C \succ^c C'$. Then

$$R_C = \bigcup_{C \succ^c C'} E_{C'} \quad \text{and} \quad I_C = R_C^*.$$

Moreover

$$E_C = \{s \approx t\}$$

if C is a ground instance $\Gamma \rightarrow \Delta, s \approx t$ of N such that (i) C is reductive for $s \approx t$, (ii) s is irreducible by R_C, (iii) $\Gamma \subseteq I_C$, and (iv) $\Delta \cap I_C = \emptyset$. In that case, we also say that C ***produces*** the equation (or rule) $s \approx t$. In all other cases, $E_C = \emptyset$. Finally, we define I to be the equality interpretation R^*, where $R = \bigcup_C E_C$ is the set of all equations produced by ground instances of clauses in N.

Instances of N that produce equations are also called *productive*. Note that a productive clause C is false in $I_C = R_C^*$, but true in $(R_C \cup E_C)^*$. The truth value of an equation can be determined by rewriting: $u \approx v \in I$ if and only if $u \Downarrow_R v$. In many cases the truth value of an equation can already be determined by rewriting with R_C. If C is true in I_C then for $D \succeq^c C$ it is also true in I_D and in I.

4.2 Superposition and Redundancy

The interpretation I will in general not be a model of N, unless N is closed under sufficiently many applications of certain inference rules. The inference system $\mathcal{S}_S^{\succ}$ which we consider in this paper is the one described in [4, 5] and is also briefly summarized in the appendix. It is based on $\succ$ and on a selection function S. By a *selection function* we mean a mapping S that assigns to each clause C a (possibly empty) multiset of negative occurrences of equations in C. The equations in $S(C)$ are called *selected.* If $S(C) = \emptyset$, then no equation is selected. Selected equations can be arbitrarily chosen and need not be maximal. Selection functions are assumed to be compatible with substitution, i.e. an occurrence of an equation is selected in C if and only if the corresponding occurrence is selected in $C\sigma$, for any substitution σ.

$\mathcal{S}_S^{\succ}$, for short $\mathcal{S}$, if $\succ$ and S are indicated by the context, consists mainly of paramodulation rules which are restricted by ordering constraints derived from $\succ$ or by selection constraints derived from S. Paramodulation affects maximal equations only, unless some atom of the antecedent of a clause is selected. If an equation is selected in (the antecedent of) a clause C, paramodulation on C always occurs into a maximal selected equation of C. An important feature is that clauses for which an equation is selected need *not* be considered for paramodulating into any other clause. They do not directly contribute to the construction of the perfect model. This is made more precise by a notion of redundancy. Redundancy is a key aspect which allows to saturate many nontrivial sets of clauses under $\mathcal{S}$ in a finite number of steps.

A ground clause C is said to be *redundant (in N)* if C is true in I_C. A clause is called redundant in N if all its ground instances are redundant in N. Redundant clauses are true in I. The interpretation I is completely determined by productive clauses, which are non-redundant instances of N.

An inference π from ground clauses is said to be *redundant (in N)* if either one of its premises is redundant in N or else its conclusion is true in I_C, where C is the maximal (the second, if the inference has two premises) premise of π. An inference from arbitrary premises is redundant in N if all its ground instances are redundant in N. We say that N is *saturated* if every ground instance of an inference from premises in N is redundant in N.

Lemma 1. *Let N be a saturated set of clauses. If an instance C of a clause in N contains a selected equation, C is not productive.*

Theorem 2. *Let N be a consistent and saturated set of clauses. Then I is the perfect model of N with respect to $\succ$.*

These two lemmas are the key to our method. The first shows that clauses C in N with selected equations do not contribute to the perfect model. In a saturated set of clauses N, C is therefore an inductive consequence of $N \setminus \{C\}$. The second lemma shows that any consistent set of clauses has a perfect model (with respect to any given complete reduction ordering). In particular, the construction of section 4.1 yields the perfect model, provided N is consistent and saturated. Fair theorem proving derivation are a means to saturate a given set of clauses, though the limit may not be reachable in a finite number of steps. All this gives hints of how to compute in the perfect model, an aspect which is made more precise below.

4.3 Saturation

The next question we address is how to construct a saturated set of clauses. The notion of redundancy is not effectively usable as it is not stable under addition or deletion of clauses.

Let N be a set of clauses and C be a ground clause (not necessarily a ground instance of N). We call C *composite* with respect to N, if there exist ground instances $C_1, \ldots, C_k$ of N such that $C_1, \ldots, C_k \models C$ and $C \succ^c C_j$, for all j with $1 \leq j \leq k$. A non-ground clause is called composite if all its ground instances are composite.

A ground inference π with conclusion B is called *composite* (with respect to N) if either some premise is composite with respect to N, or else there exist ground instances $C_1, \ldots, C_k$ of N such that $C_1, \ldots, C_k \models B$ and $C \succ^c C_j$, for all j with $1 \leq j \leq k$, where C is the maximal premise of π. A non-ground inference is called composite if all its ground instances are composite.

Lemma 3. *For any saturated and consistent set of clauses N compositeness with respect to N implies redundancy with respect to N—for clauses as well as for inferences. Moreover, compositeness is stable under addition of clauses to N and stable under deletion of* composite *clauses from N.*

A *theorem proving derivation* is a (finite or countably infinite) sequence $N_0, N_1, N_2, \ldots$ of sets of clauses such that either

(*Deduction*) $N_{i+1} = N_i \cup \{C\}$ and $N_i \models C$, or

(*Deletion*) $N_{i+1} = N_i \setminus \{C\}$ and C is composite with respect to $N_i \cup \{C\}$.

The set $N_\infty = \bigcup_j \bigcap_{k \geq j} N_k$ is called the *limit* of the derivation. Clauses in N_∞ are called *persisting*.

Deduction adds clauses that logically follow from given clauses; deletion eliminates composite clauses. Simplification can be modeled as a sequence of deduction steps followed by a deletion step.

A theorem proving derivation is called *fair* if every inference in $\mathcal{S}$ from premises in N_∞ is composite with respect to $\bigcup_j N_j$.

A fair derivation can be constructed, for instance, by systematically adding conclusions of non-composite inferences in $\mathcal{S}$. As the maximal premise of a ground inference is always greater with respect to $\succ^c$ than its conclusion, the inference becomes composite as soon as the conclusion has been added.

A set of clauses N is called *complete* if all inferences from N are composite with respect to N. A complete set of clauses that does not contain the empty clause is saturated.

Lemma 4. *Let $N = N_0, N_1, N_2, \ldots$ be a fair theorem proving derivation. If N is inconsistent then the empty clause is contained in $\bigcup_j N_j$. Otherwise N and N_∞ are logically equivalent, and N_∞ is complete (and hence saturated).*

4.4 Computing in Perfect Models

The rewrite system R which defines I is canonical, hence constitutes a decision procedure for equality in the perfect model, provided the one-step rewrite relation $\Rightarrow_R$ is computable. This need not be the case in general, not even for finite and complete sets of clauses. However, if the set of clauses is such that matching a ground term against the maximal term of a clause always results in a reductive ground clause, one can use a recursive algorithm to decide the word problem for I.

A clause $C = \Gamma \rightarrow \Delta$ is called *universally reductive* if either the succedent Δ is empty, or else Δ can be written as $\Delta', s \approx t$ such that (i) all variables of C also occur in s, (ii) $C\sigma$ is reductive for $s\sigma \approx t\sigma$, for all ground substitutions σ. A set N of clauses is called universally reductive if any clause in N is either universally reductive, or else contains a selected atom.

Lemma 5. *Suppose $\succ$ is decidable, and let N be a saturated, finite and universally reductive set of clauses. Then it is decidable whether a ground equation $s \approx t$ is valid in the perfect model for N.*

An obvious consequence of the above lemma is the decidability of validity in the perfect model for ground clauses in the indicated case.

5 Proof by Consistency

Inductive validity and redundancy are equivalent concepts. If $C = \Gamma \rightarrow \Delta$ is an inductive consequence of a consistent and saturated set N of clauses then the logically equivalent clause $\top \approx \top, \Gamma \rightarrow \Delta$, with $\top$ being a new constant of a new sort and maximal with respect to $\succ$, is redundant in N. Conversely, if C is redundant in N is is true in I_C and in I. In this case N and $N \setminus \{C\}$ have the same perfect model. Therefore C is an inductive consequence of $N \setminus \{C\}$. Inductive theorem proving is proving redundancy, and vice versa. Lemmas 3 and 1 are the basis of the technique that we are going propose in this section. In simple cases a conjecture (or some derived clause) can be eliminated by a direct proof of compositeness using specific techniques such contextual rewriting, cf. section 5.5. Otherwise one may attempt to make the clause become redundant in the limit of a saturation process. Closing conjectures by type predicates for their variables is a technical device to translate a non-ground clause into an equivalent one with a non-empty antecedent from which an equation can be selected for superposition. Selecting a type predicate of a variable corresponds to selecting that variable as the induction variable.

5.1 Type Predicates

For each sort s in Σ we add a predicate symbol gnd_s, and for each operator $f : s_1, \ldots, s_n \rightarrow s$ in Σ a clause

$$G(f) = \mathsf{gnd}_{s_1}(x_1), \ldots, \mathsf{gnd}_{s_n}(x_n) \rightarrow \mathsf{gnd}_s(f(x_1, \ldots, x_n)).$$

gnd_s is called the *type predicate* for s and $G(f)$ the *type clause* for f. A ground term $t = f(t_1, \ldots, t_n)$ over Σ uniquely determines a ground instance

$$\mathsf{gnd}(t_1), \ldots, \mathsf{gnd}(t_n) \rightarrow \mathsf{gnd}(f(t_1, \ldots, t_n))$$

of the clause $G(f)$, which we will denote by $G(t)$. For a signature Σ we define $G(\Sigma)$ as the set of all $G(f)$ where f is an operator in Σ. The union of N and $G(\Sigma)$ will be denoted by N^g, while the extended signature will be denoted by Σ^g. By R^g and I^g we denote the set of rewrite rules and interpretation, respectively, constructed from N^g according to section 4.1. $G(\Sigma)$ encodes the notion of a "ground term", i.e., an atom $\mathsf{gnd}(t)$ is provable if and only if t is equal (modulo N) to a ground term. We assume that the given complete reduction ordering over Σ is arbitrarily extended to a complete reduction ordering over the extended signature Σ^g. Such an extension always exists.

Lemma 6. *Let t be a ground term over Σ of form $f(t_1, \ldots, t_n)$. Then $\mathsf{gnd}(t)$ is true in $(R^g_{G(t)} \cup E^g_{G(t)})^*$ and hence in I^g.*

Lemma 7. *Let N be a saturated set of clauses. Then N^g is also saturated.*

Lemma 8. *Let C be a ground instance of a clause in N^g. R^g_C restricted to rules not containing any type predicates is equal to R_C.*

Lemma 9. *Let C be a ground instance of a clause in N. Then C is true in I^g if and only if C is true in I.*

Let $C = \Gamma \rightarrow \Delta$ be a Σ-clause and $var(C) = \{x_1, \ldots, x_n\}$. Then we define $G(C)$ to be the Σ^g-clause $\mathsf{gnd}(x_1), \ldots, \mathsf{gnd}(x_n), \Gamma \rightarrow \Delta$. $G(C)$ is called the *closed form* of C. For a set of clauses H, we define $G(H)$ as $\{ G(C) \mid C \in H \}$.

Lemma 10. *C is true in I if and only if $G(C)$ is true in I^g.*

Let us summarize the contents of this section. If we have a saturated set of clauses N, we can transform it to N^g, which is still saturated. A clause C is valid in I if and only if C is valid in N^g if and only if $G(C)$ is valid in N^g. For inductive theorem proving we may hence we may use N^g in place of N, and we may replace conjectures H by the closed forms $G(H)$.

5.2 Inductive Theorem Proving Derivations

From now on we shall assume to be given a consistent, finite, complete and universally reductive set of clauses N with perfect model I. By an *inductive theorem proving derivation* for N we mean a finite or countably infinite sequence $H_0, H_1, \ldots$ of sets of clauses C such that

(*Deduction*) $H_{i+1} = H_i \cup \{C\}$ and $N \cup H_i \models C$, or

(*Deletion*) $N_{i+1} = N_i \setminus \{C\}$ and C is composite with respect to $N \cup H_i \cup \{C\}$

The set $H_\infty = \bigcup_j \bigcap_{k \geq j} H_k$ is called the *limit* of the derivation. Clauses in H_∞ are called *persisting*.

An inductive theorem proving derivation is called *fair* if every inference by selective equality resolution on a clause in H_∞ and every inference by selective superposition of a clause in N on a clause in H_∞ is composite with respect to $\bigcup_j H_j$.

Given a selection function S, an inductive theorem proving derivation is called *failed*, if there exists a ground clause in $\bigcup_j H_j$ which is false in I (failure with "disproof"), or else there exists a clause in H_∞ for which no equation is selected by S (failure with "don't know").

Theorem 11. *Let $H_0, H_1, \ldots$ be a fair inductive theorem proving derivation for N.*

(i) If the derivation is non-failed then the clauses in H_0 are inductive theorems of N, i.e., valid in I.

(ii) If the derivation fails with "disproof", then H_0 is not valid in I.

Proof. (i) If $H_0, H_1, \ldots$ is fair and non-failed, the sequence $N \cup H_0, N \cup H_1, \ldots$ is a theorem proving derivation (with respect to S). Inferences with premises all in N are composite in N as N is complete. Inferences with at least one premise in H_∞ are composite in $N \cup H_\infty$ as the derivation is fair and as there is no clause in H_∞ for which no equation is selected by S. Therefore the sequence $N \cup H_0, N \cup H_1, \ldots$ is a fair theorem proving derivation with limit $N \cup H_\infty$, hence $N \cup H_\infty$ is complete. As the clauses in H_∞ have selected equations, they are not productive. Therefore the perfect models of N and $N \cup H_\infty$ (and hence $N \cup H_0$) are identical.

(ii) follows immediately from the soundness of deduction. □

Note that the fairness requirement for inductive derivations does not imply the need for computing any non-linear inferences with premises all in N or with two premises both not in N. To achieve refutation completeness of the method the production of non-ground clauses with an empty antecedent has to be avoided. Using type predicates is a technique to achieve this goal.

5.3 Refutation Completeness

Retutation completeness in this context means to avoid failure with "don't know" in inductive theorem proving derivations. For that purpose we can assume without loss of generality that the given theory presentation N includes all type predicates and type clauses, i.e., $N = \tilde{N}^g$, as this does not affect completeness, consistency, finiteness, and universal reductivity. Similarly, we can assume the initial set of conjectures H_0 to be closed. Suppose that we only admit selection functions which always select some type atom of form $\mathsf{gnd}(x)$, x a variable, and only such atoms, if there are any in a given clause. Furthermore assume that in an inductive theorem proving the only deductions one makes are by selective equality resolution on a clause in $\bigcup_j H_j$ or by selective superposition of a clause in N on a clause in $\bigcup_j H_j$. Then any clause in $\bigcup_j H_j$ is closed, hence either is a ground clause, for which validity is decidable, or else contains a selected equation, Therefore failure with "don't know" is impossible.

In practice, however, one would not want to restrict selection functions to always select type predicates only, nor to limit deductions to what is required by fairness. Otherwise one might not detect situations in which the antecedent is false for certain substitutions. Also simplification, e.g., by demodulation, would not be allowed. Failure with "don't know" will only occur in extreme cases cases anyway, and there are other ways of achieving refutation completeness.

Superposition of a type clause on a selected type atom in a conjecture is a principal kind of inferences to be computed in inductive theorem proving derivations. The more type clauses one can be prove redundant, the less such inferences one needs to consider. This problem will be addressed in the next section.

5.4 Sufficient Completeness

A type clause for a function f is redundant, if the function is sufficiently completely defined with respect to the remaining functions. More generally, we consider subsignatures Σ_B, called *base signatures*, of the given signature Σ. A set of Σ-clauses N (together with an ordering $\succ$) is called *sufficiently complete* with respect to Σ_B if for any Σ-ground term s there exists a Σ_B-ground term t such that $s \approx t$ is true in the perfect model I of N. Again, this property is defined with respect to the perfect model rather than all minimal models. Furthermore, we assume that ground terms over Σ_B are smaller in the reduction ordering than ground terms containing function symbols not in Σ_B.

Lemma 12. *N is sufficiently complete with respect to Σ_B if and only if the type clauses $G(f)$, for f in $\Sigma \setminus \Sigma_B$, are redundant.*

Corollary 13. *Let N be a saturated set of clauses, and let S be a selection function such that each clause in $G(\Sigma \setminus \Sigma_B)$ either contains a selected equation if its antecedent is nonempty or else is redundant. Then N is sufficiently complete with respect to Σ_B if and only if N^g is saturated.*

The inductive proof procedure outlined above can be used to prove sufficient-completeness by starting with the theory $N \cup G(\Sigma_B)$ and proving the inductive validity of the clauses in $G(\Sigma \setminus \Sigma_B)$. If this is successful in finite time, the result is a complete presentation $N \cup G(\Sigma_B) \cup G'$ where all clauses in G' have a selected atom. In later inductive proofs no inferences from G', in particular no inferences from any type clause $G(f)$, with f in $\Sigma \setminus \Sigma_B$, need to be computed.

With these remarks, the reader may want to take another look at the proof of sufficient-completeness of addition that we presented in section 2.

Proving the sufficient completeness of a function definition is a particular case in which a lemma is produced that makes subsequent proofs go through or at least makes them more efficient. In general, clauses which have been proved may be kept and used for proving compositeness in a subsequent inductive proof, without the need of superposing such a lemma on some conjecture.

5.5 Proofs of Compositeness

Inductive validity is reduced to proving compositeness of certain clauses that are derived from the conjectures. For the method to be applicable in practice one needs to have powerful methods available for verifying compositeness. Moreover, a failure in proving compositeness often gives an indication of what kind of lemma would be required in order to make the proof go through. It is here that generalization and lemma suggestion techniques should be incorporated. In this paper we shall only scratch the surface by making a few technical remarks on the subject related to orderings and first-order clauses.

In proofs of compositeness, one may assume that instances of the given theory presentation N are smaller (with respect to $\succ^c$) than any "new" clause that is introduced during an inductive theorem proving derivation (including the initial conjectures). Formally this can be justified by assuming that a new clause $\Gamma \to \Delta$

actually represents the logically equivalent clause $\top \approx \top, \Gamma \rightarrow \Delta$, where $\top$ is a new constant of a new sort which is the maximal term with respect to $\succ$.

For first-order clauses a technique for obtaining compositeness proofs, called *contextual reductive rewriting* in [2, 3], that has proven to be useful in practice. Let C be a clause and let N be a set of clauses. N_C denotes the set of instances C' of N such that $C \succ^c C'$. Let τ be a skolemizing substitution, i.e., a substitution that replaces variables by new constants. Let $C = \Gamma, u[l\sigma] \approx v \rightarrow \Delta$ (or $C = \Gamma \rightarrow \Delta, u[l\sigma] \approx v$) be a clause in N. Suppose there exists an instance $D\sigma$ of a clause $D = \Lambda \rightarrow \Pi, l \approx r$ in N such that (i) $l\sigma \succ r\sigma$, (ii) $C \succ^c D\sigma$, (iii) $N_C\tau \models \Gamma\tau \rightarrow A\tau$ for all equations A in $\Lambda\sigma$ and $N_C\tau \models A\tau \rightarrow \Delta\tau$ for all equations A in $\Pi\sigma$. Then C can be contextually rewritten to $\Gamma, u[r\sigma] \approx v \rightarrow \Delta$ (or $\Gamma \rightarrow \Delta, u[r\sigma] \approx v$). After this the clause C becomes composite in $N \cup \{\Gamma, u[r\sigma] \approx v \rightarrow \Delta\}$ and may be eliminated.

We may also do several steps of contextual rewriting in a row; in this case the bound on the complexity is provided by the first clause. If we eventually arrive at a clause that is composite, we have proved that the first clause is composite. Or we may use the method to prove inferences composite; in this case the bound is provided by the maximal premise of the inference.

More liberal ways of rewriting where a term is sometimes replaced by a larger term (as long as the instances of clauses which are involved in the rewriting are still sufficiently small) are suggested by the "rippling-out" method of [8] and can be extended without problems to our framework.

6 Conclusion

We have described a method of proof by consistency for first-order clauses with equality. Inductive theorem proving was defined as proving validity in the perfect model of a theory. We have built on methods for saturating sets of clauses and shown that inductive validity and redundancy are equivalent concepts. Selection strategies for superposition provide means to make clauses become redundant in the limit of a successful saturation process. For this idea to always be applicable an explicit closing of clauses by type predicates has been suggested As a side-effect our method allows for proofs of sufficient-completeness of function definitions, a property that is essential in other contexts too, e.g., for hierarchical specifications. We have shown that the concepts of proof by consistency for the purely equational case can be appropriately extended retaining most of their characteristic properties.

We have implemented this method for the restricted case of horn clauses in the CEC system for conditional equational completion [10] and some encouraging, initial practical experience has been made.

References

1. Leo Bachmair. Proof by consistency in equational theories. In *Proc. 3rd IEEE Symp. on Logic in Computer Science*, pages 228–233, Edinburgh, July 1988.
2. Leo Bachmair and Harald Ganzinger. On restrictions of ordered paramodulation with simplification. In *Proc. 10th Int. Conf. on Automated Deduction*, Kaiserslautern, July 1990. Springer LNCS 449.

3. Leo Bachmair and Harald Ganzinger. Completion of first-order clauses with equality by strict superposition. In *Proc. 2nd Int. Workshop on Conditional and Typed Rewriting Systems*, Montreal, June 1990. Springer LNCS 516.
4. Leo Bachmair and Harald Ganzinger. Perfect model semantics for logic programs with equality. In *Proc. 8th Int. Conf. on Logic Programming*. MIT Press, 1991.
5. Leo Bachmair and Harald Ganzinger. Rewrite-based equational theorem proving with selection and simplification. Technical Report MPI-I-91-208, Max-Planck-Institut für Informatik, Saarbrücken, August 1991.
6. Eddy Bevers and Johan Lewi. Proof by consistency in conditional equational theories. In *Proc. 2nd Int. Workshop on Conditional and Typed Rewriting Systems*, Montreal, June 1990. Springer LNCS 516.
7. Robert S. Boyer and J. Strother Moore. *A Computational Logic*. Academic Press, New York, 1979.
8. A. Bundy, F. van Harmelen, A. Smail, and A. Ireland. Extensions to the rippling-out tactic for guiding inductive proofs. In *Proc. 10th Int. Conf. on Automated Deduction*, pages 132–146, Kaiserslautern, July 1990. Springer LNCS 449.
9. Laurent Fribourg. A strong restriction of the inductive completion procedure. In *Proc. 13th Int. Coll. on Automata, Languages and Programming*, pages 105–115, Rennes, France, July 1986. Springer LNCS 226.
10. H. Ganzinger and R. Schäfers. System support for modular order-sorted Horn clause specifications. *Proc. 12th Int. Conf. on Software Engineering*, Nice, pages 150–163, 1990.
11. Stephen J. Garland and John V. Guttag. Inductive methods for reasoning about abstract data types. In *Proc. 15th Annual ACM Symp. on Principles of Programming Languages*, pages 219–228, San Diego, January 1988.
12. Gérard Huet and Jean-Marie Hullot. Proofs by induction in equational theories with constructors. *Journal of Computer and System Sciences*, 25:239–266, 1982.
13. Jean-Pierre Jouannaud and Emmanuel Kounalis. Proofs by induction in equational theories without constructors. In *Proc. Symp. on Logic in Computer Science*, pages 358–366, Cambridge, Mass., June 1986.
14. Stephane Kaplan and Marianne Choquer. On the decidability of quasi-reducibility. *EATCS Bulletin*, 28:32–34, 1986.
15. David R. Musser. On proving inductive properties of abstract data types. In *Proc. 7th Annual ACM Symp. on Principles of Programming Languages*, pages 154–162, Las Vegas, January 1980.
16. Fernando Orejas. Theorem proving in conditional-equational theories. Draft.
17. Peter Padawitz. Inductive expansion: A calculus for verifying and synthesizing functional and logic programs. *Journal of Automated Reasoning*, 7(1):27–103, March 1991.
18. David A. Plaisted. Semantic confluence tests and completion methods. *Information and Control*, 65:182–215, 1985.
19. T. C. Przymusinski. On the declarative semantics of deductive databases and logic programs. In J. Minker, editor, *Foundations of Deductive Data Bases and Logic Programming*, pages 193–216. Morgan Kaufmann Publishers, Los Altos, 1988.
20. Jürgen Stuber. Inductive theorem proving for horn clauses. Master's thesis, Universität Dortmund, April 1991.

Reduction Techniques for First-Order Reasoning

Francois Bronsard
Uday S. Reddy

Department of Computer Science
The University of Illinois at Urbana-Champaign
Urbana, IL 61801.
Net: {bronsard,reddy}@cs.uiuc.edu

1 Introduction

The objective of this paper is to develop reduction techniques for first-order clausal reasoning inspired by the techniques of term-rewriting. We define a notion of *reduction* (similar to rewriting) and a procedure for *completion* of first-order clausal theories.

Our method for clausal reasoning closely follows the Knuth-Bendix method for equational reasoning. (See [4] for a survey of the latter). Given a set of clauses $\mathcal{T}$ (called the *theory*) and a ground clause Γ, the problem is to decide $\mathcal{T} \models \Gamma$. (We call this the "clausal word problem"). Evidently, the problem is equivalent to the unsatisfiability of $\mathcal{T} \cup \bar{\Gamma}\theta$, where $\bar{\Gamma}\theta$ is the set of skolemized negations of the literals in Γ (called the *goal clauses*).

- We choose a well-founded order $>$ over literals (with certain additional properties), and
- complete $\mathcal{T}$ by adding certain "critical" resolvents of $\mathcal{T}$.
- In the process, the existing clauses of $\mathcal{T}$ can be deleted (or simplified) if they can be proved from other clauses of $\mathcal{T}$ by a *reductive* proof as described below.

When a complete set of clauses $\mathcal{T}_C$ is found, the validity of the clause Γ can be decided by carrying out a reductive proof from the set $\bar{\Gamma}$. Γ is a consequence of $\mathcal{T}$ if and only if the empty clause $\perp$ is derivable from $\bar{\Gamma}$ by a reductive proof. (This seems superficially different from the equational method, but note that rewriting $t = u$ to an identity can be equivalently viewed as rewriting the inequality $t \neq u$ to a contradiction). More generally, the reductive proof from $\bar{\Gamma}$ can be interleaved with the completion process to obtain a semi-decision procedure.

What is a *reductive proof*? Intuitively it is a proof where the complexity of the goal is decreased with each inference step. For ground proofs (or subproofs) this can be obtained by the use of *ordered cut* (ordered ground resolution) where cut is restricted to largest literals. For nonground proofs, we define a specialized version of the resolution inference (more specifically, of the hyper-resolution inference) called *reductive deduction* (or *reduction* for short) where selected literals of one or more goal clauses $\Gamma_1, \ldots, \Gamma_k$ are matched against all the *maximal* literals of a theory clause.

The reduction inference is similar to rewriting in that the goal literals are *pattern matched* against the literals of a theory clause (not unified), and, secondly, the conclusion, Γ' of this hyper-resolution is smaller than the largest goal premise by the chosen order. Moreover, if the goal clauses are ground, then Γ', the derived clause, is ground as well. The major difference from rewriting is that the derived clause Γ' is not equivalent to $\Gamma_1, \ldots, \Gamma_k$; it is only a consequence. Thus, multiple reductive steps are possible from a given goal clause (or set of goals clauses) and the proof is non-linear in the sense that all these different reductions might be necessary for the proof. Notice, however, that reductive proofs are "theory-linear", i.e., all inferences use at most one theory clause.

This is not of course the first application of term-rewriting ideas to first-order reasoning. Hsiang and Rusinowitch [6], and Bachmair [1] describe an *ordered resolution* technique where resolution is restricted to maximal literals (after instantiation). These works do not involve a notion of reduction. Zhang and Kapur [8] use a contextual rewriting-based reduction technique for simplifying/deleting clauses in the theory. However, this reduction technique is not complete for refuting goals. This work inspired our previous work reported in [3], where we proposed *contextual deduction* as the appropriate generalization of contextual rewriting for refuting goals. While we were able to show that it was a complete refutation method using "saturated" theories (theories closed under critical resolution), it was not known whether it was a complete refutation method using *complete* theories. (The difference is that complete theories need not contain simplifiable clauses). Independently, Nieuwenhuis and Orejas [7] developed a reduction technique and a completion procedure, but the reduction was defined for single literals (as opposed to our multiple-literal reduction). Thus, it could be a complete refutation method only if it was *nondeterministic* (it would have to "guess" the instantiation of a theory clause). The present work may be seen as a combination of the strengths of [3] and [7]. It uses contextual deduction (or, equivalently, multiple-literal reduction) to achieve determinacy and uses the techniques of [7] to show completeness with respect to complete theories.

2 Definitions

An *atomic formula* (*atom* for short) is a predicate symbol applied to a tuple of terms. A *literal* is a positive or negative atom. A *clause* is a set of literals interpreted as their disjunction. We use $A, B, C, \ldots$ to denote literals and write clauses as $A_1, \ldots, A_k$ (without enclosing braces). $\Gamma, \Delta, \ldots$ range over clauses. The empty clause is denoted by $\perp$. If Γ is a clause, $\bar{\Gamma}$ denotes the set of clauses $\{(\neg A) \mid A \in \Gamma\}$.

Consider the following inferences on clauses:

Resolution	Res	$\dfrac{\neg A', \Gamma \quad A, \Delta}{\Gamma\sigma, \Delta\sigma}$	if $\sigma = \mathrm{mgu}(A, A')$
Factoring	Fac	$\dfrac{A, A', \Gamma}{A\sigma, \Gamma\sigma}$	if $\sigma = \mathrm{mgu}(A, A')$
Cut	Cut	$\dfrac{\neg A, \Gamma \quad A, \Delta}{\Gamma, \Delta}$	
Instantiation	$\mathrm{I}_{\mathcal{T}}$	$\dfrac{\Gamma}{\Gamma\sigma}$	where $\Gamma \in \mathcal{T}$

It is well-known that a set of clauses $\mathcal{T}$ is unsatisfiable if and only if $\mathcal{T} \vdash \bot$ using Res and Fac. In this work, we rely instead on the equivalent characterization that $\mathcal{T}$ is unsatisfiable if and only if $\mathcal{T} \vdash \bot$ using Cut and I.

We adopt a well-founded order $\succ$ on literals with the following properties:
- preservation of the sign, *i.e.*, if $A \succ B$ then $\neg A \succ B$, $A \succ \neg B$, and $\neg A \succ \neg B$,
- stable, *i.e.*, if $A \succ B$ then $A\sigma \succ B\sigma$,
- transparent, *i.e.*, if $A \succ B$ then $V(A) \supseteq V(B)$,
- complete, *i.e.*, $\succ$ can be extended to an order $>$ that is total over ground literals.

All these properties are standard (e.g., [6], [1]) except for transparency. This property ensures groundness of goal clauses without which unification would be needed, and this, in turn, would make the reduction process potentially infinite.

A cut inference as described above is said to be *ordered* if $\neg A$ and A are the largest literals in $\neg A, \Gamma$ and A, Δ respectively. The results of [6] and [1] show that, for an unsatisfiable set $\mathcal{G}$ of ground clauses, there is a proof $\mathcal{G} \vdash \bot$ using ordered cut inferences.

A resolution inference as described above is said to be *critical* if $\neg A'$ and A are *maximal* in their respective premise clauses. Note that critical inferences are *ordered*, as defined in [6], but not conversely. The stronger condition of criticality is used to ensure groundness of goal clauses.

A *reductive deduction* (*reduction* for short) using a theory $\mathcal{T}$ is the inference

$$\mathrm{R}_{\mathcal{T}} \quad \frac{\neg A_1\sigma, \Gamma_1 \quad \cdots \quad \neg A_k\sigma, \Gamma_k}{\Delta\sigma, \Gamma_1, \ldots, \Gamma_k}$$

if there is a clause (or a factor of a clause) $A_1, \ldots, A_k, \Delta$ from $\mathcal{T}$ such that $A_1, \ldots, A_k$ is the set of *maximal literals* of the clause. Note that, by the transparency property of $\succ$, $V(A_1, \ldots, A_k) \supseteq V(\Delta)$. So, pattern matching is adequate to compute σ and $\sigma = \mathrm{mgu}(\langle A_1\sigma, \ldots, A_k\sigma\rangle, \langle A_1, \ldots, A_k\rangle)$. Moreover, if the premises of a reduction are ground then so is the conclusion.

The number of premises of an inference step is called its *arity*. For example, a Cut has arity 2, an I has arity 1 and a reduction $\mathrm{R}_{\mathcal{T}}$ like the one presented above has arity k. In the following presentation we limit ourselves to inferences of arity 3 or less. The discussion easily generalizes to higher arities.

A set of ground clauses $\mathcal{G}$ is said to be *reductively refutable* (using $\mathcal{T}$) if $\mathcal{G} \vdash \bot$ using $\mathrm{R}_{\mathcal{T}}$ and Ordered Cut inferences. A clause Γ is said to be provable by reductive refutation if $\bar{\Gamma}$ is reductively refutable.

Example 1 *Consider the following simple example. The theory contains two clauses*

$$\{P, \neg Q, R\} \textit{ and } \{\neg P, R\}$$

and we are given the goal $\{Q\}, \{\neg R\}$. *If the ordering is* $R \succ Q \succ P$ *then the following reductive derivation is possible:*

$$\dfrac{\dfrac{\dfrac{\neg R}{P, \neg Q}\; R \textit{ with } P, \neg Q, R \qquad Q}{P}\,\text{Cut} \qquad \dfrac{\neg R}{\neg P}\; R \textit{ with } \neg P, R}{\bot}\,\text{Cut}$$

If, instead, the ordering is $P \succ R \succ Q$, *then applying the completion procedure to the theory would generate the complete theory*

$$\{P, \neg Q, R\},\ \{\neg P, R\},\ \textit{and } \{\neg Q, R\}$$

Now the goal has the reductive refutation

$$\dfrac{\dfrac{\neg R}{\neg Q}\; R \textit{ with } \neg Q, R \qquad Q}{\bot}\,\text{Cut}$$

Example 2 *Although we do not address here the incorporation of equational reasoning in our approach, such an incorporation represents our final goal. Incorporating equational reasoning would allow the following example. Consider the following set of clauses expressing some simple properties of the greatest common divisor function (gcd)*

$$\begin{array}{l}
\{\neg(K * X = B * A),\ \neg(A = gcd(X * A, B * A)),\ divide(X, A)\} \\
\{divide(X, B),\ \neg(prime(X)),\ gcd(X, B) = 1\} \\
\{gcd(X * Z, Y * Z) = gcd(X, Y) * Z\} \\
\{\neg(Y = 1),\ X = X * Y\}
\end{array}$$

Consider the goal

$$\{prime(a)\}, \{\neg(divide(a, c))\}, \{\neg(divide(a, b))\}, \{k * a = b * c\}$$

and the complexities $= > * > gcd > divide > prime$.

The completion procedure applied to the above set of clauses would generate in particular the following clause

$$\{\neg(K * X = B * A),\ \neg(1 = gcd(X, B)),\ divide(X, A)\}$$

as a result of simplifying the first clause with the third clause and resolving the result with the fourth clause. Now we can prove our goal by the following reductive refutation

$$\cfrac{\cfrac{k * a = b * c}{divide(a,c) \vee \neg(1 = gcd(a,b))}\,R}{divide(a,c) \vee divide(a,b) \vee \neg(prime(a))}\,R$$

Three simple Cut inference between this last goal clause and the three goal clauses: $\{prime(a)\}, \{\neg(divide(a,c))\}, \{\neg(divide(a,b))\}$ *would complete the refutation.*

3 Completion

The following procedure is similar in spirit to Bachmair's "basic completion" for the equational case. The procedure maintains two sets of clauses $\mathcal{T}$ and $\mathcal{R}$. The clauses in the former set can be simplified and deleted, while the clauses in the latter set can be used for reduction. The procedure involves the following state transitions.

Orient $\quad \dfrac{(\mathcal{T} \cup \{\Gamma\}, \mathcal{R})}{(\mathcal{T}, \mathcal{R} \cup \{\Gamma\})}$

Deduce $\quad \dfrac{(\mathcal{T}, \mathcal{R})}{(\mathcal{T} \cup \{\Gamma\}, \mathcal{R})}$ if Γ is a critical resolvent of $\mathcal{R}$

Delete $\quad \dfrac{(\mathcal{T} \cup \{\Gamma\}, \mathcal{R})}{(\mathcal{T}, \mathcal{R})}$ if Γ is provable by reductive refutation using $\mathcal{R}$

Let us point out that Delete subsumes the classical deletion strategies like tautology deletion and subsumption. We discuss more general notions of simplification in Section 4.

A completion *derivation* is a sequence $(\mathcal{T}, \emptyset), (\mathcal{T}_1, \mathcal{R}_1), (\mathcal{T}_2, \mathcal{R}_2), \ldots$ such that $(\mathcal{T}_i, \mathcal{R}_i) \vdash (\mathcal{T}_{i+1}, \mathcal{R}_{i+1})$ for each i. The set of *persisting* clauses is defined by $\mathcal{T}^\infty = \bigcup_i \bigcap_{j \geq i} \mathcal{T}_j$, $\mathcal{R}^\infty = \bigcup_i \bigcap_{j \geq i} \mathcal{R}_j$. A derivation is said to be fair if $\mathcal{T}^\infty = \emptyset$ and $\bigcup_i \mathcal{T}_i$ includes all the critical resolvents of $\mathcal{R}^\infty$. Note that, if a derivation is fair and $\mathcal{R}^\infty$ is finite, then $\mathcal{R}^\infty = \mathcal{R}_i$ and $\mathcal{T}_i = \emptyset$ for some i. Such a set denotes a "complete" set of clauses and its word problem becomes decidable.

The completeness result for the completion procedure is the following:

Theorem 1 *If* $\mathcal{T} \models \Gamma$ *then* Γ *is provable by reductive refutation, using the* $\mathcal{R}^\infty$ *of a basic completion derivation from* $(\mathcal{T}, \emptyset)$.

We prove this theorem using the approach of proof transformation. $\mathcal{T} \models \Gamma$ implies that there is an ordered refutation of $\mathcal{T} \cup \bar{\Gamma}$ using Cut and I steps. We show, in the remainder of the section, that this ordered proof can be transformed into a reductive refutation using $\mathcal{R}^\infty$.

3.1 Intermediate inferences

First, we add some inference rules for theoretical purposes:

$$\text{MCut}\quad \frac{\neg A_1, \Gamma_1 \quad \cdots \quad \neg A_k, \Gamma_k \quad A_1, \ldots, A_k, \Delta}{\Delta, \Gamma_1, \ldots, \Gamma_k}$$

$$\text{Con}\quad \begin{array}{c} [\bar{\Gamma}] \\ \vdots \\ \dfrac{\bot}{\Gamma} \end{array}$$

The inference rule MCut (multiple cut) allows one to cut multiple literals of a clause at once. It is related to the hyper-resolution inference. The inference rule Con (contradiction) expresses the classical technique of proof by contradiction (brackets indicate discharged hypotheses). The role of Con in our framework is the same as that of Case Analysis in [7].

Note that MCut is not symmetric. The rightmost premise, called the *major premise*, plays a special role, while the other premises, called the *minor premises*, can appear in any order. Although, a Cut step could be seen as a special case of MCut of arity 2, we intend MCut steps to be used only as intermediate states leading to reduction steps. So, we will maintain a clear distinction between MCuts and Cuts.

The notion of ordered proof in the presence of MCut must be made relative to Cuts. We say that a proof is ordered if the Cut steps are ordered and pairs of steps, of the form Cut-MCut and MCut-Cut, show a reduction of complexities.

3.2 Proof transformation

Initially, we have an ordered proof of $\bar{\Gamma} \vdash \bot$ using $I_{\mathcal{T}}$ and Cut inferences. The transitions of the completion procedure induce proof transformations that introduce R, $I_{\mathcal{R}}$, MCut, or Con inferences, but will ultimately produce a proof with only R and ordered Cut steps, i.e. a reductive proof.

The transformations corresponding to the transitions of the completion procedure are the following: (we use the symbol $\Rightarrow$ to denote proof transformation)

- Orient changes $I_{\mathcal{T}}$ steps to $I_{\mathcal{R}}$ steps:

$$\frac{\Gamma}{\Gamma\sigma} I_{\mathcal{T}} \quad \Longrightarrow \quad \frac{\Gamma}{\Gamma\sigma} I_{\mathcal{R}} \tag{1}$$

- Deduce (adding resolvent) replaces $I_{\mathcal{R}}$ steps followed by Cut or MCut with $I_{\mathcal{T}}$ steps. There are three transformations corresponding to that transition.

$$\frac{\dfrac{\neg A', \Gamma'}{\neg A\sigma, \Gamma} I_{\mathcal{R}} \qquad \dfrac{A, \Delta}{A\sigma, \Delta\sigma} I_{\mathcal{R}}}{\Gamma, \Delta\sigma} \text{Cut} \quad \Longrightarrow \quad \frac{\Gamma'\tau, \Delta\tau}{\Gamma, \Delta\sigma} I_{\mathcal{T}} \tag{2}$$

where $\Gamma'\tau, \Delta\tau$ is the critical resolvent of $\neg A', \Gamma'$ and A, Δ.

$$\dfrac{\dfrac{\neg A_1\sigma \qquad \dfrac{A_1, A_2}{A_1\sigma, A_2\sigma}\,\mathrm{I}_{\mathcal{R}}}{A_2\sigma}\,\mathrm{MCut} \qquad \dfrac{\neg A_2', \Pi}{\neg A_2\sigma, \Pi\sigma}\,\mathrm{I}_{\mathcal{R}}}{\Pi\sigma}\,\mathrm{Cut} \tag{3}$$

$$\Longrightarrow \quad \dfrac{\neg A_1\sigma \qquad \dfrac{A_1\tau, \Pi\tau}{A_1\sigma, \Pi\sigma}\,\mathrm{I}_{\mathcal{T}}}{\Pi\sigma}\,\mathrm{MCut}$$

where $A_1\tau, \Pi\tau$ is the critical resolvent of A_1, A_2 and $\neg A_2', \Pi$.

$$\dfrac{\dfrac{\neg A_1\sigma \qquad \dfrac{A_1, A_2}{A_1\sigma, A_2\sigma}\,\mathrm{I}_{\mathcal{R}}}{A_2\sigma}\,\mathrm{MCut} \qquad \dfrac{\neg A_3\sigma \qquad \dfrac{\neg A_2', A_3, \Pi}{\neg A_2\sigma, A_3\sigma, \Pi\sigma}\,\mathrm{I}_{\mathcal{R}}}{\neg A_2\sigma, \Pi\sigma}\,\mathrm{MCut}}{\Pi\sigma}\,\mathrm{Cut} \tag{4}$$

$$\Longrightarrow \quad \dfrac{\neg A_1\sigma \qquad \neg A_3\sigma \qquad \dfrac{A_1\tau, A_3\tau, \Pi\tau}{A_1\sigma, A_3\sigma, \Pi\sigma}\,\mathrm{I}_{\mathcal{T}}}{\Pi\sigma}\,\mathrm{MCut}$$

where $A_1\tau, A_3\tau, \Pi\tau$ is the critical resolvent of A_1, A_2 and $\neg A_2', A_3, \Pi$.

- Delete replaces $\mathrm{I}_{\mathcal{T}}$ steps with Con steps:

$$\dfrac{\Gamma}{\Gamma\sigma}\,\mathrm{I}_{\mathcal{T}} \quad \Longrightarrow \quad \dfrac{\begin{matrix}[\bar{\Gamma}\sigma] \\ \vdots\ \alpha \\ \bot\end{matrix}}{\Gamma\sigma}\,\mathrm{Con} \tag{5}$$

  where α is obtained from the proof of $R \cup \bar{\Gamma}\theta \vdash \bot$ by appropriately modifying the substitutions. Since θ is a skolemizing substitution, replacing θ by σ and modifying the other substitutions appropriately is possible. Moreover, each reduction step in the proof of $R \cup \bar{\Gamma}\theta \vdash \bot$ is replaced in α by an instantiation step ($\mathrm{I}_{\mathcal{R}}$) followed by an MCut step.

The above proofs transformations could be used to show the *compatibility* of the completion procedure with the proof transformation relation. That is, if α is a proof using $(\mathcal{T}, \mathcal{R})$ and $(\mathcal{T}, \mathcal{R}) \vdash (\mathcal{T}', \mathcal{R}')$ by basic completion, then there is a proof α' using $(\mathcal{T}', \mathcal{R}')$ such that $\alpha \Longrightarrow^* \alpha'$. The exact proof of the compatibility of the completion procedure is given in Lemma 4.

These transformations are not sufficient to obtain reductive proofs. We also need proof transformations to introduce or eliminate Con, $\mathrm{I}_{\mathcal{R}}$, and MCut steps. These transformations are given below.

- MCut steps are introduced by

$$\dfrac{\neg A\sigma, \Gamma \quad \dfrac{A, \Delta}{A\sigma, \Delta\sigma}\,\mathrm{I}_{\mathcal{R}}}{\Gamma, \Delta\sigma}\,\mathrm{Cut} \quad \Longleftrightarrow \quad \dfrac{\neg A\sigma, \Gamma \quad \dfrac{A, \Delta}{A\sigma, \Delta\sigma}\,\mathrm{I}_{\mathcal{R}}}{\Gamma, \Delta\sigma}\,\mathrm{MCut} \qquad (6)$$

- An $\mathrm{I}_{\mathcal{R}}$ step followed by an MCut that cuts all the maximal literals of the clause can be transformed into an R step as follows:

$$\dfrac{\neg A_1\sigma \quad \neg A_2\sigma \quad \dfrac{A_1, A_2, \Delta}{A_1\sigma, A_2\sigma, \Delta\sigma}\,\mathrm{I}_{\mathcal{R}}}{\Delta\sigma}\,\mathrm{MCut} \quad \Longrightarrow \quad \dfrac{\neg A_1\sigma \quad \neg A_2\sigma}{\Delta\sigma}\,\mathrm{R} \qquad (7)$$

  To simplify the presentation we will assume that this transformation is postponed until no other transformations are applicable. Therefore, none of the other transformations needs to consider the case of R steps.

- To modify the arity of an MCut we use the following transformation.

$$\dfrac{\dfrac{\neg A_1 \quad A_1, A_2, \Gamma}{A_2, \Gamma}\,\mathrm{MCut} \quad \neg A_2}{\Gamma}\,\mathrm{Cut} \quad \Longleftrightarrow \quad \dfrac{\neg A_1 \quad \neg A_2 \quad A_1, A_2, \Gamma}{\Gamma}\,\mathrm{MCut} \qquad (8)$$

- A Con step can be eliminated by merging the refutation leading to it with the proof step following it. This requires the transformation

$$\dfrac{\neg A, \Gamma \quad \dfrac{\begin{array}{c}[\neg A]\,[\bar{\Pi}] \\ \vdots\,\alpha \\ \bot\end{array}}{A, \Pi}\,\mathrm{Con}}{\Gamma, \Pi}\,\mathrm{Cut} \quad \Longrightarrow \quad \dfrac{\begin{array}{c}[\bar{\Gamma}]\,[\bar{\Pi}] \\ \vdots\,\alpha' \\ \bot\end{array}}{\Gamma, \Pi}\,\mathrm{Con} \qquad (9)$$

  where α' is obtained from α by applying the following transformation to inference steps involving the $\neg A$'s in α:

$$\dfrac{[\neg A] \quad A, \Delta}{\Delta}\,\mathrm{Cut} \quad \Longrightarrow \quad \dfrac{[\bar{\Gamma}] \quad \dfrac{\neg A, \Gamma \quad A, \Delta}{\Gamma, \Delta}\,\mathrm{Cut}}{\Delta}\,\mathrm{Cut\ (several)}$$

  A similar transformation is applicable if the Con step is a minor premise of an MCut step.

After the transformation 5 a Con step might become the major premise of an MCut, in which case the following transformation is applicable.

$$\dfrac{\neg A_1, \Gamma_1 \qquad \dfrac{\begin{array}{c}[\neg A_1]\,[\neg A_2]\,[\bar{\Gamma}_2]\\ \vdots\,\alpha \\ \bot\end{array}}{A_1, A_2, \Gamma_2}\,\text{Con}}{\Gamma_1, A_2, \Gamma_2}\,\text{MCut} \quad \Longrightarrow \quad \dfrac{\begin{array}{c}[\bar{\Gamma}_1]\,[\neg A_2]\,[\bar{\Gamma}_2]\\ \vdots\,\alpha' \\ \bot\end{array}}{\Gamma_1, A_2, \Gamma_2}\,\text{Con} \tag{10}$$

α' is then obtained from α by transformations like the one given earlier or the following one

$$\dfrac{[\neg A_1] \quad \neg B \quad A_1, B, \Pi}{\Pi}\,\text{MCut} \Longrightarrow \dfrac{[\bar{\Gamma}_1] \qquad \dfrac{\neg A_1, \Gamma_1 \quad \neg B \quad A_1, B, \Pi}{\Gamma_1, \Pi}\,\text{MCut}}{\Pi}\,\text{Cut (several)}$$

Finally we rely strongly on the notion that proofs can be ordered using simple *commutation rules* as shown in [1]. These transformation rules are

$$\dfrac{\dfrac{\neg A_1 \quad A_1, A_2, \Gamma}{A_2, \Gamma}\,\text{Cut} \qquad \neg A_2}{\Gamma}\,\text{Cut} \quad \Longleftrightarrow \quad \dfrac{\neg A_1 \qquad \dfrac{A_1, A_2, \Gamma \quad \neg A_2}{A_1, \Gamma}\,\text{Cut}}{\Gamma}\,\text{Cut} \tag{11}$$

$$\begin{array}{c} \dfrac{\dfrac{\neg A_1 \quad A_1, A_2}{A_2}\,\text{Cut} \qquad A_3 \quad \neg A_2, \neg A_3, \Pi}{\Pi}\,\text{MCut} \\[2ex] \Longleftrightarrow \quad \dfrac{\neg A_1 \qquad \dfrac{A_1, A_2 \quad A_3 \quad \neg A_2, \neg A_3, \Pi}{A_1, \Pi}\,\text{MCut}}{\Pi}\,\text{Cut} \end{array} \tag{12}$$

3.3 Proof reduction

One purpose of proof transformations is to replace I steps, which are nondeterministic in choosing the instantiations, by deterministic R steps. To distinguish between the two kinds of instantiations, we mark each ground literal in a proof with a tag, as in A^t, where t is either $\top$ or $\bot$. Tagged literals A^t are compared with each other by comparing the corresponding pairs $\langle t, A\rangle$ in lexicographic order.

A proof is marked as follows. The literals in the conclusion of an ¡i step are tagged with $\top$ unless

- the I step is followed by an MCut, in which case all the literals cut against $\bot$-tagged literals receive the tag $\bot$, or

- the I step uses a rule clause from $\mathcal{R}$ and it is followed by an MCut that cuts all the maximal literals of the instantiated clause. In this case, all the literals receive the tag $\bot$.

For Con steps, we must assume that the transformations introducing or modifying Con steps mark permanently the literals involved in such transformations. Thus, the literals introduced by Con steps receive as tags whatever permanent marks were given to them. The literals of goal clauses are tagged with $\bot$ and, finally, Cut and MCut steps propagate tags from premises to conclusions.

Treat a marked proof α as a tree of inferences. With each inference i, we associate a complexity $C(i)$ as a multiset of tagged literals. Inference-complexities are compared by the multiset order and proof-complexities are compared by the recursive path order. This order on proofs is denoted $>_P$. The complexities of inferences are as follows.

$$\frac{A^t, \Gamma \quad \neg A^{t'}, \Delta}{\Gamma, \Delta}\,\text{Cut} \qquad \{\max(A^t, \neg A^{t'})\}$$

$$\frac{\neg A_1^{t_1}, \Gamma_1 \quad \cdots \quad \neg A_k^{t_k}, \Gamma_k}{\Gamma_1, \ldots, \Gamma_k, \Delta}\,\text{R} \qquad \{A_1^{t_1}, \ldots, A_k^{t_k}\}$$

$$\frac{A_1', \ldots, A_n'}{A_1^{t_1}, \ldots, A_n^{t_n}}\,\text{I}_{\mathcal{T}} \qquad \{\max\{A_1^{t_1}, \ldots, A_n^{t_n}, \bot^{\top}\}\}^3$$

$$\frac{A_1', \ldots, A_n'}{A_1^{t_1}, \ldots, A_n^{t_n}}\,\text{I}_{\mathcal{R}} \qquad \{\max_i A_i^{t_i}\}$$

$$\frac{\neg A_1^{s_1} \cdots A_k^{s_k} \qquad \dfrac{A_1', \ldots, A_k', \ldots, A_n'}{A_1^{t_1}, \ldots, A_k^{t_k}, \ldots, A_n^{t_n}}\,\text{I}}{A_{k+1}^{t_{k+1}}, \ldots, A_n^{t_n}}\,\text{MCut} \qquad \{\max(A_i^{s_i}, A_i^{t_i})\}_{i=1,\ldots,k} \cup C(I)$$

$$\frac{\begin{matrix}[\bar{\Gamma}] \\ \vdots\,\alpha \\ \bot\end{matrix}}{\Gamma}\,\text{Con} \qquad \{\bot^{\bot}\}$$

This defines the proof ordering $>_P$. The *proof reduction* relation $\Longrightarrow_P$ is defined as the intersection of the proof transformation $\Longrightarrow^+$ and the proof ordering $>_P$. Evidently, $\Longrightarrow_P^+$ is a well-founded order. Theorem 1 is proved by induction on $\Longrightarrow_P^+$. First, note the following lemmas:

Lemma 2 *For every marked proof α using $(\mathcal{T}, \mathcal{R})$, there is a proof α' such that $\alpha \Longrightarrow_P^* \alpha'$ and all MCuts in α' are* simple, *i.e., they resolve only $\bot$-tagged literals.*

Lemma 3 *For every proof α using $(\mathcal{T},\mathcal{R})$, there is an ordered proof α' such that $\alpha \Longrightarrow_P^* \alpha'$.*

Lemma 2 uses transformations 6 and 8 while Lemma 3 uses the commutation transformations 11 and 12.

Lemma 4 (Compatibility of Completion with $>_p$) *If $(\mathcal{T},\mathcal{R}) \vdash^* (\mathcal{T}',\mathcal{R}')$ then for each proof α using $(\mathcal{T},\mathcal{R})$ there is a proof α' using $(\mathcal{T}',\mathcal{R}')$ such that $\alpha = \alpha'$ or $\alpha \Longrightarrow_P \alpha'$.*

Proof: The two transitions that force a modification of proofs are Orient and Delete.

1. Orient : The transformation 1 is clearly a reduction: $\{A^t\}^3 >_P \{A^t\}$.
2. Delete : The transformations 5 and 10 are applicable after a Delete steps, as follows.

$$\dfrac{\begin{array}{c}\vdots\,\beta\\ \neg A_1^\perp,\Delta\end{array} \qquad \dfrac{A_1',A_2',\Gamma}{A_1^\perp,A_2^\top,\Gamma^t}\,I_{\mathcal{T}}}{\Delta,A_2^\top,\Gamma^t}\,\text{MCut} \quad\Longrightarrow\quad \dfrac{\begin{array}{c}\vdots\,\beta\\ \neg A_1^\perp,\Delta\end{array} \qquad \dfrac{\begin{array}{c}[\neg A_1^\perp]\,[\neg A_2^\top]\,[\bar{\Gamma}^t]\\ \vdots\,\alpha'\\ \bot\end{array}}{A_1^\perp,A_2^\top,\Gamma^t}\,\text{Con}}{\Delta,A_2^\top,\Gamma^t}\,\text{MCut}$$

$$\Longrightarrow\quad \dfrac{\begin{array}{c}[\bar{\Delta}]\,[\neg A_2^\top]\,[\bar{\Gamma}^t]\\ \vdots\,\alpha''\\ \bot\end{array}}{\Delta,A_2^\top,\Gamma^t}\,\text{Con}$$

Assuming $A_2^\top$ is the largest literal in $A_1^\perp,A_2^\top,\Gamma^t$, the complexity of the initial proof is $\{A_1^\perp\}\cup\{A_2^\top\}^3(C(\beta),\{A_2^\top\}^3)$. We show that both α and α' have smaller complexity than this.

α' is a reductive refutation from the goal clauses $\{\neg A_1^\perp\},\{\neg A_2^\top\}$, and $\bar{\Gamma}$. Since $\neg A_2^\top$ is the largest literal among them, every cut in α' would have complexity at most $\{A_2^\top\}$, every $I_{\mathcal{R}}$ step complexity at most $\{A_2^\top\}$ and every MCut step complexity less than $\{A_2^\top\}^3$.

In the worst case, we have an MCut of the form

$$\dfrac{[\neg A_2^\top] \qquad \begin{array}{c}\vdots\,\gamma\\ \neg A_3^\top\end{array} \qquad \dfrac{A_2'',A_3'',\Pi''}{A_2^\top,A_3^\top,\Pi}\,I_{\mathcal{R}}}{\Pi}\,\text{MCut}$$

with complexity $\{A_2^\top,A_3^\top,A_2^\top\}(C(\gamma),\{A_2^\top\})$ where A_3 is necessarily smaller than A_2. But, this complexity is smaller than $\{A_2^\top\}^3$.

The proof α'' is the same as α' except that every use of the hypothesis $\neg A_1^{\perp}$ is replaced by the proof β followed by several cuts to eliminate the literals of Δ. None of the literals in $\neg A_1^{\perp}, \Delta$ can be larger than $A_2^{\top}$ (if not, the initial proof is not ordered), so each of these cuts has complexity smaller than $\{A_2^{\top}\}$.

Hence, the complexity of the final proof $\{\perp^{\perp}\}(C(\alpha''))$ is smaller than that of the initial proof. □

Our proof reduction relation cannot show that any Con step in an arbitrary non-reductive proof can be reduced. However, it maintains invariants over proofs involving Con steps such that we can show that non-reductive proofs respecting these invariants can be reduced.

Proposition 5 *The proof reduction relation maintains the following invariants. Consider a Con step with a clause Γ as its conclusion and a refutation α as its premise, then*

1. *Γ is a premise of a Cut step or a minor premise of an MCut step,*
2. *a goal clause of the form $[\neg A^t]$ in α can only occur in Cut steps where it is resolved against a clause of the form A^{t_2}, Δ, with $t \succeq t_2$, and*
3. *the goal clauses from $\bar{\Gamma}$ occur only once along any path of the proof α.*

Lemma 6 *If a proof α_0 contains a Con step, then there is a proof α_0' such that $\alpha_0 \Longrightarrow_P \alpha_0'$.*

Proof: Suppose the Con step has the clause A^t, Π as its conclusion, the refutation α as its premise, and that it is followed by a Cut resolving the literal A (a similar reasoning applies if the Con step is followed by an MCut step). The complexity of the Cut step is then $\{A^{t_0}\}$ with $t_0 \succeq t$. The transformation 9 is applicable and we must show that it corresponds to a reduction of complexity.

By Proposition 5, $[\neg A^t]$, if it occurs in α, can only occur in Cut steps where it is resolved against clauses of the form A^{t_2}, Δ with $t \succeq t_2$. Hence, such a Cut step has the complexity $\{A^t\}$. Proposition 5 also indicates that the subproof whose conclusion is A^{t_2}, Δ does not use $[\neg A^t]$, so it is unaffected by the transformation 9. Moreover, since the proof is ordered, all steps in α following such a Cut step have complexity $<_P \{A^t\}$. Similarly, the new Cut steps created by the transformation 9 must have a complexity $<_P \{A^{t_0}\}$. Therefore, by the properties of the rpo ordering, we can show that transformation 9 corresponds to the following reduction of complexity (β is the subproof whose conclusion is the second premise of the Cut)

$$\{A^{t_0}\}(C(\beta), \{\perp^{\perp}\}C(\alpha)) >_P \{\perp^{\perp}\}C(\alpha')$$

□

Lemma 7 *If a proof α is not reductive, there exists α' such that $\alpha \Longrightarrow_P \alpha'$.*

Proof: The previous lemmas have shown that any proof that is not ordered, contains Con steps, or contains non-simple MCut steps can be reduced. Thus, the desired result will follow if we can show that the presence of I steps also induces possible reductions.

- If α contains a $I_{\mathcal{T}}$ step, then the transformation 1 is applicable and corresponds to a reduction.
- Consider the case of a $I_{\mathcal{R}}$ step followed by a Cut step. Since the proof is ordered, this Cut step must resolve the largest literal occurring in the conclusion of the $I_{\mathcal{R}}$ step. If this literal is resolved in the Cut step with a $\perp$-tagged literal, then the transformation 6 used from left to right corresponds to a reduction:

  $$\{A^{\top}\sigma\}(C(\beta), \{A^{\top}\sigma\}) >_P \{A^{\perp}\sigma\} \cup \{B^{\top}\sigma\}(C(\beta), \{B^{\top}\sigma\})$$

  where $B\sigma$ is the second largest literal in the conclusion of the $I_{\mathcal{R}}$ step.

  If the literal $A\sigma$ is resolved with a $\top$-tagged literal, and the other premise of the Cut is also the conclusion of an $I_{\mathcal{R}}$ step, then the transformation 2 is applicable and again corresponds to a reduction.

  If the other premise of the Cut is not the conclusion of an $I_{\mathcal{R}}$ step, then there must be some inner I step, since only I steps introduce $\top$-tagged literals, and some transformation must be applicable to it.
- Consider the case of a $I_{\mathcal{R}}$ step followed by an MCut. If the MCut step cuts all the maximal literals of the premise of the $I_{\mathcal{R}}$, then the transformation 7 is applicable. However, this transformation is postponed until no other transformation is applicable.

  If the MCut is followed by a Cut that resolves a maximal literal of the premise of the $I_{\mathcal{R}}$ step with a $\perp$-tagged literal then the transformation 8 (from left to right) is applicable, inducing the following reduction of complexities

  $$\begin{array}{l}\{A_2^{\top}\}(\{A_1^{\perp}\} \cup \{A_2^{\top}\}(C(\beta), \{A_2^{\top}\}), C(\gamma)) \\ \quad >_P \;\; \{A_1^{\perp}, A_2^{\perp}\} \cup \{B^{\top}\}(C(\beta), C(\gamma), \{B^{\top}\})\end{array}$$

  where B is the largest literal, after A_1 and A_2, in the conclusion of the $I_{\mathcal{R}}$ step.

  If the MCut is followed by a Cut that resolves a maximal literal of the premise of the $I_{\mathcal{R}}$ step with a $\top$-tagged literal then either the transformation 3 or 4 is applicable, or some inner I step can be transformed. □

Lemma 8 *If α is a non-reductive proof using $(\mathcal{T}_i, \mathcal{R}_j)$ then there exists a $j \geq i$ such that $\alpha \Longrightarrow_P \alpha'$ where α' is a proof using some $(\mathcal{T}_j, \mathcal{R}_j)$.*

Proof: By lemma 7 we know that there must be some transformation applicable to a non-reductive proof. Only four transformations (1), (2), (3),and (4) might require a delay until a later stage of the completion procedure. In these four cases, the fairness condition insure that in some state $j \geq i$ we will, accordingly, realize the Orient transition or generate the appropriate critical resolvent. By lemma 4, at that stage we will either have a simpler proof or the same proof, in which case the appropriate transformation can be executed. □

4 Simplification

The general notion of simplification of clauses can be expressed by the following state transition (which subsumes Delete):

$$\text{Simplify} \quad \frac{(\mathcal{T} \cup \{\Gamma\}, \mathcal{R})}{(\mathcal{T} \cup \{\Gamma'\}, \mathcal{R})} \quad \begin{array}{l}\text{if } \Gamma \succ \Gamma',\ \Gamma\theta\ (\theta \text{ a skolemizing substitution})\\ \text{has a reductive proof using } \mathcal{R} \cup \{\Gamma'\}, \text{ and } \Gamma' \text{ is}\\ \text{provable using } \{\Gamma'\} \cup \mathcal{R} \cup \mathcal{T}.\end{array}$$

Unfortunately, this definition of Simplify is too general to preserve the refutational completeness of our method, at least relatively to our current proof technique. However, simpler versions of this rule can be used without modifying our completeness proof.

A simple and useful version of Simplify is the following rule:

$$\text{Simplify} \quad \frac{(\mathcal{T} \cup \{\Gamma\}, \mathcal{R})}{(\mathcal{T} \cup \{\Gamma'\}, \mathcal{R})} \quad \begin{array}{l}\text{if } \Gamma' \subset \Gamma \text{ and } \Gamma' \text{ has a reductive refutation}\\ \text{using } \mathcal{R} \cup \{\Gamma\}.\end{array}$$

Note that if $\Gamma' \subset \Gamma$ then Γ has a trivial reductive refutation using Γ'. It is easy to show that Lemma 4 still hold with this definition of Simplify. Let us add that in theory we only have to show that Γ' is provable using $\mathcal{R} \cup \mathcal{T} \cup \{\Gamma\}$. However, for efficiency, we simply check that Γ' is reductively provable using $\mathcal{R} \cup \{\Gamma\}$.

Finally, in the context of equational reasoning we would add the classical simplification rule

$$\text{Simplify} \quad \frac{(\mathcal{T} \cup \{\Gamma\}, \mathcal{R})}{(\mathcal{T} \cup \{\Gamma'\}, \mathcal{R})} \quad \text{if } \Gamma \text{ rewrites to } \Gamma' \text{ using the theory } R.$$

5 Conclusion

We have outlined a completion procedure for first-order clausal reasoning with inspiration from completion procedures for equational reasoning. We believe this to be a useful method for many practical applications involving clausal reasoning where the theory stays constant and is used repeatedly for proving many goals. Program synthesis (see [3]) is such an application. We have implemented a prototype completion procedure in Prolog and used it for simple examples. We are currently involved in integrating it with the Focus program derivation system.

One important problem not touched upon in this work is the incorporation of equational reasoning. With the addition of equality, this method becomes viable for conditional equations, and it would extend the currently known completion procedures for them [2, 5].

References

[1] L. Bachmair. Proof normalization for resolution and paramodulation. In N. Dershowitz, editor, *Rewriting Techniques and Applications*, pages 15–28. Springer-Verlag, Berlin, 1989. (Lecture Notes in Comp. Science, Vol 355).

[2] Leo Bachmair and Harald Ganzinger. Completion of first order clauses with equality. In S. Kaplan and M. Okada, editors, *Conditional and Typed Rewriting Systems — Second International CTRS Workshop*, pages 162–180. Springer-Verlag, 1991.

[3] F. Bronsard and U. S. Reddy. Conditional rewriting in Focus. In S. Kaplan and M. Okada, editors, *Conditional and Typed Rewriting Systems — Second International CTRS Workshop*, pages 2–13. Springer-Verlag, Berlin, 1991.

[4] N. Dershowitz and J.-P. Jouannaud. Rewrite systems. In J. van Leeuwen, editor, *Handbook of Theoretical Computer Science B: Formal Methods and Semantics*, chapter 6, pages 243–320. North-Holland, Amsterdam, 1990.

[5] H. Ganzinger. A completion procedure for conditional equations. *J. Symbolic Computation*, 11:51–81, 1991.

[6] Jieh Hsiang and Michaël Rusinowitch. A new method for establishing refutational completeness in theorem proving. In J. H. Siekmann, editor, *Proceedings of the Eighth International Conference on Automated Deduction*, pages 141–152, Oxford, England, July 1986. Vol. 230 of *Lecture Notes in Computer Science*, Springer, Berlin.

[7] R. Nieuwenhuis and F. Orejas. Clausal rewriting. In S. Kaplan and M. Okada, editors, *Conditional and Typed Rewriting Systems — Second International CTRS Workshop*, pages 246–258. Springer-Verlag, 1991.

[8] H. Zhang and D. Kapur. First-order theorem proving using conditional rewrite rules. In E. Lusk and R. Overbeek, editors, *9th Intern. Conf. on Automated Deduction*, pages 1–20. Springer-Verlag, 1988.

Conditional Term Rewriting and First-Order Theorem Proving†

David A. Plaisted, Geoffrey D. Alexander, Heng Chu, and Shie-Jue Lee
Department of Computer Science
University of North Carolina at Chapel Hill
Chapel Hill, NC 27599-3175
e-mail: {plaisted | alexande | chu}@cs.unc.edu

Abstract. We survey some basic issues in first-order theorem proving and their implications for conditional term-rewriting systems. In particular, we discuss the propositional efficiency of theorem proving strategies, goal-sensitivity, and the use of semantics. We give several recommendations for theorem proving strategies to enable them to properly treat these issues. Although few current theorem provers implement these recommendations, we discuss the clause linking theorem proving method and its extension to equality and semantics as examples of methods that satisfy most or all of our recommendations. Implicit induction theorem provers meet our recommendations, to some extent. We also discuss correctness and efficiency issues involved in the use of semantics in first-order theorem proving. Finally, we discuss issues of efficiency and semantics in conditional term-rewriting.

1 Introduction

We argue that term-rewriting and conditional term-rewriting research should be increasingly concerned with larger theorem proving issues as the difficulty of the problems addressed increases. In particular, search complexity, goal-sensitivity, and semantics ought to be emphasized more. Search complexity is important because many common theorem proving strategies have severe propositional inefficiencies. These same inefficiencies appear in term-rewriting-based theorem provers. Goal sensitivity means that the inferences performed should depend on the particular theorem being proved, and not only on general axioms. This is necessary when there are many axioms, to reduce the irrelevant inferences. Typical term-rewriting approaches are based on Knuth-Bendix completion [KB70] or unfailing completion [BDP89] for which superpositions are performed in a manner usually independent of the theorem. Of course, it's often a good idea to perform some inferences that are goal-sensitive and some that are not. By semantics we mean the use of a first-order structure that is a model of the axioms used; such structures can help to focus the search and are commonly used by human mathematicians. It seems difficult to imagine a human mathematician proving a theorem about group theory without considering some examples of groups during the proof process, for example. These concerns

† This research was partially supported by the National Science Foundation under grant CCR-9108904; Fachbereich Informatik, Universitaet Kaiserslautern, Kaiserslautern, Germany; and the Max-Planck-Institut fuer Informatik, Saarbruecken, Germany

become even more relevant since term-rewriting based strategies are increasingly being extended to first-order logic; examples include refinements of paramodulation [HR91, BG90, BGLS91] and implicit induction methods for full first-order logic.

We believe that these basic problems need to be addressed before theorem provers can become significantly more powerful. New methods such as implicit induction and its variants perform better on some problems than existing methods, but for hard problems these new methods face the same barriers as existing methods. There is also a need for progress in other areas, such as the treatment of equality and mathematical induction, but the importance of these problems is widely recognized and good progress is being made.

2 Propositional behavior of theorem proving strategies

We now discuss the propositional inefficiency of many common theorem proving strategies. In [Pla92], we analyze the search complexity of a number of different theorem proving strategies. For this, we describe the search space as a tree, where each node of the tree represents a state of the search, and the descendents of a node represent alternative possible successor states of a state. For example, for resolution, each state is a set of clauses, and each state has one descendent, obtained by doing all possible resolutions according to some specified strategy. For model elimination, each state is a single chain, and the descendents of a state are the chains obtained by permissible operations from the chain. We are interested in the total size (number of nodes) of the search space, and also in other measures of its size; for details, see [Pla92]. There has been some work analyzing the size of proofs in various inference systems, but to our knowledge not much in the size of the search space, which is more relevant for efficiency considerations. Of course, if the smallest proof has exponential size, then the search space must be exponential also.

We consider easy sets of clauses to distinguish strategies more finely than the results of [Hak85], who showed that the smallest resolution proofs of some propositional formulas are exponential. We consider propositional Horn sets and related sets of clauses. We show that even for such sets of clauses, most strategies perform poorly, namely, the size of the search space is exponential. Those strategies that perform well in terms of the size of the search space are not goal sensitive, except for a few exceptions; these tend to be back chaining strategies that use caching. Examples of these are model elimination and MESON with caching of unit lemmas [AS92], the simplified problem reduction format [Pla82], the modified problem reduction format [Pla88], and clause linking [LP90a]. We show in [Pla92] that similar results apply to SLD resolution, used as the semantics of Prolog, implying some severe inefficiencies in Prolog execution for pathological programs. We also show that the usual extensions of term-rewriting-based strategies to full first-order logic [HR91, BG90, BGLS91] (and others) have similar inefficiencies in the size of the search space. This implies that new approaches are necessary to obtain efficient extensions of rewriting-based theorem proving strategies to full first-order logic. In practice, resolution and similar strategies perform well on Horn clauses because the problems are simple enough that the lack of goal-sensitivity is not a problem and methods (like hyper-resolution) are used that have good search behavior for Horn clauses. However, for highly non-Horn problems and for problems for which goal-sensitivity is important, such methods are inefficient.

3 Semantics

Next we consider the use of semantics by theorem provers. By semantics we mean the use of a first-order structure to guide the search for a proof. For example, when proving a theorem about groups, the structure used might be a group. Assuming unsorted structures for the moment, a structure I has a *domain* D and to each function symbol f associates a function $f^I : D^n \rightarrow D$ for suitable n. To each predicate symbol P there is associated a relation P^I on D^n for some n. For example, if the structure were a group, then the domain would be the set of elements of the group, some operator f would be interpreted so that f^I is the product operation on the group, and $=$ would be interpreted as the equality relation. These f^I and P^I need to be represented in some way, possibly as computer programs for computing f^I and P^I. Also, there needs to be some method for guiding the search using this structure.

This approach can be applied to refutational theorem proving. To show that a theorem R follows from some general axioms A, we show that $A \wedge \neg R$ is unsatisfiable, which is done by Skolemizing (eliminating existential quantifiers) and converting to clause form S. The conversion to clause form is satisfiability-preserving, so that $\models (R \supset A)$ iff S is unsatisfiable. Since the syntax of S is much simpler than that of A and R, this approach is in some ways more suitable for computer implementation than reasoning directly in first-order logic. Now, S is a set of clauses representing the conjunction of the clauses in the set. We can express S as $S_1 \cup S_2$ where the clauses in S_1 come from skolemizing A and those in S_2 come from skolemizing $\neg R$. Suppose we have a first-order structure M which satisfies the axioms A. It follows that M (suitably modified to account for Skolem functions) also satisfies S_1. Therefore, to get a goal-sensitive theorem proving strategy, we need a way to focus attention on the clauses that are not satisfied by M.

Another way to use semantics is to detect false conjectures. For example, suppose we have a set A of axioms of arithmetic. Then we know that from A we cannot prove $0 = s(0)$ (assuming arithmetic is consistent), and we need not attempt such a theorem or lemma. In general, we can detect unprovable lemmas by showing that they are not satisfied by a model M of the axioms. The reason is that if $M \models A$ and $A \vdash B$ (B is derivable from A) then $M \models B$, assuming the logic used is sound. However, we can often test whether $M \models B$ by straightforward evaluation techniques. For example, we can evaluate 0 and $s(0)$ to different elements of the domain of M, that is, to different integers, so we know that $\neg(M \models 0 = s(0))$, and so $\neg(A \vdash B)$. Thus B is not derivable from A, and we know that there is no proof of B from A without searching for one. Detecting this can reduce the search performed by a theorem prover and increase efficiency.

3.1 Correctness of models

There is some concern about verifying that M is a model of A. We will discuss this below. However, note that this is not a correctness issue, but rather an efficiency issue. Some semantics-based strategies work for any structure, whether it is a model of A or not, but the efficiency may depend on the model. Even if the completeness of the method depends on having a model of A, the correctness of the proof obtained only depends on the correct application of the rules of inference. Therefore, assuming the rules of inference of the underlying system are applied correctly, the use of an incorrect model will not lead to incorrect proofs.

We now return to the correctness issue. Given a model M and a set A of (first-order) axioms, the problem is to check that M satisfies A. For our purposes, we can assume that A is a set of first-order clauses. This means that all quantifiers are universal and existential quantifiers have been replaced by Skolem functions. Finite models can be checked by exhaustive testing. To test if a formula $A[x_1, ..., x_n]$ is satisfied by a model M with finite domain D, we only have to look at all the ways of interpreting x_i as d_i for $d_i \in D$. This gives us a finite set of structures I extending M by interpreting the variables of A as elements of D. We need to verify that all such I satisfy A. This assumes that the variables are implicitly universally quantified. We call this the ***universal satisfiability test.*** Often we want to do the test with the variables regarded as implicitly existentially quantified; then we want to know whether one of the structures I satisfies A, that is, is there some way of assigning elements of D to the variables x_i making A true? We call this the ***existential satisfiability test.*** For these tests, it is convenient to consider formulas $A[d_1, ..., d_n]$ in which the variables x_i have been replaced by elements d_i of the domain. Then, M universally satisfies A if M satisfies all such formulae $A[d_1, ..., d_n]$, and M existentially satisfies A if M satisfies some such formula $A[d_1, ..., d_n]$. Since $A[d_1, ..., d_n]$ is a ground formula with some domain elements in it, this formula can be evaluated using the interpretations f^M of the function symbols f and the interpretations P^M of the predicate symbols P. This test can be made more efficient by backtracking; sometimes we can detect that A is not satisfied by the model when only a subset of the x_i have been assigned elements of the domain. Note that finite models are useful because there are finite models of groups and rings and related mathematical structures. This means that simple models are available for theorems about groups and rings and related structures. For models with an infinite domain, the exhaustive testing approach cannot be applied; this approach can show that M does not satisfy A but not that M satisfies A when D is infinite (for the universal satisfiability test). Therefore, for such models, we need to prove that $M \models A$ in some more powerful logical system, possibly Peano arithmetic, set theory, or some system with mathematical induction but less powerful than Peano arithmetic. For example, if the domain of M is the real numbers and the functions f^M are all linear and the P^M are all either $=$, $>$, or $<$, then we can use the sup-inf procedure [Ble75] or some similar procedure for the existential satisfiability test. This test can also be generalized to the case where the functions f^M are piecewise linear. Implicit induction may be useful here because it can be mechanized to a great extent. Another way to verify models is to use mathematical knowledge; we have a considerable knowledge of the axioms that are true in various commonly used models, such as the real numbers. This knowledge can be given to the theorem prover. The task of verifying that $M \models A$ is made easier by the fact that the same models and axioms are often reused.

In order to guarantee that such common knowledge is correctly used, it can be stored in tables having three parts: The axioms A, the model M, and a proof that $M \models A$ in some sufficiently powerful logical system. Then, in order to verify that $M \models A$, it is only necessary to check the proof that $M \models A$. Even though this proof may be in a more powerful logical system, proof checking is still much easier than proof finding and it would not be difficult to write such a proof checker.

3.2 Generation of models

There are also systematic ways to generate models from other models. For example, if there is a homomorphism h from one logical system L_1 to another system L_2, and M_2 is a model of L_2, then it is often possible to construct a model M_1 of L_1 from M_2 and the homomorphism h. For this, it is necessary for h to have certain properties; for example, axioms of L_1 should map onto formula of L_2 satisfied by M_2. An interesting idea is to use abstractions as such homomorphisms; for a discussion of this, see [GW]. Another method is to use the model intersection property. For Horn sets, equational systems, and equational Horn sets, we have the *model intersection property*, which says that if $M_1 \models A$ and $M_2 \models A$ then there is a model M_3 such that $M_3 \models A$, where M_3 is defined so that if P is a predicate symbol and t_i are terms, then $M_3 \models P(t_1, ..., t_n)$ iff $M_1 \models P(t_1, ..., t_n)$ and $M_2 \models P(t_1, ..., t_n)$. We call M_3 the *intersection* of M_1 and M_2. This model intersection property is useful for generating models, and also has an implication for efficiency. It turns out that if M_1 and M_2 have finite domains D_1 and D_2, respectively, then $D_3 = D_1 \times D_2$ can be taken as the domain of M_3. Now, the size of D_3 is the product of the sizes of D_1 and D_2. However, testing if $M_3 \models P(t_1, ..., t_n)$ can be done by testing if $M_1 \models P(t_1, ..., t_n)$ and $M_2 \models P(t_1, ..., t_n)$. Thus the time to test if $M_3 \models P(t_1, ..., t_n)$ is the sum of the time to test for M_1 and M_2. This is interesting, because models with larger domains tend to give better guidance in theorem proving, since there are fewer "accidental" formulas true, but models with smaller domains have an easier satisfiability test. The use of the model intersection property allows us to get models with large domains that are easy to test. Of course, we can intersect more than two models in the same way.

3.3 Implementations of models

There is another issue, and that is the difference between an abstract model and its procedural realization. An example of an abstract model is the integers; it is an abstract mathematical object. Its procedural realization would be procedures for computing addition and subtraction and other functions and predicates, as well as a procedure to perform the existential satisfiability test. We would like to verify that the procedures related to the model are also correct. Of course, theorem provers should also be verified and usually they are not. This is partly due to the fact that it is possible to check a proof independently of the correctness of the theorem prover. Verification of the realization of a model is made easier by the fact that the same models are re-used even more than axioms are. In our semantic clause linking prover, the only models we have used so far are models with a finite domain and piecewise linear models over the reals. This has been enough to prove all the theorems of interest to us so far. For example, we can get a model of a small subset of set theory by taking the domain as the natural numbers and interpreting set intersection as minimum, set union as maximum, and the powerset operation as adding one. The empty set is interpreted as zero. Some other set operations (like set difference) also have natural interpretations. This model is even piecewise linear, so the sup-inf procedure can be used as a satisfiability test. This is enough to prove simple properties of union, intersection, complement, powerset, and related operations, though of course it is not a model of all the axioms of set theory. We are also interested in finding simple models for temporal logic. For this, we can note that it is sufficient to have a systematic way to translate a formula A into a formula B such that $M \models A$ iff $M' \models B$, where M' is easier to handle. For example, we can eliminate the temporal connectives in favor of their first-order logic definitions in terms of time, and then use a linear model. It seems that in some cases, a sorted logic might permit simpler models than an unsorted logic; this might be a good argument for using sorted logic in some cases, since a different domain can be used for each sort. Thus, one can accomplish much with a small number of models, and verification is not a major a problem. In general, however, a declarative representation of a model (as a set of equations, say) is easier to verify than a procedural one, but probably slower for computation. Some kind of compilation might help to get the advantages of both approaches.

4 Recommendations for a theorem prover

The above considerations lead to some specific recommendations for a theorem prover. The clause linking strategy in itself is not as significant as the general approach. The recommendations that we suggest are the following:

1. The theorem prover should use case analysis to deal with propositional connectives.
2. The theorem prover should not combine literals from different clauses in the same structures, or should do so only rarely.
3. The theorem prover should use semantics.
4. The inferences should be goal-sensitive, or at least, a large percentage of them should be goal-sensitive.

By case analysis we mean that the cases (L is true) and (L is false) should be considered independently, when attempting to decide if a formula involving a literal L is satisfiable. This approach is used for example in the Davis-and-Putnam satisfiability test [DP60]. For Horn clauses or sets of clauses for which unit resolution is sufficient, no case analysis is needed because unit resolution does not combine literals from different clauses. However, in general, a combination of unit resolution and case analysis can be used to handle the propositional structure of a theorem, and this is essentially the approach used by Davis and Putnam's method. Note that resolution combines literals from different clauses into a single clause, except in unit resolution, and so (non-unit) resolution violates recommendation 2.

We have implemented three provers embodying most of these recommendations: clause linking [LP89], its extension to equality [AP92], and its extension to semantics (described below), and we hope to combine the latter two eventually. The same philosophy can be extended to non-clausal theorem proving and non-refutational theorem proving, but then the concept of combining literals has to be modified. Note that the process of combining literals can happen in subtle ways; Prolog does not explicity combine literals from different clauses but it does so implicity, because the current state of the search can be expressed as a sequence of choices of literals from more than one clause in the Prolog program. The same comment applies to conditional term rewriting. For this kind of literal combination, caching can eliminate the combinatorial explosion, but only for Horn clauses.

4.1 Implicit induction

We now comment on implicit induction and its relationship to the above criteria. Actually, implicit induction in its later versions, using the ideas of linear completion and inductively complete positions, has many of these features. The method is goal sensitive, since all inferences are derived from the equation to be proved. Something like case analysis is used to consider the different instantiations of the variables in an equation to be proved; each instantiation is considered separately, and the proof attempt for one instantiation does not interfere with the proof attempt for another instantiation. For example, we may attempt to show that $\forall x P(x)$ by showing that $P(0)$ and $P(s(Y))$. These two proofs are done independently. Something like semantics is used to detect false equations, since they result in identifying distinct constructor terms in the base theory, or something similar. For pure equational theories, there is no combination of literals from different clauses. However, extensions of implicit induction to first-order logic will probably encounter the same combinatorial problems as most theorem proving strategies, unless they are specifically designed to embody the above four criteria. Even for Horn clauses or conditional term-rewriting systems, there can be a subtle combination of literals, and some kind of caching is useful to eliminate this. Finally, the use of semantics is not quite the same as using an explicity model to detect false lemmas; possibly the use of semantics in implicit induction can be strengthened.

4.2 Clause linking

We now describe clause linking (hyper-linking), which is a refutational clause-form theorem proving method for first-order logic. Suppose S is a set of clauses. The clause linking strategy performs unifications between literals of clauses in S to instantiate these clauses little by little. If $C = L_1, ..., L_m$ is a clause in S, and $M_1, ..., M_n$ are literals from clauses in S, with variables renamed to avoid conflicts, and Θ is a most general substition such that L_i and M_i are complementary for all i, then we call $C\Theta$ a *hyper-instance* of C. We call the literals M_i *electrons*. If R is a set of clauses, let $H(R)$ be the set of hyper-instances of clauses in R. Let S_0 be the set S of input clauses. Let S_1 be $S_0 \cup H(S_0)$. In general, let S_i be $S_{i-1} \cup H(S_{i-1})$. (Actually, it is enough to let S_i be $H(S_{i-1})$, but when restrictions on hyper-linking are added, such as the support strategies mentioned below, we may also need S_{i-1}.) The hyper-linking strategy computes this sequence $\{S_0, S_1, S_2, ...\}$ of sets of clauses. At each stage, S_i is a set of (instances of) clauses; no connections between clauses are remembered. If R is a set of clauses, let $Gr(R)$ be R with all variables replaced by the same constant symbol c. A propositional unsatisfiability test similar to Davis and Putnam's method is applied to the sets $Gr(S_i)$ of ground clauses. If the propositional test returns an answer of "unsatisfiable" for some such $Gr(S_i)$ then we know that the original S was unsatisfiable. This is complete for first-order logic. In [LP90a], we give experimental evidence that this prover avoids some of the propositional inefficiencies of common theorem proving strategies such as resolution and model elimination. There, we show that this prover has good performance on propositional calculus problems and logic puzzles. In [LP90b], we consider the application of this prover to logic puzzles and show how the propositional decision procedure can be used to find a model of the axioms, in some cases. We also show that this prover performs well on constraint satisfaction problems. In [LP90c], we show how this prover performs well on simple set theory problems and simple temporal logic theorems. One disadvantage of this prover is that one cannot do subsumption deletion; if a clause C subsumes clause D, it still may be necessary to retain clause D. However, we have shown that D can be deleted in this case if C is a unit clause, which is the most common case, reducing the handicap of retaining instances. Also, D can be deleted if C is a proper subset of D. In fact, if C subsumes D, D can be deleted unless C is a non-unit clause and D is an instance of C. (This was partially observed by Wayne Snyder.) Therefore, the restriction on subsumption deletion is not as severe as it first appears. Another disadvantage of clause linking is that the information transfer is slower than for resolution, since clauses are never combined into larger structures. This is partially compensated for by the fact that the hyper-linking operation unifies all of the literals of a clause with their electrons at the same time, instead of just the negative literals, as in hyper-resolution. This makes the information transfer faster since more literals are involved in the unification. Hyper-linking can be made goal-sensitive by choosing a proper support strategy; these support strategies are somewhat analogous to those for resolution. Thus hyper-linking satisfies all of the above recommendations except the third one, specifying that semantics be used.

4.3 Clause linking with semantics

As stated, clause linking performs case analysis in the propositional unsatisfiability test, avoids combining literals from different clauses, and is goal sensitive (with the proper support strategy). However, it does not use semantics. We modified clause linking to use semantics and obtained some impressive results on problems that were difficult or impossible for our previous provers, and for most other theorem provers as well. The semantics prover preserves the philosophy of reducing first-order logic to propositional calculus and applying a propositional decision procedure. It also does not combine literals from different clauses. However, instead of alternating between hyper-linking and a propositional unsatisfiability test, the two operations are combined and run in parallel. This modified method requires that in addition to a set S of clauses, a model M be specified and a procedure be given to perform an existential satisfiability test for M. Instead of using unification to instantiate the variables, as in clause linking, the existential satisfiability test for M is used. Also, this method requires that the signature contain at least one individual constant symbol. It would be interesting to see if a similar method can be devised that uses unification to instantiate the variables. The semantic hyper-linking method operates as follows:

```
T <- {};
I <- a first-order structure;
done <- false;
while (not done) do begin
        if I |= S then done <- true
        else    D <- a ground instance of a clause in S
                such that not (I |= D);
                T <- T U {D};
                if T is unsatisfiable then done <- true else
                I <- a first-order structure such that I |= T
                fi
        fi
end;
if T is unsatisfiable then [[S is unsatisfiable]] else
        [[S is satisfiable]] fi;
```

Here a structure I can be specified in any way whatever. However, typically the initial I will be a model M of the general axioms. Thus the first clause D found will be a ground instance of a clause obtained from the negation of the particular theorem. Each time a new I is chosen, I is modified in a minimal way so that I is still as close to M as possible. For two structures I and J, we can define $d(I, J)$ to be the set of atoms A such that either ($I \models A$ and $\neg(J \models A)$) or ($J \models A$ and $\neg(I \models A)$), where an atom is a predicate symbol followed by a list of terms. The model M is given initially, and at each step, a new I is chosen so that $d(I, M)$ is minimal subject to the condition that $I \models T$. (In fact, $d(I, M)$ is finite for all I chosen.) Thus, when D is chosen so that $\neg(I \models D)$, D will be related to M and T. Since all elements of T were chosen previously, and therefore are related to the theorem, D will also be related to the theorem. This prevents instances of general axioms from entering the set T unless they are related to the theorem, giving the method a goal-sensitive behavior. Also, a Davis-and-Putnam-like method is used to test unsatisfiability of T. This method therefore satisfies all of the four recommendations given above for a general theorem proving strategy.

4.4 Completeness of clause linking with semantics

We now give a sketch of a completeness proof for this method. The completeness proof is independent of the way in which the interpretations I are chosen. Suppose we have a complexity ordering $<$ on on clauses such that for a clause C, there are only finitely many clauses D such that $\neg(C < D)$. Thus, for any clause C, almost all clauses D are bigger than C in the ordering. For example, we can say that $C < D$ if C is shorter than D, written as a character string. We assume that whenever a ground instance D is chosen such that $\neg(I \models D)$, one is chosen that is minimal in the complexity ordering. With this restriction, the method is complete.

Theorem. The semantic hyper-linking method is complete, regardless of how the structures I are chosen, if the ground clauses D are chosen minimal in the complexity ordering.

Proof. The proof is fairly simple. Suppose to the contrary. Then there is an unsatisfiable set S of clauses such that the method never terminates. Thus there is an infinite sequence $I_1I_2I_3...$ of structures chosen, and an infinite sequence $D_1D_2D_3...$ of ground instances of clauses in S chosen such that $\neg(I_i \models D_i)$ and such that if $i < j$ then $I_j \models D_i$, by the way I_j is chosen. Now, let $A_1A_2A_3...$ be an enumeration of the ground atoms in the Herbrand base for S. Each interpretation I corresponds to a way of choosing truth values for the A_j. Construct an interpretation J in the following way: $J \models A_1$ if there are infinitely many i such that $I_i \models A_1$, and $J \models \neg A_1$ otherwise; in that case, there are infinitely many i such that $I_i \models \neg A_1$. In general, we choose whether $J \models A_j$ as follows: Say that two structures J_1 and J_2 are *identical for set B of atoms* if for all atoms A in B, $J_1 \models A$ iff $J_2 \models A$. We choose J so that $J \models A_j$ if there are infinitely many i such that I_i and J are identical for $A_1, A_2, ..., A_{j-1}$ and such that $I_i \models A_j$. Also, $J \models \neg A_j$ otherwise; in that case, there are infinitely many i such that I_i and J are identical for $A_1, A_2, ..., A_{j-1}$ and such that $I_i \models \neg A_j$. Now, since S is unsatisfiable, there must be a ground instance D of a clause in S, such that $\neg(J \models D)$. Suppose D contains literals $L_1...L_n$ and with the negation signs removed, these are the atoms $A_{a_1}, ..., A_{a_n}$. Let b be the maximum of the a_i. Then all atoms in D occur among $A_1...A_b$. Thus, any interpretation identical with J for these atoms will falsify D. However, by the way J is chosen, there are infinitely many such interpretations I that are identical with J for the atoms $A_1...A_b$. Call this set of interpretations $I(b)$, and note that this set is infinite. Thus there are infinitely many i (those for which $I_i \in I(b)$) such that $\neg(I_i \models D)$. For each such I, a clause D' is chosen such that $\neg(I \models D')$. Thus, there are infinitely many such D'. However, each such D' is chosen as minimal in the complexity ordering subject to the condition that $\neg(I \models D')$. There are only finitely many clauses altogether that are not strictly larger than D in this ordering. Thus, after a finite number of choices of D' for the interpretations in $I(b)$, D itself will be chosen. Thus D is D_i for some i. However, for $j > i$, $I_j \models D_i$. This implies that there are only finitely many I such that $\neg(I \models D)$. This contradicts what was said above, that there are infinitely many such I, since $I(b)$ is infinite, and if $I_i \in I(b)$ then $\neg(I_i \models D)$. This completes the proof.

Note the importance of the complexity measure in the proof. In practice, we prefer clauses that are small, although the complexity measure can be modified to weight different function symbols differently, to prefer certain term structures, and so on. This choice of a weight function gives a natural way to incorporate heuristics into semantic clause linking, although we would prefer to obtain proofs without having to be careful in the choice of a weight function. A problem with this method is how to find the clauses D given I. The general idea is to use an existential satisfiability test to instantiate the variables little by little and reject unpromising instantiations. This can be done using the fact that if a clause C is not existentially unsatisfiable in I (in other words, I satisfies the universal closure of C), then C cannot have any ground instances D such that $\neg(I \models D)$. Of course, it would be complete but inefficient just to enumerate all the ground instances of all the clauses in S and test them one by one. With this prover, we obtained a number of good results; we obtained fairly fast proofs of the intermediate value theorem, wos15, wos19 and wos20 combined, wos21, and am8. The theorems wos15, wos19, wos20, and wos21 are about groups and rings. The theorems wos19 and wos20 together state that subgroups of index two are normal. Also, am8 is a simplified form of the theorem that a continuous function attains its maximum value on a closed interval. The intermediate value theorem states that if f is continuous and $a < b$ and $f(a) < 0$ and $f(b) > 0$ then there exists an x between a and b such that $f(x) = 0$. We also obtained a proof of gcd, a theorem about properties of the greatest common divisor function, but it took about 50,000 seconds. Most of these theorems are difficult for almost all existing theorem provers. These theorems are well known in the automated deduction community; more details about the contents of these theorems can be obtained in a number of places, including [WB87] and [Lee90]. The searches for proofs were noteworthy in that little or no guidance was given except for the model. The models we have considered so far are finite models, linear models, and piecewise linear models. In general, semantic clause linking works best when the terms occurring in the desired proof are small, the clauses are non-Horn clauses, or the proofs are long. If the terms are large, the clauses are Horn clauses, and the proofs are short, then resolution and similar methods perform reasonably well compared to clause linking with semantics; examples of such problems are the implicational propositional calculus problems of [MW92].

4.5 Clause linking with equality

We also extended clause linking to equality [AP92], using Brand's modification method [Bra75] and a restricted form of term-rewriting. This combination has a number of similarities to the basic paramodulation of [BGLS91, NR92]. Brand's modification is similar to the flattening operation used in logic programming with equality. For example, a clause $P(f(x))$ would be tranformed into $\neg(f(x) = y) \vee P(y)$. In general, the transformed clauses have negative literals of the form $\neg f(x_1 ... x_n) = y$ and positive literals of the form $f(x_1 ... x_n) = y$ and literals of the form $P(x_1 ... x_n)$ for non-equality predicates, where the x_i are variables. This transformation Tr has the property that S together with the equality axioms is unsatisfiable iff $Tr(S) \cup \{x = x\}$ is unsatisfiable. We apply clause linking directly to $Tr(S) \cup \{x = x\}$, with rewriting added to reduce clauses to normal form. We also allow equations to be used as rewrite rules, if the instances used are orientable, as in unfailing completion. For this to work, we have to restrict the rewriting method in a way similar to blocked basic paramodulation [BGLS91]; for our method, this means that in a literal of the form $f(s_1 ... s_n) = t$ or of the form $\neg f(s_1 ... s_n) = t$, a rewrite cannot be applied at the top level of the term $f(s_1 ... s_n)$ except in restricted cases. This combination works reasonably well, and can obtain problems involving equality that are beyond the ability of clause linking without any special equality methods. We are currently working on adding some kind of constrained rewriting to deal with associative-commutative operators.

We are interested in combining clause linking with equality and semantics at the same time, since this would increase the effectiveness for equality problems and would at the same time permit the use of semantics, goal-sensitivity, and efficient propositional decision procedures. One way to do this is just to apply the semantic prover to the transformed set $Tr(S) \cup \{x = x\}$. However, this would not permit rewriting (demodulation) to be done. If we add rewriting, then we probably need to restrict it in some way, as in the current extension of clause linking to equality. Instead of considering all ground instances as in semantic clause linking, we may only need to consider those in which the variables are replaced by terms in normal form. This may make the search for ground instances more efficient. Also, we would like to add mathematical induction. For us, the most direct way seems to be to use the general induction schema, which is stated in terms of a predicate P and a well-founded ordering $<$ and gives conditions under which $P(x)$ is true for all x, using well-founded induction based on the ordering $<$. We would need some way to choose P and $<$. Once these are specified, and a first-order axiomatization of $<$ is given, induction is reduced to a first-order statement which our prover can handle. This approach has the advantage that semantics, goal-sensitivity, and efficient propositional decision procedures could be combined with induction. This may avoid some of the inefficiencies of current implicit and explicit induction approaches when dealing with Boolean connectives, especially for non-Horn problems.

4.6 Specialized decision procedures

It is also possible to combine semantic hyper-linking with specialized decision procedures. This seems natural, since semantic hyper-linking constructs sets of ground clauses, and specialized decision procedures typically apply most easily to ground clauses. The advantage of using specialized decision procedures is that it is not necessary to explicitly include the axioms of a specialized theory, and the search is often more efficient because methods adapted to the theory can be used instead of general first-order methods. There are a number of ways in which specialized decision procedures can be incorporated into semantic clause linking. We will only mention one, which seems fairly natural. Suppose we have a specialized theory $\mathcal{T}$. This may be axiomatized by a finite set of axioms, or it may not be. Let us define $\models_{\mathcal{T}}$ so that $\models_{\mathcal{T}} A$ iff $\mathcal{T} \models A$ and we define $\models_{\mathcal{T}}$ so that $A \models_{\mathcal{T}} B$ iff for all first-order structures I satisfying $\mathcal{T}$, if $I \models A$ then $I \models B$. Then, we can adapt semantic hyper-linking to specialized decision procedures by replacing $\models$ everywhere by $\models_{\mathcal{T}}$. This includes replacing the test for satisfiability of T by a test for $\mathcal{T}$-satisfiability of T. However, it's not clear how to retain goal-sensitivity with this approach, since it's not easy to see in general how to obtain a structure I satisfying $\mathcal{T}$ such that $d(I, M)$ is minimal and $I \models \mathcal{T}$. For finitely axiomatizable theories, the axioms may have to be used to construct M. Also, it may be possible to develop methods specialized to various theories for the purpose of generating such an I and preserving goal-sensitivity.

5 Conditional term rewriting

We now consider the application of these ideas to conditional term-rewriting systems. We note that conditional term-rewriting systems are like Horn clause, and the recursive evaluation of conditions by rewriting is similar to Prolog's depth-first search. For example, the Prolog clause p :- q,r can be translated into the conditional rewrite rule $p \rightarrow true$ if $q = true \wedge r = true$. Therefore the same propositional inefficiencies can occur for conditional term-rewriting as for Prolog and resolution. To avoid this, some kind of caching is needed, so that the same rewrites are not done over and over again. For this, we also need to cache failures as well as successes in rewriting, since it can take a lot of work in a conditional system just to detect that a term is in normal form.

We also would like to use semantics. This seems most relevant for conditional narrowing, since there may be many ways a variable can be instantiated and semantics can be used to reject bad instantiations (that is, instantiations that cannot lead to a proof). Suppose we are rewriting a term t using a conditional rewrite rule of the form $r \rightarrow s$ if C where C is a condition involving further equalities. Suppose r unifies with a subterm of t with Θ as a most general unifier. Suppose M is a model of the conditional rewrite rules and equality axioms. Then if $\neg(M \models (r\Theta = s\Theta))$, we know that this attempted rewrite will fail, and we may save some work this way. Similarly, if we are attempting to prove a theorem $u = v$ by conditional narrowing, we can reject a potential instantiation Θ if $\neg(M \models (u\Theta = v\Theta))$.

Note that there may be a number of active subgoals at the same time. Each equation in a condition $C\Theta$ as above can be considered as a subgoal that must be satisfied before the rewrite can occur. We satisfy an equation by showing that the left and right-hand sides are equal using term-rewriting. For example, if we are attempting to show that $u = v$ and we use a conditional rule $r \rightarrow S$ if C, then all the (instantiated) equations in C become subgoals. Suppose $r' = s'$ is one of these equations in C. When rewriting r', another conditional rule may be used; it's conditions then become additional subgoals that must be solved. We may generate subgoals using a number of conditional rewrite rules recursively, and all of these conditions may be active at the same time. Suppose $e_1, ..., e_k$ are the active conditions. Then we can reject the current rewrite unless $M \models e_1 \wedge ... \wedge e_k$. For example, the conditions $x = 1$ and $x = 2$ are satisfiable separately but not jointly.

This approach requires us to combine more than one condition in order to test whether $M \models e_1 \wedge ... \wedge e_k$. Unfortunately, this kind of combination is what leads to search inefficiency of theorem proving methods. Caching does not avoid this problem, since a given rewrite can be invoked by many different paths of subgoaling, and there can be many combinations of e_i to consider for a given term r. We don't have a full answer; an obvious possibility is just to test the current subgoal independently of the earlier ones, and have a weaker test. Another possibility is to instantiate each subgoal to a ground subgoal as it is generated, and then perform the semantics test. This of course requires us to examine many possible ground instantiations of each subgoal. Probably only normalized substitions need to be considered, which may help some. Also, the model may help to guide the instantiation. This is essentially the method used in semantic clause linking (with some refinements). However, the process of generating these ground instantiations without using unification may be expensive, especially for large terms.

Another idea is to use a general first-order theorem prover instead of conditional term-rewriting to deal with conditional systems. This will work, but has the disadvantage that a general first-order theorem prover cannot always detect non-theorems. With a confluent conditional term-rewriting system, one can detect non-theorems as well as theorems.

6 Existing applications of semantics

We now consider some other existing applications of these ideas. The geometry theorem prover of Gelernter [GHL63] is an early use of semantics. Bledsoe [Ble77] recognized the general importance of semantics in theorem proving a number of years ago. The recent paper [Pro92] shows how something very much like semantics can increase the efficiency of an implicit induction theorem prover by detecting that certain proposed induction hypotheses are false. The use of constraints [Com90, KKR90] can be considered as using semantics since the constraints are evaluated in a fixed model. Another idea is to detect true subgoals; such subgoals may not need to be proved. For example, we may use a specialized decision procedure to detect true statements about arithmetic. In this case, the correctness of the model is necessary to insure the correctness of the proof. Avenhaus and Becker [AB92] use semantics to enable the use of efficient representations of integers and for the construction of simple termination orderings for term-rewriting systems. In the system of Bachmair and Ganzinger [BG90], certain paramodulations are only necessary if a certain equation is true in the well-founded model of the axioms; a system for testing this might reduce the search space. There have of course been many other applications of semantics which we cannot mention here.

References

[AB92] J. Avenhaus and K. Becker. Conditional rewriting modulo a built-in algebra. Technical Report SEKI-Report SR 92-11, Universitaet Kaiserslautern, March 1992.

[AP92] G. Alexander and D. Plaisted. Proving equality theorems with hyper-linking. In *Proceedings of the 11th International Conference on Automated Deduction*, pages 706–710, 1992. system abstract.

[AS92] Owen Astrachan and M. Stickel. Caching and lemma use in model elimination theorem provers. In D. Kapur, editor, *Proceedings of the Eleventh International Conference on Automated Deduction*, 1992.

[BDP89] Leo Bachmair, N. Dershowitz, and D. Plaisted. Completion without failure. In Hassan Ait-Kaci and Maurice Nivat, editors, *Resolution of Equations in Algebraic Structures 2: Rewriting Techniques*, pages 1–30, New York, 1989. Academic Press.

[BG90] Leo Bachmair and Harold Ganzinger. On restrictions of ordered paramodulation with simplification. In Mark Stickel, editor, *Proceedings of the 10th International Conference on Automated Deduction*, pages 427–441, New York, 1990. Springer-Verlag.

[BGLS91] L. Bachmair, H. Ganzinger, C. Lynch, and W. Snyder. Basic paramodulation and basic strict superposition. submitted, 1991.

[Ble75] W. W. Bledsoe. A new method for proving certain presburger formulas. In *Proc. of the 3 rd IJCAI*, pages 15–21, Stanford, CA, 1975.

[Ble77] W. W. Bledsoe. Non-resolution theorem proving. *Artificial Intelligence*, 9:1–35, 1977.

[Bra75] D. Brand. Proving theorems with the modification method. *SIAM J. Comput.*, 4:412–430, 1975.

[Com90] H. Comon. Solving inequations in term algebras. In *Proceedings of 5th IEEE Symposium on Logic in Computer Science*, pages 62–69, 1990.

[DP60] M. Davis and H. Putnam. A computing procedure for quantification theory. *Journal of the Association for Computing Machinery*, 7:201–215, 1960.

[GHL63] H. Gelernter, J.R. Hansen, and D.W. Loveland. Empirical explorations of the geometry theorem proving machine. In E. Feigenbaum and J. Feldman, editors, *Computers and Thought*, pages 153–167. McGraw-Hill, New York, 1963.

[GW] F. Giunchiglia and T. Walsh. A theory of abstraction. *Artificial Intelligence.* to appear.

[Hak85] A. Haken. The intractability of resolution. *Theoretical Computer Science*, 39:297–308, 1985.

[HR91] J. Hsiang and M Rusinowitch. Proving refutational completeness of theorem-proving strategies: the transfinite semantic tree method. *J. Assoc. Comput. Mach.*, 38(3):559–587, July 1991.

[KB70] D.E. Knuth and P.B. Bendix. Simple word problems in universal algebras. In *Computational Problems in Abstract Algebra*, pages 263–297. Pergamon, Oxford, U.K., 1970.

[KKR90] C. Kirchner, H. Kirchner, and M. Rusinowitch. Deduction with symbolic constraints. *Revue Francaise d'Intelligence Artificielle*, 4(3):9 – 52, 1990.

[Lee90] S.-J. Lee. *CLIN: An Automated Reasoning System Using Clause Linking.* PhD thesis, University of North Carolina at Chapel Hill, 1990.

[LP89] S.-J. Lee and D. Plaisted. Theorem proving using hyper-matching strategy. In *Proceedings of the 4th International Symposium on Methodologies for Intelligent Systems*, pages 467–476, 1989.

[LP90a] S.-J. Lee and D. Plaisted. Eliminating duplication with the hyper-linking strategy. *Journal of Automated Reasoning*, 1990. to appear.

[LP90b] S.-J. Lee and D. Plaisted. New applications of a fast propositional calculus decision procedure. In *Proceedings of the 8th Biennial Conference of Canadian Society for Computational Studies of Intelligence*, pages 204–211, 1990.

[LP90c] S.-J. Lee and D. Plaisted. Reasoning with predicate replacement. In *Proceedings of the 5th International Symposium on Methodologies for Intelligent Systems*, 1990.

[MW92] W. McCune and L. Wos. Experiments in automated deduction with condensed detachment. In *Proceedings of the 11th International Conference on Automated Deduction*, pages 209–223, July 1992.

[NR92] R. Nieuwenhuis and A. Rubio. Theorem proving with ordering constrained clauses. In *Proceedings of the 11th International Conference on Automated Deduction*, pages 477–491, July 1992.

[Pla82] D. Plaisted. A simplified problem reduction format. *Artificial Intelligence*, 18:227–261, 1982.

[Pla88] D. Plaisted. Non-Horn clause logic programming without contrapositives. *Journal of Automated Reasoning*, 4:287–325, 1988.

[Pla92] D. Plaisted. The search efficiency of theorem proving strategies. Draft, 1992.

[Pro92] M. Protzen. Disproving conjectures. In *Proceedings of the 11th International Conference on Automated Deduction*, pages 340–354, July 1992.

[WB87] T.C. Wang and W.W. Bledsoe. Hierarchical deduction. *Journal of Automated Reasoning*, 3:35–77, 1987.

Decidability of Regularity and Related Properties of Ground Normal Form Languages

Gregory KUCHEROV

Centre de Recherche en Informatique de Nancy
CNRS and INRIA-Lorraine
Campus Scientifique, BP 239
F-54506 Vandœuvre-lès-Nancy
France
email: kucherov@loria.fr

Mohamed TAJINE

Université Louis Pasteur
Centre de Recherche en Informatique
7, rue René Descartes
F-67084 Strasbourg
France
email: tajine@dpt-info.u-strasbg.fr

Abstract

We study language-theoretical properties of the set of reducible ground terms with respect to a given rewriting system. As a tool for our analysis we introduce the property of finite irreducibility of a term with respect to a variable and prove it to be decidable. It turns out that this property generalizes numerous interesting properties of the language of ground normal forms. In particular, we show that testing regularity of this language can be reduced to checking this property. In this way we prove the decidability of the regularity of the set of ground normal forms, the problem mentioned in the list of open problems in rewriting [2]. Also, the decidability of the existence of an equivalent ground term rewriting system and some other results are proved.

1 Introduction

Although the term rewriting formalism has been studied for many years, very little is known about language-theoretical properties of term sets induced by term rewriting systems. In particular, in this paper we focus our interest on the properties of the set of reducible ground terms $Red(\mathcal{R})$ and the set of ground normal forms $NF(\mathcal{R})$, where $\mathcal{R}$ is an ordinary (non-conditional, non-equational) term rewriting system. It turns out that these sets, regarded as tree languages [4], have a very interesting structure that is worth to be studying. In particular, it is useful to relate this structure with classical notions from the formal language theory. Such results would not only be of theoretical importance but could also have a practical interest in the application domains of term rewriting systems.

As a tool for our analysis in this paper we introduce a property of finite irreducibility of a term by a given term rewriting system $\mathcal{R}$ with respect to a variable. It means that if we consider the set of all irreducible ground instances of the term, the set of different ground terms substituted for the variable is finite. In this paper we show that we can effectively bound the depth of these ground terms. In other words, given a term t and a variable x, we give a bound depending on $\mathcal{R}$, t, and x such that if t is finitely irreducible with respect to x and $\delta(t)$ is an irreducible ground instance of t, then the depth of $\delta(x)$ is smaller than this bound. Using the bound we reduce the property of finite irreducibility to that of ground reducibility which is known to be decidable [9, 13, 1]. In this way we show the decidability of the finite irreducibility property.

It turns out that this property is closely related to many interesting properties of the language of ground normal forms (or its complement, the set of reducible ground terms). In particular, we show that for a rewriting system $\mathcal{R}$, if the set of reducible ground terms $Red(\mathcal{R})$ is regular then every non-linear term t in $\mathcal{R}$ is finitely irreducible by $\mathcal{R} \setminus \{t\}$ with respect to all its non-linear variables.[1] On the other hand, we show that the latter condition is implied by another one, namely the existence of a left-linear term rewriting system $\mathcal{L}$ such that $Red(\mathcal{R}) = Red(\mathcal{L})$. This "linearizability property" was studied by one of the authors in [12] and proved equivalent to the regularity of $Red(\mathcal{R})$. In this paper we give a shorter proof of this result that uses a well-known Ramsey's theorem. Combining these results with the decidability of finite irreducibility, we prove the decidability of the regularity of ground normal form languages. This solves the problem 7 of the list of open problems in rewriting [2] (see also [5]).

Using similar ideas we show some other decidability results. In particular, we prove that it is decidable whether a term rewriting system has an equivalent ground term rewriting system with the same set of reducible ground terms. The difference with the previous problem is that *all* variables and not only the non-linear ones must be substituted by finite sets of ground terms. Also, we show how the results obtained imply the decidability of finiteness of the set of ground normal form, the result proved previously in [9, 13].

2 Preliminaries

We use standard basic notions of term rewriting system theory. $T_\Sigma(X)$ stands for the set of (finite, first-order) terms over a finite signature Σ and an enumerable set of variables X. $\mathcal{V}ar(t) \subset X$ is the set of variables in $t \in T_\Sigma(X)$. T_Σ denotes the set of ground terms over Σ that will also be naturally treated as finite labeled trees. For $t \in T_\Sigma(X)$, $\mathcal{P}os(t)$ denotes the set of positions in t defined in the usual way as sequences of natural numbers and $\mathcal{VP}os(t) = \{\pi \in \mathcal{P}os(t) | t|\pi \in X\}$ is the set of variable positions in t. By ε we denote the empty sequence that corresponds to the

[1] In this paper we identify a term rewriting system with the set of left-hand sides of the rules

root position. For $\pi_1, \pi_2 \in \mathcal{P}os(t)$, $\pi_1 \succeq \pi_2$ iff π_2 is a prefix of π_1, and $\pi_1 \succ \pi_2$ iff $\pi_1 \succeq \pi_2$ and $\pi_2 \not\succeq \pi_1$. $\pi \cdot \tau$ denotes the concatenation of π and τ. As usual, for $\pi \in \mathcal{P}os(t)$, $t|_\pi$ is a subterm of t at π and $t[\pi \leftarrow s]$ is the result of replacement of $t|_\pi$ by s in t.

A variable $x \in \mathcal{V}ar(t)$ is said to be linear in t if there exists only one position $\pi \in \mathcal{P}os(t)$ such that $t|_\pi = x$ and is said to be non-linear in t otherwise. A term $t \in T_\Sigma(X)$ is linear if all its variables are linear and is non-linear otherwise.

Given $\tau \in \mathcal{P}os(t)$, $|\tau|$ stands for the length of τ. For $t \in T_\Sigma(X)$ and $S \subseteq T_\Sigma(X)$, the depth of t and S, denoted by $\|t\|$ and $\|S\|$, is defined by $\|t\| = max\{|\tau| | \tau \in \mathcal{P}os(t)\}$ and $\|S\| = max\{\|t\| \,| t \in S\}$. Substitutions and ground substitutions are defined in the usual way.

We will deal with ordinary term rewriting systems defined as a finite set of rules $t \to s$, where $t, s \in T_\Sigma(X)$ and $\mathcal{V}ar(s) \subseteq \mathcal{V}ar(t)$. The only property of term rewriting systems we will be concerned with in this paper is their "reduction power", that is the set of reducible ground terms. For a rewriting system $\mathcal{R}$, a ground term $g \in T_\Sigma$ is ($\mathcal{R}$-)reducible if there exist $\pi \in \mathcal{P}os(t) \setminus \mathcal{VP}os(t)$ and a rule $t \to s$ in $\mathcal{R}$ such that $g|_\pi$ is a ground instance of t, that is $g = \delta(t)$ for some ground substitution δ. Thus, we will identify a term rewriting system $\mathcal{R}$ with the set of its left-hand sides and we will freely mix up term rewriting systems and finite term sets. For example, we will say "linear rewriting system" or simply "linear set" instead of "left-linear rewriting system".

Given a term rewriting system $\mathcal{R}$, a term $t \in T_\Sigma(X)$ is called ($\mathcal{R}$-)ground reducible (or ground reducible by $\mathcal{R}$) iff for each ground substitution δ, $\delta(t)$ is ($\mathcal{R}$-)reducible. By $Gr(\mathcal{R})$ we denote the set of ground instances of $\mathcal{R}$ and by $Red(\mathcal{R})$ the set of $\mathcal{R}$-reducible ground terms. Thus, t is $\mathcal{R}$-ground reducible iff $Gr(\{t\}) \subseteq Red(\mathcal{R})$. $NF(\mathcal{R})$ stands for the set of $\mathcal{R}$-irreducible ground terms (ground normal forms), i.e. $NF(\mathcal{R}) = T_\Sigma \setminus Red(\mathcal{R})$.

3 Finite Irreducibility, Regularity, and Linearity

A well-known problem related to the properties of $Red(\mathcal{R})$ is the ground reducibility problem that was proved decidable by several authors in the mid eighties. It consists in testing whether all ground instances of a given term t are reducible by a rewriting system $\mathcal{R}$, or, formally, if $Gr(\{t\}) \subseteq Red(\mathcal{R})$ holds. From the language-theoretical point of view the ground reducibility problem is equivalent to the inclusion problem for the languages $Red(\mathcal{R})$.

Theorem 1 (generalized ground-reducibility problem) *[9, 13, 1] It is decidable whether $Red(\mathcal{R}_1) \subseteq Red(\mathcal{R}_2)$ for given arbitrary term rewriting systems $\mathcal{R}_1, \mathcal{R}_2$.*

Another known result is the decidability of finiteness of $NF(\mathcal{R})$. Note that $Red(\mathcal{R})$ is always infinite provided that $\mathcal{R}$ is not empty and T_Σ is infinite.

Theorem 2 *[9, 10] For an arbitrary rewriting system $\mathcal{R}$, it is decidable whether $NF(\mathcal{R})$ is finite.*

In this paper we are concerned with the regularity property of $Red(\mathcal{R})$ (equivalently, $NF(\mathcal{R})$), where $Red(\mathcal{R})$ is regarded as a tree language. More specifically, our objective is to prove the decidability of regularity of $Red(\mathcal{R})$. First we recall the definitions of finite tree automaton and regular tree language. The following definition of tree automaton follows [4].

Definition 1 • *Given a signature Σ, a bottom-up tree automaton $\mathcal{A}$ is a finite Σ-algebra $A = (Q, \Sigma)$, where elements of the finite carrier Q are called* states, *together with a distinguished subset $Q_{fin} \subset Q$ of* final states.

- *The language of ground terms $L \subseteq T_\Sigma$* recognized by $\mathcal{A}$ *is defined by $L = \{t \in T_\Sigma | t^A \in Q_{fin}\}$, where t^A denotes the interpretation of t in the algebra A. In what follows we denote t^A by $\mathcal{A}(t)$.*

- *A language $L \subseteq T_\Sigma$ is called* regular *(or recognizable) iff there exists an automaton $\mathcal{A}$ over Σ that recognizes L.*

In this section we aim to express the regularity property of $Red(\mathcal{R})$ in terms of other "more syntactic" properties that are easier to test. In particular, we introduce a property of *finite irreducibility* of a term with respect to a variable.

Definition 2 *Let $\mathcal{R}$ be a rewriting system and $t \in T_\Sigma(X)$. t is said to be* infinitely irreducible by $\mathcal{R}$ with respect to a variable $x \in \mathcal{V}ar(t)$ *iff there exists an infinite sequence of distinct $\mathcal{R}$-irreducible ground instances $\{\delta_1(t), \delta_2(t), \delta_3(t), \ldots\} \subseteq NF(\mathcal{R})$ such that*

- *for every $i \geq 1$, $\delta_i(t)$ contains no proper subterm that is an instance of t,*
- *the set $\{\delta_1(x), \delta_2(x), \delta_3(x), \ldots\}$ is infinite.*

Otherwise t is called finitely irreducible by $\mathcal{R}$ with respect to x.

A property similar to infinite irreducibility was called transnormality in [11]. The first condition in the definition can be interpreted by treating t as a special rewrite rule that can be applied to every position of $\sigma(t)$ but the root position. To give a simple example, suppose $\Sigma = \{f, h, a\}$, $\mathcal{R} = \{h(x)\}$, and $t = f(y, y)$. Then t is finitely irreducible with respect to y, but this would not be the case if the first condition had been dropped. It should be noted that this condition is purely technical, and the results of this paper concerning the analysis of finite irreducibility property (cf section 4) are valid both with and without this condition. In fact, the condition is natural for some application but is not needed for the others (cf section 5). In the sequel we will assume this condition to be present and we will mention explicitly when it is not taken into account.

Now we relate the property of regularity of the set of reducible ground terms to that of finite irreducibility. It is the latter property that we will actually test afterwards. The results of the rest of the section are based on the results proved by one of the authors in [12] without using explicitly the finite irreducibility property. Besides, we present here another proof of the main proposition (theorem 4 below) that uses a more general technique.[2] In particular, the well-known theorem of Ramsey is used.

We need the following technical lemma.

[2]We find the difference important, although it concerns the technical part of the proof and not the main idea. The full proofs of [12] can be found in *Kucherov, G.A., On relationship between term rewriting systems and regular tree languages, Rapport de Recherche, 1273, INRIA, 1990.*

Lemma 1 *Let $t \in T_\Sigma(X)$. Let $g_1, g_2, g_3 \in T_\Sigma$, and π be a position that belongs to $\mathcal{P}os(g_1), \mathcal{P}os(g_2)$, and $\mathcal{P}os(g_3)$. If $g_1[\pi \leftarrow g_2|_\pi]$, $g_1[\pi \leftarrow g_3|_\pi]$, $g_2[\pi \leftarrow g_3|_\pi]$ are instances of t, then g_2 is also an instance of t.*

Proof: We proceed by case analysis of the location of π with respect to t.

Suppose that π is "outside" t, that is there exists $\tau \in \mathcal{VP}os(t)$ such that $\tau \preceq \pi$. Assume that $t|_\tau = x$ and $\pi = \tau \cdot \nu$. If x is linear in t, then since $g_2[\pi \leftarrow g_3|_\pi]$ is an instance of t, then g_2 is also an instance of t. If x is non-linear in t, then since $g_1[\pi \leftarrow g_2|_\pi]$ and $g_1[\pi \leftarrow g_3|_\pi]$ are both instances of t, we conclude that $g_1[\pi \leftarrow g_2|_\pi]|_\tau = g_1[\pi \leftarrow g_3|_\pi]|_\tau$ and therefore $g_2|_\pi = g_3|_\pi$. As soon as $g_2[\pi \leftarrow g_3|_\pi]$ is an instance of t, g_2 is also an instance of t.

Now suppose that π is "within" t, that is $\pi \in \mathcal{P}os(t) \setminus \mathcal{VP}os(t)$. Since $g_1[\pi \leftarrow g_2|_\pi]$ is an instance of t, then $g_2|_\pi$ is an instance of $t|_\pi$. Denote $\mathcal{VP}os_\pi(t) = \{\tau | \tau \in \mathcal{VP}os(t),\ \pi \prec \tau\}$. If there is no non-linear variable $x \in \mathcal{V}ar(t)$ located at a position in $\mathcal{VP}os_\pi(t)$ as well as at a position in $\mathcal{VP}os(t) \setminus \mathcal{VP}os_\pi(t)$, then since $g_2[\pi \leftarrow g_3|_\pi]$ is an instance of t, g_2 is also an instance of t. Now assume that for some variable $x \in \mathcal{V}ar(t)$, there exist $\tau_1 \in \mathcal{VP}os_\pi(t), \tau_2 \in \mathcal{VP}os(t) \setminus \mathcal{VP}os_\pi(t)$ such that $t|_{\tau_1} = t|_{\tau_2} = x$. We prove that $g_2|_{\tau_1} = g_2|_{\tau_2}$. Since $g_1[\pi \leftarrow g_2|_\pi]$ and $g_1[\pi \leftarrow g_3|_\pi]$ both are instances of t, we conclude that $g_1|_{\tau_2} = g_2|_{\tau_1}$ and $g_1|_{\tau_2} = g_3|_{\tau_1}$. Hence, $g_2|_{\tau_1} = g_3|_{\tau_1}$. Since $g_2[\pi \leftarrow g_3|_\pi]$ is an instance of t, then $g_2|_{\tau_2} = g_3|_{\tau_1}$ and hence $g_2|_{\tau_1} = g_2|_{\tau_2}$. Since x, τ_1, τ_2 were chosen arbitrary, we conclude that g_2 is an instance of t. □

The lemma above is a generalization of lemma A.7 from [9]. It plays an important role as it gives a link between reasoning on terms and combinatorial reasoning, the latter being necessary for proving the results of this paper.

In the combinatorial part of the proof below we use the following "infinite version" of the well-known Ramsey theorem (see [6, page 16]).

Theorem 3 (Ramsey's theorem, infinite version) *Let I be an infinite set and n a natural number. Denote by $\mathcal{P}_n(I)$ the set of all n-element subsets of I and assume that $\mathcal{P}_n(I) = P_1 \uplus P_2$. Then there exists an infinite subset $J \subset I$ such that either $\mathcal{P}_n(J) \subseteq P_1$ or $\mathcal{P}_n(J) \subseteq P_2$.*

Now we are in position to prove the main result of this section.

Theorem 4 *For a rewriting system $\mathcal{R}$, if the set $Red(\mathcal{R})$ is a regular tree language, then every non-linear term $t \in \mathcal{R}$ is finitely irreducible by $\mathcal{R}\setminus\{t\}$ with respect to all its non-linear variables $x \in \mathcal{V}ar(t)$.*

Proof: By contradiction, assume that there exists a non-linear term $t \in \mathcal{R}$ and a non-linear variable $x \in \mathcal{V}ar(t)$ such that t is infinitely irreducible by $\mathcal{R}\setminus\{t\}$ with respect to x. By definition, there exists an infinite set of ground substitutions $\{\delta_1, \delta_2, \ldots\}$ such that for every i, $i \geq 1$, $\delta_i(t)$ is irreducible by $\mathcal{R}\setminus\{t\}$ and does not contain an instance of t as a proper subterm, and the set $\{\delta_1(x), \delta_2(x), \ldots\}$ is infinite. Without loss of generality we assume that for $i, j \geq 1$, $i \neq j$, $\delta_i(x) \neq \delta_j(x)$.

Assume now that $\mathcal{A}$ is an automaton that recognizes $Red(\mathcal{R})$. Since the set of states Q of $\mathcal{A}$ is finite, and $\{\delta_1(x), \delta_2(x), \ldots\}$ is infinite, assume without loss of generality that there exists $q_0 \in Q$ such that $\mathcal{A}(\delta_i(x)) = q_0$ for all i, $i \geq 1$.

Let π be a position of x in t. Denote $g_{ij} = \delta_i(t)[\pi \leftarrow \delta_j(x)]$. Since $\mathcal{A}(\delta_i(x)) = \mathcal{A}(\delta_j(x))$, then $\mathcal{A}(\delta_i(t)) = \mathcal{A}(g_{ij})$ and thus g_{ij} must be reducible. We now contradict this by proving that there exist $i, j \geq 1$, $i \neq j$ such that g_{ij} is not reducible. Actually, the statement we will prove is much stronger. We show that there exists an infinite subset of indexes $\bar{J}$ such that g_{ij} is not reducible for every $i, j \in \bar{J}$, $i < j$.

We observe that since $\delta_i(x) \neq \delta_j(x)$ for $i \neq j$, and x is non-linear in t, then g_{ij} is not an instance of t. Also, g_{ij} can be potentially reducible only at a position above π. Consider a position τ, $\tau \prec \pi$ and a rule $s \in \mathcal{R}$. We assume that $s \neq t$ whenever $\tau = \varepsilon$. It follows from lemma 1 that if for distinct i_1, i_2, i_3 the terms $g_{i_1 i_2}, g_{i_1 i_3}, g_{i_2 i_3}$ are reducible by s at τ, then $g_{i_2 i_2} = \delta_{i_2}(t)$ is also reducible by s at τ and this would contradict the assumption that every $\delta_i(t)$ is reducible only by t at root. Therefore, given s and τ as above, among every three pairs $(i_1, i_2), (i_1, i_3), (i_2, i_3)$ there exists at least one such for which the corresponding term g_{ij} is not reducible by s at τ.

Now we apply theorem 3 with $n = 2$ and I being the set of natural numbers. We identify uniquely every pair (i, j), $i < j$ with the 2-element subset $\{i, j\}$, and we split the set of all such pairs into two subsets P_1, P_2 in the following way. A pair (i, j) belongs to P_2 iff g_{ij} is reducible by s at τ, otherwise it belongs to P_1. By theorem 3 there exists an infinite set of indexes J such that either $\mathcal{P}_2(J) \subseteq P_1$ or $\mathcal{P}_2(J) \subseteq P_2$. By the above remark, the latter alternative is impossible even for a 3-element set J. Thus, we get an infinite set of indexes J such that g_{ij} is irreducible by s at τ for all $i, j \in J, i < j$. Since the number of rules in $\mathcal{R}$ and the number of positions in t are finite, by applying theorem 3 iteratively for every $s \in \mathcal{R}, \tau \in \mathcal{P}os(t)$ (except for $s = t, \tau = \varepsilon$), we finally get an infinite set of indexes $\bar{J}$ such that g_{ij} is $\mathcal{R}$-irreducible for all $i, j \in \bar{J}$, $i < j$. Thus, we get a contradiction with the fact that every g_{ij} belongs to the language recognized by $\mathcal{A}$. □

Now we show that for a given rewrite system $\mathcal{R}$, the language $Red(\mathcal{R})$ is regular if and only if there exists a linear rewriting system $\mathcal{L}$ such that $Red(\mathcal{R}) = Red(\mathcal{L})$. The "if part" follows immediately from the following theorem proved in [3].

Theorem 5 ([3]) *If $\mathcal{L}$ is a linear rewriting system, then $Red(\mathcal{L})$ is a regular tree language.*

Thus, we concentrate on proving the existence of a linear system $\mathcal{L}$ equivalent to $\mathcal{R}$ provided that $Red(\mathcal{R})$ is regular. Moreover, we show that $\mathcal{L}$ may be constructed by instantiating the non-linear variables of $\mathcal{R}$.

Definition 3 *Given finite sets $\mathcal{R} \subseteq T_\Sigma(X)$, $\mathcal{L} \subseteq T_\Sigma(X)$, $\mathcal{L}$ is said to be an* instantiation *of $\mathcal{R}$ iff every term of $\mathcal{L}$ is an instance of some term of $\mathcal{R}$. We call $\mathcal{L}$* linear (resp. ground,...) instantiation *of $\mathcal{R}$ iff $\mathcal{L}$ is a linear (resp. ground,...) set in addition.*

In other words, $\mathcal{L}$ is an instantiation of $\mathcal{R}$ iff $\mathcal{L}$ can be obtained by instantiation or deletion of terms of $\mathcal{R}$. Furthermore, if $\mathcal{L}$ is a linear instantiation of $\mathcal{R}$ then every non-linear term in $\mathcal{R}$ either is deleted or has its non-linear variables substituted by a finite number of ground terms. The following lemma relates the condition of finite irreducibility and the existence of an equivalent linear instantiation.

Lemma 2 *For a rewriting system $\mathcal{R}$, there exists a linear instantiation $\mathcal{L}$ of $\mathcal{R}$ such that $Red(\mathcal{R}) = Red(\mathcal{L})$ then every non-linear term $t \in \mathcal{R}$ is finitely irreducible by $\mathcal{R}\backslash\{t\}$ with respect to all its non-linear variables $x \in \mathcal{V}ar(t)$.*

Proof: By contradiction, if a non-linear term $t \in \mathcal{R}$ is infinitely irreducible by $\mathcal{R}\backslash\{t\}$ with respect to some non-linear variable $x \in \mathcal{V}ar(t)$, then there is an infinite sequence

$\{\delta_1(t), \delta_2(t), \delta_3(t), \ldots\}$ of ground terms reducible only by t and only at the root position, and the set $\{\delta_1(x), \delta_2(x), \delta_3(x), \ldots\}$ is infinite. Clearly, if we replaced x by any finite set of ground terms, infinitely many terms from $\{\delta_1(t), \delta_2(t), \delta_3(t), \ldots\}$ would become irreducible, and thus the set of reducible ground terms would be changed. Also, t cannot be deleted from $\mathcal{R}$. Thus, an equivalent linear instantiation does not exist and we have a contradiction. □

We note that the converse of the lemma above does not hold [8]. The reason is that the linearizability condition is a property of whole rewriting systems that cannot be decomposed into a sum of local conditions imposed on each of their terms. For example, suppose $\mathcal{R} = \{h(f(x,y)), f(f(x,y),z), f(x,f(y,z)), f(x,x), f(h(x),h(x))\}$, and the signature consists of f, h, and a constant a. It is easy to see that $Red(\mathcal{R}) = T_\Sigma \setminus NF(\mathcal{R})$, where $NF(\mathcal{R}) = \{f(h^i(a), f(h^j(a)) | i \neq j\}$. The term $f(x,x)$ can be replaced by $f(a,a)$ without changing $Red(\mathcal{R})$. Also, $f(h(x),h(x))$ can be simply dropped since it is subsumed by $f(x,x)$. However, both terms cannot be instantiated simultaneously, and thus the system cannot be linearized as a whole.

Thus, linearizability of a system is a stronger condition than finite irreducibility of its elements with respect to the non-linear variables. According to theorem 4, the latter condition is also implied by the regularity of the language of reducible ground terms. The following theorem proves equivalent the regularity and the linearizability.

Theorem 6 ([12]) *For a rewriting system $\mathcal{R}$, $Red(\mathcal{R})$ is a regular tree language iff there exists a linear instantiation $\mathcal{L}$ of $\mathcal{R}$ such that $Red(\mathcal{R}) = Red(\mathcal{L})$.*

Proof: The "if part" follows immediately from theorem 5. Now suppose that $Red(\mathcal{R})$ is regular. Take a non-linear term $t \in \mathcal{R}$ and a non-linear variable $x \in \mathcal{V}ar(t)$. By theorem 4, t is finitely irreducible by $\mathcal{R} \setminus \{t\}$ with respect to x. By definition 2 we can transform t into $\sigma_1(t), \ldots, \sigma_n(t)$ by replacing x by a finite number of ground terms such that if $\delta(t)$ is a ground instance of t and no instance of t is a proper subterm of $\delta(t)$, then $\delta(t)$ is either reducible by $\mathcal{R} \setminus \{t\}$ or is an instance of $\sigma_i(t)$ for some $i, 1 \leq i \leq n$. Hence, it is easy to see that replacing t by $\sigma_1(t), \ldots, \sigma_n(t)$ does not affect the set of reducible ground terms. On the other hand, the transformation eliminates one non-linear variable. By iterating this transformation for each non-linear term and each non-linear variable we obtain a linear system $\mathcal{L}$ that satisfies the theorem. □

Finally, we note that if we have a procedure for testing finite irreducibility and computing a corresponding set of replacement terms, then the above theorem gives an effective procedure for testing the existence of $\mathcal{L}$ and computing it. Checking finite irreducibility is the subject of the following section.

4 Substitution Bound and Main Result

In this section we prove the decidability of the property of finite irreducibility with respect to a variable. Given a rewriting system $\mathcal{R}$, a term t, and a variable $x \in \mathcal{V}ar(t)$, we give a bound on the depth of $\delta(x)$, where $\delta(t)$ is irreducible and t is finitely irreducible with respect to x. Using this bound we reduce the property of finite irreducibility to that of ground reducibility that is known to be decidable.

The technique that we use to construct the bound is similar to that of [9]. The idea is to give a bound such that if $\delta(x)$ exceeds the bound, then a larger substitution

σ can be constructed such that $\sigma(t)$ is irreducible. Obviously, in this way we get an infinite number of such substitutions. In [9] the attention is focused on constructing another bound that is in a sense complementary to ours. Its meaning is exactly the opposite: if $\delta(x)$ exceeds the bound then a *smaller* substitution σ exists such that $\sigma(t)$ is irreducible. However, the possibility of constructing a bound in the sense of this paper was indicated in [9] too, and the idea of the construction was sketched. In section 6 we will make further remarks on the relation between the two bounds.

Assume that we are given a term rewriting system $\mathcal{R}$, a term $t \in T_\Sigma(X)$ and a variable $x \in \mathcal{V}ar(t)$. Now we give a number $B(\mathcal{R}, t, x)$ that we use in the proof of the main theorem below. Actually, we will show that it bounds the depth of $\delta(x)$ where $\delta(t)$ is an $\mathcal{R}$-irreducible instance of t, and t is finitely irreducible by $\mathcal{R}$ with respect to x.

Let $card(\mathcal{R})$ denote the number of terms in $\mathcal{R}$, $maxarity(\Sigma)$ the maximal arity of function symbols in Σ, $nocc(x,t)$ and $depth(x,t)$ respectively the number and the maximal depth of occurrences of x in t. Suppose $C(\mathcal{R}) = 3\times(card(\mathcal{R})\times\|\mathcal{R}\|)!$, $D(\mathcal{R}) = C(\mathcal{R}) \times maxarity(\Sigma)^{\|\mathcal{R}\|}$, $A(\mathcal{R}, t, x) = D(\mathcal{R}) \times (card(\mathcal{R}) \times nocc(t, x) \times depth(x, t))$, $B(\mathcal{R}, t, x) = \|\mathcal{R}\| \times A(\mathcal{R}, t, x)$. Now we prove the following main theorem.

Theorem 7 *Let a term rewriting system $\mathcal{R}$, a term $t \in T_\Sigma(X)$ and a variable $x \in \mathcal{V}ar(t)$ be given. A number $B(\mathcal{R}, t, x)$ can be computed that verifies the following condition. If there exists a substitution δ such that $\delta(t)$ is $\mathcal{R}$-irreducible and $\|\delta(x)\| > B(\mathcal{R}, t, x)$, then there exists a substitution σ such that $\sigma(t)$ is irreducible and $\|\sigma(x)\| > \|\delta(x)\|$.*

We prove that the number $B(\mathcal{R}, t, x)$ defined in the above formula satisfies the theorem. Before giving the proof we give the following technical proposition. The proof is omitted because of space limitations.

Proposition 1 *Let $g \in T_\Sigma$, $t \in T_\Sigma(X)$, and g is not an instance of t.*

(i) Let $\pi \in \mathcal{P}os(g)$ and $|\pi| \geq \|t\|$. If $g_1 \in T_\Sigma$, $g' = g[\pi \leftarrow g_1]$, and $\|g'\| \geq \|g\| + \|t\|$, then g' is not an instance of t.

(ii) Let $\pi_1, \ldots, \pi_n \in \mathcal{P}os(g)$ and $|\pi_i| \geq \|t\|$ for every $i, 1 \leq i \leq n$. Assume that $g|_{\pi_1} = \ldots = g|_{\pi_n}$. Assume that $g_1, g_2 \in T_\Sigma$, $g' = g[\pi_1 \leftarrow g_1, \ldots, \pi_n \leftarrow g_1], g'' = g[\pi_1 \leftarrow g_2, \ldots, \pi_n \leftarrow g_2]$. If both g' and g'' are instances of t, then $g_1 = g_2$.

Now we are ready to give the proof.

Proof of theorem 7: Consider the term $\delta(x)$ and a path of maximal length in it. Since it is longer than $B(\mathcal{R}, t, x)$, find on this path $A(\mathcal{R}, t, x) + 1$ occurrences $\tau_0, \ldots, \tau_{A(\mathcal{R},t,x)}$ such that $\tau_{i-1} \prec \tau_i$ and $|\tau_i| - |\tau_{i-1}| = \|\mathcal{R}\|$ for every $i, 1 \leq i \leq A(\mathcal{R}, t, x)$. For $i, j, 1 \leq i < j \leq A(\mathcal{R}, t, x)$, denote by δ_{ij} the substitution defined by $\delta_{ij}(x) = \delta(x)[\tau_j \leftarrow \delta(x)|_{\tau_i}]$ and $\delta_{ij}(y) = \delta(y)$ for $y \neq x$. Since τ_i, τ_j belong to this longest path, $\|\delta_{ij}(x)\| > \|\delta(x)\|$. We show that a substitution σ verifying the theorem can be chosen among δ_{ij}. Similar to [9] we distinguish global and local reducibility.

Definition 4 *A term $g \in T_\Sigma$ is said to be locally (respectively globally) reducible with respect to a position $\tau \in \mathcal{P}os(g)$ iff it is reducible at some position $\pi \prec \tau$ such that $|\tau| - |\pi| \leq \|\mathcal{R}\|$ (respectively $|\tau| - |\pi| > \|\mathcal{R}\|$).*

The proof consists of three parts. In the first part we show that each $\delta_{ij}(x)$ is not globally reducible with respect to τ_j. In the second part we select a subset of pairs (i,j) such that $\delta_{ij}(t)$ is not reducible at any position preceding a position of x in t. Finally, we prove that among the remaining pairs there exists a pair (i,j) such that $\delta_{ij}(x)$ is not locally reducible with respect to τ_j. Clearly, these three parts cover all possibilities for $\delta_{ij}(t)$ to be reducible and prove the theorem.

Part 1. Consider a pair (i,j), $1 \le i < j \le A(\mathcal{R},t,x)$ and a position $\nu \prec \tau_j$ such that $|\tau_j| - |\nu| > \|\mathcal{R}\|$. Denote $g = \delta(x)|_\nu$ and $g' = \delta_{ij}(x)|_\nu$. Since ν, τ_j are both on the longest path and $|\tau_j| - |\tau_i| \ge \|\mathcal{R}\|$, then $\|g'\| - \|g\| \ge \|\mathcal{R}\|$. Since g is irreducible, then $g \notin Gr(\mathcal{R})$ and by proposition 1(i), $g' \notin Gr(\mathcal{R})$. Since ν and (i,j) were chosen arbitrary, we conclude that for each (i,j), $1 \le i < j \le A(\mathcal{R},t,x)$, $\delta_{ij}(x)$ is not reducible at any $\nu \prec \tau_j$, $|\tau_j| - |\nu| > \|\mathcal{R}\|$, i.e. is not globally reducible with respect to τ_j.

Part 2. Take a position $\nu \in \mathcal{V}\mathcal{P}os(t)$ of x in t and consider any position $\tau \prec \nu$. Note that the subterm $t|_\tau$ may have several occurrences of x. Take $s \in \mathcal{R}$. Let $\pi = \nu \cdot \tau_1$. Since $\delta_{ij}(x)|_\pi \neq \delta_{ik}(x)|_\pi$ for $j \neq k$, by proposition 1(ii) for every $i \ge 1$, there exists at most one $j > i$ such that $\delta_{ij}(t)$ is reducible by s at τ. Since the number of positions of x in t is $nocc(t,x)$ and the depth of any of them is bounded by $depth(x,t)$, there are at most $nocc(t,x) \times depth(x,t)$ possible values of τ. Consequently, given i, $i \ge 1$, there are at most $card(\mathcal{R}) \times nocc(t,x) \times depth(x,t)$ indexes j, $j > i$ such that $\delta_{ij}(t)$ is reducible at some $\tau \in \mathcal{P}os(t)$ by some $s \in \mathcal{R}$.

Now we construct a subsequence $\{l_1, \ldots, l_{D(\mathcal{R})}\} \subset \{1, \ldots, A(\mathcal{R},t,x)\}$ through the following "diagonalization procedure". Take $l_1 = 1$. Delete from the sequence $\{2, \ldots, A(\mathcal{R},t,x)\}$ those j for which $\delta_{l_1 j}(t)$ is reducible at some $\tau \in \mathcal{P}os(t)$. By the above remark, we have deleted at most $card(\mathcal{R}) \times nocc(t,x) \times depth(x,t)$ numbers. Take l_2 to be the smallest element in the resulting sequence, and apply the same deleting procedure to the rest of it. By iterating this procedure $D(\mathcal{R})$ times we construct a subsequence $\{l_1, \ldots, l_{D(\mathcal{R})}\} \subset \{1, \ldots, A(\mathcal{R},t,x)\}$. Note that since at every step we delete at most $card(\mathcal{R}) \times nocc(t,x) \times depth(x,t)$ indexes, and $A(\mathcal{R},t,x) = D(\mathcal{R}) \times (card(\mathcal{R}) \times nocc(t,x) \times depth(x,t))$, the procedure can be applied $D(\mathcal{R})$ times and therefore is correctly defined. Note finally that by construction for every $l', l'' \in \{l_1, \ldots, l_{D(\mathcal{R})}\}$, $l' < l''$, the term $\delta_{l'l''}(t)$ is not reducible at any $\tau \in \mathcal{P}os(t)$ by any $s \in \mathcal{R}$.

Part 3. Now we observe that the positions $\tau_{l_1}, \ldots, \tau_{l_{D(\mathcal{R})}}$ have at most $maxarity(\Sigma)^{\|\mathcal{R}\|}$ different suffixes of the length $\|\mathcal{R}\|$, and hence among $l_1, \ldots, l_{D(\mathcal{R})}$ there are at least $C(\mathcal{R})$ indexes $k_1, \ldots, k_{C(\mathcal{R})}$ such that for some ρ, $|\rho| = \|\mathcal{R}\|$, $\tau_{k_i} = \tau'_{k_i} \cdot \rho$ for every i, $1 \le i \le C(\mathcal{R})$. Let $\pi_n^{(m)}$, $n \in \{k_1, \ldots, k_{C(\mathcal{R})}\}$, $1 \le m \le \|\mathcal{R}\|$, be the positions defined by $\pi_n^{(m)} \prec \tau_n$, $|\tau_n| - |\pi_n^{(m)}| = m$. We prove that there exists a pair $k', k'' \in \{k_1, \ldots, k_{C(\mathcal{R})}\}, k' < k''$ such that $\delta_{k'k''}(x)$ is not reducible at any $\pi_{k''}^m$, $1 \le m \le \|\mathcal{R}\|$ by any $s \in \mathcal{R}$, which also means that $\delta_{k'k''}(x)$ is not locally reducible with respect to $\tau_{k''}$. The proof of this is similar to that of theorem 4 but we use the "finite version" of Ramsey's theorem.

Theorem 8 (Ramsey's theorem, finite version) *Let a finite set I and natural numbers N, n be given. Let $A_1, \ldots, A_N$ be natural numbers, and $A_i \ge 2$, $1 \le$*

$i \leq N$. Denote by $\mathcal{P}_n(I)$ the set of all n-element subsets of I and assume that $\mathcal{P}_n(I) = P_1 \uplus \ldots \uplus P_N$. Then there exists a number $R(A_1, \ldots, A_N; n)$ such that if I contains at least $R(A_1, \ldots, A_N; n)$ objects, then there exist i, $1 \leq i \leq N$ and a subset $J \subseteq I$ such that J contains at least A_i objects, and $\mathcal{P}_n(J) \subseteq P_i$.

$R(A_1, \ldots, A_N; n)$ are called Ramsey numbers. We are going to apply the theorem with $n = 2, A_1 = \ldots = A_N = 3$. It is known [6] that the numbers $R_N = R(\underbrace{3, \ldots, 3}_{N}; 2)$, $N \geq 2$ satisfy the recurrence relation $R_2 = 6$, $R_N \leq N \times (R_{N-1} - 1) + 2$. Hence, $R_N \leq 3 \times N!$.

Now we observe that for a given $s \in \mathcal{R}$ and m, $1 \leq m \leq \|\mathcal{R}\|$, if $k', k'', k''' \in \{k_1, \ldots, k_{C(\mathcal{R})}\}$ and $k' < k'' < k'''$, then either $\delta_{k'k''}(x)$ is not reducible at $\pi^m_{k''}$ by s, or $\delta_{k'k'''}(x)$ is not reducible at $\pi^m_{k'''}$ by s, or $\delta_{k''k'''}(x)$ is not reducible at $\pi^m_{k'''}$ by s. Otherwise by assuming $g_1 = \delta(x)|_{\pi^m_{k'''}}$, $g_2 = \delta(x)|_{\pi^m_{k''}}$, $g_3 = \delta(x)|_{\pi^m_{k'}}$, and applying lemma 1, we would conclude that $\delta(x)$ is reducible at $\pi^m_{k''}$ by s which is a contradiction.

Now we apply Ramsey's theorem. We identify uniquely every pair (k', k''), $k', k'' \in \{k_1, \ldots, k_{C(\mathcal{R})}\}$, $k' < k''$ with the 2-element subset $\{k', k''\}$, and we split the set of all such pairs into $(card(\mathcal{R}) \times \|\mathcal{R}\| + 1)$ subsets $P_0, P_1, \ldots, P_{card(\mathcal{R}) \times \|\mathcal{R}\|}$ in the following way. Each P_i, $1 \leq i \leq card(\mathcal{R}) \times \|\mathcal{R}\|$ is one-to-one associated with a pair s, m, $s \in \mathcal{R}, 1 \leq m \leq \|\mathcal{R}\|$. The way we distribute the pairs among P_i is the following. If P_i corresponds to a pair s, m, then a pair (k', k'') belongs to P_i iff $\delta_{k'k''}(x)$ is reducible by s at $\pi^m_{k''}$. If there are several possibilities to place (k', k''), we choose any of them. If there are no s, m as above such that $\delta_{k'k''}$ is reducible by s at $\pi^m_{k''}$, we place (k', k'') into P_0. If no pair is finally placed in P_0, then since $C(\mathcal{R}) = 3 \times (card(\mathcal{R}) \times \|\mathcal{R}\|)!$, by theorem 8 there exists a 3-element subset $J \subseteq \{k_1, \ldots, k_{C(\mathcal{R})}\}$ such that $\mathcal{P}_2(J) \subseteq P_i$ for some i, $1 \leq i \leq card(\mathcal{R}) \times \|\mathcal{R}\|$. But this contradicts the above remark. Therefore, there exists at least one pair k', k'' such that $\delta_{k'k''}(x)$ is not reducible by any $s \in \mathcal{R}$ at any $\pi^m_{k''}$, $1 \leq m \leq \|\mathcal{R}\|$. Thus, $\delta_{k'k''}(x)$ is not locally reducible with respect to $\tau_{k''}$. This completes the proof.

□

We remark that many ideas in the proof above were borrowed from [9]. Moreover, a slightly simpler proof of part 3 can be given which uses a minor modification of lemma 5.4 from [9]. However, we have preferred to give a longer proof not only in order to make the paper self-contained, but also because it uses Ramsey's theorem which embodies complex combinatorial reasoning of [9]. Also, it was interesting for us to discover that the same "Ramsey's theorem technique" is applicable for proving both principal results of this paper - theorem 4 and theorem 7. We believe that Ramsey's theorem, being a very powerful combinatorial result, can be very fruitful in proving this kind of properties of term sets.

The following corollary adapts theorem 7 to the first condition in the definition of finite irreducibility.

Corollary 1 *Let a term rewriting system $\mathcal{R}$, a term $t \in T_\Sigma(X)$ and a variable $x \in \mathcal{V}ar(t)$ be given. A number $B(\mathcal{R}, t, x)$ can be computed that verifies the following condition. If there exists a substitution δ such that $\delta(t)$ is $\mathcal{R}$-irreducible, $\delta(t)$ has no proper subterm that is an instance of t, and $\|\delta(x)\| > B(\mathcal{R}, t, x)$, then there exists a substitution σ such that $\sigma(t)$ is irreducible, $\|\sigma(x)\| > \|\delta(x)\|$, and $\sigma(t)$ has no proper subterm that is an instance of t.*

Proof: The proof of theorem 7 remains valid but we have to insert t into $\mathcal{R}$ and to treat it as a "special" rewrite rule that cannot be applied at the root position. This particularity is relevant only to part 2. It is easy to see that the proof of part 2 still works. The only modification is that if $\tau = \varepsilon$, then every rule of $\mathcal{R}$ but t is potentially applicable. Thus, we have one less possibility of reduction and even more freedom in choosing a suitable subsequence.

We have to correct obviously the bound B by taking $card(\mathcal{R})+1$ instead of $card(\mathcal{R})$ and $\|\mathcal{R} \cup \{t\}\|$ instead of $\|\mathcal{R}\|$. □

In the rest of the paper we will assume that $B(\mathcal{R}, t, x)$ denotes the bound modified according to the proof of corollary 1 unless the contrary is explicitly stated.

Theorem 7 allows us to prove the decidability of finite irreducibility of a term with respect to a variable. Note that the proof uses the ground reducibility property that is known to be decidable [13, 9, 1].

Theorem 9 *It is decidable whether given a rewriting system $\mathcal{R}$, a term $t \in T_\Sigma(X)$ is finitely $\mathcal{R}$-irreducible with respect to a variable $x \in \mathcal{V}ar(t)$.*

Proof: Compute $B(\mathcal{R}, t, x)$ and compute all instances $\sigma_1(t), \dots, \sigma_K(t)$ such that for every i, $1 \le i \le K$,

- $\sigma_i(x) \in T_\Sigma$, and $\sigma_i(y) = y$ for every $y \neq x$,
- $\sigma_i(t)$ is $\mathcal{R}$-irreducible,
- $\|\sigma_i(x)\| \le B(\mathcal{R}, t, x)$.

We show now that t is finitely $\mathcal{R}$-irreducible with respect to x iff t is ground reducible by $\mathcal{R} \cup \{\sigma_1(t), \dots, \sigma_K(t)\}$. We use an easy observation that t is ground reducible if and only if for every ground instance $\delta(t)$ in which no proper subterm is an instance of t, $\delta(t)$ is reducible.

Assume that t is finitely $\mathcal{R}$-irreducible with respect to x. Let $\delta(t)$ be a ground instance of t that has no instance of t as its proper subterm. If $\|\delta(x)\| \le B(\mathcal{R}, t, x)$, then by construction of σ_i, $\delta(t)$ is either reducible by $\mathcal{R}$, or is an instance of $\sigma_i(t)$ for some i, $0 \le i \le K$. If $\|\delta(x)\| > B(\mathcal{R}, t, x)$, then $\delta(t)$ is reducible since otherwise t cannot be finitely irreducible. Thus, $\delta(t)$ is reducible by $\mathcal{R} \cup \{\sigma_1(t), \dots, \sigma_K(t)\}$ and therefore t is ground reducible by $\mathcal{R} \cup \{\sigma_1(t), \dots, \sigma_K(t)\}$.

Conversely, assume that t is ground reducible by $\mathcal{R} \cup \{\sigma_1(t), \dots, \sigma_K(t)\}$. Consider a ground instance $\delta(t)$ and suppose it contains no proper subterm that is an instance of t. Assume that $\|\delta(x)\| > B(\mathcal{R}, t, x)$. $\delta(t)$ is not reducible by $\sigma_1(t), \dots, \sigma_K(t)$ at any position different from ε since $\delta(t)$ has no proper subterm that is an instance of t. On the other hand, since $\|\sigma_i(x)\| \le B(\mathcal{R}, t, x)$, $\delta(t)$ cannot be an instance of $\sigma_1(t), \dots, \sigma_K(t)$. Therefore, $\delta(t)$ is reducible by $\mathcal{R}$. This proves that t is finitely irreducible by $\mathcal{R}$ with respect to x. □

The proof of theorem 9 gives a decision procedure for testing finite irreducibility of t with respect to x. It consists in computing all substitutions σ with $\sigma(x) \in T_\Sigma$, $\sigma(x) \leq B(\mathcal{R}, t, x)$, and $\sigma(y) = y$ for $y \neq x$, then selecting out those for which $\sigma(t)$ is $\mathcal{R}$-reducible, and then checking if t is ground reducible by $\mathcal{R} \cup \{\sigma_1(t), \ldots, \sigma_K(t)\}$, where $\sigma_1, \ldots, \sigma_K$ are the remaining substitutions.

It should be noted that theorem 9 is still valid if the first condition in the definition 2 is dropped. The decision procedure is now the following. At first, we compute $B(\mathcal{R}, t, x)$ according to theorem 7. Then we compute all instances $\sigma_1(t), \ldots, \sigma_K(t)$ such that for every i, $1 \leq i \leq K$,

- $\sigma_i(y) = y$ for every $y \neq x$,
- $\sigma_i(t)$ is $\mathcal{R}$-irreducible,
- $\sigma_i(x)$ contains variables, $\|\sigma_i(x)\| = B(\mathcal{R}, t, x)$, and the height of each variable in $\sigma_i(x)$ is exactly $B(\mathcal{R}, t, x)$.

The following statement is trivial. t is finitely $\mathcal{R}$-irreducible with respect to x iff for every i, $1 \leq i \leq K$, $\sigma_i(t)$ is ground reducible by $\mathcal{R}$. This gives the decision procedure.

5 Decidability Results

In this section we apply the above results to the analysis of several interesting properties of the set of ground normal forms.

5.1 Regularity of *Red*($\mathcal{R}$)

The decidability of the regularity of the set of reducible ground terms follows naturally from the results of the previous sections.

Theorem 10 *It is decidable whether given a rewriting system $\mathcal{R}$, the set Red($\mathcal{R}$) is a regular tree language.*

Proof: The result follows from theorem 9. A decision procedure implied by the proof of theorem 6 is the following. Starting from the initial system $\mathcal{R}$, transform it by iterating the following procedure. Take a non-linear term $t \in \mathcal{R}$ and a non-linear variable $x \in \mathcal{V}ar(t)$. Compute the bound $B(\mathcal{R} \setminus \{t\}, t, x)$. Substitute x by all ground terms not deeper than $B(\mathcal{R} \setminus \{t\}, t, x)$ and select out those instances that are $\mathcal{R}$-irreducible. If $\sigma_1(t), \ldots, \sigma_K(t)$ are the resulting terms, check if t is ground reducible by $\mathcal{R} \cup \{\sigma_1(t), \ldots, \sigma_K(t)\} \setminus \{t\}$. If this is the case, proceed with the system $\mathcal{R} \cup \{\sigma_1(t), \ldots, \sigma_K(t)\} \setminus \{t\}$.

If all ground reducibility tests succeed, then *Red*($\mathcal{R}$) is regular, otherwise *Red*($\mathcal{R}$) is not regular. Note that in the first case the system is finally transformed into a linear one. □

It should be noted that the decision procedure defined in the proof of theorem 6 treats *subsequently* each non-linear term and each non-linear variable in it. The straightforward way of applying the procedure makes the size of the system enormous. The reason is that every time we instantiate a variable the depth of the rewriting system increases and formally we have to correct the bound. However, this correction is unnecessary and this follows from the following general observation [8]. Assume that we are given a rewriting system $\mathcal{R}$, a term t, and a variable $x \in \mathcal{V}ar(t)$. Let N be a number that can be taken to verify the finite irreducibility of t by $\mathcal{R}$ with respect to x. It means that N can be taken as a value of $B(\mathcal{R}, t, x)$ in theorem 7. If instead of $\mathcal{R}$ we now consider another rewriting system $\mathcal{R}'$ such that $Red(\mathcal{R}) = Red(\mathcal{R}')$, then the same bound N can be taken as a value of $B(\mathcal{R}', t, x)$. Also, we remark that the algorithm instantiates one variable at a time and does not affect the others. In particular, the depth of the variables in $\mathcal{R}$ is always bounded by the depth of the initial system. Taking these two arguments into account, we conclude that if we take the bound to be $max\{B(\mathcal{R}\setminus\{t\}, t, x) | t \in \mathcal{R}, x \in \mathcal{V}ar(t)\}$, then it can be used throughout the whole run of the algorithm. Obviously, such a bound can be computed by the formula given before theorem 4 where $depth(x, t)$ and $nocc(x, t)$ are replaced respectively by $\|\mathcal{R}\|$ and the maximal number of occurrences of a variable in a term in $\mathcal{R}$.

Furthermore, from the possibility of using a single bound it follows that we can also instantiate all the non-linear variables *simultaneously*. In this way we construct a system $\mathcal{L}$ by replacing the non-linear variables by ground terms of depth smaller than the bound and then check if each term from $\mathcal{R}$ is $\mathcal{L}$-ground reducible. This is equivalent to $Red(\mathcal{R}) \subseteq Red(\mathcal{L})$ (cf theorem 1). Note that $\mathcal{L}$ is always linear.

Obviously, since $NF(\mathcal{R}) = T_\Sigma \setminus Red(\mathcal{R})$, it is also decidable if the set of ground normal forms is regular.

5.2 Existence of an equivalent ground system

The results of the previous sections allow us to prove decidable the problem of whether given a rewriting system, there exists an equivalent ground rewriting system.

Theorem 11 *It is decidable whether given a rewriting system $\mathcal{R}$, there exists a finite ground rewriting system $\mathcal{G} \subseteq T_\Sigma$ such that $Red(\mathcal{R}) = Red(\mathcal{G})$.*

Proof: Assume that a finite system $\mathcal{G} \subseteq T_\Sigma$ exists such that $Red(\mathcal{R}) = Red(\mathcal{G})$. Consider the set $\mathcal{G}'$ of subterms of terms in $\mathcal{G}$ that are reducible by $\mathcal{R}$ and have no proper subterms reducible by $\mathcal{R}$. Clearly, $\mathcal{G}' \subseteq Gr(\mathcal{R})$. On the other hand, $Red(\mathcal{G}') = Red(\mathcal{G})$, and hence, $Red(\mathcal{G}') = Red(\mathcal{R})$. Thus, we can always assume that $\mathcal{G} \subseteq Gr(\mathcal{R})$, that is, every term of $\mathcal{G}$ is a ground instance of a term of $\mathcal{R}$.

It is easy to see now that the existence of a finite set $\mathcal{G} \subseteq Gr(\mathcal{R})$ such that $Red(\mathcal{G}) = Red(\mathcal{R})$ implies the finite irreducibility of every $t \in \mathcal{R}$ by $\mathcal{R}\setminus\{t\}$ with respect to *every* variable $x \in \mathcal{V}ar(t)$. By analogy with the previous subsection, we can test the existence of $\mathcal{G}$ by iterating the following procedure while the system contains non-ground terms. Take $t \in \mathcal{R}$ and $x \in \mathcal{V}ar(t)$. Compute the bound $B(\mathcal{R}\setminus\{t\}, t, x)$. Substitute x by all ground terms not deeper than $B(\mathcal{R}\setminus\{t\}, t, x)$ and select out those instances that are $\mathcal{R}$-irreducible. If $\sigma_1(t), \ldots, \sigma_K(t)$ are the resulting terms, check if t is ground reducible by $\mathcal{R} \cup \{\sigma_1(t), \ldots, \sigma_K(t)\} \setminus \{t\}$. If this is the case, iterate the procedure with the system $\mathcal{R} \cup \{\sigma_1(t), \ldots, \sigma_K(t)\} \setminus \{t\}$. □

Thus, testing the existence of a finite ground system equivalent to $\mathcal{R}$ is equivalent to testing the existence of an equivalent finite ground instantiation of $\mathcal{R}$, and is done by checking the finite irreducibility of the terms in $\mathcal{R}$ with respect to *all* variables. Note the only difference with the previous case: testing the regularity of $Red(\mathcal{R})$ is testing the existence of an equivalent finite *linear* instantiation of $\mathcal{R}$, and is done by checking the finite irreducibility of the terms in $\mathcal{R}$ with respect to the ***non-linear*** variables. All the comments from the previous subsection concerning the bound and the strategy of applying the decision procedure are valid for this case too.

5.3 Finiteness of $NF(\mathcal{R})$

It is known that the finiteness of the set of ground normal forms is a decidable property [9, 13]. However, it is interesting to see that this is a very particular case of the above results.

Theorem 12 *It is decidable whether given a rewriting system $\mathcal{R}$, the set of ground normal forms $NF(\mathcal{R})$ is finite.*

Proof: The finiteness of $NF(\mathcal{R})$ can be expressed as the finite irreducibility of the degenerate term $t = x$ by $\mathcal{R}$ with respect to x, where the definition of finite irreducibility (definition 2) is taken without the first condition. By the remark at the end of the previous section, this property is decidable. □

From the construction of B it follows that $\|NF(\mathcal{R})\|$ is bounded by $3 \times \|\mathcal{R}\|^2 \times maxarity(\Sigma) \times (card(\mathcal{R}) \times \|\mathcal{R}\|)!$ in the case when $NF(\mathcal{R})$ is finite.

Obviously, if $NF(\mathcal{R})$ is finite, then $Red(\mathcal{R})$ is regular. Moreover, if $NF(\mathcal{R})$ is finite, then there exists a finite ground system $\mathcal{G}$ such that $Red(\mathcal{R}) = Red(\mathcal{G})$ ($\|\mathcal{G}\|$ can be bounded by $\|NF(\mathcal{R})\| + 1$). Consequently, the decidable properties we have considered in this section induce the following classification of rewriting systems. Note that every class is strictly embedded into the one below.

$$\{\mathcal{R} | NF(\mathcal{R}) \text{ is finite}\}$$
$$\cap$$
$$\{\mathcal{R} | \text{ there exists a finite } \mathcal{G} \subseteq T_\Sigma \text{ such that } Red(\mathcal{G}) = Red(\mathcal{R})\}$$
$$\cap$$
$$\{\mathcal{R} | Red(\mathcal{R}) \text{ is regular}\}$$
$$\cap$$
$$\text{all term rewriting systems}$$

6 Concluding Remarks

Throughout the paper we identified term rewriting systems with the sets of left-hand sides, and we considered two rewriting system equivalent if they had the same set of reducible ground terms. The results of section 5 allow us to transform, if this is at all possible, a rewriting system $\mathcal{R}$ into an equivalent "good" (linear or ground) rewriting system $\mathcal{L}$ by substituting some variables by ground terms. We remark that this instantiation can be extended to the right-hand sides of $\mathcal{R}$. Moreover, if $\mathcal{R}$ is convergent, the system we obtain is equivalent to $\mathcal{R}$ in the classical sense, i.e. it generates the same equivalence relation on T_Σ. More precisely, if $\mathcal{R}$ is convergent and $\mathcal{L}$ is an instantiation of $\mathcal{R}$ such that $Red(\mathcal{R}) = Red(\mathcal{L})$, then $\leftrightarrow^*_{\mathcal{R}} = \leftrightarrow^*_{\mathcal{L}}$ on T_Σ.

During the work on this paper we came to know of the work of D.Hofbauer and M.Huber [7]. Using the approach of test sets, they proved independently that the existence of an equivalent linear rewriting system (and therefore the regularity of the ground normal form language) can be effectively tested. Also, recently we became aware that similar results were obtained by S.Vágvölgyi and R.Gilleron. They also attacked the problem of decidability of regularity of ground normal form languages using a very similar approach combining the results of [9] and [12].

References

[1] H. Comon. *Unification et disunification. Théories et applications.* Thèse de Doctorat d'Université, Institut Polytechnique de Grenoble (France), 1988.

[2] N. Dershowitz, J.-P. Jouannaud, and J.W. Klop. Open problems in rewriting. In R. V. Book, editor, *Proceedings 4th Conference on Rewriting Techniques and Applications, Como (Italy)*, volume 488 of *Lecture Notes in Computer Science*, pages 445–456. Springer-Verlag, 1991.

[3] J.H. Gallier and R. V. Book. Reductions in tree replacement systems. *Theoretical Computer Science*, 37:123–150, 1985.

[4] F. Gécseg and M. Steinby. *Tree automata.* Akadémiai Kiadó, Budapest, Hungary, 1984.

[5] R. Gilleron. Decision problems for term rewriting systems and recognizable tree languages. Research Report IT 200, Laboratoire d'Informatique Fondamentale de Lille, 1990.

[6] R.L. Graham, B.L. Rothschild, and J.H Spencer. *Ramsey theory.* John Wiley and Sons, 1980.

[7] D. Hofbauer and M. Huber. Computing linearizations using test sets. In *Proceedings of the 3rd International Workshop on Conditional Term Rewriting Systems*, 1992. this volume.

[8] D. Hofbauer and M. Huber. Joint discussions, 1992.

[9] D. Kapur, P. Narendran, and H. Zhang. On sufficient completeness and related properties of term rewriting systems. *Acta Informatica*, 24:395–415, 1987.

[10] E. Kounalis. Pumping lemmas for tree languages generated by rewrite systems. In *Fifteenth International Symposium on Mathematical Foundations of Computer Science, Banská Bystrica (Czechoslovakia)*, Lecture Notes in Computer Science. Springer-Verlag, 1990.

[11] E. Kounalis. Testing for inductive (co)-reducibility. In A. Arnold, editor, *Proceedings 15th CAAP, Copenhagen (Denmark)*, volume 431 of *Lecture Notes in Computer Science*, pages 221–238. Springer-Verlag, May 1990.

[12] G. A. Kucherov. On relationship between term rewriting systems and regular tree languages. In R. V. Book, editor, *Proceedings 4th Conference on Rewriting Techniques and Applications, Como (Italy)*, volume 488 of *Lecture Notes in Computer Science*, pages 299–311. Springer-Verlag, April 1991.

[13] D. Plaisted. Semantic confluence tests and completion methods. *Information and Control*, 65:182–215, 1985.

Computing Linearizations Using Test Sets

Dieter Hofbauer* Maria Huber**

CRIN and INRIA-Lorraine
Campus Scientifique, BP 239, F - 54506 Vandœuvre-lès-Nancy, France
email: {hofbauer, huber}@loria.fr

Abstract. Often non left-linear rules in term rewriting systems can be replaced by a finite set of left-linear ones without changing the set of irreducible ground terms. Using appropriate test sets, we can always decide if this is possible and, in case it is, effectively perform such a transformation. We thus can also decide if the set of irreducible ground terms is a regular tree language, using a result of Kucherov.

1 Introduction

Finite test sets as a tool for reasoning about term rewriting systems have widely been used wherever the set of irreducible ground terms plays a crucial role, so e.g. for checking sufficient completeness or proving inductive equalities using the concept of ground reducibility. Plaisted [16] and Kapur, Narendran, Zhang [9] have shown the decidability of ground reducibility in the general case by computing finite sets of ground terms (see Comon [3] for a different approach). Much smaller test sets - containing also non-ground terms - have been obtained for left-linear systems; see among others [7], [8], [2], and for a less restricted class of rewriting systems [1]. The test sets used in this paper are inspired by the ones presented by Kounalis [10] for the general case. His approach, however, turned out to be not correct; after the second author in [6] gave a construction of test sets having additional properties strong enough for using them within ground reducibility tests, Kounalis [11] has also rectified the older version.

In this paper we show how the test sets in [6], which have originally been designed for deciding ground reducibility, can be used to effectively replace non left-linear rules by finite sets of left-linear ones, if this is possible without changing the set of irreducible ground terms. For this purpose we use a (single) deduction rule; it looks for non left-linear rewrite rules whose non-linear variables are "idle", i.e., for rewrite rules that can be safely replaced by a finite number of instances where the non-linear variables are substituted by ground terms. If the deduction rule is no longer applicable we have either reached a left-linear system with the same set of irreducible ground terms as the initial system or, in case there is still some non-linear rewrite rule, we know that such a system does not exist. This (nondeterministic) process terminates after at most k steps where k is the

* supported by M.E.N.
** supported by M.R.T.

number of non left-linear rules in the initial system. Therefore it can be used to decide whether a linearization is possible.
Kucherov has shown in [12] that a finite linearization exists if and only if the set of irreducible ground terms is a regular tree language. Deciding the linearizability is therefore equivalent to deciding if the set of irreducible ground terms is regular. In the case of a regular set, the left-linear set of rules determined by our transformation can easily be used to give a regular grammar (see Gallier, Book [4]). Similar results have been independently obtained by Kucherov and Tajine [13] and, as we heard recently, by Vágvölgyi and Gilleron [17].

The paper is organized as follows: Section 3 contains our definition of test sets where the notion of "typical terms" plays a key role; proofs are omitted and can be found in [5]. Some fundamental properties of linearizations are discussed in section 4; finally our linearization procedure is presented in section 5.

2 Preliminaries

Since right hand sides of rewrite rules are irrelevant as far as only reducibility is concerned, we identify rewrite systems with the set of its left hand sides. Throughout the paper we assume signatures Σ to be manysorted. $\mathcal{T}_\Sigma(X)$ ($\mathcal{T}_\Sigma$) denotes the set of (ground) terms over Σ where X is an infinite set of variables. $Pos(t) \subseteq \mathbb{N}^+$ is the set of positions in a term t as usual; λ denotes the root position. $t|_u$ is the subterm in t at $u \in Pos(t)$; $t|_U = \{t|_u \mid u \in U\}$ for sets of positions U. $Pos_X(t) = \{u \in Pos(t) \mid t|_u \in X\}$ are the variable positions in t; we will also use $Pos_\Sigma(t) = Pos(t) \setminus Pos_X(t)$ and $Var(t) = t|_{Pos_X(t)}$.
For $u \in \mathbb{N}^+$, $|u|$ denotes the length of u, thus $|\lambda| = 0$. Subterms occuring at positions of length 1 in a term are called its principals. For $u, v \in \mathbb{N}^+$ we use $u \leq v$ if $uw = v$ for some $w \in \mathbb{N}^*$, $u < v$ if $u \leq v$ and $u \neq v$, $u|v$ if neither $u \leq v$ nor $v \leq u$, $u - v = w$ if $u = vw$. The depth $|t|$ of t is defined by $|x| = 0$ for $x \in X$, $|t| = max\{|u| + 1 \mid u \in Pos_\Sigma(t)\}$ otherwise.
Substitutions σ have finite domain $dom(\sigma) = \{x \in X \mid x\sigma \neq x\}$. The restriction of σ to $Y \subseteq X$ is denoted by $\sigma|_Y$ (i.e., $x\sigma|_Y = x\sigma$ for $x \in Y$, $x\sigma = x$ otherwise). t is an instance of s if $t = s\sigma$ for some substitution σ; t is a variant of s if t is an instance of s and vice versa.
$Ground(t)$ denotes the set of all ground instances of t. Let R be a set of terms; t is R-reducible if some subterm of t is an instance of some term in R, otherwise t is R-irreducible. $Red_R(t)$ ($Nf_R(t)$) is the set of all R-(ir)reducible ground instances of t, Red_R (Nf_R) the set of all R-(ir)reducible ground terms. t is ground reducible by R if $Ground(R) \subseteq Red_R$.

A variable x is non-linear in t if it occurs at more than one position in t; $Pos_{nl}(t)$ denotes the set of all positions of non-linear variables in t; we use $Var_{nl} = t|_{Pos_{nl}(t)}$. A term is (non-)linear if $Var_{nl}(t)$ is (non-)empty. A set of terms is linear if it contains only linear terms. A substitution σ is linear on $Y \subseteq X$ if $y\sigma$ is linear for all $y \in Y$ and $y\sigma$ and $y'\sigma$ are variable disjoint for $y, y' \in Y$, $y \neq y'$.

Let t, s be unifiable terms. t is called non-linear w.r.t. s if $s|_u$ is ground for some $u \in Pos_{nl}(t) \cap Pos(s)$, or if there is a substitution σ such that t and $s\sigma$ are unifiable, $Pos(s\sigma) \supseteq Pos(t)$, and $s\sigma|_u \neq s\sigma|_v$ for some $u, v \in Pos_X(t)$ for which $t|_u = t|_v$. Otherwise t is linear w.r.t. s. For example, $f(a,a)$ and $f(x,y)$ are both linear w.r.t. $f(x,x)$ (as linear terms are always linear w.r.t. unifiable terms), whereas $f(x,x)$ is non-linear w.r.t. $f(a,a)$ and $f(x,y)$ (as non-linear terms are never linear w.r.t. linear terms). $f(x,x)$ and $f(g(x),g(x))$ are linear w.r.t. each other. This generalizes the notion of restrictedness given in [14], where t has to be an instance of s. Indeed, according to the above definition, $t\sigma$ is linear w.r.t. t iff σ is linear on $Var(t)$. Moreover, t is linear w.r.t. $t\sigma$ iff $x\sigma$ is non-ground for all non-linear variables x in t. t is linear iff t is linear w.r.t. x iff t is linear w.r.t. all s, s unifiable with t. For sets R and S, R is called linear w.r.t. S if each term in R is linear or is linear w.r.t. some term in S.

Lemma 1 *Suppose s is an instance of r. Then there is an instance q of r, which has s as an instance, and which is linear w.r.t. both r and s.*

Proof For $s = r\sigma$ choose $q = r\sigma|_V$ where $V = \{x \in Var_{nl}(r) \mid x\sigma \in T_\Sigma\}$. Now $s = q\sigma$. q is linear w.r.t. r since $\sigma|_V$ is linear on $Var(r)$; q is linear w.r.t. $s = q\sigma$ since $x\sigma$ is non-ground for all $x \in Var_{nl}(q) = \{x \in Var_{nl}(r) \mid x\sigma \notin T_\Sigma\}$. □

Given a set $T \subseteq T_\Sigma(X)$, $t\tau$ is called a T-instance of t if for all $x \in Var(t)$, $x\tau$ is
- a variant of some term in T, and
- variable disjoint to t and to all $y\tau$, $y \in Var(t)$, $y \neq x$.

For linear T this makes T-instances $t\tau$ (and even $t\tau|_Y$ for all $Y \subseteq Var(t)$) linear w.r.t. t. In order to guarantee the finiteness of the set of all T-instances of t in case of a finite T, we additionally assume T-instances to be "minimal" w.r.t. some fixed well-ordering $\succeq_X$ on X. $t\tau$ is a minimal T-instance of t, if no $t\tau\{x \mapsto y\}$ is a T-instance of t for $x \in Var(t\tau)$, $y \in X$, $x \succ y$.

For a position u in t and a term $l \in R$, the pair (u, l) is an R-reducer for t if $t|_u$ is an instance of l. For terms t_1, t_2 and $U \subseteq Pos(t_1)$, we write $t_1 \subseteq_R^U t_2$ if every R-reducer (u, l) for t_1, $u \in U$, is also an R-reducer for t_2. U is omitted in case of $U = Pos(t_1)$. Let $\simeq_R^U$ denote the equivalence relation induced by the partial order $\subseteq_R^U$ and define $\subset_T^U = \subseteq_R^U \setminus \supseteq_R^U$. For a single position u we use $\simeq_R^u$ instead of $\simeq_R^{\{u\}}$ for short. Note that for substitutions σ we always have $t \subseteq_R^U t\sigma$.

3 Typical Terms and Test Sets

Test sets are one of the most popular tools for performing ground reducibility tests and they are used world-wide, or at least they should be. Given a test set T for R, ground reducibility of t by R can be checked as follows:

t is ground reducible by R iff all T-instances $t\tau$ of t are R-reducible.

For this purpose, T has to be a finite representation of the (possibly infinite) set of R-irreducible ground terms. To ensure ground reducibility of t in case all T-instances $t\tau$ are reducible, it suffices that each irreducible ground term is represented as an instance of a test set term, i.e., that the test set is "complete".

Definition 1 (complete) *A set T is* complete w.r.t. R *if $Nf_R \subseteq Ground(T)$.*

To guarantee the other direction, we would like each T-instance $t\tau$ to have a ground instance $t\tau\gamma$ such that $t\tau\gamma \simeq_R t\tau$. Now, if a T-instance $t\tau$ is irreducible, it has an irreducible ground instance, implying that t itself is not ground reducible. For linear sets R, this can be achieved if T meets two additional requirements. First, T is not allowed to contain terms which are (ground) reducible. Second, all terms in T have to be "expanded" to a certain extend. Various kinds of expansion criteria are used in all test set approaches; they all are refinements of "tops", a concept introduced in [15].

Definition 2 (expanded) *t is* expanded w.r.t. l *if t and l are not unifiable, or $Pos_\Sigma(l) \subseteq Pos_\Sigma(t)$. t is* expanded w.r.t. R *(*w.r.t. all proper subterms of R*) if t is expanded w.r.t. all terms $l \in R$ (all terms $l|_u$, $l \in R$, $u \in Pos(l)$, $u \neq \lambda$).*

Lemma 2 (linear reducers) *Let R be linear.*
(1) If t is expanded w.r.t. R, then $t\sigma \simeq^\lambda_R t$ for all substitutions σ.
(2) Let all terms in T be expanded w.r.t. all proper subterms of R, let $t\tau$ be a T-instance of t.

- *$t\tau\sigma \simeq_R t\tau$ for all substitutions σ where $\tau\sigma$ is R-irreducible.*
- *If additionally no term in T is ground reducible, then there is a ground instance $t\tau\gamma$ of $t\tau$ such that $t\tau\gamma \simeq_R t\tau$.*

Example 1 Consider $R_1 = \{h(h(x)), h(g(x))\}$ and $R_2 = R_1 \cup \{f(x,x)\}$ over signature $\Sigma = \{a :\rightarrow s_1, g, h : s_1 \rightarrow s_1, f : s_1 s_1 \rightarrow s_2, i : s_2 \rightarrow s_2\}$. A finite test set for R_1 is $\{a, h(x), g(x), f(x,y), i(z)\}$. This, however, is no test set for R_2, since, e.g., the T_1-instance $f(h(x), h(a))$ of $f(y, h(a))$ is R_2-irreducible, but ground reducible by R_2, i.e. has no ground instance with the same reducers.

Thus, if R is linear, complete sets T which contain no ground reducible term, and which are expanded w.r.t. all proper subterms of R, can already be used for a ground reducibility test. In the non-linear case, however, this of course is not sufficient. All non-ground terms in our test sets have not only to be not ground reducible but have to be "typical" for their "problem positions". Problem positions in a term t are essentially those which correspond to non-linear variables in some rule l such that for some (non-empty, ground) context c, $c[t]$ is not reducible by l, but probably some ground instance of $c[t]$ is. Typicality of t then guarantees the existence of infinitely many irreducible ground instances $t\gamma$ of t such that the depths of subterms in $t\gamma$ at problem positions in t can differ arbitrarily. This ensures that $c[t]$ has a ground instance $c[t\gamma]$ with exactly the same reducers as $c[t]$.
In example 1, λ is a problem position in $t = h(x)$ w.r.t. R_2 because $c[t] = f(h(x), h(a))$ is unifiable with $f(x,x) \in R_2$. Whereas $f(h(x), h(a))$ is no instance of $f(x,x)$, since $h(x)|_\lambda \neq h(a)$, all ground instances $t\gamma$ of t, where γ is R_2-irreducible, are reducible by $f(x,x)$, as $h(x)\gamma|_\lambda = h(a)$. Therefore $h(x)$ is not allowed to be in a test set for R_2; $h(x)$, having only one R_2-irreducible ground instance, is not typical at position λ.

Definition 3 (problem positions) *The* problem positions in t w.r.t. R *are the set* $P_R(t) = \bigcup_{u \in N^+} P_R^u(t)$, *where* $P_R^u(t) =$
$\{v \in Pos(t) \mid t|_v \notin T_\Sigma$ *and* $\exists l \in R\colon uv \in Pos_{nl}(l)$ *and* t *is unifiable with* $l|_u\}$.

Note that $P_R(t) = P_R^\lambda(t) = \emptyset$ for linear R or for ground terms t. If R is not linear and t is not ground, then $P_R(t)$ contains always λ.

Definition 4 (typical) *For* $Y \subseteq Var(t)$, t *is* Y-typical w.r.t. R *if there exists a ground substitution* η *with domain* $Var(t) \setminus Y$ *and a family* $(G_y)_{y \in Y}$ *of infinite sets of ground terms such that* $t\eta\gamma \in Nf_R$ *for all substitutions* γ *where* $y\gamma \in G_y$ *for all* $y \in Y$.
For linear t *and* $U \subseteq Pos(t)$, t *is called* U-typical w.r.t. R *if there is a set* $Y \subseteq Var(t)$ *such that* $Var(t|_u) \cap Y \neq \emptyset$ *for all* $u \in U$ *and* t *is* Y*-typical w.r.t.* R.

Especially, t is $\emptyset$-typical iff t has at least one irreducible ground instance (i.e., it is not ground reducible) and t is $\{\lambda\}$-typical iff t has infinitely many irreducible ground instances. Y-typicality of a term t is equivalent to the existence of an irreducible ground instance of t where all variables in Y are substituted by terms whose depths exceed a certain – computable – bound.

Lemma 3 (bound $b(R)$**)** *Given a finite set* $R \subseteq T_\Sigma(X)$, *we can compute a bound* $b(R) \in \mathbb{N}$ *such that for all* $t \in T_\Sigma(X)$ *and all* R*-irreducible ground substitutions* γ *with domain* $Var(t)$ *the following holds:*

(1) For all $y \in Var(t)$ *with* $|y\gamma| \geq b(R)$ *there is an infinite set* $G \subseteq T_\Sigma$ *such that* $t\{y \mapsto g\}\gamma \subseteq_R t\gamma$ *for all* $g \in G$.
(2) More generally, for all $Y \subseteq Var(t)$ *with* $|y\gamma| \geq b(R)$ *for all* $y \in Y$ *there is a family* $(G_y)_{y \in Y}$ *of infinite sets of ground terms such that* $t\eta \subseteq_R t\gamma$ *for all substitutions* η *where* $y\eta \in G_y$ *for all* $y \in Y$ *and* $y\eta = y\gamma$ *for all* $y \notin Y$.

Corollary 1 (deciding typicality) t *is* Y*-typical iff there is a ground substitution* γ *such that* $t\gamma \in Nf_R(t)$ *and* $|y\gamma| \geq b(R)$ *for all* $y \in Y$.

If all terms in test sets meet the requirement "typical at problem positions" – besides being expanded –, then each test set instance $t\tau$ of a term t has a ground instance with the same reducers; as discussed above, this enables their use within a ground reducibility test. There are even infinitely many ground instances of $t\tau$ with this property if τ contains at least one typical term.

Lemma 4 (reducers) *Let* $t \in T_\Sigma(X)$, $R \subseteq T_\Sigma(X)$.

(1) If t *is expanded and* $P_R^\lambda(t)$*-typical w.r.t.* R, *then there exists a ground instance* $t\gamma$ *of* t *such that* $t\gamma \simeq_R^\lambda t$.
(2) Let T *be a linear set such that all* $s \in T$ *are expanded w.r.t. all proper subterms of* R *and are* $P_R(s)$*-typical w.r.t.* R, *let* $t\tau$ *be a* T*-instance of* t.
  (i) If σ *is a substitution such that* $x\tau\sigma$ *is* $P_R(x\tau)$*-typical w.r.t.* R *for all* $x \in Var(t)$, *then* $t\tau\sigma \simeq_R t\tau$.
  (ii) There exists a ground instance $t\tau\gamma$ *of* $t\tau$ *such that* $t\tau\gamma \simeq_R t\tau$.
  (iii) Let $Y \subseteq Var(t)$ *and* $y\tau$ *be* λ*-typical for all* $y \in Y$. *Then for each* $n \in \mathbb{N}$ *there is a ground instance* $t\tau\gamma$ *of* $t\tau$ *such that* $t\tau\gamma \simeq_R t\tau$ *and* $|y\tau| \geq n$ *for all* $y \in Y$.

We now have introduced all properties which are crucial for test sets. Note that Nf_R is always an infinite test set for R.

Definition 5 (test set) *$T \subseteq T_\Sigma(X)$ is a* test set *for R if T is linear and*
- *T is complete w.r.t. R,*
- *all terms in T are expanded w.r.t. all proper subterms of R,*
- *all $s \in T$ are $P_R(s)$-typical.*

If R is linear, then $P_R(s) = \emptyset$ for all $s \in T$, thus $P_R(s)$-typicality just requires that s is not ground reducible.

Example 2 (Example 1 cont'd) Let $R_3 = R_2 \cup \{f(g(x), g(y))\}$ and $R_4 = R_3 \cup \{i(f(x,x))\}$ over Σ. The finite test set $\{a, h(x), g(x), f(x,y), i(z)\}$ for R_1 is no test set for R_2 and R_3. Here we can choose $\{a, h(a), g(x), f(x,y), i(z)\}$. A test set for R_4 can be obtained by additionally replacing $f(x,y)$ by its instances $f(a,x), f(x,a), f(h(a),x)$, and $f(x,h(a))$. Note that $P_{R_4}(f(x,y)) = \{\lambda, 1, 2\}$, $f(x,y)$ is $\{1\}$-typical and $\{2\}$-typical, but not $\{1,2\}$-typical w.r.t. R_4.

A finite test set exists for all finite sets R. A simple algorithm for computing test sets can be given using corollary 1. It allows to find a "brute force" test set by simply enumerating terms of bounded depth. Choose T as a subset (where variants are omitted in order to make T finite) of

$$\{\, cut(t, d_R) \mid t \in T_\Sigma(X) \text{ is linear and not ground reducible by } R, \\ |t| \leq b(R) + d_R \text{ and } |u| = b(R) + d_R \text{ for all } u \in Pos_X(t) \,\}$$

where d_R is the maximal depth of terms in R and $cut(t, d_R)$ denotes (unique up to variants) a linear term such that t is an instance of $cut(t, d_R)$, $t = cut(t, d_R)\sigma$ say, variables in $cut(t, d_R)$ occur only at depth d_R, and $x\sigma$ is not ground for all variables x in $cut(t, d_R)$. Certainly much better algorithms exist; we do not elaborate on this point here, however.

The following results show how finite test sets can be used to decide ground reducibility and typicality of terms. They are immediate consequences of lemma 3, lemma 4, and the completeness of test sets.

Theorem 1 (ground reducibility test) *Let T be a test set for R. Then t is ground reducible w.r.t. R iff all T-instances $t\tau$ of t are R-reducible.*

Theorem 2 (typicality test) *Let T be a finite test set for R, let $Y \subseteq Var(t)$. Then t is Y-typical w.r.t. R iff there is a T-instance $t\tau$ of t such that $t\tau$ is R-irreducible and $y\tau$ is $\{\lambda\}$-typical w.r.t. R for all $y \in Y$.*

Later we will have to know that terms with a specific typicality are contained in all test sets for a given set.

Lemma 5 *Let $t \in T_\Sigma(X)$, $Y \subseteq Var(t)$, let T be finite and complete w.r.t. R. If there is an R-irreducible ground substitution γ such that $|y\gamma| \geq b(R)$ for all $y \in Y$, then there is a T-instance $t\tau$ of t such that $t\tau \subseteq_R t\gamma$ and $y\tau$ is λ-typical for all $y \in Y$.*

We also will have to compare different test sets for the same set, or more generally for different sets with the same reducible ground terms. Those test sets, indeed, are very similar.

Lemma 6 (different test sets) *Let $R_1, R_2 \subseteq T_\Sigma(X)$, $Red_{R_1} = Red_{R_2}$, and let T_1 (T_2) be a finite test set for R_1 (R_2). Then for all t and all T_1-instances $t\tau_1$ there is a T_2-instance $t\tau_2$ such that $t\tau_1$ and $t\tau_2$ are unifiable, with most general unifier σ say, and for all $x \in Var(t)$, if $x\tau_1$ is U-typical for $U \subseteq Pos(x\tau_1)$, then $x\tau_1\sigma$ is U-typical. Consequently*
(1) If R_1 is non-linear and $x\tau_2$ is ground for $x \in Var(t)$, then $x\tau_1$ is ground.
(2) $t\tau_1\sigma \simeq_{R_1} t\tau_1$ and $t\tau_2 \sqsubseteq_{R_1} t\tau_1$.
(3) $t\tau_2 \sqsubseteq_{R_2} t\tau_1$ if additionally for all $l_2 \in R_2 \setminus R_1$ there is a (non-linear) $l_1 \in R_1$ and a ground substitution γ with domain $Var_{nl}(l_1)$ such that $l_1\gamma = l_2$.

4 Linearizations and Idle Variables

In this section we answer the question for which sets R there is a *linear* set S with the same set of reducible ground terms, i.e., with $Red_R = Red_S$. Of course, choosing $S = Ground(R)$, such a linear set always exists; we therefore concentrate on finding *finite* sets S in the following. Now a first basic observation is that a linear set S with $Red_R = Red_S$ exists only if such a set S can be found as a *linear instantiation* of R, i.e., as a set consisting of linear instances of terms in R only.

Definition 6 (instantiation, linearization) *A set R^* is an* instantiation of *R if each term in R^* is an instance of some term in R; it is a* linear instantiation *if it is a linear set. A linear instantiation R^* of R is called* linearization of *R if $Red_{R^*} = Red_R$.*

Example 3 (Example 2 cont'd) A finite linearization of both R_3 and R_4 is $\{h(h(x)), h(g(x)), f(a,a), f(h(a), h(a))\}$, whereas R_2 has no finite linearization.

Lemma 7 *Assume $Red_R = Red_S$ for $R, S \subseteq T_\Sigma(X)$. Then there is an instantiation I of both R and S which is linear w.r.t. both R and S such that $Red_I = Red_R$. Moreover, if R and S are finite, then I is finite.*

Proof Let T be a test set for R and define $inst(S, T, R) =$
$\{ s\tau \mid s \in S,\ s\tau$ a T-instance of s, all principals of $s\tau$ are R-irreducible $\}$.
We first show that $I' = inst(S, T, R)$ is an instantiation of both R and S, that it is linear w.r.t. S, and $Red_{I'} = Red_R$. Obviously, I' is an instantiation of S, and it is linear w.r.t. S since T is linear. To show that I' is also an instantiation of R we use theorem 1: each $s \in S$ is ground reducible w.r.t. S, thus ground reducible w.r.t. R from $Red_R = Red_S$. Now by theorem 1 all T-instances $s\tau$ are R-reducible; hence all terms in I' are instances of terms in R.
Since I' is an instantiation of S, $Red_{I'} \subseteq Red_S$. To show $Red_S \subseteq Red_{I'}$ consider some $g \in Red_S$; let $s\gamma$ be an innermost redex in g w.r.t. S. $Red_R = Red_S$ implies that $s\gamma$ is also an innermost redex w.r.t. R in g; thus by completeness of T

w.r.t. R, $s\gamma = s\tau\eta$ for some T-instance $s\tau$ and a ground substitution η. As $s\tau\eta$ is innermost w.r.t. R, all principals of $s\tau$ are R-irreducible, hence $s\tau \in I'$ and $g \in Red_{I'}$.
Applying lemma 1 to R and its instantiation I', we obtain a set Q which is an instantiation of R, which has I' as an instantiation, and which is linear w.r.t. both R and I'. From $Red_{I'} \subseteq Red_Q \subseteq Red_R$ and $Red_{I'} = Red_R$ we get $Red_Q = Red_R$. Now choose $I = inst(Q, T_S, S)$ where T_S is a test set for S; as shown above, I is an instantiation of both S and Q, is linear w.r.t. Q, and $Red_I = Red_S$. By transitivity, I is an instantiation of R (via Q), and I is linear w.r.t. R (via Q and I').
If R is finite, a finite test set T for R exists; then finiteness of S implies finiteness of I' (cf. the remark in the preliminaries). Hence Q and I are finite, too. □

Since finite test sets can be effectively computed for all finite sets, the proof of lemma 7 provides a way to construct such a set I from given finite sets R and S. Now as an immediate consequence, for a (finite) set R there is a (finite) linear set with the same reducible ground terms if and only if R has a (finite) linearization; note that a set which is linear w.r.t. a linear set is linear itself.

Corollary 2 *Suppose R is finite. Then a finite set S with $Red_R = Red_S$ exists iff R has a finite linearization. Moreover, given a finite R and a finite linear set S with $Red_R = Red_S$, a finite linearization of R can be computed.*

Linearizations enjoy a nice property which makes them easy to handle: they can be approximated locally. Suppose there is a linearization of R; then each $t \in R$ can be replaced by a finite set of some of its linear instances without changing the set of reducible ground terms. In other words, a linearization of R does not exist if there is a (non-linear) term $t \in R$ that can not be linearized "locally", i.e., $Red_R \neq Red_{R \setminus \{t\} \cup t^*}$ for all linear instantiation t^* of $\{t\}$. On the other hand, given R, $t \in R$, and a set t^* such that $Red_R = Red_{R \setminus \{t\} \cup t^*}$, we know that $R \setminus \{t\} \cup t^*$ has a linearization if R has one.

Lemma 8 (linearizing step by step) *If R has a finite linearization and $S \subseteq R$, then*

- *there is a finite linear instantiation S^* of S such that $Red_R = Red_{R \setminus S \cup S^*}$,*
- *for all finite S^* with $Red_R = Red_{R \setminus S \cup S^*}$, $R \setminus S \cup S^*$, if finite, has a finite linearization.*

Proof Let R^* be a finite linearization of R. Define $S^* = \{s\sigma \mid s \in S,\ s\sigma \in R^*\}$. As R^* is finite and linear, S^* is so. $Red_{R \setminus S \cup S^*} \subseteq Red_R$ holds since S^* is an instantiation of S. In order to show $Red_R \subseteq Red_{R \setminus S \cup S^*}$ note that R^*, being an instantiation of R, is an instantiation of $R \setminus S \cup S^*$ as well, thus $Red_{R^*} \subseteq Red_{R \setminus S \cup S^*}$. Using $Red_{R^*} = Red_R$ we get $Red_R \subseteq Red_{R \setminus S \cup S^*}$.
Furthermore for each finite set $R \setminus S \cup S^*$ with $Red_{R^*} = Red_R = Red_{R \setminus S \cup S^*}$, $R \setminus S \cup S^*$ has a finite linearization by corollary 2. □

Note, however, that there are sets R whose non-linear terms can all be linearized "locally" w.r.t. R, whereas R has no finite linearization. Consider e.g.

$R = \{f(x,x), f(g(x),g(x)), f(f(x,y),z), f(x,f(y,z))\}$ over $\Sigma = \{f,g,a\}$. All terms in R can be safely replaced by finite linear instantiations separately - for $f(x,x)$ we can choose $\{f(a,a)\}$, for $f(g(x),g(x))$ the empty set -, whereas there is no finite linearization of the entire set R.

Lemma 8 enables an incremental approach to linearization: Starting with a (finite) set R, replace some term t in R by a finite linear instantiation t^* of $\{t\}$ if this does not affect the set of reducible ground terms; if for some t no such linear instantiation exists, then there is no finite linearization of R. Continuing with this process (starting with $R \setminus \{t\} \cup t^*$), we either reach a linearization of R or we finally know for sure that R cannot be linearized. Thus we can look for a linearization step by step looking in each step for a "local linearization" t^*.
But how to decide if such a t^* exists and, if so, how to construct it? A first observation is that a finite linear instantiation of a term t consists always of a set of instances of t, where all non-linear variables are substituted by ground terms. Thus "local linearizability" can easily be characterized using the notion of *idle* variables. These are variables which are rather idle, as their job, as far as only reducibility of ground terms is concerned, could be done as well by a finite set of ground terms.

Definition 7 (idle variables) $Y \subseteq Var(t)$ *is* idle in t w.r.t. R *if there exists a finite set Γ of ground substitutions with domain Y such that* $Red_{R \cup \{t\}} = Red_{R \cup t\Gamma}$.

Often we will call a single variable y idle meaning $\{y\}$ instead. For example, given $\Sigma = \{a,f\}$, x is idle in $f(x)$ w.r.t $\emptyset$, since $Red_{\{f(x)\}} = Red_{\{f(a)\}}$. Note that $Red_{R \cup \{t\}} = Red_{R \cup t\Gamma}$ iff $Red_{\{t\}} \subseteq Red_{R \cup t\Gamma}$ iff $Red_{\{t\}} \cap Nf_R \subseteq Red_{t\Gamma}$.

Example 4 (Example 2 cont'd) x is not idle in $f(x,x)$ w.r.t. R_2. x is idle in $f(x,x)$ w.r.t. R_3, however, as $R_3 = \{h(h(x)), h(g(x)), f(x,x), f(g(x),g(y))\}$ and $\{h(h(x)), h(g(x)), f(a,a), g(h(a),h(a)), f(g(x),g(y))\}$ reduce the same ground terms.

Lemma 9 *A finite linear instantiation t^* of $\{t\}$ with $Red_{R \cup \{t\}} = Red_{R \cup t^*}$ exists iff $Var_{nl}(t)$ is idle in t w.r.t. R.*

Proof Let $t^* = \{t\sigma_1, \ldots, t\sigma_n\}$ be a finite linear instantiation of $\{t\}$ such that $Red_{R \cup \{t\}} = Red_{R \cup t^*}$; define $\gamma_i = \sigma_i|_{Var_{nl}(t)}$ for $1 \leq i \leq n$ and $\Gamma = \{\gamma_1, \ldots, \gamma_n\}$. Clearly $Red_{R \cup t^*} \subseteq Red_{R \cup t\Gamma} \subseteq Red_{R \cup \{t\}}$, hence $Red_{R \cup \{t\}} = Red_{R \cup t^*}$ implies $Red_{R \cup \{t\}} = Red_{R \cup t\Gamma}$. Since t^* is linear, Γ is a set of ground substitutions with domain $Var_{nl}(t)$, i.e., $Var_{nl}(t)$ is idle in t w.r.t. R.
The other direction holds since $t\Gamma$ is always a finite linear instantiation of $\{t\}$ for finite sets of ground substitutions Γ with domain $Var_{nl}(t)$. □

In order to characterize idleness, we will use the following notations:
$red(t,Y,R,T) = \{\, t\tau|_Y \mid t\tau \text{ a } T\text{-instance of } t,\ t \text{ is } R\text{-irreducible},\ t\tau \simeq_{R \cup \{t\}} t \,\}$,
$red(t,Y,R) = red(t,Y,R,T_\Sigma)$.

It will be shown that Y is idle in t w.r.t. R if and only if $red(t, Y, R)$ is finite; if it is finite, $\Gamma = \{\gamma \mid t\gamma \in red(t, Y, R)\}$ is just such a finite set of ground substitutions we are looking for in order to verify idleness of Y. Finiteness of $red(t, Y, R)$ can be decided by computing the finite set $red(t, Y, R, T)$, where T is a finite test set for $R \cup \{t\}$. We first prove some (rather technical) basic properties of $red(T, Y, R)$ and $red(t, Y, R, T)$, and establish some connections between them. Note that $red(t, Y, R, T) = \{\, t\tau|_Y \mid t\tau$ is an R-irreducible T-instance of t whose principals are $\{t\}$-irreducible$\}$. Moreover, $red(t, Y, R, T)$ is an instantiation of $\{t\}$, and it is finite if T is finite.

Lemma 10 (properties of red) *Let $t \in \mathcal{T}_\Sigma(X), Y \subseteq Var(t)$, $R, T \subseteq \mathcal{T}_\Sigma(X)$.*

(1) $Red_{R\cup\{t\}} = Red_{R\cup red(t,Y,R)}$, and $Red_{R\cup\{t\}} = Red_{R\cup red(t,Y,R,T)}$ if T is complete w.r.t. $R \cup \{t\}$ and $t \notin X$.

(2) $red(t, Y, R)$ is an instantiation of $t\Theta|_Y$ for all sets of substitutions Θ with $Red_{R\cup\{t\}} = Red_{R\cup t\Theta}$ and $Y\Theta$ and t variable disjoint. Especially

(i) $red(t, Y, R)$ is an instantiation of $red(t, Y, R, T)$ if T is complete w.r.t. $R \cup \{t\}$ and $t \notin X$,

(ii) $red(t, Y, R) \subseteq t\Gamma$ for all sets of ground substitutions Γ with domain Y and $Red_{R\cup\{t\}} = Red_{R\cup t\Gamma}$,

(3) If T is a test set for $R\cup\{t\}$, then each term in $red(t, Y, R, T)$ has an instance in $red(t, Y, R)$.

(i) $red(t, Y, R, T) = red(t, Y, R)$ if additionally $Y\tau$ is ground for all $t\tau \in red(t, Y, R, T)$ and $t \notin X$.

(ii) $red(t, Y, R, S) = red(t, Y, R)$ for all sets $S \subseteq \mathcal{T}_\Sigma$ where $Red_{R\cup\{t\}} = Red_{R\cup red(t,Y,R,S)}$.

Proof (1) Let $g \in Red_{\{t\}} \setminus Red_R$ and $t\gamma$ be an innermost $\{t\}$-redex in g, i.e., $t\gamma$ is a subterm of g and no proper subterm of $t\gamma$ is an instance of t. As $t\gamma$ is R-irreducible and no proper subterm of $t\gamma$ is $\{t\}$-reducible, we conclude $t\gamma \simeq_{R\cup\{t\}} t$, thus $t\gamma|_Y \in red(t, Y, R)$ and $g \in Red_{red(t,Y,R)}$.
If $t \notin X$, then γ is $R \cup \{t\}$-irreducible. For a complete set T, $t\gamma$ is an instance of a T-instance $t\tau$ of t. From $t\gamma|_Y \in red(t, Y, R)$ we know $t \simeq_{R\cup\{t\}} t\gamma$. Now $t \subseteq_{R\cup\{t\}} t\tau \subseteq_{R\cup\{t\}} t\gamma$ implies $t \simeq_{R\cup\{t\}} t\tau$, thus $t\tau|_Y \in red(t, Y, R, T)$. As the subterm $t\gamma$ of g is an instance of $t\tau|_Y$, g is in $Red_{red(t,Y,R,T)}$.
(2) Let $t\gamma|_Y \in red(t, Y, R)$, where $t\gamma$ is an R-irreducible ground instance of t whose principals are $\{t\}$-irreducible. As $t\gamma \in Red_{\{t\}} \setminus Red_R$ and $Red_{\{t\}} \setminus Red_R \subseteq Red_{t\Theta}$, $t\gamma$ is an instance of some term t' in $t\Theta|_Y$; since $Y\Theta$ and t are variable disjoint, $t\gamma|_Y$ is an instance of t' as well. (2)(i) follows directly from (1), (2)(ii) is trivial.
(3) Let $t\tau$ be a T-instance of t, $t\tau|_Y \in red(t, Y, R, T)$. By lemma 4 (2)(ii), there is a ground instance $t\tau\gamma$ of $t\tau$ such that $t\tau\gamma \simeq_{R\cup\{t\}} t\tau$. Now $t\tau \simeq_{R\cup\{t\}} t$ implies $t\tau\gamma \simeq_{R\cup\{t\}} t$ and thus $t(\tau\gamma)|_Y \in red(t, Y, R)$. $t(\tau\gamma)|_Y$ is an instance of $t\tau|_Y$ since $t\tau$ and t are variable disjoint. (3)(i) follows directly from this and (2)(i). $red(t, Y, R) \subseteq red(t, Y, R, S)$ in (3)(ii) is a consequence of (2)(ii). By definition, $red(t, Y, R, S) \subseteq red(t, Y, R)$ if $S \subseteq \mathcal{T}_\Sigma$. Since every $S \supseteq Nf_{R\cup\{t\}}$ is complete w.r.t. $R \cup \{t\}$, by (1) we have $Red_{R\cup\{t\}} = Red_{R\cup red(t,Y,R,S)}$. □

Lemma 11 *Y is idle in t w.r.t. R iff $red(t, Y, R)$ is finite.*

Proof If Γ is a finite set of ground substitutions with domain Y and $Red_{R\cup\{t\}} = Red_{R\cup t\Gamma}$, then $red(t, Y, R) \subseteq t\Gamma$ by lemma 10 (2)(ii), thus $red(t, Y, R)$ is finite. For the other direction we observe that $red(t, Y, R) = t\Gamma$ for some set of ground substitutions Γ with domain Y; if $red(t, Y, R)$ is finite, then Γ is finite, too. By lemma 10 (1), $Red_{R\cup\{t\}} = Red_{R\cup red(t,Y,R)}$, hence Y is idle in t w.r.t. R. □

Using this result, it is easy to show that Y is idle in t w.r.t. R if and only if all $y \in Y$ are idle in t w.r.t. R.
Our test set approach for deciding the idleness of variables, and thus local linearizability, is based on the following result.

Theorem 3 (decidability of idleness) *Let $R \subseteq \mathcal{T}_\Sigma(X)$, $t \in \mathcal{T}_\Sigma(X) \setminus X$, $Y \subseteq Var(t)$, let T be a finite test set for $R \cup \{t\}$. Then the following are equivalent:*
(1) Y is idle in t w.r.t. R.
(2) $red(t, Y, R)$ is finite.
(3) $|y\gamma| < b(R \cup \{t\})$ for all $t\gamma \in red(t, Y, R)$, $y \in Y$.
(4) For all $t\tau \in red(t, Y, R, T)$ and all $y \in Y$, $y\tau$ is not λ-typical w.r.t. $R \cup \{t\}$.

Proof (1) ⇔ (2) is lemma 11, (3) ⇒ (2) is trivial.
(2) ⇒ (4): Let $y\tau$ be λ-typical where $t\tau$ is a T-instance of t, $t\tau \simeq_{R\cup\{t\}} t$, t R-irreducible, $y \in Y$. Define $\Gamma = \{\tau\gamma \mid t\tau\gamma$ a ground instance of $t\tau$, $t\tau\gamma \simeq_{R\cup\{t\}} t\tau\}$; by lemma 4 (2)(iii), $\{y\gamma \mid \gamma \in \Gamma\}$ is infinite. Since $t\tau \simeq_{R\cup\{t\}} t$, we have $t\gamma \simeq_{R\cup\{t\}} t$, thus $t\gamma|_{\{y\}} \in red(t, \{y\}, R)$, for all $\gamma \in \Gamma$. Hence $red(t, \{y\}, R)$, and for that $red(t, Y, R)$, is infinite.
(4) ⇒ (3): Let $t\gamma'$ be a ground instance of t, t R-irreducible, such that $t\gamma' \simeq_{R\cup\{t\}} t$, let $\gamma = \gamma'|_{\{y\}}$ and $|y\gamma| \geq b(R \cup \{t\})$ for some $y \in Y$. By lemma 5 there is a T-instance $t\tau$ of t where $t\tau \subseteq_{R\cup\{t\}} t\gamma'$ and $y\tau$ is λ-typical w.r.t. $R \cup \{t\}$. As $t\tau \subseteq_{R\cup\{t\}} t\gamma' \simeq_{R\cup\{t\}} t$ implies $t\tau \simeq_{R\cup\{t\}} t$, we get $t\tau|_{\{y\}} \in red(t, \{y\}, R, T)$. □

Especially, Y is idle in a non-linear term t if and only if $Y\tau$ is ground for all $t\tau$ in $red(t, Y, R, T)$; this is because all non-ground terms in a test set for a non-linear set are λ-typical. As a consequence, we have $red(t, Y, R) = red(t, Y, R, T)$ in this case.

Corollary 3 *Let T be a test set for $R \cup \{t\}$, $t \notin X$, where each term in $T \setminus \mathcal{T}_\Sigma$ is λ-typical. If $red(t, Y, R)$ is finite, then $red(t, Y, R) = red(t, Y, R, T)$.*

Proof If $red(t, Y, R)$ is finite, and each non-ground term in T is λ-typical, theorem 3, (2)⇒(4), implies that $Y\tau$ is ground for all $t\tau \in red(t, Y, R, T)$. Thus $red(t, Y, R) = red(t, Y, R, T)$ by lemma 10 (3)(i). Note that the proof of (2)⇒(4) in theorem 3 makes no use of the finiteness of T. □

Now we come back to linearizations. As stated in lemma 9, local linearizability of a term is equivalent to the idleness of its non-linear variables. In order to combine the results obtained so far, we use the notations

$$lin(t, R, T) = red(t, Var_{nl}(t), R, T) \text{ and } lin(t, R) = lin(t, R, \mathcal{T}_\Sigma).$$

Obviously, $lin(t, R)$ is a linear instantiation of $\{t\}$, and $Red_{R\cup\{t\}} = Red_{R\cup lin(t,R)}$ by lemma 10 (1). We even know by lemma 10 (2)(ii) that $lin(t, R) \subseteq t\Gamma|_{Var_{nl}(t)}$ for all linear instantiations $t\Gamma$ of $\{t\}$ with $Red_{R\cup\{t\}} = Red_{R\cup t\Gamma}$; in this sense, $lin(t, R)$ is minimal among all local linearizations of t.

Corollary 4 (local linearization) *Let $R \subseteq T_\Sigma(X)$, let t be non-linear, T be a finite test set for $R \cup \{t\}$. Then the following are equivalent:*
(1) There is a finite linear instantiation t^* of $\{t\}$ with $Red_{R\cup\{t\}} = Red_{R\cup t^*}$.
(2) $lin(t, R)$ is a finite linear instantiation of $\{t\}$ with $Red_{R\cup\{t\}} = Red_{R\cup lin(t,R)}$.
(3) $lin(t, R) \subseteq \{ t\gamma \mid \gamma$ a ground substitution with domain $Var_{nl}(t)$, $|y\gamma| < b(R \cup \{t\})$ for all $y \in Var_{nl}(t)$ $\}$.
(4) $lin(t, R) = lin(t, R, T)$.
(5) $lin(t, R, T)$ is linear.
(6) $lin(t, R, T)$ is a finite linear instantiation of $\{t\}$ with $Red_{R\cup\{t\}} = Red_{R\cup lin(t,R,T)}$.

Proof (1) is equivalent to the idleness of $Var_{nl}(t)$ in t w.r.t. R by lemma 9, thus by lemma 11 to the finiteness of $lin(t, R)$, i.e., to (2). (2)⇒(3) is a direct consequence of theorem 3, (2)⇒(3). Clearly (3) implies finiteness of $lin(t, R)$, thus (2). (2) implies (4) by corollary 3, (4) implies (2) since $lin(t, R, T)$ is always finite if T is finite.
(2)⇔(5): $lin(t, R, T)$ is linear iff for all $t\tau \in lin(t, R, T)$ and all $y \in Var_{nl}(t)$, $y\tau$ is ground, i.e., since T is a test set for a non-linear set, $y\tau$ is not λ-typical. By theorem 3, (4)⇔(2), this holds if and only if $lin(t, R)$ is finite.
From (2) we conclude (6) using (4); (6)⇒(5) is trivial. □

This shows that a finite linear instantiation t^* of $\{t\}$ with $Red_{R\cup\{t\}} = Red_{R\cup t^*}$ exists if and only if $lin(t, R, T)$ is linear, where T is any finite test set for $R\cup\{t\}$. Furthermore, if such sets t^* exist, then $lin(t, R, T)$ is always one of them. Since $lin(t, R, T)$ can easily be computed from a finite test set T, and since linearity of a finite set is trivially decidable, we can decide if a term t can be linearized locally in the context of a finite set R. As discussed above, we consequently can even decide, using lemma 8, if a finite set can be linearized entirely, and - in case it can - compute a linearization.

Corollary 5 *It is decidable if a finite set R has a finite linearization, and, in case it has, such a linearization can be computed.*

Corollary 4 provides another approach for computing linearizations without using test sets; this will be sketched briefly. Computing the finite set $G = \{t\gamma \mid \gamma$ a ground substitution, $dom(\gamma) = Var_{nl}(t)$, $|y\gamma| < b(R \cup \{t\})$ for $y \in Var_{nl}(t)\}$, we can decide whether there is a local linearization of t in the context of R as follows. If $Red_{R\cup\{t\}} = Red_{R\cup G}$, then obviously G is a local linearization of t. Otherwise, $lin(t, R)$ is no subset of G, since always $Red_{R\cup\{t\}} = Red_{R\cup lin(t,R)}$; thus by corollary 4, (1)⇒(3), t can not be linearized locally. $Red_{R\cup\{t\}} = Red_{R\cup G}$ can be checked by deciding if t is ground reducible w.r.t. $R \cup G$. Kucherov and Tajine [13] are using this approach for deciding local linearizability.

Using test sets, the decision procedure becomes easier: neither a ground reducibility test nor a computation of the bound $b(R \cup \{t\})$ is visible. This is possible because test sets are complete (which guarantees $Red_{R\cup\{t\}} = Red_{R\cup lin(t,R)}$), and because knowledge about the bound is already incorporated in the test set terms via the typicality requirement.

5 Computing Linearizations Using Test Sets

In the following we will use a *deduction rule* to describe and discuss different approaches for computing linearizations. This deduction rule operates on pairs (N, L) where N and L are finite sets of non-linear and linear terms respectively. In each step one of the non-linear terms is removed from N and replaced by a finite set of linear terms which is added to L. Given a set $R_0 = N_0 \cup L_0$, we non-deterministically apply the rule (LIN) defined below starting with (N_0, L_0) until a normal form is reached. If this normal form contains only linear terms, then its second component is a finite linearization of R_0. In case there is still some non-linear term, i.e., the first component of the normal form is non-empty, we know that no finite linearization of R_0 exists. Depending on the set T, the deduction rule (LIN) is defined as

$$\frac{(\,N\dot{\cup}\{l\},\ L\,)}{(\,N,\ L\cup lin(l, N\cup L, T)\,)} \qquad \text{if } lin(l, N\cup L, T) \text{ is linear.} \qquad (LIN)$$

We write $\vdash^{l,T}_{LIN}$ (or just $\vdash^{T}_{LIN}$) for the (one step) deduction relation using (LIN) with set T and non-linear rule l; its transitive and reflexive closure is denoted by $\models^{*\,l,T}_{LIN}$ ($\models^{*\,T}_{LIN}$). Some properties of this deduction relation are easy to prove using lemma 10 (1).

Lemma 12 (properties of LIN**)** *Let N, L be finite sets of non-linear and linear terms respectively, let $(N, L) \vdash^{T_1}_{LIN} \cdots \vdash^{T_n}_{LIN} (N', L')$. Then L' is linear; $N' \cup L'$ is an instantiation of $N \cup L$ and it is finite if all T_i $(1 \le i \le n)$ are finite. $Red_{N\cup L} = Red_{N'\cup L'}$ if all T_i $(1 \le i \le n)$ are complete (w.r.t. $N \cup L$).*

As suggested by corollary 4, the set T can be choosen as a finite test set for $N \cup \{l\} \cup L$. We use $(N, L) \vdash^{local}_{LIN} (N', L')$ iff $(N, L) \vdash^{T}_{LIN} (N', L')$ for some finite test set T for $N \cup L$.

Theorem 4 (linearizing using test sets) *Let N_0, L_0 be finite non-linear and linear sets respectively, $R_0 = N_0 \cup L_0$, let $(N_0, L_0) \models^{*\,local}_{LIN} (\bar{N}, \bar{L})$, and let $(\bar{N}, \bar{L})$ be in normal form w.r.t. $\vdash^{\bar{T}}_{LIN}$ for some finite test set $\bar{T}$ for $\bar{N} \cup \bar{L}$.*
If $\bar{N} = \emptyset$, then L is a finite linearization of R_0, otherwise R_0 has no finite linearization.

Proof (1) follows from lemma 12 as all test sets used are finite and complete for $N_0 \cup L_0$. To show (2) assume $l \in \bar{N}$. Since $\vdash^{\bar{T}}_{LIN}$ is not applicable to $(\bar{N}, \bar{L})$, $lin(l, \bar{N}\backslash\{l\} \cup \bar{L}, \bar{T})$ is non-linear. By lemma 4, l has no finite linear instantiation

l^* such that $Red_{\bar{N}\cup\bar{L}} = Red_{\bar{N}\setminus\{l\}\cup\bar{L}\cup l^*}$. Now lemma 8 implies that R_0 has no finite linearization. □

As $\vdash_{LIN}^{local}$ terminates (after at most $|N_0|$ steps) we are able to decide linearizability of R_0, and in case R_0 is linearizable $\bar{L}$ is a linearization of R_0.
Obviously, the major disadvantage with this procedure is that a new test set has to be computed for each deduction step. It can be shown, however, that it is possible to keep on using the same test set, namely a finite test set for the initial set R_0, during the whole (LIN)-deduction.

Lemma 13 *Let N_0, L_0 be finite sets of non-linear and linear terms respectively, let $(N_0, L_0) \vdash^{*\,local}_{LIN} (N, L) \vdash_{LIN}^{local} (N\setminus\{l\}, L \cup lin(l, N\setminus\{l\} \cup L, T))$. Then for all finite test sets T_0 for $N_0 \cup L_0$ we have $lin(l, N\setminus\{l\} \cup L, T_0) = lin(l, N\setminus\{l\} \cup L, T)$.*

Proof Let N_0, L_0, N, L, l, and T be as above (i.e., $l \in N$ is non-linear, T is a finite test set for $N \cup L$, and $lin(l, N\setminus\{l\} \cup L, T)$ is linear), let T_0 be a finite test set for $N_0 \cup L_0$. We use $lin_0 = lin(l, N\setminus\{l\} \cup L, T_0)$ and $lin = lin(l, N\setminus\{l\} \cup L, T)$ for short.
$lin_0 \subseteq lin$: Let $l\tau_0|_{Var_{nl}(l)} \in lin_0$. For the T_0-instance $l\tau_0$ of l we choose a T-instance $l\tau$ according to lemma 6. Assume $l\tau|_{Var_{nl}(l)} \notin lin$, i.e., $l \subset_{N\cup L} l\tau$. Applying lemma 6(3) ($N_0 \cup L_0$ corresponds to R_1, $N \cup L$ to R_2), we get $l\tau \subseteq_{N\cup L} l\tau_0$ and therefore $l \subset_{N\cup L} l\tau_0$, which implies $l\tau_0|_{Var_{nl}(l)} \notin lin_0$, a contradiction. Thus $l\tau|_{Var_{nl}(l)} \in lin$. Since lin is linear, $x\tau$ is ground for all $x \in Var_{nl}(l)$, thus $x\tau_0$ is ground, too, by lemma 6 (1). Now $x\tau$ and $x\tau_0$ are unifiable by lemma 6, thus $x\tau_0 = x\tau$. Hence $l\tau_0|_{Var_{nl}(l)} = l\tau|_{Var_{nl}(l)} \in lin$.
$lin \subseteq lin_0$: Let $l\tau|_{Var_{nl}(l)} \in lin$. For the T-instance $l\tau$ of l we choose a T_0-instance $l\tau_0$ according to lemma 6. Assume $l\tau_0|_{Var_{nl}(l)} \notin lin_0$, i.e., $l \subset_{N\cup L} l\tau_0$. By lemma 6(2) ($N \cup L$ corresponds to R_1, $N_0 \cup L_0$ to R_2), $l\tau_0 \subseteq_{N\cup L} l\tau$ and therefore $l \subset_{N\cup L} l\tau$, which implies $l\tau|_{Var_{nl}(l)} \notin lin$, a contradiction. Thus $l\tau_0|_{Var_{nl}(l)} \in lin_0$, and $l\tau_0|_{Var_{nl}(l)} \in lin$ from $lin_0 \subseteq lin$. If $x\tau_0$ would not be ground for some $x \in Var_{nl}(l)$, then $l\tau_0|_{Var_{nl}(l)}$ would not be linear, contradicting the linearity of lin. Thus $x\tau_0$ is ground for all $x \in Var_{nl}(l)$, and therefore $x\tau$ is ground, too, by lemma 6 (1). Now $x\tau_0 = x\tau$, as $x\tau_0$ and $x\tau$ are unifiable according to lemma 6. Hence $l\tau|_{Var_{nl}(l)} = l\tau_0|_{Var_{nl}(l)} \in lin_0$. □

Corollary 6 *Let T_0 be a finite test set for $N_0 \cup L_0$. Then $(N_0, L_0) \vdash_{LIN}^{l_1,local} \dots \vdash_{LIN}^{l_n,local} (N, L)$ iff $(N_0, L_0) \vdash_{LIN}^{l_1,T_0} \dots \vdash_{LIN}^{l_n,T_0} (N, L)$.*

Proof By induction on n; the case $n = 0$ is trivial. Suppose $(N_0, L_0) \vdash^{l_1,local} \dots \vdash^{l_{n-1},local} (N', L') \vdash^{l_n,local} (N, L)$. Lemma 13 implies $(N', L') \vdash^{l_n,T_0} (N, L)$, the induction hypothesis yields $(N_0, L_0) \vdash^{l_1,T_0} \dots \vdash^{l_{n-1},T_0} (N'L')$.
Now assume $(N_0, L_0) \vdash^{l_1,T_0} \dots \vdash^{l_{n-1},T_0} (N', L') \vdash^{l_n,T_0} (N, L)$. Since T_0 is complete w.r.t. $N' \cup L'$ and $l_n^* = lin(l_n, N'\setminus\{l_n\} \cup L', T_0)$ is linear, l_n^* is a finite linear instantiation of $\{l_n\}$ with $Red_{N'\cup L'} = Red_{N'\setminus\{l_n\}\cup L'\cup l_n^*}$, using lemma 12. Hence by corollary 4 $lin(l_n, N'\setminus\{l_n\} \cup L', T)$ is linear for all finite test sets T for $N' \cup L'$. The induction hypothesis yields $(N_0, L_0) \vdash^{l_1,local} \dots \vdash^{l_{n-1},local} (N'L')$, thus by

lemma 13 $l_n^* = lin(l_n, N'\backslash\{l_n\} \cup L', T)$, hence $(N', L') \vdash^{l_n, local} (N, L)$. □

Combining theorem 4 and corollary 6, we finally arrive at the result discussed above, showing that computing one test set is sufficient for the linearization of an entire set.

Theorem 5 (linearizing using one test set) *Let N_0, L_0 be finite sets of non-linear and linear terms respectively, $R_0 = N_0 \cup L_0$, let T_0 be a finite test set for R_0, let $(\bar{N}, \bar{L})$ be a normal form of (N_0, L_0) w.r.t. $\vdash_{LIN}^{T_0}$.*
If $\bar{N} = \emptyset$, then $\bar{L}$ is a finite linearization of R_0, otherwise R_0 has no finite linearization.

References

1. R. Bündgen. *Term completion versus algebraic completion.* Dissertation, Universität Tübingen (1991).
2. R. Bündgen and W. Küchlin. Computing ground reducibility and inductively complete positions. *Proc. 3rd RTA, LNCS* Vol. 355 (1989), 59–75.
3. H. Comon. *Unification et disunification. Théories et applications.* Thèse de Doctorat d'Université, Institut Polytechnique de Grenoble (1988).
4. J. H. Gallier and R. V. Book. Reductions in tree replacement systems. *Theoretical Computer Science* 37 (1985), 123–150.
5. D. Hofbauer and M. Huber. Typical terms and test sets. Unpublished (1992).
6. M. Huber. Testmengen für Grundreduzierbarkeit: Konstruktionen, Komplikationen, Korollare. Diplomarbeit, Technische Universität Berlin (1991).
7. J.-P. Jouannaud and E. Kounalis. Proof by induction in equational theories without constructors. *Proc. 1st LICS* (1986), 358–366.
8. D. Kapur, P. Narendran, and H. Zhang. Proof by induction using test sets. *Proc. 8th CADE,* LNCS Vol. 230 (1986), 99–117.
9. D. Kapur, P. Narendran, and H. Zhang. On sufficient completeness and related properties of term rewriting systems. *Acta Informatica* 24 (1987), 395–415.
10. E. Kounalis. Testing for inductive (co)-reducibility. *Proc. 15th CAAP,* LNCS Vol. 431 (1990), 221–238.
11. E. Kounalis. Testing for the ground (co)-reducibility property in term rewriting systems. Revised version (1991), to appear.
12. G. A. Kucherov. On relationship between term rewriting systems and regular tree languages. *Proc. 4th RTA,* LNCS Vol. 488 (1991), 299–311.
13. G. A. Kucherov and M. Tajine. Decidability of regularity and related properties of ground normal form languages. This volume (1992).
14. J.-L. Lassez and K. Marriott. Explicit representation of terms defined by counter examples. *Journal of Automated Reasoning* 3 (1987), 301–317.
15. T. Nipkow and G. Weikum. A decidability result about sufficient completeness of axiomatically specified abstract data types. *Proc. 6th GI Conf.,* LNCS Vol. 145 (1983), 257–268.
16. D. A. Plaisted. Semantic confluence tests and completion methods. *Information and Control* 65(2/3) (1985), 182–215.
17. S. Vágvölgyi and R. Gilleron. For a rewrite system it is decidable whether the set of irreducible ground terms is recognizable. Unpublished (1992).

Proving Group Isomorphism Theorems
(Extended Abstract)

Hantao Zhang*
Department of Computer Science
The University of Iowa
Iowa City, IA 52242
hzhang@cs.uiowa.edu

Abstract

We report the first computer proof of the three isomorphism theorems in group theory. The first theorem, the easiest of the three, was considered by Larry Wos as one of challenging problems for theorem provers. The technique we used is conditional completion which consists of one simplification rule called *contextual rewriting* and one inference rule called *clausal superposition*. Conditional completion works on conditional equations made from clauses and is a powerful method for clause-based theorem proving with equality.

1 Introduction

In this extended abstract, we present a case study of a conditional completion procedure which is an extension of the Knuth-Bendix completion procedure [5] for conditional equations [6, 7, 3, 2, 10]. The used conditional completion procedure can be viewed as a theorem proving method with good redundance control because it consists of a powerful simplification rule called contextual rewriting and a very restricted inference rule called clausal superposition (see [10] for the formal definitions of these two rules). We implemented the conditional completion procedure in the Rewrite Rule Laboratory (*RRL*) [4].

The problems of our case study consist of the three isomorphism theorems in group theory. Our attention on the group isomorphism problems was brought up by Larry Wos's thought-provoking book: ***Automated Reasoning: 33 Basic Research Problems*** [8]. Research Problem 8 of the book asks what inference rule, if any, effectively performs for set theory as paramodulation does for equality. The first isomorphism theorem (Test Problem 5) was chosen by Wos for testing possible solutions of this research problem. To challenge the people who may attack this problem, Wos wrote [8, pp. 128]:

> "If one were able to suggest a general approach that succeeds in proving various theorems of this type in a reasonable amount of computer time, starting with a fixed database of axioms and a set of lemmas, then we would consider the achievement most notable, whether or not the approach relies on an inference rule for set theory."

Because of the above wording, we thought in the beginning of our experiment that the first isomorphism theorem must be a tough one. In fact, a machine proof of a lemma related to the

*Partially supported by the National Science Foundation Grants no. CCR-9009414, INT-9016100 and CCR-9202838.

first isomorphism theorem was recently published in Journal of Automated Reasoning [9]. When we input a formulation of the first isomorphism theorem to *RRL*, surprisely, *RRL* produced a proof in seconds. Encouraged by this result, we continued to prove, successfully, the second and the third isomorphism theorems. If Wos's judgement on this test problem is right, then our experimental results really demonstrate that conditional completion is a powerful method to theorem proving with equality.

Because of the space restriction, we only briefly introduce the three group isomorphism theorems and omitted the details of the proofs (which is provided in the full version of the paper). Instead, we discuss the problems and issues related to the proofs of these theorems.

2 The Three Group Isomorphism Theorems

A ***group*** G is an algebraic structure $(G, *)$, where (i) G is a set of elements closed under the binary operation $*$, (ii) there exists a ***unit element*** $e \in G$ such that for any $x \in G$, $x * e = x$, (iii) for any $x \in G$, there exists an ***inverse*** $y \in G$ such that $x * y = e$. H is called a ***subgroup*** of $(G, *)$ if $H \subseteq G$ and $(H, *)$ is a group.

Given two groups $(G, *)$ and $(H, \circ)$, a function $h : G \rightarrow H$ is said to be a ***homomorphism*** if $h(x_1 * x_2) = h(x_1) \circ h(x_2)$ for any $x_1, x_2 \in G$. If h is one-to-one and onto, we say G is ***isomorphic*** to H. The ***kernel*** of h is the set $ker(h) = \{x \in G \mid h(x) = 0\}$, where 0 is a unit of $(H, \circ)$.

Let $(G, *)$ be a group. If $A \subset G$ and $B \subset G$, then AB denotes the set $\{x * y \mid x \in A, y \in B\}$. If $x \in G$ and $S \subset G$, $xS = \{x * y \mid y \in S\}$ is called a ***left coset*** of S. Similarly, Sx denotes a ***right coset*** of S.

A subgroup N of G is ***normal*** in G (denoted by $N \rhd G$) if $xN = Nx$ for any $x \in G$. In this case, we use G/N to denote the set of all cosets of N (i.e., $G/N = \{xN \mid x \in G\}$), with the operation f defined on G/N: $f(xN, yN) = (x * y)N$ for any $x, y \in G$. It is relatively easy for *RRL* to prove that f is well-defined, i.e., for any $x_1, x_2, y_1, y_2 \in G$, $x_1N = x_2N$ and $y_1N = y_2N$ imply $(x_1 * y_1)N = (x_2 * y_2)N$.

Here are the three famous isomorphism theorems in group theory:

- **The First Isomorphism Theorem** Let h be a homomorphism from the group G onto the group H. Then $ker(h) \rhd G$ and $G/ker(h)$ is **isomorphic** to H. Conversely, let $N \rhd G$. Then there exists a homomorphism h' from G onto G/N, defined by $h'(g) = gN$, with $ker(h') = N$.

- **The Second Isomorphism Theorem** Let G be a group, $A \rhd G$, and B a subgroup of G. Then $(A \cap B) \rhd B$ and AB/A is **isomorphic** to $B/(A \cap B)$.

- **The Third Isomorphism Theorem** Let G be a group and let $H \rhd G$, $N \rhd G$, and N a subgroup of H. Then $H/N \rhd G/N$ and $(G/N)/(H/N)$ is **isomorphic** to G/H.

3 Problems and Issues

In this section, we discuss two problems which we feel that their solutions are important to the efficiency of the proofs of the three group isomorphism theorems. One problem concerns about specification of the theorems in the first order logic; the other concerns about the control of the conditional completion procedure.

3.1 Choose Right Specification

For any theorem, there may exist several ways to specify it in a first-order logic. The choice of specification may have little impact to the eye of the human but is important to theorem provers. The following small tricks are used in our specification and they facilitate the proving process and improve the readability of the proofs.

- To avoid the predicates like $x \in G$, we used a typed (i.e., multi-sort) language for formulas. For instance, to say that G is closed under $*$, we declare the arity of $*$ as $* : G, G \rightarrow G$. Similarly, to say h is a homomorphism from $(G, *)$ to $(H, \circ)$, we declare the arity of h as $h : G \rightarrow H$ and input the equation $h(x * y) = h(x) \circ h(y)$ without quantifying on x and y. By using a multi-sort language, we may reduce the number of literals in a clause, thus reduce the search space in general.
- When requiring to define a set, instead of giving a construction of a set (which is hard to do), we define this set by giving its membership function. For instance, the kernel K of $h : G \rightarrow H$ is defined as $(x \in K) \equiv (h(x) = 0)$, where 0 is the unit of H.
- In order to avoid the redundance caused by converting formulas into clauses, and to take advantage of rewriting, some non-clausal formulas like $(x \in K) \equiv (h(x) = 0)$ are used as a rewrite rule $(x \in K) \rightarrow (h(x) = 0)$ in the proofs. We consider that the left-hand side of this kind of rules is just a shorthand for the right-hand side.
- Instead of proving directly the equivalence of two sets A and B, we split the proof into two parts: For any x, (i) $x \in A$ implies $x \in B$ and (ii) $x \in B$ implies $x \in A$ (both of them are in clausal form). However, once these two sets are proved to be equivalent, we simply assume that $A = B$ and replace one by the other at will.

3.2 Control of the Completion Procedure

The input of each subproblem of the three isomorphism theorems contain the axiomatization of a group $(G, *)$:

$$\begin{array}{llll} (1) & x * e & = & x & \text{(* } e \text{ is a right unit of } G. \text{ *)} \\ (2) & x * i(x) & = & e & \text{(* } i(x) \text{ is a right inverse of } x \text{ in } G. \text{ *)} \\ (3) & (x * y) * z & = & x * (y * z) & \text{(* } * \text{ is associative. *)} \end{array}$$

When the above equations, together with the precedence order $i > * > e$, are input to *RRL*, *RRL* runs the Knuth-Bendix completion procedure [5] and generates a set of ten rewrite rules [5].

However, if we mix the above three equations with other conditional equations and input them to the completion procedure, the completion procedure often wastes too much time to compute critical pairs with some rules which are generated from the above three equations but are not the members of the final set of ten rules. We found in our experiment that dividing the completion into two phases helps a lot: In the first phase, we run the completion procedure on the above three equations (*RRL* takes only 0.4 seconds to complete on a Sun Sparcstation 1 with 16 Megabytes main memory). In the second phase, we input conditional equations special to each subproblem. Note that we did not use any set-of-support strategy in the proofs.

We also found that the use of lemmas is very critical. Some lemmas can be easily proved to be true, but they cannot be easily generated by the completion procedure. Providing one such lemma by the user could speed a proof by several hundred factors. However, more user-provided lemmas means less automatic power of the prover. This is explained in details in the full version of the paper.

problem	time (second)	lemmas by user
(1a), (1b), (1c)	0.09, 0.11, 0.20	1
(2a), (2b), (2c)	0.20, 0.48, 0.30,	2
(3a)	0.42	1

Table 1: Computer time of the three isomorphism theorems.

4 Discussion about the Proofs

The first isomorphism theorem requires to prove the following:

- (1a) If h is a homomorphism from G to H, then $ker(h) \rhd G$.
- (1b) If h is a homomorphism from G to H, then $G/ker(h)$ is isomorphic to H.
- (1c) If $N \rhd G$ and $h_3(x) = xN$, then h_3 is a homomorphism from G onto G/N and $ker(h_3) = N$.

Problem (1a) is Test Problem 4 in [8]. A computer proof of this problem is reported in [9] by Yuan Yu using the Boyer-Moore theorem prover [1]. Yu's proof, which needs a lot of assistance from the user, assumes that groups are finite (we do not need this assumption). Our proof of (1a) is obtained instantly without any assistance from the user.

Problem (1b) is the so-called ***first isomorphism theorem*** in [8]. It is so because (1b) is the major part of the whole theorem.

Let G be a group, $A \rhd G$, and B a subgroup of G. The second isomorphism theorem requires to show that

- (2a) $(A \cap B) \rhd B$.
- (2b) AB is a subgroup of G and $A \rhd AB$.
- (2c) AB/A is isomorphic to $B/(A \cap B)$.

Instead of directly proving (2c), we prove that there is a homomorphism h_4 from AB onto $B/(A \cap B)$, with $ker(h_4) = A$. If this is done, by the first isomorphism theorem, $A \rhd AB$, and AB/A and $B/(A \cap B)$ are isomorphic.

Let G be a group and let $H \rhd G$, $N \rhd G$, and N a subgroup of H. Like the proof of (2c), instead of directly proving the third isomorphism theorem, we prove that there is a homomorphism h_5 from (G/N) onto G/H, with $ker(h_5) = H/N$. If this is done, by the first isomorphism theorem, $H/N \rhd G/N$, and $(G/N)/(H/N)$ and G/H are isomorphic. Hence, the proof of the third isomorphism theorem is reduced to the following statement:

- (3a) There is a homomorphism h_5 from (G/N) onto G/H, with $ker(h_5) = H/N$.

Table 1 records the cumulated computer time for each of these theorems. The times are measured in AKCL on a Sun Sparcstation 1 with 16 Megabytes main memory and exclude the time for obtaining a set of ten rules from equations (1)-(3) in the previous section. The table also gives the number of intermediate lemmas designed by the user (each lemma must be proved to be true before it can be added in the system).

The proofs of these theorems are not deep; for most of them, contextual rewriting is sufficient to do the job. Interestingly, many problems in group theory are of this type. In fact, besides the three isomorphism theorems, we are able to prove theorems about automorphisms and products of groups, and homomorphisms of rings.

Strictly speaking, our proofs of the three isomorphism theorems are not fully automatic. For instance, when requiring to prove the existence of a homomorphism from a set to another, the user has to explicitly construct a function and then lets the theorem prover to check that the constructed function is really a homomorphism. Also, in the proofs of the second and the third isomorphism theorems, the first isomorphism theorem is used by the user, not by the prover. However, all the lemmas and theorems input to *RRL* are automatically proved by *RRL*. We think that a fully automatic proof of these theorems in the sense that a homomorphism can automatically generated is beyond the power of today's theorem provers.

Our experiment does not provide any solution to Wos' Research Problem 8, which asks what inference rule effectively performs for set theory. Our experiment only demonstrates again the power of conditional completion in solving problems with clauses and equality.

References

[1] Boyer, R.S., Moore, J S.: (1979) A Computational Logic. Academic Press, New York.

[2] Ganzinger, H.: (1987) A completion procedure for conditional equations. In Kaplan, S., Jouannaud, J.-P. (eds) Proc. of Conditional Term Rewriting Systems. Lecture Notes in Computer Science, vol. 308. pp. 62-83.

[3] Kaplan S.: (1984) Fair conditional term rewriting systems: unification, termination and confluence. Technical Report, LRI, Orsay.

[4] Kapur, D., Zhang, H.: (1989) An overview of RRL: Rewrite Rule Laboratory. In: Dershowitz, N. (ed.): Proc. of the third international conference on rewriting techniques and its applications. Lecture Notes in Computer Science 355, Springer. pp. 513-529.

[5] Knuth, D., Bendix, P.: (1970) Simple word problems in universal algebras. In: Leech, (ed.) Computational problems in abstract algebra. New York: Pergamon Press, pp. 263-297.

[6] Lankford, D.S.: (1979) Some new approaches to the theory and applications of conditional term rewriting systems. Report MTP-6, Dept. of Mathematics, Lousiana Tech University, Ruston, LA.

[7] Remy, J.L.:(1982) Etudes des systemes reecriture conditionelles et applications aux types abstraits algebriques. These d'etat, Universite de Nancy I, Nancy, France.

[8] Wos, L.R.: (1988) Automated reasoning: 33 basic research problems. New Jersey: Prentice Hall.

[9] Yu, Y.: (1990) Computer proofs in group theory. J. of Automated Reasoning **6** 251-286.

[10] Zhang, H., Kapur, D.: (1988) First-order logic theorem proving using conditional rewrite rules. In: Lusk, E., Overbeek, R., (eds.): Proc. of 9th international conference on automated deduction. Lecture Notes in Computer Science 310, Springer, pp. 1-20

Semigroups Satisfying $x^{m+n} = x^n$

Nachum Dershowitz*
Department of Computer Science

University of Illinois
Urbana, IL 61801
U.S.A.
nachum@cs.uiuc.edu

Hebrew University
Jerusalem 91904
Israel
nachum@cs.huji.ac.il

Abstract

We summarize recent results on semigroups satisfying the identity $x^{m+n} = x^n$, for $n \geq 0$ and $m \geq 1$, and some rewrite techniques that have contributed to their investigation.

1 Introduction

Ninety years ago, Burnside [1902] posed the question whether every group satisfying the identity $x^m = 1$, and having a finite number of generators, is finite. In 1969, Brzozowski (see the list of open questions in [Brzozowski, 1980]) conjectured that the congruence classes on words generated by $x^{n+1} = x^n$, are all regular sets. Recently, McCammond [1991] extended this conjecture to all semigroups satisfying $x^{m+n} = x^n$ and investigated the decidability of their word problems. These conjectures have been the topic of recent research, which we summarize here.

Consider the set A^* of finite words over some *finite* alphabet A containing at least two letters, and suppose we identify certain repetitious words. (The case $|A| = 1$ is patently uninteresting.) Specifically, a word of the form $ux^{m+n}v$, where x is any subword repeated contiguously $m+n$ times ($n \geq 0$, $m \geq 1$), is equivalent to the shorter word ux^nv. Let $\sim_{m,n}$ denote this congruence on words. In other words, we are looking at the algebras $A^*/\sim_{m,n}$, with finite generating set A, and with a binary juxtaposition operation that satisfies the axiom of associativity, $(xy)z = x(yz)$, as well as $x^{m+n} = x^n$. The different cases are portrayed in Table 1.

We are interested in the following three questions:

1. Does $\sim_{m,n}$ have finite index (finitely many congruence classes)? In other words: Is the algebra $A^*/\sim_{m,n}$ finite?

2. Is each of the congruence classes in $A^*/\sim_{m,n}$ a regular (recognizable, rational) set?

*This work was supported in part by a Lady Davis fellowship at the Hebrew University and by the U. S. National Science Foundation under Grants CCR-90-07195, CCR-90-24271, and INT-90-16958.

	$n=0$	1	2	3	4	5	$\cdots$
$m=1$	$x=1$	$xx=x$	$x^3=xx$	$x^4=x^3$	$x^5=x^4$	$x^6=x^5$	$\cdots$
2	$xx=1$	$x^3=x$	$x^4=xx$	$x^5=x^3$	$x^6=x^4$	$x^7=x^5$	$\cdots$
3	$x^3=1$	$x^4=x$	$x^5=xx$	$x^6=x^3$	$x^7=x^4$	$x^8=x^5$	$\cdots$
4	$x^4=1$	$x^5=x$	$x^6=xx$	$x^7=x^3$	$x^8=x^4$	$x^9=x^5$	$\cdots$
5	$x^5=1$	$x^6=x$	$x^7=xx$	$x^8=x^3$	$x^9=x^4$	$x^{10}=x^5$	$\cdots$
6	$x^6=1$	$x^7=x$	$x^8=xx$	$x^9=x^3$	$x^{10}=x^4$	$x^{11}=x^5$	$\cdots$
7	$x^7=1$	$x^8=x$	$x^9=xx$	$x^{10}=x^3$	$x^{11}=x^4$	$x^{12}=x^5$	$\cdots$
8	$x^8=1$	$x^9=x$	$x^{10}=xx$	$x^{11}=x^3$	$x^{12}=x^4$	$x^{13}=x^5$	$\cdots$
9	$x^9=1$	$x^{10}=x$	$x^{11}=xx$	$x^{12}=x^3$	$x^{13}=x^4$	$x^{14}=x^5$	$\cdots$
$\vdots$	$\vdots$	$\vdots$	$\vdots$	$\vdots$	$\vdots$	$\vdots$	$\vdots$
115	$x^{116}=1$	$x^{117}=x$	$x^{118}=xx$	$x^{119}=x^3$	$x^{120}=x^4$	$x^{121}=x^5$	$\cdots$
$\vdots$	$\vdots$	$\vdots$	$\vdots$	$\vdots$	$\vdots$	$\vdots$	$\vdots$

Table 1: Semigroups satisfying $x^{m+n} = x^n$.

3. Is the congruence $\sim_{m,n}$ on A^* decidable? In other words: Is the free word (identity) problem for the variety defined by (associativity and) $x^{m+n} = x^n$ recursively solvable?

The top-left case ($m = 1$, $n = 0$) of Table 1 is a trivial algebra. The next case of the top row ($m, n = 1$), the (free) idempotent semigroups, are called *(free) bands*. The algebras $x^m = 1$ in the first column are called *Burnside groups.* (They are groups, since every element of A has an inverse x^{m-1}.) By extension, the rest of the algebras have been called *Burnside algebras.* We might call the first row *Brzozowski semigroups*, since he proposed the question whether their equivalent words form regular sets.

Before proceeding, it is important to realize that two words may be congruent, though neither contains an instance of x^{m+n}. For example, from $xxx = xx$ one can derive $yxyxxyxyxxyxy = yxyxxyxy$. The point is that the equation $x^{m+n} = x^n$ on strings has critical pairs with itself (modulo associativity).

A general approach to these three questions is the following: Construct an infinite convergent (terminating, and confluent) string-rewriting system (semi-Thue system) for the theory by looking at longer and longer ground instances of the axiom. If any sufficiently long word is reducible by some rule, then the algebra is finite. If only finitely many rules are needed to rewrite all elements of an equivalence class to normal form, then under certain conditions the class is regular. If there is an effective way of generating a finite set of rules for reducing any given word to normal form, then the word problem is decidable.

For notation and concepts related to rewriting, see [Dershowitz and Jouannaud, 1990].

2 Finiteness

Burnside [1902] showed that the groups $x^m = 1$, for $m = 1, 2, 3$, are finite and asked whether the same was true for all m. Sanov [1940] proved $x^4 = 1$ to be finite and Hall [1957] showed the same for exponent 6. In 1968, Novikov and Adian, in a series of papers, showed that there are *infinite* Burnside groups for all odd $m \geq 4381$; this negative result was extended to odd $m \geq 665$ in Adian's monograph [1979] on the subject (to which [Adian, 1977] serves as an introduction). More recently, the bound was improved to 115, and the conjecture was shown false for all m greater than 2^{13} [Adian, personal communication]. The question remains open for $m = 5, 7, 8, 9, \ldots, 113, 114, 116, 118, \ldots, 2^{13}$. Much work has been done on this and related questions; it is still an active area of research.

Green and Rees [1952] proved that $x^m = 1$ is finite if and only if the semigroup $x^{m+1} = x$ is. So columns 1 and 2 of Table 1 have identical finiteness properties. In particular, bands ($xx = x$) are finite. In fact, the conditional equation

$$C(y) \subseteq C(x) = C(z) \quad \Rightarrow \quad xyz \;=\; xz$$

captures all the infinitely many critical pairs derivable from the defining axiom $xx = x$, where $C(x)$ denotes the set of letters in x. It applies to any subword xyz such that the letters in x and z are the same and include all those in y. See [Howie, 1976, Chap. IV] and [Siekmann and Szabó, 1982]. Finitely generated bands are finite, since there are only finitely many words not equivalent to a shorter word. To see this, suppose a word w contains n letters and is of length $2^{n+1} - 1$. We show, by induction on n, that it must contain an instance of xyz with $C(y) \subseteq C(x) = C(z)$; hence, by the above equation, it is equivalent to a shorter word. Let x be the shortest prefix of w containing all n letters, and z the shortest such suffix. If x and z overlap, or if $w = xz$, then one must be of length at least 2^n and have a subword of length $2^n - 1$ containing only $n - 1$ letters, which, by the inductive hypothesis, can be replaced with a smaller word. If, on the other hand, $w = xyz$, with y non-empty, then $C(y) \subseteq C(x) = C(z)$ and $xyz = xz$. A similar, but more complicated, argument in [Green and Rees, 1952] establishes that the algebra is precisely of size

$$\sum_{k=1}^{n} \binom{n}{k} \prod_{i=1}^{k-1} (k - i + 1)^{2^i}$$

which, asymptotically, looks more like $(n/e)^{n^2}$ than like the n^{2^n} one gets by the above simplistic reasoning.

The other algebras of the table are all infinite, since there exist infinitely many square-free words (words not of the form $uxxv$). Since neither side of the axiom $x^{m+n} = x^n$, for $n \geq 2$, can apply to any of these words, each square-free word is in a different class, of which there are infinitely many. It was Axel Thue [1912] who first constructed an infinite square-free word over a three letter alphabet, as well as an infinite cube-free word in a binary alphabet. The infinite sequence of square-free words in Figure 1 is due to Aršon [Aršon, 1937]: Each word is obtained from the previous by substituting $a \mapsto abc$, $b \mapsto bca$, and $c \mapsto cab$ for letters in odd positions, and $a \mapsto cba$, $b \mapsto acb$, and $c \mapsto bac$, for letters in even positions. To obtain a sequence of cube-free words over two letters, 0 and 1, one can apply the map: $a \mapsto 01$, $b \mapsto 010$, and $c \mapsto 0110$. For details, see [Adian, 1979, Chap. I].

Since there are only finitely many square-free words over a binary alphabet, this argument does not work for $n = 2$ and $|A| = 2$. Nevertheless, Brzozowski, Culik and

a
abc
$abc\ acb\ cab$
$abc\ acb\ cab\ cba\ cab\ acb\ cab\ cba\ bca$
⋮

Figure 1: Square-free words.

[0]
[001]
[001 001 100]
[001 001 100 001 001 100 100 001 001]
⋮

Figure 2: Congruence classes for $x^3 = xx$.

Gabrielian [1971] showed that the congruence induced by $x^3 = xx$ has infinitely many classes, as shown in Figure 2, where each class is obtained from the previous representative by applying the morphism $0 \mapsto 001$, $1 \mapsto 100$. No sequence of applications of the axiom can equate the representative elements of distinct classes.

When the algebra is finite, ordered completion (see [Hsiang and Rusinowitch, 1987; Dershowitz, 1992; Bachmair and Dershowitz, 199?]) can be used to generate its multiplication table. One computes critical pairs with the axioms, and, at the same time, normal forms of successively larger and larger words. This works provided one can determine when sufficiently many rules have been generated from the axioms for words of any given length. For example, for bands ($xx = x$) and $A = \{a, b\}$, one starts with $a \to 1$, $b \to 2$, $ba \to 3$, and $ab \to 4$. Since aab has two normal forms, 14 and 4, we get $14 \to 4$. Eventually, one gets the six elements in Table 2 (as predicted by the formula on the previous page). Pedersen [1988] performed some experiments with such a method.

	1	2	3	4	5	6
1	1	4	6	4	4	6
2	3	2	3	5	5	3
3	3	5	3	5	5	3
4	6	4	6	4	4	6
5	3	5	3	5	5	3
6	6	4	6	4	4	6

Table 2: The free band on two generators.

3 Regularity

If an algebra is finite, then each equivalence class is regular, since each congruence class can be identified with a state, congruent prefixes being interchangeable. By the same token, if all prefixes (or all suffixes, let alone all subwords) of words in a particular class belong to a finite number of classes, then that class is regular.

Brzozowski, Culik, and Gabrielian [1971] showed that each of the classes in Figure 2 is regular which lent support to the conjecture that such is the case for all equivalence classes for the semigroup varieties $x^{n+1} = x^n$. Imre Simon, in unpublished notes (see [Brzozowski, 1980]), contributed to this problem. The first solution, for $n \geq 5$, was by de Luca and Varricchio [1990] who believed their method could be extended to $n = 4$. McCammond [1991] generalized the question to all Burnside semigroups $x^{m+n} = x^n$, and—taking a different approach—solved it for all $m \geq 1$ and $n \geq 6$. Most recently, do Lago [1992] (for his Master's thesis) refined the approach of de Luca and Varricchio, showing regularity for all $m \geq 1$ and $n \geq 4$, and leaving hope that the method applies to $n = 3$, too.

The following combinatorial result from Fine and Wilf [1965] is essential to obtaining these results: If w is a word with periods p and q (that is, if $u'w = u^p$ and $v'w = v^q$ for some suffixes u' of u and v' of v), then w also has a period $gcd(p,q)$, the greatest common divisor of its two periods—provided w is of length at least $p+q-gcd(p,q)$. This is used in [do Lago, 1992] to show that all the critical pairs $l \rightarrow r$, have a special form: the right-hand (shorter) side r is a suffix of the left side l ($l = ur$), as well as a prefix of l ($l = rv$), the remainder of which (v) is of the form w^m, for the given m, where w is the *shortest* periodic suffix of r. For example, the reduced critical pair, $(01)^2(10101)^2 \rightarrow (01)^2(10101)$, formed from the instances $(01)^3 \rightarrow (01)^2$ and $(10101)^3 \rightarrow (10101)^2$ of the axiom (for $m = 1$ and $n = 2$), is of the desired form. In general, for the critical pairs to have this form, $n \geq 4$ is required. By analyzing the structure of derivations (with the closure set of reduced critical pairs of this form), it can be shown that the set of normal forms of subwords of elements of any one class is finite, establishing regularity. See [de Luca and Varricchio, 1990; do Lago, 1992].

4 Decidability

When an algebra is finite, there is a finite (unconditional) rewrite system to decide its word problem, that is, validity of ground equations over a finite set of generators. For bands, for example, one need only include a rule $w \rightarrow w'$, where w' is the shortest word equivalent to w, whenever w is of length up to $2^{n+1}-1$ and $w \neq w'$. (See [Benninghofen *et al.*, 1987, Chap. II].) Longer rules have reducible left-hand sides, and contribute nothing. Siekmann and Szabó [1982] give a simple decision procedure for free bands using the following conditional string-rewriting system:

$$\begin{array}{rrcl} & xx & \rightarrow & x \\ C(y) \subseteq C(x) = C(z) \mid & xyz & \rightarrow & xz \end{array}$$

for which they give a proof of the convergence (that is, termination and confluence). There are infinitely many "square-free" words to which the first rule does not apply, but the second does. (An extension of this system, for the join of bands and commutative semigroups, is given in [Nordahl, 1992].)

Even the infinite Burnside groups, odd $n \geq 665$, $m = 1$, have decidable free word problems (see [Adian, 1979, Chap. VI]). The word problem is also decidable for finitely presented groups whose relations are all of the form $w^n = 1$, with n is sufficiently large [Adian, 1979, Preface].

The identity problems for all the cases known to be regular, namely $n \geq 4$, $m \geq 1$, are similarly decidable, since there is an effective way of constructing just the rewrite rules up to the size needed to map an element of A^* to its normal form, rather than generate the whole, infinite system for the theory. See [de Luca and Varricchio, 1990] and [do Lago, 1992].

5 Discussion

Much of the work we have described considers *production* rules $r \rightarrow l$, rather than *reduction* rules $l \rightarrow r$, as we have. In these papers, termination of reduction is invariably based simply on word length. The notions of local and global confluence do play an important role in the work on regularity of the Burnside semigroups. The notion of critical pair is also central, but less explicit. (The "closure under reductions" of [do Lago, 1992], for example, is exactly closure under critical pair generation.)

There are many other questions about semigroups (let alone richer algebraic structures) to which rewriting techniques have been applied. For example, Ehrenfeucht, Haussler, and Rozenberg [1983] give the following generalization of the Myhill-Nerode Theorem: A subset of a semigroup S is regular if and only if it closed with respect to some well-quasi-order $\preceq$ on S that has the replacement property: $x \preceq y$ implies $uxv \preceq uyv$ for all (empty or nonempty) words x, y, u, v. (A well-quasi-order $\preceq$ is a reflexive-transitive binary relation that has no infinite descending sequences $s_1 \succ s_2 \succ \cdots$ and no infinite antichains of incomparable elements.) From this, it follows that a language (over a finite alphabet) is regular if and only if it is produced by a string-rewriting relation $\rightarrow$ whose derivation relation $\rightarrow^*$ is a well-quasi-order of A^* [Ehrenfeucht *et al.*, 1983; de Luca and Varricchio, 1992]. Higman's Lemma is generalized in [Ehrenfeucht *et al.*, 1983] to show that certain productions give a well-quasi-order. See de Luca and Varricchio [1992] for additional applications of rewrite relations and well-quasi-orders to regular languages.

See Benninghofen, Kemmerich, and Richter's [1987] monograph and Book's [1987] survey for various applications of rewriting to questions of formal languages and decidability in semigroups. They also point out the limitations of the rewriting approach (see, for example, [Squier, 1987]).

Other applications of rewriting to the investigation of semigroups includes the use of ordered completion—and the introduction of new operators—by Pedersen [1989] to construct new decision procedures for some one-relation monoids. Decidability for one-relation Burnside varieties has not been investigated.

Acknowledgement

I thank Sergei Adian for enlightening conversations, Gregory Kucherov and John Pedersen for their comments, and Michaël Rusinowitch for his encouragement.

References

[Adian, 1977] Sergei I. Adian. Classifications of periodic words and their applicaion in group theory. In J. L. Mennicke, editor, *Proceedings of a Workshop on Burnside Groups*, pages 1–40, Bielefeld, Germany, 1977. Vol. 806 of *Lecture Notes in Mathematics*, Springer-Verlag, Berlin.

[Adian, 1979] Sergei I. Adian. *The Burnside Problem and Identities in Groups*. Springer-Verlag, Berlin, 1979. Translated from the Russian.

[Aršon, 1937] S. E. Aršon. Proof of the existence of n-valued infinite asymmetric sequences. *Matematicheskii Sbornik*, 2(44):769–779, 1937.

[Bachmair and Dershowitz, 199?] Leo Bachmair and Nachum Dershowitz. Equational inference, canonical proofs, and proof orderings. *J. of the Association for Computing Machinery*, 199? To appear. Available as Technical Report DCS-R-92-1746, Department of Computer Science, University of Illinois, Urbana, IL.

[Benninghofen *et al.*, 1987] Benjamin Benninghofen, Susanne Kemmerich, and Michael M. Richter. *Systems of Reductions*, volume 277 of *Lecture Notes in Computer Science*. Springer, Berlin, 1987.

[Book, 1987] Ronald V. Book. Thue systems as rewriting systems. *J. Symbolic Computation*, 3(1&2):39–68, February/April 1987.

[Brzozowski, 1980] Janusz Brzozowski. Open problems about regular languages. In R. Book, editor, *Formal Language Theory: Perspectives and Open Problems*, pages 23–47. Academic Press, New York, 1980.

[Brzozowski *et al.*, 1971] Janusz Brzozowski, Karl Culik II, and A. Gabrielian. Classification of non-counting events. *J. of Computer and System Sciences*, 5:41–53, 1971.

[Burnside, 1902] W. Burnside. On an unsettled question in the theory of discontinuous groups. *Quarterly J. of Pure and Applied Mathematics*, 33:230–238, 1902.

[Dershowitz, 1992] Nachum Dershowitz. Rewriting methods for word problems. In M. Ito, editor, *Words, Languages & Combinatorics (Proceedings of the International Colloquium, Kyoto, Japan, August 1990)*, pages 104–118, Singapore, 1992. World Scientific.

[Dershowitz and Jouannaud, 1990] Nachum Dershowitz and Jean-Pierre Jouannaud. Rewrite systems. In J. van Leeuwen, editor, *Handbook of Theoretical Computer Science B: Formal Methods and Semantics*, chapter 6, pages 243–320. North-Holland, Amsterdam, 1990.

[Ehrenfeucht *et al.*, 1983] Andrzej Ehrenfeucht, David Haussler, and G. Rozenberg. On regularity of context-free languages. *Theoretical Computer Science*, 27(3):311–332, December 1983.

[Fine and Wilf, 1965] N. J. Fine and M. S. Wilf. Uniqueness theorems for periodic functions. *Proceedings of the American Mathematical Society*, 16:109–114, 1965.

[Green and Rees, 1952] J. A. Green and D. Rees. On semigroups in which $x^r = x$. *Proceedings of the Cambridge Philosophical Society*, 48:35–40, 1952.

[Hall, 1957] M. Hall, Jr. Solution of the Burnside problem for exponent 6. In *Proceedings of the National Academy of Sciences of the USA*, volume 43, pages 751–753, 1957.

[Howie, 1976] J. M. Howie. *An Introduction to Semigroup Theory*. Academic Press, London, 1976.

[Hsiang and Rusinowitch, 1987] Jieh Hsiang and Michaël Rusinowitch. On word problems in equational theories. In T. Ottmann, editor, *Proceedings of the Fourteenth EATCS International Conference on Automata, Languages and Programming*, pages 54–71, Karlsruhe, West Germany, July 1987. Vol. 267 of *Lecture Notes in Computer Science*, Springer-Verlag, Berlin.

[do Lago, 1992] Alair Pereira do Lago. On the Burnside semigroups $x^n = x^{n+m}$. In I. Simon, editor, *Proceedings of the First Latin American Symposium on Theoretical Informatics*, pages 329–343, São Paulo, Brazil, April 1992. Vol. 583 of *Lecture Notes in Computer Science*, Springer-Verlag, Berlin.

[de Luca and Varricchio, 1990] Aldo de Luca and Stefano Varricchio. On non counting regular classes. In M. Paterson, editor, *Proceedings of the Seventeenth International Colloquium on Automata, Languages and Programming*, Warwick, June 1990. EATCS. Vol. 443 of *Lecture Notes in Computer Science*, Springer-Verlag, Berlin; to appear in *Theoretical Computer Science*.

[de Luca and Varricchio, 1992] Aldo de Luca and Stefano Varricchio. Some regularity conditions based on well quasi-orders. In I. Simon, editor, *Proceedings of the First Latin American Symposium on Theoretical Informatics*, pages 356–371, São Paulo, Brazil, April 1992. Vol. 583 of *Lecture Notes in Computer Science*, Springer-Verlag, Berlin.

[McCammond, 1991] J. McCammond. The solution to the word problem for the relatively free semigroups satisfying $t^a = t^{a+b}$ with $a \geq 6$. *Intl. J. of Algebra and Computation*, 1:1–32, 1991.

[Nordahl, 1992] Thomas E. Nordahl. On the join of the variety of all bands and the variety of all commutative semigroups via conditional rewrite rules. In M. Ito, editor, *Words, Languages & Combinatorics (Proceedings of the International Colloquium, Kyoto, Japan, August 1990)*, pages 365–372, Singapore, 1992. World Scientific.

[Pedersen, 1988] John Pedersen. Computer solution of word problems in universal algebra. In M. Tangora, editor, *Computers in Algebra*, pages 103–128. 1988. Vol. 111 of *Lecture Notes in Pure and Applied Mathematics*, Marcel-Dekker, New York.

[Pedersen, 1989] John Pedersen. Morphocompletion for one-relation monoids. In N. Dershowitz, editor, *Proceedings of the Third International Conference on Rewriting Techniques and Applications*, pages 574–578, Chapel Hill, NC, April 1989. Vol. 355 *Lecture Notes in Computer Science*, Springer, Berlin.

[Sanov, 1940] I. N. Sanov. Solution of the Burnside problem for exponent 4. *Učen. Zap. Leningrad Univ.*, 10:166–170, 1940.

[Siekmann and Szabó, 1982] Jorg Siekmann and P. Szabó. A Noetherian and confluent rewrite system for idempotent semigroups. *Semigroup Forum*, 25(1/2):83–110, 1982.

[Squier, 1987] Craig Squier. Word problems and a homological finiteness condition for monoids. *J. of Pure and Applied Algebra*, 49:201–217, 1987.

[Thue, 1912] Axel Thue. Über die gegenseitige Lage gleicher Teile gewisser Zeichenreichen. *Norske Videnskabssellskabets Skrifter I Mat. Nat. Kl.*, 1:1–67, 1912.

Could Orders Be Captured by Term Rewriting Systems?

Sergei G. Vorobyov

CRIN & INRIA-Lorraine
Technopole de Nancy-Brabois, Campus Scientifique
B.P. 101, 54602 Villers-lès-Nancy, Cedex, France
e-mail: Serge.Vorobyov@loria.fr

Abstract. Imposing restrictions on the structure of rewrite rules, it is interesting to know the resulting loss of expressiveness in terms of ability to capture some natural properties. Can transitivity be described by rewrite rules of the suitable form, as $(x \le y) \& (y \le z) \rightarrow (x \le z)$? It turns out that none decidable quantifier-free theory of similarity type $\{ \le \}$ validating transitivity of $\le$ can be axiomatized by means of a finitely terminating AC-rewrite rule system of the same similarity type, provided that conjunction is the only logical connective permitted in left-hand sides of non-logical rewrite rules. Therefore, some simple decidable theories admit no simple rewrite axiomatizations. This result applies, for example, to universal fragments of theories of linear, discrete, and dense orders. We conclude by demonstrating a simple canonical axiomatization of quantifier-free theory of partial order based on addition of one auxiliary predicate symbol and formula preprocessing.

1 Introduction

During last years rewrite rule systems and rewriting techniques have been drawing increasing attention as an elegant tool with diverse applications in automated theorem proving, program verification, algebraic specifications and simplification, functional and logic programming, see [HO80, DJ90].

Whenever axioms of a theory T can be transformed into a finite set of uniquely terminating oriented replacement rules R_T, the theory T can be easily decided by a simple normal form reduction: to test T-equivalence of formulas it suffices to compute their R_T- irreducible forms and check them for coincidence. Accordingly, we say that a rewrite rule system axiomatizes a logical theory iff all valid formulas of the theory and only these formulas are reducible to *true*.

Generally speaking, a rewrite axiomatization and a necessary transformation exist for every decidable theory, since, for example, the corresponding Turing machine can be encoded into a uniquely terminating rewrite rule system. It would be the end of the story, but in the framework of rewriting we are more interested in straightforward "*orient axioms into rewrite rules and make a completion*" transformations, without awkward reduction to Turing machines or other encodings using, say, Diophantine polynomials.

Given a decidable theory the problem is: whether there exists a *natural* uniquely terminating rewrite rule system axiomatizing it? There is a long tradition of *natural* classification of rule-based systems (grammars) according to the structure of their rules, e.g., right-linear, context-free, etc. Unfortunately, very little is known on similar classification of rewrite rule systems and expressiveness of corresponding classes. Therefore, it would be useful to obtain some general permissive or prohibitive principles leaving no doubt that there exists or does not exist an axiomatizing system of a given sort provided a theory possesses or lacks some properties. These properties could be formulated in terms of expressive power, i.e., an ability to capture some simple and well-known logical theories, e.g., of natural numbers, lists, graphs, orders, etc.

In [Vor88] we obtained one result of this sort. Call a rewrite rule system *context-free* iff all its non-logical rules do not contain logical connectives in left-hand sides. Logical rules describe tautological transformations, while non-logical allow domain-dependent. For example,

$$\alpha \,\&\, (\,\beta \vee \gamma\,) \rightarrow \alpha \,\&\, \beta \vee \alpha \,\&\, \gamma$$

is a logical rule, but

$$(x \leq y) \,\&\, (y \leq z) \rightarrow (x \leq z)$$

is not; both rules are *non-context-free*. Note, that pure PROLOG is context-free. Context-freedom is reasonable and desirable property, since it significantly simplifies the problems of matching and unification. It turns out, however, that no interesting (i.e., containing all true numeric inequalities) fragment of quantifier-free Presburger arithmetic can be axiomatized by a uniquely and finitely terminating context-free rewrite rule system, usual or conditional, finite or infinite, see [Vor88]. Adding auxiliary symbols does not help, cf., [KN85].

In this paper we present another negative result on classification and expressive power of rewriting systems. We prove that using non-logical rewrite rules of similarity type $\{\leq\}$ and of restricted form admitting only conjunctions in left-hand sides gives possibility to axiomatize *none* quantifier-free decidable theory containing transitivity.

To state this result in more simple terms, imagine a *thinking engine* able to infer conclusions about transitive relation between variables, clever enough to

- conduct arbitrary purely logical inferences, and
- capture several number of atomic relations at a time (i.e., a conjunction) conducting non-logical ones.

We show that every Artificial Intelligence with the IQ like this is too weak to cope with transitivity. It should either be contradictory, or fall into infinite loops.

In [Vor89] we demonstrated one possible way to overcome the barrier of low expressive power of term rewriting systems based on building-in decision algorithms and special means of inductive inference into conditional rewrite rule systems.

The paper is organized as follows. In Section 2 we give necessary definitions and examples, in Section 3 we state and prove the abovementioned result, and in Section 4 we demonstrate a simple two-rule canonical axiomatization of quantifier-free theory of partial order based on addition of one auxiliary predicate symbol and formula preprocessing.

2 Basic Definitions

2.1 The Language

Definition 1 (Quantifier-Free Order Formulas). Let V be a set of variables and $\leq$ be a binary predicate symbol with the intended meaning "*less or equal*". The set of *quantifier-free order formulas*, later referred to as *QFO-formulas*, is defined as the least set satisfying the following properties:

1. **0** and **1** are atomic *QFO*-formulas (*false* and *true* respectively);
2. expressions of the form $(x \leq y)$ are *atomic QFO-formulas* for all variables x and y from V;
3. all atomic *QFO*-formulas are *QFO-formulas*;
4. if ϕ and ψ are *QFO*-formulas then so are $(\phi + \psi)$ and $(\phi * \psi)$ (where $+$ and $*$ stand for "*exclusive or*" and "*conjunction*").

Let $Var(\phi)$ denote the set of variables occurring in ϕ. Sometimes we will freely use $x = y$ as an abbreviation for $x \leq y$ & $y \leq x$, and $x < y$ as an abbreviation for $x \leq y$ & $\neg x = y$. □

Remark (On Boolean-Ring Connectives). We have chosen the Boolean-ring connectives $*$ (multiplication) $+$ (exclusive or), **1** (*true*), and **0** (*false*) instead of the standard logical connectives &, $\vee$, $\supset$, $\Leftrightarrow$, $\neg$. The latter can be introduced by the well-known rewrite rules, see [Hsi85]:

(a) $x \,\&\, y \rightarrow x * y$;
(b) $x \vee y \rightarrow x * y + x + y$;
(c) $x \supset y \rightarrow x * y + x + \mathbf{1}$;
(d) $x \Leftrightarrow y \rightarrow x + y + \mathbf{1}$;
(e) $\neg x \rightarrow x + \mathbf{1}$

We will discuss the reasons of this choice later on. □

Convention. In what follows we use the common rules to omit parentheses, keeping in mind that $*$ binds stronger than $+$. We also preassume associative and commutative properties of both operations. Sometimes will write *true* instead of **1**. □

Definition 2 (Theory, Quantifier-Free Theory of Order).

- A *theory* is an arbitrary set of formulas.
- A theory is called *quantifier-free* iff it consists of *universal* formulas of the form $\forall x_1, \ldots, \forall x_n\ \Phi(x_1, \ldots, x_n)$, where the quantifier prefix $\forall x_1, \ldots, \forall x_n$ (usually omitted) contains universal quantifiers only, and the formula $\Phi(x_1, \ldots, x_n)$ contains no quantifiers.
- For an arbitrary theory T, let $Th^{\forall}(T)$ denote the quantifier-free fragment of T, i.e.,

$$T^{\forall}(T) = \{ \Phi \mid \Phi \in T \text{ and } \Phi \text{ is universal} \}.$$

- The *quantifier-free theory of order, $Ord^{\forall}$* for short, is any decidable quantifier-free theory of similarity type $\{\leq\}$ validating two properties:

$$Ord^{\forall} \models x \leq y \,\&\, y \leq z \supset x \leq z$$

and

$$Ord^{\forall} \models x = y \,\&\, x \leq z \supset y \leq z,$$

i.e., transitivity and substitution property of the $\leq$ predicate. □

Remark (On Substitution Property and Genericity of $Ord^{\forall}$)

- The substitution property easily follows from transitivity if we suppose that $Ord^{\forall}$ is closed w.r.t. logical inferences. In this case the second condition on $Ord^{\forall}$ can be omitted. We will not go deep in these details here, since the substitution property does not seem very restrictive.
- Note that $Ord^{\forall}$ is a generic name for a decidable theory capable to express transitivity of $\leq$, introduced to make the result of this paper more general. $Ord^{\forall}$ may designate, e.g., any of the theories mentioned in the following example. □

Examples of Theories of Order

1. Consider the theory of *linearly ordered sets* axiomatized by the following set of formulas OR:

$$\forall x \forall y (x \leq y \ \vee\ y \leq x),$$

$$\forall x \forall y \forall z (x \leq y \,\&\, y \leq z \ \supset\ x \leq z).$$

2. Let DO be the set of formulas obtained from OR by adding

$$\exists x \forall y (y \leq x \supset x = y),$$

$$\forall x \exists y (x < y \,\&\, \forall z (x < z \supset y \leq z)),$$

$$\forall x \forall y (y < x \supset \exists z (z < x \,\&\, \forall w (z < w \supset x \leq w))).$$

These axioms express the existence of the least element, the immediate successor, and predecessor respectively. The set DO axiomatizes the *theory of discrete linear order with the first element.*

3. Let DNO be OR plus two axioms stating the existence of intermediates and unboundedness:

$$\forall x \forall z \exists y (x < z \supset x < y \,\&\, y < z),$$

$$\forall x \exists y \exists z (y < x \,\&\, x < z),$$

DNO axiomatizes the theory of *dense linear order without endpoints.*

All the above theories, thus all their unquantified subtheories, $Th^{\forall}$(OR), $Th^{\forall}$(DO), $Th^{\forall}$(DNO), are decidable, see [Rab77]. □

2.2 Rewrite Rule Systems, Axiomatizability

Assuming familiarity with basic notions of the rewrite rule theory (see, e.g., [HO80, DJ90]) we will briefly remind necessary definitions. These definitions are equally applicable to formulas of any similarity type (signature), e.g., to QFO-formulas.

Definition 3 (Rewrite Rule System). A *rewrite rule system* is a set R of oriented replacement rules of the form

$$l \rightarrow r,$$

where l and r are formulas satisfying the *variable restriction*

$$Var(r) \subseteq Var(l).$$

The formula l is called the *left-hand side* of the rule, and r is called its *right-hand side.* □

The definitions of the reduction relation between formulas $\rightarrow_R$ as well as its reflexive-transitive closure $s \rightarrow_R^*$ and R-normal (or irreducible) form are straightforward and can be found elsewhere. Of course, associativity and commutativity of $+$ and $*$ should be taken into account.

Definition 4 (Finite and Unique Termination). We say that a rewrite rule system R is *canonical* iff it is both *finitely* and *uniquely* terminating, i.e.,

1. there are no infinite reduction chains of the form

$$t_0 \rightarrow_R t_1 \rightarrow_R t_2 \rightarrow_R \cdots$$

2. whenever $t \rightarrow_R^* t_1$ and $t \rightarrow_R^* t_2$, there exists a formula t_3 such that

$$t_1 \rightarrow_R^* t_3 \text{ and } t_2 \rightarrow_R^* t_3.$$

Definition 5 (Soundness, Completeness, Axiomatizability). Let T be a theory. We say that a rewrite rule system R

- is *T-sound* iff

$$\phi \rightarrow_R^* \psi \text{ implies } \phi + \psi + \mathbf{1} \in T,$$

(in other words, ϕ and ψ are equivalent modulo T);
- *axiomatizes* a theory T iff R is T-sound, and

$$\phi \in T \text{ implies } \phi \rightarrow_R^* \mathit{true},$$

(i.e., all theorems of T could be proved by reduction to *true*).

We say that a theory T is *axiomatizable* iff there exists a canonical rewrite rule system axiomatizing T. □

Remark (On Soundness and Structure Restrictions).

1. If we had allowed non-sound systems, any theory of similarity type Σ would have been axiomatizable by means of the system

$$\{\phi \rightarrow \mathit{true} \mid \phi \textit{ is a formula of similarity type } \Sigma\}.$$

2. Moreover, if we had not restricted somehow the structure of rewrite rules, any theory T closed w.r.t. replacements of equals by equals might have been axiomatized by means of the system

$$\{\phi \rightarrow \mathit{true} \mid \phi \in T\}.$$

Obviously, systems like these are practically useless and not very interesting. Therefore, we should extract "reasonable" classes of rewrite rule systems ,e.g., finite or well-structured otherwise, and investigate the problem of axiomatizability for these classes. □

Definition 6 (Logical and Non-Logical Rules). A rewrite rule $l \rightarrow r$ is called *logical* iff $l \Leftrightarrow r$ is a Boolean tautology, and *non-logical* otherwise. □

Example (Canonical Rewrite System for Boolean Algebra). The following associative and commutative system **BA**, consisting of logical rules only, axiomatizes the ∀-theory of Boolean algebras, see [Zhe27, Hsi85]:

(a) - (e) *Rules (a) through (e) from Section 2.1*
(f) $1 * x \rightarrow x;$
(g) $x * x \rightarrow x;$
(h) $0 * x \rightarrow 0;$
(i) $0 + x \rightarrow x;$
(j) $x * x \rightarrow 0;$
(k) $x * (y + z) \rightarrow x * y + x * z.$ □

Remarks (On Boolean-Ring Subsystem BA).

- The main reason why we have chosen the Boolean-ring connectives as basic is the following: there does not exist a canonical rewrite rule system axiomatizing propositional tautologies in the basis { &, ∨, ¬ }, see [Hsi85]).
- It is reasonable to suppose that each axiomatizing rewrite system contains something like **BA** as a subsystem to be able to axiomatize at least propositional tautologies. A guess like this leads to no loss of generality, since a logical subsystem may be considered empty. Consequently, we will assume that every system aspiring to axiomatize some theory consists of logical and non-logical parts.
- We propose to classify rewrite rule systems imposing restrictions on the structure of left-hand sides of non-logical rewrite rules. □

2.3 *N*-Systems

Which classes of rewrite rule systems could be considered "reasonable" or easy feasible? In [Vor88] we investigated the expressive power of *context-free* rewrite rule systems (CF-systems for short). CF-systems consist of logical rules **BA** and non-logical rules that contain no Boolean connectives in left-hand sides. For example,

$$x \leq y \rightarrow x < y \vee x = y$$

is CF-rule, but

$$x \leq y \,\&\, y \leq z \rightarrow x \leq z$$

is not. CF-systems possess nice structural properties and may be considered reasonable, because reductions in these systems are easy feasible: there is no necessity to take associativity, commutativity, and other properties of Boolean connectives into account when reducing in a CF-system. It is the validity of normal forms that should be decided in **BA**, but not necessarily by means of direct **BA**– reductions: resolution-like methods can be applied as well. Roughly speaking, CF-systems separate purely logical and non-logical assumptions and reasoning.

J.Hsiang [Hsi85] considered a more general class of N-rewrite rule systems admitting conjunctions in left-hand sides of rewrite rules. Owing to more simple matching and unification, both reduction and critical pair/superposition algorithms become considerably easier being restricted to the case of N-rules, as compared with the general case.

Definition 7 (N-System).

- An *N-formula* is a Boolean product (conjunction) of atomic formulas.
- An *N-rule* is a rewrite rule of the form

$$l \to r_1 + \ldots + r_n ,$$

  where l, r_1, ..., r_n are N-formulas.
- A rewrite rule system R is an *N-rewrite rule system* iff $R \setminus \mathbf{BA}$ consists of N-rules only. □

Remarks (On N-Systems and BA).

1. It is rather difficult, at least by means of a computer, to reduce using a system more general than N-system, because distributivity of $*$ w.r.t. $+$ should be taken into account. This complicates the problem of redex searching. On the other hand, reducing in an N-system is much more simple, see [Hsi85].
2. Since provability of at least propositional tautologies is the necessary condition of axiomatizability, we should allow logical non-N-rules, e.g., distributivity of **BA**, alongside with the pure N-rules. □

Now we are at the point to pose the main problem we are interested in.

The Problem of N-Axiomatizability.

- Stated in simple terms: can transitivity be captured by N-systems?
- More precisely: whether there exists a decidable theory $Ord^{\forall}$ axiomatizable by means of a canonical or even finitely terminating N-rewrite rule system of similarity type $\{ \leq \}$? □

3 Main Result

In this Section we will give a negative answer to the problem of axiomatizability of $Ord^{\forall}$ by means of finitely terminating N-systems.

Theorem. *None theory $Ord^{\forall}$ can be axiomatized by a finitely terminating N-rewrite rule system of similarity type $\{ \leq \}$.*

Proof. The proof consists of two lemmas. The first describes and restricts the form of right-hand sides of finitely terminating rewrite rule systems. The second demonstrates a particular formula that is true in every $Ord^{\forall}$, but is irreducible in any $Ord^{\forall}$-sound canonical system.

Lemma 8. *If an N-system of similarity type $\{\ \leq\ \}$ is finitely terminating then for any rule $l \rightarrow r_1 + \ldots + r_n$ of this system every N-formula r_i occurring in the right-hand side r contains fewer atoms than the left-hand side l.* □

Proof. If the statement of the lemma had been false the formula

$$(x \leq x) * \ldots * (x \leq x),$$

containing as many equal atoms as the left-hand side of the corresponding violating rule would have given an infinite reduction chain. □

Now we will prove that the following simple formula Φ, which should be valid in $Ord^{\forall}$ by assumption, cannot be reduced to *true* whatever $Ord^{\forall}$-sound and finitely terminating N-rewrite rule system we would choose. Let Φ be the formula

$$(x \leq y) * (y \leq x) * (x \leq z) + (x \leq y) * (y \leq x) * (y \leq z) + \mathbf{1},$$

which simply states, rephrased in terms of usual equality and equivalence, that

$$(x = y) * (x \leq z) \Leftrightarrow (x = y) * (y \leq z),$$

i.e., ordinary substitution property of the $\leq$ predicate that should be valid in every $Ord^{\forall}$ theory by assumption, cf., Definition 2.

It suffices to prove that neither of the summands of Φ is reducible at all in an arbitrarily chosen finitely terminating N-rewrite rule system This fact follows directly from Lemma 8 and Lemma 9 below.

Lemma 9. *The formula*

$$\Psi = (x \leq y) * (y \leq x) * (x \leq z)$$

is not equivalent w.r.t. $Ord^{\forall}$ to a Boolean sum of atomic formulas and two-atomic products of inequalities depending on variables x, y, and z. □

Remarks:

1. Therefore, any possible finitely terminating system should be $Ord^{\forall}$-unsound, hence, cannot axiomatize $Ord^{\forall}$.
2. Lemma 9 shows that equality (or loop in a graph representing a conjunction of inequalities between variables) cannot be captured by N-systems.
3. Due to Lemma 8 there is no need to consider three-atomic products, four-atomic, etc. □

Proof of Lemma 9. Suppose, on the contrary, that a Boolean sum Π of different atoms and two-atomic products equivalent to Ψ does exist. In what follows, a pair of a model of $Ord^{\forall}$ and a valuation of variables will be referred to as an ***interpretation***. The arguments below contain simple statements on the existence of particular interpretations.

We can assume without the loss of generality that:

1. Π is in the **BA**-normal form;
2. Π does not contain reflexive atoms, i.e., $(x \leq x)$, which could be replaced by **1** and then **BA**-simplified if necessary;
3. Π does not contain **1** as a summand; otherwise the equivalence of Ψ and Π, i.e., $\Psi + \Pi + 1$, would be false in an interpretation validating reflexive atoms exclusively;
4. Π contains no atomic summand; otherwise the equivalence of Ψ and Π, i.e., $\Psi + \Pi + 1$, would be false in an interpretation that validates only this particular atomic summand and all reflexive atoms.

Henceforth, the only remaining possibility is that Π could be a sum of different two-atomic products. We will show, however, that this is not the case. This will conclude the proof.

There exist two different possibilities.

Let us suppose first, that there is a product $p * q$ in Π which implies in $Ord^{\forall}$ none consequence different from p, q, and $p * q$, i.e., itself. Then we can easily choose an interpretation $\mathcal{I}$ that makes Π true and Ψ false. Hence, this case is excluded.

Let us consider the last remaining possibility. Let every product $p*q$ occurring in Π imply in $Ord^{\forall}$ a consequence different from both p and q. There are $3! = 6$ possible cases for three variables x, y, z:

(1) $(x \leq y) * (y \leq z)$ *implies* $(x \leq z)$
(2) $(x \leq z) * (z \leq y)$ *implies* $(x \leq y)$
(3) $(y \leq x) * (x \leq z)$ *implies* $(y \leq z)$
(4) $(y \leq z) * (z \leq x)$ *implies* $(y \leq x)$
(5) $(z \leq x) * (x \leq y)$ *implies* $(z \leq y)$
(6) $(z \leq y) * (y \leq x)$ *implies* $(z \leq x)$

Let $\mathcal{I}_i$ ($1 \leq i \leq 6$) be an interpretation that validates only reflexive atoms and the atoms occurring in (i) above. Then $\mathcal{I}_i$ gives a counterexample to the equivalence of Ψ and Π, i.e., $\Psi + \Pi + \mathbf{1}$. This concludes the proof of Lemma 2 and the proof of the Main theorem. □

4 Adding Auxiliary Symbol and Preprocessing Formulas

There exist several possibilities of increasing the expressive power. First, we can allow more complex left-hand sides. But this is highly undesirable, because it complicates rewriting. Second, there might exist some natural refinements of rewriting capturing the transitivity quite naturally. Third, we can add auxiliary symbols playing the role of "memory", and to perform some formula preprocessing. Let us consider this last possibility.

Let Φ be an unquantified formula of similarity type $\{\leq\}$ and let Φ^* denote a $\mathbf{BA} \cup \{(x \leq x) \to \mathbf{1}\}$-normal form of Φ:

$$\Phi^* = \sum_{i=1}^{m} p_i \;, \tag{1}$$

where p_i are conjunctions of inequalities between different variables.

Definition 10 (Closure). We define a *closure* $\widehat{\Phi^*}$ of a $\mathbf{BA} \cup \{(x \leq x) \to \mathbf{1}\}$-normalized formula Φ^* as follows. Let ? be a new binary predicate symbol and $\overline{x}^i = \{x^i_1, \ldots, x^i_{k_i}\}$ be all variables occurring in p_i from (1). Each formula $\widehat{p_i}$, called the *closure* of p_i, is the conjunction of p_i with *all* atoms of the form $u?v$ such that (a) $u, v \in \overline{x}^i$, (b) $u \leq v$ does not occur in p_i, and (3) u is different from v. Then the closure of Φ is defined as

$$\widehat{\Phi^*} = \sum_{i=1}^{m} \widehat{p_i} \;.$$

Definition 11 (The System R_{po}). Consider the AC rewriting system R_{po} consisting of $\mathbf{BA}$ and the following two rules:

$$\text{(Refl)} \qquad (x \leq x) \quad \to \quad \mathbf{1} \;,$$

$$\text{(Trans)} \qquad (x \leq y) * (y \leq z) * (x?z) \to (x \leq y) * (y \leq z) * (x \leq z) \;. \;\square$$

Obviously, R_{po} is finitely terminating, because the number of "?" symbols decreases at each rewrite step. In general, R_{po} is not confluent, since, e.g.,

$$(x \leq x) * (x \leq z) * (x?z)$$

could be reduced either to

$$(x \leq z) * (x?z)$$

(by (Refl)), or to

$$(x \leq x) * (x \leq z) * (x \leq z)$$

(by (Trans)). But, by the closure construction, the combination

$$\ldots(x \leq z) * (x?z)\ldots$$

is impossible in $\widehat{\Phi^*}$, and we have the following

Theorem (Completeness of R_{po}). *For any quantifier-free formula Φ of similarity type $\{\leq\}$ the R_{po}-normal form of its closure $\widehat{\Phi^*}$ is unique and is equal to* **1** *once $Th^{\forall}(PO) \models \Phi$.* □

Proof. The unique termination follows from the fact that the rules (Refl) and (Trans) superpose only in

$$(a \leq a) * (a \leq b) * (a?b) \ ,$$

which is impossible neither in $\widehat{\Phi^*}$, nor in any formula R_{po}-reducible from it.

Let Φ be valid in all models of partial order. We first claim that any of its $\mathbf{BA} \cup \{(x \leq x) \rightarrow \mathbf{1}\}$-normal forms are of the form

$$\Phi^* = \sum_{i=1}^{2k} p_i + \mathbf{1} \ ,$$

where p_i are conjunctions of inequalities between different variables. In fact, if it contained no $+\ \mathbf{1}$ member, it would be false in the discrete model of PO where different elements are incomparable and different variables are interpreted by different elements. Next, if it contained an even number of summands, it would be false in the one-element model of PO.

Our second claim is that Φ^* could be represented as

$$\Phi^* = \sum_{i=1}^{k} p_i + \sum_{i=1}^{k} q_i + \mathbf{1} \ , \qquad (2)$$

where all p_i and q_i are pairwise PO-equivalent.

Suppose not, then we can assume that all equivalent pairs of p_i and q_i are erased from (2). This does not influence the PO-validity of Φ^*. After this simplification we have $k > 0$ in (2), otherwise we immediately get a contradiction.

Denote $Cons_{po}(p_i)$ the set of inequalities between different variables being the PO-consequences of p_i. Let us choose p_{i_0} (or q_{i_0}) with the property "*$Cons_{po}(p_{i_0})$ is minimal as compared with the corresponding sets of other p_i's and q_i's*". Note that for different p_i or q_i in (2) their sets $Cons_{po}(\ldots)$ do not coincide; otherwise we would have equivalent pairs. Now we can easily construct an interpretation satisfying only $Cons_{po}(p_{i_0})$, which contradicts the validity of Φ.

So, we can suppose that in (2) all p_i and q_i are pairwise equivalent modulo PO. It remains to prove that for any p_i and different x, y, if $PO \models p_i \supset x \leq y$, then $x \leq y$ occurs in the R_{po}-normal form of its closure $\widehat{p_i}^*$ (and similarly for

q_i). In fact, if this is true, then $\widehat{p_i}^*$ and $\widehat{q_i}^*$ do coincide and by idempotence $p_i + q_i \longrightarrow^*_{R_{op}} 0$ which is needed.

We proceed by induction on the length of shortest possible proof of $p_i \supset x \leq y$ in PO. The basic case is trivial: if $x \leq y$ occurs in p_i then it is not discarded during R_{po}-normalization, hence appears in $\widehat{p_i}^*$. Let the last step of the proof be achieved by applying the transitivity law to $x \leq w$ and $w \leq y$. The inductive hypothesis applies to $x \leq w$ and $w \leq y$. At the same time $x?y$ should necessarily present in $\widehat{p_i}$ by construction, and it could not be discarded before; otherwise it would exist a shorter proof of $x \leq y$. But then the rule (Trans) applies. This finishes the proof. □

5 Concluding Remarks

Our main aim in this paper was to draw attention to the problem of measuring expressive power of term rewriting systems imposing structural restrictions on the form of the rules. The research along these lines might give insights for simple and straightforward generalizations of term rewriting systems able to axiomatize naturally simple theories like $Ord^{\forall}$, and to elaborate subtle criteria to measure expressiveness of rewriting.

References

[DJ90] N. Dershowitz and J.-P. Jouannaud. Rewrite systems. In J. van Leuven, editor, *Handbook of Theoretical Computer Science*. Elsevier Science Publishers North-Holland, 1990.

[HO80] G. Huet and D. Oppen. Equations and rewrite rules: A survey. In R. V. Book, editor, *Formal Language Theory: Perspectives and Open Problems*, pages 349–405. Academic Press, New York, 1980.

[Hsi85] J. Hsiang. Refutational theorem proving using term rewriting systems. *Artificial Intelligence*, 25(1):255–300, 1985.

[KN85] D. Kapur and P. Narendran. A finite Thue system with decidable word problem and without equivalent finite canonical system. *Theoretical Computer Science*, 35:337–344, 1985.

[Rab77] M.O. Rabin. Decidable theories. In J. Barwise, editor, *Handbook of Mathematical Logic*, pages 595–629. North-Holland Pub. Co., 1977.

[Vor88] S.G. Vorobyov. On the arithmetic inexpressiveness of term rewriting systems. In *Proceedings 3rd IEEE Symposium on Logic in Computer Science, Edinburgh (UK)*, pages 212–217, 1988.

[Vor89] S. G. Vorobyov. Conditional rewrite rule systems with built-in aritnmetic and induction. In N. Dershowitz, editor, *Proceedings 3rd Conference on Rewriting Techniques and Applications, Chapel Hill (North Carolina, USA)*, volume 355 of *Lecture Notes in Computer Science*, pages 492–512. Springer-Verlag, April 1989.

[Zhe27] I. I. Zhegalkin. The technique of calculation of statements in symbolic logic. *Mathematical Sbornik*, 34:9–28, 1927. In russian.

A Categorical Formulation for Critical-Pair/Completion Procedures

Karel Stokkermans *
RISC-Linz
Johannes-Kepler Universität
A-4040 Linz, Austria
email: kstokker@risc.uni-linz.ac.at

August 14, 1992

Abstract

A novel representation of so-called critical-pair/completion procedures is introduced. This representation, by the means of category theory, serves to give a treatment procedures at as general a level as possible. First we give an overview of the fundamental notions in a systematic, axiomatic way, and then translate these notions into a categorical framework. The elements of the given universe (for instance terms, or polynomials) serve as objects in a category baptized *CPC*. Its arrows are the reductions in the universe (for instance rewrite rules, or reductions defined by polynomials from a given ideal). This category can alternatively be obtained by free generation from the underlying graph based on the reduction rules.

Substitutions (or multipliers) and embeddings into superpatterns, typical operations for all completion procedures, are viewed as functors onto the category itself.

The basic concepts involved in critical-pair/completion algorithms are modeled by well-known categorical concepts, to form a basis for the analysis of the differences and essential correspondences between the various critical-pair/completion algorithms.

1 Introduction

In this paper we present a categorical formulation of so-called critical-pair/completion procedures (*CPC-procedures* for short). An introduction to CPC-procedures, along with first ideas for a systematic, axiomatic formulation for them, was given in [Buc85]. The main instances of CPC-procedures are the Gröbner basis algorithm in polynomial ideal theory (cf. [Buc65]), the Knuth-Bendix completion procedure for term rewriting systems (cf. [KB67]), and the resolution procedure in automated theorem proving (cf. [Rob65]).

A general framework for CPC-procedures will certainly not result in a computational speed-up, but it will provide insight in the minimal requirements for the critical-pair/completion approach to be relevant. It also may give some clues how to modify and simplify concrete programs.

The principal thought behind the categorical approach is the use of abstraction to arrive at a unified formalism, and then to clarify the notions and their relationships. Notions can be characterized by universal properties, and then one can go down from this metalevel to concrete applications, exploiting the gained insight. For an introduction to the language of categories and the basic notions and results in category theory we refer to [Mac71, LS86]. As an illustration for the application of category theory to the clarification of mathematical concepts, we would like to mention the presentation of combinatory logic and typed λ-calculus by means of cartesian closed categories, cf. [Lam80, LS86].

We will now present a short discussion of other categorical approaches to CPC-procedures, a more extensive evaluation can be found in [Sto91].

Almost all categorical approaches to CPC-procedures dealt with the completion of term rewriting systems. [Joh87] developed a 2-categorical approach to term rewriting. It was pursued further in [Joh91], where, apparently independent from earlier work by Benson, string rewriting systems are modeled by

*sponsored by the Austrian Ministry of Science and Research (BMWF), ESPRIT BRA 3125 "MEDLAR"

2-categories. Then, the approach is extended by modeling linear term rewriting systems by 3-categories. One major problem is that the multiple appearance of the same function symbol is hard to cover in this framework. Nevertheless, it may prove possible to carry out the axiomatization suggested in [Buc85], by patterns, multipliers and replacements in such 3-categories.

Independently, in [Ben75] rewriting systems determine so-called categories of derivations. The categories obtained are free strict monoidal categories, simple forms of 2-categories. Morphisms in the category generalize the notion of "derivation tree". Here also, only string rewriting is modeled.

Finally, [Hue86] defines the *computation category* associated with a regular term rewriting system as its derivation category, quotiented by the permutation equivalence. It has as objects the terms, and as arrows between terms the permutation class of parallel derivations from the domain to the codomain. This theory of derivations was first developed in the framework of λ-calculus in [Lév78], and adapted to the framework of regular term rewriting systems in [HL79]. A more general framework of combinatory reduction systems allowing binding operators and providing a general theory encompassing λ-calculus was investigated in [Klo80].

The condition of regularity of the term rewriting system essentially stipulates that it is left-linear (no variables occur twice in the left hand side of a rule) and does not contain critical pairs. This enables one to show confluence.

Since this is not what we are after (such conditions may serve as criteria within the framework to be developed for distinguishing between certain instances of critical-pair/completion procedures but certainly not as preconditions for the framework itself), we present an extension of the concepts of *derivation* and *computation* category by giving (categorically) axiomatic formulations of the basic notions indicated in [Buc85], namely pattern, multiplier, and a suitable notion of replacement. The multiplier and replacement (here: embedding) notions are used to express the *encompassment* quasiordering (cf. [DJ90]), called *containment* in [Hue81].

Section 3 axiomatizes the prerequisites for CPC-procedures and the basic notions involved in them and explains which notions in the three main CPC-procedures correspond to the notions introduced in the general framework, section 4 is concerned with the construction of the categorical framework, into which the basic notions are translated in section 5. Finally, section 6 provides an outlook to future work.

2 Overview of the Critical-Pair/Completion Procedures to be Modeled

This section consists of algorithmic specifications for the three main instances of CPC-algorithms and a general algorithmical framework for such procedures. What consists the completion (of the set of reduction rules, such that confluence can be guaranteed) in these procedures should be clear from the algorithmic descriptions below. For the (general) concept of critical pairs we give a short intuitive and graphical explanation. A definition will follow in section 3. This general description should enable the reader to understand how the critical pairs are built in the individual algorithms.

In general, a critical pair is constructed from two reduction rules $p_1 : l_1 \rightarrow r_1$ and $p_2 : l_2 \rightarrow r_2$. One looks for a "most general common instance" for the left hand sides of the two rules, say L, and then reduces L according to both rules, obtaining $p_1(L)$ and $p_2(L)$ respectively as reducts. These two reducts form a critical pair. They are equivalent with respect to the equivalence relation induced by the reduction relation. Hence, to obtain a unique normal form $NF(L)$ for L, they should either *be* equal, or *made* equal, which is enforced by adding a new reduction rule.

Pictorially,

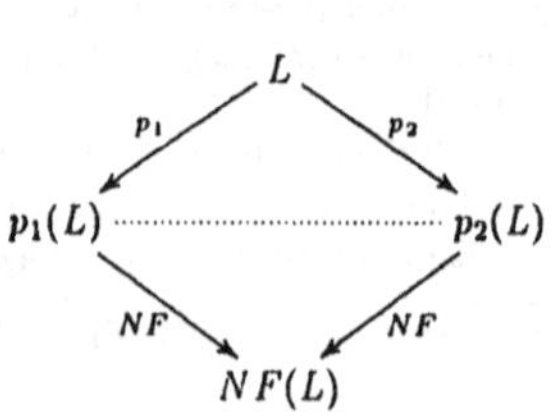

where the NF arrows are derived from the normal form algorithm in the desired, completed situation in which confluence and termination are guaranteed.

2.1 The Knuth-Bendix Procedure

We present a simple version of the Knuth-Bendix algorithm, as described in [Klo90]. In this algorithm, a critical pair is generated from a most general superposition situation (cf. section 3) between the left hand sides of two rewrite rules.

Input: an equational specification (Σ, E), and a reduction ordering $>$ on $Ter(\Sigma)$ (i.e. a program that computes $>$).

Output: a complete TRS R such that for all $s, t \in Ter(\Sigma)$: $s =_R t \Leftrightarrow (\Sigma, E) \vdash s = t$.

Algorithm:

```
R := ∅
while E ≠ ∅ do
   choose an equation s = t ∈ E;
   reduce s and t to respective normal forms s' and t' with respect to R;
   if s' ≡ t' then
      E := E − {s = t}
   else
      if s' > t' then
         α := s'; β := t'
      else if t' > s' then
         α := t'; β := s'
      else
         failure
      fi;
      CP := {P = Q|(P,Q) is a critical pair between the rules in R and α → β};
      R := R ∪ {α → β};
      E := E ∪ CP − {s = t}
   fi
od;
success
```

2.2 The Gröbner Basis Algorithm

The following description is taken from [Buc83b]. The critical pairs in this algorithm arise from unifying two polynomials in the given basis by taking the least common multiple of their leading monomials, and taking the difference the polynomials constructed by the corresponding multiplications (this is the *S*Polynomial in the algorithm below).

Definition 1 (Gröbner Basis) *F is called a* Gröbner basis *iff for all g, h_1, h_2 in a given polynomial ring $K[x_1, \ldots, x_n]$:*
if h_1 and h_2 are normal forms of g modulo F then $h_1 = h_2$.

Input: a finite subset F of the polynomial ring $K[x_1, \ldots, x_n]$.

Output: a finite subset G of the polynomial ring $K[x_1, \ldots, x_n]$, such that Ideal(F) = Ideal(G) and G is a Gröbner basis.

Algorithm:

```
G := F;
CP := {(f1, f2)|f1, f2 ∈ G, f1 ≠ f2};
while CP ≠ ∅ do
   (f1, f2) := a pair in CP;
   CP := CP − {(f1, f2)};
   h := SPolynomial(f1, f2);
```

h' := the normal form of h modulo G;
if $h' \neq 0$ **then**
$CP := CP \cup \{(g, h') | g \in G\}$;
$G := G \cup \{h'\}$
fi
od

2.3 The Resolution Procedure

The following presentation is taken from [Ric83]. In this algorithm the critical pairs are formed by unifying a positive and a negative literal from two different clauses (these two literals form the critical pair), and adding their resolvent (consisting of all remaining literals in the two clauses, with the unifying substitution taken into account) to the original set of clauses.

Input: a set of clauses F in first-order predicate logic.

Output: a (semi-)decision whether F is unsatisfiable.

Algorithm:
while F has not been proven unsatisfiable and new clauses can be added **do**
(c_1, c_2) := a pair in F;
$\text{Res}((c_1, c_2)) := \bigcup\{L | L \in c_1 \vee L \in c_2\} - \bigcup\{L | (L \in c_1 \wedge \neg L \in c_2) \vee (L \in c_2 \wedge \neg L \in c_1)\}$;
if $\text{Res}((c_1, c_2)) = \Box$ **then**
F is *unsatisfiable*
else $F := F \cup \text{Res}((c_1, c_2))$
fi
od

2.4 The CPC Frame Algorithm

We now give a framework for all CPC-procedures. How to fit the three procedures into it should be clear from the algorithmic skeletons given above. The set G is the set generated during the completion of the set of original patterns (cf. section 3) P. The set CP keeps track of all critical pairs still to be considered. The set $cp(f, g)$ denotes all critical pairs between the patterns f and g, the set $cp(f, G)$ all critical pairs arising from computing $cp(f, g)$ for all $g \in G$.

Input: a set P of patterns in a universe T and an ordering $>$ on T.

Output: an extended set of patterns such that the reduction relation induced by the patterns allows a canonical normal form algorithm.

Algorithm:
$G := P$;
$CP := \bigcup_{f,g \in G} cp(f, g)$;
while $CP \neq \emptyset$ **do**
(s, t) := an element in CP;
$CP := CP - \{(s, t)\}$;
$(s, t) := (\text{Sim}(s), \text{Sim}(t))$;
if $s \neq t$ **then**
analyze (s, t);
l := left hand side of rule to be added; r := right hand side of rule to be added;
$h := (l, r)$;
$CP := CP \cup \bigcup cp(h, G)$;
$G := G \cup \{h\}$.
fi
od

Sim is a simplification algorithm, computing (weak) normal forms of elements of the universe T with respect to the reduction relation generated by the patterns in P (and any additional patterns added during the algorithm). In the subalgorithm *analyze* the procedure tests whether a reduction from Sim(s) to Sim(t) or vice versa can be added without violating the noetherianity of the reduction relation.

3 Axiomatic Preliminaries

In this section we describe the prerequisites for the application of CPC-procedures and the basic notions involved with them.

3.1 Prerequisites

In a situation where we can apply CPC-procedures, we have a structure $(T, >, \rightarrow_P, M, E)$. Below, we describe the components of this quintuple, and give their basic properties.

The universe T. The universe T describes the domain from which the elements on which we will apply the CPC-procedure are taken. For the Knuth-Bendix algorithm the elements of T are the terms in the term algebra $Ter(\Sigma)$, for a given signature Σ.

Objects. The objects are the polynomials in $K[x_1, \ldots, x_n]$. in the Gröbner basis algorithm T consists of the polynomials in $K[x_1, \ldots, x_n]$, and in the resolution procedure T consists of the clauses of first-order predicate logic (boolean algebra).

Ordering on T. The universe T must be equipped with an ordering, either partial or total. In the Knuth-Bendix case, the reduction ordering $<$ on $Ter(\Sigma)$ is given in advance as a parameter to the completion procedure. Note that a different ordering may influence the outcome of the completion. In the case of the Gröbner basis algorithm, we stipulate that the ordering (inferred from a total ordering on the power products) is total. In fact, we require one more property, related to the reduction, to obtain the concept of ***admissible orderings***, cf. e.g. [Buc83a]. A common choice is the total degree ordering. In the resolution algorithm, the clauses are ordered lexically, cf. [Rob65].

The pattern set P and the generated reduction relation $\rightarrow_P \subset T \times T$. In the terminology of [Buc85], we have a set of *patterns* that generate the reduction relation. For the Knuth-Bendix algorithm, the patterns are derived from the equations in E respecting the ordering relation: if $s = t \in E$ then if $s > t$ we have the pattern $s \rightarrow_P t$, and if $t > s$ we have the pattern $t \rightarrow_P s$. In the Gröbner basis algorithm the patterns are formed from the elements of F as follows. If f is some polynomial in F, we have the pattern $LM(f) \rightarrow_P LM(f) - f$, where $LM(f)$ is the leading monomial of f. In the resolution procedure the patterns (s, t) are obtained from the clauses in F. Any clause $\{L_1, \ldots, L_m, M_1, \ldots, M_n\}$ (where the L_i are positive and the M_j negative literals) can be used to form $m + n$ patterns, depending on which of the literals is taken as left hand side s.

For practical applications, the set P of all pairs (l, r) such that $l \rightarrow_P r$ is finite (this corresponds to a finite number of generators for an ideal in the Gröbner basis algorithm, a finite number of rewrite rules in Knuth-Bendix, and a finite set of clauses in resolution). Of course, nothing stops us from allowing an infinite set of such pairs in the theoretical setting, but in general that raises additional termination problems.

In order to be a true *reduction* relation, $\rightarrow_P$ has to fulfill the following property:

$$\text{(Ax1)} \quad x \rightarrow_P y \Rightarrow x > y \quad P\text{-reduction}$$

The transitive and reflexive closure of $\rightarrow_P$ will be denoted by $\rightarrow_P^*$.

The multiplying operation. In all three CPC-algorithms, we have a concept of ***multiplier***, which essentially expresses substituting on an atomic term, that can later be incorporated into a composite term by a replacement (see below).

In the Knuth-Bendix algorithm the multipliers correspond to all variable substitutions on the terms. In the Gröbner basis algorithm the multipliers correspond to all multiplications by monomials (which includes coefficients, monomials with an empty power product). In the resolution procedure the multipliers are the variable substitutions of first order predicate logic.

Given a suitable notion of multiplier for the algorithm under consideration, we can define:

Definition 2 (Multipliers) *We define* $M = \{\mu : T \to T\}$ *as the set of all* multipliers.

We stipulate that M contains the identity operation and preserves the ordering. We also demand that the composition of two multiplications again is a multiplication.

(Ax2)	$\mathrm{id}_T \in M$	**identity**
(Ax3)	$x > y \Rightarrow (\forall \mu)\, \mu x > \mu y$	**order preservation**
(Ax4)	$\mu, \nu \in M \Rightarrow \mu \circ \nu \in M$	**composition in M**

The operation of embedding. The notions corresponding to *place* (as in [Buc85]) for the three algorithms considered are as follows. In the Knuth-Bendix algorithm the places are the positions where subterms can occur. They can be coded by sequences of positive integers using the Dewey decimal notation. This notation describes the path from the outermost symbol to the head of the subterm at that position. In the Gröbner basis algorithm the places are the positions where terms can be added, i.e. they are coded by the power products less or equal the leading power product of the polynomial in question. In this case the embedders, operations based on the concept of places as explained in section 3, are only used for the definition of the generalized reduction relation $\to_G$, they are not important for the determination of critical pairs. Any labeling unambiguously distinguishing the (linearly ordered) places of the power product will do. In the resolution procedure the places are just the positions in a clause where literals occur, they may be labeled successively.

In all three cases we can code the places in a given term, e.g. by natural numbers. We will assume that a concept of places in a term and their numbering is given for all CPC-procedures.

For a term t, its set of places is denoted by $P(t)$. We denote the subterm occurring at place p of term t by $t|_p$, and the result of replacing this subterm by another term s is denoted by $t[s]_p$. The set of all terms u that are the same as t except at place p, i.e. all terms u such that for some term s, $u[s]_p = t$, is called the *context* of the replacement. This set of terms is denoted by $u[\sqcup]_p$.

Now, for every context $u[\sqcup]_p$ we define a function on T that embeds a term t at the empty place in the context.

Definition 3 (Embedders) $E_{up} : T \to T$ *is defined by*

$$E_{up}(t) := u[t]_p.$$

If $p \notin P(u)$ *then* $E_{up}(t) := u$. *We also define the set of all embedding functions:*

$$E = \{E_{up} | u \in T, p \in P(u)\}.$$

Elements of E *will be called* embedders. *In the following, we will use the symbol* η *to denote an embedder.*

E contains the identity operation on T (namely embedding in an empty context), and the embedders η in E should preserve the order. Finally, E is closed under composition of embedders with the appropriate domain and codomain.

(Ax5)	$\mathrm{id}_T \in E$	**identity**
(Ax6)	$x > y \Rightarrow (\forall \eta)\eta x > \eta y$	**order preservation**
(Ax7)	$\eta, \theta \in E, \mathrm{Cod}(\theta) = \mathrm{Dom}(\eta) \Rightarrow \eta \circ \theta \in E$	**composition in E**

From M and E, we define the relation G expressing that the second object is more general than the first.

Definition 4 (Generalizers)

$$(x, y) \in G \text{ iff } (\exists \mu \in M \exists \eta \in E)\, y = \eta(\mu(x)).$$

In the context of reduction, $(x, y) \in G$ means that if x can be reduced by a pattern $p \in P$, then the same pattern p can be used to reduce y.

The extended reduction relation $\rightarrow_G$. We now define an extension of the reduction relation generated by $\rightarrow_P$.

Definition 5 (Generalized Reduction) *$\rightarrow_G$ will denote the reduction relation generated by $\rightarrow_P$ and the following two rules:*

$$\begin{array}{lll} \text{(Gen1)} & x \rightarrow_P y \Rightarrow (\forall \mu) \mu x \rightarrow_G \mu y & M\textbf{-extension} \\ \text{(Gen2)} & x \rightarrow_P y \Rightarrow (\forall \eta) \eta x \rightarrow_G \eta y & E\textbf{-extension} \end{array}$$

The two rules Gen1 and Gen2 correspond to the *multiplier* and *replacement rules* in [Buc85], or the *fully invariant property* and *replacement property* in [DJ90], respectively. In the case of Gröbner bases the generated reduction relation is not compatible with embedding (so Gen2 would not hold), only *semi-compatible* (cf. [BL82]), i.e. that if $x \rightarrow_P y$ then ηx and ηy have a common normal form. Conceiving the Gröbner basis algorithm as completion modulo the actual representation of polynomials (i.e., not insisting on the usual distributive normal form, but viewing polynomials as finite sums of monomials) would overcome this problem, as shown in [MS91].

Lemma 1

$$\text{(g-Red)} \quad x \rightarrow_G y \Rightarrow x > y \quad G\textbf{-reduction}$$

Proof If $x \rightarrow_G y$, then by definition 5 one of the following three cases must hold.

1. $x \rightarrow_P y$. Then by Ax1 $x > y$.
2. There exists a μ in M and x', y' in T such that $x = \mu x'$, $y = \mu y'$, and $x' \rightarrow_P y'$. Then by Ax1 $x' > y'$, and by Ax3 $x > y$.
3. There exists an η in E and x', y' in T such that $x = \eta x'$, $y = \eta y'$, and $x' \rightarrow_P y'$. Then by Ax1 $x' > y'$, and by Ax6 $x > y$.

□

Definition 6 ($\rightarrow_G^*$) *Finally, we denote the transitive and reflexive closure of $\rightarrow_G$ by $\rightarrow_G^*$.*

The quintuple $(T, >, \rightarrow_P, M, E)$ will be used in our definition of the base category *CPC* in section 4.

3.2 Basic Notions

We define the basic concepts of CPC-procedures within the framework $(T, >, \rightarrow_P, M, E)$.

Weak Normal Form. Intuitively, $y \in T$ is a *weak normal form* for $x \in T$, if x reduces to y, and it is not possible to reduce y any further. The weakness expresses that uniqueness is not guaranteed. In other words, y is a minimal element with respect to $\rightarrow_G^*$. Formally:

$$\text{(WNF)} \quad \text{WNF}(x, y) :\Leftrightarrow x \rightarrow_G^* y \wedge \neg(\exists z) y \rightarrow_G z$$

Normal Form. Intuitively, $y \in T$ is the *normal form* of $x \in T$ if y is a weak normal form of x and y is uniquely determined. Formally:

$$\text{(NF)} \quad y = \text{NF}(x) :\Leftrightarrow \text{WNF}(x, y) \wedge [(\forall z)\text{WNF}(x, z) \Rightarrow z = y]$$

We will denote the normal form of an element of the universe $t \in T$ by $\underline{t}$.

Termination Condition. The completion of the set of patterns P is finished when all terms have a (unique) normal form. We therefore have:

$$\text{(TC)} \quad (\forall t \in T \exists \underline{t} \in T)\, \underline{t} = \mathrm{NF}(t)$$

Completion Step. Completion is the creation of a set of generating arrows such that for all objects (any) normal form algorithm yields a (unique) normal form. A single completion step consists of adding, if necessary and possible, one new reduction pattern. The idea behind the categorical setup is to formulate the completion step as the formation of a limit in the categorical sense. Here we give an algorithmic description of the form of such a completion step. (It is a further specification of the analysis step in the CPC Frame algorithm.)

Precondition:	$(\exists x)\mathrm{WNF}(x, x_1) \wedge \mathrm{WNF}(x, x_2) \wedge x_1 \neq x_2$
Action:	**if** $x_1 > x_2$ **then** $\rightarrow_P := \rightarrow_P \cup \{(x_1, x_2)\}$
	elif $x_2 > x_1$ **then** $\rightarrow_P := \rightarrow_P \cup \{(x_2, x_1)\}$
	elif *failure*
Continue	with the extended relation $\rightarrow_P$

Most General Superposition Situation. In order to find critical pairs, a CPC-algorithm computes ***most general superposition situations*** between two patterns (in fact, the same pattern may be used to be superposed on itself). Such situations occur when there are two multipliers (μ_1, μ_2) such that, for some patterns $l_1 \rightarrow_P r_1$, $l_2 \rightarrow_P r_2$, $\mu_2(l_2)$ occurs as a subterm in $\mu_1(l_1)$. To capture the ***most general*** demand, we insist that if two other multipliers (μ_3, μ_4) can be found such that $\mu_4(l_2)$ occurs as a subterm in $\mu_3(l_1)$ ***at the same place*** u, we can find a multiplier exposing (μ_3, μ_4) as a special instance for (μ_1, μ_2). (Note that multipliers do not change the arrangement of places of subterms within a term.) Formally, we have the following definition.

Definition 7 (Most General Superposition Situation) *A* most general superposition situation *for two patterns* $p_1 = l_1 \rightarrow_P r_1$ *and* $p_2 = l_2 \rightarrow_P r_2$ *is a quadruple* (η, u, μ_1, μ_2) *such that:* $\mu_1 l_1 = \eta_{\mu_1(l_1),u}(\mu_2(l_2))$, *and for all* (μ_3, μ_4) *such that* $\mu_3 l_1 = \eta_{\mu_3(l_1),u}(\mu_4(l_2))$ *there exists a* μ_5 *such that* $\mu_3(l_1) = \mu_5(\mu_1(l_1))$ *and* $\mu_4(l_2) = \mu_5(\mu_2(l_2))$.

In the case of the Gröbner Basis algorithm, the embedder $\eta_{\mu_1(l_1),u}$ is always taken to be the identity embedder of the whole pattern l_1 at the root position u of the term l_1. Then, we arrive at the simpler concept of ***most general unifier.***

Most General Unifier. In the case of Gröbner bases we only need to compute the least common multiple of the leading monomials of the two patterns under consideration. In the general formulation, this amounts to say that we only have to look at a most general overlapping by multipliers on two different patterns.

Definition 8 *A* most general unifier *for two patterns* $p_1 = l_1 \rightarrow_P r_1$ *and* $p_2 = l_2 \rightarrow_P r_2$ *is a pair of multipliers* (μ_1, μ_2) *such that:* $\mu_1 l_1 = \mu_2 l_2$ *and for all* (μ_3, μ_4) *such that* $\mu_3 l_1 = \mu_4 l_2$ *there exists a* μ_5 *such that* $\mu_3 l_1 = \mu_5 \mu_1 l_1$ *(and therefore* $\mu_4 l_2 = \mu_5 \mu_2 l_2$*).*

Note that most general unifiers need not be unique. A most general unifier will correspond to a (weak) coproduct in categorical terms.

Critical Pair. From most general superposition situations we can compute critical pairs, which play an essential role in bringing down the complexity of CPC-procedures.

A critical pair is computed by reducing the most general superposition situation for two patterns by these patterns. The two terms obtained form the critical pair. The pattern to be added in the completion step is then established by reducing the critical pair to (weak) normal forms.

Deviating from ordinary terminology, we will find it convenient to refer to the unreduced critical pair as ***provisional critical pair***, and to the reduced forms used for computing the new pattern as ***critical pair***.

Consequently, we get the following definition.

Definition 9 (Critical Pair) *A* critical pair *for two patterns $p_1 = l_1 \rightarrow_P r_1$ and $p_2 = l_2 \rightarrow_P r_2$ is a pair of objects $(\underline{s}, \underline{t}) \in T \times T$ such that $s = \mu_1 r_1$ and $t = \eta_{\mu_1(l_1),u}(\mu_2(r_2))$ is a most general superposition situation for the two patterns, and $\underline{s}$ and $\underline{t}$ are (weak) normal forms of s and t, respectively. For such a pair of objects (s,t) we will use the term* provisional critical pair.

Two examples.

For the Knuth-Bendix case, suppose we have two patterns (rewrite rules) $p_1 = a.(b.c) \rightarrow (a.b).c$ and $p_2 = a.a^{-1} \rightarrow e$. Then we have $l_1 = a.(b.c)$, $r_1 = (a.b).c$, $l_2 = a.a^{-1}$, $r_2 = e$, $\mu_1 = b^{-1}/c$ (read: b^{-1} substituted for c), $\mu_2 = b/a$, $s = (a.b).b^{-1}$, $u = 2$, $t = \eta_{a.(b.b^{-1}),2}(e) = a.e$, and, in the presence of a third pattern $p_3 = a.e \rightarrow a$, $\underline{t} = a$, whereas s cannot be reduced any further. So the critical pair is $((a.b).b^{-1}, a)$.

For Gröbner bases, take two polynomials $p_1 = xy - x^2$ and $p_2 = y^2 - x$. With lexicographic ordering we obtain as respective left and right hand sides $l_1 = xy$, $r_1 = x^2$, $l_2 = y^2$, $r_2 = x$, $\mu_1 = y$ (read: multiplying by y), $\mu_2 = x$, $s = x^2y$, $u = 1$, $t = \eta_{xy^2,1}(x^2) = x^2$ (note that the embedder is always the identity operator in the case of Gröbner bases), and, because of p_1, $\underline{s} = x^3$. We can't reduce t further, so the critical pair is (x^3, x^2).

4 Constructing the Categorical Framework

We start with the construction of a basic category $\mathcal{CPC}$, within which all the defining notions can be modeled. The universe of a CPC-procedure yields the objects for $\mathcal{CPC}$, and the reduction relation induces the arrows. Finally, it is shown that the operations of multiplication and embedding fulfill the functor properties (for functors from $\mathcal{CPC}$ onto itself).

4.1 Construction of the Category $\mathcal{CPC}$

Definition 10 ($\mathcal{CPC}$) *$\mathcal{CPC}$ is constructed as follows:*

- *The objects are the elements in the universe T.*
- *The arrows in $\mathcal{CPC}$ are generated from the patterns in P. We take the transitive and reflexive closure of $\rightarrow_P$ and extend it with the reductions that can be carried out because of the multiplier and replacement rules. In other words, the arrows are based on the $\rightarrow_G^*$ relation. So, there is an arrow with as domain an $x \in T$ and as codomain a $y \in T$ if and only if $x \rightarrow_G^* y$. Because of this, there is always at most one arrow between two objects, i.e. no distinction is made between different sequences of reductions leading to $x \rightarrow_G^* y$.*

Theorem 1 *$\mathcal{CPC}$ is a category.*

Proof
We have to check the defining properties for categories.

1. Identity morphism for objects: this is defined by the reflexiveness of the defining relation $\rightarrow_G^*$.
2. Composition of arrows: this is defined by the transitivity of $\rightarrow_G^*$.
3. The identity and associativity rules trivially hold because if there are two arrows between given objects A and B, they are the same by definition.

□

Alternatively, we may construct $\mathcal{CPC}$ by taking the discrete category consisting of the objects from the universe and the identity arrows, and adding the generating patterns defining $\rightarrow_P$ as indeterminates.

Proposition 1 *The category $\mathcal{CPC}$ can equivalently be described by constructing the category freely generated by the graph generated by the reduction relation $\rightarrow_G$, where the objects are the objects of the universe, and all arrows with common domain and codomain are identified.*

Proof Direct from the definition of freely generated categories in [LS86] and the identification of all the arrows with identical source and target.
□

4.2 Constructing Functors on $\mathcal{CPC}$

We now define two types of functors from $\mathcal{CPC}$ to $\mathcal{CPC}$.

- *Multiplier* functors (MUL) corresponding to the *multipliers*.
- *Embedder* functors (EMB) corresponding to the *embedders*.

Definition 11 (Multiplier Functors) *Every* multiplier μ *defines a* multiplier functor $\text{MUL}(\mu)$ *from* $\mathcal{CPC}$ *to* $\mathcal{CPC}$ *such that*

- *for every object* $A \in \mathcal{CPC}$ $\text{MUL}(\mu)(A) = \mu(A)$, *and*
- *for every arrow* $\phi : A \to B$ $\text{MUL}(\mu)(\phi) = \mu(\phi) : \mu(A) \to \mu(B)$. *Rule* Gen1 *guarantees that this arrow exists.*

Definition 12 (Embedder Functors) *Every* embedder η *defines an* embedder functor $\text{EMB}(\eta)$ *from* $\mathcal{CPC}$ *to* $\mathcal{CPC}$ *such that*

- *for every object* $A \in \mathcal{CPC}$ $\text{EMB}(\eta)(A) = \eta(A)$, *and*
- *for every arrow* $\phi : A \to B$ $\text{EMB}(\eta)(\phi) = \eta(\phi) : \eta(A) \to \eta(B)$. *Rule* Gen2 *guarantees that this arrow exists.*

Theorem 2 *All functors* $\text{MUL}(\mu)$ *and* $\text{EMB}(\eta)$ *fulfill the functor properties.*

Proof
Trivial, since all required arrows exist because of Gen1 and Gen2, respectively, and are uniquely determined by the definition of $\mathcal{CPC}$.
□

5 Discussion of the Categorical Model

In this section we formulate the relevant notions of CPC-algorithms in categorical terms, starting from the category $\mathcal{CPC}$, but invoking categories derived from it for some of the concepts.

5.1 Weak Normal Form

In the categorical model, a weak normal form of an object A is an object B such that there exists an arrow $f : A \to B$ and such that there are no arrows with domain B except the identity arrow id_B. In arrow notation,

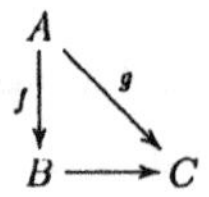

implies $B = C$ (or $g = f$).

5.2 Normal Form

The canonical normal form of an object A is characterized as a terminal object in the comma category $\mathcal{CPC} \uparrow A$.

To be precise, the canonical normal form $\underline{A}$ of the object A is the unique element fulfilling, for all B, the commutativity of

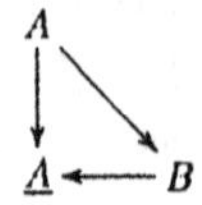

Lemma 2 *If the reduction system is strongly normalizing, any object A has as its normal form the object $\underline{A}$ characterized by being the terminal object in the comma category $\mathcal{CPC} \uparrow A$.*

Proof If $\underline{A}$ is the canonical normal form of A, it clearly holds that for any object B such that A reduces to B, B can be reduced to $\underline{A}$. This fact yields the required diagram in $\mathcal{CPC} \uparrow A$, where the uniqueness of the arrow from $A \rightarrow B$ to $A \rightarrow \underline{A}$ is induced by the uniqueness of the arrow $B \rightarrow \underline{A}$ in $\mathcal{CPC}$.

Conversely, if $\underline{A}$ is the terminal object in $\mathcal{CPC} \uparrow A$, we know that any object B such that A reduces to B has a reduction to $\underline{A}$. On the condition that all reduction sequences are finite (i.e. the reduction relation $\rightarrow_G$ is strongly normalizing), this implies that $\underline{A}$ is the canonical normal form of A.

□

5.3 Strongly Normalizing

Initially, the reduction relation is supposed to be strongly normalizing in a CPC-algorithm, and the algorithm ends with failure if the strong normalizability cannot be preserved. (In the case of Gröbner bases this never is a problem.)

In a more general setting, one might attempt to drop this limitation, in which case the comma categories $\mathcal{CPC} \uparrow A$ may have infinitely many objects. (One can easily show, using König's lemma, that the strong normalizability of the reduction relation is equivalent with the finiteness of $\mathcal{CPC} \uparrow A$ for all A.)

5.4 Most General Superposition Situation

The categorical characterization of most general superposition situations is based on that of most general unifiers. Essentially, one takes the left hand side of a fixed pattern and tries to unify the left hand side of any other pattern (possibly the same) with any of its subterm. If this is possible, one only considers a most general pair of multipliers (for the precise definition, cf. section 3).

Hence, a most general superposition situation is given by an embedder functor EMB and two multiplier functors MUL1, MUL2 such that the pair (MUL1,MUL2) forms a most general unifier in the (product) category of multipliers on $\mathcal{CPC}$. For the precise definition, we first give the (simpler) case of plain most general unifiers (as used in the Gröbner basis algorithm) and then generalize to the most general superposition situation case.

5.5 Most General Unifier

A most general unifier of two objects (left hand sides of patterns) is a weak coproduct in the category of multiplier functors on $\mathcal{CPC}$. First, we define this *category of multipliers* $\mu\mathcal{CPC}$.

Definition 13 ($\mu\mathcal{CPC}$) *$\mu\mathcal{CPC}$ is constructed as follows:*

- *The objects are the elements in the universe T, the same objects as in $\mathcal{CPC}$.*
- *The arrows in $\mu\mathcal{CPC}$ are generated by the multiplier functors MUL(μ). I.e., for all A, B in $\mathcal{CPC}$, there is an arrow MUL : $A \rightarrow B$ if and only if MUL(A) $= B$, for a multiplier MUL in M.*

Lemma 3 *$\mu\mathcal{CPC}$ is a category.*

Proof

1. Identity morphism for objects: this is defined by Ax2: the set of all multipliers contains the identity operation, so MUL(id_T) yields the identity arrow for all objects.
2. Composition of arrows: this is defined by Ax4: the compositions of two multipliers is again a multiplier, so for two arrows MUL1 : $A \rightarrow B$ and MUL2 : $B \rightarrow C$, the composition MUL2 ∘ MUL1 : $A \rightarrow C$ is defined by the multiplier MUL2 ∘ MUL1.
3. The identity and associativity rules: If any two arrows have identical source and target, they are trivially identifiable, as no internal properties of source and target are involved.

□

Now, a most general unifier is a weak coproduct in this category.

Lemma 4 *The most general unifier of two objects A, B in T (so objects in $\mathcal{CPC}$, and $\mu\mathcal{CPC}$) is the weak coproduct of A and B in $\mu\mathcal{CPC}$.*

Proof The defining property of a most general unifier (namely that any other unifier is just a special instance of it, attainable by an intermediate multiplier (here MUL5)) is exactly what is expressed by the weak coproduct diagram for $\mu\mathcal{CPC}$, below. (The fact that such a coproduct is not uniquely determined in general explains the use of the notion *weak coproduct* here.)

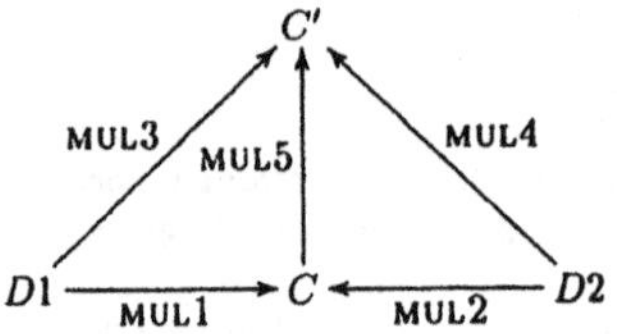

□

From this, we derive the definition of a *most general superposition situation*. We consider the product category $\mu\mathcal{CPC} \times \mu\mathcal{CPC}$, for which we have:

Lemma 5 *The most general superposition situation of two objects A, B in T (so objects in $\mathcal{CPC}$, and $\mu\mathcal{CPC}$) is given by a triple of functors in $\mathcal{CPC}$ (EMB, MUL1, MUL2) such that EMB $\circ$ MUL1 = MUL2 and the pair (MUL1, MUL2) fulfills the property that for every pair (MUL3, MUL4) for which EMB $\circ$ MUL3 = MUL4 there is a pair (MUL5, MUL5) such that the following diagram in $\mu\mathcal{CPC} \times \mu\mathcal{CPC}$ commutes:*

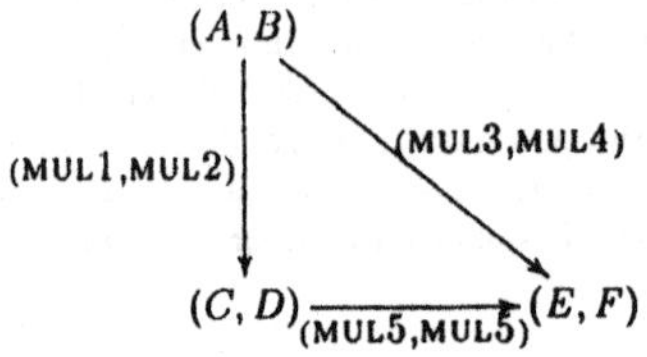

where (C, D) = (MUL1(A), EMB(MUL1(A))) *and* (E, F) = (MUL3(A), EMB(MUL3(A))).

Proof Directly from the definition of most general superposition situations in section 3.

□

5.6 Critical Pair

A provisional critical pair for $D1$ and $D2$ with respect to the patterns $f1$ and $f2$ is given by the pair (A, B), a critical pair consists of Sim(A) and Sim(B). The computation of the critical pair is exemplified by the following diagram.

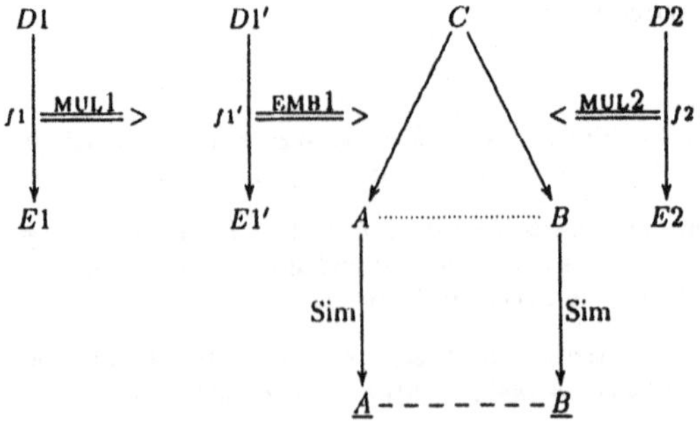

The provisional critical pair can categorically be obtained from $D1, D2$ and the patterns $f1$ and $f2$ by computing on the arrows of $\mathcal{CPC}$. To formulate this, the ***extended category of multipliers*** $\mu^*\mathcal{CPC}$ is defined.

Definition 14 ($\mu^*\mathcal{CPC}$) *$\mu^*\mathcal{CPC}$ is constructed as follows:*

- *The objects are the arrows in $\mathcal{CPC}$.*
- *The arrows in $\mu^*\mathcal{CPC}$ are generated by the multiplier functors* MUL(μ). *I.e., for all $f: A \rightarrow B, g: C \rightarrow D$ in $\mathcal{CPC}$, there is an arrow* MUL $: f \rightarrow g$ *if and only if* MUL$(f) = g$, *for a multiplier* MUL *in M.*

Lemma 6 *$\mu^*\mathcal{CPC}$ is a category.*

Proof
The argument is completely analogous to lemma 3 (treating $\mu\mathcal{CPC}$).
□

In this extended category of multipliers, a provisional critical pair of two patterns $f1 : D1 \rightarrow E1$ and $f2 : D2 \rightarrow E2$ is obtained from computing the most general superposition situation of $D1$ and $D2$ in $\mu\mathcal{CPC} \times \mu\mathcal{CPC}$ and then applying the defining embedder and multipliers EMB, MUL1, MUL2 on the arrow $f1, f2$, objects in $\mu^*\mathcal{CPC}$. From the provisional critical pair (cf. the corresponding definition in section 3) the real critical pair is found by calculating the weak normal forms of codomains EMB(MUL1($E1$)) and MUL2($E2$) in $\mathcal{CPC}$.

5.7 Arrow Addition

This is the basic operation on the category $\mathcal{CPC}$, corresponding to a completion step, and consists of adding an arrow (new basic pattern) Sim(A) $\rightarrow$ Sim(B) or vice versa (compatibility with >), cf. Polynomial Categories, [LS86]. Denoting such an indeterminate arrow with domain Sim(A) and codomain Sim(B) by X, this step consists of constructing the polynomial category $\mathcal{CPC}[X]$. This is equivalent to the construction of the category freely generated by the graph (on T) generated by the basic patterns from $\rightarrow_P$ and the additional reduction Sim(A) $\rightarrow$ Sim(B).

The whole completion process hence consists of successive constructions of polynomial categories, until the category is completed, i.e. fulfills the property that all objects have canonical normal forms.

5.8 The Completed Category $\underline{\mathcal{CPC}}$

The completion process terminates successfully if and only if all objects of $\mathcal{CPC}$ have uniquely determined normal forms. I.e., no infinite reductions exist and for all objects A, if there are arrows $f : A \rightarrow B$ and $g : A \rightarrow C$ then we can find arrows $h : B \rightarrow D$ and $k : C \rightarrow D$. The normal form itself is given by the pseudopushout of f and g in $\mathcal{CPC}$, where a pseudopushout is defined as an ordinary pushout in which the arrow in the limiting diagram is reversed.

Theorem 3 *The completion of $\mathcal{CPC}$ is terminated successfully, creating the completed category $\underline{\mathcal{CPC}}$, if all pseudopushouts exist (in $\underline{\mathcal{CPC}}$).*

Proof:
The confluence of the reduction relation is expressed by the fact that for any two reductions $f : A \rightarrow^*_G B$ and $g : A \rightarrow^*_G C$ there exists an object D and arrows $h : B \rightarrow^*_G D$, $k : C \rightarrow^*_G D$ in the completed category $\underline{\mathcal{CPC}}$ such that the corresponding diagram commutes (i.e., $h \circ f = k \circ g$).

The uniqueness of the normal form is then equivalent to the statement that for any D', h', k' making the pushout diagram of f and g commutative, we can find an arrow $l : D' \rightarrow^*_G D$ (taking D as the normal form). This last arrow is exactly inverse to the arrow defining a ***pushout***, hence the label ***pseudopushout***.

The above discussion can be summarized in the following commutative diagram.

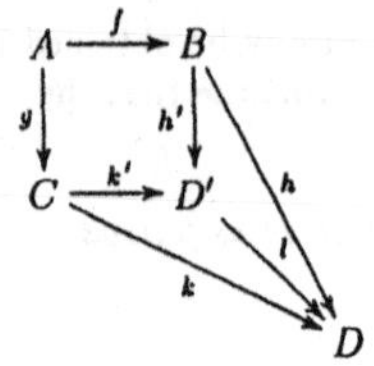

□

6 Final Remarks

In parallel to the establishment of the categorical framework and the clarification of the relations between the various CPC-algorithms and the conditions for the algorithm(s) to be effective, we are carrying out an implementation of the categorical CPC-algorithm. The implementational work is done in Standard ML ([Wik87]). It need not be stressed that this implementation will not lead to a more efficient way of computing Gröbner bases or completed term rewriting systems. Rather, it will provide us with insight in the minimal conditions for successful termination, failure, or non-termination, in a clear and well-established mathematical language.

I would like to thank Bruno Buchberger and Jochen Pfalzgraf for many helpful remarks and discussions.

References

[Ben75] David B. Benson. The basic algebraic structures in the categories of derivations. *Information and Control*, 28:1–29, 1975.

[BL82] Bruno Buchberger and Rudiger Loos. Algebraic simplication. *Computing*, Supplement 4:11–43, 1982. Appeared as report CAMP 82-12.0, RISC, Linz, Austria.

[Buc65] Bruno Buchberger. *Ein Algorithmus zum Auffinden der Basiselemente des Restklassenringes nach einem nulldimensionalen Polynomideal.* PhD thesis, Universität Innsbruck, 1965.

[Buc83a] Bruno Buchberger. A critical-pair/completion algorithm in reduction rings. CAMP-Technical Report 83-21.0, Universität Linz, 1983.

[Buc83b] Bruno Buchberger. Gröbner bases: An algorithmic method in polynomial ideal theory. In N.K. Bose, editor, *Recent Trends in Multidimensional Systems Theory*, chapter 6, pages 184–232. Reidel, Dordrecht, The Netherlands, 1983.

[Buc85] Bruno Buchberger. Basic features and development of the critical-pair/completion procedure. In Jean-Pierre Jouannaud, editor, *Proc. First Int. Conf. RTA, LNCS 202*, pages 1–45, Dijon, 1985. Springer Verlag.

[DJ90] Nachum Dershowitz and Jean-Pierre Jouannaud. Rewriting systems. In J. van Leeuwen, editor, *Handbook of Theoretical Computer Science*, pages 243–320. Elsevier Publishers, Amsterdam, 1990.

[HL79] Gérard Huet and Jean-Jacques Lévy. Call by need computations in non-ambiguous linear term rewriting systems. Rapport Laboria 359, IRIA, August 1979.

[Hue81] Gérard Huet. A complete proof of correctness of the Knuth-Bendix completion algorithm. *Journal of Computer and System Sciences*, 23(1):11–21, 1981.

[Hue86] Gérard Huet. Formal structures for computation and deduction. Working material for lectures at the International Summer School on Logic of Programming and Calculi of Discrete Design in Marktoberdorf, Germany, July 29 – August 10 1986.

[Joh87] Michael Johnson. *Pasting Diagrams in n-Categories with Applications to Coherence Theorems and Categories of Paths.* PhD thesis, University of Sydney, 1987.

[Joh91] Michael Johnson. Linear term rewriting systems are higher dimensional string rewriting systems. In C.M.I. Rattray and R.G. Clark, editors, *The Unified Computation Laboratory*, pages 3–12. Oxford University Press, 1991.

[KB67] Donald E. Knuth and Peter B. Bendix. Simple word problems in universal algebras. In J. Leech, editor, *Computational Problems in Abstract Algebra*, pages 263–298, Oxford, 1967. Pergamon Press. Appeared 1970.

[Klo80] Jan Willem Klop. *Combinatory Reduction Systems.* PhD thesis, Mathematisch Centrum, Amsterdam, 1980.

[Klo90] Jan Willem Klop. Term rewriting systems. Technical Report CS-R9073, Centrum voor Wiskunde en Informatica, Amsterdam, 1990.

[Lam80] Joachim Lambek. From λ-calculus to cartesian closed categories. In James Paul Seldin and James Roger Hindley, editors, *To H. B. Curry: Essays on Combinatory Logic, Lambda Calculus and Formalism*, pages 375–402. Academic Press, 1980.

[Lév78] Jean-Jacques Lévy. Réductions correctes et optimales dans le λ-calcul. Thèse d'Etat, Université de Paris VII, 1978.

[LS86] Joachim Lambek and P.J. Scott. *Introduction to Higher-Order Categorical Logic*, volume 7 of *Cambridge Studies in Advanced Mathematics.* Cambridge University Press, 1986.

[Mac71] Saunders MacLane. *Categories for the Working Mathematician*, volume 5 of *Graduate Texts in Mathematics.* Springer Verlag, 1971.

[MS91] Aart Middeldorp and Mirjana Starčević. A rewrite approach to polynomial ideal theory. Technical Report CS-R9160, Centrum voor Wiskunde en Informatica, Amsterdam, 1991.

[Ric83] Elaine Rich. *Artificial Intelligence.* McGraw-Hill, London, 1983.

[Rob65] John Alan Robinson. A machine-oriented logic based on the resolution principle. *Journal of the ACM*, 12(1):23–41, 1965.

[Sto91] Karel Stokkermans. A categorical approach to critical-pair/completion procedures. Technical Report 91-58.0, RISC-Linz, J. Kepler University, Linz, Austria, Europe, December 1991. Literature Survey and Work Plan for Ph.D Thesis.

[Wik87] Åke Wikström. *Functional Programming Using Standard ML.* Prentice Hall International, Hemel Hempstead, United Kingdom, 1987.

Trace Rewriting Systems*

Yabo Wang†
David Lorge Parnas

Communications Research Laboratory
Department of Electrical and Computer Engineering
McMaster University
Hamilton, Ontario, Canada L8S 4K1

ABSTRACT We first describe a finite state machine based module interface specification method — trace assertion method. Then, viewing trace assertions in a trace specification as defining an equivalence relation among traces, we define a trace rewriting system from the specification. Such a trace rewriting system resembles some aspects of string rewriting, membership conditional rewriting and priority rewriting. We prove that a *proper* trace rewriting system is both terminating and confluent and compare trace rewriting systems with term rewriting systems.

1. Introduction

Software modules, just as other engineering products, should have precise specifications, i.e., precise descriptions of the externally observable behaviour of the modules. To abstractly specify the requirements on each module's interface, we view modules as "black-boxes" and specify them by the trace specification method [12] (a concise summary of this method can be found in [11]).

In the trace method, abstract data objects implemented by software modules are viewed as finite state machines. For a deterministic object, the externally observable behaviour is completely determined by the input history — the traces of the object. An object's behaviour can thus be specified in terms of assertions about traces. To make a trace specification a practical engineering reference document, so that a user can quickly find relevant information without searching through the whole document, trace assertions in trace specifications are rigidly organized (using a "one element extension of canonical traces" approach which will be described in Section 2).

Given a trace specification S, by interpreting the trace assertions in S as rewriting rules, S can be viewed as defining a (trace) rewriting system. Thus, the externally observable behaviour of the module specified by S can be simulated by the rewriting system. We shall see that the rigid structure on trace assertions in trace specifications, originally introduced for easy information accessing, also plays a crucial role in guaranteeing some very important properties of trace rewriting systems such as confluence and termination.

---

†. An off campus Ph.D student of Queen's University, Kingston, Ontario

*. This research is supported by the Telecommunications Research Institute of Ontario.

In this paper, we restrict ourselves to trace specifications for modules that implement a single, deterministic abstract data object. In Section 2, we introduce the trace specification method and the underlying semantics. We then formally define trace rewriting system in Section 3. We prove that a special type of trace rewriting systems — *proper* trace rewriting systems are confluent and terminating in Sections 4. The conclusions, discussions, and some future research topics are presented in Section 5.

2. Trace Specification Method

In the trace specification method [12], each "information hiding" [10] module is viewed as implementing one finite state machine (hereafter referred to as an *object*).

An *event (of interest)* of an object is either an invocation of module access program or a change of input variable value. Events are denoted by *event expressions* of the form $E(a_1, ..., a_k)$, where E is an event class identifier (e.g., program name) and a_i's are the actual values of event arguments†.

Each object can be in a finite number of *states*. The state of an object can only be changed by the occurrence of one of its events of interest. For an object A in its initial state, the occurrence of a sequence of discrete events of A, say, $e_{i_1}, e_{i_2}, ..., e_{i_m}$, will leave the object in some state S. The string $E_{i_1}.E_{i_2}.\cdots.E_{i_m}$, called a *trace* of A, is used to denote the sequence of events and the state S, where E_{i_j} is the event expression denoting the event e_{i_j} and "." is the string concatenation operator.

Since the length of a trace can be arbitrarily long (but must be finite), there are an infinite number of possible traces for any object, but an object can only have a fixed number of distinct states. Therefore, it must be the case that many traces denote the same state and hence, "denotes the same stateas" (symbolically "$\equiv$") is an equivalence relation over the set of all possible traces of an object. For an object of *n* states, "$\equiv$" partitions the trace set into *n equivalence classes.* In each such class, a single trace, called the *canonical trace*, can be chosen to *represent* that class.‡

Since we only consider deterministic machines, it is obvious that if two traces T1 and T2 denote the same state, then T1.S and T2.S, where S is an arbitrary trace, will denote the same state. This fact is formally stated as an axiom:

$$(\forall\, T1, T2, S)(T1 \equiv T2 \leftrightarrow T1.S \equiv T2.S) \qquad \langle 2.1 \rangle$$

For an object, only the values of its output variables are considered as *externally observable.* An object's externally observable behavior is completely defined if the value of its output variables are defined for any trace. The latter can be achieved by (1) defining an *output function o* which maps any canonical trace to the vector of output values and (2) defining a *reduction function r* which maps any possible trace to its representative canonical trace: for the object in a state denoted by a given trace T, the vector of output values is determined by $o(r(T))$.

---

†. We assume that each event, e.g, a program invocation, has a fixed number of arguments.

‡. In theory, the choice of the canonical trace can be arbitrary. In practice, there are "good" and "bad" choices — see the discussion in **Conclusions** (6).

Once the set of canonical traces has been chosen, defining *o* is straightforward and hence will not be further discussed in this paper. Defining the function *r*, however, requires a more structured approach because *r*'s domain, the set of all possible traces, is infinite.

The reduction function *r* we intend to define should have the property that any two traces T1 and T2 are in the same equivalent class if and only if *r* maps them to the same canonical trace, i.e.,

$$(\forall\, T1, T2)(T1 \equiv T2 \leftrightarrow r(T1) = r(T2)) \qquad \text{<2.2>}$$

Also, since *r* maps a canonical trace to itself, it is idempotent, i.e.,

$$(\forall\, T)(r(r(T)) = r(T)) \qquad \text{<2.3>}$$

For a given trace $E_{i_1}.E_{i_2}.\cdots.E_{i_{m-1}}.E_{i_m}$, by <2.3>, we have

$$r(r(E_{i_1}.E_{i_2}.\cdots.E_{i_{m-1}}.E_{i_m})) = r(E_{i_1}.E_{i_2}.\cdots.E_{i_{m-1}}.E_{i_m}) \qquad \text{<2.4>}$$

hence, by <2.2> and <2.4>

$$r(E_{i_1}.E_{i_2}.\cdots.E_{i_{m-1}}.E_{i_m}) \equiv E_{i_1}.E_{i_2}.\cdots.E_{i_{m-1}}.E_{i_m} \qquad \text{<2.5>}$$

Similarly,

$$r(E_{i_1}.E_{i_2}.\cdots.E_{i_{m-1}}) \equiv E_{i_1}.E_{i_2}.\cdots.E_{i_{m-1}} \qquad \text{<2.6>}$$

by <2.1> and <2.6>

$$r(E_{i_1}.E_{i_2}.\cdots.E_{i_{m-1}}).E_{i_m} \equiv E_{i_1}.E_{i_2}.\cdots.E_{i_{m-1}}.E_{i_m} \qquad \text{<2.7>}$$

Then, by <2.5> and <2.7>

$$r(E_{i_1}.E_{i_2}.\cdots.E_{i_{m-1}}.E_{i_m}) \equiv r(E_{i_1}.E_{i_2}.\cdots.E_{i_{m-1}}).E_{i_m} \qquad \text{<2.8>}$$

By <2.2> and <2.4>,

$$r(E_{i_1}.E_{i_2}.\cdots.E_{i_{m-1}}.E_{i_m}) = r(r(E_{i_1}.E_{i_2}.\cdots.E_{i_{m-1}}).E_{i_m}) \qquad \text{<2.9>}$$

Consequently,

$$\begin{aligned} r(E_{i_1}.E_{i_2}.\cdots.E_{i_{m-1}}.E_{i_m}) &= r(r(E_{i_1}.E_{i_2}.\cdots.E_{i_{m-1}}).E_{i_m}) \\ &= r(r(r(E_{i_1}.E_{i_2}.\cdots.E_{i_{m-2}}).E_{i_{m-1}}).E_{i_m}) \\ &= \dots\dots \\ &= r(r(r(\cdots r(r(r(\_).E_{i_1}).E_{i_2}).\cdots.E_{i_{m-1}}).E_{i_m}) \end{aligned} \qquad \text{<2.10>}$$

where "_", called the *empty trace*, is the default canonical trace denoting the *initial state* of every object, i.e., r(_) = _ by default.

Therefore, *r* can be recursively defined: for *every* canonical trace T extended by *any* event expression E, (i.e., for each trace of the form T.E, where T is a canonical trace), we define its equivalent canonical trace T_c by

$$r(T.E) \triangleq T_c \qquad \text{<2.11>}$$

Using an infix notation $=_r$, the above definition can be expressed as a *trace assertion*†

$$T.E =_r T_c \qquad \text{<2.12>}$$

†. For historical reasons, in [12], such a trace assertion is written as $T.E \equiv T_c$

This method of defining trace equivalence classes by single event extension of canonical traces was first explicitly stated in [6]†. The advantage of defining an object's externally observable behavior this way is that, in effect, such trace assertions define the state transition function of the object: when the object is in the state denoted by the canonical trace T, the occurrence of the event described by E changes the object into the state denoted by the canonical trace T_c.

We use an integer stack object as an example. For the canonical trace PUSH(1).PUSH(2).PUSH(3) (denoting the stack in the state of containing the three elements 1, 2, and 3), the effect of a POP event on the object in this state can be described by:

$$PUSH(1).PUSH(2).PUSH(3).POP =_r PUSH(1).PUSH(2)$$

which states that PUSH(1).PUSH(2).PUSH(3).POP and PUSH(1).PUSH(2) denote the same state of the stack object (containing the elements 1, and 2) and the latter is the canonical trace; or operationally, the event POP changes the stack from the state denoted by PUSH(1).PUSH(2).PUSH(3) to the state denoted by PUSH(1).PUSH(2).

Such a trace assertion can be generalized to an *assertion schema* that describes a set of trace assertions. The format of an assertion schema is the same as a trace assertion except that either side of "$=_r$" is a *trace pattern*, i.e., a parameterized trace where in some positions of event expressions and arguments, *variables* are used to denote a set of event expressions and arguments, respectively. Such a trace pattern defines a set of trace instances – the traces that can be obtained by replacing all the event expression and argument variables by event expressions and event arguments, respectively. More formally, let $\chi(T)$ denote the set of variables appearing in the trace pattern T and θ denote a variable replacement operation. $\theta(T)$ is a trace instance defined by the pattern T. The assertion schema below describes the effect of POP on any stack containing three integer elements:

$$PUSH(x_1).PUSH(x_2).PUSH(x_3).POP =_r PUSH(x_1).PUSH(x_2)$$

This schema can be further generalized so that the effect of POP on any non-empty stack can be described by

$$[T1.PUSH(x)].POP =_r T1$$

where expression in "[...]" is a canonical trace pattern defining the subset of canonical traces — traces end with PUSH(x) for some x.

The most general trace assertion schema is a conditional schema of the format:

$$P(\chi(T.E)) :: T.E =_r T_c \qquad <2.13>$$

where

- T.E, often collectively denoted as T_L, called the LHS (left hand side) of the assertion schema, consists of a canonical trace pattern T (any instance of T is a canonical trace) extended by an *event schema E* (an event class identifier with its parameterized arguments);

---

†. For each equivalence trace class, we choose exactly one *canonical trace* as the representative while in [6] it is possible to have more than one *normal trace* as representatives.

- $P(\chi(T.E))$, or $P(\chi(T_L))$, denotes a predicate, called the *condition* of the assertion schema such that, all the *free* variables in P must be in T_L; and
- T_c, called the RHS (right hand side) of the assertion schema, is a trace pattern such that all the variables in T_c must appear in $\chi(T_L)$ and when $\theta(P(\chi(T_L))) =$ **true**, $\theta(T_c)$ is a canonical trace instance.

In such an assertion schema, $P(\chi(T_L))$, together with T, defines a subset of canonical traces: T defines the pattern of the subset (e.g., the orders of the event expressions, the positions of the variables with respect to constants, etc.) and P defines some extra constraints on $\chi(T_L)$. When $\chi(T_L)$ is empty or there is no constraint on $\chi(T_L)$, $P(\chi(T_L)) =$ **true** and can be omitted. In the following, when no confusion may arise, we often abbreviate $P(\chi(T_L))$ as P. We put all the assertion schemas with the same extending event schema E, e.g., POP, or PUSH(x), in a tabular format as shown below†

$$\begin{array}{|c|}\hline P_1 \\ \hline \cdots \\ \hline P_n \\ \hline \end{array} \;::\; \begin{array}{|c|}\hline T_1 \\ \hline \cdots \\ \hline T_n \\ \hline \end{array}\,.\,E \quad \equiv_r \quad \begin{array}{|c|}\hline T_{c1} \\ \hline \cdots \\ \hline T_{cn} \\ \hline \end{array}$$

where each P_i is a condition which, together with T_i, defines a subset CS_i of the canonical trace set CS.

In a trace module specification, to guarantee such tables completely define the reduction function *r* described above, the trace method requires that

(1) there be one such a table for each event schema E; and

(2) in each table, $CS_1, CS_2, \ldots, CS_n$ be a partitioning of CS, i.e.,

- $CS_i \cap CS_j = \emptyset$, for each $0 \leq i, j \leq n$ and $i \neq j$; and
- $CS_1 \cup CS_2 \cup \ldots \cup CS_n = CS$.

The second requirement can be equivalently expressed as: for any canonical trace T, there is one and only one row k in each of the tables such that $T \in CS_k$.

We refer to trace specifications satisfying (1) and (2) as *proper* and hereafter, we only consider proper trace specifications.

3. Trace Rewriting

In a trace specification, the set of trace assertion schemas (cf. <2.13>) together with the "inference rules" <2.1> – <2.3> completely define a trace equivalence relation. This approach has been used in equational systems for a long time; there, a set of equations together with the equational inference rules are used to define an equivalence relation among terms. An equational system is often simulated by means of term rewriting where equations are applied to one direction only [7,8]. This leads us to consider simulating trace specifications by interpreting trace assertion schemas as trace rewriting rules. Actually, a trace schema $P(\chi(T_L)) :: T_L \equiv_r T_R$ has already been viewed as a con-

---

†. In [12], the tables are in a different style: there is no explicit separations between the conditions and the canonical trace patterns (the conditional symbols "::" are omitted).

ditional rewriting rule in Section 2: any trace T that is an instance of T_L and satisfies $P(\chi(T_L))$ can be rewritten to a trace which is an instance of T_R (with the variables in T_R being replaced by their corresponding trace elements in T). Such a rule is expressed as $P(\chi(T_L)) :: T_L \Rightarrow T_R$.

Note that these rules obey the following two restrictions on rewriting rules [8]:

(1) no LHS of any assertion contains only a single variable; and

(2) for each assertion, all the variables appearing in the RHS also appear either at the LHS or in the condition.

The set of trace assertion schemas in a trace specification can be therefore viewed as a rewriting system. Formally, we define trace rewrite systems as follows.

Definition 3.1 (*free traces, ground traces*): Let

(1) $\mathbf{E}$ be a non-empty, finite set of elements called *event names*;

(2) $\mathbf{C}$ be a denumerable set of elements called *constants* and $\mathbf{C} \cap \mathbf{E} = \emptyset$;

(3) $\mathbf{V}$ be a denumerable set of elements called *variables* and $\mathbf{V} \cap \mathbf{E} = \emptyset, \mathbf{V} \cap \mathbf{C} = \emptyset$; and

(4) α: $\mathbf{E} \rightarrow \mathbf{N}$ be an *arity function* which assigns to each element of $\mathbf{E}$ a natural number (we denote $\mathbf{E}_i = \{e \in \mathbf{E} \mid \alpha(e) = i\}$).

Then, the *free trace set* $\mathbf{T}(\mathbf{E}, \mathbf{V} \cup \mathbf{C}, \alpha)$ is constructed as follows:

(i) the distinguished element "_", the *empty trace*, is in $\mathbf{T}(\mathbf{E}, \mathbf{V} \cup \mathbf{C}, \alpha)$, i.e, $\_ \in \mathbf{T}(\mathbf{E}, \mathbf{V} \cup \mathbf{C}, \alpha)$;

(ii) all the elements in $\mathbf{V} \cup \mathbf{C}$ are also in $\mathbf{T}(\mathbf{E}, \mathbf{V} \cup \mathbf{C}, \alpha)$, i.e., $\mathbf{T}(\mathbf{E}, \mathbf{V} \cup \mathbf{C}, \alpha) \supseteq (\mathbf{V} \cup \mathbf{C})$;

(iii) for each $e \in \mathbf{E}_k$ ($k \geq 0$) and $a_i \in (\mathbf{V} \cup \mathbf{C})$ ($0 \leq i \leq k$), $e(a_1, a_2, \ldots, a_k) \in \mathbf{T}(\mathbf{E}, \mathbf{V} \cup \mathbf{C}, \alpha)$, (when $e \in \mathbf{E}_0$, e() is written as e);

(iv) if $T_1, T_2 \in \mathbf{T}(\mathbf{E}, \mathbf{V} \cup \mathbf{C}, \alpha)$, then, $T_1 \cdot T_2 \in \mathbf{T}(\mathbf{E}, \mathbf{V} \cup \mathbf{C}, \alpha)$; and

(v) these are the only members of $\mathbf{T}(\mathbf{E}, \mathbf{V} \cup \mathbf{C}, \alpha)$.

A subset of the free traces, $\mathbf{T}(\mathbf{E}, \mathbf{C}, \alpha)$ (i.e., free traces without variables), will be called as *ground traces*. ❑

Definition 3.2 (*lengths of ground traces*):

(1) length(_) = 0; and

(2) $\text{length}(E_{i_1}.E_{i_2}.\cdots.E_{i_n}) = n$. ❑

Definition 3.3 (*subtraces*): *Subtraces* of a given ground trace $T = E_{i_1}.E_{i_2}.\cdots.E_{i_n}$ are the sequences of events in

$T_s = \{E_{i_j}.\cdots.E_{i_k} \mid \text{where } 1 \leq j \leq k \leq n\} \cup \{\_\}$

A subtrace $E_{i_j}.\cdots.E_{i_k}$ of T can be denoted as $T]_{j-1}^{k}$, and the empty trace _ as $T]_0^0$. A subset of T_s, $T]_0^k$, where $0 \leq k \leq n$, are called the *prefix subtraces* of T and often simply denoted as $T]^k$. ❑

It is clear that $T]^{\text{length}(T)} = T$, and $T]_0^0$, or $T]^0$ (i.e., _), is a prefix subtrace of any trace.

Definition 3.4 (*prefix subtrace replacement*): *A prefix subtrace replacement* $T[<j> \leftarrow S]$ replaces the prefix subtrace $T]^j$ of the ground trace T by the ground trace

S. More formally, assuming $T = E_{i_1}.\cdots.E_{i_j}.\cdots.E_{i_n}$, and $S = E_{k_1}.\cdots.E_{k_m}$, then $T[<j> \leftarrow S] = E_{k_1}.\cdots.E_{k_m}.E_{i_{j+1}}.\cdots.E_{i_n}$. ❑

Note that if $j = \text{length}(T)$, then $T[<j> \leftarrow S] = S$.

Let **P** be a set of predicate symbols and denote $\mathbf{P}_i = \{p \in \mathbf{P} \mid p \text{ has } i \text{ parameters}\}$.

Definition 3.5 (*substitution, matching*): A *substitution* θ is a mapping on $\mathbf{T}(\mathbf{E} \cup \mathbf{P}, \mathbf{V} \cup \mathbf{C}, \alpha)$ such that:

(1) $\theta(x) = t \in \mathbf{T}(\mathbf{E}, \mathbf{C}, \alpha)$ for all $x \in X$, where X is a finite subset of **V**;
(2) $\theta(y) = y$ for all $y \in (\mathbf{V} - X)$;
(3) $\theta(c) = c$ for all $c \in \mathbf{C}$;
(4) $\theta(e(a_1,...a_n)) = e(\theta(a_1),\theta(a_2),\ldots,\theta(a_n))$ for $e \in \mathbf{E}_n$ $(1 \leq n)$, and $\theta(e) = e$ for $e \in \mathbf{E}_0$;
(5) $\theta(p(b_1,...b_n)) = p(\theta(b_1),\theta(b_2),\ldots,\theta(b_n))$ for $p \in \mathbf{P}_n$ $(1 \leq n)$, and $\theta(p) = p$ for $p \in \mathbf{P}_0$; and
(6) $\theta(T_1 \cdot T_2) = \theta(T_1) \cdot \theta(T_2)$.

A free trace S *matches* a ground trace T if there exists a substitution θ, such that $\theta(S) = T$. ❑

Definition 3.6 (*trace rule*): Given a set of free traces $\mathbf{T}(\mathbf{E}, \mathbf{V} \cup \mathbf{C}, \alpha)$, a *trace rule* is a triple $<P, T_L, T_R>$ of one predicate and two free traces such that $T_L \notin \mathbf{V}$ and any variable in T_R must also occurs in T_L or P. ❑

Definition 3.7 (*traces defined by a trace specification*): A trace module specification S defines a set of free traces, $\mathbf{T}(\mathbf{E}, \mathbf{V} \cup \mathbf{C}, \alpha)$, such that the set can be constructed from

E = {the union of all the event names listed in the events syntax tables of S};
C = {the union of the domains of all the event arguments in S};
V = {any denumerable set of elements that
(a) includes all variables named in S; and
(b) $\mathbf{V} \cap \mathbf{E} = \emptyset$, $\mathbf{V} \cap \mathbf{C} = \emptyset$}; and
α is the function which assigns each event name in **E** a non-negative integer that is the number of arguments the event takes. ❑

Definition 3.8 (*trace rules defined by a trace specification*): A trace module specification S defines a set of trace rules $\Rightarrow$ such that if $P :: T_L =_r T_R$ is an assertion schema in S, then $<P, T_L, T_R> \in \Rightarrow$. ❑

Definition 3.9 (*trace rewriting system defined by a trace specification*): A trace module specification S defines a *trace rewriting system* $R = <\mathbf{T}(\mathbf{E}, \mathbf{V} \cup \mathbf{C}, \alpha), \Rightarrow>$ where $\mathbf{T}(\mathbf{E}, \mathbf{V} \cup \mathbf{C}, \alpha)$ is the set of free traces defined by S, and $\Rightarrow$ is the set of trace rules defined by S. ❑

In the following, we simply refer R as a trace rewriting system when the trace specification S which defines R is not in our concern.

Definition 3.10 (*redex, redex point, rpsr triple, mrpsr*): In a trace rewriting system $<\mathbf{T}(\mathbf{E}, \mathbf{V} \cup \mathbf{C}, \alpha), \Rightarrow>$, for a trace $T \in \mathbf{T}(\mathbf{E}, \mathbf{V} \cup \mathbf{C}, \alpha)$, if there exist a $j \geq 0$, a substitution θ, and a rule $<P, T_L, T_R> \in \Rightarrow$ such that

$$T^{|j} = \theta(T_L) \quad \text{and} \quad \theta(P(\chi(T_L))) = \textbf{true}$$

then, $T^{|j}$ is called a ***redex (reducible expression)*** of T, j is called a ***redex point*** of T, and the triple (j, θ, <P, T_L, T_R>) is called a ***rpsr (redex point-substitution-rule) triple*** of T. A rpsr triple (j, θ, <P, T_L, T_R>) is called a ***max rpsr triple (mrpsr)*** of T, if j is the maximum redex point of T. The set of mrpsr's of T is denoted as *mrpsr*(T). ❑

In Section 4, we will prove that in a proper trace rewriting system, a ground trace T has a unique mrpsr triple.

Definition 3.11 (*trace rewriting relation*): Given a trace rewriting system R = <**T**(**E**, **V** ∪ **C**, α), ⇨>, the set of trace rules ⇨ generates a ***trace rewriting relation***, denoted $\rightarrow_R$, on the ground traces **T**(**E**, **C**, α) such that:

(1) for a trace T and each rpsr triple (j, θ, <P, T_L, T_R>) ∈ mrpsr(T),
if $T_L \neq T_R$, then T $\rightarrow_R$ T[<j> ← θ(T_R)]; and

(2) these are the only members of $\rightarrow_R$. ❑

When T $\rightarrow_R$ T', T is often said *can be rewritten* to T' in one rewriting step.

Definition 3.12 (*canonical form*): Given a trace rewriting relation $\rightarrow_R$, for a ground trace T in R, if there exists a rewriting sequence $T \rightarrow_R T_1 \rightarrow_R T_2 \rightarrow_R \cdots \rightarrow_R T_x$, but T_x can not be further rewritten, then T_x is called a ***canonical form*** of T in R. ❑

Note that, in Definition 3.11, the condition $T_L \neq T_R$ eliminates the useless rewriting of the form T $\rightarrow_R$ T which otherwise would be generated by the trace assertion schema of the form P :: T $=_r$ T. Such a schema is necessary in trace specifications to completely define the reduction function *r*.

Therefore, by Definition 3.12, for a trace T, if (j, θ, <P, T_L, T_R>) ∈ mrpsr(T) and $T_L = T_R$, then T is a canonical form since there does not exist any T' to which T can be rewritten.

Definition 3.13 (*proper trace rewriting system*): A trace rewriting system R is ***proper*** if the trace specification S which defines R is proper. ❑

4. Properties of Proper Trace Rewriting Systems

We first prove the following lemma dealing with the size of mrpsr set:

Lemma (*Single Triple Lemma*): *In a proper trace rewriting system R, for any trace T ≠ _, mrpsr(T) contains exactly one rpsr triple.*

Proof: (1) We first prove that mrpsr(T) is not empty. Since T ≠ _, we can assume $T = E_{i_1}.\cdots.E_{i_n}$, where n > 0.

According to Definition 3.10, all the rpsr triples contained in the set mrpsr(T) have the same redex point, the maximum one. Since R is proper, there must exist a trace rule <P, T_L, T_R> ∈ ⇨ and a substitution θ such that θ(P(χ(T_L))) = **true** and $\_.E_{i_1} = \theta(T_L)$. Therefore, T has rpsr triples.

Due to the fact that the length of T is finite, T has a maximum redex point. By Definition 3.10, T has max rpsr triples. Thus, mrpsr(T) is not empty.

(2) We now prove. by contradiction, that mrpsr(T) contains no more than one rpsr tri-

ple. Suppose that mrpsr(T) contains more than one rpsr triple. Then there exist two rpsr triples (j, θ, <P, T_L.E, T_R>) and (j, θ', <P', T_L'.E', T_R'>), such that

$$\theta(P(\chi(T_L))) = \textbf{true}, \theta(T_L) = T|^j \text{ and } \theta'(P'(\chi(T_L'))) = \textbf{true}, \theta'(T_L') = T|^j.$$

This would mean that in S, which is the trace specification defines R, $T|^{j-1}$ is both in the canonical trace subset defined by P and T_L and in the subset defined by P' and T_L'. Therefore, S is not proper and neither is R. This contradicts the assumption that R is a proper rewriting system. ❑

For the relationship between the canonical traces in trace specifications and canonical forms in trace rewriting systems, we have:

Theorem (*Canonical Form Theorem*): *Let R be the proper trace rewriting system defined by a proper trace specification S. Then a ground trace is a canonical form in R if and only if it is a canonical trace in S.*

Proof: For the special case, we show that the empty trace, _, is both a canonical form in R and canonical trace in S. By default, _ is the canonical trace denoting the initial state of the object specified by S. Since there is no trace schema in S and hence no trace rule in R with the LHS being the empty trace, it can not be rewritten in R and therefore is a canonical form in R.

For the general cases:

(1) For the "if" part. Let T_c be a non-empty and ground canonical trace in S. Since the reduction function *r* maps a canonical trace to itself, there is a trace schema of the form P :: T $=_r$ T in S and hence a trace rule of the form P :: T ⇨ T in R, where $\theta(T) = T_c$ for some substitution θ. Thus, T_c's max redex point j = length(T_c), and the only rpsr triple in mrpsr(T_c) is (j, θ, <P, T, T>). However, such a triple does not satisfy the condition $T_L \neq T_R$ in Definition 3.11. Therefore, T_c can not be rewritten in R and hence is a canonical form.

(2) For the "only if" part. Let T_c be a non-empty and ground canonical form in R, i.e., T can not be rewritten in R. By Definition 3.11, there are two possibilities: either (a) mrpsr(T_c) is empty, or (b) for all rpsr triples (j, θ, <P, T_L, T_R>) ∈ mrpsr(T_c), it is the case that $T_L = T_R$. By the *Single Triple Lemma*, we know mrpsr(T_c) is not empty. Therefore, (b) is true, i.e., T_c is mapped to itself by the reduction function *f* and hence is a canonical trace in S. ❑

Definition 4.1 (*reduced rewriting system*): A conditional trace rewriting system R is *reduced* if for each rewriting rule P(χ(T_L)) :: T_L ⇨ T_R in R and each substitution θ, θ(P(χ(T_L))) = **true** implies θ(T_R) is a canonical form. ❑

Theorem (*Reduced System Theorem*): *A proper trace rewriting system R is a reduced rewriting system.*

Proof: Let S be the proper trace specification which defines R. A trace rule P(χ(T_L)) :: T_L ⇨ T_R is in R if and only if the trace assertion schema P(χ(T_L)) :: T_L $=_r$ T_R is in S. For each substitution θ, when θ(P(χ(T_L))) = **true**, by <2.13>, we know θ(T_R) is a canonical trace. By the *Canonical Form Theorem*, θ(T_R) is a canonical form in R. ❑

Two of the most extensively studied properties of rewriting systems are the confluence property and the termination property.

Definition 4.2 (*confluent rewriting system*): Let $\rightarrow_R$ be the rewriting relation generated by a rewriting system R and $\overset{*}{\rightarrow}_R$ is the transitive-reflexive closure of $\rightarrow_R$. R is said to be <u>*confluent*</u> if T $\overset{*}{\rightarrow}_R$ T_1 and T $\overset{*}{\rightarrow}_R$ T_2 implies there exists a T' such that $T_1 \overset{*}{\rightarrow}_R$ T' and $T_2 \overset{*}{\rightarrow}_R$ T'. ❑

Definition 4.2 states that in a confluent rewriting system, any term having canonical forms has a unique canonical form.

Definition 4.3 (*terminating rewriting system*): Let $\rightarrow_R$ be the rewriting relation generated by a rewriting system R. R is said to be *(finitely) terminating* if for any ground term T, it has no possible infinite sequence (or path) of rewritings

$$T \rightarrow_R T_1 \rightarrow_R T_2 \rightarrow_R T_3 \rightarrow_R \ldots\ldots$$ ❑

In other words, in a terminating rewriting system, every ground term has canonical forms.

A confluent and terminating rewriting system has the property that any ground term T has one and only one canonical form. In such a system, the order in which the rewriting rules are applied has no effect on the rewriting results. The question of whether or not a rewriting system is confluent or terminating is, in general, undecidable [3, 7].

In the following, we prove that a proper trace rewriting system has both the confluence and termination properties.

Theorem (*Confluence Theorem*): *A proper trace rewriting system R is confluent.*

Proof: First, we know the empty trace is a canonical form in R and hence can not by rewritten.

By the *Single Triple Lemma*, in R, for any non-empty ground trace T, mrpsr(T) contains exact one rpsr triple. Therefore, if such a T is not a canonical form, then there exists a *unique* T_1 such that $T \rightarrow T_1$. By the transitivity of this "uniqueness", T has a unique rewriting path. By Definition 4.2, R is (trivially) confluent. ❑

Before we prove the termination property of proper trace rewriting systems, we introduce the concept of "well ordered set":

Definition 4.4 (*well ordered set*): A set W is $\succ$-well-ordered if $\succ$ is an irreflexive and transitive relation on W such that for any $X \in W$, there is no infinite sequence of elements

$$X \succ X_1 \succ X_2 \succ \ldots\ldots$$ ❑

Theorem (*Termination Theorem*): *A proper trace rewriting system R is terminating.*

Proof: We prove the termination property based on the following theorem:

> **Theorem** [9]: *A rewrite system R over a set of expressions S is terminating if, and only if, there exists a $\succ$-well-ordered set W and a mapping τ from S to W such that*
>
> $$u \rightarrow_R v \quad \textit{implies} \quad \tau(u) \succ \tau(v)$$
>
> *for all expressions u and v in S.*

Obviously, the set of natural numbers, **NAT**, is well-ordered by the "greater than" relation, >.

Denoting the maximum redex point of a ground trace T as maxr(T), we define

$$\tau: \ \mathbf{T(E, C, \alpha)} \ \longrightarrow \ \mathbf{NAT}$$

as the function such that for a ground trace T, .

$$\tau(T) = \begin{cases} 0, & \text{if T is a canonical trace} \\ \text{length}(T) - \text{maxr}(T) + 1 & \text{otherwise} \end{cases}$$

Now, we prove that, for any two ground traces M and N, M $\longrightarrow$ N implies $\tau(M) > \tau(N)$.

Let $M = E_{i_1}.\cdots.E_{i_k}.\cdots.E_{i_n}$ and mrpsr(M) = {(j, θ, <P, T_L, T_R>)} (by the *Single Triple Lemma*, we know mrpsr(M) contains only one rpsr triple). Then,

$$M|^j = E_{i_1}.\cdots.E_{i_j} = \theta(T_L),\ \theta(P) = \mathbf{true},\ \text{and } N = M[<j> \longleftarrow \theta(T_R)]$$

Since R is proper and $\theta(T_R)$ is an instance of the RHS of a trace rule, by the *Reduced System Theorem*, $\theta(T_R)$ is a canonical form and hence a canonical trace. There are two cases:

(1) If j = length(M) = n, then N = $\theta(T_R)$, i.e., N is a canonical trace, and hence

$$\tau(N) = 0. \qquad \text{<4.1>}$$

On the other hand, we have

$$\tau(M) = \text{length}(M) - j + 1 = 1. \qquad \text{<4.2>}$$

(2) If j < n, then $N = \theta(T_R).E_{i_{j+1}}.\cdots.E_{i_n}$ and

$$\tau(M) = \text{length}(M) - j + 1 = n - j + 1. \qquad \text{<4.3>}$$

Since $\theta(T_R)$ is a canonical trace and R is proper, there must exist a trace rule in R such that its LHS can be matched to $\theta(T_R).E_{i_{j+1}}$. This implies

$$\text{maxr}(N) \geq \text{length}(\theta(T_R)) + 1.$$

And hence

$$\text{maxr}(N) > \text{length}(\theta(T_R)).$$

On the other hand,

$$\text{length}(N) = \text{length}(\theta(T_R)) + [n - j] = \text{length}(\theta(T_R)) + n - j.$$

Therefore,

$$\begin{aligned} \tau(N) &= \text{length}(N) - \text{maxr}(N) + 1 \\ &< \text{length}(N) - [\text{length}(\theta(T_R))] + 1 \\ &= [\text{length}(\theta(T_R)) + n - j] - [\text{length}(\theta(T_R))] + 1 \\ &= n - j + 1. \end{aligned} \qquad \text{<4.4>}$$

By <4.1> – <4.4>, in both cases, we have $\tau(M) > \tau(N)$. Therefore, R is terminating. ❑

Corollary: *In a proper trace rewriting system R, a non-empty ground trace T of length m can be rewritten to its canonical form in no more than m rewriting steps.*

Proof: From the above proofs we observe that if T is not canonical form, it has a unique rewriting path $T \rightarrow_R T_1 \rightarrow_R T_2 \rightarrow_R T_3 \rightarrow_R \ldots T_n$. But, using the same τ definition, we have

$$m \geq \tau(T) > \tau(T_1) > \tau(T_2) > \tau(T_3) \ldots > \tau(T_n) = 0.$$

Therefore, $n \leq m$. ❑

5. Conclusions

There are many similarities between the trace rewriting systems (TraceRS) and the traditional term rewriting systems (TermRS). Almost all the terminology defined in Section 3 is from the literature on TermRS and the definitions are also very similar to their counterparts in the TermRS literature. Furthermore, TraceRS resembles Toyama's membership conditional rewriting [14]. Comparing TraceRS with TermRS, we offer the following observations:

(1) Trace Rewriting versus String Rewriting. TraceRS is different from, and, more powerful (in the sense of expressive power) than, string rewriting [2]: the conditions in trace assertion schemas can describe more complex traces whereas terms in most string rewriting systems can only be string instances.

(2) System Structure. A TermRS usually consists of a set of rewriting rules rarely having any structure; whereas a trace specification consists of a set of rigidly structured trace assertions (structured by canonical traces). The structure of trace assertions in a trace specification also facilitates the consistency and completeness checks of trace specifications: by the definition given in [5], a proper trace specification is always consistent and sufficiently complete.

(3) The "Word Problem". Since a proper TraceRS is both terminating and confluent, the "word problem" is decidable: for any two traces M and N (of length m, and n, respectively) in a proper TraceRS, the problem "is M ≡ N?", or "do M and N denote the same state" can be solved in m + n rewriting steps since in at most m + n rewriting steps, M and N can be rewritten to their representative canonical trace(s).

(4) System Construction/Modification. The trace specification method provides a step by step guide for a specifier to construct a proper specification and hence a terminating and confluent TraceRS; whereas when a TermRS is constructed, the termination and confluence properties are solely depend on the talent of a designer. The proof of either termination or confluence property of a TermRS is often rule dependent, and any change in the rewriting rules (e.g., modifying the theory they define) may destroy the properties and the proof has to start all over again. For this reason, it is not convenient to use general TermRS's in an environment in which the system design is constantly changing. On the other hand, the rigid structure in TraceRS makes changes easier and if the revised specification is proper, the confluence and termination properties are certain.

(5) Implementation. Intuitively, trace rewriting replaces the *longest* canonical trace and the following event expression of a trace by its representative canonical trace. It is very similar to the priority term rewriting [1] except that instead of syntactically assigning priority to rewriting rules, we use the length of matching prefix as the priority. While introducing priorities complicates the definition of rewriting relation, it makers implementation easier: given a trace $T = E_{i_1}.\cdots.E_{i_n}$, we can use a "current maximum redex point" pointer to scan T from left to right and do rewrite whenever a prefix of T is matched to the LHS of a trace rule.

TraceRS is particularly convenient in simulating an object's externally observable behavior: initially, the object is set to its initial state denoted by the empty trace _. When an event of the object occurs, the object will be in the state denoted by the canonical traces derived from the rewriting rules. Any externally observable behaviour of the object, e.g., the returned value of a program call, the values of output variables, etc. can be derived from the canonical trace (by the output function *o*).

(6) Canonical Trace/Term Concept. To successfully write a proper trace specification, the most important and creative step is to choose the canonical trace set. The wisely chosen canonical traces can greatly simplify the specification. This step should not be viewed as an obstacle of the method. Many papers in algebraic specifications implicitly use the canonical term idea [4, 5, 15]. Roughly speaking, events chosen to construct canonical traces correspond to van Horebeek's *constructors* [15], Guttag's *C constructors* [5], and SPECIAL's *primitive O-operators* [13], which have to be identified and used to guide the constructing of specifications, although they are not explicitly used to structure a specification.

Topics for Further Study. The unrestricted use of first order predicates in describing trace patterns enables us to construct more concise and expressive specifications and hence TraceRS's. On the other hand, lack of restrictions makes the properness check (of trace specifications and TraceRS's) more difficult and requires more powerful matching algorithms. We need to study all the three aspects together in order to find a reasonably powerful trace language and yet keep the properness check and matching algorithm feasible.

The work described here can be further extended in two directions: (i) The current trace rewriting is a single module and single level rewriting: any values of the objects not implemented by the module are considered as constants since no rules of the module can rewrite them. We can extend the trace rewriting to many module and multiple level rewriting such that the "constants" are traces of other lower level modules. We conjecture that such an extended rewriting system (with certain restrictions) still has both termination and confluence properties; (ii) In [12], two other types of specifications are also defined: multiple objects module specifications and non-deterministic module specifications. Trace rewriting systems must be extend to handle the two types of specifications.

References

[1] Baeten, J. C. M., Bergstra, J. A., and Klop, J. W. "Term Rewriting Systems with Priorities" *Proceedings of the Second International Conference on Rewriting Techniques and Applications*, France, Lecture Notes in Computer Science 256, pp. 83-94, 1987.

[2] Book, R. V. "Thue Systems as Rewriting Systems" *Journal of Symbolic Computation*, **3**, pp. 39-68, 1987.

[3] Dershowitz, N. "Termination of Rewriting" *Journal of Symbolic Computation*, **3**,

pp. 69-116, 1987.

[4] Goguen, J. A., Thatcher, J. W., and Wagner, E. "An Initial Algebra Approach to the Specification and Implementation of Abstract Data Types" *Current Trends in Programming Methodology* IV, R. T. Yeh (ed.), pp. 80-184, Prentice-Hall, 1978.

[5] Guttag, J., and Horning, J. J. "The Algebraic Specification of Abstract Data Types" *Acta Informatica* 10, pp. 27-52, 1978.

[6] Hoffman, D. "The Specification of Communication Protocols" *IEEE Transactions on Computers*, Vol. C-34, No. 12, pp. 1102-1113, December 1985.

[7] Huet, G., Oppen, D. C. "Equations and Rewrite Rules – A Survey" *Formal Language Theory: Perspectives and Open Problems*, pp. 349-405, R. Book (ed.), Academic Press, 1980.

[8] Klop, J. W. "Term Rewriting Systems" *Handbook of Logic in Computer Science*, Chapter 6, S. Abramsky, D. Gabby and T. Maibaum (eds.), Oxford University Press, 1992.

[9] Manna, Z, and Ness, S. "On the Termination of Markov Algorithms" *Proceedings of 3rd International Conference on System Sciences*, pp. 789-792, Hawaii, January 1970.

[10] Parnas, D. L. "Information Distributions Aspects of Design Methodology" *Proceedings of IFIP Congress 1971*, pp. 26-30, 1972.

[11] Parnas, D. L., and Madey, J. "Functional Documentation for Computer Systems Engineering (Version 2)" *CRL Report* 237, Telecommunications Research Institute of Ontario (TRIO), McMaster University, September 1991.

[12] Parnas, D. L., and Wang, Y. "The Trace Assertion Method of Module Interface Specification" *Technical Report* 89-261, Telecommunications Research Institute of Ontario (TRIO), Queen's University, 1989. (Available on request from the address shown in title page.)

[13] Robinson, L., and Roubine, O. "SPECIAL – A Specification and Assertion language" *Technical Report* CSL-46, Computer Science Laboratory, Stanford Research Institute, 1977.

[14] Toyama, Y. "Confluent Term Rewriting Systems with Membership Conditions" *Conditional Terms Rewriting Systems*, Lecture Notes in Computer Science 308, pp. 229-241, 1987.

[15] van Horebeek, I., Lewi, J., and Bevers, E. "An Exception Handling Method for Constructive Algebraic Specifications" *Software Practice and Experience*, Vol. 18, No. 5, pp. 443-438, May 1988.

A Calculus for Conditional Inductive Theorem Proving

Ulrich Fraus

Bavarian Research Center for Knowledge Based Systems (FORWISS)
University of Passau, Innstr. 33, W-8390 Passau, Germany
E-mail: fraus@forwiss.uni-passau.de

Abstract. We describe a calculus for proofs of conditional theorems over algebraic specifications. The main principle used in these proofs is "natural" induction over the structure of terms. Hereby we have to deal with conditional induction hypotheses. The correctness of our calculus can be proved by use of the Natural Deduction Calculus. (This proof is omitted in this short version.)

1 Inductive Theorem Proving

Inductive reasoning is a widespread method for proving universally quantified theorems over specifications. Two main directions are distinguished in the field of inductive theorem proving. The first one is the "natural" induction as used in the well known Boyer-Moore-Prover [1]. For this kind of proving the generation and application of induction hypotheses are explicit proof steps.

The other direction is the inductive proof by completion [3] (sometimes also called inductionless induction). The idea hereby is to show that it is not possible to derive any contradiction from the specification enriched by the theorem. In this case we do not have explicit induction hypotheses.

The calculus described in this paper works with the "natural" induction approach. We think that this approach is more similar to human reasoning and so it is easier to control and understand such proofs.

Both approaches allow the (sometimes restricted) usage of conditional axioms in the specifications. But the theorems to be proved have to be unconditional equations. Now we want to go one step further and prove true conditional theorems. That is, the implication need not be coded in a boolean function symbol like:

```
impl( equal(x,1), equal(mult(x,y),x) ) = true
```

In our calculus we want to deal with conditional theorems like:

```
x = 1   =>   mult(x,y) = y
```

Maybe at the first look this seems to be just a different syntactic form of writing the theorem. But this new form of true conditional theorems gives us two big advantages. The first one is that now the conditions are real equations which can be seen as additional axioms for proving the correctness of the conclusion. An important difference between the conditions and lemmas is that the conditions share the variables with the conclusion.

The second advantage is that an induction hypothesis is now a simple equation which is connected with a set of equations (its conditions). These connected conditions can now be verified themselves as independent proofs. This makes inductive theorem proving more powerful.

2 Definitions

Before showing the calculus we need to give some basis and definitions.

- A *specification* consists of a signature $\Sigma = (S, F)$ and a set of axioms A. S is a set of sorts and F is a set of function symbols. The expression $W_\Sigma(X)$ denotes the set of all terms with variables from a set X. For further details see [5].
- A *substitution* σ is a mapping with a finite domain $dom(\sigma) \subseteq X$ from the set of variables to the set of terms. The application of a substitution σ to a term t will be written in postfix notation as $t\sigma$. A substitution which maps its domain to variable-free terms is called *ground-substitution*. If σ is ground then $t\sigma$ is called a *(s-)instance* of t.
- The theorems that we want to prove with our calculus are *conditional equations* of the form $C \Rightarrow e$. Hereby C is a conjunction of unconditional equations (the *conditions*) and e is an unconditional equation (the *conclusion*). Each conditional equation should be proved with the help of an exactly defined set of conditional induction hypotheses H. A theorem together with such a set constitutes a *goal*. For such a goal we write:

$$C \Rightarrow e \text{ when } H$$

- We need a set of hypotheses because we have many nested induction proofs. In our calculus the *induction hypotheses* have the form $(\eta : P \Rightarrow l = r)$, where η is a substitution, P are the conditions and l and r are two terms of the same sort. We need the substitution η because we have to record all global substitutions which took place since to generation of this hypothesis. This protocol is necessary to decide if an application of a hypothesis is valid, i.e. it is smaller than the original goal w.r.t. the induction ordering used (see rule 6 of the calculus).
- A set of substitutions $\{\tau_1, \ldots, \tau_m\}_{x,s}$ for one variable x of sort s is called *complete set* if for each variable-free term $t \in W_\Sigma(\emptyset)$ of sort s the following holds:

$$\exists i \in \{1, \ldots, m\} : \exists \text{ground-substitution } \sigma\text{: } A \vdash t = x\tau_i\sigma$$

This means that every variable-free term t is equivalent (modulo the set of axioms A) to an instance of one $x\tau_i$. Therefore, a complete set constitutes a complete case analysis.
- The expression *var(t)* denotes the set of all variables which occur in the term t. We can extend this notation to equations and conjunctions of equation.
- The notation *VAR(C,e)* is an abbreviation for var(C) ∪ var(e).
- The used induction ordering > has to be a stable, monotone and noetherian strict-ordering on terms [2].

3 The Calculus

Our calculus for inductive proofs of conditional theorems contains nine rules.

3.1 Complete Case Analysis

• a complete case analysis for an arbitrary term (is added to the conditions).

$$\frac{\begin{array}{c} C \cup (t = x\tau_1) \Rightarrow e \text{ when } H \\ \cdots \\ C \cup (t = x\tau_n) \Rightarrow e \text{ when } H \end{array}}{C \Rightarrow e \text{ when } H} \qquad \textit{(Rule 1)}$$

if $\{\tau_1, \ldots, \tau_n\}_{x,s}$ is a *complete set* for the variable x, if $x \notin VAR(C, e)$ and if $\forall i \in \{1, \ldots, n\}: var(x\tau_i) \cap VAR(C, e) = \emptyset$ (i.e. x and the variables in τ_i are „new" variables). $t \in W_\Sigma(X)$ is an arbitrary term of sort s and $var(t) \subseteq VAR(C, e)$ (i.e. no „new" variables in t)

Example: A possible case analysis in the conditions: *iszero(x) = true* and *iszero(x) = false*

- a complete case analysis for one variable of the goal.

$$\frac{\begin{array}{c} C\tau_1 \Rightarrow e\tau_1 \text{ when } \{ (\sigma_1\tau_1 : c_1 \Rightarrow l_1 = r_1), \ldots, (\sigma_m\tau_1 : c_m \Rightarrow l_m = r_m) \} \\ \ldots \\ C\tau_n \Rightarrow e\tau_n \text{ when } \{ (\sigma_1\tau_n : c_1 \Rightarrow l_1 = r_1), \ldots, (\sigma_m\tau_n : c_m \Rightarrow l_m = r_m) \} \end{array}}{C \Rightarrow e \text{ when } \{ (\sigma_1 : c_1 \Rightarrow l_1 = r_1), \ldots, (\sigma_m : c_m \Rightarrow l_m = r_m) \}} \quad \textit{(Rule 2)}$$

if $\{\tau_1, \ldots, \tau_n\}_{x,s}$ is a *complete set* for the variable x, if $x \in VAR(C, e)$ and if $\forall i \in \{1, \ldots, n\}: var(x\tau_i) \cap VAR(C, e) = \emptyset$ (i.e. x is a variable occurring in the goal and all the variables in τ_i are „new").

Example: A complete case analysis for the variable x: *[x\zero]* and *[x\succ(y)]*

3.2 Generation of a new induction hypothesis

The conditional equation $C \Rightarrow l = r$ can form a new induction hypothesis. Hypotheses can be applied (see rule 6) only to smaller instances. This is controlled by the substitution.

$$\frac{C \Rightarrow l = r \text{ when } (H \cup \{ (\mathrm{id} : C \Rightarrow l = r) \})}{C \Rightarrow l = r \text{ when } H} \quad \textit{(Rule 3)}$$

if $var(r) \cup var(C) \subseteq var(l)$. (This is necessary because when testing if a hypothesis application is valid (see Rule 6) only the left side is examined.)

The actual conditional equation is declared as a new induction hypothesis. Using the identical substitution id guarantees that the hypothesis cannot be applied to the equation it was generated from. First there have to be some global substitutions (see rule 2).

Example: This is equivalent to say in a mathematical proof: "Let us try to prove something for n+1 while assuming that we know it is true for n."

3.3 Conditional Rewriting

- a rewriting step in the conditions

$$\frac{C[u \leftarrow r\tau] \Rightarrow e \text{ when } H \qquad \{\ \} \Rightarrow P_1\tau \text{ when } H \quad \cdots \quad \{\ \} \Rightarrow P_n\tau \text{ when } H}{C \Rightarrow e \text{ when } H} \quad \textit{(Rule 4)}$$

if $C/u = l\tau$ and $(P_1 \wedge \ldots \wedge P_n \Rightarrow (l = r)) \in R$.

We have empty conditions when validating the preconditions, because the rewrite step is applied to the conditions C. This is necessary to avoid cycles during conditional rewriting.

• a rewriting step in the conclusion

$$\frac{C \Rightarrow e\,[u \leftarrow r\tau] \text{ when } H \qquad C \Rightarrow P_1\tau \text{ when } H \quad \cdots \quad C \Rightarrow P_n\tau \text{ when } H}{C \Rightarrow e \text{ when } H} \qquad \textit{(Rule 5)}$$

if $e/u = l\tau$ and $(P_1 \wedge \ldots \wedge P_n \Rightarrow (l = r)) \in R$.

3.4 Application of an induction hypothesis

$$\frac{C \Rightarrow e\,[u \leftarrow r\tau] \text{ when } H \qquad C \Rightarrow P_1\tau \text{ when } H \quad \cdots \quad C \Rightarrow P_n\tau \text{ when } H}{C \Rightarrow e \text{ when } H} \qquad \textit{(Rule 6)}$$

if $e/u = l\tau$ and $(\eta{:}P_1 \wedge \ldots \wedge P_n \Rightarrow l = r) \in H$ and $l\tau < l\eta$.

A conditional hypothesis from the set H can be applied to the conclusion only if the application of substitution τ results in a smaller term than the substitution η which is stored for this hypothesis. Therefore, both substitutions are applied to the left side of the conclusion.

Example: For the goal $\{\ \} \Rightarrow f(y) = 0$ when $\{\,(\,[x\backslash succ(y)] : \{\ \} \Rightarrow f(x) = 0)\,\}$ one can apply the induction hypothesis, because $l\tau < l\eta$ holds (i.e $f(x)\,[x\backslash y] < f(x)\,[x\backslash succ(y)]$).
The substitution connected with the hypothesis states that $[x\backslash succ(y)]$ was the only global substitution applied since the generation of the hypothesis.

3.5 Application of a condition

Any condition contained in C can be applied to the conclusion.

$$\frac{C \Rightarrow e\,[u \leftarrow r] \text{ when } H}{C \Rightarrow e \text{ when } H} \qquad \textit{(Rule 7)}$$

if $e/u = l$ and either $l = r$ or $r = l$ is an equation which is contained in the conditions C. For this rule we do not have to apply a substitution, because the variables in the conclusion and in the conditions are shared.

Example: When we have the goal $(x = 1 \Rightarrow mult(x, y) = y)$ when H then we can apply $x = 1$ to the conclusion and get $(x = 1 \Rightarrow mult(1, y) = y)$ when H. We need no substitution because x is the same variable in the condition and the conclusion.

3.6 Trivial conclusion

$$\frac{}{C \Rightarrow (t = t) \text{ when } H} \qquad \textit{(Rule 8)}$$

Trivial conclusions (i.e. both sides are syntactically identical) are always valid.

3.7 Unsatisfiable condition

If there is a contradiction in the conditions then the goal is valid, because *false* implies every conclusion. All recognizable contradictions are described by the predicate UNSAT.

$$\frac{}{c_1 = d_1 \wedge \ldots \wedge c_n = d_n \Rightarrow e \text{ when } H} \qquad \textit{(Rule 9)}$$

if UNSAT(c_i, d_i) is valid for any $i \in \{1, \ldots, n\}$, i.e if one of the conditions contains a contradiction. UNSAT is a binary predicate which is *not part of* this calculus!

Example: If UNSAT(zero, succ(x)) is assumed to be true then
(zero = succ$(x) \Rightarrow$ add(2, 2) = 5) when H is a valid goal.

By selecting UNSAT we can change the strength of our calculus.

4 Proof of the Correctness

A *derivation step* in our calculus is defined in the usual way. A *derivation* is a finite or infinite sequence of derivation steps.

Without giving all the details here, we define that a goal $(C \Rightarrow e)$ when H is valid if the following expression is true:

$\forall$ground substitutions σ with dom(σ) = VAR(C, e):

$$(A \cup \{P\sigma' \Rightarrow l\sigma' = r\sigma' \mid ((l\sigma' < l\eta\sigma) \wedge (\eta: P \Rightarrow l = r) \in H)\}) \vdash \quad C\sigma \Rightarrow e\sigma$$

where A is the set of axioms of a specification Σ, $<$ is a strict, stable and noetherian ordering on terms and $\vdash$ is the Natural Deduction Calculus.

The expression above means that every instance of the conditional equation $C \Rightarrow e$ can be deduced from the axioms and from a set of σ'-instances of the induction hypotheses. Those σ'-instances are determined by comparing the σ'-instance of the left hand side and the term which results from applying the recorded substitution η and the ground substitution σ to the left hand side.

For our proof of correctness we assume that the Natural Deduction Calculus is correct. Then we prove the correctness of our calculus w.r.t. the Natural Deduction Calculus (see also the expression above). Our proof is structured in two parts. First we have to show that every derivation step of our calculus does not influence the correctness. This means that if the goals over the line are valid then the goal below the line is valid too. The second part of our correctness proof is to show that every sequence of derivation steps constitutes a correct derivation.

The first is done by a case analysis over all rules of our calculus. The second is (very easily) done by an induction over the length of the sequence. These proofs are not included in this paper, because of space reasons. The first one especially is too long.

5 Short Example

We assume a specification of the Boolean values (with constructor functions T and F for true and false) and of the natural numbers (with constructor functions *zero* and *succ()*). We use the two function *odd* and *even*:

function even, odd: (Nat) Bool
even(zero) $\rightarrow T$
even(succ(zero)) $\rightarrow F$
even(succ(succ(x))) $\rightarrow$ even(x)
odd(zero) $\rightarrow F$
odd(succ(zero)) $\rightarrow T$
odd(succ(succ(x))) $\rightarrow$ odd(x)

Now we want to prove the theorem $\forall x\colon\ \mathrm{even}(x)=F \Rightarrow \mathrm{odd}(x)=T$. For the predicate UNSAT we only use $\mathrm{UNSAT}(T,F)$, i.e. we use the classical logic. (Off course the search for a proof in this calculus is done bottom-up.) The 4th block uses the first three blocks for a case analysis:

$$\dfrac{\dfrac{}{T=F \Rightarrow \mathrm{odd}(\mathrm{zero})=T \text{ when } \{x\backslash \mathrm{zero}: \mathrm{even}(x)=F \Rightarrow \mathrm{odd}(x)=T\}}\ R9}{\mathrm{even}(\mathrm{zero})=F \Rightarrow \mathrm{odd}(\mathrm{zero})=T \text{ when } \{x\backslash \mathrm{zero}: \mathrm{even}(x)=F \Rightarrow \mathrm{odd}(x)=T\}}\ R4$$

$$\dfrac{\dfrac{}{\mathrm{even}(\mathrm{succ}(\mathrm{zero}))=F \Rightarrow T=T \text{ when } \{x\backslash \mathrm{succ}(\mathrm{zero}): \mathrm{even}(x)=F \Rightarrow \mathrm{odd}(x)=T\}}\ R8}{\mathrm{even}(\mathrm{succ}(\mathrm{zero}))=F \Rightarrow \mathrm{odd}(\mathrm{succ}(\mathrm{zero}))=T \text{ when } \{x\backslash \mathrm{succ}(\mathrm{zero}): \mathrm{even}(x)=F \Rightarrow \mathrm{odd}(x)=T\}}\ R5$$

$$\dfrac{\dfrac{\dfrac{}{\mathrm{even}(y)=F \Rightarrow T=T \text{ when } \{\ldots\}}\ R8 \qquad \dfrac{\dfrac{}{\mathrm{even}(y)=F \Rightarrow F=F \text{ when } \{\ldots\}}\ R8}{\mathrm{even}(y)=F \Rightarrow \mathrm{even}(y)=F \text{ when } \{\ldots\}}\ R7}{\mathrm{even}(y)=F \Rightarrow \mathrm{odd}(y)=T \text{ when } \{x\backslash \mathrm{succ}(\mathrm{succ}(y)): \mathrm{even}(x)=F \Rightarrow \mathrm{odd}(x)=T\}}\ R6}{\mathrm{even}(\mathrm{succ}(\mathrm{succ}(y)))=F \Rightarrow \mathrm{odd}(\mathrm{succ}(\mathrm{succ}(y)))=T \text{ when } \{x\backslash \mathrm{succ}(\mathrm{succ}(y)): \mathrm{even}(x)=F \Rightarrow \mathrm{odd}(x)=T\}}\ R4,R5$$

$$\dfrac{\dfrac{\begin{array}{c}\mathrm{even}(\mathrm{zero})=F \Rightarrow \mathrm{odd}(\mathrm{zero})=T \text{ when } \{x\backslash \mathrm{zero}: \mathrm{even}(x)=F \Rightarrow \mathrm{odd}(x)=T\}\\ \mathrm{even}(\mathrm{succ}(\mathrm{zero}))=F \Rightarrow \mathrm{odd}(\mathrm{succ}(\mathrm{zero}))=T \text{ when } \{x\backslash \mathrm{succ}(\mathrm{zero}): \mathrm{even}(x)=F \Rightarrow \mathrm{odd}(x)=T\}\\ \mathrm{even}(\mathrm{succ}(\mathrm{succ}(y)))=F \Rightarrow \mathrm{odd}(\mathrm{succ}(\mathrm{succ}(y)))=T \text{ when } \{x\backslash \mathrm{succ}(\mathrm{succ}(y)): \mathrm{even}(x)=F \Rightarrow \mathrm{odd}(x)=T\}\end{array}}{\mathrm{even}(x)=F \Rightarrow \mathrm{odd}(x)=T \text{ when } \{\mathrm{id}: \mathrm{even}(x)=F \Rightarrow \mathrm{odd}(x)=T\}}\ R2}{\mathrm{even}(x)=F \Rightarrow \mathrm{odd}(x)=T \text{ when } \{\ \}}\ R3$$

6 Conclusion

Conditional theorems as described above are something very common in software verification. In most cases the conditions are restrictions of the universally quantified variables. We have already examined some examples where the function-coded implication is too weak for doing an inductive proof. This is because the induction hypothesis is too complex to be applied to the terms where we need it.

The calculus described above is structured enough to allow the possibility of system support. An implementation of our calculus is currently under development. It is based on the implementation of an unconditional theorem prover [4]. A prototype of the conditional prover may be available in summer 1992.

References

1. R.S. Boyer, J.S. Moore: A Computational Logic. Academic Press, 1979.
2. N. Dershowitz: Termination of Rewriting. Journal of symbolic Computation 3, 1987.
3. D. Duffy: Principles of Automated Theorem Proving. Wiley, 1991.
4. U. Fraus, H. Hußmann: A Narrowing-Based Theorem Prover. Proceedings of the IMA Conference, July 1990, University of Stirling, Oxford University Press, 1992.
5. M. Wirsing, e.a.: On hierarchies of abstract data types. Acta Informatica 20, 1983.

Implementing Contextual Rewriting

Hantao Zhang*
Department of Computer Science
The University of Iowa
Iowa City, IA 52242
hzhang@cs.uiowa.edu

Abstract

Contextual rewriting as a generalization of conditional rewriting has been studied in different forms. We show that contextual rewriting is a powerful simplification rule for the first-order theorem proving with equality and preserves the refutational completeness of many reasoning systems. After comparing definitions of contextual rewriting by Boyer-Moore, Remy, Ganzinger and Zhang-Kapur, we show that the definition of Zhang-Kapur is more suitable for implementation because it is natural to organize the context as a ground term rewriting system. We provide a detailed procedure for simplying clauses using contextual rewriting. We also provide a solution on how to handle variables which appear only in the condition of a conditional rewrite rule.

1 Introduction

Contextual rewriting, as a more general case of the traditional conditional rewriting [11, 8] has been studied by different researchers, such as Boyer and Moore [1], Remy [12], Ganzinger [6] and Zhang and Kapur [14]. To simplify a term t by contextual rewriting, in addition to a set R of conditional rewrite rules, we also have a set C of equations called the ***context*** of t. Under the assumption that the equations in C are true, C and R together rewrite t to another term. If C is empty, the contextual rewriting is identical to the traditional conditional rewriting. The research on contextual rewriting has been focused on (i) how to obtain the context of a term and (ii) how to use contexts in the rewriting.

We have implemented a version of contextual rewriting as described in [14] in the theorem prover *RRL*[1] to support both deductive and inductive theorem proving. For deductive theorem proving, the contextual rewriting is a very powerful simplification rule which subsumes strictly several simplification rules such as tautology deletion, subsumption and demodulation. Contextual rewriting helped us to obtain automatic proofs of a number of problems which are challenge for resolution theorem provers. For instance, we obtained the first proof (to the best of our knowledge) of SAM's lemma on lattice theory when the axioms are formulated using equality. We also obtained the first computer proofs of the three famous isomorphism theorems in group theory. For inductive theorem proving, because of contextual rewriting, RRL is faster than Boyer-Moore theorem prover [15] to prove one theorem.

For the application of contextual rewriting to be effective, we must have an efficient implementation of contextual rewriting. In this paper, we show that contextual rewriting is a powerful

*Partially supported by the National Science Foundation Grants no. CCR-9009414, INT-9016100 and CCR-9202838.

[1] *RRL* (the Rewrite Rule Laboratory) is a theorem proving environment for experimenting with existing reasoning algorithms as well as for developing new reasoning algorithms based on rewriting paradigm [9].

simplification rule for the first-order theorem proving with equality and preserves the refutational completeness of many reasoning systems. We compare different definitions of contextual rewriting in regard to its efficiency and power. We show how to implement the definition of Zhang-Kapur [14] efficiently and provide a solution to the problem of handling variables which appear only in the condition of a rewrite rule.

2 Contextual Rewriting

In this section, after introducing some basic concepts, we formally define the so-called contextual rewriting. We then show how to simplify clauses by contextual rewriting and explain the role played by the context.

2.1 Preliminaries

Rewriting is usually defined in terms of rewrite rules. Rewrite rules can be easily made from equations as well as from general formulas. The rewrite rules used in our approach are made from general formulas or clauses. For simplicity, we make the following assumptions:

1. We assume that all quantifiers (both existential and universal) in a formula are already removed using skolemization and the result is converted into a set of clauses, even though our discussion does not depend upon the input being in the clausal form.

2. Since we will frequently use the equality relation, and since every nonequality atom A can be written as $A = \mathsf{T}$ (T and F denote the Boolean constants ***true*** and ***false***), without loss of any generality, we assume that ***every atom is an equality (atom)***. A negative literal is assumed to be an nonequality. A literal is either an equality or an nonequality. We write a literal as $t \rightleftharpoons s$ which denotes either $t = s$ or $t \neq s$.

3. We assume that a reduction ordering (which is a well-founded partial ordering on terms and is stable under instantiation and replacement) is used to make rewrite rules from clauses. When a precedence relation on operators is given, Dershowitz' recursive path ordering [4] is used to extend the precedence relation from operators to terms.

4. A term t is said to be ***maximal*** in a clause $\mathbf{c}$ if for any term t' of $\mathbf{c}$, $t' \not\succ t$; Similarly, a literal L is said to be ***maximal*** in $\mathbf{c}$ if for any literal L' of $\mathbf{c}$, $L' \not\succ_l L$, where $(s_1 \rightleftharpoons t_1) \succ_l (s_2 \rightleftharpoons t_2)$ iff $\{s_1, t_2\} \succ\!\!\succ \{s_2, t_2\}$.

5. A clause $L_1 \vee L_2 \vee \cdots \vee L_n$ can be always written as a conditional equation:

$$lhs(L_i) = rhs(L_i) \textbf{ if } \neg L_1 \wedge \cdots \wedge \neg L_{i-1} \wedge \neg L_{i+1} \wedge \cdots \wedge \neg L_n,$$

   or a conditional rewrite rule

$$lhs(L_i) \rightarrow rhs(L_i) \textbf{ if } \{\neg L_1, ..., \neg L_{i-1}, \neg L_{i+1}, ..., \neg L_n\},$$

   where, $lhs(L_i) = a_i$ and $rhs(L_i) = b_i$ if L_i is an equality $(a_i = b_i)$, or $lhs(L_i) = (a_i = b_i)$ and $rhs(L_i) = \mathsf{F}$ if L_i is $(a_i \neq b_i)$.

Example 2.1 Given the following two ground clauses

$$(\mathbf{c}_1) \quad (p(a) = \mathsf{T}) \vee (f(a) = a) \vee (g(a) \neq a),$$
$$(\mathbf{c}_2) \quad (p(a) \neq \mathsf{T}) \vee (f(a) \neq g(a)).$$

Different rules can be made from the same clause under different orderings. A few examples are illustrated in Table 1. □

ORDERING		REWRITE RULES
$p \succ f \succ g \succ a$	$(\mathbf{r}_1)$	$p(a) \to \mathsf{T}$ **if** $\{(f(a) \neq a), (g(a) = a)\}$
	$(\mathbf{r}_2)$	$(p(a) = \mathsf{T}) \to \mathsf{F}$ **if** $\{(f(a) = g(a))\}$
$f \succ g \succ p \succ a$	$(\mathbf{r}'_1)$	$f(a) \to a$ **if** $\{(g(a) = a), (p(a) \neq \mathsf{T})\}$
	$(\mathbf{r}'_2)$	$(f(a) = g(a)) \to \mathsf{F}$ **if** $\{(p(a) = \mathsf{T})\}$
$g \succ f \succ p \succ a$	$(\mathbf{r}''_1)$	$(g(a) = a) \to \mathsf{F}$ **if** $\{(f(a) \neq a), (p(a) \neq \mathsf{T})\}$
	$(\mathbf{r}''_2)$	$(f(a) = g(a)) \to \mathsf{F}$ **if** $\{(p(a) = \mathsf{T})\}$

Table 1: Rewrite rules from clauses: an example

To make the study of these rewrite rules simple, we may require that each rewrite rule be *terminating*, i.e., for each rule $l \to r$ **if** $\{p_1, ..., p_n\}$, $l \succ r$ and $l \succ p_i$. This requirement is too restrictive in practice. Instead, we may require that $l \succ r$ and $(l = r)$ be maximal in $\{p_1, ..., p_n\}$.

2.2 A Formal Definition

We first define a congruence relation called "constant congruence" and then give a formal definition of "contextual rewriting". In the next section, we illustrate the use of contextual rewriting by examples.

Definition 2.2 The *constant congruence* $\simeq_C$ generated by a set C of pure (i.e., unconditional) equations is the minimal equivalence relation satisfying the following properties: (i) For any equation $t_1 = t_2$ of C, $t_1 \simeq_C t_2$, and (ii) if $t_1 \simeq_C t_2$, then $f(...t_1...) \simeq_C f(...t_2...)$.

This relation is different from the *equational congruence relation* usually used in the literature since in the constant congruence $\simeq_C$, a variable is treated as a constant. That is, no instantiations for variables are allowed in the constant congruence relation.

Definition 2.3 (contextual rewriting $\to_{C,R}$) Given a set C of equations and a set R of (conditional) rewrite rules, a rewriting under the context C is recursively defined as follows: A term t is rewritten to t' by contextual rewriting using R and C, denoted as $t \to_{C,R} t'$, if either

- $t \simeq_C t'$ or
- there exist a subterm position p of t, a rewrite rule $l \to r$ **if** $\{p_1, p_2, ..., p_k\}$ in R and a substitution σ such that
  1. $\sigma(l) = t/p$,
  2. for each p_i, $1 \le i \le k$, $\sigma(p_i) \to^*_{C,R} \mathsf{T}$,
  3. $t' = t[p \leftarrow \sigma(r)]$.

We say $l \to r$ **if** $\{p_1, p_2, ..., p_k\}$ is *applicable* on t if the conditions (1) and (2) are satisfied.

Example 2.4 Let R contain the definitions of $+$, $*$ (*multiplication*), *gcd* on natural numbers, including the rule

$$(\mathbf{r}_1) \quad gcd(u * v, u) \to u,$$

which states that the gcd of any number and its multiplication is the number itself. Let the context C be $\{z = (x * y)\}$ and the term be $gcd(z, x)$. Then $gcd(z, x) \to_{C,R} gcd(x * y, x)$ because $z \simeq_C (x * y)$. Applying rule $\mathbf{r}_1$ on $gcd(x * y, x)$, we get x. That is, $gcd(z, x) \to_{C,R} gcd(x * y, x) \to_{C,R} x$ by two steps of the context rewriting. □

2.3 Simplification of Clauses

We have just illustrated how rewrite rules can be obtained from clauses and how to perform rewriting using these rewrite rules. Now we show how to simplify a clause using the rewrite rules made from other clauses.

In order to explain how to apply contextual rewriting, we prefer to reformulate each clause into two parts:

1. a literal of the clause which is going to be simplified, called the ***focus literal***;
2. the ***context*** created from all the literals of the clause except the focus literal.

Suppose the focus literal is L_1 in $L_1 \vee L_2 \vee ... \vee L_n$; the ***context*** for simplifying L_1 consists of all the equality representations of the negated remaining literals in the clause. That is, the context can be written as

$$\{lhs(\neg L_i) = rhs(\neg L_i) \mid 2 \leq i \leq n\}.$$

Using the above context and a set of rewrite rules made from other clauses, we may simplify L_1 using the contextual rewriting. We illustrate this idea by the following examples.

Example 2.5 (continuation of Example 2.4) The clause to be simplified is

$$(\mathbf{c}) \quad (z \neq x * y) \vee (gcd(z, x) = x).$$

Let the focus literal be $gcd(z, x) = x$. The only remaining literal is $(z \neq x * y)$ and the equality representation of its negation is $\{z = x * y\}$, which will serve as the ***context*** of $gcd(z, x) = x$. As shown in Example 2.4, $gcd(z, x)$ is rewritten as x. That is, $\mathbf{c}$ is simplified to $(z \neq x * y) \vee (x = x)$, a trivial clause. □

Example 2.6 Given two rewrite rules:

$$\begin{array}{lrcl} (\mathbf{r}_1) & hasteeth(x) & \rightarrow & \mathsf{T} \text{ if } (istiger(x) = \mathsf{T}), \\ (\mathbf{r}_2) & isanimal(x) & \rightarrow & \mathsf{T} \text{ if } (istiger(x) = \mathsf{T}). \end{array}$$

and a clause to be simplified:

$$(\mathbf{c}) \quad (istiger(y) \neq \mathsf{T}) \vee (isanimal(y) \neq \mathsf{T}) \vee (hasteeth(y) \neq \mathsf{T}).$$

To simplify $(hasteeth(y) \neq \mathsf{T})$, the remaining literals are $(isanimal(y) \neq \mathsf{T})$ and $(hasteeth(y) \neq \mathsf{T})$, and the equality representations of their negation are

$$C = \{(istiger(y) = \mathsf{T}), (isanimal(y) = \mathsf{T})\},$$

which serves as the context of $(hasteeth(y) \neq \mathsf{T})$. We match the left side of $(\mathbf{r}_1)$ to $hasteeth(y)$ to obtain the substitution $\sigma = \{x \leftarrow y\}$. The condition of $(\mathbf{r}_1)$ under σ is $\{istiger(y) = \mathsf{T}\}$, which can be simplified to T by the context C, i.e, by the constant congruence $\simeq_C$. So $(\mathbf{r}_1)$ is applicable and $hasteeth(y)$ is reduced to T. Similarly, $isanimal(y)$ can also be reduced to T by $(\mathbf{r}_2)$. After removing $\mathsf{T} \neq \mathsf{T}$ from the simplified clause, we obtain a unit clause, i.e., $(istiger(y) \neq \mathsf{T})$. □

By definition, demodulation is a special case of contextual rewriting. Contextual rewriting is also more general than many conditional rewritings proposed in the literature. We show below that contextual rewriting subsumes both ***subsumption*** and ***tautology elimination***, two simplification rules widely used in clausal theorem proving.

Subsumption has been proven to be a powerful simplification rule for clausal theorem proving. Our contextual rewriting is more general than subsumption in that if a clause $\mathbf{c}$ is discarded because of being subsumed by another clause, then $\mathbf{c}$ can be rewritten to a trivial clause by contextual rewriting.

Lemma 2.7 *If* $\mathbf{c}$ *subsumes* $\mathbf{c}'$*, then a rule made from* $\mathbf{c}$ *can reduce* $\mathbf{c}'$ *to a trivial clause using contextual rewriting.*

Proof. Let $\mathbf{c} = L_1 \vee L_2 ... \vee L_m$ and $\mathbf{c}' = M_1 \vee M_2 \vee ... \vee M_n$, such that $\sigma\{L_1, L_2, ..., L_m\} \subset \{M_1, M_2, ..., M_n\}$ for some substitution σ. For simplicity, let us assume $\sigma(L_1) = M_1$ and all literals in $\mathbf{c}'$ are equalities (nonequalities can be handled similarly). The rule made from $\mathbf{c}$ is

$$(\mathbf{r}_1) \quad L_1 \rightarrow \mathsf{T} \textbf{ if } \{\neg L_2, ..., \neg L_m\}.$$

To simplify $\mathbf{c}'$, we take M_1 as the focus literal and let the context be

$$C = \{lhs(\neg M_i) = rhs(\neg M_i) \mid 2 \leq i \leq n\} = \{M_i = \mathsf{F} \mid 2 \leq i \leq n\}.$$

L_1 matches M_1 since $\sigma(L_1) = M_1$. Since $\mathbf{c}$ subsumes $\mathbf{c}'$, each $\sigma(L_j)$, $2 \leq j \leq m$, is equal to some M_i. So, using the context C, each $\sigma(L_j)$, $2 \leq j \leq m$, is equal to F. In other words, the instance of the condition of $(\mathbf{r}_1)$ under the substitution σ is equal to T. Thus $(\mathbf{r}_1)$ is applicable and can reduce M_1 to T. In this way, $\mathbf{c}'$ is reduced to $\mathsf{T} \vee M_2 \vee ... \vee M_n$, a trivial clause. Thus $\mathbf{c}$ can be discarded safely. □

A ***tautology*** is a clause in which both a literal and its negation are presented. Since a tautology is true in any interpretation and cannot contribute to a refutational proof, tautologies are often deleted from a proof system and this deletion is called ***tautology elimination***.

Constant congruence can be used to transform a tautology into a trivial clause. That is, suppose the clause $\mathbf{c}$ contains both $s = t$ and $s \neq t$. When $s = t$ is the focus literal, the negation of $s \neq t$, i.e., $s = t$, serves as one of the equalities in the context. Obviously, $s = t$ will be rewritten by the context to $t = t$ (we assume that $x = x$ is built-in). Thus, $\mathbf{c}$ becomes a trivial clause.

2.4 Role of Constant Congruence

The reader may wonder what role the constant congruence relation has played in the simplification of clauses. The clause given in Example 2.5 is

$$(\mathbf{c}) \quad (z \neq x * y) \vee (gcd(z, x) = x),$$

which can be reformulated as

$$(\mathbf{c}') \quad (z \neq x * y) \vee (gcd((x * y), x) = x)$$

using the constant congruence relation. That is, taking $\{z = (x * y)\}$ as the context C, we have $(gcd(z, x) = x) \simeq_C (gcd((x * y), x) = x)$. By abuse of notation, we write $\mathbf{c} \simeq_C \mathbf{c}'$ to denote that $\mathbf{c}'$ is obtained from $\mathbf{c}$ using the constant congruence.

Now the question is: Without the constant congruence, can $\mathbf{c}'$ be derived from $\mathbf{c}$ using some rules of inference? The answer is yes. In fact, $\mathbf{c}'$ can be derived by paramodulating an instance of the following tautology $(u = v) \vee (u \neq v)$ (this instance can be obtained by paramodulating functionally reflexive axioms into the symmetry axiom of equality) into the clause $\mathbf{c}$.

For the above example, an instance of the tautology is

$$(z = (x * y)) \vee (z \neq (x * y)),$$

which may be paramodulated into $\mathbf{c}$ at z and the result is $\mathbf{c}'$.

This holds for any two clauses $\mathbf{c}$ and $\mathbf{c}'$ if $\mathbf{c} \simeq_C \mathbf{c}'$. Note that the relation $\simeq_C$ is symmetric, that is, $\mathbf{c} \simeq_C \mathbf{c}'$ implies $\mathbf{c}' \simeq_C \mathbf{c}$. This symmetry property allows us to keep one clause (either $\mathbf{c}$ or $\mathbf{c}'$, not both) in the system, since if a proof can be derived from $\mathbf{c}$, then the same proof can be derived from $\mathbf{c}'$ (by some paramodulations of the equality axioms).

In many problems, we have clauses like

$$(\mathbf{c}_1) \quad (member(x, cons(y, z)) = \mathsf{T}) \vee (x \neq y).$$

where $(x \neq y)$ is the negation of an equality with both arguments being variables. Using the constant congruence, the above clause can be reformulated into:

$$(\mathbf{c}_1') \quad (member(x, cons(x, z)) = \mathsf{T}) \vee (x \neq y).$$

This clause can be further simplified to

$$(\mathbf{c}_2) \quad (member(x, cons(x, z)) = \mathsf{T})$$

by a resolution between $(x \neq y)$ and $(u = u)$. There is no doubt that $\mathbf{c}_2$ is more effective than $\mathbf{c}_1$ in a resolution theorem prover because $\mathbf{c}_2$ is a unit clause. Note also that $\mathbf{c}_1$ can be derived from $\mathbf{c}_2$ and the equality axioms. Hence it is safe to replace clauses like $\mathbf{c}_1$ by clauses like $\mathbf{c}_2$ in a resolution theorem prover. This can be done systematically using the constant congruence relation. We believe that the constant congruence relation is an effective additional way to handle the equality axioms, and can increase the power of the reasoning systems in which the equality is built in.

3 Refutational Completeness

The most significant value of a simplification rule is that if a clause gets simplified, then the original clause can be thrown away without violating the completeness of the system. We show below that contextual rewriting is such a simplification rule.

We will assume the standard definitions of an ***interpretation***, and an interpretation ***satisfying*** or ***falsifying*** a formula or a set of formulas, as well as a formula or a set of formulas being ***satisfiable*** or ***unsatisfiable***. An interpretation is ***consistent*** if no atom under this interpretation has more than one value. Every interpretation must satisfy the equality axioms, that is, $\Pi(t = t) = \mathsf{T}$ and $\Pi(f(...t_1...)) = \Pi(f(...t_2...))$ if $\Pi(t_1 = t_2) = \mathsf{T}$.

The following lemma asserts that for any interpretation Π which satisfies every input formula, a clause and its reduced form always have the same value under Π.

Lemma 3.1 (soundness of rewriting $\rightarrow_{C,R}$) *Let R be a set of rewrite rules and a clause* $\mathbf{c}$ *be rewritten (once) to* $\mathbf{c}'$ *by R. If there exists an interpretation* Π *such that each rewrite rule of R is interpreted to* T, *then* $\Pi(\mathbf{c}) = \Pi(\mathbf{c}')$.

Using Lemma 3.1, we can show easily that if a set of clauses is unsatisfiable and one of the clauses is reducible by the rewrite rules made from the rest, we can safely replace the reducible clause by its reduced form because the new set of clauses is unsatisfiable, too. This property allows us to use contextual rewriting in any reasoning program employing refutational strategies.

Theorem 3.2 *Given a set S of clauses and* $\mathbf{c} \in S$, *if* $\mathbf{c}$ *is rewritten to* $\mathbf{c}'$ *by the rules made from other clauses in S, then S is unsatisfiable if and only if* $S - \{\mathbf{c}\} \cup \{\mathbf{c}'\}$ *is unsatisfiable.*

Proof. Let $S' = S - \{\mathbf{c}\} \cup \{\mathbf{c}'\}$. If S' is unsatisfiable, since $\mathbf{c}'$ is a consequent of S, S must be unsatisfiable, too.

Suppose S is unsatisfiable but S' is not. We have an interpretation Π such that each clause of S' is interpreted to T. By Lemma 3.1, $\Pi(\mathbf{c}) = \Pi(\mathbf{c}')$, so S is satisfiable, too. This is a contradiction. □

For those who are not convinced by the above theorem and who are familiar with the technique of semantic trees for establishing refutational completeness, we provide below an alternative proof.

Let $\mathcal{P}$ be a set of inference rules whose refutational completeness can be established by the technique of semantic trees. Moreover, let us assume that only terminating rules are allowed in the contextual rewriting, and the Herbrand base $\{a_i\}_{i<\lambda}$, where λ is an ordinal, is ordered by the reduction ordering $\succ$ used in the contextual rewriting. That is, $a_i \succ a_j$ implies $i \succ j$.

Let us first make the meaning of "established by the technique of semantic tree" explicit.

Given S, an inconsistent set of clauses, let $\mathcal{P}(S)$ be the smallest set of clauses containing S and closed by $\mathcal{P}$, a set of inference rules. By the Herbrand theorem, there exists an inconsistent finite set G of ground instances of clauses in $\mathcal{P}(S)$. A ***semantic tree*** with respect to G is a binary tree of height n, where a_n is the maximal atom appearing in G (n is an ordinal and may be transfinite), such that (i) the left out-edge of each internal node at level i (assume the root node is at level 1) is labeled with $a_i = \mathsf{F}$; (ii) the right out-edge of each internal node at level i is labeled with $a_i = \mathsf{T}$.

For each node t at level k in a semantic tree, let the labels of the edges in the path from the root to t be $\Pi_t = \{a_i = \delta_i \mid 1 \leq i < k\}$, where $\delta_i \in \{\mathsf{T}, \mathsf{F}\}$. Then Π_t is a partial interpretation of G. In fact, associating with each node t of a semantic tree, there exists a unique (partial) interpretation Π_t. Obviously, if t is the root node, then Π_t is the empty interpretation; if $k = n$, then Π_t is a complete interpretation of G.

Because G is inconsistent, by the Herbrand theorem, we can assign a clause $\mathbf{c} \in G$ to each leaf node of a semantic tree, such that $\mathbf{c}$ is false in the interpretation associated with the leaf node. A semantic tree in which every leaf node is assigned such an equation is said to be ***closed***.

We say that the refutational completeness of $\mathcal{P}$ can ***be established by the technique of semantic tree***, if whenever a closed semantic tree has more than one node, some edges of the tree can be deleted and the resulted new tree is still closed (by assigning some clauses, which are generated by $\mathcal{P}$ from the clauses assigned to the leaves, to the new leaves.)

Note that most resolution strategies can be established by this technique. Various restricted paramodulation strategies proposed in [7], can be established by this technique, too.

Theorem 3.3 ***If the refutational completeness of a set $\mathcal{P}$ of inference rules can be established by the technique of semantic trees, then the contextual rewriting preserves the refutational completeness of $\mathcal{P}$.***

Proof. During the proof process without the contextual rewriting, we are able to show that a closed semantic tree can be shrunk with the help of the inference rules in $\mathcal{P}$. To prove that the contextual rewriting does not affect the completeness of $\mathcal{P}$, it is sufficient to show that a closed semantic tree with more than one node can be always shrunk when only irreducible (by contextual rewriting) clauses are assigned to a closed tree, that is, we restrict the use of $\mathcal{P}$ to irreducible clauses.

If we can show that an irreducible clause can always be assigned to every leaf node of a closed tree, then the same proof process to shrink a closed tree can be used. The following lemma asserts this is the case.

Lemma 3.4 ***Suppose $\sigma(\mathbf{c}) \in G$, where $\mathbf{c} \in \mathcal{P}(S)$, is the clause assigned to a leaf node t of a closed semantic tree T. If $\mathbf{c}$ is reduced (by the contextual rewriting) to $\mathbf{c}_0$ by a set of clauses $\{\mathbf{c}_1, ..., \mathbf{c}_m\}$. Then one of the clauses in $\{\sigma(\mathbf{c}_0), \sigma(\mathbf{c}_1), ..., \sigma(\mathbf{c}_m)\}$ is false in Π_t, the interpretation associated with t.***

The proof of the above lemma is simple: If one of clauses in $\{\sigma(\mathbf{c}_1),...,\sigma(\mathbf{c}_m)\}$ is false in Π_t, then we are done. If every clause in $\{\sigma(\mathbf{c}_1),...,\sigma(\mathbf{c}_m)\}$ is true in Π_t, by lemma 3.1, $\sigma(\mathbf{c}_0)$ has the same truth value as $\sigma(\mathbf{c})$ in Π_t. Thus $\sigma(\mathbf{c}_0)$ is false in Π_t.

Using the above lemma, when $\sigma(\mathbf{c})$, the clause assigned to the leaf node t of T, is reducible, we can replace it by $\mathbf{c}' \in \{\sigma(\mathbf{c}_0), \sigma(\mathbf{c}_1),...,\sigma(\mathbf{c}_m)\}$ such that $\mathbf{c}'$ is false in Π_t. This process cannot continue forever because only terminating rewrite rules are allowed. Finally, every clause assigned to T will be irreducible. At this stage, we can shrink T with the help of $\mathcal{P}$, as in the case when the contextual rewriting is not used. □

4 Comparison with Related Work

Conditional rewriting, as a generalization of unconditional rewriting or demodulation, was studied in different forms by many researchers, see, for example, [1], [11], and [12]. The rewriting studied by Kaplan [8] represents the classical conditional rewriting which is similar to contextual rewriting when the context C is empty in each rewriting. Contextual rewriting is more powerful than the classical conditional rewriting because it uses an additional context. We point out that our contextual rewriting is not equivalent to the union of the classical conditional rewriting and the constant congruence relation. Actually, let $\rightarrow_R$ denote the classical conditional rewriting relation, then $(\simeq_C \cup \rightarrow_R)^* \subset \rightarrow^*_{C,R}$, but not $\rightarrow^*_{C,R} \subset (\simeq_C \cup \rightarrow_R)^*$.

In the literature on conditional rewritings, the formalisms proposed by Boyer and Moore [1], Remy [12] and Ganzinger [6] are closely related to the contextual rewriting discussed here.

4.1 Boyer-Moore's Rewriting and Cross-Fertilization

Our definition of contextual rewriting is influented most by the simplification process and the cross-fertilization process of Boyer-Moore theorem prover [1], even though the word "context rewriting" is not used in [1] and the description of the rewriting in [1] is very informal.

The idea of using the remaining literals in a clause as the context to simplify one literal was first described in [1]. In the simplification process of the Boyer-Moore theorem prover, a literal is simplified under the assumption that the remaining literals are F (we negate each literal first before obtaining an equality representation). In other words, the context of a literal comes from the negation of the rest literals of the clause. In [1] it neither says how C is used nor that the variables of C must be treated as constants.

In the cross-fertilization process, an equality in the condition is used to reformulate other literals. Let us quote the definition of "cross-fertilization" from [1, pp. 146-7]:

> In order to use `(EQUAL s' t')` in
>
> ```
> (IMPLIES (EQUAL s' t')
>          (EQUAL s (h t'))),
> ```
>
> we can substitute `s'` for `t'` in the other literals of the clause and delete the hypothesis. But rather than substitute for all occurrences of `t'` in `(EQUAL s (h t'))`, we prefer to substitute just for those in `(h t')`. That is, if we have decided to use `(EQUAL s' t')` by substituting the left-hand side for the right, and one of the places into which we substitute is itself an equality, `(EQUAL s (h t'))`, related to `(EQUAL s' t')` by induction, then we substitute only into the right-hand side. We call this "cross-fertilization".

In term of our terminology, the above paragraph describes a restricted (or heuristic) use of reformulation by constant congruence. When an equality in the context is used to simplify other literals, this equality is deleted from the clause in the Boyer-Moore theorem prover. In particular, cross-fertilization is used only if an induction has been done as a prior step in the proof attempt, and no user-supplied rules affect the cross-fertilization process. For more detail about "cross-fertilization", please see [1]. For our case, we cannot delete any literal unless it is trivial (i.e. $t \neq t$).

We believe that our contribution is that we combined rewriting and cross-fertilization processes together, and gave a formal description of the idea under the form of contextual rewriting. We also emphasized that the constant congruence should be used for the equalities in the context. More importantly, we show that this rewriting can apply to deductive theorem proving when viewing each clause as a (conditional) rewrite rule.

4.2 Remy's Contextual Rewriting

To the best of our knowledge, the name of contextual rewriting first appeared in Remy's thesis [12]. In Remy's definition, a *context* is a Boolean expression.

Definition 4.1 (Remy's definition of $\rightarrow_{C,R}$) Given a Boolean expression C and a set R of (conditional) rewrite rules, a rewriting under the context C is defined as follows: A term t is reduced to t' by R under the context C, denoted as $t \rightarrow_{C,R} t'$, if there exist a subterm position p of t, a rewrite rule $l \rightarrow r$ **if p** in R and a substitution σ such that

1. $\sigma(l) = t/p$,
2. $(C \Rightarrow \sigma(\mathbf{p})) \rightarrow_R^* \mathsf{T}$, where $\rightarrow_R$ is the classical conditional rewriting and $\Rightarrow$ denotes the logical implication,
3. $t' = t[p \leftarrow \sigma(r)]$.

From the above definition, it is easy to see that Remy's definition does not use the context to reformulate a term; the context of a term t is used only to establish the condition of each applicable rewrite rule. We use an example to illustrate the difference.

Example 4.2 Suppose we are given a rewrite rule $\mathbf{r} : rem(u*v, u) \rightarrow 0$ **if** $(u \neq 0)$ and a conditional equation

$$\mathbf{c} : rem(y * z, x) = 0 \text{ if } \{(x * y = y * z), (x \neq 0)\}$$

to be simplified. Let t be $rem(y * z, x)$ in $\mathbf{c}$. When we try to rewrite t, we take the condition of $\mathbf{c}$, i.e., $C = \{(x * y = y * z), (x \neq 0)\}$, as the *context* of t. In Remy's definition, the context is $(x * y = y * z) \wedge (x \neq 0)$.

At first, t is reformulated by C to $rem(x * y, x)$; this step is not allowed in Remy's definition.

Next, we apply the left-hand side of $\mathbf{r}$ to $rem(x*y, x)$, with the matching substitution $\sigma = \{u \leftarrow x, v \leftarrow y\}$. Using our definition, we need to show that the condition of $\sigma\mathbf{r}$, i.e., $x \neq 0$, is rewritten (by $\rightarrow_{C,R}$) to T. Using Remy's definition, we need to show that $((x*y = y*z) \wedge (x \neq 0)) \Rightarrow (x \neq 0)$ is rewritten (by $\rightarrow_R$) to T. Both conditions can be easily satisfied, so the term $rem(x * y, x)$ is replaced by the right-hand side of $\sigma\mathbf{r}$, i.e., 0. In short, the term t is simplified by $\mathbf{r}$ with C to 0. □

The contextual rewriting method proposed by Zhang and Remy in [16] is different from Remy's definition: The context given in [16] is also a Boolean expression; this context is not used for simplification but is enriched when a conditional rewrite rule is applied (i.e., the condition of a rewrite rule is appended into the context).

4.3 Ganzinger's Contextual Rewriting

Ganzinger's definition of contextual rewriting [6] pays particular attention to the fact that the variables in a context are treated as constants. To simplify a term t in the context C by a set R of rewrite rules, a skolem substitution θ (which replaces each variable of t and C by a new constant) is first applied to t and C, then $\theta(t)$ is rewritten by the new rewrite system $R \cup \theta C$ using the classical conditional rewriting. That is, $t \rightarrow_{C,R} t'$ if and only if $\theta(t) \rightarrow_{R \cup \theta(C)} \theta(t')$.

For instance, in Example 4.2, if R contains only $\mathbf{r}$ and $\theta = \{x \leftarrow s_1, y \leftarrow s_2, z \leftarrow s_3\}$, then $\theta(rem(y * z, x))$ is reduced to 0 by

$$\{\mathbf{r},\ \ s_2 * s_3 \rightarrow s_1 * s_2,\ \ (s_1 = 0) \rightarrow \mathsf{F}\}.$$

Ganzinger's definition is simple and easy to understand. However, the use of the context C in this definition is limited. At first, $\rightarrow_{\theta(C)}$ is not equivalent to $\simeq_C$. Even though $\theta(C)$ is a canonical rewrite system, because $\theta(C)$ is used together with R, the ordering to orient $\theta(C)$ is restricted and $R \cup \theta(C)$ may be not canonical even if both R and $\theta(C)$ are.

Example 4.3 Suppose that the rule $rem(gcd(x, y), y) \rightarrow 0$ is in R. To rewrite $rem(x, y)$ to 0 with $C = \{gcd(x, y) = x\}$, we have

$$rem(x, y) \simeq_C rem(gcd(x, y), y) \rightarrow_{C,R} 0.$$

If C is represented by a canonical system, we cannot have $x \rightarrow_C gcd(x, y)$ or $\theta(x) \rightarrow_C \theta(gcd(x, y))$ for any skolem substitution θ. □

We also point that a direct implementation of Ganzinger's definition will have the following drawback: We have to frequently skolemize terms and update the set of rewrite rules because different terms have different contexts. In our definition, the context of a term t can be treated as a ground rewrite system private to t.

5 Implementing Contextual Rewriting

For the application of contextual rewriting to be effective, we must have an efficient implementation of contextual rewriting. In this section, we show how to implement contextual rewriting efficiently and how to control contextual rewriting. We also provide a solution to the problem of handling variables which appear only in the condition of a rewrite rule.

5.1 A Procedure for Simplifying Clauses

To implement the contextual rewriting $\rightarrow_{C,R}$, at first, we have to implement the constant congruence relation $\simeq_C$. Because each ground rewrite system admits a canonical rewrite system, we can use this canonical system to obtain a unique normal form of any term with respect to $\simeq_C$ only. This can be easily done using the Knuth-Bendix completion procedure for ground equations [10]. By the well-know Knuth-Bendix' theorem, to complete a ground rewrite system, it is sufficient to mutually reduce each equation in the system and the result will be a canonical system, which serves as a decision procedure for $\simeq_C$.

We describe below how the constant congruence relation and the contextual rewriting are used in the procedure of clause-simplifying, which is actually implemented in *RRL* [9].

Suppose $\mathbf{c}$ is a clause $L_1 \vee \cdots \vee L_n$ to be simplified. Let $C = \{lhs(\neg L_i) = rhs(\neg L_i) \mid 1 \le i \le n\}$, the equality representation of the negated literals in $\mathbf{c}$. The following processes are performed on C:

1. **Rule making:** We organize $C = \{l_i = r_i\}$ as a set of "ground" (i.e., assuming all variables are constants) rewrite rules $\{l_i \rightarrow r_i\}$ such that $l_i \succ r_i$ for a reduction ordering $\succ$.

   In practice, we use Dershowitz' recursive path ordering by assuming a total ordering on variable names and every function is greater than variables. One exception with this method is that, when $(t = x) \in C$, where the variable x does not appear in t, we make a rewrite rule $x \rightarrow t$ instead of $t \rightarrow x$, if x can be made be greater than any function or variable of t in the precedence.

2. **Reformulating by context:** We still refer each rewrite rule $(l_i \rightarrow r_i) \in C$ as a literal $(l_i \neq r_i)$. Literals in C are used to rewrite each other until the set is inter-reduced. When a literal is subject to rewriting, it serves as the focus literal and the rest as the context.

   This reformulating process will always terminate because $l_i \succ r_i$ by a well-founded ordering. By Knuth-Bendix's result [10], a terminating and inter-reduced ground rewrite system is canonical. Because every term has a unique normal form in a canonical rewrite system, given a context C, we can easily decide whether two terms can be reformulated to the same term by $\simeq_C$.

3. **Elimination of irrelevants:** If C contains $t \neq t$, we just return T because C becomes a trivial clause. Otherwise, we delete any literal from C whose rule-form is $t \neq t$ or $x \rightarrow t$, where x is a variable. The removal of the latter is sound because after reformulating, x does not appear in any other literals and the literal $x \neq t$ can be resolved off with the unit clause $(x = x)$ without affecting any other literals.

   At the end of this process, we make a clause $\mathbf{c}'$ from these literals and it has been proved in [13] that the two clauses, $\mathbf{c}'$ and $\mathbf{c}$, are logically equivalent.

4. **Rewriting:** If a set R of (conditional) rewrite rules (made from other clauses) is available, we rewrite each literal L_i in C in turn by R while using the remaining literals of C as the context of L_i. If L_i is rewritten to a new literal, this new literal replaces L_i in C and simplifies, if possible, the remaining literals of C.

5. **Reporting result:** During processes (2) and (4), if a literal is reformulated by the context or rewritten by R to T, the procedure stops and this result is returned. If a literal is rewritten to F, this literal is discarded. Otherwise, if no further rewriting is possible in (4), a new clause (may be an empty clause F or a trivial clause T) is formulated from C and is returned.

We warn the reader that the above implementation of contextual rewriting is not complete, as illustrated by Example 4.3. We learned recently from Kapur and Narendran[2] that the following problem is NP-complete:

Instance: A conditional rewriting system R, a context C, and a term t.

Question: Does there exist a term t' such that $t \not\simeq_C t'$ and $t \rightarrow_{C,R} t'$?

Because of this result, it is unlikely that we will have a complete and efficient implementation of contextual rewriting. Hence, the use of ground completion for the context seems to be an efficient partial implementation of contextual rewriting.

---

[2] Private communication.

5.2 Control of Contextual Rewriting

There are three sources of nontermination in contextual rewriting: (i) the nontermination of $\simeq_C$; (ii) the right-hand side of a rewrite rule is not smaller than its left-hand side; (iii) the condition of a rewrite rule is not smaller than the left-hand side of its head.

For case (i), we may use a ground canonical rewrite system for $\simeq_C$. As seen in the previous section, this results in an efficient but incomplete implementation of contextual rewriting. For case (ii), the techniques developed for unfailure completion procedures can be adopted to handle this problem.

For case (iii), we restrict the ***depth***[3] of simplification to a small number, say 2. For instance, suppose we are given the following rules:

$$\begin{aligned} \mathbf{r}_1: &\ a_0 \rightarrow b_0 \textbf{ if } a_1 = b_1 \\ \mathbf{r}_2: &\ a_1 \rightarrow b_1 \textbf{ if } a_2 = b_2 \\ \mathbf{r}_3: &\ a_2 \rightarrow b_2 \textbf{ if } a_3 = b_3 \end{aligned}$$

To simplify the clause $a_0 = b_0 \vee a_2 \neq b_2$ to T by contextual rewriting, a depth 2 simplification is needed. To simplify $a_0 = b_0 \vee a_3 \neq b_3$ to T, a depth 3 simplification is needed. If the depth is limited to 2, then $a_0 = b_0 \vee a_3 \neq b_3$ cannot be simplified to T.

The use of contextual rewriting is optional in deductive theorem proving, so whether a clause is reducible or not affects only the computer time to obtain a proof; it does not affect the completeness of other inference rules. In inductive theorem proving, we wish that contextual rewriting is as powerful as possible. To compensate the loss of the power by the depth restriction, the user can use ***bridge lemmas***. For the above example, a bridge lemma will be $\mathbf{r}_4 : a_1 \rightarrow b_1$ **if** $a_3 = b_3$. When this bridge lemma is added (after it is proved to be true), the clause $a_0 = b_0 \vee a_3 \neq b_3$ can be simplified to T by a depth 2 simplification.

Every definition of rewriting induces a reduction relation $\rightarrow$ over terms as follows: $t_1 \rightarrow t_2$ if and only if t_1 is rewritten to t_2. Let $\rightarrow^*$ be the transitive and reflexive closure of $\rightarrow$. An important property regarding the relation $\rightarrow$ is ***confluence***. $\rightarrow$ is said to be ***confluent*** (or Church-Rosser) if and only if for any terms t_1 and t_2, $t_1 \leftrightarrow^* t_2$ implies there exists a term t_3 such that $t_1 \rightarrow^* t_3$ and $t_2 \rightarrow^* t_3$.

When the input consists of a set R of pure equations and the context C is empty, contextual rewriting is identical to unconditional rewriting or demodulation. In this case, $\leftrightarrow_R^*$, the transitive, reflexive and symmetric closure of $\rightarrow_R$ is equivalent to the equational congruence generated by R. If $\rightarrow_R$ is canonical, then $\rightarrow_R$ provides an effective decision procedure for the relation $\leftrightarrow_R^*$.

If R consists of a set of conditional rewrite rules, let $\rightarrow_{C,R}$ be the reduction relation induced by the contextual rewriting. As pointed out in [14], contextual rewriting is not confluent in general and even if it is confluent and terminating, the semantics of the rewriting relation is not clear. In [2] and [3], Bronsard and Reddy have generalized contextual rewriting as a restricted inference rule called contextual deduction and they proved that contextual deduction is refutational complete under certain conditions.

Currently ***RRL*** does not support automatic techniques to solve the confluence problem of contextual rewriting. The user can ***freeze*** (or ***disenable***) some rewrite rules to avoid undesirable rewriting or tells ***RRL*** explicitly through an interactive interface to use a particular rewrite rule.

[3]Suppose initially the ***depth*** of the simplification is 0. If the current simplification is at ***depth d***, and a conditional rewrite rule is applied, then the simplification of the conditions of that rule is at ***depth*** $d + 1$.

5.3 Handling Extra Variables in the Premises of Rules

In many cases, when a rewrite rule is made from a clause, the condition of the rule contains some variables which do not appear in the head of the rule. $less(x,y) \rightarrow \mathsf{T}$ **if** $\{(less(x,z) = \mathsf{T}, less(z,y) = \mathsf{T}\}$ is such an example. It is impossible to use such rules for contextual rewriting as defined previously.

In many applications, we wish to use such rules to simplify other clauses. For instance, during an experiment of proving a version of Ramsey's theorem in graph theory [15], we encount rules like

$$(\mathbf{r}_1)\quad member(x,z) \rightarrow \mathsf{F} \text{ if } \{member(x,y) \neq \mathsf{T}, subsetp(z,y) = \mathsf{T}\},$$

where y does not appear in the head. Suppose we also have a rule say $(\mathbf{r}_2)$ $subsetp(a,b) \rightarrow true$ and wish to prove that $(\mathbf{c}) : (member(u,a) \neq \mathsf{T}) \vee (member(u,b) = \mathsf{T})$. Note that $\mathbf{c}$ can be easily proved by resolution (by negating $\mathbf{c}$ first). Here, we wish to prove it by contextual rewriting.

The method we implemented to handle clauses, in which there is not a single literal containing all the variables, is illustrated by solving the above example.

Instead of $(\mathbf{r}_1)$, we consider a variant of $\mathbf{r}_1$, i.e.,

$$(\mathbf{r}_1')\quad \{member(x,z) \neq \mathsf{T}, member(x,y) = \mathsf{T}\} \text{ if } \{subsetp(z,y) = \mathsf{T}\}.$$

Now, every variable in the condition appears in the head. Another variant of $\mathbf{r}_1$ is

$$(\mathbf{r}_1'')\quad \{member(x,z) \neq \mathsf{T}, subsetp(z,y) \neq \mathsf{T}\} \text{ if } \{member(x,y) \neq \mathsf{T}\}.$$

To apply $(\mathbf{r}_1')$ to a clause, we need to

1. find a substitution σ such that both $\sigma(member(x,z)) \neq \mathsf{T}$ and $\sigma(member(x,y)) = \mathsf{T}$ are literals of that clause;

2. verify that the condition $\sigma(subsetp(z,y)) \rightarrow_{C,R} \mathsf{T}$.

It is not difficult to check that the clause $(\mathbf{c})$ above, i.e., $(member(u,a) \neq \mathsf{T}) \vee (member(u,b) = \mathsf{T})\}$, can be rewritten to a trivial one.

A formal description of the above two steps can be easily done and is omitted here. Because the idea used in (1) is the same as the subsumption rule, we call this method ***generalized subsumption***.

5.4 Some Experimental Results

Contextual rewriting has been used to support deductive theorem proving and inductive theorem proving methods in *RRL*. For deductive theorem proving, the method has been used successfully on a number of problems which appear hard for resolution theorem provers. For instance, to the best of our knowledge, we do not know any proof of SAM's lemma on lattice theory, when the axioms are formulated using equality. Our method successfully produced an automatic proof of this theorem in less than 30 seconds on a Sun 3/60 workstation.

In the proof of SAM's lemma, the deletion of the input clause (4) may illustrate the power of contextual rewriting. The axiomatization of modular lattice theory is as follows, plus the properties that *min* and *max* are commutative and associative.

$$
\begin{array}{lll}
(1) & min(x,x) = x & /*\text{ idempotence }*/ \\
(2) & max(x,x) = x & \\
(3) & min(x,z) = x \vee (max(x,y) \neq z) & /*\text{ absorption }*/ \\
(4) & max(x,z) = x \vee (min(x,y) \neq z) & \\
(5) & min(0,x) = 0 & \\
(6) & max(x,0) = x & \\
(7) & min(1,x) = x & \\
(8) & max(x,1) = 1 & \\
(9) & max(x, min(y,z)) = min(y, max(x,z)) \vee min(x,y) = x &
\end{array}
$$

The involved rewrite rules are:

$$
\begin{array}{llll}
(r1) & min(x,x) & \rightarrow & x, \\
(r3) & min(x, max(x,y)) & \rightarrow & x, \\
(r9) & max(x, min(y,z)) & \rightarrow & min(y, max(x,z)) \text{ if } min(x,y) = x.
\end{array}
$$

The literal $max(x,z) = x$ in clause (4) was first reformulated as $max(x, min(x,y)) = x$ by the context $\{z = min(x,y)\}$, which is made from the other literal of (4). Next, we can apply rule (r9) on $max(x, min(x,y))$, since the condition of (r9) can be established by rule (r1). The result of this rewriting is $min(x, max(x,y))$, which can be further reduced to x by rule (r3). In short, the first literal of clause (4) is reduced to $(x = x)$, or T. Clause (4) becomes trivial and can be thrown away safely. This also indicates the fact that the axiomatization of modular lattice theory is not minimal since one of the axioms can be derived from the others. Note that neither demodulation nor subsumption can remove clause (4).

References

[1] Boyer, R.S., Moore, J S.: (1979) A Computational Logic. Academic Press, New York.

[2] Bronsard, F., Reddy, U.S.: (1991) Conditional rewriting in Focus. In Kaplan, S., Okada, N. (eds): Procs. of Second International CTRS Workshop. pp. 2-13. Springer-Verlag, Berlin.

[3] Bronsard, F., Reddy, U.S.: (1992) Reduction techniques for first-order reasoning. In Rusinowitch, M., Remy, J.L.: (eds): Procs. of Third International CTRS Workshop.

[4] Dershowitz, N.: (1982) Ordering for term-rewriting systems. Theoretical Computer Science 17(3), 279-301.

[5] Dershowitz, N., Jouannaud J.P.: (1990) Rewrite systems. In J. van Leeuwen, editor, *Handbook of Theoretical Computer Science B: Formal Methods and Semantics*, chapter 6, pages 243–320, North-Holland, Amsterdam.

[6] Ganzinger, H.: (1987) A completion procedure for conditional equations. In Kaplan, S., Jouannaud, J.-P. (eds) Proc. of Conditional Term Rewriting Systems. Lecture Notes in Computer Science, vol. 308. pp. 62-83.

[7] Hsiang, J., Rusinowitch, M.: (1986) A new method for establishing refutational completeness in theorem proving. In: Siekmann, J.H. (ed.): Proc. of 8th conference on automated deduction, Lecture Notes in Computer Science 230, Springer, pp. 141-152

[8] Kaplan S.: (1984) Fair conditional term rewriting systems: unification, termination and confluence. Technical Report, LRI, Orsay,

[9] Kapur, D., Zhang, H.: (1989) An overview of RRL: Rewrite Rule Laboratory. In: Dershowitz, N. (ed.): Proc. of the third international conference on rewriting techniques and its applications. Lecture Notes in Computer Science 355, Springer. pp. 513-529.

[10] Knuth, D., Bendix, P.: (1970) Simple word problems in universal algebras. In: Leech, (ed.) Computational problems in abstract algebra. New York: Pergamon Press. pp. 263-297.

[11] Lankford, D.S.: (1979) Some new approaches to the theory and applications of conditional term rewriting systems. Report MTP-6, Dept. of Mathematics, Lousiana Tech University, Ruston, LA.

[12] Remy, J.L.: (1982) Etudes des systemes reecriture conditionelles et applications aux types abstraits algebriques. These d'etat, Universite de Nancy I, Nancy, France.

[13] Zhang, H.: (1988) Reduction, superposition and induction: automated reasoning in an equational logic. Ph.D. Thesis, Department of Computer Science, Rensselaer Polytechnic Institute, Troy, NY.

[14] Zhang, H., Kapur, D.: (1988) First-order logic theorem proving using conditional rewrite rules. In: Lusk, E., Overbeek, R., (eds.): Proc. of 9th international conference on automated deduction. Lecture Notes in Computer Science 310, Springer, pp. 1-20.

[15] Zhang, H., Hua, X.: (1992) Proving Ramsey theorem by cover-set induction: a case and comparison study. Presented at Second International Symposium on Artificial Intelligence and Mathematics. Fort Lauderdale, Florida.

[16] Zhang, H., Remy, J.L.: (1985) Contextual rewriting. In Jouannaud, J.P. (ed.): Proc. of 1st International Conference on Rewriting Techniques and Applications, Lecture Notes in Computer Science 202, Springer. pp. 1-20.

Confluence of Terminating Membership Conditional TRS *

Junnosuke Yamada
NTT Research Laboratories
Seika-cho, Soraku-gun, Kyoto 619-02 Japan
e-mail:jun@nttlab.ntt.jp

abstract

Membership conditional term rewriting systems (MCTRSs) are term rewriting systems in which applications of rewriting rules are restricted by membership conditions. In this paper, a sufficient condition (the critical pair lemma) for terminating MCTRSs is proposed. To show the condition, the notion of contextual rewriting discussed by Zhang-Rémy is extended and a new technique to divide membership conditions into simpler subconditions is introduced. Furthermore, when membership conditions of MCTRSs are divided into infinite subconditions with some well-founded structures, their confluence are shown by induction on the structures. Finally, a completion algorithm for MCTRSs is proposed on the basis of this sufficient condition and a completion of McCarthy's 91-function is demonstrated.

1. Introduction

The membership conditional TRS (MCTRS) is an extension of the TRS proposed by Toyama[15] Rules of MCTRSs are rewriting rules of TRSs incorporating some membership conditions on variables in them, and they only reduce the terms which fulfill their membership conditions. Notice that whether or not membership conditions hold is assumed to be separated from reductions in MCTRSs. This makes MCTRSs promising to study the general properties of extended TRSs [1,5,10,13] as abstract TRSs separating the interactions between conditions and rules [10]. Among of them meta-rewriting systems was independently proposed by H. Kirchner[10] and is very similar to MCTRSs. MCTRSs are also closely connected to the approach to simulate order sorted rewriting by many sorted ones, as they have no restrictions on sort decreasingness and regularity which are indispensable for order sorted rewritings.

One of the most important properties of TRSs is confluence. This paper investigates the confluence of MCTRSs and clarifies its sufficient condition. The conditions of confluence of TRSs differ according to whether or not the TRS is terminating [2,6]. Here a sufficient condition of terminating MCTRSs is treated (as for non-terminating MCTRSs, refer to [15,16]).

It is well-known that a terminating unconditional TRS is confluent if and only if its every critical pair converges [6]. To have some corresponding results for MCTRSs, some modification to the notions of overlap and critical pair are necessary as with other extended TRSs [10,13]. In this paper, a sufficient condition for the confluence of unconditional TRSs is extended to MCTRSs by adapting the contextual rewriting introduced by Zhang-Rémy[22] to the reduction of terms with membership conditions (called contextual terms), and introducing the technique to

* *Presented in CTRS92 at Pont-à-Mousson, 8-10 July 1992. A preliminary version appeared in Trans. IEICEJ J-74-D-I (1991), pp.666-674, in Japanese.*

divide membership conditions into simpler subconditions. Furthermore, when membership conditions of MCTRSs are divided into infinite subconditions with some well-founded structures, their confluence are shown by induction on the structures. Finally a completion algorithm of MCTRSs is proposed.

Sections 2 and 3 briefly prepare necessary notions and symbols of TRSs and MCTRSs, respectively. In section 4, contextual rewriting and the split of membership conditions are introduced. Section 5 shows a sufficient condition for confluence, and applys the condition for the inductively proven case. In section 6, a completion algorithm based on this condition is proposed, and the proposed algorithm is exemplified by demonstrating a property of McCarthy's 91-function.

2. Term Rewriting Systems (TRS)

In this section, basic notions and symbols of TRSs are briefly explained. The reader can refer to [2,6] for details.

$T = T(F,V)$ is a set of terms composed of a set of function symbols F and a infinite set of variables V. The syntactical identity of terms is indicated by $\equiv$. $Var(t)$ is the set of all the variables appearing in t, and a term t is called a ground term if $Var(t) = \emptyset$.

A term t is denoted by $t(x_1, \cdots, x_n)$ to specify all the variables $Var(t) = \{x_1, \cdots, x_n\}$ in it, and $t(s_1, \cdots, s_n)$ is the term obtained by substituting a term s_i for x_i in term t.

The occurrences $\mathcal{O}(t)$ of t, the subterm t/u of t at $u \in \mathcal{O}(t)$, the substitution $t[u \leftarrow s]$ and so on are defined completely same as [6].

A rewriting rule (or simply, rule) on T is a pair of terms (l, r) such that $Var(l) \supset Var(r)$ and $l \notin V$. A set of rules is denoted by $\triangleright$, and $(l, r) \in \triangleright$ by $l \triangleright r$. A term t reduces to a term t' by a rule $l \triangleright r$ at $u \in \mathcal{O}(t)$, if there are a substitution θ, an occurrence $u \in \mathcal{O}(t)$ such that $t/u \notin Var(t)$, $t/u \equiv l\theta$, and $t' \equiv t[u \leftarrow r\theta]$, and is denoted by $t \to t'$. The subterm $s \equiv t/u$ is called a redex in t of the rule. $\to^*$ is the transitive reflexive closure of $\to$.

A term rewriting system (TRS) is defined as follows.

Definition 2.1. A *term rewriting system* $R = (T, \to)$ is a structure of a binary relation $\to$ on the set of objects $T = T(F,V)$ defined by a set of (rewriting) rules $\triangleright$. The set of rules $\triangleright$ and the TRS R defined by $\triangleright$ are identified if not confusing.

A term t is a normal form, if there is no term s such that $t \to s$. A term t' is called a normal form of a term t, if $t \to^* s$ and s is a normal form. Two terms t_1 and t_2 converge, if there is a term s such that $t_1 \to^* s$ and $t_2 \to^* s$.

Hereafter any two rules are assumed to have no common variable by renaming variables appropriately.

Two rules $l \triangleright r$ and $l' \triangleright r'$ overlap, if there are an occurrence $u \in \mathcal{O}(l)$ and a substitution θ such that $l/u \notin V$ and $l\theta/u \equiv l'\theta$.

A critical pair of two overlapping rules is defined as follows[6].

Definition 2.2. Let two rules $l_1 \triangleright r_1$, $l_2 \triangleright r_2$ overlap at an occurrence $u \in \mathcal{O}(l_1)$. The pair of terms $\langle P, Q \rangle$ such that

$$P \equiv l_1\theta[u \leftarrow r_2\theta], \quad Q \equiv r_1\theta$$

is the *critical pair* of the two rules at the occurrence u where θ is the most general unifier of $l_1/u \notin V$ and l_2.

The two properties below are the most important for the investigation of TRSs [6].

Definition 2.3. A TRS $R = (T, \rightarrow)$ is *terminating*, if every reduction in R terminates, i.e., there is no infinite reduction sequence $t_1 \rightarrow t_2 \rightarrow t_3 \rightarrow \cdots$ in R where $t_i \in T$.

Definition 2.4. A TRS $R = (T, \rightarrow)$ is *confluent*, if there exists $s \in T$ such that $t_1 \rightarrow^* s$, $t_2 \rightarrow^* s$ for every $t, t_1, t_2 \in T$ such that $t \rightarrow^* t_1$, $t \rightarrow^* t_2$. R is *locally confluent*, if there exists $s \in T$ such that $t_1 \rightarrow^* s$, $t_2 \rightarrow^* s$ for every $t, t_1, t_2 \in T$ such that $t \rightarrow t_1$, $t \rightarrow t_2$.

A TRS is *complete* if it is terminating and confluent. In complete TRSs, every term has necessarily an unique normal form.

As for terminating TRSs, the following lemma on the relationship between confluence and local confluence is well-known [6].

Theorem 2.5. A terminating TRS R is confluent if and only if R is locally confluent.

The next theorem holds for unconditional TRSs[6].

Theorem 2.6. A terminating TRS R is confluent if and only if every critical pair of R converges.

3. Membership Conditional TRS

The membership conditional TRS (MCTRS) is a kind of conditional TRS proposed by Toyama [15]. Every rewriting rule of MCTRSs is a rewriting rule with membership conditions on the variables in the left hand side, and is applied only when the conditions hold. For example, a rule $f(x) \triangleright 0 : x \,\varepsilon\, \{\text{even}\}$ of a MCTRS results in a reduction $f(2) \rightarrow 0$, but no reduction of $f(3)$. MCTRSs are known to be effective for in describing the systems in which types, values, semantics of variables are restricted[15,16], and also in treating infinite number of rules as one[10].

Definition 3.1. A pair of a term t and some membership conditions $(x_1, \cdots, x_n) \,\varepsilon\, S_1 \times \cdots \times S_n$ on all the variables $\{x_1, \cdots, x_n\} \supset Var(t)$,

$$t : (x_1, \cdots, x_n) \,\varepsilon\, S_1 \times \cdots \times S_n$$

is called a *contextual term* (*c-term*), where each S_i is a non-empty subset of T. $(x_1, \cdots, x_n) \,\varepsilon\, S_1 \times \cdots \times S_n$ is called the *context* of the c-term along with Zhang-Rémy[22]. The context is abbreviated by c and the c-term by $t : c$. A *MC-rule* $l \triangleright r : c$ is the pair of a (rewriting) rule $l \triangleright r$ and membership conditions c on the variables in its left hand side l.

Henceforth, a membership condition on a term t to a set S is denoted by $t \,\varepsilon\, S$, and any membership condition can be allowed if syntactically correct. Correspondingly, $\in$ means to belong to some set mathematically. This means, a membership condition holds on some appropriate model. Moreover for every set S, $t \in S$ is assumed to be decidable.

Definition 3.2. A substitution θ is a *c-substitution* from $t : (x_1, \cdots, x_m) \,\varepsilon\, S_1 \times \cdots \times S_m$ to $t' : (y_1, \cdots, y_n) \,\varepsilon\, S'_1 \times \cdots \times S'_n$, if

(1) $t' \equiv t\theta$, and,
(2) $x_i\theta(s'_1, \cdots, s'_n) \in S_i$ holds for all $(s'_1, \cdots, s'_n) \in S'_1 \times \cdots \times S'_n$, for every $i = 1 \cdots m$.

By a MC-rule $l \triangleright r : (x_1, \cdots, x_n) \varepsilon S_1 \times \cdots \times S_n$ a term t reduces to a term t', if for some subterm t/u in t, there is a substitution θ such that $t/u \equiv l\theta$, $t' \equiv t[u \leftarrow r\theta]$ and $x_i\theta \in S_i$ for every x_i, and is denoted by $t \rightarrow t'$. Then, a set of MC-rules $\triangleright$ defines a relation $\rightarrow$ on the set of terms T, and defines a membership conditional TRS as below.

Definition 3.3. A *membership conditional TRS* (*MCTRS*) is a TRS defined by a set of MC-rules.

Also in MCTRSs, any two MC-rules have no common variable by renaming variables appropriately.

A MCTRS is *terminating* or (*locally*) *confluent* if the relation $\rightarrow$ defined by its set of MC-rules $\triangleright$ is terminating or (locally) confluent, respectively. From now on, only terminating MCTRSs are considered.

An example of a MCTRS and its reductions follow:
Example 3.4. Let $F = \{eq, d, +, s, 0\}$ and $F' = \{+, s, 0\}$ be sets of function symbols. The following MCTRS

$$\begin{cases} x + 0 \triangleright x \quad : x \varepsilon \mathsf{N} & (1) \\ x + s(y) \triangleright s(x+y) \quad : (x, y) \varepsilon \mathsf{N}^2 & (2) \\ d(x) \triangleright x + x \quad : x \varepsilon T(F') & (3) \\ eq(x, x) \triangleright x \quad : x \varepsilon T(F') & (4) \end{cases}$$

defines addition $+$, double d, and equal eq on the set of natural numbers $\mathsf{N} = T(\{s, 0\})$ where 0 is the constant and s the successor function. In this MCTRS, there is a reduction sequence

$$eq(d(0), d(0)) \rightarrow eq(0+0, d(0)) \rightarrow eq(0+0, 0+0) \rightarrow 0+0 \rightarrow 0.$$

Notice that $eq(d(0), d(0))$ cannot directly reduce to $d(0)$ using the MC-rule (3). This is because the membership condition $d(0) \varepsilon T(F')$ does not hold and MC-rule (3) cannot be applied.

The following notion necessary later must be defined.
Definition 3.5. A set of terms $S(\subset T)$ is *closed* for a MCTRS $R = (T, \rightarrow)$, if for every $t \in S$ its normal form is also in S*. A MCTRS R is *closed*, if every set of membership conditions in $\triangleright$ of R is closed.

Remark. MCTRSs might be contradictory according to their membership conditions, but only consistent MCTRSs will be treated in this paper. Several sufficient conditions for MCTRSs not to be contradictory can found in [15].

4. Contextual Rewriting

In MCTRSs, the notions of overlap and critical pair must be modified. For example, the following two MC-rules

---

* This condition can be relaxed as follows: "$\forall t, s \in S[[t \rightarrow s] \Rightarrow \exists s' \in S[s \rightarrow^* s']]$." The results in this paper hold under this relaxed definition.

$$f(x) \triangleright 0 : x \;\varepsilon\; \{\text{even}\}, \quad f(x) \triangleright 1 : x \;\varepsilon\; \{\text{odd}\}$$

overlap without their membership conditions, but they do not overlap as MC-rules because they reduce no same single term simultaneously. To discuss overlap and critical pairs only from MC-rules, the notion of contextual rewriting introduced by Zhang-Rémy[22] is adapted to our case. Notice that the contextual rewriting in this paper is different from that in [22]. This is because [22] treats only the contexts of Boolean values, and the contexts in this paper are membership conditions. Therefore the definitions of contextual rewriting differ.

Reductions (rewritings) of c-terms by MC-rules are defined as follows, based on substitutions between c-terms.

Definition 4.1. A c-term $t : c$ is *contextually reducible* (*c-reducible*) by a MC-rule $l \triangleright r : (x_1, \cdots, x_n) \;\varepsilon\; S_1 \times \cdots \times S_n$, if there exists a c-substitution θ' from $l : (x_1, \cdots, x_n) \;\varepsilon\; S_1 \times \cdots \times S_n$ to $t' : c$ for the subterm $t' \equiv t/u$ of $t : c$ at some occurrence $u \in \mathcal{O}(t)$. Then $t : c$ *c-reduces* to $s : c \equiv t[u \leftarrow r\theta'] : c$, and is denoted by $t : c \twoheadrightarrow s : c$. In this definition, $x_i\theta' \in S_i$ must hold under the assumption that the variables appearing in $x_i\theta'$ are restricted by the context c.

$\twoheadrightarrow^*$ denotes the reflexive transitive closure of $\twoheadrightarrow$. A c-term $t : c$ is a *$\twoheadrightarrow$-normal form*, if there is no c-term $t' : c$ such that $t : c \twoheadrightarrow t' : c$. A c-term $t' : c$ is a *$\twoheadrightarrow$-normal form* of $t : c$ if $t : c \twoheadrightarrow^* t' : c$ and $t' : c$ is a $\twoheadrightarrow$-normal form. Two c-terms $t_1 : c$ and $t_2 : c$ with a common context *c-converge*, if there is a c-term $s : c$ such that $t_1 : c \twoheadrightarrow^* s : c$ and $t_2 : c \twoheadrightarrow^* s : c$.

Next, the relationship between contextual terms and terms without contexts is investigated.

Definition 4.2. A term t is an *instance* of c-term $s : (x_1, \cdots, x_n) \;\varepsilon\; S_1 \times \cdots \times S_n$, if $t \equiv s\theta$ by a substitution θ such that $x_i\theta \in S_i$ for each x_i.

Lemma 4.3. For every c-term $s : c$, there exist an instance t and a substitution θ such that $t \equiv s\theta$. Conversely arbitrary term t is an instance of some c-term $s : c$.

This claims a correspondence between reductions between terms and c-reductions between c-terms.

Lemma 4.4. Let a c-term $s : c$ c-reduce to a c-term $s' : c$, and $t \equiv s\theta$ (θ is a substitution) be an instance of $s : c$. Then there is an instance $t' \equiv s'\theta$ of $s' : c$ such that $t \to t'$. Namely the diagram bellow commutes:

$$\begin{array}{ccccc} & s : c & \twoheadrightarrow & s' : c & \\ & \downarrow & & \downarrow & \\ s\theta \equiv & t & \to & \exists t' & \equiv s'\theta \end{array}$$

Proof. Let $s : c \equiv l\tilde{\theta} : c \twoheadrightarrow r\tilde{\theta} : c \equiv s' : c$ by a MC-rule $l \triangleright r : \tilde{c}$. As $t' \equiv r\tilde{\theta}\theta$ is clearly an instance of $s' : c$, $l\tilde{\theta}\theta \to r\tilde{\theta}\theta$ by $l \triangleright r : \tilde{c}$, i.e., $y\tilde{\theta}\theta \in \tilde{S}$ must be proven for every condition $y \;\varepsilon\; \tilde{S}$ of $\tilde{c}$. As $l\tilde{\theta} : c \twoheadrightarrow r\tilde{\theta} : c$ by $l \triangleright r : \tilde{c}$, $y\tilde{\theta} \in \tilde{S}$ for all $(s_1, \cdots, s_n) \in S_1 \times \cdots \times S_n$ where $c = (x_1, \cdots, x_n) \;\varepsilon\; S_1 \times \cdots \times S_n$. On the other hand $l\tilde{\theta}\theta$ is an instance of $l\tilde{\theta} : c$, $t_j \in S_j$ for every t_j in $\theta = \{\cdots, x_j/t_j, \cdots\}$. Thus $y\tilde{\theta}\theta \in \tilde{S}$. □

From this lemma, a TRS in which membership conditions can be treated together with terms is defined for each MCTRS.

Definition 4.5. Let $R = (T, \rightarrow)$ be a MCTRS, and $\tilde{T}$ the set of all the c-terms whose instances are some terms in T. The *covering CTRS* $\tilde{R} = (\tilde{T}, \twoheadrightarrow)$ of R is a TRS where the c-reduction $\twoheadrightarrow$ is defined on $\tilde{T}$ by the MC-rules $\triangleright$ of R. Trivially every MCTRS has a covering CTRS.

To consider overlap and critical pairs between MC-rules in MCTRSs, unification between two c-terms must be clarified first. For that purpose, the membership conditions attached to the variables must also be considered in finding an unification substitution.

Definition 4.6. Let $t_1 : c_1 \equiv t_1 : (x_1, \cdots, x_m) \varepsilon S_1 \times \cdots \times S_m$, $t_2 : c_2 \equiv t_2 : (x_{m+1}, \cdots, x_{m+n}) \varepsilon S_{m+1} \times \cdots \times S_{m+n}$ be two c-terms with no common variables, $\theta = \{\cdots, x_j/s_j, \cdots\}$ the unifier of t_1 and t_2 where $\{x_1, \cdots, x_{m+n}\} \supset \bigcup Var(s_i)$. $t_1 : c_1$ and $t_2 : c_2$ are *c-unifiable*, if there exist sets $S'_1, \cdots, S'_{m+n}$ such that

$$s_j \in S_j \text{ for } 1 \leq \forall j \leq m+n \Leftrightarrow (x_1, \cdots, x_{m+n}) \in S'_1 \times \cdots \times S'_{m+n}. \qquad (*)$$

If $t_1 : c_1$ and $t_2 : c_2$ are c-unifiable, a pair (θ, c) of the substitution θ and the membership conditions $c = (x_1, \cdots, x_{m+n}) \varepsilon S'_1 \times \cdots \times S'_{m+n}$ is called their *c-unifier*. Furthermore (θ, c) is called the *most general c-unifier* when θ is the most general unifier of t_1 and t_2.

To have a c-unification substitution, for each substitution $x_j/s_j(x_1, \cdots, x_{m+n})$, the sets S'_i satisfying $(*)$ under $x_1 \in S_1, \cdots, x_{m+n} \in S_{m+n}$ must be determined. Generally, such S'_i can only be determined in limited cases.

For example, we restrict ourselves to a very simple case such that contexts can be generated from $(*)$ using intersection and difference between sets. Even in this very simple case, it is necessary to check the c-unifiability of two c-terms, so all the sets in the resulting contexts must be decided to be non-empty. For this decision, all the sets in the membership condition of two c-unifiable c-terms are limited only recursive sets.

In this paper, the c-unifiers satisfying $(*)$ are assumed to be constructable, and more detailed observations can be found in appendix.

Definition 4.7. Any two MC-rules

$$l_1 \triangleright r_1 : (x_1, \cdots, x_m) \varepsilon S_1 \times \cdots \times S_m,$$
$$l_2 \triangleright r_2 : (x_{m+1}, \cdots, x_{m+n}) \varepsilon S_{m+1} \times \cdots \times S_{m+n}$$

c-overlap or are *c-overlapping* at an occurrence $u \in \mathcal{O}(l_1)$ such that $l_1/u \notin V$, if there exists a c-unifier of the two c-terms l_1/u and l_2 attached with each membership conditions in MC-rules.

Here the c-critical pair of two c-overlapping MC-rules can be defined:

Definition 4.8. The *contextual critical pair* (*c-critical pair*) of two MC-rules

$$l_1 \triangleright r_1 : (x_1, \cdots, x_m) \varepsilon S_1 \times \cdots \times S_m,$$
$$l_2 \triangleright r_2 : (x_{m+1}, \cdots, x_{m+n}) \varepsilon S_{m+1} \times \cdots \times S_{m+n}$$

c-overlapping at $u \in \mathcal{O}(l_1)$ is a triple $\langle P, Q\rangle : c$ of two terms and membership conditions,

$$P \equiv l_1\theta[u \leftarrow r_2\theta], \quad Q \equiv r_1\theta, \text{ and } c = (x_1, \cdots, x_{m+n}) \varepsilon S'_1 \times \cdots \times S'_{m+n}$$

where (θ, c) is the most general c-unifier of c-terms l_1/u and l_2 with each membership conditions.

An example of a c-critical pair between two MC-rules

$$f(f(y)) \triangleright h(y) : y \ \varepsilon \ g(T) \cup h(T) \quad \text{and} \quad f(g(z)) \triangleright g(z) : z \ \varepsilon \ T$$

is illustrated. These two MC-rules overlap by a substitution $\{y/g(z)\}$ as unconditional rules when their membership conditions are ignored. Then some condition on z such that $z \ \varepsilon \ T$ and $g(z) \ \varepsilon \ g(T) \cup h(T)$ (= { the set of the terms in the form of $g(\cdots)$ or $h(\cdots)$}) must be found to overlap as MC-rules. In this example, only the latter condition has to be considered, and we can restrict ourselves to a sub-condition $z \ \varepsilon \ T$ such that $g(z) \ \varepsilon \ g(T) \cup h(T)$. Thus a condition $z \ \varepsilon \ T$ and a c-critical pair $\langle f(g(z)), h(g(z))\rangle : z \ \varepsilon \ T$ are easily obtained.

The notion of c-reductions between c-terms make it possible to investigate behaviors of MCTRSs without considering all the reduction relations generated by MC-rules. That is, an infinite number of reduction relations can be realized only through a finite number of MC-rules.

Example 4.9. A MCTRS

$$\begin{cases} x + 0 \triangleright x & : x \ \varepsilon \ \mathsf{N} & (1) \\ x + s(y) \triangleright s(x) + y & : (x, y) \ \varepsilon \ \mathsf{N}^2 & (2) \\ x + s(y) \triangleright s(x + y) & : (x, y) \ \varepsilon \ \mathsf{N}^2 & (3) \end{cases}$$

defines addition + on the set of natural numbers $\mathsf{N} = T(\{s, 0\})$. Two MC-rules (2) and (3) c-overlap in the covering CTRS, and there are two c-reductions.

$$\begin{array}{ccc} & x + s(y) : (x, y) \ \varepsilon \ \mathsf{N}^2 & \\ \swarrow & & \searrow \\ s(x + y) : (x, y) \ \varepsilon \ \mathsf{N}^2 & & s(x) + y : (x, y) \ \varepsilon \ \mathsf{N}^2 \end{array}$$

The resulting two c-terms are c-normal forms and never c-converge, even though the corresponding MCTRS is locally confluent.

To avoid this phenomenon caused by the difference between $\twoheadrightarrow$ and $\rightarrow$, c-reduction is extended by splits of membership conditions as introduced below. By this extension, the two c-terms in the last example converge.

Definition 4.10. Let $c = (x_1, \cdots, x_n) \ \varepsilon \ S_1 \times \cdots \times S_n$ a context and $S_1 \times \cdots \times S_n = S_1^1 \times \cdots \times S_n^1 \cup \cdots \cup S_1^k \times \cdots \times S_n^k$ be a union of sets. Then c ***splits*** or has a ***split*** into $c_1, \cdots, c_k$ where $c_j = S_1^j \times \cdots \times S_n^j$. This is denoted by $c = c_1 \cup \cdots \cup c_k$ or simply by $c = \cup c_j$.

Definition 4.11. Two c-terms $t_1 : c$ and $t_2 : c$ ***c-converge with split***, if there is a split $c = c_1 \cup \cdots \cup c_k$ such that $t_1 : c_i$ and $t_2 : c_i$ c-converge for all i.

Soon after theorem 5.2, it will be illustrated how the (ground) confluence of MCTRS in example 4.9 is derived from application of split to two c-terms $s(x+y) : (x, y) \ \varepsilon \ \mathsf{N}^2$ and $s(x) + y : (x, y) \ \varepsilon \ \mathsf{N}^2$ which do not c-converge.

In covering CTRSs, as their reduction $\twoheadrightarrow$ between c-terms preserves their contexts, $\twoheadrightarrow$ is a weaker relation than $\rightarrow$ of the underlying MCTRSs. Then the confluence of a terminating covering CTRS does not necessarily imply confluence of its underlying MCTRS. By extending c-reduction by split introduced in the next section, this difficulties will be solved.

5. Confluence of MCTRS

In this section, a sufficient condition for the confluence of MCTRSs will be clarified. The rules of MCTRSs have contexts (condition parts), it becomes necessary to consider c-critical pairs in their covering CTRSs. Consequently, it is shown to be a sufficient condition for confluence that every c-critical pair c-converges with split. Additionally, the condition is extended for the cases with c-critical pairs with infinite splits.

To show the confluence of terminating closed MCTRSs, first, we prepare the next lemma on the c-convergence of contextual critical pairs in their covering CTRSs and local confluence of MCTRSs.

Lemma 5.1. Let $R = (T, \rightarrow)$ be a terminating closed MCTRS, $\tilde{R} = (\tilde{T}, \twoheadrightarrow)$ its covering CTRS. If every c-critical pair $\langle P, Q\rangle : c$ of $\tilde{R}$ c-converges with split, R is locally confluent.

Proof. Let a term t reduce to different terms t_1, t_2 by applying two MC-rules (r1) $l_1 \triangleright r_1 : c_1$, (r2) $l_2 \triangleright r_2 : c_2$ on redexes t/u, t/v where $u, v \in \mathcal{O}(t)$ in R, respectively.

The proof separates into the following two cases according to the relative positions of u and v.

Case 1: t/u and t/v are not a subterm of each other.

Applying (r2) to t_1/v, (r1) to t_2/u, an identical term is obtained.

Case 2: Either t/u or t/v is a subterm of the other.

Without loss of generality, t/v can be assumed to be a subterm of t/u, then only a subterm t/u of t has to be taken into account. Then there are two subcases.

a. $l_1/(v/u) \notin V$.

In this subcase, two terms $t_2/u, t_1/u$ are obtained by some common c-substitution θ applied to the two terms $P : c$, $Q : c$ respectively where $\langle P, Q\rangle : c$ is a c-critical pair of $\tilde{R}$.

From the local confluence with split, for some split $c = \cup c_i$ there is a c-term $s'_1 : c_i$ such that $P : c_i \twoheadrightarrow^* s'_i : c_i$ and $Q : c_i \twoheadrightarrow^* s'_i : c_i$ for each i. Moreover for some i_0, t_1/u and t_2/u are instances of $Q : c_{i_0}$ and $P : c_{i_0}$ obtained by some c-substitution θ_{i_0}, respectively. By lemma 4.4, there exists a term $t' \equiv t[u \leftarrow s'_{i_0}\theta_{i_0}]$ and $t_j \rightarrow^* t'$ for $j = 1, 2$ as diagram below.

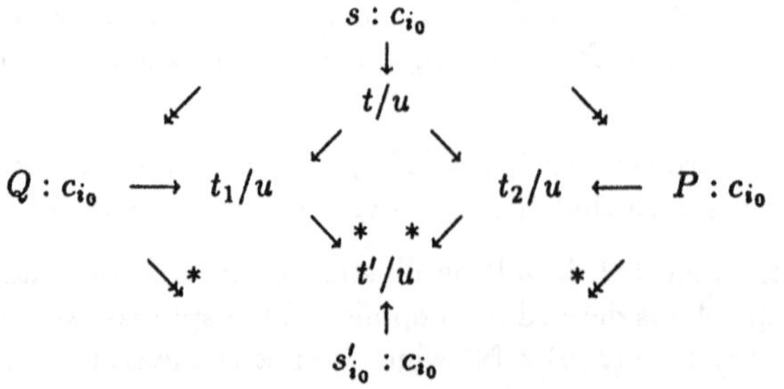

b. $l_1/(v/u) \in V$ or $v/u \notin \mathcal{O}(l_1)$.
The left hand side of (r1) has a variable $x \equiv l_1/(w/u)$ such that $u \leq w \leq v$. Let $w'/u, w''/u, \cdots, w'''/u$ be the occurrences of x in l_1 other than w/u. As R is terminating and closed, $t_2/(w/u)$ reduces to a normal form satisfying the membership condition $x \ \varepsilon \ S$ in (r1). Similarly $t_2/(w'/u), t_2/(w''/u), \cdots, t_2/(w'''/u)$ reduce to the same normal forms as $t_2/(w/u)$. Now the resulting term is reducible by (r1), and let the term reduced by (r1) be t'/u.

Let $v'/u, v''/u, \cdots, v'''/u$ be the occurrences of x in the right hand side of (r1). Reductions of $t_1/(v'/u)$, $t_1/(v''/u)$,$\cdots$,$t_1/(v'''/u)$ to their normal forms as $t/(w/u)$ results in t'/u.

Note that, if x does not appear in the right hand side of (r1), $t_1 \equiv t'$. □

This proof is parallel to that in [6], but subcase 2b comes into consideration first for MCTRSs. Because, in unconditional TRSs, the variable x in subcase 2b has a membership condition $x \ \varepsilon \ T(F, V)$.

Remark. Subcase 2b is called *variable overlap* case. Because left hand side of one rule occurrs below a variable occurrence of left hand side of another rule. Although unification between c-terms depends on which kind of sets are allowed in membership conditions. In this paper, the details of unification/matching are not mentioned, since our aim is to clarify abstract properties of MCTRSs by separating membership conditions and reductions.

On the other hand, there are some works by H. Kirchner and M. Hermann[11] and H. Comon[2,3] in which sets are restricted to many(order) sorted algebras and regular tree languages. In [3], variable overlap cases are analyzed by using the properties of regular tree languages and some complicated unification/matching mechanism in second order framework.

Immediately from this lemma, the following sufficient condition for the confluence of MCTRSs holds.

Theorem 5.2. **(Main Theorem)** Let R be a closed terminating MCTRS, $\tilde{R}$ its covering CTRS. If every c-critical pair of $\tilde{R}$ c-converges with split, R is confluent.

Proof. Locally confluent by lemma 5.1, and then clear from lemma 2.5 as R is terminating. □

Now example 4.9 is reconsidered. Take a split $\mathsf{N} \times \mathsf{N} = \mathsf{N} \times \{0\} \cup \mathsf{N} \times \{1, 2, \cdots\}$. Then there appear two convergent diagrams for $y \ \varepsilon \ \{0\}$ and $y \equiv s(s(y')) \ \varepsilon \ \{1, 2, \cdots\}$:

$$\begin{array}{ccccc} & & x + s(0) : x \ \varepsilon \ \mathsf{N} & & \\ & \swarrow & & \searrow & \\ s(x) + 0 : \cdots & & & & s(x + 0) : \cdots \\ & \searrow & & \swarrow & \\ & & s(x) : x \ \varepsilon \ \mathsf{N} & & \end{array} \qquad \begin{array}{ccccc} & & x + s(s(y')) : (x, y') \ \varepsilon \ \mathsf{N}^2 & & \\ & \swarrow & & \searrow & \\ s(x) + s(y') : \cdots & & & & s(x + s(y')) : \cdots \\ & \searrow & & \swarrow & \\ & & s(s(x + y')) : (x, y') \ \varepsilon \ \mathsf{N}^2 & & \end{array}$$

Thus the MCTRS in example 4.9 is (ground) confluent by theorem 5.2.

Definition 4.10 of finite split can be naturally extended to infinite split. and theorem 5.2 can be applicable for the cases in which infinite splits occur.

Definition 5.3. Let $\tilde{R}$ be a covering cTRS of a MCTRS R. A c-critical pair $\langle P, Q\rangle : c$ of $\tilde{R}$ is *inductively convergent* if it satisfies the following condition: there are a infinite split $c = \bigcup c_i$ and a well-founded ordering $<$ on c_i's such that
(1) $\langle P, Q\rangle : c_0$ c-converges, and
(2) if $\langle P, Q\rangle : c_j$ c-converges for $\forall j \leq i$, then $\langle P, Q\rangle : c_{i+1}$ c-converges, where c_i's are assumed to be reordered so as to fulfill $c_i < c_{i+1}$.

Definition 5.4. Let $R = (T, \rightarrow)$ be a TRS. R is *inductively confluent* if for every t, t_1, t_2 such that $t \rightarrow^* t_i$, there exists a term s such that $t\theta \rightarrow^* t_i\theta \rightarrow^* s\theta$ for every ground substitution θ.

By these definitions, and well-foundedness of the sets of closed terms:
Corollary 5.5. Let R be a closed terminanting MCTRS, every sets in membership conditions in R be ground. If every contextual critical pair in its covering cTRS $\tilde{R}$ is inductively convergent, R is inductively confluent.

By this corollary, the inductive confluence of the next MCTRS is proven. **Example 5.6.**

$$\begin{cases} x + 0 \triangleright x \quad : x \in \mathsf{N} & (1) \\ x + s(y) \triangleright s(x+y) \quad : (x, y) \in \mathsf{N}^2 & (2) \\ f(x,y,z) \triangleright (x+y)+z \quad : (x,y,z) \in \mathsf{N}^3 & (3) \\ f(x,y,z) \triangleright x+(y+z) \quad : (x,y,z) \in \mathsf{N}^3 & (4) \end{cases}$$

This MCTRS has two pairs of overlapping rules (1),(2) and (3),(4), and it is clear that the former converges. On the other hand, the latter cannot converge, even though $f(x, y, z)$ always calculates the unique sum of three natural numbers x, y, z. Induction proceeds as follows: the base case of $z = 0$ and the induction hypothesis have the diagrams below respectively, where $c_n = \mathsf{N} \times \mathsf{N} \times \{k \in \mathsf{N} \mid k \leq n\}$:

$$\begin{array}{ccc} & f(x,y,0) : c_0 & \\ \swarrow & & \searrow \\ (x+y)+0 : c_0 & & x+(y+0) : c_0 \\ \searrow & & \swarrow \\ & x+y : c_0 & \end{array} \qquad \begin{array}{ccc} & f(x,y,z) : c_n & \\ \swarrow & & \searrow \\ (x+y)+z : c_n & & x+(y+z) : c_n \\ \searrow^{*} & & {}^{*}\swarrow \\ & \exists t : c_n & \end{array}$$

Now our induction step is to show the c-convergence of the below diagram:

$$\begin{array}{ccccc} (x+y)+s(z) : c_{n+1} & \twoheadleftarrow & f(x,y,s(z)) : c_{n+1} & \twoheadrightarrow & x+(y+s(z)) : c_{n+1} \\ \downarrow & & & & \downarrow \\ s(\underline{(x+y)+z}) : c_{n+1} & & & & x+s(y+z) : c_{n+1} \\ & & & & \downarrow \\ & & & & s(\underline{x+(y+z)}) : c_{n+1} \end{array}$$

Then the underlined two terms reduce to an identical term by the hypothesis, because $(x, y, z) \in c_n$ then $(x, y, s(z)) \in c_{n+1}$. Thus the ground confluence of this MCTRS has been proven inductively.

6. Completion of MCTRSs

Based on the theorem 5.2, a completion algorithm for MCTRSs (or, MC-equalities' when regarding MC-rules as unoriented) can be designed as for unconditional TRSs[12] and other extensions of TRS[5,8]. Then, a completion algorithm for

MCTRS is proposed, and the algorithm exemplified by demonstrating a property of McCarthy's 91-function.

A set of MC-equalities E and a reduction ordering $\succ$ are given as inputs of algorithm. Additionally, selection of MC-equality $t = s : c$ from E assumed to satisfy the *fairness hypothesis*[2], i.e., every equality in E is necessarily chosen in finitely many steps.

Completion Algorithm

procedure $cmp(E, R, \succ)$

loop

if $E = \emptyset$ **then** $return(R)$

choose $t = s : c \in E$

find some finite split$^{(**)}$ $c = \bigcup c_i$

$t'_i : c_i := \twoheadrightarrow_R$normal form of $t : c_i$

$s'_i : c_i := \twoheadrightarrow_R$normal form of $s : c_i$

if $t'_i \not\succ s'_i \wedge t'_i \not\prec s'_i$ for some i **then** $return(fail)$

if $t_i \equiv s_i$ for all i **then** $E = E - \{t = s : c\}$

else

$$R^{new} := \left\{ l_i \triangleright r_i : c_i \;\middle|\; \begin{array}{l} t'_i \not\equiv s'_i \\ l_i \equiv t'_i, r_i \equiv s'_i \text{ if } t'_i \succ s'_i \\ l_i \equiv s'_i, r_i \equiv t'_i \text{ if } t'_i \prec s'_i \end{array} \right\}$$

$$R^{eq} := \left\{ \begin{array}{r} l \triangleright r : c \\ \in R \end{array} \;\middle|\; \begin{array}{l} l : c' \text{ or } r : c' \text{ is } \twoheadrightarrow_{R^{new}} \\ \text{reducible for some } c' \subset c \end{array} \right\}$$

$E^{eq} := \{l = r : c \mid l \triangleright r : c \in R^{eq}\}$

$R := R \cup R^{new} - R^{eq}$

$E := E - \{t = s : c\} \cup E^{eq} \cup CP\langle R, R^{new}\rangle$

Where $CP\langle R, R^{new}\rangle$ is the set of all the c-critical pairs generated between MC-rules in R and R^{new}. For membership conditions on the same variables $c = (x_1, \cdots, x_n) \in S_1 \times \cdots \times S_n$, $c' = (x_1, \cdots, x_n) \in S'_1 \times \cdots \times S'_n$, $c' \subset c$ if $S'_i \subset S_i$ for all i.

In this completion algorithm *cmp*, membership conditions must actually split in (**). A such practical method is to repeat finitely manys split and c-reductions. In the example 6.2 of the 91-function appearing later, only the set of natural numbers less than or equal to 100 or its subsets split. Then only the finitely many splits occur, and this *cmp* works. Even a terminating MCTRS may have infinite splits, and this completion may run forever. In such a case, while some method of finding the inductive structure of splits is necessary, it is known that there exists no such a complete method. Methods in automated theorem proving are helpful for such a purpose, but they are beyond this paper. For such methods, the reader can refer [9,14].

During completion by *cmp* rules which do not satisfy closedness may appear. But, by restricting closedness to $\forall s \in S \forall t \in T[s \rightarrow t \Rightarrow t \in S]$, *cmp* can manipulate any sets of *closed* MC-equalities.

The next theorem shows the completeness of this algorithm.

Theorem 6.1. **(Completness of *cmp*)** Let E be an arbitrary set of MC-equalities and $\succ$ a reduction ordering. If this algorithm terminates and a MCTRS R is obtained, then R is complete, i.e., terminating and confluent. Moreover $=_E \equiv \sim_R$ holds, where $=_E$ and $\sim_R$ are equivalence relations generated by $=$ of E and $\rightarrow$ of R, respectively.

Proof. $E_i, R_i, \cdots$ denote $E, R, \cdots$ in the i-th loop. First, $=_{E_i} \cup \sim_{R_i} \equiv =_{E_{i+1}} \cup \sim_{R_{i+1}}$ is shown by the induction on i. This is trivial for $i = 1$, and the case of $i+1$ is proven assuming the case of i. Clearly $=_{E_i} \cup \sim_{R_i} \subset =_{E_{i+1}} \cup \sim_{R_{i+1}}$. As $\sim_{R_i} \cup =_{\{t=s:c\}} \supset =_{CP\langle R_i, R_i^{new}\rangle}$, $=_{E_i} \cup \sim_{R_i} \supset =_{E_{i+1}} \cup \sim_{R_{i+1}}$. Then $=_E \equiv \sim_R$ if the *cmp* terminates at the n-th loop with $E_n = \emptyset, R_n = R$. Termination of the resulting R is guaranteed by $\succ$, and R is locally confluent because every critical pair of R converges. Thus R is complete. □

By this completion algorithm *cmp*, the inductionless induction for unconditional TRSs [7,17] is also possible for MCTRSs. That is, if axioms are described as a complete MCTRS R, whether or not a MC-equality $t = s : c$ is a theorem of the axioms can be determined by completing $R \cup \{t = s : c\}$. If the completion terminates with success, $t = s : c$ is a theorem under the given axioms R.

Example 6.2. McCarthy's 91-function f is defined recursively by

$$f(x) = \text{ if } x \leq 100 \text{ then } f(f(x+11)) \text{ else } x - 10$$

and has the following property:

$$f(x) = 91 \text{ for } \forall x \leq 100.$$

This definition and property of f are formulated in two MC-equalities (1),(2) and a MC-equality (3) respectively:

$$E\!: \begin{cases} f(x) = x - 10 \quad : 101 \leq x & (1) \\ f(x) = f(f(x+11)) \quad : x \leq 100 & (2) \\ f(x) = 91 \quad : x \leq 100 & (3) \end{cases}$$

For simplicity, membership conditions are expressed by inequalities or equalities whose elements satisfy. For example, $x \leq 100$ means $x \in \{n \in \mathsf{N} \mid n \leq 100\}$, $x = 91$ does $x \in \{91\}$. The successor s and the predecessor p are abbreviated by $+$ and $-$ respectively, $x - 10$ means $p^{10}(x)$ and $x + 11$ does $s^{11}(x)$. Also the rules $s(p(x)) \triangleright x : x \in \mathsf{N}$ and $p(s(x)) \triangleright x : x \in \mathsf{N}$ are assumed and used freely.

Property (3) is demonstrated by the completion algorithm *cmp* for MCTRSs, i.e., $E = \{(1),(2),(3)\}$ is shown to be complete. Notice that algorithm *cmp* can complete $\{(1),(2)\}$, so the prerequisite for inductionless induction holds. Hereafter, set of MC-rules is denoted by R, and reduction ordering $\succ$ is defined from the precedence $f \gg s, p \gg 0$ on the function symbols.

When MC-equalities (1) and (3) (or in reverse order) are chosen from E, the following MCTRS is obtained:

$$R\!: \begin{cases} f(x) \triangleright x - 10 \quad : 101 \leq x & (1') \\ f(x) \triangleright 91 \quad : x \leq 100 & (3') \end{cases}$$

Finally the MC-equality (2) is selected from E as a candidate for a new MC-rule, then the left hand side of (2) reduces as $f(x) : x \leq 100 \twoheadrightarrow 91 : x \leq 100$ and the right hand side does as below by MC-rules (1′) and (3′).

$$\begin{array}{ccccc} f(f(x+11)) : x \leq 100 & \twoheadrightarrow & f(x+1) : 90 \leq x \leq 100 & \twoheadrightarrow & x-9 : x = 100 \\ \downarrow & & \downarrow & & \downarrow \\ f(91) : x \leq 89 & & 91 : 90 \leq x \leq 99 & & 91 : x = 100 \\ \downarrow & & & & \\ 91 : x \leq 89 & & & & \end{array}$$

This shows, both sides of (2) reduce to an identical term, and $E = \emptyset$. Namely, completion using *cmp* has been successful.

Consequently, the definition of the 91-function and its property are described in a MCTRS, and the property has been proven by completing the MCTRS using *cmp*.

This f can be defined not by a MCTRS but by an unconditional TRS below:

$$R' : \begin{cases} f(f(s^{11}(0))) \triangleright f(0) \\ f(f(s^{11}(s(0)))) \triangleright f(s(0)) \\ f(f(s^{11}(s(s(0))))) \triangleright f(s(s(0))) \\ \vdots \\ f(f(s^{11}(s^{100}(0)))) \triangleright f(s^{100}(0)) \\ f(s^{101}(x)) \triangleright s^{91}(x) \end{cases}$$

This R' has many rules and its completion is complicated and obscure. Compared to R' and its completion, MCTRS is capable of much simpler completion.

7. Conclusion

This paper has clarified a sufficient condition for the confluence of terminating MCTRSs. This condition has been obtained by a new extension to MCTRS of the notion of contextual rewriting introduced by Zhang-Rémy[22]. Additionally, the notion of *split* of membership conditions has been invented. For MCTRSs whose confluence is difficult to show for their membership conditions, split has enabled to demonstrate their confluence by decomposing them into smaller confluent sub-MCTRSs. It has also been shown possible to extend finite split to infinite split with some well-founded structures. Finally a completion algorithm of MCTRS has been proposed and a famous property of McCarthy's 91-function was demonstrated utilizing the proposed completion algorithm.

MCTRSs seem to be promising to consider the relationship between unconditional TRSs and various extensions of them as conditional TRSs [1], meta rewriting systems by H. Kirchner [10] and many(order) sorted rewriting systems by J. Goguen, G. Smolka et al.[5,13]. The technique of split will also enable to handle order sorted rewritings as many sorted ones.

Acknowledgement

The author is very grateful to Dr. Yoshihito Toyama of NTT Research Laboratories for many helpful discussions and invaluable suggestions. He thanks to Dr. H. Comon for fruitful discussions. He also thanks to Dr. R. Treinen and Prof. J. -P. Jouannaud for inspiring him of using regular tree languages as sets in membership conditions and informing him of the works by Dr. Comon.

References

[1] Bergstra, J. and Klop, J. : "Conditional Rewrite Rules", J. of Comp. and Sys. Sci. 32, pp.323-362 (1986).

[2] Comon, H. : "Equational Formulas in Order Sorted Formulas", Springer LNCS 443, pp.674-688 (1990).

[3] Comon, H. : "Completion of Rewrite Systems with Membership Conditions", Springer LNCS 623, pp.392-404 (1992).Draft for full version, May 1992.

[2] Dershowitz, N. and Jouannaud, J.-P. : "Rewrite Systems", in Handbook of Theoretical Computer Science , Vol.B, N.-Holland (1990).

[5] Gnaedig, I., Kirchner, C. and Kirchner, H. : "Equational Completion in Order Sorted Algebras", TCS 72, pp.169-202 (1990).

[6] Huet, G. : "Confluent Reductions : Abstract Properties and Applications to Term Rewriting Systems", J. of ACM 27, pp.797-821 (1980).

[7] Huet, G. and Hullot, J.-M. : "Induction in Equational Theories with Constructors", J. of Comp. and Sys. Sci. 25, pp.239-266 (1982).

[8] Kaplan, S. : "Simplifying Conditional Term Rewriting Systems : Unification, Termination and Confluence", J. of Symbolic Computation 4, pp.295-334 (1987).

[9] Kapur, D., Narendran, P. and Zhang, H. : "Proof by Induction Using Test Sets", Springer LNCS 230, pp.99-117 (1985).

[10] Kirchner, H. : "Schematization of Infinite Sets of Rewrite Rules Generated by Divergent Completion Processes", TCS 67, pp.303-332 (1989).

[11] Kirchner H. and Hermann, M. : "Meta-rule Synthesis from Crossed Rewrite Systems", Springer LNCS 516, pp.143-154 (1991).

[12] Knuth, E. and Bendix, B. : "Simple Word Problems in Universal Algebras", in Computational Problems in Abstract Algebra, ed. J. Leech, pp.263-297, Pergamon Press (1970).

[13] Nutt, W., Smolka, G., Goguen, J., and Meseguer, J. : "Order Sorted Computation", Proc. the Colloq. on Resolution of Equations in Algebraic Structures (1987).

[14] Thiel, J.-J. : "Stop Losing Sleep over Incomplete Data Type Specification", Proc. 11th ACM POPL, pp.76-82 (1984).

[15] Toyama, Y. : "Term Rewriting Systems with Membership Conditions", Springer LNCS 308, pp.228-244 (1988).

[16] Toyama, Y.: "Membership Conditional Term Rewriting Systems", Trans. IEICEJ, E72, pp.1224-1229 (1989).

[17] Toyama, Y. : "How to Prove Equivalence of Term Rewriting Systems without Induction", TCS 90(1991), pp. 369-390.

[18] Yamada, J. : "Membership Conditional TRS with Infinite Sets in Conditions", IPSJ SIG Notes 90-SF-35-6 / IEICEJ Tech. Rep. COMP90-6 (1990).

[19] Yamada, J. : "On the Confluence of Terminating Membership Conditional TRS", IEICEJ Trans. IEICEJ J-74-D-I (1991), pp.666-674, in Japanese.

[20] Yamada, J. : "Membership Conditional TRS with Regular Tree Automata", ICOT PAR-WG, Feb. 1992.

[21] Yamada, J. and Toyama, Y. : "Confluence of Membership Conditional TRS", IEICEJ Tech. Rep. COMP89-8 / IPSJ SIG Notes 89-SF-28-8 (1989).

[22] Zhang, H. and Rémy, J.-L. : "Contextual Rewriting", Springer LNCS 202, pp.46-62 (1985).

Appendix.

In this appendix, some detailed investigation on the sets in membership conditions is executed.

First, c-unification is taken into consideration in a very abstract fashion. Let $\theta = \{\cdots, x_j/s_j, \cdots\}$ be a unifier of t_1 and t_2 of two c-terms $t_1 : c_1 \equiv t_1 : (x_1, \cdots, x_m) \;\varepsilon\; S_1 \times \cdots \times S_m$ and $t_2 : c_2 \equiv t_2 : (x_{m+1}, \cdots, x_{m+n}) \;\varepsilon\; S_{m+1} \times \cdots \times S_{m+n}$ with no common variables. Next compose sets $S'_1, \cdots, S'_{m+n}$ such that

$$s_j \in S_j \text{ for } 1 \leq \forall j \leq m+n \Leftrightarrow (x_1, \cdots, x_{m+n}) \in S'_1 \times \cdots \times S'_{m+n}. \qquad (*)$$

That is, for each substitution $x_j/s_j(x_1,\cdots,x_{m+n})$ the sets S'_i satisfying $(*)$ under $x_1 \in S_1,\cdots,x_{m+n} \in S_{m+n}$ must be determined.

Generally, such S'_i can only be determined in limited cases, but the sets S'_i in $(*)$ can be easily determined in the following cases.

1° If the left hand side of $(*)$ is of the form of $s_j(\cdots,x_i,\cdots) \in T(F,V)$, then $S'_i = S_i$.

2° If the left hand side of $(*)$ is in the form of $s_j(\cdots,x_i,\cdots) \in T(F)$ and $S_i \subset T(F)$ for every i in the right hand side, then $S'_i = S_i$.

3° If $(*)$ is $s(x) \in S \Leftrightarrow x \in S'$ and either S or S' is finite, S' can be determined by checking for all the elements in the finite set. (This is the case of the 91-function in example 6.2.)

In these cases above, the c-most general unifier can be easily established, and our examples are within the above cases.

Next, the relationship between the primitive operations in MCTRSs and the set operations is observed. Notice that primitives in MCTRSs are matching, replacement, unification between c-terms and split is newly added to them.

The followings are set operations which must be equipped for realization of primitives in MCTRSs. The sets in membership conditions have to be decidable and closed for the them.

(1) Decision of emptiness.
(2) Intersection($\cap$) and set difference($-$).
(3) From the membership condition on a term, construct the membership conditions on the variables in it. Namely, find $S'_1,\cdots,S'_n$ such that $t \equiv f(x_1,\cdots,x_n) \in S \Leftrightarrow (x_1,\cdots x_n) \in S'_1 \times \cdots \times S'_n$, i.e., to solve $f^{-1}(S)$ for an n-ary function symbol f.
(4) Concatenation, i.e., construct the set $\{f(s_1,\cdots,s_n) \mid s_1 \in S_1,\cdots,s_n \in S_n\}$ from given an n-ary function symbol f and sets $S_1,\cdots,S_n$.

First of all, as membership conditions make sense only when every set in them are not empty, (1) is indispensable. C-reduction is based on c-matching, and c-matching necessitates concatenation (4) and decision of inclusion between sets. Noticing that inclusion can be deduced to intersection, set-difference and (1), we obtain (2) and (4). Split consists of intersection and concatenation with check of non-emptiness, then (2) and (4) are required.

Unification is decomposed into the primitives as follows. Solve $s_j^{-1}(S_j)$ and have the condition $x_i \in S_i^{(j)}$ on variable for every j by (3). Next take $S_i \cap \bigcap_j S_i^{(j)}$ for each i and obtain the contexts for c-unifier by (1), and emptiness must be certified in every primitive operation. Thus, (2) and (3) must exist.

It is well-known that the class of regular tree languages or equivalently finite bottom-up tree automata(FBTA) is the smallest class closed under above primitive operations. Then they can be models of sets in membership conditions, and [20] showed how the primitives of MCTRSs are realized by manipulating labels and transition rules of FBTA, and also illustrated the completion of 91-function looking on sets of natural numbers as FBTA. Split was also constructed by directling handling FBTA for its simplicity and importance in our *cmp* in it, so split may be added to the above primitives.

H. Kirchner and M. Herman[10,11] tried to regard membership conditions as many sorted algebra and H. Comon[2,3] fully exercised the MCTRSs whose sets are modeled by FBTA. The class of reglar tree laguages is still wider than required and more smaller class for the model of sets in CTRSs is desired.

Completeness and Confluence of Order-Sorted Term Rewriting

Lars With

Technische Universität Berlin, Germany, Franklinstr. 28/29, FR 6–2, 1000 Berlin 10.
E-mail: lars@cs.tu-berlin.de

Abstract. We show various results for order-sorted term rewriting.
We extend the result that order-sorted rewriting is complete if the rewrite system is compatible and confluent [17] by proving that order-sorted rewriting is complete if we consider the semantical sorts of the terms, i.e., we have confidence in the rewrite steps because they replace equals by equals.
In the order-sorted case the critical pair theorem [13] holds only if the rewrite system is weakly sort decreasing [16]. We show that every rewrite system can be made weakly sort decreasing by extending the expressibility of the signature by adding term declarations.
Unification with term declarations is undecidable in general [16]. We give a condition for the decidability of unification with term declarations that is considerably more general than the conditions given in [16]. We furthermore solve two open problems stated in [16]: unification in linear signatures is decidable and thus regularity of linear signatures, too.

1 Introduction

Order-sorted term rewriting is used as operational semantics for order-sorted specifications, i.e., order-sorted equational logic [6, 5, 17, 8, 12, 20, 21]. Order-sorted equational logic is the continuation of the extension of unsorted to many-sorted equational logic.

In the unsorted case there is no influence of the semantics of the equations on the sort structure, because there is none. In the many-sorted case the semantics does not influence the sort structure either, if equations between terms of different sorts are not allowed, which is the usual assumption [7, 15]. In the order-sorted case things are more complicated. An ordered sort structure makes no sense if we do not allow equations between terms of two sorts where one sort is a subsort of the other. With such equations, for example $a \equiv b$, where a is of sort A, b is of sort B, and B is a subsort of A, it is possible that the value of a term, here a, is of a certain sort, here B, in every model of the equation, although the term syntactically is not of this sort (a is syntactically not of sort B). Such an equation may also logically imply an equation that is not syntactically well formed. Assuming a unary function f that maps B to B, denoted by the function declaration $f(B)\colon B$, the above equation $a \equiv b$ implies $f(a) \equiv f(b)$ but $f(a)$ is not well formed.

A more natural example is the equation $\mathsf{square}(x_{\mathsf{int}}) \equiv x_{\mathsf{int}} * x_{\mathsf{int}}$ where the square-function maps int to nat, $*$ maps int $\times$ int to int, and nat is a subsort of int. The core of the problem in this example is that $x_{\mathsf{int}} * x_{\mathsf{int}}$ is of sort nat in every model of a proper specification of $*$, but it is not possible to express this fact by a function declaration. Thus the semantical sort structure may differ from the syntactical one.

This difference has also significant consequences on order-sorted term rewriting, since the semantics of a rewrite rule $l \rightarrow r$ is the semantics of the equation $l \equiv r$. It is possible that the left hand side of an order-sorted rewrite rule matches with a subterm of a term but the application of the rule yields a term that is not well-formed. The application of the rewrite rule $b \rightarrow a$ to $f(b)$ in the example above would yield the ill formed term $f(a)$. Rewrite rules are called *compatible* if they never yield an ill formed term. The common approach is not to allow incompatible rewrite steps [17, 16, 12].

One fundamental theorem in the unsorted case is that term rewriting is complete [10]. This does not hold in the order-sorted case: Consider the extension of the example above with a constant symbol b' of sort B and the rewrite system $R = \{b \rightarrow a, a \rightarrow b'\}$. Then $f(b) = f(b')$ in every model of $E_R = \{b \equiv a, a \equiv b'\}$ but not $f(b) \stackrel{*}{\leftrightarrow} f(b')$ because $f(a)$ is not a syntactically well formed term.

Smolka et al. give a completeness theorem [17] essentially saying that order-sorted rewriting is complete if the rewrite system is compatible and confluent. However, the confluence of a rewrite system does not say anything about its power. Thus this restriction does not seem adequate. The key is that compatibility is not strong enough to guarantee that every rewrite step that is necessary for completeness is possible.

In section 3 we give a completeness theorem saying that order-sorted rewriting is complete without any restriction if we have confidence to the rewrite steps because they replace equals by equals.

Another fundamental theorem in the unsorted case is the critical pair theorem [9]. This theorem says that the rewrite relation is locally confluent if and only if all critical pairs converge. In the unsorted case the "critical" pairs of "overlaps" where the subterm of one left hand side is a variable always converge: For example $f(b)$ can be rewritten with the rewrite rule $b \rightarrow a$ to $f(a)$ and with $f(x) \rightarrow x$ to b. Because $f(x) \rightarrow x$ is still applicable to $f(a)$ we can rewrite $f(a)$ to a. The application of $b \rightarrow a$ to b also yields a. Thus the pair $f(a)$ and b converges and by definition does not form a critical pair. But consider the following order-sorted variant of this example: Let a be of sort A, b and x of sort B and f a mapping from A to A. Then $f(b)$ can again be rewritten to $f(a)$ and b. But the rewrite rule $f(x) \rightarrow x$ is not applicable to $f(a)$ since x is of sort B and can not be substituted by a. The key to this problem is that $b \rightarrow a$ is not sort decreasing.

Thus in [17] a critical pair theorem is given that is restricted to sort decreasing rewrite systems and in [16] this theorem is strengthened to weakly sort decreasing systems. *Weakly sort decreasing* means that if t can be rewritten in one step to t' then t' either is of smaller sort than t or can be rewritten in one or more steps to a term of smaller sort than t. This is a considerable restriction because there is a large class of rewrite systems that are not weakly sort decreasing. This class contains practical examples like $\text{square}(x_{\text{int}}) \rightarrow x_{\text{int}} * x_{\text{int}}$ (s. above). Furthermore a Knuth-Bendix completion procedure [13] has to orient equations with respect to weak sort decreasingness [4]. This is an important problem since orientation w.r.t. to the termination order might not be weakly sort decreasing or an equation might not be weakly sort decreasing in both directions.

We tackle this problem by extending the expressibility of the signature. We add so-called term declarations [16] like $x_{\mathrm{int}} * x_{\mathrm{int}}$: nat. In section 4 we give a theorem saying that every rewrite system can be made weakly sort decreasing by adding term declarations in a suitable way.

Unfortunately unification with term declarations is undecidable [16] in general. In section 5 we give a theorem saying that unification with term declarations is decidable if there are no unsolved critical overlaps between term declarations. Such an overlap is solved if the resulting term declaration is part of the signature. We further give a theorem saying that every linear term declaration can be replaced by function declarations with auxiliary sorts. This solves two open problems stated in [16]: Unification in linear signatures is decidable and thus regularity of linear signatures is decidable, too.

We at most give sketches of the proofs in this paper. The full proofs can be found in [22, 23].

2 Preliminaries

2.1 Syntax

An *order-sorted signature* Σ is a quadruple $(\mathcal{S}, \sqsubseteq, \mathcal{F}, \mathcal{V})$:

- $\mathcal{S}$ is a finite set of sort symbols,
- $\sqsubseteq$ is a set of *subsort declarations* $s \sqsubseteq s'$ with $s, s' \in \mathcal{S}$, such that the reflexive and transitive closure of $\sqsubseteq$ (denoted by $\sqsubseteq$ as well) is antisymmetric,
- $\mathcal{F}$ is a set of *function declarations* $f(s_1, \ldots, s_n): s$ with $s_1, \ldots, s_n, s \in \mathrm{S}_\Sigma$ (if $n = 0$ the brackets are omitted), where $\mathrm{S}_\Sigma := \mathcal{S}$[1],
- $\mathcal{V}$ is a *set of variables* with $V \cap \mathcal{F} = \emptyset$ that can be decomposed into pairwise disjoint infinite but enumerable sets $\mathcal{V}_s$ with $s \in \mathrm{S}_\Sigma$.

We denote the set of all function symbols of Σ by F_Σ. We denote the set of all $s' \in \mathcal{S}$ with $s \sqsubseteq s'$ by $[s, \infty]$, the set of all *lower bounds* of $S \subseteq \mathcal{S}$ relative to $\sqsubseteq$ by $\mathrm{Lbs}(S)$. We extend the partial order on the sort symbols to sets of sort symbols: $S \sqsubseteq S'$ if and only if $S \supseteq S'$. We denote a variable $x \in \mathcal{V}_s$ by x_s. In the following we assume an order-sorted signature $\Sigma = (\mathcal{S}, \sqsubseteq, \mathcal{F}, \mathcal{V})$.

The set $\mathcal{T}_s^\Sigma$ of all well sorted Σ*-terms of sort* $s \in \mathrm{S}_\Sigma$ over Σ is inductively defined in the following way:

1. if $s' \sqsubseteq s$ then $x_{s'} \in \mathcal{T}_s^\Sigma$,
2. if $f(s_1, \ldots, s_n): s' \in \mathcal{F}$, $t_i \in \mathcal{T}_{s_i}^\Sigma$ and $s' \sqsubseteq s$ then $f(t_1, \ldots, t_n) \in \mathcal{T}_s^\Sigma$

and the set of all well sorted Σ*-terms over* Σ is $\mathcal{T}^\Sigma := \bigcup_{s \in \mathrm{S}_\Sigma} \mathcal{T}_s^\Sigma$. If we do not care about the arity and the argument sorts of a function symbol we get the following set of F_Σ-terms $\mathcal{T}^{\mathrm{F}_\Sigma}$:

1. $x_s \in \mathcal{T}^{\mathrm{F}_\Sigma}$

[1] This indirection is necessary because we later use S_Σ as a parameter for the syntactic definitions, i.e. change the definition of S_Σ.

2. if $f \in F_\Sigma$ and $t_i \in \mathcal{T}^{F_\Sigma}$ then $f(t_1, \ldots, t_n) \in \mathcal{T}^{F_\Sigma}$

If Σ and thus F_Σ is clear from the context we drop the prefix F_Σ. We denote the set of all *sorts of a Σ-term t* by $\mathrm{Sorts}_\Sigma(t) := \{s \in S_\Sigma \mid t \in \mathcal{T}_s^\Sigma\}$.

An order-sorted signature Σ is *regular* if every Σ-term t has a least sort, i.e. there is an $s \in \mathrm{Sorts}_\Sigma(t)$ such that $s \sqsubseteq s'$ for every $s' \in \mathrm{Sorts}_\Sigma(t)$. We abbreviate $\mathrm{Sorts}_\Sigma(t) \sqsubseteq \mathrm{Sorts}_\Sigma(t')$ by $t \sqsubseteq t'$.

A well sorted *Σ-substitution* σ is a mapping from $\mathcal{V}$ to $\mathcal{T}^\Sigma$ satisfying the following conditions:

1. $\sigma(x) \sqsubseteq x$ for all $x \in \mathcal{V}$,
2. $\sigma(x) = x$ for almost all $x \in \mathcal{V}$.

A Σ-substitution is uniquely extended to an endomorphism from $\mathcal{T}^{F_\Sigma}$ to $\mathcal{T}^{F_\Sigma}$. The application of the extension of a substitution σ to a term t is denoted by $t\sigma$. We represent a substitution σ by a set of replacements $\{x_1 \leftarrow t_1, \ldots, x_n \leftarrow t_n\}$ with $\sigma(x_i) = t_i$ for $i = 1, \ldots, n$ and $\sigma(x) = x$ for all $x \notin \{x_1, \ldots, x_n\}$. The *restriction* $\sigma/_V$ of a Σ-substitution σ to $V \subseteq \mathcal{V}$ is defined by: $\sigma/_V(x) := \sigma(x)$ for all $x \in V$, and $\sigma/_V(x) := x$ for all $x \notin V$. The functional *composition* of two Σ-substitutions σ and θ is denoted by $\sigma \circ \theta$ (first θ, then σ). A Σ-substitution σ *subsumes* a Σ-substitution θ relative to a set of variables $V \subseteq \mathcal{V}$ ($\sigma \preceq_V \theta$) if there exists a Σ-substitution δ with $(\delta \circ \sigma)/_V = \theta/_V$.

A well sorted *Σ-equation* (*F_Σ-equation*) is an ordered pair of Σ-terms (F_Σ-terms) $t \equiv t'$. We denote the set of all variables occurring in a term t by $\mathrm{Vars}(t)$.

A Σ-substitution σ is called a *Σ-unifier* of a set of F_Σ-equations E if $t\sigma = t'\sigma$ for all $t \equiv t' \in E$. We denote the set of all Σ-unifiers of a set of F_Σ-equations E by $\mathrm{Unif}(E)$. A Σ-unifier σ of a set of F_Σ-equations E is a *most general Σ-unifier (Σ-mgu)* of E relative to a finite set of variables $V \subseteq \mathcal{V}$ with $V \supseteq \mathrm{Vars}(E)$, if $\sigma \preceq_V \theta$ for every Σ-unifier θ of E. A set of Σ-unifiers U of a set of F_Σ-equations E is *minimal* relative to a finite set of variables $V \subseteq \mathcal{V}$ with $V \supseteq \mathrm{Vars}(E)$ if for all $\sigma, \theta \in U$ the subsumption $\sigma \preceq_V \theta$ implies $\sigma = \theta$; U is *complete* relative to V if for every Σ-unifier θ of E there exists $\sigma \in U$ with $\sigma \preceq_V \theta$. We abbreviate *minimal complete set of Σ-unifiers* by *Σ-mcsu*. We say that Σ-unification is

- *unitary*, if for every set of F_Σ-equations E with $\mathrm{Unif}(E) \neq \emptyset$ there is an Σ-mgu of E relative to $\mathrm{Vars}(E)$
- *finitary*, if for every set of F_Σ-equations E with $\mathrm{Unif}(E) \neq \emptyset$ there is a finite Σ-mcsu of E relative to $\mathrm{Vars}(E)$
- *infinitary*, if it is not finitary.

A well sorted *Σ-rewrite rule* is an ordered pair of Σ-terms $l \rightarrow r$ with $l \notin \mathcal{V}$ and $\mathrm{Vars}(r) \subseteq \mathrm{Vars}(l)$. The *occurences* of a term t ($\mathrm{Occ}(t)$) are defined as usual as sequences of positive integers. We denote the *subterm of a Σ-term t at the occurence u* with t/u and use $t[u \leftarrow t']$ to denote the term obtained from t by replacing the subterm t/u at occurence u by t'. A finite set of Σ-rewrite rules R (*Σ-rewrite system*) defines the following *Σ-rewrite relation*:[2]

$$t \rightarrow_{\Sigma,R} t[u \leftarrow r\sigma]$$

[2] We denote this rewrite relation by $\rightarrow_{\Sigma,R}$ since we later define a new rewrite relation, which is more powerful.

if t is a Σ-term and there is a rewrite rule $l \rightarrow r \in R$ and an occurence $u \in \mathrm{Occ}(t)$ with $t/u = l\sigma$ for a Σ-substitution σ and $t[u \leftarrow r\sigma]$ is a Σ-term. We call this process Σ-rewriting. A Σ-rewrite system R is called *Σ-compatible* if the following holds: let $t' = t[u \leftarrow r\sigma]$ for a rewrite rule $l \rightarrow r$ and an occurence $u \in \mathrm{Occ}(t)$ with $t/u = l\sigma$ for a Σ-substitution σ, then t' is a Σ-term if t is a Σ-term. We denote the set of Σ-equations given by a Σ-rewrite system R by $E_R := \{t \equiv t' \mid t \rightarrow t' \in R\}$.

Since we later define other rewrite relations, we give the following definitions for binary relations in general. We denote the reflexive and transitive closure of a binary relation $\rightarrow$ by $\stackrel{*}{\rightarrow}$ and the reflexive, symmetric and transitiv closure by $\stackrel{*}{\leftrightarrow}$. Two elements x and y *converge* $(x \downarrow y)$ if there is a z with $x \stackrel{*}{\rightarrow} z$ and $y \stackrel{*}{\rightarrow} z$. We call a relation $\rightarrow$ *locally confluent* if $x \rightarrow y$ and $x \rightarrow z$ implies $x \downarrow z$, *confluent* if $x \stackrel{*}{\rightarrow} y$ and $x \stackrel{*}{\rightarrow} z$ implies $x \downarrow z$.

2.2 Semantics

To avoid semantical problems we assume in the following that every signature is fully inhabited i.e. that for every $s \in S_\Sigma$ there is a Σ-term t of sort s with $\mathrm{Vars}(t) = \emptyset$. We use non-overloaded semantics here as in [18], where a comparison between overloaded and non-overloaded semantics is given. We assume the value of a function to be undefined, if the function is applied to a tuple of arguments that is not in the domain of the function. Furthermore we assume all functions to be strict.

With these assumptions the validity of a F_Σ-equation can be defined in the same way as the validity of a Σ-equation. If $t \equiv t'$ is valid in a Σ-algebra $\mathcal{A}$ we also say that $\mathcal{A}$ *satisfies* $t \equiv t'$ $(\mathcal{A} \models t \equiv t')$ or that $\mathcal{A}$ is a *Σ-model* of $t \equiv t'$. We extend this to sets of equations as usual. A set of F_Σ-equations E *implies* a F_Σ-equation $t \equiv t'$ $(E \models t \equiv t')$, iff every Σ-model of E is also a Σ-model of $t \equiv t'$. We abbreviate $E \models t \equiv t'$ by $t =_E t'$.

3 Completeness

One fundamental theorem in the unsorted case is that rewriting is complete [10]. The rewrite relation $\rightarrow_{\Sigma,R}$ is not complete since $\stackrel{*}{\leftrightarrow}_{\Sigma,R}$ is not a congruence relation on $\mathcal{T}^{\mathrm{F}_\Sigma}$. The solution is to define a new rewrite relation that allows every possible rewrite step: A Σ-rewrite system R defines the following *R-rewrite relation*:

$$t \rightarrow_{\Sigma,R} t[u \leftarrow r\sigma]$$

if t is a F_Σ-term and there is a rewrite rule $l \rightarrow r \in R$ and an occurence $u \in \mathrm{Occ}(t)$ with $t/u = l\sigma$ for a Σ-substitution σ. We call this process R-rewriting. This rewrite relation is complete without any restriction, since $\stackrel{*}{\leftrightarrow}_{\Sigma,R}$ is a congruence relation on $\mathcal{T}^{\mathrm{F}_\Sigma}$.

Theorem 1 *Let R be a rewrite system. Then $t =_{E_R} t'$ if and only if $t \stackrel{*}{\leftrightarrow}_{\Sigma,R} t'$.*

Hence with this rewrite relation the important completeness result generalizes nicely from unsorted to order-sorted term rewriting. This is a strong indication that this rewrite relation is the right choice.

4 Confluence

Another fundamental theorem in the unsorted case is the critical pair theorem [9]. This theorem says that the rewrite relation is locally confluent if and only if all critical pairs converge. In the unsorted case the critical pairs of a rewrite system are finitely representable because most general unifiers exist for every unifiable set of equations. In the order-sorted case unification in non regular signatures is not finitary. But if a term is of two or more uncomparable sorts then in every algebra these sorts have a non empty intersection: at least the value of this particular term. Thus we can make every signature regular signature adding function declarations on intersection sorts.

Therefore we have to change our definitions of syntax and semantics slightly: We denote the set $\bigcup_{s\in S}[s,\infty]$ by $[S,\infty]$. We redefine[3]

$$\mathrm{S}_\Sigma := \{[S,\infty] \mid S \subseteq \mathcal{S}, \mathrm{Lbs}(S) \neq \emptyset\}.$$

We represent $[S,\infty]$ by the smallest set S' with $[S',\infty] = [S,\infty]$. Furthermore we abbreviate sets of sort symbols $\{s\}$ by s and $S_1 \cup \cdots \cup S_n$ by $\{S_1,\ldots,S_n\}$.[4] The interpretation of a set of sort symbols S is the intersection of the interpretations of the sort symbols in S. If we consider a lower case s possibly with index as a set of sort symbols, all other definitions and denotations remain unchanged.

Let Σ be a non regular signature. The ***regulated signature*** is constructed from Σ in the following way: For every subset

$$F = \{f(s_{11},\ldots,s_{1n})\colon s_1,\ldots,f(s_{m1},\ldots,s_{mn})\colon s_m\}$$

of $\mathcal{F}$ with $\mathrm{Lbs}(\{s_{1i},\ldots,s_{mi}\}) \neq \emptyset$ for $i = 1,\ldots,n$ we add a new function declaration

$$f(\{s_{11},\ldots,s_{m1}\},\ldots,\{s_{1n},\ldots,s_{mn}\})\colon \{s_1,\ldots,s_n\}$$

to $\mathcal{F}$, and to ensure that $\{s_1,\ldots,s_n\} \in \mathrm{S}_\Sigma$ a new sort symbol $\langle\{s_1,\ldots,s_n\}\rangle$ to $\mathcal{S}$, the subsort declarations

$$\langle\{s_1,\ldots,s_n\}\rangle \sqsubseteq s_i$$

for $i = 1,\ldots,n$ to $\sqsubseteq$ and new sets of variables $\mathcal{V}_{\langle\{s_1,\ldots,s_n\}\rangle}$ and $\mathcal{V}_{\{s_1,\ldots,s_n\}}$ to $\mathcal{V}$. The obtained signature is "equivalent" to Σ in a very strong sense, because we only add sorts and subsort relations to the signature that do exist semantically anyway. We use the notion of a conservative extension similar to [3].

Theorem 2 *Let Σ be a non regular signature and E be a set of F_Σ-equations. Then there is a regular signature Σ' such that (Σ', E) is a conservative extension of (Σ, E).*

---

[3] For notational reasons we prefer not to explicitly add sets like $\{A,B\}$ to $\mathcal{S}$.

[4] This is quite a long chain of abbreviations and denotations. So let's have a look at a short example: Let $\mathcal{S} = \{A,B,C\}$ and $\sqsubseteq = \{C \sqsubseteq A, C \sqsubseteq B\}$. Then C is an abbreviation for $\{C\}$, $\{C\}$ is an abbreviation for $[\{C\},\infty]$ and also for $[\{B,C\},\infty]$. Furthermore $[\{B,C\},\infty]$ denotes $[B,\infty] \cup [C,\infty] = \{A,B,C\}$. Hence altogether C denotes $\{A,B,C\}$.

Another difference to the unsorted case is that in the order-sorted case the "critical" pairs of "overlaps" where the subterm of one left hand side is a variable, do not always converge. This problem comes from rewrite rules that are not sort decreasing. Thus in [16] a critical pair theorem is given that is restricted to weakly sort decreasing rewrite systems. A binary relation $\rightarrow$ on $\mathcal{T}^{\mathrm{F}_\Sigma}$ is called *weakly Σ-sort decreasing* if $t \rightarrow t'$ implies that there is a Σ-term t'' with $t' \xrightarrow{*} t''$ and $t'' \sqsubseteq t$.

4.1 Rewriting Based on E-Semantical Sorts

Again, the key to the problem is that the semantical sort structure differs from the syntactical one. We define rewriting based on semantically well sorted substitutions:

We denote the *E-semantical sorts* of a term t relative to a set of Σ-equations E by

$$\mathrm{Sorts}_E(t) := \bigcup_{t' =_E t} \mathrm{Sorts}_\Sigma(t')$$

and abbreviate $\mathrm{Sorts}_E(t) \sqsubseteq \mathrm{Sorts}_E(t')$ by $t \sqsubseteq_E t'$. A *(Σ, E)-substitution* σ is a mapping from $\mathcal{V}$ to $\mathcal{T}^{\Sigma,E}$ satisfying the following conditions:

1. $\sigma(x) \sqsubseteq_E x$ for all $x \in \mathcal{V}$,
2. $\sigma(x) = x$ for almost all $x \in \mathcal{V}$.

A Σ-rewrite system R defines the following *E_R-rewrite relation*:

$$t \Rightarrow_{\Sigma,R} t[u \leftarrow r\sigma]$$

if there is a rewrite rule $l \rightarrow r \in R$ and an occurence $u \in \mathrm{Occ}(t)$ with $t/u = l\sigma$ for a (Σ, E_R)-substitution σ.

With this rewrite relation we obtain a theorem that is similar to the unsorted case.

Theorem 3 *Let R be a rewrite system. Then $\Rightarrow_{\Sigma,R}$ is locally confluent if and only if all critical pairs obtained by (Σ, E)-substitutions of R converge.*

The proof is similar to the proof of the critical pair theorem in the unsorted case. This theorem is mainly of theoretical interest because in general it is undecidable whether a substitution is semantically well sorted. Nevertheless this theorem demonstrates that order-sorted rewriting based on syntactical sorts is too a weak calculus.

4.2 Sort Decreasingness by Construction

We approximate the semantical sort structure in such a way that a rewrite system becomes weakly sort decreasing. We do this by extending the expressibility of order-sorted signatures. We add so called term declarations [16] like $x_{\mathrm{int}} * x_{\mathrm{int}} : \mathrm{nat}$.

An *order-sorted signature Σ with term declarations* is a quintuple $(\mathcal{S}, \sqsubseteq, \mathcal{F}, \mathcal{V}, \mathcal{D})$, where $(\mathcal{S}, \sqsubseteq, \mathcal{F}, \mathcal{V})$ is an order-sorted signature and $\mathcal{D}$ is a set of term declarations $t : s$ with $t \in \mathcal{T}^{(\Sigma, \sqsubseteq, \mathcal{F}, \mathcal{V})}$ and $s \in \mathrm{S}_\Sigma$. For technical reasons in the following we assume

$$\mathcal{D} \supseteq \{ f(x_{1_{s_1}}, \ldots, x_{n_{s_n}}) : s \mid f(s_1, \ldots, s_n) : s \in \mathcal{F} \}.$$

A term declaration $t : s \in \mathcal{D}$ is called *redundant* if $s \in \mathrm{S}_{(\mathcal{S}, \sqsubseteq, \mathcal{F}, \mathcal{V}, \mathcal{D} \setminus \{t:s\})}(t)$. The set of all well sorted *Σ-terms of sort $s \in \mathrm{S}_\Sigma$* is inductively defined in the following way:

1. if $s' \sqsubseteq s$ then $x'_s \in \mathcal{T}_s^\Sigma$
2. if $t{:}\,s' \in \mathcal{D}$ and $s' \sqsubseteq s$ then $t \in \mathcal{T}_s^\Sigma$
3. if $t \in \mathcal{T}_s^\Sigma$, $t' \in \mathcal{T}_{s'}^\Sigma$ and $s' \sqsubseteq s''$, then $t\{x_{s''} \leftarrow t'\} \in \mathcal{T}_s^\Sigma$, where $t\{x_{s''} \leftarrow t'\}$ is constructed from t by simultaneously replacing[5] every occurence of $x_{s''}$ in t by t'.

All other definitions and notations are the same as for order-sorted signatures without term declarations.

We want to add term declarations to the signature such that a rewrite system becomes weakly sort decreasing. Hence at first we need sufficient conditions for a rewrite system to be weakly sort decreasing. Therefore it is useful to know exactly what determines the sort of a term. In [16] a proposition is given saying that a term t is of sort s if and only if $t'{:}\,s \in \mathcal{D}$ with $t'\sigma = t$ for a Σ-substitution σ. Thus it is critical for sort decreasingness if an instance of a term declaration can be rewritten:

Let $t{:}\,s$ be a term declaration, $l \rightarrow r \in R$, $u \in \mathrm{Occ}(t)$ with $t/u \notin \mathcal{V}$ and σ a Σ-substitution such that $(t/u)\sigma = l\sigma$. Then $(t\sigma, (t[u \leftarrow r])\sigma)$ is called *Σ-critical sort relation* obtained by Σ-superposition of $t{:}\,s$ to $l \rightarrow r$ at u with σ.

A Σ-critical sort relation (t, t') is *Σ-satisfied*, if $t' \sqsubseteq t$, and *weakly Σ-satisfied*, if there is a Σ-term t'' with $t' \overset{*}{\rightarrow}_{\Sigma,R} t''$ and $t'' \sqsubseteq t$.

We give a criterion for weak Σ-sort decreasingness that is considerably more general than the conditions in [16]. In [16] also critical sort relations obtained by superposition at variable positions are considered, for every critical sort relation (t, s) the term t has to be of a sort smaller than s, and the condition is only sufficient in combination with local confluence and termination.

Theorem 4 *Let Σ be a regular signature and R be a Σ-rewrite system. Then $\rightarrow_{\Sigma,R}$ is weakly Σ-sort decreasing if and only if all Σ-critical sort relations of R are weakly Σ-satisfied.*

The restriction to regular signatures is not crucial since every non regular signature with term declarations can be transformed into an equivalent regular one. This can be done because of considerations similar to those for signatures without term declarations. Thus theorem 2 also holds for signatures with term declarations.[6]

Because of theorem 4 for every rewrite system we can add term declarations to the signature, such that the rewrite system becomes weakly sort decreasing. We simply have to add the term declaration $t{:}\,s$ if (t, s) is a critical sort relation that is not weakly satisified. The obtained signature is equivalent to the former one in a very strong sense, since it only approximates the semantical sorts.

Theorem 5 *Let R be a rewrite system for an order-sorted signature Σ. Then there is a signature Σ', such that (Σ', E_R) is a conservative extension of (Σ, E_R) and R is weakly Σ'-sort decreasing.*

---

[5] Think of $\{x''_s \leftarrow t'\}$ as an unsorted substitution. The problem is to define well sorted terms without well sorted substitutions.

[6] However, we need unification with term declarations to make a signature with term declarations regular. For the discussion and solutions to this problem we refer the reader to section 5.

5 Unification in signatures with term declarations

Unfortunately unification with term declarations is undecidable in general [16]. The most general condition given in [16] that ensures that unification with term declarations is decidable and finitary is that signatures have to be regular and *almost elementary*, i.e., in every term declaration $f(t_1, \ldots, t_n)$ the terms t_i are either ground or variables and no variable occurs twice. We give conditions that are considerably more general.

In [16] unification rules in the style of [14] are given for regular signatures with term declarations. Since theorem 2 also holds for signatures with term declarations this is not a crucial restriction. The following unification rule is the critical one for termination of the unification rules:

- *term declaration*

$$\frac{E \uplus \{x \equiv f(p_1, \ldots, p_n)\}}{E \cup \{x \equiv f(p_1, \ldots, p_n), p_1 \equiv q_1, \ldots, p_n \equiv q_n\}}$$

if $x \notin \text{Vars}(f(p_1, \ldots, p_n))$ and $f(q_1, \ldots, q_n): s \in \mathcal{D}$ with $s \sqsubseteq \text{Sorts}_\Sigma(x)$

This rule is crucial for termination because it introduces new terms (q_i). The following example demonstrates in which cases the term declaration rule really is critical for termination:

Suppose we want to unify the equation

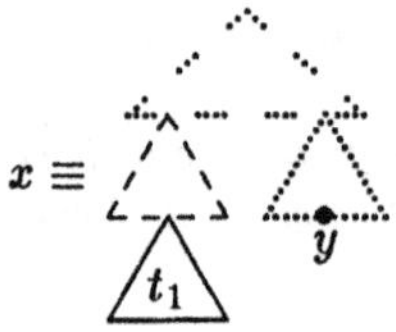

(1)

and we have apply the term declaration rule with the following term declaration

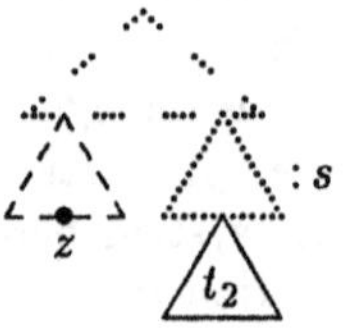

(2)

with $s \sqsubseteq \text{Sorts}_\Sigma(x)$, because otherwise we could not substitute the right hand side for x. Then we get two new equations

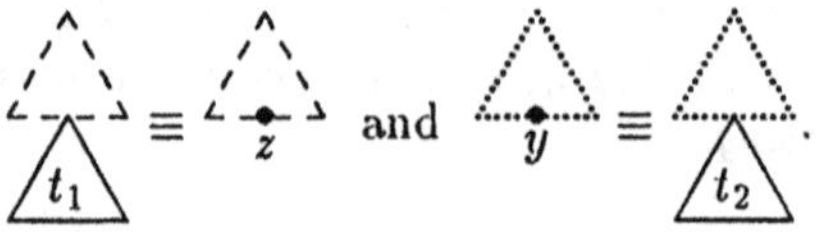

(3)

The first of the two new equations can be reduced to

This equation does not cause any problem for termination, since t_1 is a subterm of the initial equation (1).

The second equation of (3) can be reduced to

$$y \equiv \triangle_{t_2} .$$

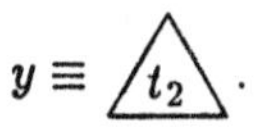

This equation may cause problems for termination, because t_2 is not a subterm of the initial equation (1). But if there occur only finitely many steps of this kind during unification, this also would not be a problem.

Thus we only have problems with termination of the unification procedure if it is possible to chain this kind of steps. Suppose we have to apply the term declaration rule with the term declaration $t_2': s'$. Then we get the equation $t_2 \equiv t_2'$. If this equation is unifiable we have an overlap between the term declaration (2) and $t_2': s'$.

Let $t_1: s_1$ and $t_2: s_2$ be two variable disjoint variants of term declarations, $u \in$ $Occ(t_1)$ with $t_1/u \notin V$ and σ a Σ-substitution such that $(t_1/u)\sigma = t_2\sigma$. Then $t_1\sigma$ is called Σ-*critical overlap between term declarations* obtained by Σ-*superposition* of $t_1: s_1$ to $t_2: s_2$ at u with σ.

Now we modify the term declaration rule: We mark the subterms of the involved term declaration and add the condition that f must not be marked. Then we can prove termination of the unification procedure by a termination function that does not count marked function symbols.[7]

Up to this point we can give a theorem saying that unfication with term declarations is decidable if there is no critical overlap between term declarations. However, the following example will show that critical overlaps between term declarations do not always destroy termination of the unification procedure:

Suppose we have the following two term declarations that do overlap

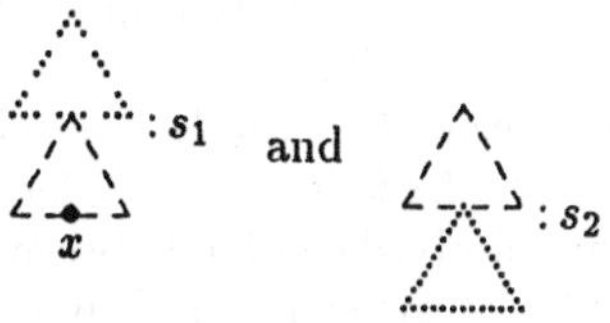

$$\triangle : s_1 \quad \text{and} \quad \triangle : s_2 \tag{4}$$

and we have to unify the following equation

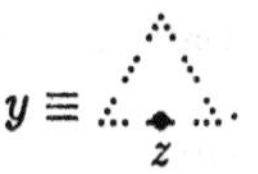

$$y \equiv \triangle_z \tag{5}$$

If we have to apply the term declaration rule with the first term declaration of (4), we obtain the following equations

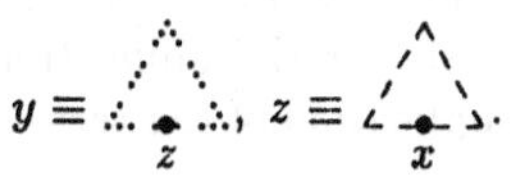

$$y \equiv \triangle_z, \quad z \equiv \triangle_x \tag{6}$$

---

[7] We also have to take care of possibly marked function symbols in all other unification rules [23].

If we then have to apply the term declaration rule with the second term declaration of (4) to the second equation of (6), we get

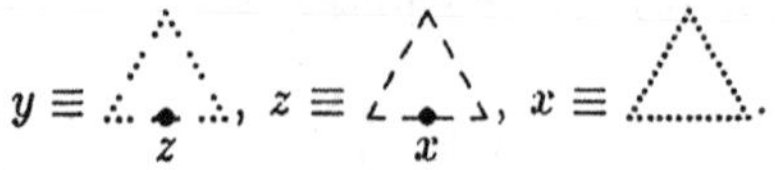

(7)

Because of the third equation of (7) we can eliminate x by substitution and obtain the following set of equations

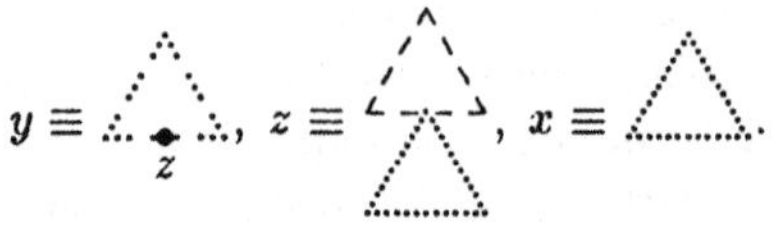

(8)

If we would use the modified term declaration rule we could not obtain (7) from (6) since

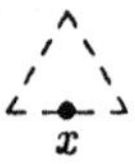

would be marked. But if the term declaration

(9)

is part of the signature, we can directly obtain (8) from (5)[8]. If the term declaration (9) is part of the signature we call the overlap between the two term declarations (4) solved:

A Σ-critical overlap $t_1\sigma$ between two term declarations $t_1: s_1$ and $t_2: s_2$ obtained by Σ-superposition at u with σ is *solved* if $t_1\sigma: s_1 \in \mathcal{D}$ for $u \neq \lambda$ or if $t_1\sigma: \{s_1, s_2\} \in \mathcal{D}$ for $u = \lambda$. Note that there is an unsolved overlap between term declarations obtained by superposition at λ if and only if the signature is not regular.

A unification step that is not possible with the modified term declaration rule can be replaced by application of the term declaration rule with the term declaration that solves the overlap. These considerations lead to the following theorem:

Theorem 6 *Let Σ be a signature. Then Σ-unification is decidable and finitary if every Σ-critical overlap between term declarations is solved.*

If we try to compute a critical overlap between two term declarations we start with an equation that only consists of subterms of term declarations on both sides. If we use the term declaration rule for the unification of such an equation and if the equations obtained by the application of the term declaration rule are unifiable,

[8] We do not obtain the third equation of (8). But x is only an auxiliary variable, since it does not occur in the initial equation (5). Thus the third equation of (8) is irrelevant for unification.

there is another critical overlap between term declarations, and so on. However, if the initial equation was unifiable, this chain must have an end. And at the end of this chain there is a critical overlap between two term declarations that can be obtained without application of the term declaration rule.

Theorem 7 *Let $\Sigma = (\mathcal{S}, \sqsubseteq, \mathcal{F}, \mathcal{V}, \mathcal{D})$ be a signature. Then there is a Σ-critical overlap between term declarations if and only if there is a $(\mathcal{S}, \sqsubseteq, \mathcal{F}, \mathcal{V})$-critical overlap between term declarations.*

This theorem gives us the possibility to compute critical overlaps between term declarations without using the term declarations therefore.

5.1 Linear Term Declarations

We further show that every linear term declaration can be replaced by function declarations with auxiliary sorts for every subterm of the linear function declarations.

Let Σ be an order-sorted signature with term declarations. Then we construct a signature without linear term declarations from Σ in the following way:

Let $t = f(t_1, \ldots, t_n) : \in \mathcal{D}$ be a linear term declaration. If $t: s$ is redundant then remove $t: s$ from $\mathcal{D}$. Otherwise do the following (let $\langle t: s, u\rangle = s'$ for every $u \in \mathrm{Occ}(t)$ with $t/u \in \mathcal{V}_{s'}$):

- for every $u \in \mathrm{Occ}(t) \setminus \{\lambda\}$ with $t/u \notin \mathcal{V}$ (let $t/u = g(p_1, \ldots, p_m)$) add $\langle t: s, u\rangle \notin \mathcal{S}$ to $\mathcal{S}$ and
$$g(\langle t: s, u \cdot 1\rangle, \ldots, \langle t: s, u \cdot m\rangle) : \langle t: s, u\rangle$$
to $\mathcal{F}$
- add $f(\langle t: s, 1\rangle, \ldots, \langle t: s, n\rangle) : s$ to $\mathcal{F}$

Because of theorem 2 the obtained signature can be transformed into a regular one. The obtained signature is equivalent to the former one in a very strong sense, since it only introduces auxiliary sorts. These auxiliary sorts interfere with the former sorts at the occurence λ of a linear term declaration only. Thus the new function declarations together with the auxiliary sorts simply imitate the effect of the linear term declarations.

Theorem 8 *Let Σ be a signature with term declarations and E be a set of F_Σ-equations. Then there is a regular signature Σ' without linear term declarations such that (Σ', E) is a conservative extension of (Σ, E).*

Thus every linear signature (a signature that contains only linear term declarations) can be transformed into an equivalent regular one without term declarations and thus into a signature where unification is decidable and finitary. A set of equations is unifiable w.r.t. the new signature if and only if it is unifiable w.r.t. the initial signature. Furthermore the new signature is effectively computable from the initial one.

This solves two open problems stated in [16]: Unification in linear signatures and regularity of linear signatures is decidable.

Theorem 9 *Σ-unification for linear signatures with term declarations is decidable.*

Corollary 10 *Regularity of linear signatures is decidable.*

The following example shows how the construction of the signature without linear term declarations solves the problems for unification in linear signatures:

Let $\mathcal{S} := \{A, B\}$, $\sqsubseteq := \{B \sqsubseteq A\}$, $\mathcal{F} = \{b: B, f(A): A\}$ and

$$\mathcal{D} = \{f(x_A): A, f(b): B, f(f(x_B)): B\}.$$

Then every Σ-mcsu of $\{x_B \equiv f(y_B)\}$ relative to $\{x_B, y_B\}$ is infinitary, since the set of Σ-substitutions σ with $B \in \mathrm{Sorts}_\Sigma(f(y_B)\sigma)$ is

$$\{\{y_B \leftarrow b\}, \{y_B \leftarrow f(b)\}, \{y_B \leftarrow f(f(b))\}, \ldots\}.$$

Let the regular signature without linear term declarations be $\Sigma' = (\mathcal{S}', \sqsubseteq', \mathcal{F}', \mathcal{D}')$. Then

$$\begin{aligned}
\mathcal{S}' = \mathcal{S} \cup \{ \quad & \langle f(b): B, 1\rangle, \langle\{B, \langle f(b): B, 1\rangle\}\rangle, \\
& \langle f(f(x_B)): B, 1\rangle, \langle\{B, \langle f(f(x_B)): B, 1\rangle\}\rangle \ \},
\end{aligned}$$

$$\begin{aligned}
\sqsubseteq' = \sqsubseteq \cup \{ \quad & \langle\{B, \langle f(b): B, 1\rangle\}\rangle \sqsubseteq B, \\
& \langle\{B, \langle f(b): B, 1\rangle\}\rangle \sqsubseteq \langle f(b): B, 1\rangle, \\
& \langle\{B, \langle f(f(x_B)): B, 1\rangle\}\rangle \sqsubseteq B, \\
& \langle\{B, \langle f(f(x_B)): B, 1\rangle\}\rangle \sqsubseteq \langle f(f(x_B)): B, 1\rangle \ \},
\end{aligned}$$

$$\begin{aligned}
\mathcal{F}' = \mathcal{F} \cup \{ \quad & b: \langle f(b): B, 1\rangle, \\
& b: \{B, \langle f(b): B, 1\rangle\}, \\
& f(\langle f(b): B, 1\rangle): B, \\
& f(B): \langle f(f(x_B)): B, 1\rangle, \\
& f(\langle f(f(x_B)): B, 1\rangle): B, \\
& f(\{B, \langle f(b): B, 1\rangle\}): \{B, \langle f(f(x_B)): B, 1\rangle\} \ \}
\end{aligned}$$

and $\mathcal{D}' = \emptyset$. Now $\{x_B \equiv f(y_B)\}$ has the Σ'-mcsu

$$\begin{aligned}
\{ \quad & \{y_B \leftarrow z_{\langle f(b):B,1\rangle}, x_B \leftarrow f(z_{\langle f(b):B,1\rangle})\} \\
& \{y_B \leftarrow z'_{\langle f(f(x_B)):B,1\rangle}, x_B \leftarrow f(z'_{\langle f(f(x_B)):B,1\rangle})\} \ \}.
\end{aligned}$$

6 Conclusions

Theorem 1 generalizes the important completeness theorem from unsorted to order-sorted rewriting by modifying the rewrite relation in a very simple way.

For a large class of rewrite systems that are not weakly sort decreasing theorem 6 and 9 yield perfect solutions to the problem of the failure of the critical pair theorem in the order-sorted case. Perfect solution means that there are not more critical pairs than those obtained by semantically well sorted substitutions.

Furthermore theorems 4, 6 and 7 yield a completion procedure for order-sorted signatures. By such a completion procedure we can possibly make a rewrite system weakly sort decreasing. During this completion we can try to complete the signature such that unification remains decidable and finitary.

While we have not considered order-sorted term rewriting modulo sets of equations the results should extend straight-forwardly to the equational case [4, 18].

7 Related Work

In [2] a fragment of second order logic is used to compute critical pairs in order to tackle the problem of an order-sorted critical pair theorem. Contrary to our approach there one has also to obey critical pairs obtained by superposition at variable positions.

In [1] a critical pair theorem is given that needs so-called sort convergence of the rewrite system. However, simple considerations show that sort convergence is equivalent to weak sort decreasingness. Also a completion procedure is given that makes a rewrite system sort convergent. However, this completion causes problems since the new rewrite system is in a sense not equivalent to the former one.

In [19] a critical pair theorem is given for range-unique signatures. For this theorem one has also to obey critical pairs obtained by superposition at variable positions. To make a signature range-unique new rewrite rules have to be added. These new rewrite rules may yield new critical pairs. There seem to be some parallels between the new critical pairs and the fact that unification in signatures with term declarations is decidable if there are no unsolved overlaps between the term declarations. Our approach has the advantage that we have not to obey critical pairs obtained by superposition at variable positions and that we can distinguish between problems that arise because of the semantical sort structure and those that arise because of the rewrite rules.

In [11] the computation of the sorts of a term is embedded into the rewriting process. This is a quite universal approach because it is possible to enrich a term with its complete semantical sorts. However, this can also be of disadvantage, since semantical sorts are undecidable in general. Hence our approach is more conservative and more sensitive because we only approximate the semantical sorts as close as necessary. The approach in [11] also has the disadvantage that the sort problems are mixed up with the rewrite problems.

Acknowledgements I wish to thank Albrecht Hoene, Dieter Hofbauer, Arfst Nickelsen and Dirk Siefkes for their very helpful support and Andreas Werner for his critical comments during the preparation of this paper.

References

1. H. Chen and J. Hsiang. Order-sorted equational specification and completion. Technical report, State University of New York at Stony Brook, 1991.
2. H. Comon. Completion of rewrite systems with membership constraints. Research report, CNRS-LNRI, 1991.
3. H. Ehrig and B. Mahr. *Fundamentals of Algebraic Specification 1*. Springer, 1985.
4. I. Gnaedig, C. Kirchner, and H. Kirchner. Equational completion in order-sorted algebras. In *Proceedings of the 13th Colloquium on Trees in Algebra and Programming (CAAP)*, number 299 in Lecture Notes in Computer Science, pages 165–184. Springer, 1988.
5. M. Gogolla. *Über partiell geordnete Sortenmengen und deren Anwendung zur Fehlerbehandlung in abstrakten Datentypen*. PhD thesis, Naturwissenschaftliche Fakultät der Technischen Universität Braunschweig, 1986.

6. J. A. Goguen, J.-P. Jouannaud, and J. Meseguer. Operational semantics for order-sorted algebra. In W. Brauer, editor, *Proceedings of the 12th International Colloquium on Automata, Languages and Programming*, number 194 in Lecture Notes in Computer Science, pages 221–231. Springer, 1985.
7. J. A. Goguen and J. Meseguer. Completeness of many-sorted equational logic. Technical Report CSLI-84-15, Center for the Study of Language and Information, Stanford University, 1984.
8. J. A. Goguen and T. Winkler. Introducing OBJ3. Technical Report SRI-CSL-88-9, SRI International, 1988.
9. G. Huet. Confluent reductions: Abstract properties and applications to term rewriting systems. *J. ACM*, 27(4):797–821, 1980.
10. G. Huet and D. C. Oppen. Equations and rewrite rules: A survey. In R. Book, editor, *Formal Languages: Perspectives and Open Problems*, pages 349–405. Academic Press, 1980.
11. C. Kirchner and H. Kirchner. Order-sorted computations in G-algebra. Technical report, Centre de Recherche en Informatique de Nancy, 1991.
12. C. Kirchner, H. Kirchner, and J. Meseguer. Operational semantics of OBJ-3. In T. Lepistö and A. Salomaa, editors, *Proceedings of the 15th International Colloquium on Automata, Languages and Programming*, number 317 in Lecture Notes in Computer Science, pages 287–301. Springer, 1988.
13. D. E. Knuth and P. B. Bendix. Simple word problems in universal algebras. In J. Leech, editor, *Computational Problems in Abstract Algebras*, pages 263–297. Pergamon Press, 1970.
14. A. Martelli and U. Montanari. An efficient unification algorithm. *ACM Trans. Prog. Lang. Syst.*, 4(2):258–282, 1982.
15. J. Meseguer and J. A. Goguen. Initiality, induction and computability. In M. Nivat and J. C. Reynolds, editors, *Algebraic Methods in Semantics*. Cambridge University Press, 1985.
16. M. Schmidt-Schauss. *Computational Aspects of an Order-Sorted Logic with Term Declarations*. Number 395 in Lecture Notes in Artificial Intelligence. Springer, 1989.
17. G. Smolka, W. Nutt, J. A. Goguen, and J. Meseguer. Order-sorted equational computation. SEKI Report SR-87-14, Fachbereich Informatik der Universität Kaiserslautern, 1987.
18. U. Waldmann. Semantics of order-sorted specifications. *Theoretical Comput. Sci.*, 94(1):1–36, 1992.
19. A. Werner. A semantic approach to order-sorted rewriting. unpublished, 1991.
20. L. With. Linear order-sorted unification. Forschungsberichte des Fachbereichs Informatik 1989-15, Technische Universität Berlin, 1989.
21. L. With. Multi-sort variables — a new concept for order-sorted unification. Forschungsberichte des Fachbereichs Informatik 1989–19, Technische Universität Berlin, 1989.
22. L. With. Completeness and confluence of order-sorted term rewriting. Forschungsberichte des Fachbereichs Informatik, Technische Universität Berlin, 1992. to appear.
23. L. With. Order-sorted unification with term declarations. Forschungsberichte des Fachbereichs Informatik, Technische Universität Berlin, 1992. to appear.

Completion for Constrained Term Rewriting Systems*

Charles Hoot
Department of Computer Science
University of Illinois at Urbana-Champaign
Urbana, IL 61801, U.S.A.
email: hoot@cs.uiuc.edu

Abstract

A framework for rewriting with membership, equality, and inequality constraints is presented. A ground completion procedure is described which modifies the standard critical pair lemma. One consequence is the desire to rewrite terms in a set represented by a grammar, which leads to strong requirements on the conditional rules. Examples of the use of a completed constrained term rewriting system are presented, including an instance of an inductive theorem.

1 Introduction

Unconditional equational theories have long been of interest in many areas of computer science. One of the milestones was the Knuth-Bendix completion procedure for taking a set of axioms and generating a rewrite system which can be used to decide the equivalence of terms [KB70] [Der89]. Unconditional equations, however, are not always successful when one wants to describe partial functions such as the arithmetic inverse over natural numbers. In the area of high level languages there are many instances of functions which are defined over limited domain.

A number of approaches have been proposed to address this issue. One approach is to use functions and test for equivalence in conditions. Another approach taken by Smolka and more recently Kirchner is to introduce sorts and declare that functions and symbols are of certain types. Ill-sorted terms are not allowed [DW91] [SNGM89]. Another approach taken recently in constraint logic programming by Frisch and others is the introduction of membership constraints [CD91], [Uri92], [Fri91]. Within the field of rewriting, Toyama proposed the use of monadic membership constraints to force some nonconfluent unconditional term rewriting systems to be confluent [Toy87]. More recent work related to this paper

*This research supported in part by the National Science Foundation under Grant CCR-90-24271.

has been performed by Comon where the introduction of second order logic is used to allow completion [Com].

In this paper a constraint framework with disjunctions of simple monadic membership constraints with equality and inequality conditions is introduced for terms, rewrite rules, grammars and theorems. The process of applying a rewrite rule and unification are explored and the results are used to design a modification of the Knuth-Bendix completion procedure for ground terms.

2 Membership Constrained Terms

A number of different kinds of constraints could be considered for conditional rewriting. This paper will only address membership constraints which are monadic in conjunction with auxiliary equality and inequality constraints. The *membership constraint* $x \in CS_x$ is satisfied iff a substitution maps x to an element of the *constraint set* CS_x. An *equality constraint* $t(x_1, \ldots, x_n) = s(x_1, \ldots, x_n)$ is satisfied by a substitution iff the terms resulting from a substitution σ are identical. Similarly, an inequality constraint $t(x_1, \ldots, x_n) \neq s(x_1, \ldots, x_n)$ is satisfied iff the terms resulting from a substitution σ are not identical. Note that this is very different from the standard meaning of equality in a condition in more traditional conditional term rewriting systems. In such a system, the condition is satisfied if the two terms are joinable in the conditional rewriting system. As a consequence, when proving termination the left hand side of the rule must be greater than each of the terms in the condition as well as the right hand side. In contrast, for the membership constrained term rewriting case, determining the satisfaction of a condition is independent of the particular rewriting system under consideration. Hence proofs of termination can proceed using the same techniques as for the unconditional case.

A *constrained term* is represented by the following notation:

$$t(x_1, \ldots, x_n) \parallel x_1 \in CS_1, \ldots, x_n \in CS_n, C(x_1, \ldots, x_n)$$

where $C(x_1, \ldots, x_n)$ is a conjunction of equality and inequality constraints and denotes the set of ground terms obtained by ground substitutions which satisfy the constraints. The conjunction of the membership constraints with the equality and inequality constraints will be called a *simple condition*. A *disjunctive condition* will also be allowed for convenience. It consists of a finite disjunction of simple conditions. For example, the constrained term

$$f(x, y) \parallel x \in \{a, b\}, y \in \{c, d\}$$

denotes the ground terms $f(a, c)$, $f(a, d)$, $f(b, c)$, and $f(b, d)$.

A condition is in *constraint normal form* if it meets all of the following criteria:

1. The term is linear in all variables.
2. The equality conditions are all of the form variable equals variable, *e.g.*, $x = y$.
3. For each equivalence class formed by the equality constraints over the variables, one variable is chosen to represent the class.

4. If two variables are equal via an equality constraint, then the constraint sets in the membership constraints for the two variables are identical.

5. The inequality conditions are all of the form variable not equals term, *e.g.*, $x \neq f(a, y, z)$.

6. The variables in the inequality constraints consist of only the representative variables from the equality constraints.

Converting a term to normal form is mostly straightforward. The equality constraints can be put into solved form with the standard method used in unification. Any equations left in the solved form as variable equals term will be substituted throughout the inequality constraints (and eventually the term itself).

For the inequality constraints, an algorithm similar to the one for equalities will be used. The main thing to note is that there is a rule: *If there is an inequality of the form $h(t_1, \ldots, t_n) \neq h(s_1, \ldots, s_n)$, replace the term with n terms. Each of the new terms inherits all of the other constraints from the original term. Each of the conditions has one new inequality constraint, namely $t_i \neq s_i$.* Since the solved form may be a disjunction, corresponding splitting will occur. Termination can be shown by a multiset ordering with one element for each disjunct (simple condition) consisting of the sum of the heights of the terms in the disjunct. References for solving unification and disunification problems include [MMR86], [KL87], [CL89].

As an example, consider the following conjunction of equality and inequality constraints:

$$f(x, y) = f(t(z), x) \wedge g(z, x, h(a, z)) \neq g(x, y, h(z, w)).$$

The equality constraints can be put into the solved form

$$E \equiv x = t(z) \wedge y = t(z).$$

After substituting for x and y in the inequalities we are left with

$$E \wedge g(z, t(z), h(a, z)) \neq g(t(z), t(z), h(z, s(w))).$$

The first application of the rules for inequalities results in a disjunction of three simple conditions

$$[E \wedge z \neq t(z)] \vee [E \wedge t(z) \neq t(z)] \vee [E \wedge h(a, z) \neq h(z, s(w))].$$

The first two inequality constraints reduce to $\top$ and $\bot$ respectively, and the third disjunct splits into two new ones. Each of the simple conditions is in solved form.

$$E \vee [E \wedge a \neq z] \vee [E \wedge z \neq s(w)].$$

This example also serves to illustrate a special case. If all of the inequality constraints for any disjunct D produced via splitting can be removed (converted into $\top$), then a substitution which satisfies one of the other disjuncts will also satisfy D. Therefore D can subsume all of the other disjuncts from the split. In the current example, this leaves just E.

Now that the equality and inequality parts are in solved form, substitutions must be performed using the equality constraints. However, this step must be performed with some caution as the following example illustrates. The condition for the term

$$t(x) \parallel x \in \{a,b\}, y \in \{a,b\}, x = h(y)$$

is satisfied by no substitution, yet the term obtained by blindly substituting and then discarding the equality constraint

$$t(h(y)) \parallel x \in \{a,b\}, y \in \{a,b\}$$

is satisfied by substitutions resulting in the ground terms $t(h(a))$ and $t(h(b))$.

To avoid this problem, the membership constraints must be modified to reflect the equality constraint. This process is called *reconciling* the variables with the equalities. Equality constraints in solved form can have one of two forms, variable = variable, or variable = term, where none of the variables occurring on the right side of an equation are also on the left side of any equation and no variable occurs on the left side of more than one equation. (The variables on the right sides of the equations will eventually be the representative variables of equivalence classes.)

Reconciling the variables consists of the following steps:

1. Let the membership constraints be $x_1 \in CS_{x_1}, \ldots, x_m \in CS_{x_m}$ and $y_1 \in CS_{y_1}, \ldots, y_n \in CS_{y_n}$
2. For each variable y_j with equality constraints of the form $x_i = y_j$ compose the sets $CS'_{y_j} = CS_{y_j} \cap \bigcap_{x_i} CS_{x_i}$.
3. While there are equality constraints left unprocessed do:
   (a) Pick an unprocessed equality $x_i = t_i(y_1, \ldots, y_n)$.
   (b) Form the set $CS'_{x_i} = CS_{x_i} \cap \{t_i(y_1, \ldots, y_n) \mid y_1 \in CS'_{y_1}, \ldots, y_n \in CS'_{y_n}\}$ (form the intersection with the composition).
   (c) Split the term into k (possibly infinite) pieces with grammars $Y_1^k, \ldots, Y_n^k$ such that
      - If $y_1 \in Y_1^k, \ldots, y_n \in Y_n^k$ then $t_i(y_1, \ldots, y_n) \in CS'_{x_i}$.
      - If $\exists y_1, \ldots, y_n$ such that $t_i(y_1, \ldots, y_n) \in CS'_{x_i}$ then there is some k such that $y_1 \in Y_1^k, \ldots y_n \in Y_n^k$.

   (d) For each of the pieces construct the following $CS'_{y_j} \leftarrow CS'_{y_j} \cap Y_j^k$.
4. In each of the resulting pieces construct the reconciled grammars. For each variable x_i construct $X_i = \{t_i(y_1, \ldots, y_n) \mid y_1 \in CS'_{y_1}, \ldots, y_n \in CS'_{y_n}\}$ with the equality constraint given by $x_i = t_i(y_1, \ldots, y_n)$. For each variable y_j construct $Y_j = CS'_{y_i}$.
5. Substitute with the equality constraints in the term, and remove them.

The term can now be put into constraint normal form by introducing a new variable for each duplicate variable in the term. Each new variable will have an associated equality constraint and membership constraint. Step 4c is the heart of the algorithm for reconciling the variables. The first condition guarantees that in each of the pieces (disjuncts) there are no dependencies between variables. *E.g.*, with $x \in \{h(a,b), h(b,a)\}, y \in \{a,b\}, z \in \{a,b\}$ and $x = h(y,z)$, the variables y and z have a dependency due to the equality condition. Note that if there is only a single variable no splitting is needed to represent dependencies. Also, if regular tree languages are used, only a finite number of splits is needed. The second condition guarantees that all of the ground instances are covered by the disjuncts.

3 Grammars

To this point the specification of the constraint sets has been left open. It is desirable that the grammars describing the constraint sets be closed under union, intersection, and composition. It will be seen later that decidability of testing the grammar for emptiness is also needed. Terms corresponding to trees accepted by some finite tree automaton are one possibility. These sets are *regular tree languages* and possess many of the same properties as their cousins, the regular languages over words. In particular, the class of regular tree languages is a boolean algebra. Therefore, it is closed under finite unions, intersections, and differences. In addition, for a given finite tree automaton $\mathcal{U}$, it is effectively decidable if an arbitrary tree t is accepted by $\mathcal{U}$, and if the language accepted by $\mathcal{U}$ is empty. Notice that the decidability of whether two tree automata $\mathcal{U}_1$ and $\mathcal{U}_2$ accept the same language follows from the closure property of differences and the decidability of a language being empty [Don70].

For the purposes of this paper an extension of the regular tree languages will be considered. This extension allows equality and inequality constraints to be posed; it is motivated by the desire to handle term rewriting systems, which typically have equality constraints. *E.g.*, the rewrite rule $h(x,x) \rightarrow g(x)$ has an implicit equality constraint which is made explicit in the equivalent rule $h(x,y) \rightarrow g(x) \parallel x = y$. The particular method used to describe the grammar will be "production" rules of the form

$$t(x_1,\ldots,x_n) \in G \Leftarrow x_1 \in G_1,\ldots,x_n \in G_n, C(x_1,\ldots,x_n).$$

This denotes that a ground instance of the term t is an element of grammar G if each of the constraints is satisfied. It is assumed throughout the rest of this paper that all equality constraints will be made explicit. Therefore, the occurrences of the variables x_1 through x_n in the term t should be distinct. If this property does not hold, it is trivial to introduce extra variables and equality constraints to make it so.

Of some interest is what effect restrictions on the constraints have on the properties of the languages representable by the grammar. If $C(x_1,\ldots,x_n)$ is empty, then the language is a regular tree language. Recent work by Bogaert [BT91] has shown that if the equality/inequality constraints are between variables on the same level, then all of the standard properties hold. With the conditions for constraints outlined in this paper previously (constraint normal form), testing for emptiness is decidable. Union is closed. This author does not know if intersection is closed, but suspects that it is not. Certainly if intersection is not closed then complement is not either, as is shown by the following grammar G and its complement:

$$\begin{aligned} t(x) \in G &\Leftarrow x \in G_x \\ t(y) \in G &\Leftarrow y \in G_y \\ t(z) \in \neg G &\Leftarrow z \in \neg G_x \cap \neg G_y. \end{aligned}$$

An iterative process for constructing a grammar is useful for showing that membership of a word in a grammar and emptiness of a grammar are decidable when the conditions are in constraint normal form. Three categories of production rules must be considered— constant terms with no membership constraints, terms consisting of a single variable, and everything else. Each iteration consists of two parts. In the first part, all of the new terms are produced. In the second part, the production rules with variables for the term are applied until a steady state is reached.

Define the *threshold* $\Lambda_{G,\mathcal{P}}$ of a grammar G with respect to a production rule $\mathcal{P}$ as follows:

$$\Lambda_{G,\mathcal{P}} = 1 + \sum_{\mathcal{I} \text{ in } \mathcal{P}} O(\mathcal{I}, G)$$

where $\mathcal{I}$ is an inequality constraint and $O(\mathcal{I}, G)$ is 1 if there is some variable x such that $x \in G$ is a membership constraint in $\mathcal{P}$ and x is in $\mathcal{I}$. $O(\mathcal{I}, G)$ is 0 otherwise. The following lemma, in combination with the iterative construction, is crucial to showing the decidability properties.

Lemma 1 *Once the number of members in an approximation to a grammar G reaches the threshold $\Lambda_{G,\mathcal{P}}$, additional members in G will not directly affect the satisfiability of the condition for $\mathcal{P}$.*

Proof: Suppose there is an approximation G^i where $|G^i| = \Lambda_{G,\mathcal{P}}$ and there is no satisfying substitution, but for a later approximation $G^j = G^i \cup \{a_1, \ldots, a_m\}$ there is a satisfying substitution. Call it $\sigma = \{x_1 \mapsto a_1, \ldots, x_m \mapsto a_m, \ldots, x_n \mapsto g_n, y_1 \mapsto t_1, \ldots y_n \mapsto t_n\}$ where the x_i have membership constraints for G, all values x_i with $i > m$ are from the original approximation G^i, and each y maps to some value from the appropriate corresponding grammar approximation. Substitute the values for $x_{m+1}, \ldots, x_n, y_1, \ldots, y_n$ into the condition. This yields a new condition with at most $\Lambda_{G,\mathcal{P}} - 1$ undetermined inequality constraints over m variables. For this new condition with respect to G^i there are $(\Lambda_{G,\mathcal{P}})^m$ possible substitutions. Consider an undetermined inequality constraint with k variables. There are $(\Lambda_{G,\mathcal{P}})^{k-1}$ ways of substituting for $k-1$ of the variables, which can rule out at most one value for the remaining variable. Thus this inequality can fail for at most $(\Lambda_{G,\mathcal{P}})^{m-1}$ substitutions. Therefore, the undetermined constraints can fail for at most $(\Lambda_{G,\mathcal{P}} - 1)(\Lambda_{G,\mathcal{P}})^{m-1}$ substitutions. Since this is strictly less than the total number of possible substitutions, there must be a satisfying substitution from G^i.

The key is that conditions like $x \in G, y \in H, x = y$ are prohibited. For such a condition there is no threshold for guaranteeing that extra terms in a grammar are not needed to determine satisfiability of a condition.

4 Rewriting Constrained Terms

Consider a constrained term with a simple condition

$$t(x_1,\ldots,x_p) \parallel x_1 \in CS_{x_1},\ldots,x_p \in CS_{x_p}, C(x_1,\ldots,x_p)$$

and the constrained rewriting rule

$$\begin{aligned} l(y_1,\ldots,y_m) \rightarrow r(z_1,\ldots,z_n) \quad \parallel \quad & y_1 \in CS_{y_1},\ldots,y_m \in CS_{y_m}, \\ & z_1 \in CS_{z_1},\ldots,z_n \in CS_{z_n}, \\ & C(y_1,\ldots,y_m,z_1,\ldots,z_n). \end{aligned}$$

What happens when the rule is applied to a subterm s of t represented as $t[s]$? To apply the rule there must be a substitution σ which maps variables from $y_1,\ldots,y_m$ into terms over $x_1,\ldots,x_p$ such that $l\sigma = s$. There must be a ground instance of the rule applicable to a ground instance of the term t. The instances of t to which the rule is applicable are

$$\begin{aligned} t(x_1,\ldots,x_p) \quad \parallel \quad & x_1 \in CS_{x_1},\ldots,x_p \in CS_{x_p}, C(x_1,\ldots,x_p), \\ & y_1 \in CS_{y_1},\ldots,y_m \in CS_{y_m}, \\ & z_1 \in CS_{z_1},\ldots,z_n \in CS_{z_n}, \\ & C(y_1,\ldots,y_m,z_1,\ldots,z_n), E(\sigma) \end{aligned}$$

where $E(\sigma)$ is a conjunction of equality constraints generated from σ. If $x \mapsto \Phi$ is in σ, then $x = \Phi$ is in $E(\sigma)$.

After rewriting, the term is

$$\begin{aligned} t[r] \quad \parallel \quad & x_1 \in CS_{x_1},\ldots,x_p \in CS_{x_p}, C(x_1,\ldots,x_p), \\ & y_1 \in CS_{y_1},\ldots,y_m \in CS_{y_m}, \\ & z_1 \in CS_{z_1},\ldots,z_n \in CS_{z_n}, \\ & C(y_1,\ldots,y_m,z_1,\ldots,z_n), E(\sigma) \end{aligned}$$

While this is correct, typically simplifications to the condition will be applied (including converting the condition to constraint normal form.) One useful simplification is to remove membership constraints for variables which occur only in the constraint. For example, in the term

$$t(x_1,\ldots,x_p) \parallel x_1 \in CS_{x_1},\ldots,x_p \in CS_{x_p}, y \in CS_y, C(x_1,\ldots,x_p)$$

the membership constraint for y can be removed as long as CS_y has at least one member. If CS_y is empty, there are no substitutions which satisfy the condition and the entire condition can be replaced by T.

Another simplification can be applied to terms of the form

$$t(x_1,\ldots,x_p) \parallel x_1 \in CS_{x_1},\ldots,x_p \in CS_{x_p}, y \in CS_{x_i}, x_i = y, C(x_1,\ldots,x_p)$$

which typically arise during rewriting. Here the two terms involving y can be removed. The constraints involving y are satisfiable iff the constraint $x_i \in CS_{x_i}$ is satisfiable.

Note that in the discussion of rewriting, the presence of constraints containing variables not in the term was ignored. This is acceptable since one can consider that the term $t(x_1,\ldots,x_p)$ need not use all of the variables $x_1,\ldots,x_p$. Under this assumption the above process correctly handles the situation. In subsequent discussions, this assumption will be made.

5 Unification of Constrained Terms

Of practical interest for the completion process and narrowing is whether two terms are unifiable. The focus will be on terms with simple conditions. The extension to the general case of disjunctive conditions can be done by forming the disjunct of the unifications of the simple conditions from the two terms in a pairwise fashion.

Consider two simple conditional terms

$$t(x_1,\ldots,x_n) \parallel x_1 \in CS_{x_1},\ldots,x_n \in CS_{x_n}, C(x_1,\ldots,x_n)$$

and

$$s(y_1,\ldots,y_m) \parallel y_1 \in CS_{y_1},\ldots,y_m \in CS_{y_m}, C(y_1,\ldots,y_m)$$

Clearly, any ground instances of the unified term for the conditional terms must also be a ground instance of the unconditional terms. The first step in unification is to perform standard unification on the unconditional terms t and s. This results in a most general unifier, σ. The unifier maps variables from a subset S of $T \equiv \{x_1,\ldots,x_m,y_1,\ldots,y_n\}$ onto terms over the remaining variables $T - S$.

The result of unifying the constrained terms is given by

$$\begin{aligned} t(x_1,\ldots,x_n)\sigma \quad \parallel \quad & E(\sigma), x_1 \in CS_{x_1},\ldots,x_n \in CS_{x_n}, C(x_1,\ldots,x_n), \\ & y_1 \in CS_{y_1},\ldots,y_m \in CS_{y_m}, C(y_1,\ldots,y_m) \end{aligned}$$

where $E(\sigma)$ converts each mapping to an equality constraint. For a substitution to be valid for the unified term, it must satisfy the conditions on the original terms as well as the new equality constraints generated in the unification.

The final step is to put the condition in constraint normal form as outlined previously. As noted, this may require a disjunctive condition. Hence unification is not unitary.

6 Ground Completion

Standard Knuth-Bendix completion attempts to find a decision procedure for equality of terms by rewriting for an equational theory. Ground completion limits the requirements for the decision procedure to deciding the equality of ground terms. The critical pair lemma gives sufficient conditions for local confluence in unconditional rewrite systems. Unfortunately, this is no longer the case with membership conditions. To show local confluence, one needs to consider three cases — non-overlapping rule applications, nested overlaps, and critical overlaps. The non-overlapping case is still trivially joinable.

The case where rules overlap at a non-variable subterm is similar to standard Knuth-Bendix completion. Given two simple conditional rules $l_1 \rightarrow r_1$ and $l_2 \rightarrow r_2$, the following procedure computes the critical pairs:

1. Rename variables to avoid name clashes and unify the lefthand side of rule 1 with a non-variable subterm of the lefthand side of rule 2 (with constraints of course).

2. Rewrite with the two rules to form two terms t_1 and t_2.

3. Generate conditional critical pairs (one for each pairing of disjuncts between the disjunctive conditions for t_1 and t_2). Note that the terms in the critical pairs are all identical and generated in the standard way. By construction, the conditions of the two rules will be satisfied by any ground terms covered by the conditions.

4. Guarantee that each of the simple conditions is in constraint normal form. (This may result in further splitting.) Eventually, a rule will be added for each critical pair which is not joinable.

Where the standard critical pair lemma breaks down is for variable subterms (nested overlap). With unconditional rules the application of rule 1 to a subterm, which is bound to a variable part of rule 2, could not prevent the application of the rule 2. Now, however, the condition of rule 2 may be violated, preventing the trivial joining of rules.

To fix the critical pair lemma, overlap at a variable subterm (rewriting within a condition) must be considered. In particular, the overlap (rule application) can be rooted at any possible subterm of a variable. The ground terms which a variable can take on are those of the substitutions which satisfy the condition of the rule. In particular, suppose there is a rule

$$l \rightarrow r \parallel y \in CS_y, x_1 \in CS_{x_1}, \ldots, x_m \in CS_{x_m}, C(y, x_1, \ldots, x_m).$$

A grammar G can be constructed which accepts the ground terms which y can assume in satisfying substitutions.

$$y \in G \Leftarrow y \in CS_y, x_1 \in CS_{x_1}, \ldots, x_m \in CS_{x_m}, C(y, x_1, \ldots, x_m).$$

The critical pair lemma for constrained rewriting must consider the applications of rules to such grammars. In particular, given a grammar G as above and a production rule $\mathcal{R}$, a new grammar G' must be generated which satisfies

$$t \in G' \text{ iff } \exists s \text{ such that } s \in G \text{ and } s \underset{\mathcal{R}}{\rightarrow} t.$$

This results in a new critical pair

$$(l, r) \parallel y \in G', x_1 \in CS_{x_1}, \ldots, x_m \in CS_{x_m}, C'(y, x_1, \ldots, x_m).$$

where the constraint C' is satisfied by a ground substitution $\sigma = \{y \mapsto t_y, x_1 \mapsto t_1, \ldots, x_m \mapsto t_m\}$ iff there exists a term s such that $\sigma' = \{y \mapsto s, x_1 \mapsto t_1, \ldots, x_m \mapsto t_m\}$ satisfies the condition $y \in G, x_1 \in CS_{x_1}, \ldots, x_m \in CS_{x_m}, C(y, x_1, \ldots, x_m)$ and $s \rightarrow_{\mathcal{R}} t_y$.

Unfortunately, if C is non-empty, the condition C' need not be equality or inequality between terms. Consider the following term

$$h(x, y) \parallel x \in G, y \in G, x = y$$

with grammar $G \equiv \{a, fa, ffa, \ldots\}$ in conjunction with the rule

$$f(x) \rightarrow g(g(x)).$$

The set of terms which are derivable in one step from G is

$$\{h(f^i g^2 f^j a, f^n a) \mid i+j=n-1, i \geq 0, j \geq 0\}$$
$$\cup \quad \{h(f^n a, f^i g^2 f^j a) \mid i+j=n-1, i \geq 0, j \geq 0\}$$

which is not finitely representable using only equality and inequality constraints on terms.

The obvious counter to this in the case of equality constraints is to force rewriting on the equal subterms to occur on all equivalent occurrences, *i.e.*, leave the equality constraints completely unchanged. Unfortunately, the single step rewrites of terms in the grammar are no longer represented and local confluence is not guaranteed.

Even this approach does not work for inequality constraints. For example, with the term

$$h(x,y) \parallel x \in G, y \in G, x \neq y$$

keeping the inequality constraint after a single rewrite with the previously defined term rewriting system would admit the ground instance $h(f(a), g(g(a)))$. Yet the only term from which this could have been derived is $h(f(a), f(a))$, which is not a ground instance of the original term.

Guideline 1 *Never allow rewriting inside a grammar G which is part of a condition which has equality or inequality constraints.*

Note that this prevents rewriting in all of the grammars on which G depends. ¿From here on, in order to follow the guideline, only rewriting of grammars without equality and inequality constraints (regular tree languages) will be considered.

In general, this guideline has the practical consequence of ruling out non-left-linear and non-right-linear rules when grammars can be rewritten. Non-right-linear rules can directly introduce an equality constraint to a grammar. Non-left-linear rules can introduce a constraint by an indirect route. When such a rule is applied to a constrained theorem, the term is split to account for all of the ground terms. It is in the split to which the rule is not applicable that a constraint is generated.

Even with these restrictions, another problem can arise if there is a cycle in the grammar— the completion process may not terminate. Consider the following conditional rewrite system:

$$h(x) \rightarrow b \quad \parallel \quad x \in G \equiv \{a, fa, ffa, \ldots\}$$
$$f(y) \rightarrow g(y).$$

The only critical pair is between the second rule and terms for the variable x. Applying the second rule to the grammar G results in a new grammar G_1 which has exactly one occurrence of the symbol g in each of the terms. The new critical pair can be oriented and the resulting rule

$$h(w) \rightarrow b \parallel w \in G_1 \equiv \{ga, fga, gfa, \ldots\}$$

added to the term rewriting system.

There is no critical pair between this and the first rule since $G \cap G_1 = \emptyset$. However there is a critical pair between the second rule and the variable w. We can generate a new grammar G_2 which has exactly two occurrences of the symbol g in each of its members. This process will go on generating grammars G_i which have exactly i occurrences of the constant g. This problem can sometimes be avoided by generating all of these consequences at once. To guarantee local confluence, all that must be assured is that all of the terms which can be generated in a single rewrite are included.

The following algorithm accomplishes that task:

1. Let G be the initial regular tree grammar, $\mathcal{R} = l \rightarrow r$ be the rewrite rule, with q the top symbol of l, and let $G_1, \ldots, G_n$ be any grammars associated with G.

2. Construct $G', G'_1, \ldots, G'_n$, equivalent regular tree grammars where each production rule has at most one constant symbol.

3. For each production $\mathcal{P}$ which has an occurrence of q, make a copy.

4. For each of the copies, promote productions of the variables as needed until the non-variable part of l is matched. If more than one production rule can be promoted and match for a given variable, make a copy for each one.

5. Rewrite each of the copies according to the rewrite rule.

6. Add each of the rewritten production rules to the grammars $G', G'_1, \ldots, G'_n$, resulting in G^1.

Lemma 2 *The grammar G^1 contains all of the terms which are a single rewrite via $\mathcal{R}$ away from terms in G.*

Proof: Consider a ground term $t[s] \in G$ such that $t[l\sigma] \rightarrow_{\mathcal{R}} t[r\sigma]$. The position of the application of the rewrite rule must correspond to one of the copies. All possible productions corresponding to a valid rewrite are generated and the rewrite rule is applied to the production. One of these must correspond to $s = l\sigma$. The rewritten production rule generates $r\sigma$. $t[r\sigma]$ is in G^1 since the productions for generating the context (excluding the top symbol) are the originals from G. The immediate context from the production and the rewritten subterm can be generated via the rewritten production rule.

Unfortunately, this does not generate all possible parallel rewrites of the rule $\mathcal{R}$ with the original term since the rewriting may introduce new grammars via intersection with any grammars associated with the rule. In addition to the extra productions generated by the application of the rewrite to the copies, there are a finite number of grammars of the form $N \cap G'_i$, where N is a grammar symbol from G' and G'_i is one of the grammars from the rule. To get a grammar G^2 where all possible pairs of parallel rewrites are represented, one needs only to apply the rule to each of the intersections. Note that if $N \cap G'_i = M$ where M is one of the grammars before rewrite, $N \cap G'_i$ may be replaced by it.

The following example shows an application of the algorithm. Consider the grammar G specified by

$$\begin{aligned} a &\in G \\ f(f(x)) \in G &\Leftarrow x \in G \end{aligned}$$

and the rewrite rule

$$f(f(f(x))) \rightarrow g(x) \parallel x \in \Sigma^*.$$

The grammar G can be converted to the equivalent grammar

$$\begin{aligned} a &\in G' \\ f(x) \in G' &\Leftarrow x \in T \\ f(x) \in T &\Leftarrow x \in G'. \end{aligned}$$

The second and third productions have occurrences of the symbol f, so make one copy of each rule. In both cases, promoting the first production rule in the grammar never results in a term which matches the left side of the rule. Only the second and third production rules are useful. The modified copies are respectively:

$$\begin{aligned} f(f(f(x))) \in G' &\Leftarrow x \in T \\ f(f(f(x))) \in T &\Leftarrow x \in G'. \end{aligned}$$

Rewriting gives the rules

$$\begin{aligned} g(x) \in G' &\Leftarrow x \in T \cap \Sigma^* \\ g(x) \in T &\Leftarrow x \in G' \cap \Sigma^*. \end{aligned}$$

But $\Sigma^* \cap X = X$ for any grammar X. Adding the rules to the original grammar gives the final result:

$$\begin{aligned} a &\in G' \\ f(x) \in G' &\Leftarrow x \in T \\ g(x) \in G' &\Leftarrow x \in T \\ f(x) \in T &\Leftarrow x \in G' \\ g(x) \in T &\Leftarrow x \in G'. \end{aligned}$$

There are no new grammars formed via intersection, so this grammar represents all possible parallel rewrites of the original terms in G.

Notice that the non-variable case does not add any new equalities which are syntactically different from existing ones. If the proof of termination ignores the condition, *i.e.*, uses standard techniques, then one can simply extend the condition of the rule. The only caveat is that the new part may affect old critical pairs, extending their conditions as well.

One special case of note is if all of the symbols in the constraint grammars are constructors. No rule can match a non-variable subterm, and the second part of the modified critical pair lemma is trivially satisfied. This also avoids the problems associated with rewriting grammars while proving constrained theorems. In addition, it has the advantage of tending to limit the number of possible rewrites of a ground term. In effect it is similar to applicative order (leftmost-innermost) evaluation.

7 Proving Constrained Theorems

The goal of the completion process is to develop a decision procedure to handle the word problem. Given two ground terms t_1 and t_2 are they equal under a given equational theory. One of the successes of unconditional systems is that these questions can be asked of terms with variables as well. One of the problems is dealing with theorems that require inductive proofs. Constrained theorems naturally express such inductive problems and arise during the constrained completion process.

Consider the following constrained term rewriting system for arithmetic on unary numbers with a modulo 2 function:

$$\begin{aligned} x+0 &\to x & \| \quad & x \in N \\ x+s(y) &\to s(x+y) & \| \quad & x \in N, y \in N \\ mod2(0) &\to 0 & & \\ mod2(s(0)) &\to s(0) & & \\ mod2(s(s(x))) &\to mod2(x) & \| \quad & x \in N \end{aligned}$$

where $N \equiv \{0, s0, ss0, \ldots\}$. The grammar is constructor based, so critical pairs arising from overlap at a variable position need not be considered. There are no regular critical pairs since no rule unifies with any non-variable subterms. Termination is easily shown by traditional methods, hence the above system is ground complete.

An example of a word problem is

$$mod2(s(0)+s(s(0))) = mod2(s(0)) + mod2(s(s(0))).$$

Since both sides derive $s(0)$, the two ground terms are equal under the equational theory.

A more interesting problem is one involving variables, such as the set of word problems denoted by

$$mod2(x) = 0 \; \| \; x \in E$$

where E is the evens $\{0, s(s(0)), s(s(s(s(0)))), \ldots\}$. The first rule for $mod2$ can be applied, resulting in the split

$$\begin{aligned} mod2(0) &= 0 \; \| \; x \in E, x = 0 \\ mod2(x) &= 0 \; \| \; x \in E - \{0\}. \end{aligned}$$

Splitting is necessary to account for the ground instances to which the rule can not be applied. Applying the rule to the split where it is valid results in

$$0 = 0 \; \| \; x \in \{0\}.$$

The membership condition for x can be dropped since there is at least one element in the set $\{0\}$ and x does not appear on either side of the equality. The left and right hand sides are both in a normal form and are equal so the theorem is proved for the first part of the split.

To the other part of the split the third rule for *mod*2 can be applied. The unification with the right hand term leads to

$$mod2(x) \parallel x \in E - \{0\}, y \in N, x = s(s(y)).$$

No further splitting is necessary to reconcile the variables x and y, leaving

$$mod2(x) \parallel x \in E - \{0\}, y \in E, x = s(s(y)),$$

which completely covers the term. Performing the substitution gives

$$mod2(s(s(y))) = 0 \parallel y \in E,$$

to which the fifth rule can be applied, leaving

$$mod2(y) = 0 \parallel y \in E.$$

At this point one of two things happens — either the prover infinitely performs applications of one rule followed by the other, essentially proving the theorem one ground substitution at a time, or it realizes that this represents the same set of word problems as the original statement of the problem, but each is smaller due to rewrite. Induction applies and the theorem is proven.

Another point that should be raised is that when grammars can be rewritten, using unification only on the term level is not enough. Consider the rewrite system

$$h(a) \rightarrow a$$
$$g(a) \rightarrow b$$

and the constrained theorem

$$g(z) = b \parallel z \in G$$

where G is the grammar

$$\begin{aligned} h(h(a)) &\in G \\ h(x) &\in G \Leftarrow x \in G. \end{aligned}$$

The rewrite system is certainly complete, yet unification with the h in the term does not cover any ground terms since $G \cap \{a\} = \emptyset$. Furthermore, unification with the variable z does not cover any ground terms either, since $G \cap \{h(a)\} = \emptyset$. But it is obvious that the theorem is true.

Rewriting the grammar once gives

$$\begin{aligned} h(a) &\in G' \\ h(x) &\in G' \Leftarrow x \in G'. \end{aligned}$$

Now rewriting at the term level we split the theorem into

$$\begin{aligned} g(z) = b &\quad\parallel\quad z \in \{h(a)\} \\ g(z) = b &\quad\parallel\quad z \in G' - \{h(a)\}. \end{aligned}$$

After the rewrite this is

$$\begin{aligned} g(a) = b &\quad\parallel\quad z \in \{a\} \\ g(z) = b &\quad\parallel\quad z \in G' - \{h(a)\}. \end{aligned}$$

The first split can be rewritten with the rule $g(a) \rightarrow b$ resulting in

$$b = b \parallel z \in \{a\}$$

which is true and there is at least one ground substitution satisfying the condition. Noting that $G = G' - \{h(a)\}$, the second split is the original constrained theorem and by induction it is proven.

8 Conclusions

Constrained term rewriting provides a natural way to express many conditional systems. In particular, it can be useful in many situations where order-sorted rewriting might be used. There are two "flavors" of conditional systems one can consider. If one allows rewriting within the condition, strong conditions are needed to guarantee that the rewritten grammars/terms are expressible. On the other hand, if the conditions can't be rewritten, the grammars are based on constructor symbols. In this case, the critical pair lemma does not need the extension to handle overlap at a variable position. Critical pairs are generated similarly to standard completion.

Successful completion results in a decision process for the equality of ground terms. Another use for this framework is the natural representation of theorems with inductive proofs, *i.e.*, proofs over a limited set of terms like natural numbers, even numbers, etc. This is of particular interest with respect to the completion process, since showing the joinability of a critical pair is an instance of such theorems.

References

[BT91] B. Bogaert and S. Tison. *Automata with Equality Tests.* Technical Report IT 207, Laboratoire d'Informatique Fondamental de Lille, USTL, Lille, France February 1991.

[Com] H. Comon. *Rewriting with Membership Constraints.* Draft, 1992.

[CD91] H. Comon and C. Delor. *Equational Formulas in Order-Sorted Algebras.* Proc. ICALP, 1990.

[CL89] H. Comon and P. Lescanne. *Equational Problems and Disunification.* J. Symbolic Computation 7, pages 371-425, 1989.

[Der89] N. Dershowitz. *Completion and Its Applications.* In H. Aït-Kaci and M. Nivat, editors, *Resolution of Equations in Algebraic Structures 2: Rewriting Techniques*, chapter 2, pages 31-86. Academic Press, New York, 1989.

[DW91] A. J. J. Dick and P. Watson. *Order-sorted Term Rewriting.* Computer Journal 34, pages 16-19, 1990.

[Don70] J. Doner. *Tree Acceptors and Some of Their Applications*, Journal of Computer and System Sciences 4, pages 406-451, 1970.

[Fri91] A. M. Frisch. *The Substitutional Framework for Sorted Deduction: Fundamental Results on Hybrid Reasoning.* Artificial Intelligence 49, pages 161-198, 1991.

[GM89] J. A. Goguen and J. Meseguer, *Completeness of Many-sorted Equational Logic.* SIGPLAN Notices 16, pages 24-32, 1981.

[KB70] D. E. Knuth and P. B.Bendix, *Simple Word Problems in Universal Algebras*, In J. Leech, editor, *Computational Problems in Abstract Algebra*, pages 263-297. Pergamon Press, Oxford, U. K. 1970.

[KL87] C. Kirchner and P. Lescanne. *Solving Disequations.* Proc. 2nd IEEE Symp.on Logic in Computer Science, Ithaca, NY, pages 347-352, 1987.

[MMR86] A. Martelli, C. Moiso, and G. F. Rossi. *An Algorithm for Unification in Equational Theories.* Proc. IEEE Symp. on Logic in Computer Science, Salt Lake City, UT, September 1986.

[SNGM89] G. Smolka, W. Nutt, J. A. Goguen and J. Meseguer. *Order-Sorted Equational Computation.* In H. Aït-Kaci and M Nivat, editors, *Resolution of Equations in Algebraic Structures, Vol 2, Rewriting Techniques*, pages 297-367, Academic Press, 1989.

[Toy87] Y. Toyama. *Confluent Term Rewriting Systems with Membership Conditions.* In S.Kaplan and J.P. Jouannaud, editors, *Lecture Notes in Computer Science, 308, Conditional Term Rewriting Systems 1987 Proceedings* pages 228-241, Springer Verlag 1987.

[Uri92] T. E. Uribe. *Sorted Unification Using Set Constraints.* In D. Kapur, editor, *Lecture Notes in Computer Science, 11th International Conference on Automated Deduction*, Springer-Verlag,1992.

Generalized Partial Computation using Disunification to Solve Constraints

Akihiko Takano

Advanced Research Laboratory, Hitachi, Ltd.
Hatoyama, Saitama 350-03, Japan.
e-mail: takano@harl.hitachi.co.jp

Abstract

Generalized Partial Computation (GPC) is a program optimization principle based on partial computation and theorem proving. Techniques in conventional partial computation make use of only static values of given data to specialize programs. GPC employs a theorem prover to explicitly utilize more information such as logical structure of programs, axioms for abstract data types, algebraic properties of primitive functions, etc. In this paper we formalize a GPC transformation method for a first-order language which utilizes a disunification procedure to reason about the program context. Context information of each program fragment is represented by a quantifier-free equational formula and is used to eliminate redundant transformation.

1 Introduction

Partial evaluation is known as a program optimization method where the knowledge about the static values of given data which would be passed to the program are used to specialize the program. The recent progress in the field of partial evaluation is striking enough to make us believe that it will continue to be a promising and powerful technique over the next decade ([6, 7]). But these advances also reveal the essential limit of its transformational power.

Generalized Partial Computation (GPC henceforth) was propose as a new paradigm to realize a more powerful transformation. The key idea of GPC is to employ a theorem prover to explicitly utilize more information such as logical structure of programs, axioms for abstract data types, algebraic properties of primitive functions, etc. GPC was first proposed in [3], and its transformational power was demonstrated through many examples in [4, 5]. We formalized a GPC transformation method for a lazy first-order functional language in [9], where the projection was used to approximate the program context in which an expression would be evaluated. There we assumed the availability of some theorem prover to check the satisfiability of the set of equations which described the program context more precisely. But no concrete form of equations nor concrete theorem proving method was given.

The main purpose of this paper is to give a detailed description about this part of the method. The GPC method proposed here gets an original program in the first-order functional language and a set of equational axioms as its input, and returns the specialized residual program which are optimized. The equational axioms capture the properties of data structures and primitive functions.

The *quantifier-free equational formula* is used to describe the program context in which an expression is evaluated. The *disunification* algorithm provides the decision procedure to check the satisfiability of the formula. Program fragments with unsatisfiable contexts correspond to the dead codes, which can be trimmed from the residual program. Our GPC transformation method eliminates the redundant transformation using this form of context information.

This paper is organized as follows. Section 2 describes our subject language which is a first-order functional language with minimal language constructs. In section 3 we formalize a GPC transformation method for the language. Section 4 shows the example application of the proposed method.

2 Language

We consider a set of mutually recursive function definitions $\{ f_i \; v_1 \ldots v_{n_i} = e_i \mid 1 \leq i \leq n \}$ and an expression to be evaluated in the context of these definitions. Expressions have the syntax given by the following grammar:

e	$::=$	v	variable
	$\mid$	$c \; e_1 \ldots e_n$	constructor application
	$\mid$	$p \; e_1 \ldots e_n$	primitive function application
	$\mid$	$f \; e_1 \ldots e_n$	function application
	$\mid$	case e_0 of $cp_1 : e_1 \mid \ldots \mid cp_m : e_m$	case term
cp	$::=$	$c \; v_1 \ldots v_n$	case pattern

In applications, $e_1 \ldots e_n$ are called the *arguments*, and in a case term, e_0 is called the *selector*, and $cp_1 : e_1$, ..., $cp_m : e_m$ are called the *branches*. The case patterns may not be nested. Methods to transform case terms with nested patterns to ones without nested patterns are well known. Each constructor c, primitive function p and user-defined function f has a fixed arity n.

3 GPC Transformation method $\mathcal{G}_\exists$

In this section we define a GPC method $\mathcal{G}_\exists$ which transforms a program in the first-order functional language above. The intended operational semantics of the language is the normal order (leftmost outermost first) graph reduction.

3.1 GPC Principle

Given a syntax tree representing an expression and the information regarding that expression, GPC transformation will involve specializing the expression using that information, and propagating the information toward the leaves of the tree to yield information about the subexpressions to be specialized. The information at the root describes the assumable constraints on the values of the variables, with the result of the propagation describing the constraints on the values of the variables in the lower structure of the expression, which are deducible from the condition at the root. The information at each node is expressed in some logical formula which is used by a theorem prover to specialize the corresponding expression. The specializing process yields a specialized expression (residual expression) equivalent to the original expression on the assumption of the attached information. GPC employs a theorem proving technique to guarantee this equivalence that utilizes additional information such as semantics of the language, axioms for abstract data types, algebraic properties of primitive functions, etc., which are rarely used in conventional partial evaluation.

In this paper the information at each node is expressed as a quantifier-free equational formula defined below. A finite set of equations E, which describes the properties of the data structures and the primitive functions, is given before the analysis. The redundant transformation can be detected at each node by checking the satisfiability of the purely existential equational formula under the set of axioms E. The disunification algorithm which was proposed by Nelson and Oppen [8] is used for this satisfiability check.

We borrow the definition about these basic concepts from [1]. $T(F, X)$ is the set of terms constructed on X and F. Here we take F as a set of all constructors and primitives but exclude user defined functions. An *equation* is an unordered pair of terms s, t denoted by $s = t$. For a finite set of equations E, the congruence $=_E$ generated by E is defined as the smallest equivalence relation on $T(F, X)$ such that, for every context $c[\ ]$, every substitution σ, and every equation $s = t \in E$, $c[s\sigma] =_E c[t\sigma]$. A *quantifier-free equational formula* (qfef for short) is a formula built on the logical connectives $\wedge$, $\vee$, $\neg$, $\top$ and whose atoms are equations. An expression $\neg(s = t)$ is also written as $s \neq t$ and $\neg\top$ is written as $\perp$. A *purely existential equational formula* is the formula of the form $\exists\vec{w} : \phi$, where ϕ is a qfef.

3.2 Rules for $\mathcal{G}_\exists$

GPC transformation method $\mathcal{G}_\exists$ is defined by the set of twelve rules shown below. Write $\mathcal{G}_\exists[\![e]\!]\,\phi$ to denote the residual expression of specializing expression e with the information ϕ.

It is easy to examine that the rules cover all possible expressions: of the five kinds of expression (variable, constructor application, primitive function application, function application, case term) the first four are covered directly, and for case terms, all five possibilities for the selector are considered.

It is required that $\mathcal{G}_\exists[\![e]\!]\,\phi = e$ whenever e satisfies the constraints ϕ. It is clear that each of the rules preserves equivalence.

To keep the information ϕ in the form of qfef, we define the tilde operation to approximate a program expression by a term in $T(F, X)$ which does not include any user defined function application. For an expression e, the term $\tilde{e}$ is got by substituting the all occurrences of function applications with fresh variables. Syntactically equivalent occurrences are replaced with the same variable.

Rule (1) has the top priority, i.e. the applicability of this rule is checked first. Here $\vec{w}$ is the set of all variables in ϕ. The inconsistency of the information could be detected by the disunification. If the attached formula ϕ is unsatisfiable, this portion of program could never contribute to calculate the result. We use the resultant code $\perp$ to represent these useless expression. It is possible to eliminate the case branch whose right hand expression is $\perp$.

In rules (2), (3), and (5), the basic form of the expression is not changed, and the components are converted recursively.

In rule (4), the satisfiability check is used to detect if the expression is definitely reduced to some constructor application form. Here $\vec{w}$ is the set of all variables in the extended formula. Since the number of constructors of the data type of the expression is finite, we can check the all possibilities systematically. The `let` construct is used in the residual program to introduce local variables.

Rule (6) introduces a new function f' to continue specializing that function application. It emulates the popular loop for program transformation: instantiate/unfold/simplify/fold. Here $FV(e)$ denotes the set of all free variables in e. This is the only source of nontermination of this GPC method. We proposed in [9] to use the memoization technique to elaborate this rule and make the transformation terminate more often.

$$\mathcal{G}_3[\![e]\!]\,\phi = \bot \quad \texttt{if } \exists\vec{w}:\phi \text{ is unsatisfiable under } E \tag{1}$$

$$\mathcal{G}_3[\![v]\!]\,\phi = v \tag{2}$$

$$\mathcal{G}_3[\![c\ e_1\ldots e_n]\!]\,\phi = c\ (\mathcal{G}_3[\![e_1]\!]\,\phi)\ldots(\mathcal{G}_3[\![e_n]\!]\,\phi) \tag{3}$$

$$\begin{aligned}\mathcal{G}_3[\![p\ e_1\ldots e_n]\!]\,\phi &= \text{let } c\ v_1\ldots v_k = p\ e_1\ldots e_n \text{ in} \\ &\qquad \mathcal{G}_3[\![c\ v_1\ldots v_k]\!]\,\phi \wedge (p\ \tilde{e_1}\ldots\tilde{e_n} = c\ v_1\ldots v_k) \\ &\qquad\quad \texttt{if } \exists\text{constructor } c \texttt{ s.t.} \\ &\qquad\qquad \exists\vec{w}:\phi\wedge(p\ \tilde{e_1}\ldots\tilde{e_n}\neq c\ v_1\ldots v_k) \text{ is unsatisfiable under } E\end{aligned} \tag{4}$$

$$= p\ (\mathcal{G}_3[\![e_1]\!]\,\phi)\ldots(\mathcal{G}_3[\![e_n]\!]\,\phi) \quad \texttt{otherwise} \tag{5}$$

$$\begin{aligned}\mathcal{G}_3[\![f\ e_1\ldots e_n]\!]\,\phi &= f'\ v_1\ldots v_k \\ &\texttt{where } f \text{ is defined as: } f\ x_1\ldots x_n \stackrel{\text{def}}{=} e \\ &\qquad \text{Define a new function } f' \text{ by:} \\ &\qquad\quad f'\ v_1\ldots v_k \stackrel{\text{def}}{=} \mathcal{G}_3[\![e[x_1:=e_1,\ldots,x_n:=e_n]]\!]\,\phi \\ &\qquad\quad \texttt{where } \{v_1\ldots v_k\} = FV(e_1)\cup\cdots\cup FV(e_n)\end{aligned} \tag{6}$$

$$\begin{aligned}&\mathcal{G}_3[\![\text{case } v \text{ of } p_1:e'_1|\ldots|p_m:e'_m]\!]\,\phi \\ &\quad= \text{case } v \text{ of } p_1:\mathcal{G}_3[\![e'_1]\!]\,\phi\wedge(v=p_1)|\ldots|p_m:\mathcal{G}_3[\![e'_m]\!]\,\phi\wedge(v=p_m)\end{aligned} \tag{7}$$

$$\begin{aligned}&\mathcal{G}_3[\![\text{case } c\ e_1\ldots e_n \text{ of } p_1:e'_1|\ldots|p_m:e'_m]\!]\,\phi \\ &\quad= \mathcal{G}_3[\![e'_i[x_1:=e_1,\ldots,x_n:=e_n]]\!]\,\phi \qquad \texttt{if } \exists\, i \texttt{ s.t. } p_i = c\ x_1\ldots x_n\end{aligned} \tag{8}$$

$$\quad= \bot \qquad \texttt{otherwise} \tag{9}$$

$$\begin{aligned}&\mathcal{G}_3[\![\text{case } p\ e_1\ldots e_n \text{ of } p_1:e'_1|\ldots|p_m:e'_m]\!]\,\phi \\ &\quad= \text{case } e \text{ of} \\ &\qquad p_1:\mathcal{G}_3[\![e'_1]\!]\,\phi\wedge(\tilde{e}=p_1)|\ldots|p_m:\mathcal{G}_3[\![e'_m]\!]\,\phi\wedge(\tilde{e}=p_m) \\ &\quad \texttt{where } \mathcal{G}_3[\![p\ e_1\ldots e_n]\!]\,\phi = e\end{aligned} \tag{10}$$

$$\begin{aligned}&\mathcal{G}_3[\![\text{case } f\ e_1\ldots e_n \text{ of } p_1:e'_1|\ldots|p_m:e'_m]\!]\,\phi \\ &\quad= \text{case } e \text{ of} \\ &\qquad p_1:\mathcal{G}_3[\![e'_1]\!]\,\phi\wedge(\tilde{e}=p_1)|\ldots|p_m:\mathcal{G}_3[\![e'_m]\!]\,\phi\wedge(\tilde{e}=p_m) \\ &\quad \texttt{where } \mathcal{G}_3[\![f\ e_1\ldots e_n]\!]\,\phi = e\end{aligned} \tag{11}$$

$$\begin{aligned}&\mathcal{G}_3[\![\text{case (case } e \text{ of } p_1:e_1|\ldots|p_n:e_n) \text{ of } p'_1:e'_1|\ldots|p'_m:e'_m]\!]\,\phi \\ &\quad= \mathcal{G}_3[\![\text{case } e \text{ of} \\ &\qquad\quad p_1:(\text{case } e_1 \text{ of } p'_1:e'_1|\ldots|p'_m:e'_m)|\ldots| \\ &\qquad\quad p_n:(\text{case } e_n \text{ of } p'_1:e'_1|\ldots|p'_m:e'_m)]\!]\,\phi\end{aligned} \tag{12}$$

In rule (8), the form of the expression tells which branch should be used, and the right hand expression of that branch is converted recursively. Rule (9) is used to eliminate the expression which is never used to calculate the result.

In rules (7), (10), and (11), we don't know which branch is used to calculate the result. The form of the case term is preserved and the components are converted recursively. The information of each branch is extended differently, which reflects the semantics of the case term: the right hand expression of each branch is selected to calculate the result if and only if the selector matches with the case pattern. Each branch information is extended with the identity which expresses that the selector matches the corresponding case pattern. Variables in the case pattern should be renamed to avoid conflict with the constant names in ϕ.

This extension of the information is the central idea of GPC and the most essential difference from conventional partial evaluation. It can be said that the whole GPC method is devised to utilize this extended information.

For rule (14), the nested case term is simplified, and the result is converted recursively.

4 Examples

In this extended abstract we have not enough room to show the detail of the example application of our method. We just report the result about the standard example of the string pattern matcher.

Starting with the naïve string pattern matcher and the simple four axioms on the list structure in [8], we get the optimized residual program which is as efficient as the KMP pattern matcher automatically.

5 Conclusion

In this paper we formalize a GPC transformation method for a first-order functional language which use the disunification procedure to reason about program contexts. GPC provides the natural setting to utilize additional information such as logical structure of programs, axioms for abstract data types, algebraic properties of primitive functions, etc. which have been rarely used in conventional partial evaluation.

For simplicity, we assume to use the disunification algorithm by Nelson and Oppen, and because of the requirement of it we adopt the qfef to represent the context information. Since the structure of the proposed GPC method put no restriction on the class of formulas and the structure of the disunification algorithm, we can freely incorporate the other modern disunification algorithms ([2]). GPC is surely one of the promising fields of application for the disunification and related researches.

References

[1] H. Comon. Disunification: a survey. In J.-L. Lassez and G. Plotkin, eds., *Computational Logic: Essays in Honor of Alan Robinson*. MIT Press, 1991.

[2] H. Comon, M. Haberstrau and J.-P. Jouannaud. Decidable Problems in Shallow Equational Theories. In *Proc. 7th IEEE Symp. Logic in Computer Science*, June 1992.

[3] Y. Futamura and K. Nogi. Generalized Partial Computation. In D. Bjørner, A. P. Ershov and N. D. Jones, eds., *Partial Evaluation and Mixed Computation*, 133-151, North-Holland, 1988.

[4] Y. Futamura and K. Nogi. Program Evaluation and Generalized Partial Computation. In *Proceedings of the International Conf. on Fifth Generation Computer Systems*, 685-692, Tokyo, 1988.

[5] Y. Futamura, K. Nogi and A. Takano. Essence of Generalized Partial Computation. *Theoretical Computer Science*, 90 : 61-79, 1991.

[6] N.D. Jones, P. Sestoft and H. Søndergaard. MIX: An Self-applicable Partial Evaluator for Experiments in Compiler Generation. *LISP and Symbolic Computation*, 2 (1) : 9-50, 1989.

[7] J. Launchbury. Projection Factorizations in Partial Evaluation. Cambridge University Press, 1991.

[8] G. Nelson and D. C. Oppen. Fast decision procedures based on congruence closure. *J. ACM*, 27 : 356-364, 1980.

[9] A. Takano. Generalized Partial Computation for a Lazy Functional Language. In *Proc. ACM Symp. on Partial Evaluation and Semantics-Based Program Manipulation. ACM SIGPLAN Notices*, 26 (9) : 1-11, June 1991.

Decidability of finiteness properties

Leszek Pacholski
Institute of Mathematics
Polish Academy of Sciences
and
Institute of Computer Science
University of Wrocław

Abstract

We are going to survey decidability results concerning finiteness of sets of terms or formulas that arise in investigations of equational bases of abstract algebras, axiomatisations of propositional logics, unifications theory, database theory, and logic programming.

In particular we are going to provide an unified treatment of finiteness properties that lead to several undecidability results, including for example the old result by Linial and Post solving a problem of Tarski and the recent theorem by Marcinkowski and Pacholski on the undecidability of the Horn clause implication problem.

The methods are based on a thorough analysis of the structure of derivation trees in "logics" based on one rule of inference.

Termination Proofs of Well-Moded Logic Programs via Conditional Rewrite Systems*

Harald Ganzinger, Uwe Waldmann

Max-Planck-Institut für Informatik,
Im Stadtwald, D-W-6600 Saarbrücken, Germany,
{hg,uwe}@mpi-sb.mpg.de

Abstract. In this paper, it is shown that a translation from logic programs to conditional rewrite rules can be used in a straightforward way to check (semi-automatically) whether a program is terminating under the prolog selection rule.

1 Introduction

In the last years, several methods have been proposed to check automatically, or at least semi-automatically, the termination of logic programs. The methods of Ullman and Van Gelder [11] and Plümer [9] are perhaps the best-known ones, however their applicability suffers from severe restrictions. A recent paper by Krishna Rao et al. [7] describes a termination proof technique for well-moded programs that works by a translation of programs to unconditional rewrite systems. We show that a translation to conditional rewrite systems is not only much easier to describe and to analyze, but moreover is able to prove programs terminating for which the method of [7] fails.

We assume the reader to be familiar with the basic concepts of logic programming and term rewrite systems, e.g., as described by Lloyd [8] and Dershowitz and Jouannaud [4].

2 Well-Moded Logic Programs

Logic programs differ from functional or imperative programs in that the flow of information is not not necessarily fixed a priori. When a predicate is called, a variable may be uninstantiated, instantiated to a non-ground term, or instantiated to a ground term. This generality does not only complicate the analysis of logic programs considerably, it is also often unnecessary in practice, since the author of a program frequently intends every predicate to be used in one fixed restricted manner. Modes can be used to describe these restrictions and thus simplify the analysis of logic programs.

* This research was funded by the German Ministry for Research and Technology (BMFT) under grant ITS 9103 and by the DFG project "Redundanz" (Az. Ga 261, SPP Deduktion). The responsibility for the contents of this publication lies with the authors.

Definition 1. A mode m of an n-ary predicate p is a function from $\{1, \ldots, n\}$ to the set $\{in, out\}$. If $m(i) = in$, i is called an *input position* of p, otherwise, i is called an *output position*. We say that a variable x occurs in an input (output) position of a literal $p(t_1, \ldots, t_n)$, if it occurs in some t_i such that $m(i) = in$ $(m(i) = out)$.

If a predicate p is used with different modes $m_1, \ldots, m_n$ in a program, we may consider each $p^{(m_i)}$ as a separate predicate; so we can assume without loss of generality that every predicate has exactly one mode.

Definition 2. Let x be a variable in a clause $A \leftarrow B_1, \ldots, B_n$. The head A is called a *producer* (*consumer*) of x, if x occurs in an input (output) position of A, conversely, a body literal B_j is called a *producer* (*consumer*) of x, if x occurs in an output (input) position of B_j.

In the following we will only consider left-to-right SLD-derivations, i.e., SLD-derivations via the so-called Prolog computation rule, which selects always the left-most literal of a query for the next resolution step.[2]

Definition 3. A clause $B_0 \leftarrow B_1, \ldots, B_n$ is called *LR-well-moded*, if every variable x in the clause has a producer B_i $(0 \leq i \leq n)$ and $i < j$ for every consumer B_j $(1 \leq i \leq n)$ of x in the body of the clause. A program is called *LR-well-moded*, if each of its clauses is LR-well-moded. An *LR-well-moded query* is an LR-well-moded clause without head.[3]

An easy proof by induction shows the following lemma.

Lemma 4. *Let P be an LR-well-moded program and let $G_0, G_1, \ldots$ be a left-to-right SLD-derivation starting with an LR-well-moded query G_0. Then all queries G_i are LR-well-moded, and the first literal of every non-empty G_i is ground on all its input positions.*

3 Quasi-Reductive Term Rewrite Systems

Quasi-reductive conditional rewrite rules were introduced by Ganzinger [5] in order to efficiently translate order-sorted specifications into conditional many-sorted equations. In the classical case of rewriting with reductive conditional rules, the conditions are only used for checking whether the rule can be applied. Quasi-reductivity is a generalization of reductivity. Conditions are now oriented; and instead of normalizing both sides of a condition equation, we rewrite only the left-hand side and match the result with the right-hand side. In this way we may compute values for variables occurring in the subsequent conditions or in the right-hand side of the rule.

[2] Our method can be easily extended to stronger notions of termination, for example by considering all permutations of the literals in the bodies of the clauses, such that the resulting clause remains well-moded.

[3] Similar restrictions can be found in Ground Prolog as described by Kluźniak [6]. We do not require, however, that the output parameters of every goal are unbound variables.

The traditional requirement that every variable in a condition or in the right-hand side of the rule must also occur in the left-hand side can be weakened.

We assume that $\succ$ is a reduction ordering on $T_\Sigma(X)$. The symbol $\succ_{st}$ denotes the transitive closure of $\succ \cup st$, where st is the strict subterm ordering. The ordering $\succ_{st}$ is well-founded and stable under substitutions.

Definition 5. A conditional rewrite rule $u_1 \to v_1, \ldots, u_n \to v_n \Rightarrow s \to t$ with $n > 0$ is called *deterministic*, if we have for every $1 \le i \le n$:

$$\mathrm{var}(u_i) \subseteq \mathrm{var}(s) \cup \bigcup_{j=1}^{i-1} (\mathrm{var}(u_j) \cup \mathrm{var}(v_j))$$

and

$$\mathrm{var}(t) \subseteq \mathrm{var}(s) \cup \bigcup_{j=1}^{n} (\mathrm{var}(u_j) \cup \mathrm{var}(v_j)).$$

Definition 6. A deterministic rewrite rule $u_1 \to v_1, \ldots, u_n \to v_n \Rightarrow s \to t$ is called *quasi-reductive*, if the following two conditions are satisfied for every substitution σ and for every $0 \le i < n$:

- if $u_j\sigma \succ v_j\sigma$ for every $1 \le j \le i$, then $s\sigma \succ_{st} u_{i+1}\sigma$,
- if $u_j\sigma \succ v_j\sigma$ for every $1 \le j \le n$, then $s\sigma \succ t\sigma$.[4]

An unconditional rewrite rule $s \to t$ is called *quasi-reductive*, if $s \succ t$. We say that a conditional rewrite system is quasi-reductive, if each of its rules is quasi-reductive.

Definition 7. Let R be a set of rewrite rules. The rewrite relation $\xrightarrow[Q]{}_R$ is defined as the smallest relation such that $w[s\sigma] \xrightarrow[Q]{}_R w[t\sigma]$ holds whenever $\rho = u_1 \to v_1, \ldots, u_n \to v_n \Rightarrow s \to t$ $(n \ge 0)$ is a rule in R, σ is a substitution on $\mathrm{var}(\rho)$, and $u_i\sigma \xrightarrow[Q]{}{}^{+}_R v_i\sigma$ for $1 \le i \le n$.

Lemma 8. *If R is a quasi-reductive rewrite system, then $s \xrightarrow[Q]{}_R t$ implies $s \succ t$.*

Proof. By induction on $\succ_{st}$. □

Algorithms that can be used to decide quasi-reductivity have been presented, for example, by Comon [3] for lexicographical path orderings, and by Tarski [10] and Collins [2] for polynomial orderings over the real numbers. A simple sufficient criterion to check quasi-reductivity for arbitrary reduction orderings is given in [5]:

Lemma 9. *Let Σ' be an enrichment of the original signature Σ, such that the ordering $\succ$ can be extended to a reduction ordering over $T_{\Sigma'}(X)$. A deterministic rewrite rule $u_1 \to v_1, \ldots, u_n \to v_n \Rightarrow s \to t$ is quasi-reductive, if there exists a sequence $h_i(x)$ of terms in $T_{\Sigma'}(X)$, $x \in X$, such that $s \succ h_1(u_1)$, $h_i(v_i) \succeq h_{i+1}(u_{i+1})$ for every $1 \le i < n$, and $h_n(v_n) \succeq t$.*

[4] We depart from [5] in that we substitute $u_j\sigma \succ v_j\sigma$ for $u_j\sigma \succeq v_j\sigma$; this yields a slightly weaker property, which is, however, sufficient for our purpose.

4 Termination of Logic Programs

We will now associate a conditional rewrite system with every logic program, such that the quasi-reductivity of the conditional rewrite system implies that any left-to-right SLD-derivation starting with an LR-well-moded query terminates.

Definition 10. Let Σ' be an enrichment of the original signature Σ. A *QR-interpretation* q is a function from predicate symbols to pairs of terms over Σ'. If p is an n-ary predicate symbol with input positions $1, \ldots, m$ and output positions $m+1, \ldots, n$ (without loss of generality), then q maps p to two terms (p_{in}, p_{out}) such that $\mathrm{var}(p_{in}) = \{x_1, \ldots, x_m\}$ and $\mathrm{var}(p_{out}) = \{x_{m+1}, \ldots, x_n\}$.

For every literal $A = p(t_1, \ldots, t_n)$ let $in(A) = p_{in}(x_1 \mapsto t_1, \ldots, x_m \mapsto t_m)$ and $out(A) = p_{out}(x_{m+1} \mapsto t_{m+1}, \ldots, x_n \mapsto t_n)$; then for each clause $c = B_0 \leftarrow B_1, \ldots, B_k$ the rewrite rule $q(c)$ is

$$in(B_1) \rightarrow out(B_1),\ \ldots,\ in(B_k) \rightarrow out(B_k) \Rightarrow in(B_0) \rightarrow out(B_0).$$

If P is a logic program, then $q(P) = \{\, q(c) \mid c \in P \,\}$.

Lemma 11. *If $c = B_0 \leftarrow B_1, \ldots, B_k$ is an LR-well-moded clause with non-empty body, then $q(c)$ is deterministic.*

Lemma 12. *If $\leftarrow B_1, \ldots, B_k$ is an LR-well-moded query, then $in(B_1)$ is ground.*

Lemma 13. *Let P be an LR-well-moded logic program such that $R = q(P)$ is quasi-reductive, and let $\leftarrow A$ be an LR-well-moded query. If there is a left-to-right SLD-refutation of $\leftarrow A$ with the computed answer θ, then $in(A) = in(A\theta) \xrightarrow[Q]{}_R out(A\theta)$.*

Proof. The lemma is proved by noetherian induction on $in(A)$ with respect to $\succ_{st}$; so assume that it holds for every query $\leftarrow A'$ with $in(A) \succ_{st} in(A')$.

Suppose that the first step of the derivation uses an appropriately renamed variant $B \leftarrow B_1, \ldots, B_n$, $n \geq 0$, of a program clause, and θ_0 is the most general unifier of A and B. The refutation $\leftarrow A \longrightarrow^+_{\mathrm{SLD}} \Box$ has the form

$$\begin{array}{ll} & \leftarrow A \\ \longrightarrow_{\mathrm{SLD}} & \leftarrow B_1\theta_0, B_2\theta_0, \ \ldots, B_n\theta_0 \\ \longrightarrow^+_{\mathrm{SLD}} & \qquad \leftarrow B_2\theta_0\theta_1, \ldots, B_n\theta_0\theta_1 \\ \vdots & \\ \longrightarrow^+_{\mathrm{SLD}} & \qquad\qquad \leftarrow B_n\theta_0\theta_1 \ldots \theta_{n-1} \\ \longrightarrow^+_{\mathrm{SLD}} & \qquad\qquad \Box \end{array}$$

where θ_i $(1 \leq i \leq n)$ is the computed answer of the refutation of $\leftarrow B_i\theta_0 \ldots \theta_{i-1}$ and θ is the restriction of $\theta_0 \ldots \theta_n$ to $\mathrm{var}(A)$.

By assumption, q maps the clause $B \leftarrow B_1, \ldots, B_n$ to the quasi-reductive rewrite rule $in(B_1) \rightarrow out(B_1),\ \ldots,\ in(B_k) \rightarrow out(B_k) \Rightarrow in(B) \rightarrow out(B)$.

We will now prove the following auxiliary statement: For all $1 \leq i \leq n$, we have $in(B_i\theta_0 \dots \theta_k) \xrightarrow[Q]{}_R out(B_i\theta_0 \dots \theta_k)$ for $i \leq k \leq n$ and $in(A\theta_0 \dots \theta_m) \succ_{st} in(B_i\theta_0 \dots \theta_m)$ for $i - 1 \leq m \leq n$.

Assume that the auxiliary statement holds for all $j < i$. As $in(B_j\theta_0 \dots \theta_k) \xrightarrow[Q]{}_R out(B_j\theta_0 \dots \theta_k)$ for every $j < i$ and $k \geq j$, we have $in(A) = in(A\theta_0 \dots \theta_m) = in(B\theta_0 \dots \theta_m) \succ_{st} in(B_i\theta_0 \dots \theta_m)$ for each $m \geq i - 1$ by quasi-reductivity. Since $in(A) \succ_{st} in(B_i\theta_0 \dots \theta_{i-1})$, we obtain $in(B_i\theta_0 \dots \theta_{i-1}\theta_i) \xrightarrow[Q]{}_R out(B_i\theta_0 \dots \theta_{i-1}\theta_i)$, and thus $in(B_i\theta_0 \dots \theta_k) \xrightarrow[Q]{}_R out(B_i\theta_0 \dots \theta_k)$ for $k \geq i$. This proves the auxiliary statement.

Since $in(B_i\theta_0 \dots \theta_n) \xrightarrow[Q]{}_R out(B_i\theta_0 \dots \theta_n)$ holds for every $1 \leq i \leq n$, we have now $in(A\theta_0 \dots \theta_n) = in(B\theta_0 \dots \theta_n) \xrightarrow[Q]{}_R out(B\theta_0 \dots \theta_n) = out(A\theta_0 \dots \theta_n)$, and this implies $in(A) = in(A\theta) \xrightarrow[Q]{}_R out(A\theta)$. □

Theorem 14. *If P is an LR-well-moded logic program such that $R = q(P)$ is quasi-reductive, then every left-to-right SLD-derivation starting with an LR-well-moded query G terminates.*

Proof. Obviously it suffices to consider queries of the form $\leftarrow A$. The lemma is proved by noetherian induction on $in(A)$ with respect to $\succ_{st}$. Let $B \leftarrow B_1, \dots, B_n$, $n \geq 0$, be a variant of a program clause that is used in the first derivation step and let θ_0 be the most general unifier of A and B. By quasi-reductivity, $in(A) = in(A\theta_0) \succ_{st} in(B_1\theta_0)$, hence the induction hypothesis shows that there is no infinite derivation starting with $B_1\theta_0$. So we have to distinguish two cases: If the derivation starting with $\leftarrow B_1\theta_0$ fails, the whole derivation starting with $\leftarrow A$ is obviously finite. Otherwise, $\leftarrow B_1\theta_0$ gets refuted with the computed answer θ_1, and by the previous lemma we know that $in(B_1\theta_0\theta_1) \xrightarrow[Q]{}_R out(B_1\theta_0\theta_1)$. Since $q(B \leftarrow B_1, \dots, B_n)$ is quasi-reductive, we can conclude that $in(A) = in(A\theta_0\theta_1) \succ_{st} in(B_2\theta_0\theta_1)$. Repeating this process, we see that the whole derivation terminates. □

5 Examples

Krishna Rao et al. [7] show that their method fails to prove the termination of the following logic program, which computes the transitive closure of a relation p.

$p(a, b) \leftarrow$
$p(b, c) \leftarrow$

$tc(x, x) \leftarrow$
$tc(x, y) \leftarrow p(x, z),\ tc(z, y)$

with the modes $p^{(in,out)}$ and $tc^{(in,out)}$. The failure is due to the fact that their translation to an unconditional rewrite system produces the non-terminating rewrite rule $tc(x) \rightarrow tc(p(x))$.[5]

[5] A further transformation of the rewrite system leads in fact to a terminating system. This, however, is not simply terminating, i. e., it cannot be proved terminating using any simplification ordering (e. g., recursive path orderings or polynomial orderings).

We choose a QR-interpretation q that maps p to the pair $(pin(x_1), pout(x_2))$ and tc to the pair $(tc(x_1), x_2)$. This yields the conditional rewrite system

$$\begin{aligned} pin(a) &\to pout(b) \\ pin(b) &\to pout(c) \\ tc(x) &\to x \\ pin(x) \to pout(z),\ tc(z) \to y \Rightarrow tc(x) &\to y \end{aligned}$$

This can be easily shown to be quasi-reductive, for example using Lemma 9 and a recursive path ordering with precedences $a \succ b \succ c \succ pout \succ tc \succ h_1 \succ pin \succ h_2$. Alternatively, quasi-reductivity can be shown directly using the polynomial interpretations $[\![pin]\!](x) = x + 1$, $[\![pout]\!](x) = x + 1$, $[\![tc]\!](x) = x + 2$, $[\![a]\!] = 4$, $[\![b]\!] = 3$, and $[\![c]\!] = 2$.

The program terminates because the second argument of p is smaller than the first one whenever a call of p succeeds. The fact that we can encode this by choosing appropriate interpretations for *pin* and *pout* is crucial for our ability to prove the termination. A similar technique is used for the *split* predicate in the following quicksort program:

$$\begin{aligned} &qs([\,],[\,]) \leftarrow \\ &qs(x.l, s) \leftarrow split(l, x, l_1, l_2),\ qs(l_1, s_1),\ qs(l_2, s_2),\ append(s_1, x.s_2, s) \end{aligned}$$

$$\begin{aligned} &split([\,], x, [\,], [\,]) \leftarrow \\ &split(x.l, y, x.l_1, l_2) \leftarrow x < y,\ split(l, x, l_1, l_2) \\ &split(x.l, y, l_1, x.l_2) \leftarrow x \geq y,\ split(l, x, l_1, l_2) \end{aligned}$$

$$\begin{aligned} &append([\,], l, l) \leftarrow \\ &append(x.l_1, l_2, x.l_3) \leftarrow append(l_1, l_2, l_3) \end{aligned}$$

with the modes $qs^{(in,out)}$, $split^{(in,in,out,out)}$, $append^{(in,in,out)}$, $<^{(in,in)}$ and $\geq^{(in,in)}$. It can be translated into the conditional rewrite system

$$\begin{aligned} qs([\,]) &\to [\,] \\ splitin(l, x) \to splitout(l_1, l_2),\ qs(l_1) \to s_1,\ qs(l_2) \to s_2,\ append(s_1, x.s_2) \to s \Rightarrow qs(x.l) &\to s \end{aligned}$$

$$\begin{aligned} splitin([\,], x) &\to splitout([\,],[\,]) \\ x < y \to true,\ splitin(l, y) \to splitout(l_1, l_2) \Rightarrow splitin(x.l, y) &\to splitout(x.l_1, l_2) \\ x \geq y \to true,\ splitin(l, y) \to splitout(l_1, l_2) \Rightarrow splitin(x.l, y) &\to splitout(l_1, x.l_2) \end{aligned}$$

$$\begin{aligned} append([\,], l) &\to l \\ append(l_1, l_2) \to l_3 \Rightarrow append(x.l_1, l_2) &\to x.l_3 \end{aligned}$$

We can prove this system to be quasi-reductive using the polynomial interpretations[6] $[\![qs]\!](x) = 2x$, $[\![splitin]\!](x, y) = x + y + 1$, $[\![splitout]\!](x, y) = x + y$, $[\![append]\!](x, y) = x + y$, $[\![.]\!](x, y) = 3x + y$, $[\![[\,]]\!] = 2$, $[\![<]\!](x, y) = x + y$, $[\![\geq]\!](x, y) = x + y$, $[\![true]\!] = 2$.

[6] All variables range over $\mathbb{N} \setminus \{0, 1\}$

The last example demonstrates that our method doesn't yield a complete criterion for proving termination. Consider the following program

$$p(x, g(x)) \leftarrow$$
$$p(x, f(y)) \leftarrow p(x, g(y))$$

with the mode $p^{(in,out)}$. This program terminates with every LR-well-moded query, but the input part remains constant during the recursive call of p. So it is impossible to translate this program into a quasi-reductive system.

6 Conclusions

We have shown how the termination of well-moded logic programs can be proved using quasi-reductive conditional rewrite rules. Compared to the translation to unconditional rewrite systems that was proposed by Krishna Rao et al. [7], our method is not only much easier to describe and to analyze,[7] it is also able to prove the termination of logic programs for which the criterion of [7] fails.

Plümer's method [9] (and even more Ullman's and Van Gelder's approach [11]) is subject to stronger syntactical restrictions than ours. It has, however, the advantage of being completely automatable, whereas our criterion can only be used semi-automatically. On the other hand, Apt and Pedreschi [1] present a method that is not only complete, but that can even be applied to non-well-moded programs. Still it seems that even partially mechanical proof checking is much more complicated for their criterion than for ours.

References

1. Krzysztof R. Apt and Dino Pedreschi. Studies in pure Prolog: Termination. Technical Report CS-R9048, Centrum voor Wiskunde en Informatica, Amsterdam, The Netherlands, September 1990.
2. George E. Collins. Quantifier elimination for real closed fields by cylindrical algebraic decomposition. In H. Brakhage, editor, *Automata Theory and Formal Languages, 2nd GI Conference*, LNCS 33, pages 134–183, Kaiserslautern, West Germany, May 20–23, 1975. Springer-Verlag.
3. Hubert Comon. Solving inequations in term algebras (extended abstract). In *Fifth Annual IEEE Symposium on Logic in Computer Science*, pages 62–69, Philadelphia, PA, USA, June 4–7, 1990. IEEE Computer Society Press, Los Alamitos, CA, USA.
4. Nachum Dershowitz and Jean-Pierre Jouannaud. Rewrite systems. In Jan van Leeuwen, editor, *Handbook of Theoretical Computer Science*, volume B: Formal Models and Semantics, chapter 6, pages 244–320. Elsevier Science Publishers B.V., Amsterdam, New York, Oxford, Tokyo, 1990.
5. Harald Ganzinger. Order-sorted completion: the many-sorted way. *Theoretical Computer Science*, 89:3–32, 1991.
6. Feliks Kluźniak. Type synthesis for Ground Prolog. In Jean-Louis Lassez, editor, *Logic Programming, Proceedings of the Fourth International Conference*, volume 2, pages 788–816, Melbourne, Australia, May 25–29, 1987. The MIT Press.

[7] For instance, compare Def. 10 with the formal description of the translation in [7], which fills a whole A4 page.

7. M. R. K. Krishna Rao, Deepak Kapur, and R. K. Shyamasundar. A transformational methodology for proving termination of logic programs. Draft, Computer Science Group, Tata Institute of Fundamental Research, Bombay, India, February 28, 1992. To appear in: Proceedings of the 5th Conference on Computer Science Logic 1991, LNCS, Springer-Verlag.
8. John Wylie Lloyd. *Foundations of Logic Programming*. Springer-Verlag, Berlin, Heidelberg, New York, London, Paris, Tokyo, second, extended edition, 1987.
9. Lutz Plümer. *Termination Proofs for Logic Programs*. Dissertation, Universität Dortmund, Abteilung Informatik, Dortmund, Germany, 1989. Short version: Termination Proofs for Logic Programs based on Predicate Inequalities, in David H. D. Warren and Peter Szeredi, eds., *Logic Programming, Proceedings of the Seventh International Conference*, Jerusalem, Israel, June 18–20, 1990, pages 634–648, The MIT Press.
10. Alfred Tarski. *A Decision Method for Elementary Algebra and Geometry*. University of California Press, Berkeley, second, revised edition, 1951.
11. Jeffrey D. Ullman and Allen Van Gelder. Efficient tests for top-down termination of logical rules. *Journal of the ACM*, 35(2):345–373, April 1988.

Logic Programs with Polymorphic Types: A condition for static type checking

Staffan Bonnier* and Jonas Wallgren

Department of Computer and Information Science
Linköping University
S-581 83 Linköping
{sbo,jwc}@ida.liu.se

Abstract. This paper describes a polymorphic type system for logic programs. Well-typedness is defined on so-called execution modules of the program, which is shown to yield better results than [7] in many cases. The system is partly implemented.

1 Introduction and background

There have been several attempts to introduce polymorphic type disciplines to logic programs (e.g. [4, 7, 9]). It is known that static type checking in the presence of polymorphism is difficult and several sufficient conditions for achieving this goal has been formulated in the literature. The classical approach of Mycroft and O'Keefe [7] defines the concept of well-typing of a definite program with respect to a priori given types of predicates and function symbols of the program. Their sufficient condition for static typing requires that for the head atom $p(t_1, ..., t_n)$ of any program clause the types of $t_1,...,t_n$ are identical (up to renaming of type variables) with the argument types of p. For example, consider the following program that for any list either finds its last element if the list is non-empty, or returns the integer 0 if the list is empty:

$p([\_|L], X) \leftarrow p(L, X).$
$p([X], X).$
$p([\,], 0).$

If the argument types of p are $\alpha\ list, \beta$ then the program does not fulfil the condition of Mycroft and O'Keefe since the type of the atom $p([\,], 0)$ is a proper instance of the general type of p. Additionally Mycroft and O'Keefe assume that the function symbols used in the program are "type preserving": this means that all type variables appearing in the argument types of the function type are supposed to appear also in the type of the result. For example the function symbol *cons* of type $\alpha \times \alpha\ list \rightarrow \alpha\ list$ is type-preserving, while the function symbol *length* of type $\alpha\ list \rightarrow int$ is not. In this paper we give a more general condition for static well-typing of polymorphically typed logic programs. As a matter of fact this work was done in the context of integrating logic programs

* Currently at École Normale Supérieure, Paris, France.

with external functional procedures [1, 2, 5]. For the case of polymorphic typing of such procedures the restriction to type-preserving functions seems very unnatural since it bans commonly used procedures, like the above mentioned length of lists. Our condition allows for (a limited) use of such procedures and generalises the Mycroft and O'Keefe approach. It applies also to logic programs using external procedures [1] but this topic is not discussed in the paper.

It is well known (see e.g. [9]) that the main reason of difficulties in static type checking of logic programs is the use of non-type-preserving function symbols in unification. In particular, every n-ary predicate of a typed program can be considered as a function symbol of type $\tau_1 \times \ldots \times \tau_n \rightarrow \mathbf{o}$, where $\mathbf{o}$ is a specific type constant reserved for this purpose. Thus, no predicate is type preserving, and unification of two well-typed atoms may lead to an ill-typed result. For example, if p is of the type $\alpha\ list \times \alpha\ list \rightarrow \mathbf{o}$ then the atoms $p([1], X)$ and $p(Y, [true])$ are well-typed but their most general unifier binds X to $[true]$ and Y to $[1]$ and the resulting atom $p([1], [true])$ is ill-typed. Mycroft and O'Keefe eliminate such cases by the requirement concerning the types of the unified atoms: one of them has to be as general as the type of p. This condition is not fulfilled by any of the atoms in our example. To achieve the condition for any unification that takes place in the execution of a program it suffices to require that the type of every head atom of any clause is as general as the type of its predicate. An alternative approach to achieve well-typedness of the result of unification of well-typed atoms is to require that the types of the corresponding arguments of the atoms are unifiable. This extends naturally to the case of unification of any terms whose principal functors are not type preserving. However, it is not so easy to impose static restriction for a logic program that guarantees satisfaction of this condition for any unification that takes place during execution of the program. In this paper we formulate such a restriction using a concept of *execution module* and we formalize the condition in terms of the notion of *strict well-typedness*. An execution module of a given program and goal is a finite set of atoms and there is only a finite number of such sets. The restriction is then to require that any execution module is strictly well-typed. An algorithm for finding execution modules of a given program and goal and an algorithm for checking well-typedness of (sets of) terms are given in the paper. They have been implemented in Prolog and give a basis for a prototype type checker, which can also be used for type inference.

The paper is organised as follows.

Section 2 contains some preliminary definitions and the definitions of types for terms. Section 3 contains the definitions of types for atoms and programs. Section 4 gives an algorithm for inferring/checking those types and Section 5, finally, concludes the paper.

2 Preliminaries

We assume that the reader is familiar with the usual definitions concerning definite programs (see e.g. [6]). We follow the syntactic approach of [7] to polymorphic typing of definite programs in that we consider types to be terms over certain stratified alphabets and we deal with type assignments to function symbols and predicates that allow us to type terms and atoms.

Thus, to construct our programs and goals we assume the following alphabets to be given:

Φ_i	-	the alphabet of *function symbols* of arity $i \geq 0$,
Π_i	-	the alphabet of *predicate symbols* of arity $i \geq 0$,
$\mathcal{V}$	-	the countably infinite alphabet of *variables*.

It is further assumed that the alphabets are pairwise disjoint. The notation Φ and Π will be used for $\bigcup_{i\geq 0} \Phi_i$ and $\bigcup_{i\geq 0} \Pi_i$ respectively. The terms, the atomic formulae, and the definite programs are constructed from these alphabets in the usual way.

Similarly we assume alphabets of type constructors and type variables to be given. Terms built from these alphabets will be called type expressions (or just types), and we will use the notation *Texp* for the set of all such.

We assume that the function symbols of our alphabet and the variables have assigned types. Formally this is covered by the notion of *type assignment*.

Definition 1. The notion of *type assignment* is defined for each of Φ and $\mathcal{V}$:

- for Φ; A mapping $\phi : \Phi \rightarrow \bigcup_{i\geq 1} Texp^i$ such that $\phi(f) \in Texp^{n+1}$ if $f \in \Phi_n$,
- for $\mathcal{V}$; A mapping $\nu : \mathcal{V} \rightarrow Texp$.

In the sequel ϕ is assumed to be fixed.

The notation $T_1, .., T_n \rightarrow T_{n+1}$ will be used for $\phi(f)$.

A function f is said to be polymorphic if $\phi(f)$ includes (type) variables. As pointed out above, a special kind of typed functions, called *type-preserving* functions, play an important role in formulating conditions for static type-checking of typed logic programs. We recall the related definitions.

Definition 2. The set L of *local* type variables of f is defined by:

$$L = (\bigcup_{1\leq i\leq n} \mathsf{tvar}(T_i)) - \mathsf{tvar}(T_{n+1})$$

where $\mathsf{tvar}(T)$ denotes the set of all type variables occuring in T.

When $L = \emptyset$, f is said to be *type preserving*.

However, our approach to static type checking of logic programs will not be based on the notion of type-preserving functions. We will instead use a notion of strictly typed term. To define it we need the following auxiliary definition.

Definition 3. A substitution σ is said to be a renaming of X *away from* Y iff σ is a renaming of X such that $\forall x \in X : \sigma x \notin Y$

A type substitution τ is *strict* for f w r t ϕ iff the restriction of τ to L (the set of local type variables of f) is a type renaming of L away from $\mathsf{tvar}(\tau T_{n+1})$, where $\phi(f) = T_1, \ldots, T_n \rightarrow T_{n+1}$.

The definition can be illustrated by the following example.

Example 1. Let $\phi(\mathtt{f}) = (\widetilde{\mathtt{pair}}\ \tilde{\mathtt{x}}\ \tilde{\mathtt{y}}) \rightarrow (\widetilde{\mathtt{list}}\ \tilde{\mathtt{x}})$. Then $\{\tilde{\mathtt{x}}/(\widetilde{\mathtt{list}}\ \tilde{\mathtt{y}}), \tilde{\mathtt{y}}/\tilde{\mathtt{x}}\}$, but neither $\{\tilde{\mathtt{y}}/(\widetilde{\mathtt{list}}\ \tilde{\mathtt{z}})\}$, nor $\{\tilde{\mathtt{x}}/\tilde{\mathtt{y}}\}$, is strict for $\mathtt{f}$.

Note that if f is type preserving, then every τ is strict for f. The definition says that a type substitution is strict for a function symbol f iff it only properly instantiates non-local type-variables, and (possibly) renames local ones. The intuition behind this concept concerns typing of a term with principal functor f, i.e. a term of the form $f(t_1, ..., t_n)$. Generally, if f is polymorphic then the type of this term is an instance of the type of f, and can be characterised by a type substitution. By distinguishing the class of strict type substitutions for f we distinguish a class of terms with principal functor f. The following definition makes this intuition more precise and generalises it.

Definition 4. Let $T \in \mathit{Texp}$ and let $t \in \mathit{Term}$. We define inductively what it means for t to be (strictly) T-*typed* by given type assignments ϕ and ν:

(BAS) $x \in \mathcal{V}$ is (strictly) T-typed by ν and ϕ iff $\nu(x) = T$.

(IND) $ft_1 \ldots t_n$ is (strictly) T-typed by ν and ϕ iff, for some $\tau \in \mathit{Tsubst}$ (which is strict for f w r t ϕ):

1. t_i is (strictly) τT_i-typed by ν and ϕ for $1 \leq i \leq n$, and
2. $\tau T_{n+1} = T$.

where $\phi(f) = T_1, .., T_n \rightarrow T_{n+1}$.

- t is (strictly) T-typed by ϕ iff t is (strictly) T-typed by ν and ϕ for some ν.
- t is (strictly) *well* typed by ϕ iff t is (strictly) T-typed by ϕ for some T.

Notice that the notion of strict well-typing is obtained from the usual concept of polymorphic typing (Cf. e.g. [7]) by restricting type substitutions to strict ones. Thus, for example, if $\phi(\mathit{length})$ is $\alpha\ \mathit{list} \rightarrow \mathit{int}$ then $\mathit{length}([1, 2])$ is well-typed by ϕ, but it is not strictly well-typed by ϕ. It would be, however, strictly well-typed if $\phi(\mathit{length})$ were $\mathit{int\ list} \rightarrow \mathit{int}$. It should be noticed that strict well-typedness implies well-typedness and that both concepts coincide for the terms constructed entirely with type preserving and/or monomorphic function symbols.

The concept of strict well-typedness is inspired by the concept of "type-general" in [4].

A fundamental property connected with the notion of strict well typing is the following:

Theorem 5. *Let t and t' be terms strictly well-typed by ϕ and let σ be their most general unifier. Then the term $t\sigma$ is strictly well-typed.*

3 Strictly well-typed programs do not go wrong

In this section we introduce a sufficient condition for static typing of definite programs.

We first extend the concept of strict well-typedness to atomic formulae. In the sequel we will treat terms and atoms uniformly, by assuming that ϕ is extended to predicate symbols so that $\phi(p) = T_1, ..., T_n \rightarrow \mathbf{o}$ where $\mathbf{o}$ is a reserved type constant.

Definition 6. A *typed term* is an object of the form $(t:T)$, where t is a term (or an atom) and T is a type expression. A set S of typed terms are said to be strictly well-typed by ν and ϕ iff there exists an instance ϕ' of ϕ and some ν such that, for each $(t:T) \in S$, t is strictly T-typed by ν and ϕ'. (Here ϕ' being an instance of ϕ means that $\phi' = \iota\phi$ for some type substitution ι).

Definition 7. Let P be a program and let G be a goal. Assume that no clause of $P \cup G$ shares variables with another clause. $P \cup G$ is strictly well-typed by ϕ iff $\{(A:\mathbf{o}) | A$ occuring in some clause in $P \cup G\}$ is strictly well-typed by ϕ.

It follows by Theorem 5 that the execution of a strictly well-typed augmented program $P \cup G$ starting with its goal G will never produce an ill-typed goal. This can be proved by induction on the number of steps of the SLD-resolution in the computation. At every step a unification of strictly well-typed atoms is performed and the resulting unifier preserves strict well-typedness of the result.

Thus, strict well-typedness of the augmented program is a sufficient static typing condition. However, this notion is still very restrictive. For example, the program P of the introduction that finds the last element of a non-empty list augmented with the goal $\leftarrow p([true, true], X)$ is not strictly well typed for any instance of the original type assignment. This is because P is strictly well-typed only when p is assigned the type *int list*, *int* $\rightarrow \mathbf{o}$. Thus our condition applies to P with an atomic goal $\leftarrow G$ only if the type of G is compatible with *int list*, *int* $\rightarrow \mathbf{o}$.

However, intuitively one expects that this program should work for any list, that is that the original type assignment for p would not be violated by the goals created during the execution. We now formulate an improved condition for static typing which would work for this example. This will be done by introducing so-called execution modules (or briefly modules) of a given typed program and by checking strict well-typedness of each module separately.

Informally, a module consists of all the atoms that may occur in one derivation of the goal. Starting with the goal in the set S'_0, we step by step build sets S_i of atoms that are examined and sets S'_i of atoms that should be examined. Finally, a set S constituting a module is obtained. A program is strictly well-typed by ϕ if every module is strictly well-typed by ϕ. The precise definitions of these concepts now follow.

Definition 8. Let P be a program where all clauses have been renamed apart and let G be an atomic goal.

Let $S_0 = \emptyset$ and $S'_0 = \{G\}$.

Let $A_n \in S'_n$ and $C_n = H_n \leftarrow B_{n,1}, \ldots B_{n,i} \in P$ such that A_n and H_n are unifiable.
If $(H_n : o) \in S_n$ then $S_{n+1} = S_n$ and $S'_{n+1} = S'_n - \{A_n\}$ else $S_{n+1} = S_n \cup \{(A_n : o), (H_n : o)\}$ and $S'_{n+1} = S'_n \cup \{B_{n,1}, \ldots B_{n,i}\} - \{A_n\}$.

The procedure stops when there is a k such that $S'_k = \emptyset$. Then S_k is a module.

The set $\mathcal{M}(P, G)$ of modules of a program is the set of all possible modules constructed in that way.

We now change the definition of strict well-typedness of a program into the following:

Definition 9. Let P be a program and let G be a goal. $P \cup G$ is said to be *strictly well-typed* by ϕ iff each $m \in \mathcal{M}(P, G)$ is strictly well-typed by ϕ.

4 Algorithm

This section gives an algorithm for checking (strict) well-typedness of modules and programs.

4.1 Modules

The well-typedness of a module can be checked using a rewriting system, the principles of which will be given here. The details may be found in [1].

The objects being rewritten are of the form $R = \langle S, E \rangle$, where S is a set of typed terms and E is a set of type equations that arises as constraints during the rewriting process.

Initially $R = \langle M, \emptyset \rangle$ where M is the module to be checked.

We say that R is *solved* iff it is of the form $\langle \{(v_1 : t_1), \ldots, ((v_j : t_j))\}, \{\alpha_1 = T_1, \ldots, \alpha_k = T_k\} \rangle$, where the v_i:s are pairwise distinct program variables, each α_i is a type variable occuring only once in $\{\alpha_1 = T_1, \ldots, \alpha_k = T_k\}$, and the t_i:s and the T_i:s are type expressions.

If it is possible to rewrite the initial R into solved form using the rules below, then the module is well-typed and the final R gives the types for the variables in the module, i.e. ν.

The rewriting rules are the following:

1. *Decomposition a:*
$\langle \{(f t_1 \ldots t_n : T)\} \uplus S,\ E \rangle \Rightarrow \langle \{(t_1 : \tau T_1), \ldots, (t_n : \tau T_n)\} \cup S, \{\tilde{x}_1 \doteq \tau \tilde{x}_1, \ldots, \tilde{x}_m \doteq \tau \tilde{x}_m, \tau T_{n+1} \doteq T\} \cup E \rangle$
If $\phi(f) = T_1, \ldots, T_n \rightarrow T_{n+1}, \tilde{x}_1, \ldots, \tilde{x}_m$ are all local variables of f, and τ is a renaming of tvar$(\phi(f))$ away from the set of type variables occurring in T, S, or E, or in $\phi(g)$ for some function symbol g.

2. *Merging*:
   $\langle\{(x:T_1)\}\uplus\{(x:T_2)\}\uplus S,\ E\rangle \Rightarrow \langle\{(x:T_1)\}\cup S,\ \{T_1 \doteq T_2\}\cup E\rangle$
   If x is a variable.
3. *Deletion*:
   $\langle S,\ \{\tilde{\mathsf{x}} \doteq \tilde{\mathsf{x}}\}\uplus E\rangle \Rightarrow \langle S,\ E\rangle$
   If $\tilde{\mathsf{x}}$ is a type variable.
4. *Decomposition b*:
   $\langle S,\ \{\tilde{c}T_1\ldots T_n \doteq \tilde{c}U_1\ldots U_n\}\uplus E\rangle \Rightarrow \langle S,\ \{T_1 \doteq U_1,..,T_n \doteq U_n\}\cup E\rangle$
5. *Replacement*:
   $\langle S,\ \{\tilde{\mathsf{x}} \doteq T\}\uplus E\rangle \Rightarrow \langle S,\ \{\tilde{\mathsf{x}} \doteq T\}\cup\{\tilde{\mathsf{x}}/T\}E\rangle$
   If $\tilde{\mathsf{x}}$ is a type-variable which occurs in E but not in T.
6. *Exchange*:
   $\langle S,\ \{T \doteq \tilde{\mathsf{x}}\}\uplus E\rangle \Rightarrow \langle S,\ \{\tilde{\mathsf{x}} \doteq T\}\cup E\rangle$
   If $\tilde{\mathsf{x}}$, but not T, is a type-variable.

The essential invariant preserved by these rules (see also [1]) is the existence (or non-existence) of a type-substitution ι such that $\{(t:\iota T)|(t:T)\in S\}$ is strictly well-typed by ϕ and $\iota T = \iota U$ for each $T \doteq U \in E$. In [1] all atoms in the program and in the goal constitutes one single module. It is proved in this case that the rewriting system has the desired properties, i.e. that it is terminating and, moreover, that the program is well-typed if the module can be rewritten into solved form. A simple corollary of those results is that this method with several modules also is sound — just treat every module as a separate program.

Now, an algorithm can be constructed according to the following ideas:

decompose1(R)=the result of repeatedly applying rule 1 to R as long as it is possible.

merge(R)=the result of repeatedly applying rule 2 to R as long as it is possible.

decompose2(E)=the result of repeatedly applying rule 4 to E as long as it is possible. If an equation with different constructors is found an error is signaled.

delete(E)=the result of applying rule 3 to E as long as it is possible.

exchange(E)=the result of applying rule 6 to E as long as it is possible.

replace(E)=the result of applying rule 5 once to E. If an equation is found where the left hand side variable occurs in the right hand side equation an error is signaled.

```
module_typing(M):
            R:=⟨for each atom a in M collect (a,o), ∅⟩;
            ⟨S,E⟩:=merge(decompose1(R));
            repeat R':=R;
                   R:=replace(exchange(delete(decompose2(R))));
            until R=R';
            return(true);
      error handling: return(false);
```

The reason for replace doing only one step is that this algorithm was developed in parallel with its implementation, and this was the most convenient formulation. Details on the algorithm and its implementation may be included in [10].

4.2 Programs

The first part of the algorithm describes how to build the modules of a program:

```
modules(P,M):
        repeat
            for each pair p = ⟨S, S'⟩ in M do
                let A be an atom in S';
                if A ∈ S
                   then S':=S' − {A}
                   else add to M k-1 copies of p
                            where there are k clauses H ← B1,...,Bn
                            such that A unifies with H;
                        for each such pair pi
                        and each such clause Hi ← Bi1,...,Bini do
                            S:=S ∪ {A, Hi};
                            S':=S' − A ∪ {Bi1,...,Bini}
        until no set changes;
        return(for each pair ⟨S, S'⟩ in M collect S);
```

The second part of the algoritm describes how to handle whole programs:

```
program_typing(P,G):
                flag:=true;
                for every module m in modules(P,{⟨∅,{G}⟩}) do
                    flag:=flag∧module_typing(m);
                return(flag);
```

4.3 Properties of the algorithm

Given the correctness of the rewriting system in Section 4.1, the following properties of the algorithm are easily proved:

Lemma 10. modules(P,G) *returns the set* $\mathcal{M}(\mathcal{P},\mathcal{G})$ *for the program* P *and goal* G.

Lemma 11. module_typing(m) *returns* true *iff there is a* ϕ' *that is an instance of* ϕ *such that* m *is strictly well-typed by* ϕ'.

Theorem 12. program_typing(P,G) *returns* true *iff program* P *with goal* G *is strictly well-typed.*

5 Conclusions

We presented a sufficient condition for static polymorphic typing of definite programs augmented with goals. The condition is based on the notion of strict well-typing and on static analysis of execution modules of a given program and goal. In contrast to the work of Mycroft and O'Keefe the condition allows for use of non-type-preserving functions. Viewing predicates as non-type-preserving functions we handle terms and atoms in a uniform way, unlike Mycroft and O'Keefe. Most examples well-typed according to the condition of Mycroft and O'Keefe also satisfy our condition. This is however not always true. For example consider the program:

$p(X) \leftarrow q([1]), q([true]).$
$q(Y).$

with the goal $\leftarrow p(X)$. and with the type assignment for the predicates $\phi(p) = \alpha\ list, \phi(q) = \beta\ list$. The types of the head atoms of both clauses are as general as the types of their predicates and the function symbols used are type preserving, so that the program satisfies the condition of Mycroft and O'Keefe. However, the execution modules are not strictly well-typed for any instance of the given type assignment.

The notion of execution module is a very simple tool for static analysis of the program. It seems that using more advanced methods of static analysis, like abstract interpretation (e.g. [8]) or dependency analysis [3] one could significantly extend the class of statically typeable programs.

References

1. Bonnier, S.
   A Formal Basis for Horn Clause Logic with External Polymorphic Functions
   Dissertation, 1992, Linköping University
2. Bonnier, S. and Małuszyński, J.
   Towards a Clean Amalgamation of Logic Programs with External Procedures
   In: Kowalski, R. A. and Bowen, K. A. (eds.)
   Proceedings of 5^{th}International Conference and Symposium of Logic Programming 1988
3. Boye, J.
   S-SLD-resolution — An Operational Semantics for Logic Programs with External Procedures
   In: Małuszyńsky, J. and Wirsing, M. (eds.)
   Proceedings of 3rd International Symposium on Programming Language Implementation and Logic Programming 1991
4. Hanus, M.
   Horn clause programs with polymorphic types: semantics and resolution
   Theoretical Computer Science 89(1991), pp 63–106
5. Kluźniak, F. and Kågedal, A.
   Enriching Prolog with S-Unification
   In: Darlington, J. and Dietrich, R. (eds.)

Proceedings of PHOENIX Seminar & Workshop on Declarative Programming 1991

6. Lloyd, J. W.
   Foundations of Logic Programming, 2nd ed.
   Springer-Verlag 1987, ISBN 3-540-18199-7
7. Mycroft, A. and O'Keefe, R. A.
   A Polymorphic Type System for Prolog
   Artificial Intelligence 23(1984), pp 295–307
8. Nilsson, U.
   Systematic Semantic Approximations of Logic Programs
   In: Deransart, P. and Małuszyński, J. (eds.)
   Proceedings of International Workshop on Programming Language Implementation and Logic Programming 1990
9. Pfennig, F.
   Tutorial on Types in Logic Programming
   Given at the 7th International Conference on Logic Programming 1990
10. Wallgren, J.
    Licentiate thesis, To appear (1992), Linköping University
    (On types and logic programming — No preliminary title yet)

Normalization by Leftmost Innermost Rewriting

Sergio Antoy*

Department of Computer Science
Portland State University
Portland, Oregon 97207-0751
antoy@cs.pdx.edu

Abstract. We define a transformation of rewrite systems that allows one to compute the normal form of a term in the source system by leftmost innermost rewriting of a corresponding term in the target system. This transformation is the foundation of an implementation technique used in an application of rewriting to software development. We also show in an example how our transformation accommodates lazy, non-deterministic computations in an eager, deterministic programming language.

1 Introduction

Our work originates from an application of rewriting to software development [1]. The application we are referring to proposes a conditional term rewriting system as an effective oracle for testing. In particular, rewriting is the computational paradigm for the implementation of a formal specification against which the run-time behavior of a concrete datatype is checked. Similar approaches, which however do not employ rewriting techniques, have been called *self-checking* or *N-voting* and have been shown effective in some situations [16].

Thus, the problem motivating our research is the implementation, under some constraints made precise later, of a rewrite system. In this note we disregard the often non-trivial details of such an implementation; rather we propose a theoretical framework which supports a relatively simple and efficient implementation technique for a class of systems adequate for the application at hand, namely, many-sorted, sufficiently-complete, constructor-based, weakly-orthogonal, term rewriting systems [4, 15]. The constraints mentioned above are the choice of the implementation's programming language and the choice of the computation's rewrite strategy.

The implementation language for our rewrite systems is C++ [19]. This choice is motivated by the fact that the programs of our application, which are coded in C++, and the rewrite systems serving as their oracles interact frequently and extensively. As a consequence, they must coexist in the same run-time environment.

The computation strategy of our application is *parallel outermost needed* [18]. This choice is motivated by the fact that the testing technique we are considering is based on the comparison of normal forms. As a consequence, a normalizing strategy is mandatory.

These constraints are interrelated and somewhat conflicting, in that an easy strategy for implementing a constructor-based term rewriting system in C++, as well as in

* Supported by the National Science Foundation grant CCR-8908565.

many other imperative procedural languages, is *leftmost innermost*. This because an innermost strategy can be easily mapped to the C++'s *call-by-value* parameter passing mechanism, while an outermost strategy would require either something similar to *call-by-name* or the handling of functions as first class objects—features which are both absent from C++.

To overcome these difficulties we have devised a transformation, τ, that takes a *source* rewrite system, $\mathcal{R}$, and produces a *target* rewrite system, $\mathcal{R}_\tau$, such that the normal form of a term t in $\mathcal{R}$ is computed by leftmost innermost rewriting of some associated term t_τ in $\mathcal{R}_\tau$.

2 Notation

We assume the reader is familiar with the basic notions of rewriting [4, 15]. We consider many-sorted [6] term rewriting systems. For each sort s, we assume an arbitrary but fixed ordering, called *standard*, among the constructors of s. Variables are denoted by upper case letters or, when anonymous, by the symbol "_". Any term referred to in this note type checks. The leading symbol or principal functor of a term t is called the *root* of t.

An *occurrence* is a path identifying a subterm in a term. For terms t and t' and occurrence o, t/o denotes the subterm of t at o, and $t[o \leftarrow t']$ denotes occurrence substitution. If $o_1, \ldots, o_k$ is a sequence of disjoint occurrences of t and $t_1, \ldots, t_k$ is a sequence of terms, then $t[o_i \leftarrow t_i]_{o_i \in \langle o_1, \ldots, o_k \rangle}$ denotes $t[o_1 \leftarrow t_1] \ldots [o_k \leftarrow t_k]$.

The textual order of the rules in the target system $\mathcal{R}_\tau$ is significant. Among the rules matching a term t, the most specific one [13] is always chosen to rewrite t.

3 Definitional Trees

We recall the concept of *definitional tree* [2], a hierarchical structure of the rules defining an operation, which is crucial to our transformation. The symbols *branch*, *rule*, and *exempt* in the next definition are uninterpreted functions.

Definition 1. *$\mathcal{T}$ is a* partial definitional tree, *or* pdt, *if and only if one of the following cases holds.*

> *$\mathcal{T} = branch(t, \bar{o}, \bar{\bar{\mathcal{T}}})$, where t is an open term called* template, *$\bar{o}$ is a sequence $o_1, \ldots, o_k$ of distinct occurrences of variables of t, for all j in $1, \ldots, k$, the sort of t/o_j has constructors $c_{j_1}, \ldots, c_{j_{k_j}}$ in standard ordering, and $\bar{\bar{\mathcal{T}}}$ is a sequence $\bar{\mathcal{T}}_1, \ldots, \bar{\mathcal{T}}_k$ of sequences of* pdts, *such that for all j in $1, \ldots, k$, $\bar{\mathcal{T}}_j = \mathcal{T}_{j_1}, \ldots, \mathcal{T}_{j_{k_j}}$ and for all i in $1, \ldots, k_j$, the template in the root of $\mathcal{T}_{j_i}$ is $t[o_j \leftarrow c_{j_i}(X_1, \ldots, X_n)]$, where n is the arity of c_{j_i} and $X_1, \ldots, X_n$ are fresh variables.*
>
> *$\bar{\mathcal{T}}_1, \ldots, \bar{\mathcal{T}}_k$ are called the* sequential components *of $\mathcal{T}$.*
>
> *$\mathcal{T} = rule(t, t')$, where t is a template and $t \to t'$ is a rewrite rule.*
>
> *$\mathcal{T} = exempt(t)$, where t is a template.*

$\mathcal{T}$ is a definitional tree *of an operation f if and only if $\mathcal{T}$ is a* pdt *with $f(X_1, \ldots, X_k)$ as a template argument.*

A discussion on definitional trees, including examples, can be found in [2]. In particular, a definitional tree exists for any operation of a weakly-orthogonal system $\mathcal{R}$, if certain "useless" rules are discarded [2, Th. 19]. Algorithms for building a definitional tree from the set of rewrite rules defining an operation are discussed in [20].

```
Procedure Gen-ξ(T : pdt) is
begin
    case T is
        when branch(t, ō, T̄̄) ⇒
            for all T̄ in T̄̄ loop
                for all T' in T̄ loop
                    Gen-ξ(T')
                end loop
            end loop
            output ξ(t) → t[o ← ξ(t/o)]o∈ō
        when rule(t, t') ⇒
            output ξ(t) → t'
    end case
end Gen-ξ
```

Fig. 1. Generation of a set of rules defining ξ from a complete definitional tree.

```
Procedure Gen-η is
    let X1, X2, ... be fresh variables
begin
    for any sort s of R loop
        for any constructor c of sort s loop
            output η(c(X1, ..., Xn)) → c(η(X1), ..., η(Xn))
        end loop
    end loop
    output η(X1) → η(ξ(X1))
end Gen-η
```

Fig. 2. Generation of the set of rules defining η.

4 Transformation

Our transformation is defined operationally by the procedures shown in Fig. 1 and 2 and has the following characteristics.

1. The source system $\mathcal{R}$ is a many-sorted [6], sufficiently-complete [8, 10], weakly-orthogonal [15], constructor-based [17] system. The target system $\mathcal{R}_\tau$ is constructor-based with specificity rule [13].
2. The set of constructors of $\mathcal{R}_\tau$ is the union of the sets of constructors and operations of $\mathcal{R}$.
3. There are only two unary, sort-preserving, overloaded operations in $\mathcal{R}_\tau$: η and ξ.
4. The rules defining ξ are generated by applying the procedure *Gen-ξ*, shown in Fig. 1, to a definitional tree of each operation of $\mathcal{R}$.
5. The rules defining η are generated by the procedure *Gen-η* shown in Fig. 2.

$samefringe(A,B) \to eqlist(flatten(A), flatten(B))$
$flatten(leaf(A)) \to cons(A, nil)$
$flatten(branch(A,B)) \to append(flatten(A), flatten(B))$
$append(nil, A) \to A$
$append(cons(A,B),C) \to cons(A, append(B,C))$
$eqlist(nil, nil) \to true$
$eqlist(nil, cons(\_,\_)) \to false$
$eqlist(cons(\_,\_), nil) \to false$
$eqlist(cons(A,B), cons(C,D)) \to and(eq(A,C), eqlist(B,D))$
$and(false,\_) \to false$
$and(\_, false) \to false$
$and(true, true) \to true$

Fig. 3. Rewrite system, adapted from [9, p.368], for computing whether two trees have the same fringe. The rules of the operation *eq* are not shown.

$\xi(samefringe(A,B)) \to eqlist(flatten(A), flatten(B))$
$\xi(flatten(leaf(A))) \to cons(A, nil)$
$\xi(flatten(branch(A,B))) \to append(flatten(A), flatten(B))$
$\xi(flatten(A)) \to flatten(\xi(A))$
$\xi(append(nil, nil)) \to nil$
$\xi(append(nil, cons(A,B))) \to cons(A,B)$
$\xi(append(nil, A)) \to A$
$\xi(append(cons(A,B),C)) \to cons(A, append(B,C))$
$\xi(append(A,B)) \to append(\xi(A), B)$
$\xi(eqlist(nil, nil)) \to true$
$\xi(eqlist(nil, cons(\_,\_))) \to false$
$\xi(eqlist(nil, A)) \to eqlist(nil, \xi(A))$
$\xi(eqlist(cons(\_,\_), nil)) \to false$
$\xi(eqlist(cons(A,B), cons(C,D))) \to and(eq(A,C), eqlist(B,D))$
$\xi(eqlist(cons(A,B), C)) \to eqlist(cons(A,B), \xi(C))$
$\xi(eqlist(A,B)) \to eqlist(\xi(A), B)$
$\xi(and(false,\_)) \to false$
$\xi(and(true, false)) \to false$
$\xi(and(true, true)) \to true$
$\xi(and(true, A)) \to and(true, \xi(A))$
$\xi(and(\_, false)) \to false$
$\xi(and(false, true)) \to false$ (*)
$\xi(and(true, true)) \to true$ (*)
$\xi(and(A, true)) \to and(\xi(A), true)$
$\xi(and(A,B)) \to and(\xi(A), \xi(B))$

Fig. 4. Output of *Gen-ξ* applied to (the definitional trees of the operations of) the system of Fig. 3. The rules marked with (*) can be eliminated, since the specificity rule makes their use impossible.

$\eta(true) \to true$
$\eta(nil) \to nil$
$\eta(leaf(A)) \to leaf(\eta(A)$
$\eta(A) \to \eta(\xi(A))$
$\eta(false) \to false$
$\eta(cons(A,B)) \to cons(\eta(A), \eta(B))$
$\eta(branch(A,B)) \to branch(\eta(A), \eta(B))$

Fig. 5. Output of *Gen-η* applied to (the sorts of) the system of Fig. 3. The rules for the sort of the elements decorating the trees are not shown.

5 Using τ

A rewrite system specifies which rewrites can be performed, but not where and when one should execute them during a computation. Computing requires some form of control referred to as a rewrite strategy. Without an appropriate strategy a computation may fail to terminate even when termination is possible. We argue that the leftmost innermost computation of $\eta(t)$ in $\mathcal{R}_\tau$, for any ground $\mathcal{R}$-term t, is "equivalent", in the sense formalized by the statements of *soundness* and *completeness* proposed below, to a normalizing computation of t in $\mathcal{R}$, that is, to a computation that terminates if at all possible.

Throughout this section, $\mathcal{R}$ denotes a many-sorted, sufficiently-complete, constructor-based, weakly-orthogonal, term rewriting systems. For any $\mathcal{R}$-term t, t_τ is a short-hand for $\eta(t)$. The symbol "$\rightarrow_{\mathcal{R}_\tau}$" denotes one leftmost innermost rewrite, since these are the only rewrites we care about in $\mathcal{R}_\tau$.

A term in $\mathcal{R}_\tau$ can be regarded as a term in $\mathcal{R}$ in which certain subterms are "marked". There are two markers: η, which calls for normalizing its argument, and ξ, which calls for rewriting its argument. Several rewrites of a marker's argument, including inner ones, are generally necessary to satisfy the marker's request.

The idea of looking at terms of $\mathcal{R}_\tau$ as if they were terms of $\mathcal{R}$ is formalized by the mapping s defined below. We call s *"stripping"*, since it strips a term of its occurrences of η and ξ. This action is sensible since η and ξ are unary and sort-preserving.

Definition 2. *The mapping* $\mathrm{s} : Ter(\mathcal{R}_\tau) \rightarrow Ter(\mathcal{R})$ *is defined by*

$$\mathrm{s}(t) = \begin{cases} \mathrm{s}(t') & \text{if } t = \xi(t') \text{ or } t = \eta(t'); \\ f(\mathrm{s}(t_1), \ldots, \mathrm{s}(t_k)) & \text{if } t = f(t_1, \ldots, t_k), \text{ for some } k\text{-ary symbol } f \text{ in } \mathcal{R}; \\ X & \text{if } t = X, \text{ for some variable } X. \end{cases}$$

Through s, we express an important property of the rules of $\mathcal{R}_\tau$. *If $l \rightarrow r$ is an $\mathcal{R}_\tau$-rule, then either:* (1) $\mathrm{s}(l) \equiv \mathrm{s}(r)$ *or* (2) $\mathrm{s}(l) \rightarrow \mathrm{s}(r)$ *is an $\mathcal{R}$-rule.* This fact can be verified from each *output* statement of the procedures *Gen-ξ* and *Gen-η*. The above property is crucial in establishing the soundness of τ. It implies that a computation in $\mathcal{R}_\tau$ defines a computation in $\mathcal{R}$. *For any $\mathcal{R}_\tau$-terms t and t', if $t \xrightarrow{*}_{\mathcal{R}_\tau} t'$, then* $\mathrm{s}(t) \xrightarrow{*}_{\mathcal{R}} \mathrm{s}(t')$. Furthermore, we can obtain an explicit rewrite sequence in $\mathcal{R}$ from a rewrite sequence in $\mathcal{R}_\tau$ by ignoring rewrites resulting from the application of an $\mathcal{R}_\tau$-rule $l \rightarrow r$ such that $\mathrm{s}(l) \equiv \mathrm{s}(r)$. This observation is crucial in establishing the completeness of τ.

We now characterize the ground redexes of $\mathcal{R}_\tau$. *For any ground $\mathcal{R}$-term t,* (1) $\eta(t)$ *is an $\mathcal{R}_\tau$-redex, and* (2) $\xi(t)$ *is an $\mathcal{R}_\tau$-redex if and only if t is $\mathcal{R}$-operation-rooted.* Claim (1) stems from the fact that there is a rule of η with left side $\eta(X)$, where X is variable. Claim (2) stems from the facts that the argument of ξ in the left side of any rule defining ξ is the pattern argument of some *pdt*, hence it is an operation-rooted term, and the pattern argument of the root node of a definitional tree of any k-ary operation f is $f(X_1, \ldots, X_k)$, where $X_1, \ldots, X_k$ are variables.

From the characterization of the redexes of $\mathcal{R}_\tau$ we establish the following fact, which is the second crucial property for proving the soundness of τ. *For any ground operation-rooted $\mathcal{R}$-term t, if $t_\tau \xrightarrow{*}_{\mathcal{R}_\tau} t'$, then t' is $\mathcal{R}_\tau$-irreducible if and only if* $\mathrm{s}(t') \equiv t'$. The proof is not immediate. It relies on the sufficient-completeness of $\mathcal{R}$ and on the fact that in any term of the leftmost innermost computation of t_τ any occurrence of ξ has an $\mathcal{R}$-operation-rooted term for argument. This fact stems from the structure of a definitional tree and the definition of *Gen-ξ*.

The soundness of τ is stated as follows. *For any ground $\mathcal{R}$-term t, if the leftmost innermost computation of t_τ terminates, then the $\mathcal{R}_\tau$-normal form of t_τ is the $\mathcal{R}$-normal form of t.* Using the previous results, it suffices to prove, by induction on the length of the leftmost innermost computation of t_τ, that the $\mathcal{R}_\tau$-normal form of t_τ is an $\mathcal{R}$-constructor term.

We now turn to the completeness of τ. That is, if the computation of a ground term t in $\mathcal{R}$ terminates, then the leftmost innermost computation of t_τ in $\mathcal{R}_\tau$ yields the same result. The crucial property is that computing $\xi(t)$ is somewhat equivalent to rewriting the redexes of a necessary set [18] of t. *For any ground operation-rooted $\mathcal{R}$-term t, if $\xi(t) \overset{*}{\rightarrow}_{\mathcal{R}_\tau} t'$ and $\mathrm{s}(t') \equiv t'$, then t' is the result of rewriting in $\mathcal{R}$ all the redexes of some necessary set of t.* The proof of this claim is non-trivial. For a related result in a different context see [2, Corollary 13 and its extension].

The completeness of τ is stated as follows. *For any ground $\mathcal{R}$-term t, if $t \overset{*}{\rightarrow}_{\mathcal{R}} t'$ and t' is an $\mathcal{R}$-normal form then $t_\tau \overset{*}{\rightarrow}_{\mathcal{R}_\tau} t'$ and t' is an $\mathcal{R}_\tau$-normal form.* The key step of the proof is showing the existence of a normal form of t_τ. This is achieved by looking at the leftmost innermost computation of $\eta(t)$ in $\mathcal{R}_\tau$ as a computation in $\mathcal{R}$, as outlined earlier, and by showing that in this computation there is no descendant of some redex of a necessary set of some term that is never rewritten [18]. This is a consequence of the fact that for any $\mathcal{R}$-term t, $\xi(t)$ has a normal form. The claim is then a consequence of the soundness of τ and the confluence of $\mathcal{R}$.

6 Optimizing ξ

A computation in $\mathcal{R}_\tau$ generally contains some rewrites that have no corresponding rewrites in a computation in $\mathcal{R}$. This happens for any $\mathcal{R}_\tau$-terms t and t' such that $t \rightarrow_{\mathcal{R}_\tau} t'$ and $\mathrm{s}(t) \equiv t'$. In some sense, this is the cost of using a simple strategy and preserving normalization.

However, this cost can be contained for certain systems called *inductively sequential* in [2]. Any operation in these systems has a definitional tree in which any *branch* node $\mathcal{N}$ has only one sequential component (see Def. 1.) In this situation, the occurrence argument of $\mathcal{N}$ is a sequence of length one, say $\langle o \rangle$, and o is the occurrence of a needed redex [11] of any instance s of the pattern argument of $\mathcal{N}$. This implies that in order to rewrite s one has to rewrite the subterm of s at o to a head normal form [2, Th. 11]. This suggests to employ a less "leashed" version of ξ denoted by σ. The rewrite rules defining σ are generated by the procedure *Gen-σ* shown in Fig. 6. An example of output is shown in Fig. 7. In an inductively sequential system $\mathcal{R}$, ξ can be replaced by σ, and accordingly the least specific rule of η is changed to $\eta(X_1) \rightarrow \eta(\sigma(X_1))$.

The key difference between σ and ξ is the following. *For all ground $\mathcal{R}$-terms t and t', if $\xi(t) \rightarrow_{\mathcal{R}_\tau} t'$, then* (1) *$\sigma(t) \rightarrow_{\mathcal{R}_\tau} t'$ if and only if the root of t' is an $\mathcal{R}$-constructor, and* (2) *$\sigma(t) \rightarrow_{\mathcal{R}_\tau} \sigma(t')$ if and only if the root of t' is not an $\mathcal{R}$-constructor.* This fact can be verified from each *output* statement of the procedure *Gen-σ*. Thus, σ can be regarded as a marker that calls for rewriting its argument to a $\mathcal{R}$-head normal form.

```
Procedure Gen-σ(T : pdt) is
      let X_1, ..., X_n be new variables
begin
      case T is
            when branch(t, ⟨o⟩, ⟨⟨T_1, ..., T_k⟩⟩) ⇒
                  for i in 1 .. k loop
                        Gen-σ(T_i)
                  end loop
                  output σ(t) → σ(t[o ← σ(t/o)])
            when rule(t, t') ⇒
                  case t' is
                        when constructor-rooted ⇒
                              output σ(t) → t'
                        when operation-rooted ⇒
                              output σ(t) → σ(t')
                        when variable ⇒
                              for any constructor c of the sort of t' loop
                                    let u = c(X_1, ..., X_n)
                                    output σ(t[o ← u]) → u
                              end loop
                              output σ(t) → σ(t')
                  end case
      end case
end Gen-σ
```

Fig. 6. Generation of a set of rules defining σ. Input trees are complete and their *branch* nodes have only one sequential component.

$\sigma(append(nil, nil)) \rightarrow nil$
$\sigma(append(nil, cons(A, B))) \rightarrow cons(A, B)$
$\sigma(append(nil, A)) \rightarrow \sigma(A)$
$\sigma(append(cons(A, B), C)) \rightarrow cons(A, append(B, C))$
$\sigma(append(A, B)) \rightarrow \sigma(append(\sigma(A), B))$

Fig. 7. Output of *Gen-σ* applied to a definitional tree of the operation *append*.

In the remainder of this section, we assume that $\mathcal{R}$ is inductively sequential and that $\mathcal{R}_\tau$ is optimized, that is, obtained from *Gen-σ* rather than *Gen-ξ*. The soundness and completeness of our transformation are preserved by this optimization.

The optimization shortens the length of the leftmost innermost computation of t_τ in $\mathcal{R}_\tau$. We express an upper bound on the length of this computation as a function of two arguments: (1) the length of the computation of t in $\mathcal{R}$ according to the strategy that rewrites only outermost needed redexes and (2) the size of the result of the computation. In the following, the symbol "$\rightarrow_{\mathcal{R}}$" denotes one outermost needed rewrite in $\mathcal{R}$ and the expression $|x|$ is a measure of x: that is, *length*, when x is a computation, or *size*, when x is a term.

We first claim a bound on rewrites of non-head normal forms to head normal forms. *For any ground operation-rooted $\mathcal{R}$-term t, if $\sigma(t) \xrightarrow{*}_{\mathcal{R}_\tau} t'$ and* $\mathrm{s}(t') \equiv t'$, *then* $|\sigma(t) \xrightarrow{*}_{\mathcal{R}_\tau} t'| < 2\,|t \xrightarrow{*}_{\mathcal{R}} t'|$. From the above result, we establish the following more general one. *For any ground $\mathcal{R}$-term t, if $t_\tau \xrightarrow{*}_{\mathcal{R}_\tau} t'$ and* $\mathrm{s}(t') \equiv t'$, *then* $|t_\tau \xrightarrow{*}_{\mathcal{R}_\tau} t'| \leq$

$2|t \xrightarrow{*}_{\mathcal{R}} t'| + |t'|$. Both claims are proved by induction on the length of a leftmost innermost computation in $\mathcal{R}_\tau$.

Note that the outermost needed computation in $\mathcal{R}$, used to bound the leftmost innermost computation in $\mathcal{R}_\tau$ may not be the shortest computation of a term.

7 Incompleteness

So far, we have not yet defined our transformation on *exempt* nodes. These nodes are found in the definitional trees of some operations of insufficiently-complete systems. The transformation of these systems is more complicated and we only outline a solution.

We introduce a new symbol, ι, in $\mathcal{R}_\tau$. Contrary to η and ξ, or σ, ι is a constructor of $\mathcal{R}_\tau$. This implies that there are no defining rules for ι; rather, rules must be given for η and ξ, or σ, to handle ι-rooted arguments.

We describe intuitively the behavior of these rules. Continuing our analogy, we look at ι as a marker that "notifies" the other operations of $\mathcal{R}_\tau$ that its argument is an $\mathcal{R}$-operation-rooted head normal form. This information is important, since in our application, an $\mathcal{R}$-normal form that contains occurrences of operations is unacceptable. Note that any such term is illegal also in many languages based on rewriting, such as ML, Hope, and Miranda.

If η or σ are applied to some ι-rooted term, the most sensible behavior seems to be aborting the computation, since in this case the result, if any, produced by the computation would contain an unacceptable subterm. The same action should be taken also when all the ξ-rooted subterms of term are rewritten to ι-rooted terms. In any other case, ι-rooted terms can be temporarily ignored. They may disappear as the computation proceeds.

8 Conclusions

We presented a transformation of many-sorted, sufficiently-complete, constructor-based, weakly-orthogonal, term rewriting systems. A rewrite strategy based on this transformation allows one to compute normal forms in the source system by leftmost innermost rewriting in the target system. This strategy is a key component for implementing algebraic specifications used to test C++ programs.

The direct implementation of algebraic specifications is often based on rewriting [5, 7, 12, 14]. In order to achieve executability and/or efficiency, most approaches constrain, sometimes severely, the specifications that can be implemented. Strong-normalization and/or non-overlapping are the most common limitations. A consequence of these limitations is a loss of expressive power or "declarativeness" which makes a specification look more like a functional program. Our approach mitigates this problem. In particular, our transformation is useful in conjunction with the efforts referenced above to extend, sometimes significantly, the class of implementable specifications.

Our work also suggests a technique to code lazy, non-deterministic computations in eager, deterministic programming languages. This possibility originates from the simplicity of both the target system and the evaluation strategy for the terms we are interested in. The appendix shows a program, executable by an eager language, obtained by transforming a classic example requiring lazy evaluation.

A promising extension of our work concerns the expressiveness of programming languages. Overlapping is not allowed in most programming languages based on rewriting—yet, it is a powerful feature. For example, our version of *samefringe* differs from that in [9] only for the overlapping in the operation *and*. This "parallel" *and* makes it possible to detect that two trees do not have the same fringe for arguments for which the original system fails to reach a conclusion. The implementation of a programming language allowing this programming style can be simply obtained from the interpreter or compiler of some eager functional language by adding a front-end that performs our transformation.

Appendix

We show an ML [3] program for computing the prime numbers with the Sieve of Erathostenes' algorithm. This program is the result of transforming the version proposed in [7, p. 52] except for the operation `force`, whose purpose in [7] is to alter the default order of evaluation. This option is neither available nor necessary in our context.

```
datatype lazy_list
  = nil
  | cons of int * lazy_list
  | ints_from of int
  | show of int * lazy_list
  | sieve of lazy_list
  | primes
  | filter of lazy_list * int
fun head_norm (ints_from A) = cons (A, ints_from (A+1))
  | head_norm (show (0, _)) = nil
  | head_norm (show (A, cons (B, C))) = cons (B, show (A-1, C))
  | head_norm (show (A, B)) = head_norm (show (A, head_norm B))
  | head_norm (sieve (cons (A, B))) = cons (A, (sieve (filter (B, A))))
  | head_norm (sieve A) = head_norm (sieve (head_norm A))
  | head_norm primes = head_norm (sieve (ints_from 2))
  | head_norm (filter (cons (A, B), C)) = if A mod C = 0
      then head_norm (filter (B, C))
      else cons (A, filter (B, C))
  | head_norm (filter (A, B)) = head_norm (filter (head_norm A, B))
fun normalize nil = nil
  | normalize (cons (A, B)) = cons (A, normalize B)
  | normalize A = normalize (head_norm A);
normalize (show (10, primes));
```

Fig. 8. ML program for computing prime numbers. The *datatype* statement declares the constructors. The functions *head_norm* and *normalize* correspond to σ and η respectively. The last line generates the first 10 primes.

References

1. S. Antoy and R. Hamlet. Automatically checking an implementation against its formal specification. In R. Selby, editor, *Irvine Software Symposium*, pages 29–48, Irvine, CA, March 1992.
2. Sergio Antoy. Definitional trees. In *ALP'92*, Volterra, Italy, September 1992. (to appear).
3. Andrew W. Appel and David B. MacQueen. Standard ML of New Jersey. In *PLILP'91*, pages 1–13, Passau, Germany, August 1991. Lect. Notes in Comp. Sci., Vol. 528.
4. N. Dershowitz and J.-P. Jouannaud. Rewrite systems. In J. van Leeuwen, editor, *Handbook of Theoretical Computer Science B: Formal Methods and Semantics*, chapter 6, pages 243–320. North Holland, Amsterdam, 1990.
5. Alfons Geser, Heinrich Hussmann, and Andreas Mück. A compiler for a class of conditional term rewriting systems. In S. Kaplan and J.-P. Jouannaud, editors, *CTRS'87*, pages 84–90, Orsay, France, July 1987. *LNCS* 308.
6. J. A. Goguen, J. W. Thatcher, and E. G. Wagner. An initial algebra approach to the specification, correctness, and implementation of abstract data types. In R. T. Yeh, editor, *Current Trends in Programming Methodology*, volume 4, pages 80–149. Prentice-Hall, Englewood Cliff, NJ, 1978.
7. Joseph A. Goguen and Timothy Winkler. Introducing OBJ3. Technical Report SRI-CSL-88-9, SRI International, Menlo Park, CA, 1988.
8. J. V. Guttag. Abstract data types and the development of data structures. *Comm. of the ACM*, 20:396–404, 1977.
9. Ellis Horowitz. *Fundamentals of Programming Languages*. Computer Science Press, Rockville, MD, second edition, 1984.
10. Gérard Huet and Jean-Marie Hullot. Proofs by induction in equational theories with constructors. *JCSS*, 25:239–266, 1982.
11. Gérard Huet and Jean-Jacques Lévy. Call by need computations in non-ambiguous linear term rewriting systems. Technical Report 359, INRIA, Le Chesnay, France, 1979.
12. Stéphane Kaplan. A compiler for conditional term rewriting systems. In P. Lescanne, editor, *RTA'87*, pages 25–41, Bordeaux, France, May 1987. *LNCS* 256.
13. J. R. Kennaway. The specificity rule for lazy pattern-matching in ambiguous term rewrite systems. In *Third European Symp. on Programming*, pages 256–270, 1990. *LNCS* 432.
14. H. Klaeren and K. Indermark. Efficient implementation of an algebraic specification language. In M. Wirsing and J. A. Bergstra, editors, *Algeraic Methods: Theory, Tools and Applications*, pages 69–90, Passau, Germany, June 1987. *LNCS* 394.
15. Jan Willem Klop. Term rewriting systems. Technical Report CS-R9073, Stichting Mathematisch Centrum, Amsterdam, The Netherlands, 1990.
16. J. C. Knight and N. G. Leveson. An experimental evaluation of the assumption of independence in multi-version programming. *IEEE Trans. on Soft. Eng.*, 12:96–109, 1986.
17. Michael J. O'Donnell. Computing in systems described by equations. Springer-Verlag, 1977. Lect. Notes in Comp. Sci., Vol. 58.
18. R. C. Sekar and I. V. Ramakrishnan. Programming in equational logic: Beyond strong sequentiality. In *Proceedings of the Fifth Annual IEEE Symposium on Logic in Computer Science*, pages 230–241, Philadelphia, PA, June 1990.
19. B. Stroustrup. *The* C++ *Programming Language*. Addison-Wesley, Reading, MA, 1986.
20. Yonggong Yan. Building definitional trees. Master's project, Portland State University, Portland, OR, March 1992.

A Strategy to Deal with Divergent Rewrite Systems

Paola Inverardi *

Istituto di Elaborazione dell'Informazione, C.N.R.

via Santa Maria 46, I-56126 Pisa, Italy

Monica Nesi †

University of Cambridge, Computer Laboratory

New Museums Site, Pembroke Street, Cambridge CB2 3QG, England

Abstract

In this paper a new approach to divergence in Knuth-Bendix completion is presented. Given a term rewriting system R, whose completion diverges, a strategy can be defined to simulate the application of the (infinitely many) rewrite rules derived from critical pairs, without attempting any completion. This is done by applying some of the rules in R also as expansion rules.

1 Motivation

Knuth-Bendix completion [13] is a well-known technique to produce a canonical term rewriting system from a given set T of equations. One of the major problems in this approach is that Knuth-Bendix completion can result in an infinite set of rewrite rules. This problem has been well studied [5, 6, 7, 17] and various approaches that try to cope with infinite sets of rules have been proposed [11, 12, 14, 18, 19, 1, 2].

In this paper we present a different way of dealing with divergent term rewriting systems. The idea is to simulate the behaviour of the infinite canonical system R_∞ by defining a rewriting relation based on the set R of rules obtained by orienting the given set T of equations according to a chosen term ordering, and using a (sub)set of the rules in R also as *expansion rules*. In other words, this means that only some of the equations in T are turned into rules, while the remaining ones are kept as equations. In this way, by properly controlling the rewriting steps, we can simulate the application of the rules derived from critical pairs by first expanding a term and then reducing it along the peak which generates a critical pair. Thus, we are able to compute the normal

*Research partially supported by Progetto Finalizzato Sistemi Informatici e Calcolo Parallelo of C.N.R., Italy.

†Research supported by Consiglio Nazionale delle Ricerche (C.N.R.), Italy.

form of a term with respect to R_∞ without performing any completion. In this respect, our approach is similar to the dynamic paramodulation proposed in a theorem proving context in [4], where given a term t, rewriting strategies are defined in order to transform t by dynamically generating rules from critical pairs. Moreover, a notion of (quasi-) peak climbing is proposed in [15] with the purpose of proving equational consequences in a set of equations E together with a convergent set of rules DE.

Our approach works if the set E of expansion rules is properly chosen. This amounts to requiring two basic properties on E: i) the set E is *cp-complete*, i.e. the application of any rule in R_∞ derived from a critical pair can be simulated; ii) the expansion process from a given term always terminates.

The rewriting strategy defined in [8] can be seen as a particular case of this approach to divergence. That strategy copes with divergence of completion of the equational theory for observational congruence over finite CCS [16, 9]. This is done by using those rules which are the sources of divergence as the set E of expansion rules and by properly constraining the expansion process, thus obtaining a deterministic and complete strategy.

In the following sections we first recall some basic definitions about term rewriting systems, and then informally introduce our approach by means of a simple example. Finally, the rewriting strategy is formally defined and shown to be correct and complete, and its applicability to some examples is discussed.

2 Term Rewriting Systems

We assume that the reader is familiar with the basic notions of term rewriting systems. Below we summarize the most relevant definitions, while we refer to [3, 5] for more details.

Let $F = \bigcup_n F_n$ be a set of function symbols, where F_n is the set of symbols of arity n. Let $T(F,X)$ be the set of (finite, first order) terms with function symbols F and variables X. An *equational theory* is any set $T = \{(s,t) \mid s,t \in T(F,X)\}$. Elements (s,t) are called *equations* and written $s = t$. Let $\sim_T$ be the smallest symmetric relation that contains T and is closed under monotonicity and substitution. Let $=_T$ be the reflexive-transitive closure of $\sim_T$.

A *term rewriting system* (TRS) or *rewrite system* R is any finite set $\{(l_i,r_i) \mid l_i,r_i \in T(F,X), Var(r_i) \subseteq Var(l_i)\}$. The pairs (l_i,r_i) are called *rewrite rules* and written $l_i \to r_i$. The *rewriting relation* $\to_R$ over $T(F,X)$ is defined as the smallest relation containing R that is closed under monotonicity and substitution. A term t *rewrites* to a term s, written $t \to_R s$ if there exists $l \to r$ in R, a substitution σ and a subterm $t|_u$ at the position u, called *redex*, such that $t|_u = l\sigma$ and $s = t[r\sigma]_u$. A term t is said to *overlap* a term t' if t unifies with a non-variable subterm of t' (after renaming the variables in t so as not to conflict with those in t'). If $l_i \to r_i$ and $l_j \to r_j$ are two rewrite rules (with distinct variables), u is the position of a non-variable subterm of l_i, and σ is a most general unifier of $l_i|_u$ and l_j, then the equation $(l_i\sigma)[r_j\sigma]_u = r_i\sigma$ is a *critical pair* formed from those rules.

Let $\stackrel{+}{\to}$ and $\stackrel{*}{\to}$ denote the transitive and reflexive-transitive closure of $\to$, respectively. A TRS R is *terminating* if there is no infinite sequence $t_1 \to_R t_2 \to_R \ldots$ in R. A TRS R is *confluent* if whenever $s\ {}_R\!\stackrel{*}{\leftarrow} t \stackrel{*}{\to}_R q$, there exists a term t' such that $s \stackrel{*}{\to}_R t'\ {}_R\!\stackrel{*}{\leftarrow} q$, while R is *locally confluent* if whenever $s\ {}_R\!\leftarrow t \to_R q$, there exists a term t' such that $s \stackrel{*}{\to}_R t'\ {}_R\!\stackrel{*}{\leftarrow} q$. A term t is *in R-normal form* if there is no term s such that $t \to_R s$. A term s is an

R-normal form of t if $t \xrightarrow{*}_R s$ and s is in R-normal form, in this case we write $t \xrightarrow{!}_R s$. A TRS R is *canonical* if it is terminating and confluent.

Let us call *complete* the completion procedure which applies inter-reduction whenever possible, and *nr-complete* a non-reducing completion procedure [5]. A TRS R is *divergent* in the ordering $\succ$ if complete(R) is infinite, while R is *weakly divergent* in the ordering $\succ$ if nr-complete(R) is infinite. A TRS R is *inherently (weakly) divergent* if R is (weakly) divergent for all orderings.

3 Up and Down along the Peaks

In this section we introduce the idea underlying our strategy for computing the normal form of a term with respect to an equational theory T which does not admit a finite canonical rewrite system.

Let us briefly recall the idea the Knuth-Bendix procedure is based on [13]. Let R be a terminating TRS obtained by directing the equations of a given equational theory T according to a chosen term ordering $\succ$. Starting from R, a canonical rewrite system can be computed, if it exists, by means of Knuth-Bendix completion. This procedure is based on the computation of critical pairs, which identify all *peak* situations, i.e. when a term can be rewritten by means of two (or more) rewrite rules.

In the above peak, the term t can be rewritten into t_i and t_j by applying the rules $l_i \rightarrow r_i$ and $l_j \rightarrow r_j$, respectively. If t_i and t_j do not converge to a common term by rewriting them in the current rewrite system, a new rule derived from a critical pair between the rules $l_i \rightarrow r_i$ and $l_j \rightarrow r_j$ is added to the current rewrite system.

The definition of our rewriting strategy is based on the idea that all peaks have to be recognized and the application of the associated critical pairs has to be simulated. Let us consider the following TRS R, where rules are oriented according to a recursive path ordering with the precedence $+ \succ f$ and the left-to-right status of $+$ (the example is taken from [5]):

$$(x+y)+z \rightarrow x+(y+z) \qquad (r1)$$
$$f(x)+f(y) \rightarrow f(x+y) \qquad (r2)$$

The completion procedure generates an infinite number of rewrite rules. Given any term $f^k(x+y)+z$, for any k, it is easy to see how the peaks which generate critical pairs are. For instance, let $k=2$. The term $f(f(x+y))+z$ is in R-normal form, but is not in R_∞-normal form. In fact, if it is expanded by applying $r2$ as an expansion rule on the subterm $f(x+y)$, it results in the term $f(f(x)+f(y))+z$, which can only be reduced by applying the reduction opposite to the previous expansion. Nevertheless, if another expansion step is performed by $r2$ on the subterm $f(f(x)+f(y))$, the resulting term

$(f(f(x)) + f(f(y))) + z$ can be reduced using $r1$ to the term $f(f(x)) + (f(f(y)) + z)$, which is smaller than the initial term in the term ordering $\succ$.

Note that the reduction we have just obtained could be performed directly by applying one of the infinitely many rules derived from critical pairs on the initial term $f(f(x + y)) + z$. In fact, it turns out that the infinite set of rules is characterized by the following pattern rule:

$$f^n(x+y) + z \rightarrow f^n(x) + (f^n(y) + z)$$

This means that we have simulated the application of a rule r' derived from a critical pair. It is worth noting that, since more than one expansion step is needed to perform the reduction given by r', the rule r' is obtained by overlapping a rule in R and a rule previously derived from critical pair. Moreover, from the divergence pattern and the above example, it is easy to see that each new rule (for $n \geq 2$) can be derived by overlapping $r2$ and a previously derived rule. This suggests using $r2$ as expansion rule.

In the next section we formalize the above notions by defining a rewriting strategy which properly controls each step, and characterize the properties that the set E of expansion rules and the expansion process must satisfy in order to guarantee correctness and completeness with respect to the infinite canonical rewrite system R_∞.

4 The Strategy

The formalization of our strategy is based on the following hypotheses: given an equational theory T, R is a terminating TRS obtained by directing the equations in T according to a chosen term ordering $\succ$; R is weakly divergent in $\succ$, and admits a canonical rewrite system R_∞ by the nr-complete completion procedure in the ordering $\succ$.

Definition 1 (cp-completeness)
Let $R = \{l_i \rightarrow r_i, 1 \leq i \leq n\}$ and $E = \{r_j \rightarrow l_j \mid l_j \rightarrow r_j \in R, 1 \leq j \leq m, m \leq n\}$ be a set of expansion rules. E is *cp-complete* if for any term t, $t \xrightarrow{*}_{R \cup E} s$ where s is the R_∞-normal form of t.

Obviously, if $E = R$, then E is cp-complete.

The strategy is composed of five inference rules in the style of [5]. The first one is *R_norm* which reduces the input term to R-normal form, and properly initializes the parameters of the inference rules. The second one is *expansion*, which rewrites a term by applying the expansion rules. The set *Exp* contains the terms obtained by expansion from those in the set *Exp* in the previous iteration. Then, *reduction* simply reduces a term in R, and the set *Red* contains the terms computed in this step. In *contraction* the reductions (if any) opposite to the previous expansion steps are applied and all the reduced terms which are smaller than the input term are recorded in the set G. Finally, *selection* selects one of the terms in G non-deterministically, and initializes the parameters for a new iteration of the inference rules.

$$\begin{array}{ll} R\_norm: & (\emptyset,\emptyset,\emptyset,t) \vdash (\{s\},\emptyset,\emptyset,s) \; if \; t \xrightarrow{!}_R s \\ Expansion: & (Exp, Red, G, t) \vdash (Exp', Red, G, t) \\ & where \; Exp' = \{t_i \mid t' \rightarrow_E t_i, \; t' \in Exp, \; t' \; expandible, \; t_i \notin Exp\} \\ Reduction: & (Exp, Red, G, t) \vdash (Exp, \{t_i \mid t' \rightarrow_R t_i, \; t' \in Exp\}, G, t) \\ Contraction: & (Exp, Red, G, t) \vdash \\ & (Exp, Red, G \cup \{t_i \mid t' \xrightarrow{!}_R t_i, \; t' \in Red, \; t_i \prec t\}, t) \\ Selection: & (Exp, Red, G, t) \vdash (\{t'\},\emptyset,\emptyset,t') \; where \; t' \in G \end{array}$$

We can now define the application of a critical pair by the following inference rule:

$$\begin{array}{ll} CP\_Application = & (Expansion; \; Reduction; \; Contraction); \\ & (Expansion; \; Reduction; \; Contraction)* \\ & if \; Exp \neq \emptyset \; and \; G = \emptyset \end{array}$$

and the whole strategy can be defined as follows:

$$Normal\_Form = R\_norm; \; (CP\_Application; \; Selection)*$$

where $r*$ means that the inference rule r is repeated as long as its applicability conditions are satisfied, and ";" means sequencing of rules.

Proposition (correctness)
Given a TRS R, a set E of expansion rules such that E is cp-complete, and any term t, if $Normal\_Form(t) = t'$, then t' is the R_∞-normal form of t.

Sketch of the proof Since E is cp-complete, it is guaranteed that all peaks are reachable by suitable expansions and reductions. We know that R is not a confluent rewrite system. Given the term t, the first step of *Normal_Form* normalizes it to its R-normal form obtaining a term s. There can be two cases:

1. s is an R_∞-normal form. The strategy will expand s and all expandible terms derived from it, without being able to compute a term smaller than s, *Exp* will be empty at a certain iteration of *CP_Application*, G is empty, and $t' = s$ will be returned.

2. s is not an R_∞-normal form. This means that there exists a sequence $s \rightarrow_{R_\infty} s' \ldots$ in R_∞, where the first reduction is not possible in R, and the corresponding rule r derives from a critical pair between two rules in R_∞. In the expansion and reduction phases all the terms obtained from s by expansion and those reduced by R are collected in the sets *Exp* and *Red*, respectively.

   - Let r be derived from a critical pair obtained by overlapping two rules $r1$ and $r2$ in R. In this case $t1$, the vertex of the peak, is a term expanded from s using $r1$ as an expansion rule, and since it is reduced to s' by $r2$, s' is collected in the set *Red* in the reduction phase. Thus, s' can be compared with s, since they are instances of the elements of the critical pair and R_∞ is obtained via the nr-complete completion. If $s \succ s''$, where s'' is the R-normal form of s', then s'' will also be in the output set G by applying *Contraction*.

- If r does not directly derive from two rules in R, i.e. at least one is a rule from a critical pair, we need to expand more than once. *CP_Application* will do other expansion steps as long as a smaller reduced term is found in G.

In both cases above the strategy will restart with a smaller term as input. This is a correct step since R_∞ is a confluent rewrite system, therefore any term in G can be chosen in order to build the sequence which reaches the R_∞-normal form.

Completeness is obtained if the expansion phase terminates, i.e. if after a finite number of expansions the set Exp is empty.

Corollary (completeness)
Let R be a TRS and E be a set of expansion rules such that E is cp-complete. Given any term t, if the expansion phase terminates and t' is the R_∞-normal form of t, then $Normal\_Form(t)$ terminates with t'.

Sketch of the proof It follows from the termination hypothesis for the expansion phase and the correctness above.

Two main points can affect our strategy, both of them related to the set E of expansion rules. The former deals with the notion of cp-completeness; here the problem is in determining which rules are to be used also as expansion rules. In general, in order to do this, one must have some knowledge of the sources of divergence. The latter concerns the termination of the expansion phase. If the relation $\rightarrow_E$ is terminating, then given any term t, the set of the terms obtained by expansion from t is finite. In this case no special condition on the expansion process is needed. It is worth noting that the termination property of $\rightarrow_E$ is satisfied by a number of interesting divergent systems, like the following one studied, among others, in [5, 11].

Let us consider the TRS R defining the theory of bands (idempotent semigroups) given by the two rules:

$$(x * y) * z \rightarrow x * (y * z) \qquad (r3)$$
$$x * x \rightarrow x \qquad (r4)$$

Here, the associativity rule $r3$ is the source of divergence and gives rise to an infinite number of divergence patterns. This problem does not have solution yet in the metarule approach described in [11] while it easily fits in our framework. In fact, a cp-complete set E of expansion rules is obtained by reversing the associativity rule $r3$ alone, therefore $\rightarrow_E$ is terminating. All the other examples of divergent systems, either forward or backward crossed systems presented in [5], such that $\rightarrow_E$ is terminating, can be dealt with in a similar way.

When $\rightarrow_E$ is not terminating, it is still possible to have a complete strategy by properly defining the *expandible* predicate, which occurs in the definition of the expansion inference rule. The purpose of this predicate is to properly constrain the possibility of expanding terms. In what follows we present an example of the kind of definition which can be given. The idea is to prevent from infinitely expanding the same subterm by identifying an already expanded subterm using a marking criterium. Actually, a similar definition has been used to guarantee the termination of the expansion process in the strategy defined in [8].

Definition 2 A term t is *expandible* if there exists an expansion rule $r \rightarrow l$ in E, a substitution σ and a subterm $t|_u$ at the position u such that $t|_u = r\sigma$, and $t|_u$ is either a non-entirely marked redex or a redex smaller (in the term ordering $\succ$) than the expanded redex it comes from. The expansion step results in the term $s = t[\underline{l\sigma}]_u$.

Let us consider the following TRS R given by only one rule (the example is taken from [5]):

$$f(g(f(x))) \rightarrow g(f(x))$$

The completion procedure generates an infinite number of rules, which can be characterized by the following pattern rule:

$$f(g^n(f(x))) \rightarrow g^n(f(x))$$

The set E coincides with R and $\rightarrow_E$ is not terminating. Therefore, some kind of criteria must be used to guarantee the termination of the expansion phase. Let us consider the above definition of expandible term and see if it is suitable for our example. In that above definition infinite expansions are prevented using a marking criterium on the expanded redex and by not allowing the expansion of an entirely marked subterm, unless this subterm is smaller in the given ordering than the redex it comes from. These two assumptions guarantee the termination of the expansion phase, thus we have to check if E is still cp-complete with this restricted notion of expandibility. This means that we have to show that no interesting peaks are cut out of the search space. The only expandible redexes can be contained in a term like $t[\ldots g(f(x))\ldots]$ which expands to $t[\ldots \underline{f(g(f(x)))}\ldots]$. At this point our criterium prevents from repeatedly expanding the subterm $\underline{g(f(x))}$, since it is not smaller than the redex it comes from (it is exactly the same term). Thus, the only terms we are not able to generate by expansion are of the following kind: $t[\ldots \underline{f(f(f(\ldots f(g(f(x))\ldots)))}\ldots]$. If these terms were not already reducible in R, they can only be rewritten by applying the reductions opposite to the previous expansions, because they cannot be vertices of peaks. In this way we have shown that E together with the new definition of expandible subterm is actually cp-complete, and the expansion phase is terminating.

Another interesting example is given by the term rewriting system specifying the greatest common divisor of two natural numbers (this is again taken from [5]):

$$x + 0 \rightarrow x \qquad (r5)$$
$$x + s(y) \rightarrow s(x + y) \qquad (r6)$$
$$gcd(x, 0) \rightarrow x \qquad (r7)$$
$$gcd(0, x) \rightarrow x \qquad (r8)$$
$$gcd(x + y, y) \rightarrow gcd(x, y) \qquad (r9)$$

The completion procedure using a recursive path ordering with the precedence $+ \succ s$ generates two infinite families of rules:

$$gcd(s^n(x + y), s^n(y)) \rightarrow gcd(x, s^n(y))$$

from $r6$ and $r9$, and

$$gcd(s^n(x), s^n(0)) \rightarrow gcd(x, s^n(0))$$

from the first infinite family and $r5$. By analysing the divergence patterns it is easy to see that only $r5$ and $r6$ need to be considered as expansion rules. Therefore, $\rightarrow_E$ is not terminating. If one uses the above defined notion of expandible term, it is possible to obtain a terminating expansion process while maintaining cp-completeness. In fact, the marking criterium does not affect the expansion by $r6$. As far as expanding by $r5$ is concerned, we can still reach, from those terms which are instances of the left-hand sides of the two divergence patterns, the terms of the form $gcd(x + s^k(0), s^k(0))$. These terms are the vertices of all the peaks which generate the rewrite rules whose application we must simulate.

5 Conclusion

In this paper we have presented a new approach to the treatment of the problem of divergence in Knuth-Bendix completion. Given an equational theory T, our approach tries to provide a decidable rewriting relation for T, which is a compromise between the full power of equational deduction and the full efficiency of the usual canonical rewrite system approach. In fact, the idea of keeping some of the equations as equations can seem a step backwards with respect to the canonical approach, but it turns out to be less critical than it can appear at a first glance. Of course, our approach is worth when the canonical one is not feasible and unfortunately this happens quite often.

Our use of the expansion rules is not blind, that is we have provided sufficient conditions for guaranteeing that the proposed strategy is correct and complete. These conditions are not difficult to be proved, but their proof often requires some knowledge of the sources of divergence, i.e. the rules that cause the divergence of the completion process. This may be seen as a strong requirement, but it is worth noting that any approach which tries to deal with the divergence problem is actually based on the knowledge of the sources of divergence. In this respect, we think that our requirements are less strict, since we do not need to know the exact sources of divergence, that is we can accept sets of expansion rules which include rules that are not really necessary. In the example in Section 3, we can choose both rules as E, even though only one is the source of divergence, since $\rightarrow_E$ is terminating. This has, of course, an impact on the efficiency of the method, but allows for an easy proof of cp-completeness.

Acknowledgements

The present version of this paper has benefited from many interesting comments and suggestions received during the workshop.

References

[1] Avenhaus J., 'Proving Equational and Inductive Theorems by Completion and Embedding Techniques', in Proceedings of *Rewriting Techniques and Applications*, Lecture Notes in Computer Science, Springer-Verlag, 1991, Vol. 488, pp. 361–373.

[2] Chen H., Hsiang J., Kong H. 'On Finite Representations of Infinite Sequences of Terms', in Proceedings of the *2nd International Workshop on Conditional and Typed Rewriting Systems*, Montreal, 1990, Lecture Notes in Computer Science, Springer-Verlag, 1991, Vol. 516, pp. 100–114.

[3] Dershowitz N., Jouannaud J.-P., 'Rewrite Systems', in *Handbook of Theoretical Computer Science, Vol. B: Formal Models and Semantics*, J. van Leeuwen (ed.), North-Holland, 1990, pp. 243–320.

[4] Gloess P.Y., Laurent J.-P. H., 'Adding Dynamic Paramodulation to Rewrite Algorithms', in Proceedings of the *5th Conference on Automated Deduction*, Lecture Notes in Computer Science, Springer-Verlag, 1980, Vol. 87, pp. 195–207.

[5] Hermann M., 'Vademecum of Divergent Term Rewriting Systems', Technical Report CRIN 88-R-082, Centre de Recherche en Informatique de Nancy, 1988.

[6] Hermann M., 'Crossed Term Rewriting Systems', Technical Report CRIN 89-R-003, Centre de Recherche en Informatique de Nancy, 1989.

[7] Hermann M., 'Chain Properties of Rule Closures', *Formal Aspects of Computing*, 1990, Vol. 2, pp. 207–225.

[8] Inverardi P., Nesi M., 'A Rewriting Strategy to Verify Observational Congruence', *Information Processing Letters*, 1990. Vol. 35, pp. 191–199.

[9] Inverardi P., Nesi M., 'On Rewriting Behavioural Semantics in Process Algebras', in Proceedings of the *2nd International Conference on Algebraic Methodology and Software Technology* AMAST '91, Iowa City, USA, 1991, (to appear in Workshop Series, Springer-Verlag).

[10] Inverardi P., Nesi M., 'On Dealing with Divergent Rewrite Systems', Technical Report B4-06, I.E.I.- C.N.R., Pisa, Italy, March 1992.

[11] Kirchner H., 'Schematization of infinite sets of rewrite rules generated by divergent completion processes', in *Theoretical Computer Science*, North-Holland, 1989, Vol. 67, pp. 303–332.

[12] Kirchner H., Hermann M., 'Meta-rule Synthesis from Crossed Rewrite Systems', in Proceedings of the *2nd International Workshop on Conditional and Typed Rewriting Systems*, Montreal, 1990, Lecture Notes in Computer Science, Springer-Verlag, 1991, Vol. 516, pp. 143–154.

[13] Knuth D., Bendix P., 'Simple word problems in universal algebra', *Computational Problems in Abstract Algebra*, J. Leech (ed.), Pergamon Press, 1970.

[14] Lange S., 'Towards a Set of Inference Rules for Solving Divergence in Knuth-Bendix Completion', in Proceedings of *Analogical and Inductive Inference*, K.P. Jantke (ed.), Lecture Notes in Computer Science, Springer-Verlag, 1989, Vol. 397, pp. 304–316.

[15] Lysne O., 'Term Rewriting Techniques for Systems based on Generator Induction', Research Report No. 163, Department of Informatics, University of Oslo, December 1991.

[16] Milner R., *Communication and Concurrency*, Prentice Hall, 1989.

[17] Sattler-Klein A., 'Divergence phenomena during completion', in Proceedings of *Rewriting Techniques and Applications*, Lecture Notes in Computer Science, Springer-Verlag, 1991, Vol. 488, pp. 374–385.

[18] Thomas M., Jantke K. P., 'Inductive Inference for Solving Divergence in Knuth-Bendix Completion', in Proceedings of *Analogical and Inductive Inference*, K.P. Jantke (ed.), Lecture Notes in Computer Science, Springer-Verlag, 1989, Vol. 397, pp. 288–303.

[19] Thomas M., Watson P., 'Solving Divergence in Knuth-Bendix Completion by Enriching Signatures', submitted to *Theoretical Computer Science*, abstract in Proceedings of the 2nd *International Conference on Algebraic Methodology and Software Technology* AMAST '91, Iowa City, USA, 1991, (to appear in Workshop Series, Springer-Verlag).

A New Approach to General E-Unification Based on Conditional Rewriting Systems

Bertrand DELSART

LIFIA, Institut IMAG
46 Avenue Felix Viallet
38031 GRENOBLE
FRANCE
e-mail : delsart@imag.fr

Abstract. Using strictly resolvent conditional rewriting presentations of equational theories leads to a new transformation rule. This rule defines an unifying framework for the existing topmost[1] approaches to E-unification. Thus the development of common formal optimizations and implementation techniques is possible. Moreover, new algorithms can be expressed with this rule. For example, presentations based on different kinds of conditions lead to E-unification algorithms the behavior of which depends on the axiom applied. We also present the main ideas of an efficient E-unification algorithm based on presentations the conditions of which contain only E-unification problems between variables.

Keywords : E-unification, conditional term rewriting systems.

Introduction

Unification problems between first-order terms has been introduced by Herbrand [15] in the empty theory. Uniqueness of the most general unifier of two terms, when it exists, has been proved by Robinson who gave an exponential algorithm [27]. According to Davis [6], Mc Ilroy had discovered a similar algorithm in 1962 and used it in a theorem prover. More efficient algorithms have then been devised. Corbin and Bidoit [5] broke the complexity down to a quadratic one by implementing terms with dags instead of trees. Huet [17] used the notion of equivalence classes to achieve an almost linear complexity. Paterson and Wegman gave a linear algorithm [25] but an important overhead is needed. These algorithms can be expressed with Martelli and Montanari's transformation rules[2][23], which have been extended in order to deal with equational theories.

Many algorithms have been devised for particular theories. More general ones can be produced by merging them when the theories are disjoint[16][28][1]. Fay[12] gave the first general algorithm, called narrowing. Hullot [18] proved the completeness in special cases but termination is not always ensured. Several

[1] applying axioms only at the root
[2] sketched in Herbrand's thesis

algorithms have been developed by restricting narrowing so as to limit search space [26][24]. Unfortunately, all these procedures require a convergent rewriting system equivalent to the theory. Gallier and Snyder[13][14] solved the general case by introducing delayed E-unification steps. They proposed a topmost approach, called Root-Rewriting, and another one, Lazy-Paramodulation, based on rewriting at every non variable position. Lazy-Paramodulation was improved by Dougherty and Johann[11] but termination and efficiency problems still remain. Kirchner's restriction to syntactic theories[20] results in an efficient algorithm for a large class of theories. More recently, Comon, Habertrau and Jouannaud solved shallow theories[4] with a similar technique.

These general E-unification algorithms, will be described in Sect. 2 as an extension to Martelli and Montanari's work. Section 3 is devoted to a new mutation rule based on strictly resolvent conditional rewriting presentations. Its expression power is established in Sect. 4.

1 Notations and Definitions

We suppose known the standard notion of *algebra* $< \mathcal{A}, \mathcal{F} >$ where $\mathcal{A}$ is a non empty set (called *carrier* of the algebra) and $\mathcal{F} = \bigcup_{n \geq 0} \mathcal{F}_n$ is a set of function symbols of given arity. The following definitions and notations are consistent those in [9][10].

$\mathcal{T}(\mathcal{F}, \mathcal{X})$ is the set of *terms* constructed on $\mathcal{F}$ and a denumerable set $\mathcal{X}$ of variables, disjoint from $\mathcal{F}$. $\mathcal{T}(\mathcal{F}, \mathcal{X})$ is the smallest set containing $\mathcal{X}$ such that $f(t_1 ..., t_n) \in \mathcal{T}(\mathcal{F}, \mathcal{X})$ when $f \in \mathcal{F}_n$ and $\forall i \in [1..n]\ t_i \in \mathcal{T}(\mathcal{F}, \mathcal{X})$. Elements of $\mathcal{F}_0$ are called *constants* and $\mathcal{T}(\mathcal{F}, \emptyset)$ is the set of *ground* terms.

A term t may be seen as a finite ordered labeled tree. Leaves can be either variables or constants and internal nodes are function symbols the arity of which is equal to the outdegree of the corresponding node. $\mathcal{V}ar(t)$ denotes the set of variables appearing in t. A *position* is the sequence of positive integers describing the path from the root to the subterm it describes. The empty sequence is written Λ. The concatenation of two sequences p and q is written $p.q$. A position p is said to be above a position q , $p \leq q$, if there is a sequence r such that $q = p.r$. The relation $p \parallel q$ denotes non comparable positions, also called disjoint. The *domain* of a term t, $\mathcal{P}os(t)$, is the set of its positions. Its *size*, $|t|$, is the cardinal of its domain. $\mathcal{V}\mathcal{P}os(t)$ is the subset of $\mathcal{P}os(t)$ corresponding to variable positions and $\mathcal{F}\mathcal{P}os(t)$ contains the non-variable ones.

The *subterm* of t at position p, $t|_p$, verifies:
- $t|_p = t$ if $p = \Lambda$
- $\forall i \in [1..n]\ f(t_1 ..., t_n)|_{i.p} = t_i|_p$
- $t|_p$ is not defined if $p \notin \mathcal{P}os(t)$

The symbol of t at position p is written $t(p)$, and $\mathcal{H}ead(t)$ can be used to denote $t(\Lambda)$. The *replacement* of the subterm of t at position p by s, $t[s]_p$, is defined by the following properties:
- $t[s]_p = s$ if $p = \Lambda$
- $\forall i \in [1..n] f(t_1 ..., t_n)[s]_{i.p} = f(t_1 ..., t_i[s]_p ..., t_n)$
- $t[s]_p$ is not defined if $p \notin \mathcal{P}os(t)$

A *substitution* is an endomorphism on $\mathcal{T}(\mathcal{F}, \mathcal{X})$. Thus, it may be defined as a subset of $\mathcal{X} \times \mathcal{T}(\mathcal{F}, \mathcal{X})$. Applying a substitution $\sigma = \{x_1 \mapsto t_1, x_2 \mapsto t_2 ...\}$ to the term t results in the term $t\sigma$ obtained by replacing each x_i in t by t_i. The *domain* of a substitution, $\mathcal{D}om(\sigma)$, is the subset of $\mathcal{X}$ verifying $x \in \mathcal{D}om(\sigma)$ if and only if $x\sigma \neq x$. Its *range* is defined by $\mathcal{V}Ran(\sigma) = \bigcup_{x \in \mathcal{D}om(\sigma)} \mathcal{V}ar(x\sigma)$. $\sigma_{|W}$ is the restriction of the substitution σ to the subset W of $\mathcal{X}$. Two substitutions σ, τ are *equal on a subset* W of $\mathcal{X}$, noted $\sigma = \tau[W]$, if and only if $\forall x \in W \; x\sigma = x\tau$. An injective substitution σ such that $x\sigma$ is a variable for each x in $\mathcal{D}om(\sigma)$ is a *renaming* substitution. Applying it to a term gives a *variant* of this term. The composition of substitutions $\tau \circ \sigma$ is written $\sigma\tau$. σ is said to be *idempotent* if and only if $\sigma\sigma = \sigma$.

A term s is an *instance* of a term t, or t is *more general* than s, written $t \geq s$, if and only if $s = t\sigma$ for some substitution σ. Similarly σ is an instance of τ on a subset W of $\mathcal{X}$, $\tau \geq \sigma[W]$, if and only if $\sigma = \tau\lambda[W]$ for some substitution λ.

A *rewriting system* is a set of pairs of terms, written $R = \{l_i \rightarrow r_i, i \in I\}$. In this paper, we will not suppose that $\mathcal{V}ar(r_i) \subseteq \mathcal{V}ar(l_i)$. It leads to the following *rewriting* relation, defined as the closure of R on $\mathcal{T}(\mathcal{F},\mathcal{X}) \times \mathcal{T}(\mathcal{F},\mathcal{X})$: a (given) term t is a rewriting of a term s by the rule $l_i \rightarrow r_i$ at the position p by the substitution σ if and only if $p \in \mathcal{P}os(s)$, $s|_p = l\sigma$, $\mathcal{D}om(\sigma) = \mathcal{V}ar(l) \cup \mathcal{V}ar(r)$ and $t = s[r\sigma]_p$ where l and r are variants of l_i and r_i composed of variables distinct from those of s and t. This relation, written $s \rightarrow^p_{l\rightarrow r,\sigma} t$ or $s \rightarrow_R t$ for short, can be verified by infinitely many terms t when $\mathcal{V}ar(r_i) \not\subseteq \mathcal{V}ar(l_i)$. Note that $\mathcal{V}Ran(\sigma) = \mathcal{V}ar(s|_p) \cup \mathcal{V}ar(t|_p)$ and σ is idempotent. As usual, '*' denotes the reflexive transitive closure of any relation and '+' the non reflexive one.

Let E be a set of pairs of terms called *oriented axioms* and written $l \approx r$. For two terms s and t, we write $s \leftrightarrow^p_{l\approx r,\sigma} t$ if and only if $s|_p = l\sigma$ and $t = s[r\sigma]_p$ (i.e. $s \rightarrow^p_{l\rightarrow r,\sigma} t$). Is is extended to $s \longleftrightarrow_E t$ if $s \leftrightarrow_{l\approx r} t$ with $l \approx r$ or $r \approx l$ being a variant of an axiom of E. The *equational theory* on E is the reflexive, transitive closure of this relation, also called equality modulo E. It is usually written $\xleftrightarrow{*}_E$, $=_E$ or just E.

All the previous definitions based on equality of terms can be extended to equality modulo E. For instance, $t \geq_E s$ if and only if $s =_E t\sigma$ for some substitution σ.

To solve an E-unification problem $s =^?_E t$, one must compute an idempotent substitution such that $s\sigma =_E t\sigma$. Such a substitution is said to be an *E-unifier* of s and t. $U\Sigma_E(s,t)$, the set of such substitutions, is called the set of unifiers of s and t under E.

In the empty theory, a unifier σ of s and t is said to be a *most general* unifier if each unifier of s and t is an instance of σ. In the general case, the set of the most general E-unifiers, noted $\mu U\Sigma_E(s,t)$, is defined by the following properties:

soundness : $\mu U\Sigma_E(s,t) \subseteq U\Sigma_E(s,t)$
completeness: $\forall \tau \in U\Sigma_E(s,t) \quad \exists \sigma \in \mu U\Sigma_E(s,t) / \sigma \geq_E \tau[W]$
minimality : $\forall \sigma, \tau \in \mu U\Sigma_E(s,t) \quad \sigma =_E \tau[W] \Rightarrow \sigma = \tau$

W is the set of variables of the problem (i.e. $\mathcal{V}ar(s) \cup \mathcal{V}ar(t)$). The following condition protects a subset Z of $\mathcal{X}$ containing W.

protection of Z: $\forall \sigma \in \mu U\Sigma_E(s,t) \qquad \mathcal{D}om(\sigma) \subseteq W$ and $Z \cap \mathcal{V}Ran(\sigma) = \emptyset$.

When this property is verified, $\mu U\Sigma_E(s,t)$ is said to be *away* from Z. E-unification algorithms should produce such a set when it exists. Unfortunately, E-unification is undecidable in the general case. Moreover, even if redundancies should be avoided, minimality is not often considered as a part of the unification procedure.

Rewriting systems can easily be extended with the introduction of conditions associated to each rule. They are written $R = \{c_i \,|\, l_i \rightarrow r_i, i \in I\}$. In this paper, conditions are restricted to conjunction of E-unification problems. The substitution introduced in a rewriting step must verify these conditions. Thus, given two terms s and t, t is a rewriting of s by the rule $c_i \,|\, l_i \rightarrow r_i$ at the position p by the substitution σ if and only if $p \in \mathcal{P}os(s)$, $s|_p = l\sigma$, $t = s[r\sigma]_p$, $\mathcal{V}ar(l) \cup \mathcal{V}ar(r) \subseteq \mathcal{D}om(\sigma) \subseteq \mathcal{V}ar(c) \cup \mathcal{V}ar(l) \cup \mathcal{V}ar(r)$ and $\sigma \in U\Sigma_E(c)$ for some variants c, l and r of c_i, l_i and r_i.

2 General E-Unification Algorithms

Dealing with the empty theory, Martelli and Montanari considered unification problems as sets of equations. In their approach, which is now the traditional one, rules are applied so as to compute a special set, called *solved form*. A set of equations is said to be in (tree) solved form when it is the equational presentation of an idempotent substitution $\sigma=\{x_1 \mapsto t_1 ..., x_n \mapsto t_n\}$, written $\underline{\sigma}$. Thus, $\underline{\sigma}=\{x_1 \stackrel{?}{=}_E t_1 ..., x_n \stackrel{?}{=}_E t_n\}$ such that: • $\forall\, 1 \leq i<j \leq n \;\; x_i \neq x_j$

• $\forall\, 1 \leq i,j \leq n \;\; x_i \notin \mathcal{V}ar(t_j)$

The other solved forms will not be described in this paper (see for example [19]). An unification algorithm consists of a finite set of transformation rules and a class of valid controls describing how to apply these rules to ensure completeness and possibly termination. For example, the six following rules terminate and are complete for the empty theory (i.e. $E=\emptyset$) whatever the control is[19][3].

Delete	$P \cup \{s \stackrel{?}{=}_E s\}$ $\Longrightarrow P$
Decompose	$P \cup \{f(s_1 ..., s_n) \stackrel{?}{=}_E f(t_1 ..., t_n)\}$ $\Longrightarrow P \cup \{s_1 \stackrel{?}{=}_E t_1 ..., s_n \stackrel{?}{=}_E t_n\}$
Conflict	If $f \neq g$ then $P \cup \{f(s_1 ..., s_n) \stackrel{?}{=}_E g(t_1 ..., t_m)\}$ $\Longrightarrow$ FAIL
Coalesce	If $x, y \in \mathcal{V}ar(P)$ and $x \neq y$ then $P \cup \{x \stackrel{?}{=}_E y\}$ $\Longrightarrow P\{x \mapsto y\} \cup \{x \stackrel{?}{=}_E y\}$
Check	If $x \in \mathcal{V}ar(s)$ and $s \neq x$ then $P \cup \{x \stackrel{?}{=}_E s\}$ $\Longrightarrow$ FAIL
Eliminate	If $s \notin \mathcal{X}$ and $x \in \mathcal{V}ar(P) - \mathcal{V}ar(s)$ then $P \cup \{x \stackrel{?}{=}_E s\}$ $\Longrightarrow P\{x \mapsto s\} \cup \{x \stackrel{?}{=}_E s\}$

To solve the general case, **Conflict** and **Check** must be replaced by other rules and concurrent application can be imposed[4]. When a convergent rewriting system R exists for the theory[5], narrowing[12] leads to a complete algorithm.

[3] in this paper, $s \stackrel{?}{=}_E t$ and $t \stackrel{?}{=}_E s$ are identical (in order to avoid possible loops)

[4] the set of unifiers is the union of the solutions given for each branch

[5] i.e. $\xrightarrow{*}_R \xleftarrow{*}_R = \xleftrightarrow{*}_E$ and $\xrightarrow{*}_R$ leads to a unique normal form

Narrow if $p \in \mathcal{F}Pos(s)$ then $P \cup \{s =^?_E t\}$
$\Longrightarrow P\sigma \cup \{s[r]_p\sigma =^?_E t\sigma\} \cup \underline{\sigma}$
where σ is the most general unifier of $s|_p$ and l
for some variant $l \rightarrow r$ of a rule of R

As for lazy-paramodulation, any theory can be considered but the unification step is replaced by a (delayed) E-unification step[14].

Lazy-Paramodulate If $p \in \mathcal{F}Pos(s)$ then $P \cup \{s =^?_E t\}$
$\Longrightarrow P \cup \{s|_p =^?_E l, s[r]_p =^?_E t\}$
where $l \approx r$ is a variant of an axiom of $E \cup E^{-1}$
such that $l \in \mathcal{X}$ or $\mathcal{H}ead(s|_p) = \mathcal{H}ead(l)$

The first complete algorithm was based on topmost paramodulations[13]:

Root-Rewriting If $s \notin \mathcal{X}$ then $P \cup \{s =^?_E t\}$
$\Longrightarrow P \cup \{s =^?_E l, r =^?_E t\}$
where $l \approx r$ is a variant of an axiom of $E \cup E^{-1}$
such that $l \in \mathcal{X}$ or $\mathcal{H}ead(s) = \mathcal{H}ead(l)$

Imitate If $x \in \mathcal{V}ar(f(s_1 \ldots, s_n))$ then $P \cup \{x =^?_E f(s_1 \ldots, s_n)\}$
$\Longrightarrow P \cup \{x_1 =^?_E s_1 \ldots, x_n =^?_E s_n\} \cup \underline{\sigma}$
where $x_1 \ldots, x_n$ are new variables and $\sigma = \{x \mapsto f(x_1 \ldots, x_n)\}$

Decomposition, imitation and application of every axioms must be tried concurrently but if neither s nor t are variables then the mutated term can be chosen. Infinite derivations introduced by **Imitate** are avoided because this rule must be applied only so as to allow mutation at positions above an occurrence of the variable in the other term. Moreover, **Lazy-Paramodulate** (resp. **Root-Rewriting**) need not be applied to $s|_p =^?_E l$ (resp. $s =^?_E l$). Dougherty and Johann[11] proved that $s|_p$ and l should have the same symbols at each position which is a non-variable occurrence in both terms and that the corresponding decompositions could be applied without losing completeness. The condition $s \notin \mathcal{X}$ in **Root-Rewriting** does not appear in Gallier and Snyder's definition. However, the deterministic procedure described in their paper uses this restriction in order to forbid Root-Rewriting on the non-trivial pair. Their imitation rule also leads to the elimination of the variable in the whole problem but it seems better to introduce the new variables only in order to allow decomposition.

As regards *syntactic theories*[6], **Decompose** is applied to $s =^?_E l$ and $r =^?_E t$[20].

Syntactic-Mutate $P \cup \{f(s_1 \ldots, s_n) =^?_E g(t_1 \ldots, t_m)\}$
$\Longrightarrow P \cup \{s_1 =^?_E u_1 \ldots, s_n =^?_E u_n\} \cup \{t_1 =^?_E v_1 \ldots, t_m =^?_E v_m\}$
if $f(u_1 \ldots, u_n) \approx g(v_1 \ldots, v_m)$ is a variant of an axiom of $E \cup E^{-1}$

As usual, decomposition and every possible mutation of one of the two terms must be done concurrently. This rule leads to the most efficient algorithm but the presentation considered must be *resolvent* , i.e. $\xleftrightarrow{+}_E \subseteq \xleftrightarrow{*}^{\neq\Lambda}_E \longleftrightarrow_E \xleftrightarrow{*}^{\neq\Lambda}_E$. It means that for each pair of equivalent terms there exists an equational proof with at most one equational step at the topmost occurrence. A similar technique was applied to *shallow theories*[4], for which there exists a presentation the axioms of which contain no variable at depth strictly greater than one. Pairs introduced

[6] a theory is syntactic if and only if it admits a finite resolvent presentation

by the variable part of a *collapsing rule*[7] are not considered by the mutation rules. Thus, decomposition will be applied as soon as possible. Moreover *subterm collapsing rules*[8] can be applied at every position strictly above the occurrences of s when s is a variable appearing in t. The equational presentation must be *cycle-syntactic*[9] but such a presentation can be generated for any shallow theory.

3 A New Formalism for Topmost Algorithms

Application of axioms at the root with lazy evaluation of the unifiers seems to be a good approach to general E-unification. Such an application generates two problems. Moreover, the previous topmost algorithms forbid mutation at the root of at least one of these problems. This section is devoted to the description of a new transformation rule based on these properties. Lazy evaluation is expressed by conditional rewriting rules. A (conditional) rewriting system R is a *rewriting presentation* of a theory E if and only if it is *sound* (i.e. $\xrightarrow{+}_R \subseteq \xleftrightarrow{*}_E$) and *adequate* (i.e. $\xleftrightarrow{+}_E \subseteq \xrightarrow{*}_R$)[10]. In order to protect the problem issued from the right part of a rule, we always choose R so that $\xrightarrow{+}_R \subseteq \longrightarrow_R \xrightarrow{*}{}^{\neq \Lambda}_R$. It means that at most one rewriting must be done before decomposing. Such a presentation is called *strictly resolvent presentation*. We will prove in Sect. 4.1 that any theory admits at least one strictly resolvent presentation. These properties lead to the following transformation rule:

Mutate	$P \cup \{s \stackrel{?}{=}_E t\}$
	$\Longrightarrow P \cup \{r\sigma \stackrel{?}{=}_E t\sigma\} \cup c\sigma \cup \underline{\sigma}$
	if $c \mid l \rightarrow r$ is a variant of a rule of R
	and σ is the most general unifier of s and l

Mutation of a variable is enabled to get the most general rule but we can choose which term will be considered when applying every possible **Mutate** to a given problem (see Corollary 5). If R is strictly resolvent then **Mutate** need not be applied to $r\sigma \stackrel{?}{=}_E t\sigma$ to ensure completeness (see Corollary 4). Thus, if r and t are not variables then $\mathcal{H}ead(r) = \mathcal{H}ead(t)$.

This is the only mutation rule of our E-unification algorithm. As usual, we need some rules to allow unification in the empty theory, like **Delete**, **Decompose**, **Coalesce** and **Eliminate** (for tree solved form). If the theory is *subterm collapse-free*[11], **Check** can be preserved and **Mutate** becomes useless when $s \in \mathcal{V}ar(t)$ or $t \in \mathcal{V}ar(s)$[12]. In the general case, **Check** is replaced by **Imitate**. Note that, as for **Imitate** , σ may be applied to P in **Mutate** but the corresponding proofs will not be given since it corresponds to applications of **Eliminate** and **Coalesce** on $\underline{\sigma}$.

---

[7] $l \rightarrow r$ such that $r \in \mathcal{X}$

[8] $l \rightarrow r$ such that $\forall \sigma\ |l\sigma| > |r\sigma|$ (called collapsing rule in [4])

[9] if $t[s]_q =_E s$ then $\exists p < q$ and $l \rightarrow r$ subterm collapsing $/ t[s]_q \xleftrightarrow{*}{}^{>p}_E \longrightarrow^p_{l \rightarrow r} \xleftrightarrow{*}_E s$

[10] our purpose is not to orient axioms, thus $\xleftrightarrow{+}_E \subseteq \xrightarrow{*}_R \xleftarrow{*}_R$ would be too general

[11] i.e. a term cannot be equal to one of its proper subterms

[12] no proof will be given since occur-check is well known

Delete, **Decompose**, **Coalesce**, **Eliminate** and **Imitate** are sound rules (see for example [19][13]). The following properties define the theoretical framework for our E-unification procedure.

Lemma 1. *If τ verifies $\underline{\sigma}$ then $\tau =_E \sigma\tau$*

Proof. If $x \in \mathcal{D}om(\sigma)$ then $\{x =_E^? x\sigma\} \subseteq \underline{\sigma}$. Thus $x\tau =_E x\sigma\tau$.
If $x \notin \mathcal{D}om(\sigma)$ then $x\sigma\tau = (x\sigma)\tau = x\tau$. □

Theorem 2 (Soundness). *If R is a rewriting presentation of E then **Mutate** is sound.*

Proof. Let τ be a solution of $P \cup \{r\sigma =_E^? t\sigma\} \cup c\sigma \cup \underline{\sigma}$. We must prove that τ is a solution of $P \cup \{s =_E^? t\}$. Now, $r\sigma\tau =_E t\sigma\tau \Rightarrow t\tau =_E r\sigma\tau$ (Lemma 1)

$$s\sigma = l\sigma \Rightarrow s\tau =_E l\sigma\tau \text{ (Lemma 1)}$$

$$c\sigma\tau \Rightarrow r\sigma\tau =_E l\sigma\tau \text{ (soundness of R)}$$

Thus, $s\tau =_E t\tau$. □

Theorem 3. ***Delete, Decompose, Coalesce, Eliminate, Imitate** and **Mutate** preserve the set of unifiers[13] when R is strictly resolvent and the control verifies the following property: if **Decompose**, **Imitate** or **Mutate** is applied on a given pair of terms then every transformation of this pair by one of these three rules must be tried concurrently.*

Sketch of proof. For each solution τ of the initial problem, we must prove that there is a solution of the generated problems more general than τ over the variables of the initial problem. In fact, we will exhibit a solution equal to τ over these variables. Since $s\tau =_E t\tau$, there is a strictly resolvent rewriting sequence from $s\tau$ to $t\tau$. If no rewriting steps are needed then **Coalesce**, **Decompose**, **Delete** and **Eliminate** ensure the completeness. If the first step is done at the topmost occurrence then there exists a conditional rule such that **Mutate** preserves the solution; otherwise **Decompose** or **Imitate** can be applied.

Proof. If $s = t$, $t \in \mathcal{X}\text{--}\mathcal{V}ar(s)$ or $s \in \mathcal{X}\text{--}\mathcal{V}ar(t)$ then **Delete**, **Eliminate** and **Coalesce** ensure the completeness. Thus we can suppose $s \neq t$, $s \notin \mathcal{X}\text{--}\mathcal{V}ar(t)$ and $t \notin \mathcal{X}\text{--}\mathcal{V}ar(s)$. Note that this hypothesis implies that one of these two terms is not a variable.

Let τ be a solution of $P \cup \{s =_E^? t\}$ and $W = \mathcal{V}ar(P) \cup \mathcal{V}ar(s) \cup \mathcal{V}ar(t)$. Since $s\tau =_E t\tau$ then $s\tau \xrightarrow{*}_R t\tau$. Thus, either $s\tau = t\tau$ or $s\tau \longrightarrow_R u \xrightarrow{*}{}^{\neq\Lambda}_R t\tau$.
• If $s\tau = t\tau$ then $t \notin \mathcal{X}$ and $s \notin \mathcal{X}$ (it would be a proper subterm of the other). Thus $\mathcal{H}ead(s) = \mathcal{H}ead(t)$ and **Decompose** leads to a problem solved by τ.
• Let us suppose that $s\tau \xrightarrow{+}{}^{\neq\Lambda}_R t\tau$
If $s,t \notin \mathcal{X}$ then $\mathcal{H}ead(s) = \mathcal{H}ead(t)$ and τ solves the decomposed problem.
If $t \in \mathcal{V}ar(s)$ then $\mathcal{H}ead(t\tau) = \mathcal{H}ead(s)$ and **Imitate** leads to a problem solved by $\theta = \tau\{x_i \mapsto t\tau|_i \forall i \in [1..arity(\mathcal{H}ead(s))]\}$, verifying $\theta_{|W} = \tau$.
Similarly, if $s \in \mathcal{V}ar(t)$ then **Imitate** ensures completeness.

[13] i.e. each unifier solves at least one of the concurrently generated problems

• The only remaining case is $s\tau \longrightarrow_{R,\Lambda} u \xrightarrow{*}{}_{R}^{\neq\Lambda} t\tau$ (s may be a variable). Let $c \,|\, l \rightarrow r$ be a variant of the first rewriting rule applied and μ the substitution used. By definition, $s\tau = l\mu$, $u = r\mu$ and μ verifies c.
We can suppose that $\mathcal{V}Ran(\mu) \cap \mathcal{D}om(\tau) = \emptyset$ since idempotency implies that $(\mathcal{V}ar(s\tau) \cup \mathcal{V}ar(t\tau)) \cap \mathcal{D}om(\tau) = \emptyset$. Moreover, $\mathcal{D}om(\mu)$ and $\mathcal{V}Ran(\tau)$ are disjoint since $\mathcal{D}om(\mu) \subseteq \mathcal{V}ar(l) \cup \mathcal{V}ar(c) \cup \mathcal{V}ar(r)$, a set of new variables.
Hence $\mu\tau = \mu[\mathcal{V}ar(l) \cup \mathcal{V}ar(c) \cup \mathcal{V}ar(r)]$ and $\mu\tau = \tau[W]$.
Thus $s\mu\tau = s\tau = l\mu = l\mu\tau$

$\forall x \in W, x\mu\tau\mu\tau = x\tau\mu\tau = x\tau\tau = x\tau = x\mu\tau$

$\forall x \in \mathcal{V}ar(c) \cup \mathcal{V}ar(l) \cup \mathcal{V}ar(r), x\mu\tau\mu\tau = x\mu\mu\tau = x\mu\tau$

As a consequence, s and l are unifiable and, given the most general unifier σ, there is a substitution θ so that $\mu\tau = \sigma\theta[W \cup \mathcal{V}ar(l) \cup \mathcal{V}ar(c) \cup \mathcal{V}ar(r)]$ and $\sigma\theta$ is idempotent.
Then $\sigma\theta = \tau[W]$

$\sigma\theta$ verifies P (since $P\tau = P\sigma\theta$)

$\sigma\theta$ verifies $r \overset{?}{=}_E t$ (since $r\sigma\theta = r\mu = u \xrightarrow{*}{}_{R}^{\neq\Lambda} t\tau = t\sigma\theta$)

$c\sigma\theta$ is true (since $c\sigma\theta = c\mu$).

Thus **Mutate** leads to a problem solved by $\sigma\theta = \tau[W]$. □

Corollary 4. ***Mutate*** *need not be applied to the problem $r\sigma \overset{?}{=}_E t\sigma$ generated by its previous application.*

Proof. It is a direct consequence of the strict resolventness of the equational proof introduced in the previous theorem. □

Note that if $r\sigma$ or $t\sigma$ is a variable then **Mutate** may become necessary if this variable is instantiated by the application of **Eliminate** on another problem.

Corollary 5. ***Mutate*** *need not be applied to the two terms of $s \overset{?}{=}_E t$.*

Proof. $s\tau =_E t\tau$ is also equivalent to $t\tau \xrightarrow{*}_R s\tau$ in Theorem 3. □

Theorem 6. *Any non transformable problem is either an insoluble problem or a solved form.*

Proof. Since a solved form always has a solution, we must prove than a non solved form can be transformed if it admits a solution. A problem is not in solved form if and only if there is a non solved pair, i.e. $s_i \overset{?}{=}_E t_i$ verifying one of the following properties:

- s_i and t_i are not in $\mathcal{X}$
- $s_i \in \mathcal{V}ar(t_i)$ or $t_i \in \mathcal{V}ar(s_i)$
- s_i and t_i are distinct variables that both appear elsewhere
- s_i is the only variable and $\exists j \neq i \,/\, s_i \in \mathcal{V}ar(s_j) \cup \mathcal{V}ar(t_j) - \mathcal{V}ar(t_i)$
- t_i is the only variable and $\exists j \neq i \,/\, t_i \in \mathcal{V}ar(s_j) \cup \mathcal{V}ar(t_j) - \mathcal{V}ar(s_i)$

In the first case, the proof of Theorem 3 implies that **Delete**, **Decompose** or **Mutate** can be applied when the problem is solvable. In the second one, the same proof implies that **Delete**, **Mutate** or **Imitate** are valid if there is a solution. In the last three cases, **Coalesce** or **Eliminate** can always be applied. □

Theorem 7 (Completeness). *When it terminates, the E-unification procedure based on the transformation rules* ***Delete, Decompose, Coalesce, Eliminate, Imitate*** *and* ***Mutate****, applied with the conditions given in the Theorem 3 (R being strictly resolvent), results in a sound and complete set of unifiers.*

Proof. It is a direct consequence of Theorem 2, Theorem 3 and Theorem 6. □

Minimality is not achieved but protection of a set of variables can easily be added. Termination cannot be ensured since the E-unification problem in undecidable. However, if the equational proof corresponding to a solution can be expressed by a finite strictly resolvent proof then there is a finite sequence of transformation rules that leads to a more general E-unifier. With the rewriting presentations described in this paper, finiteness of an equational proof implies the existence of such a rewriting sequence.

4 Expression Power of Strictly Resolvent Presentations

The new formalism introduced in Sect. 3 contains the part of the control that is common to **Root-Rewriting** and **Syntactic-Mutate**. The conditions can be used to refine the control, thus leading to different algorithms.

4.1 Simulating Root-Rewriting

The most important point is that any theory $E=\{l_i \approx r_i\}$ can be represented by at least the conditional rewriting system $R=\{x \stackrel{?}{=}_E l_i \mid x \to r_i\} \cup \{x \stackrel{?}{=}_E r_i \mid x \to l_i\}$.

Proof. Adequacy: if $s \leftrightarrow^{p}_{l \approx r, \sigma} t$ then $s \to^{p}_{x \stackrel{?}{=}_E l \mid x \to r, \{x \mapsto s|_p\}\sigma} t$. Thus $\xrightarrow{+}_R \supseteq \xleftrightarrow{+}_E$.

Soundness: if $s \to^{p}_{x \stackrel{?}{=}_E l \mid x \to r, \sigma} t$ then $s|_p = x\sigma$, $t = s[r\sigma]_p$ and $x\sigma =_E l\sigma$.

Since $l =_E r$ by definition of R, $s|_p =_E t|_p$. Thus $s =_E t$ and $\xrightarrow{+}_R \subseteq \xleftrightarrow{+}_E$. □

Moreover, R is strictly resolvent. In fact, the first rewriting step corresponds to the rightmost paramodulation at the root in the equational proof. The condition describes the proof leading to the paramodulated term and decomposition can then be applied since it is the last paramodulation at the root.

With this presentation, the E-unification procedure behaves like Gallier and Snyder's topmost algorithm. Proofs are build from left to right if $x \stackrel{?}{=}_E r \mid x \to l$ is applied on the term t. If mutation is applied on s, then the equational proofs are build in the reversed order.

4.2 Simulating Syntactic-Mutate

$R=\{\{X \stackrel{?}{=}_E U\} \mid f(X) \to g(V), \{Y \stackrel{?}{=}_E V\} \mid g(Y) \to f(U) \ldots\}$ is a rewriting presentation of the theory $E=\{f(U) \approx g(V) \ldots\}$, where capital letters denote lists of arguments (their equality is the conjunction of the equality of their components).

Proof. Adequacy can easily be proved.

As for soundness, if $s \rightarrow^{p}_{\{X =^{?}_{E} U\} \| f(X) \rightarrow g(V), \sigma} t$ then $s|_p = f(X\sigma)$, $t = s[g(V\sigma)]_p$ and $X\sigma =_E U\sigma$. Thus $f(X\sigma) =_E f(U\sigma) =_E g(V\sigma)$ and $s =_E t$ for the same reasons as in Sect. 4.1. □

In addition, if the equational presentation was resolvent, then R is strictly resolvent. The first rewriting step corresponds to the unique equational step at the topmost occurrence and the conditions describe the paramodulations steps which must be done before. Thus, **Mutate** can also simulate **Syntactic-Mutate**.

4.3 Theoretical results on collapsing theories

This new formalism is also interesting for common theoretical research, like dealing more efficiently with collapsing rules. When occur-check fails, **Imitate** generates a loop. However, in Gallier and Snyder's approach, **Imitate** is needed only to apply axioms above an occurrence of the initial variable. This is a consequence of the following theorem. It also proves that, after the root-rewriting step, **Eliminate** or **Coalesce** can be applied on the problem $x =^{?}_{E} r$ and **Root-Rewriting** need not be applied on the other problem (but it may be less efficient).

Theorem 8. *Let σ be a solution of $x =^{?}_{E} t$ so that $t|_p = x$, for each presentation R of the theory, there is a position q strictly above p and a substitution τ more general than σ over the variables of t so that $t\tau \xrightarrow{*}{}^{>q}_{R} \longrightarrow^{q}_{c|l \rightarrow r} \xrightarrow{*}{}^{\|q}_{R} x\tau$*

Proof. Since $t[x]_p\sigma =_E x\sigma$, there must be at least one axiom applied on the set of positions $\{(n{*}p).q \ / \ n \geq 0 \text{ and } \Lambda \leq q < p\}$, where $n{*}p$ is the concatenation of n sequences of p.

Let $(n{*}p).q$ be the topmost position verifying this criterion, $c\,|\,l \rightarrow r$ be the first rule applied at this position and θ the corresponding substitution.
By definition, $t\sigma|_{(n*p).q} \xleftrightarrow{*}{}^{\neq\Lambda}_{E} l\theta$ and $t\sigma[r\theta]_{(n*p).q} =_E x\sigma$.
Now, since it is the topmost position, $\forall p' < (n{*}p).q \ \ t\sigma|_{p'} \xleftrightarrow{*}{}^{\neq\Lambda}_{E} x\sigma|_{p'}$.
Therefore, $t\sigma|_q \xleftrightarrow{*}{}^{\neq\Lambda}_{E} x\sigma|_q = t\sigma|_{p.q} \xleftrightarrow{*}{}^{\neq\Lambda}_{E} x\sigma|_{p.q} \cdots x\sigma|_{((n-1)*p).q} = t\sigma|_{(n*p).q}$.
This implies that $t\sigma|_q \xleftrightarrow{*}{}^{\neq\Lambda}_{E} l\theta$.

Let τ be $\{x \mapsto x\sigma[r\theta]_q\}\sigma$, $\tau =_E \sigma$ since $r\theta =_E l\theta =_E x\sigma|_q$. Moreover, there is a rewriting sequence from $t\tau$ to $t\sigma$ done under the occurrences of x in t. In addition, $t\sigma \rightarrow^{q}_{c|l \rightarrow r, \theta} t\sigma[r\theta]_q \xrightarrow{*}{}^{\|q}_{R} x\sigma[r\theta]_q = x\tau$. Thus, delaying the initial steps which are not done below q leads to the wanted rewriting sequence. □

Thus, **Imitate** must only allow mutation strictly above a variable position. Moreover, the presentation used need not be strictly resolvent. However, lazy unification of the arguments is necessary (i.e. $t|_q =^{?}_{E} l$ without mutation at the root).

This result does not hold in a non-lazy approach[14]. However, since no rules are applied above a previous rewriting step in a strictly resolvent proof, we conjecture that applying rules under a variable position is useless with strictly resolvent presentations. We proved it only when any rewriting sequence is equivalent

[14] for instance, unification of x and $f(x,x)$ in $E = \{f(h(a), h(b)) \approx h(a), a \approx b\}$

to a strictly resolvent one in which the highest step is done at the same position. It is a direct consequence of the previous theorem: the proof introduced is replaced by a strictly resolvent the first step of which is done at position q. Since this property is verifyed by all the presentations we ever used, this improvement was used in our implementation.

We are presently studying the application of axioms at non-variable positions when occur-check fails. The previous theorem implies[15] the completeness of the following mutation rule when every rule of an adequate rewriting presentation R is applied concurrently at every position q above a chosen occurrence of x.

Var-Mutate	$P \cup \{x \overset{?}{=}_E t[x]_p\}$
	$\Longrightarrow P \cup \{t[r]_q \overset{?}{=}_E x\} \cup \{t\vert_q \overset{?}{=}_E l\}$
	if $q < p$ and $l \rightarrow r$ is a variant of a rule of R
	so that $l \in \mathcal{X}$ or $\mathcal{H}ead(l) = \mathcal{H}ead(t\vert_q)$

In addition, **Decompose** can be applied to $t|_q \overset{?}{=}_E l$ when l is not a variable. When dealing with cycle-syntactic theories, completeness is preserved even if only the subset of subterm-collapsing rules is considered. We are trying to restrict the set of rules considered in the general case. Unfortunately, even rules like $x \rightarrow f(x)$ are needed for some theories[16].

4.4 Adding an axiom to a syntactic theory

We have proved that **Mutate** is more general than the other topmost rules. Moreover, since the conditions are associated to the rules, the behavior may depend upon the applied rule. Thus, for example, a syntactic theory can easily be merged with a non-syntactic one if they are disjoint. Note also that, for instance, we can use the conditions to enable mutation of the generated pair $r \overset{?}{=}_E t$ for a given rule ($c \,|\, l \rightarrow r$ becomes $c \cup \{x \overset{?}{=}_E r\} \,|\, l \rightarrow x$). This can be useful to extend a syntactic theory. Let R be a strictly resolvent presentation of a syntactic theory E_0 (as described in Sect. 4.2). Let E be $E_0 \cup \{l \approx r\}$

Theorem 9. $R \cup \{\{x \overset{?}{=}_E l, y \overset{?}{=}_E r\} \,|\, x \rightarrow y, \{x \overset{?}{=}_E l, y \overset{?}{=}_E r\} \,|\, y \rightarrow x\}$ *is a strictly resolvent presentation of E.*

Proof. Soundness: since $l =_E r$, every substitution σ verifies $l\sigma =_E r\sigma$. Thus, if σ verifies the conditions, $x\sigma =_E y\sigma$ by transitivity.

Adequacy: if $s \overset{*}{\longleftrightarrow}_E u \rightarrow^{\Lambda}_{l \rightarrow r, \sigma} v \overset{*}{\longleftrightarrow}_E t$ then $s \rightarrow^{\Lambda}_{\{x \overset{?}{=}_E l, y \overset{?}{=}_E r\} \| x \rightarrow y, \{x \mapsto s, y \mapsto t\}\sigma} t$. Thus, unifiers found by application of $l \approx r$ are not lost (application of $r \rightarrow l$ or mutation below the root is similar).

Strict resolventness: the previous construction deals with topmost application of $l \approx r$. Since R is strictly resolvent, we just have to consider application of $l \approx r$ before and below a topmost rewriting in R. Now every rule of R looks like $c \,|\, f(x_1 \ldots, x_n) \rightarrow d$. Moreover, if $s \rightarrow^{i.p}_{\{x \overset{?}{=}_E l, y \overset{?}{=}_E r\} \| x \rightarrow y} u \rightarrow^{\Lambda}_{c \| f(x_1 \ldots, x_n) \rightarrow d, \sigma} t$ then $s \rightarrow^{\Lambda}_{c \| f(x_1 \ldots, x_n) \rightarrow d, \{x_i \mapsto s|_i\}\sigma} t$. Therefore any rewriting proof leads to a strictly resolvent rewriting sequence. □

[15] when allowing any rewriting after the first one at position q

[16] for instance to solve $h(x) \overset{?}{=}_E x$ in $E = \{f(x) \approx x, f(h(x)) \approx x\}$

The corresponding E-unification algorithm can be expressed as follow: if topmost applications of the new axiom are needed then introduce them without decomposition else apply the unique topmost step of the resolvent proof.

A deterministic algorithm should add the fact that the rightmost application of the new axiom at the root can be chosen, thus restricting the mutations of $r \overset{?}{=}_E t$ to conditional rules of R. In fact, a more complex presentation ensures this behavior. Let $R_{l \to r}$ be $\{x \overset{?}{=}_E l \,|\, x \to r\} \cup \{c\sigma \cup \{x \overset{?}{=}_E l\sigma\} \,|\, x \to d\sigma$ so that $c \,|\, g \to d \in$ R and σ is the most general unifier of r and $g\}$.

Theorem 10. *$R \cup R_{l \to r} \cup R_{r \to l}$ is a strictly resolvent presentation of E.*

Proof. This presentation corresponds to the rightmost application of the axiom immediately followed by the unique topmost application of an axiom of R. □

More generally, algorithms are build by searching conditions that ensure the existence of equivalent strictly resolvent proofs. More complex presentations can be used to improve efficiency of the E-unification procedure.

4.5 Presentations with the Strictest Conditions

Since unification is used to recognizes the left parts, efficiency mainly depends on the complexity of the conditions. Some theories admit a strictly resolvent presentation the conditions of which include only E-equalities between variables. Such presentations lead to the most efficient E-unification algorithms.

For instance, Transitivity[17] with Commutativity of R can be expressed by the following strictly resolvent presentation:

(1) $xRy \to yRx$
(2) $\{x \overset{?}{=}_E z\} \,|\, (xRy)*(zRt) \to (xRy)*(yRt)$
(3) $\{x \overset{?}{=}_E t\} \,|\, (xRy)*(zRt) \to (xRy)*(yRz)$
(4) $\{y \overset{?}{=}_E z\} \,|\, (xRy)*(zRt) \to (xRy)*(xRt)$
(5) $\{y \overset{?}{=}_E t\} \,|\, (xRy)*(zRt) \to (xRy)*(xRz)$
(6) $\{x \overset{?}{=}_E z, x \overset{?}{=}_E t\} \,|\, (xRy)*(zRt) \to (xRy)*(yRy)$
(7) $\{y \overset{?}{=}_E z, y \overset{?}{=}_E t\} \,|\, (xRy)*(zRt) \to (xRy)*(xRx)$

It must be compared to the presentation simulating the syntactic approach:

(1') $\{x_1 \overset{?}{=}_E x, x_2 \overset{?}{=}_E y\} \,|\, x_1 R x_2 \to yRx$
(2') $\{x_1 \overset{?}{=}_E xRy, x_2 \overset{?}{=}_E yRz\} \,|\, x_1 * x_2 \to (xRy)*(xRz)$
(3') $\{x_1 \overset{?}{=}_E xRy, x_2 \overset{?}{=}_E yRy\} \,|\, x_1 * x_2 \to (xRy)*(xRx)$

The delayed E-unification problems based on the commutativity of R are efficiently solved in the strictly resolvent presentation.

However, useless bindings of the variables may occur when one of the arguments of '*' is a variable. We have implemented an algorithm which detects the following redundancy:

(2)≡(4),(3)≡(5) and (6)≡(7) when the first argument is a variable
(2)≡(3) and (4)≡(5) when the second one is a variable

With these optimizations, the search space is at most equal to the syntactic one.

---

[17] $\{(xRy)*(yRz) \approx (xRy)*(xRz)\}$

Figure 1 shows the behavior of these sets of transformation rules for a simple term. The same terms are obtained by application of (4),(5) and (7) in one step and the last redundancy is avoided.

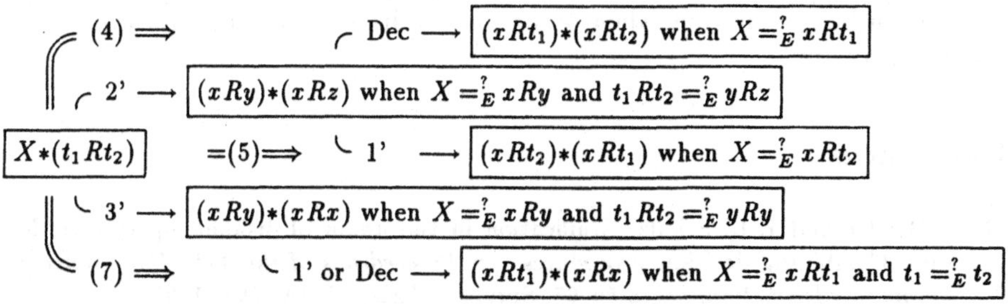

Fig. 1. transformation of $X*(t_1 Rt_2)$

We have developed an E-unification algorithm based on presentations with these strictest conditions [7][8]. We have implemented a completion algorithm that generates these presentations for any theory. Even if completeness must be restricted in order to ensure its termination, the efficient computation of a representative subset of unifiers is specially useful when using E-unification in automated deduction. We are presently experimenting this approach with the theorem provers of our inference laboratory [2]. Some features, such as a greater number of rules with respect to Gallier and Snyder's Root-Rewriting, could be seen as flaws of this presentation. However, it is compensated by a more selective application (since most of the unification is not delayed). Moreover, a lot of redundant branches are avoided thanks to the tests done (only once) in the completion step. For instance, trivial loops are avoided and cycle detection is not necessary to ensure efficiency. In addition, we have implemented an efficient narrowing module whose complexity is far less than linear in the number of rules. In fact, rewriting concurrently by many rules seems better than sequential rewriting: improvements like efficient strategy, cycle detection and problem sharing are easier since the search tree is wider but shallower.

Conclusion

We have presented a new approach to general E-unification. It leads to algorithms which depend mainly upon the chosen conditional rewriting presentation of the theory. The classic topmost approaches can be simulated and even extended since the behavior of the mutation rule may vary with the axiom applied.

Interesting practical results have been obtained when we compute the simplest presentations with respects to the conditions. Optimizations and adequate data structures are currently experimented in order to reduce the branching rate.

However, non termination of the algorithm used to produce such presentations leads to a loss of completeness in several cases. We are trying to mix the different strategies to preserve completeness (with more complex conditions). Using recurrence domains[3] or meta-rules[22][21] could be useful to build efficient complete algorithms. We are also studying the replacement of the imitation rule by a mutation below the root when occur-check fails to deal more naturally with collapsing axioms.

References

1. F. Baader and K.U. Schulz. Unification in the union of disjoint equational theories: Combining decision procedures. In *Proceedings of the 11th Conference on Automated Deduction*, pages 50–65. Springer-Verlag, LNAI 607, 1992.
2. T. Boy de la Tour, R. Caferra, and G. Chaminade. Some tools for an inference laboratory (ATINF). In *Proceedings of the 9th Conference on Automated Deduction*, pages 744–745. Springer-Verlag, LNCS 310, 1988.
3. H. Chen, J. Hsiang, and H.-C. Kong. On finite representation of infinite sequences of terms. In *Proceedings of the 2nd international Conditional and Typed Rewriting Systems workshop*, pages 100–114. Springer-Verlag, LNCS 516, 1990.
4. H. Comon, K. Haberstrau, and J.-P. Jouannaud. Decidable problems in shallow equational theories. In *Proceeding of the 7th IEEE Symposium on Logic in Computer Science*, 1991.
5. J. Corbin and M. Bidoit. A rehabilitation of Robinson's unification algorithm. In R. Mason, editor, *Proceedings of the IFIP'83*, pages 909–914, 1983.
6. M. Davis. The prehistory and early history of automated deduction. In J. Siekmann and G. Wrighston, editors, *Automation of Reasoning 1, Classical Papers on Computational Logic 1957-1966*. Springer-Verlag, 1983.
7. B. Delsart. General E-unification : A realistic approach. Presented at UNIF'91, 1991. (extended abstract).
8. B. Delsart. Efficient E-unification algorithms. Technical report, INPG, 1992.
9. N. Dershowitz and J.-P. Jouannaud. *Hanbook on Theoretical Computer Science*, volume B, chapter 6 : Rewrite Systems, pages 243–320. J. van Leeuwen (ed.), 1990.
10. N. Dershowitz and J.-P. Jouannaud. Notations for rewriting. *EATCS*, 43:162–172, 1991.
11. D. Dougherty and P. Johann. An improved general E-unification method. In *Proceedings of the 10th Conference on Automated Deduction*, pages 261–275. Springer-Verlag, LNCS 449, 1990.
12. M. Fay. First order unification in equational theories. In *Proceedings of the 4th Workshop on Automated Deduction*, pages 162–167, 1979.
13. J.H. Gallier and W. Snyder. A general complete E-unification procedure. In *Proceeding of the 2nd Conference on Rewrite Techniques and Applications*, pages 216–227. Springer-Verlag, LNCS 256, 1987.
14. J.H. Gallier and W. Snyder. Complete sets of transformations for general E-unification. *Theoretical Computer Science*, 67(2,3):203–260, 1989.
15. J. Herbrand. Recherche sur la théorie de la démonstration, 1930. Thèse de doctorat, Université de Paris.
16. A. Herold. Combination of unification algorithms. In *Proceedings of the 8th Conference on Automated Deduction*, pages 450–469. Springer-Verlag, LNCS 230, 1986.

17. G. Huet. Résolution d'équations dans les langages d'ordre 1,2,...,∞. Thèse d'Etat, Université de Paris, 1976.
18. J.-M. Hullot. Canonical forms and unification. In *Proceedings of the 5th Conference on Automated Deduction*, pages 318–334. Springer-Verlag, LNCS 87, 1980.
19. J.-P. Jouannaud and C. Kirchner. Solving equations in abstract algebras: A rule-based survey of unification. Technical Report 561, LRI, 1990.
20. C. Kirchner. Computing unification algorithms. In *Proceedings of the first IEEE Symposium on Logic in Computer Science*, pages 206–216, 1986.
21. H. Kirchner. Schematization of infinite sets of rewrite rules generated by divergent completion processes. *Theoretical Computer Science*, 67:303–332, 1989.
22. H. Kirchner and M. Hermann. Meta-rule synthesis for crossed rewrite systems. In *Proceedings of the 2nd international Conditional and Typed Rewriting Systems workshop*, pages 143–154. Springer-Verlag, LNCS 516, 1990.
23. A. Martelli and U. Montanari. An efficient unification algorithm. *ACM Transaction on Programming Languages And Systems*, 4(2):258–282, 1982.
24. W. Nutt, P. Rety, and G. Smolka. Basic narrowing revisited. *Journal of Symbolic Computation*, 7(3,4):295–318, 1989.
25. M.S. Paterson and M.N. Wegman. Linear unification. *Journal of Computer Systems, Sciences*, 16:158–167, 1978.
26. P. Rety. Improving basic narrowing. In *Proceedings of the 2nd Conference on Rewriting Techniques and Applications*, pages 226–241. Springer-Verlag, LNCS 256, 1987.
27. A. Robinson. A machine-oriented logic based on the resolution principle. *Journal of the Association for Computing Machinery*, 12(1):23–41, 1965.
28. M. Schmidt-Schauß. Unification in a combination of arbitrary disjoint equational theories. Seki report sr-87-16, University of Kaiserslauterm, 1987.

An Optimal Narrowing Strategy for General Canonical Systems*

Alexander Bockmayr[1], Stefan Krischer[2], Andreas Werner[3]

[1] MPI Informatik, Im Stadtwald, D-6600 Saarbrücken, bockmayr@mpi-sb.mpg.de
[2] CRIN & INRIA-Lorraine, B.P.239, F-54506 Vandœuvre-lès-Nancy, krischer@loria.fr
[3] SFB 314, Univ. Karlsruhe, D-7500 Karlsruhe 1, werner@ira.uka.de

Abstract. Narrowing is a universal unification procedure for equational theories defined by a canonical term rewriting system. In its original form it is extremely inefficient. Therefore, many optimizations have been proposed during the last years. In this paper, we introduce a new narrowing strategy, called LSE narrowing. LSE narrowing is complete for arbitrary canonical systems. It is optimal in the sense that two different LSE narrowing derivations cannot generate the same narrowing substitution.

1 Introduction

Narrowing is a complete unification procedure for any equational theory that can be defined by a canonical term rewriting system [Fay79, Hul80]. In its original form, it is extremely inefficient [Boc86]. Therefore, many optimizations have been proposed during the last years [Hul80, RKKL85, Fri85, Her86, Rét87, NRS89, BGM88, Ech88, Rét88, Boc88, You88, Pad88, Höl89, DG89, You91].

In this paper, which unifies and simplifies our previous results in [KB91] and [Wer91], we introduce a new narrowing strategy, called *LSE narrowing*, and its normalizing variant, called *normalizing LSE narrowing*. Both strategies are *complete*. Moreover, they are *optimal* in the sense that two different (normalizing) LSE narrowing derivations cannot generate the same narrowing substitution. These results hold for *arbitrary* canonical term rewriting systems. We do not have to impose any additional restriction on the rewrite system such as constructor discipline [Fri85] or left-linearity and non-overlapping left-hand sides [You88, You91, DG89].

The organization of the paper is as follows. After some preliminaries in Section 2, we recall in Section 3 the basic idea of narrowing and state the well-known lifting lemma of Hullot [Hul80] which establishes a fundamental relationship between rewriting and narrowing derivations. In Section 4, we discuss basic narrowing and left-to-right basic narrowing and illustrate by an example that left-to-right narrowing derivations need not generate leftmost-innermost rewriting derivations. In Section 5, we introduce the new narrowing strategy *LSE narrowing*. We show that there is a

* The first author's work was supported by the German Ministry for Research and Technology (BMFT) under grant ITS 9103. The second author's work was supported by the Centre National de Recherche Scientifique (CNRS), departement Science pour l'Ingenieur. The third author's work was supported by the Deutsche Forschungsgemeinschaft as part of the SFB 314 (project S2).

one-to-one correspondence between LSE narrowing derivations and left reductions, which are a special form of leftmost-innermost rewriting derivations. Using this correspondence, we can give very simple proofs of the completeness of LSE narrowing and the optimality property that no narrowing substitution can be generated twice. In Section 6, we present the normalizing form of LSE narrowing. The same results hold as in the non-normalizing case. The proofs, however, are much more complicated and have to be omitted for lack of space. Finally, in Section 7, we present some empirical results which illustrate the benefits of the new strategy.

2 Preliminaries

We recall briefly some basic notions that are needed in the sequel. More details can be found in the survey of [HO80].

$\Sigma = (S, F)$ denotes a *signature* with a set S of sort symbols and a set F of function symbols together with an arity function.

A *Σ-algebra* A consists of a family of non-empty sets $(A_s)_{s \in S}$ and a family of functions $(f^A)_{f \in F}$ such that if $f : s_1 \times \ldots \times s_n \rightarrow s$ then $f^A : A_{s_1} \times \ldots \times A_{s_n} \rightarrow A_s$.

X represents a family $(X_s)_{s \in S}$ of countably infinite sets X_s of *variables* of sort s. $T(F, X)$ is the Σ-algebra of *terms* with variables over Σ.

For a term $t \in T(F, X)$, $Var(t)$, $Occ(t)$, and $FuOcc(t)$ denote the set of *variables, occurrences* and *non-variable occurrences* in t respectively. The root of a term is denoted by the empty occurrence ϵ. An occurence ω is a *prefix* of an occurrence ω', $\omega \preceq \omega'$, iff there exists $v \in \mathcal{N}^*$ such that $\omega' = \omega.v$. We denote by t/ω the *subterm* of t at position $\omega \in Occ(t)$ and by $t[\omega \leftarrow s]$ the term obtained from t by *replacing* the subterm t/ω with the term $s \in T(F, X)$.

A *substitution* is a mapping $\sigma : X \rightarrow T(F, X)$ which is different from the identity only for a finite subset $Dom(\sigma)$ of X. We do not distinguish σ from its canonical extension to $T(F, X)$. $Im(\sigma) \stackrel{\text{def}}{=} \bigcup_{x \in Dom(\sigma)} Var(\sigma(x))$ is the set of *variables introduced* by σ. If σ is a substitution and V is a set of variables then the substitition $\sigma|_V$ with

$$\sigma|_V(x) \quad = \quad \begin{cases} \sigma(x) & \text{if } x \in Dom(\sigma) \cap V \\ x & \text{else} \end{cases}$$

is called *restriction* of σ on V.

A *syntactic unifier* of two terms s, t is a substitution σ such that $\sigma(s) = \sigma(t)$. A *most general syntactic unifier* of s and t is a unifier σ of s and t with $Dom(\sigma) \cap Im(\sigma) = \emptyset$ such that for any other unifier τ of s and t there exists a substitution λ with $\lambda \circ \sigma = \tau$.

A binary relation $\rightarrow$ on a Σ-algebra A is *Σ-compatible* iff $t_1 \rightarrow v_1, \ldots, t_n \rightarrow v_n$ implies $f^A(t_1, \ldots, t_n) \rightarrow f^A(v_1, \ldots, v_n)$ for all $t_i, v_i \in A_{s_i}$ and all $f : s_1 \times \ldots \times s_n \rightarrow s$ in F. By $\stackrel{*}{\rightarrow}$ we denote the reflexive-transitive closure of $\rightarrow$. A *congruence* is a Σ-compatible equivalence relation.

An *equation* is an expression of the form $s \doteq t$ where s and t are terms of $T(F, X)$ belonging to the same sort. A *system of equations* G is an expression of the form $s_1 \doteq t_1 \wedge \ldots \wedge s_n \doteq t_n, n \geq 1$ with equations $s_i \doteq t_i, i = 1, \ldots, n$.

Let E be a set of equations. The *equational theory* $\equiv_E$ associated with E is the smallest congruence $\equiv$ on $T(F,X)$ such that $\sigma(l) \equiv \sigma(r)$ for all equations $l \doteq r$ in E and all substitutions σ.

The *E-subsumption preorder* $\leq_E$ on $T(F,X)$ is defined by $s \leq_E t$ iff there is a substitution $\sigma : X \rightarrow T(F,X)$ with $t \equiv_E \sigma(s)$. For two substitutions $\sigma, \tau : X \rightarrow T(F,X)$ and a set of variables V we say $\sigma \leq_E \tau\ [V]$ iff there is a substitution λ with $\tau(x) \equiv_E \lambda(\sigma(x))$ for all $x \in V$.

A *rewriting rule* is an expression of the form $l \rightarrow r$ with terms $l, r \in T(F,X)$ of the same sort such that $Var(r) \subseteq Var(l)$ and $l \notin X$. The rule is *regular* iff $Var(l) = Var(r)$. A term rewriting system R is a set of rewriting rules.

The *reduction relation* $\rightarrow_R$ associated with R is defined as follows: $s \rightarrow_R t$, more precisely $s \rightarrow_{[v,l\rightarrow r,\tau]} t$, iff there is an occurrence $v \in Occ(s)$ and a rule $l \rightarrow r$ in R such that there exists a substitution $\tau : X \rightarrow T(F,X)$ with $\tau(l) = s/v$ and $t = s[v \leftarrow \tau(r)]$. R is *confluent* iff for any terms s, t_1, t_2 with $s \rightarrow_R t_1$ and $s \rightarrow_R t_2$ there exists a term u with $t_1 \rightarrow_R u$ and $t_2 \rightarrow_R u$. R is *noetherian* iff there exists no infinite chain $t_1 \rightarrow_R t_2 \rightarrow_R \ldots \rightarrow_R t_n \rightarrow_R \ldots$. R is *canonical* iff R is confluent and noetherian.

A term t is *irreducible* or *normalized* iff there exists no term u such that $t \rightarrow_R u$. Otherwise t is called *reducible*. A substitution σ is *normalized* iff for any $x \in X$ the term $\sigma(x)$ is irreducible. If R is canonical then there exists for any term t a unique irreducible term $t\downarrow$ such that $t \overset{*}{\rightarrow}_R t\downarrow$. $t\downarrow$ is called the *normal form* of t. For any two terms s, t, we have $s \equiv_E t$ iff $s\downarrow = t\downarrow$.

3 Narrowing: The Basic Idea

Narrowing provides a complete E-unification procedure for any equational theory E that can be defined by a canonical term rewrite system.

Definition 1. Let E be a set of equations. A system of equations G

$$s_1 \doteq t_1 \wedge \ldots \wedge s_n \doteq t_n, n \geq 1,$$

is called *E-unifiable* iff there exists a substitution $\sigma : X \rightarrow T(F,X)$ such that

$$\sigma(s_1) \equiv_E \sigma(t_1), \ldots, \sigma(s_n) \equiv_E \sigma(t_n).$$

The substitution σ is called an *E-unifier* of G.

A set $cU_E(G)$ of substitutions is called a *complete set of E-unifiers of G* iff

- every $\sigma \in cU_E(G)$ is a E-unifier of G
- for any E-unifier τ of G there is $\sigma \in cU_E(G)$ such that $\sigma \leq_E \tau\ [Var(G)]$
- for all $\sigma \in cU_E(G) : Dom(\sigma) \subseteq Var(G)$.

$cU_E(G)$ is called *minimal* iff it satisfies further the condition

- for all $\sigma, \sigma' \in cU_E(G) : \sigma \leq_E \sigma'\ [Var(G)]$ implies $\sigma = \sigma'$.

Narrowing allows to find complete sets of E-unifiers for equational theories E that can be defined by a canonical term rewrite system R.[4] The basic idea is as follows. Suppose we want to R-unify a system of equations $s_1 \doteq t_1 \wedge \ldots \wedge s_n \doteq t_n$. This means that we have to find a substitution σ such that

$$\sigma(s_1) \equiv_R \sigma(t_1), \ldots, \sigma(s_n) \equiv_R \sigma(t_n). \tag{1}$$

Since R is a canonical term rewriting system this is equivalent to

$$\sigma(s_1)\downarrow = \sigma(t_1)\downarrow, \ldots, \sigma(s_n)\downarrow = \sigma(t_n)\downarrow. \tag{2}$$

If the problem has a solution σ, then either σ is a syntactic unifier of G, which can be computed by standard unification, or σ does not syntactically unify G. In this case the system of equations $\sigma(G)$ must be reducible by R since otherwise it would be impossible to have (2). The idea is now to lift the rewriting derivation $\sigma(G) \rightarrow \ldots \rightarrow \sigma(G)\downarrow$ on the *unknown* instance $\sigma(G)$ of G to a narrowing derivation $G \rightsquigarrow_{\delta_1} \ldots \rightsquigarrow_{\delta_n} G_n$ on the given system G such that the last system of equations G_n is syntactically unifiable with most general unifier τ and $\tau \circ \delta_n \circ \ldots \circ \delta_1 \leq_R \sigma \; [Var(G)]$.

Definition 2 (Narrowing). Let R be a term rewriting system. A system of equations G is *narrowable* to a system of equations G' with *narrowing substitution* σ,

$$G \rightsquigarrow_{[v, l \rightarrow r, \sigma]} G',$$

iff there exist a non-variable occurrence $v \in Occ(G)$ and a rule $l \rightarrow r$ in R[5] such that G/v and l are syntactically unifiable with most general unifier σ and $G' = \sigma(G)[v \leftarrow \sigma(r)]$.

A *narrowing derivation* $G_0 \rightsquigarrow^*_\sigma G_n$ with *narrowing substitution* σ is a sequence of narrowing steps $G_0 \rightsquigarrow_{\delta_1} G_1 \rightsquigarrow_{\delta_2} \cdots \rightsquigarrow_{\delta_n} G_n$, $n \geq 0$, where $\sigma = (\delta_n \circ \ldots \circ \delta_1) \mid_{Var(G)}$. The narrowing substitution leading from G_i to G_j, for $0 \leq i \leq j \leq n$, will be denoted by

$$\lambda_{i,j} \stackrel{\text{def}}{=} \delta_j \circ \ldots \circ \delta_{i+1}.$$

In particular, $\lambda_{i,i} = id$, for $i = 0, \ldots, n$.

A *narrowing strategy* $\mathcal{S}$ is a property of narrowing derivations. We say that $\mathcal{S}$-*narrowing is complete* iff for any canonical term rewriting system R and any system of equations G the set of all substitutions σ such that there exists a $\mathcal{S}$-narrowing derivation $G = G_0 \rightsquigarrow_{\delta_1} \cdots \rightsquigarrow_{\delta_n} G_n, n \geq 0$, such that G_n is syntactically unifiable by a most general unifier τ and $\sigma = \tau \circ \delta_n \circ \ldots \circ \delta_1 \mid_{Var(G)}$, is a complete set of R-unifiers of G.

In order to treat syntactical unification as a narrowing step, we introduce a new rule

$$x \doteq x \rightarrow true,$$

[4] We associate with every rule $l \rightarrow r$ in R the equation $l \doteq r$ in E.

[5] We always assume that $Var(l) \cap Var(G) = \emptyset$.

where x denotes a variable. $t \doteq t' \rightsquigarrow_{\delta} true$ holds if and only if t and t' are unifiable with most general unifier δ. This additional rule is called ϵ-*rule* (since it only can be applied at occurrence ϵ) and affects neither confluence nor termination. Obviously, σ is solution of G if and only if $\sigma(G)$ can be reduced by the rules in R and the ϵ–rule to the trivial system $true \wedge \ldots \wedge true$.

Now we are able to formulate the fundamental relationship between rewriting and narrowing derivations that will provide the basis for most of the proofs in this paper.

Proposition 3 (Hullot 80). *Let R be a term rewriting system and let G be a system of equations. If μ is a normalized substitution and V a set of variables such that $Var(G) \cup Dom(\mu) \subseteq V$, then for every rewriting derivation*

$$H_0 \stackrel{\text{def}}{=} \mu(G) \quad \rightarrow_{[v_1, l_1 \rightarrow r_1, \tau_1]} \quad H_1 \quad \ldots \quad \rightarrow_{[v_n, l_n \rightarrow r_n, \tau_n]} \quad H_n \tag{3}$$

there exist a normalized substitution λ and a narrowing derivation

$$G_0 \stackrel{\text{def}}{=} G \quad \rightsquigarrow_{[v_1, l_1 \rightarrow r_1, \delta_1]} \quad G_1 \quad \ldots \quad \rightsquigarrow_{[v_n, l_n \rightarrow r_n, \delta_n]} \quad G_n \tag{4}$$

using the same rewrite rules at the same occurrences such that

$$\mu = \lambda \circ \delta_n \circ \ldots \circ \delta_1 \quad [V] \quad \textit{and} \tag{5}$$

$$H_i = (\lambda \circ \delta_n \circ \ldots \circ \delta_{i+1})(G_i), \quad i = 0, \ldots, n. \tag{6}$$

Conversely, if $\mu \stackrel{\text{def}}{=} \lambda \circ \delta_n \circ \ldots \circ \delta_1$ then there exists for any narrowing derivation (4) *and any substitution λ a rewriting derivation* (3) *such that* (6) *holds.*

Proposition 4. *A narrowing strategy $\mathcal{S}$ is complete iff for any canonical term rewriting system R, any system of equations G, and any normalized R-unifier μ of G there exists a rewriting derivation $\mu(G) \xrightarrow{*} true \wedge \ldots \wedge true$ such that the corresponding narrowing derivation is a $\mathcal{S}$-derivation.*

Putting the two preceding propositions together, we get immediately the completeness of naive narrowing.

Theorem 5. *Narrowing is complete.*

4 Left-to-Right Basic Narrowing

If μ is a normalized R-unifier of the system of equations G, then *any* rewriting derivation $\mu(G) \xrightarrow{*} true \wedge \ldots \wedge true$ can be lifted to a narrowing derivation $G \rightsquigarrow^{*}_{[\sigma]} true \wedge \ldots \wedge true$ such that the narrowing substitution σ subsumes μ. It follows that there are as many narrowing derivations generating (a generalization of) the same solution μ as there are different normalizations of $\mu(G)$. This is one of the main reasons for the inefficiency of narrowing. Another point is that narrowing may generate non-normalized solutions, which are also redundant.

In order to improve the efficiency of narrowing one can introduce a normalization strategy for $\mu(G)$ and then consider only those narrowing derivations which correspond to rewriting derivations that follow this strategy.

4.1 Basic Narrowing

A first step in this direction was Hullot's *basic narrowing* [Hul80]. A *basic narrowing derivation* is obtained by lifting an *innermost* rewriting derivation on $\mu(G)$, where μ denotes a normalized substitution, to the narrowing level. Note that naive innermost narrowing is not complete.

Definition 6 (Basic Narrowing). The sets $B_i, i = 0, \ldots, n$, of *basic occurrences* in a narrowing derivation

$$G_0 \rightsquigarrow_{[v_1,l_1\rightarrow r_1,\delta_1]} G_1 \rightsquigarrow_{[v_2,l_2\rightarrow r_2,\delta_2]} \cdots \rightsquigarrow_{[v_n,l_n\rightarrow r_n,\delta_n]} G_n$$

are inductively defined as follows

- $B_0 \stackrel{\text{def}}{=} FuOcc(G_0)$
- $B_i \stackrel{\text{def}}{=} (B_{i-1} \setminus \{v \in B_{i-1} \mid v \succeq v_i\}) \cup \{v_i.v \mid v \in FuOcc(r_i)\}, i > 0.$

For a *basic narrowing derivation* we require that $v_i \in B_{i-1}$, for all $i = 1, \ldots, n$.

While original narrowing considers any non-variable occurrence in the goal, basic narrowing discards those occurrences which have been introduced by the narrowing substitution of a previous narrowing step.

Since, in canonical systems, an innermost normalization of $\mu(G)$ always exists, Proposition 4 implies that basic narrowing is complete.

4.2 Left-to-Right Basic Narrowing

In 1986, A. Herold showed that it is possible to restrict the set of narrowing occurrences further without loosing completeness. After a narrowing step $G \rightsquigarrow_{[v,l\rightarrow r,\sigma]} G'$, we may discard also those narrowing occurrences which are strictly left of v [Her86].

Definition 7. An occurrence ω is *strictly left* of an occurrence ω', $\omega \lhd \omega'$ (resp. $\omega' \rhd \omega$) iff there exist occurrences o, v, v' and natural numbers i, i' such that $i < i', \omega = o.i.v$ and $\omega' = o.i'.v'$.

Definition 8 (Left-to-Right Basic Narrowing). The sets $LRB_i, i = 0, \ldots, n$, of *left-to-right basic occurrences* in a narrowing derivation

$$G_0 \rightsquigarrow_{[v_1,l_1\rightarrow r_1,\delta_1]} G_1 \rightsquigarrow_{[v_2,l_2\rightarrow r_2,\delta_2]} \cdots \rightsquigarrow_{[v_n,l_n\rightarrow r_n,\delta_n]} G_n$$

are inductively defined as follows

- $LRB_0 \stackrel{\text{def}}{=} FuOcc(G_0)$
- $LRB_i \stackrel{\text{def}}{=} (LRB_{i-1} \setminus \{v \in LRB_{i-1} \mid v \succeq v_i \text{ or } v \lhd v_i\}) \cup \{v_i.v \mid v \in FuOcc(r_i)\}, i > 0.$

For a *left-to-right basic narrowing derivation* we require that $v_i \in LRB_{i-1}$, for all $i = 1, \ldots, n$. Sometimes, we will use the abbreviation $LRB(U, v, l \rightarrow r)$ for the set of occurrences $U \setminus \{\omega \in U \mid \omega \succeq v \text{ or } \omega \lhd v\}) \cup \{v.\omega \mid \omega \in FuOcc(r_i)\}$.

Herold showed that narrowing derivations corresponding to leftmost-innermost normalizations of $\mu(G)$, for a normalized substitution μ, are left-to-right basic. This implies immediately the completeness of left-to-right basic narrowing. However, the converse of Herold's result is not true. A left-to-right basic narrowing derivation need not generate a leftmost-innermost rewriting derivation.

Example 1. Consider the rule

$$\pi_1 : z * 0 \to 0.$$

Starting with the term $(y*x)*x$ there are two left-to-right basic narrowing derivations

$$\begin{array}{llll} (y * x) * x & \rightsquigarrow_{[1,\pi_1,\{x \leftarrow 0, z \leftarrow y\}]} & 0 * 0 \rightsquigarrow_{[\epsilon,\pi_1]} & 0 \\ (y * x) * x & \rightsquigarrow_{[\epsilon,\pi_1,\{x \leftarrow 0, z \leftarrow y * 0\}]} & & 0 \end{array}$$

There is an obvious redundancy. In both derivations, the narrowing substitution $\{x \leftarrow 0\}$ and the derived term 0 are the same.

The reduction corresponding to the second narrowing derivation is not leftmost-innermost, since $y * 0$ can be reduced.

$$(y * 0) * 0 \quad \to_{[\epsilon,\pi_1]} \quad 0$$

5 LSE Narrowing

Our aim is now to introduce a new narrowing strategy which has the property that the corresponding rewriting derivations are always leftmost-innermost.

We start by refining the notion of a leftmost-innermost derivation. Leftmost-innermost derivations are not unique. If the rewrite system has overlapping left-hand sides, then it may happen that two different rules are applicable at the same occurrence. In order to eliminate this indeterminism we assume that the rules in our rewrite system are numbered and require that only rules with minimal rule number are applied during normalization.

Definition 9 (Left Reduction). A reduction step $t \to_{[v,\pi,\sigma]} t'$ is called a *left reduction step* iff

- all subterms t/ω with ω strictly left of v are in normal form ("leftmost")
- all proper subterms of t/v are in normal form ("innermost")
- t/v cannot be reduced by a rule whose number π' is strictly less than π ("minimal rule number").

A rewriting derivation is a *left reduction* iff all steps are left reduction steps.

Proposition 10. *For all terms t there exists a unique left reduction to the normal form of t.*

Now we will show how reducibility tests which are performed after a narrowing step can be used to obtain a one-to-one correspondence between narrowing derivations and left reductions.

Definition 11 (LSE Narrowing). In a narrowing derivation

$$G_0 \rightsquigarrow_{[v_1,\pi_1,\delta_1]} G_1 \rightsquigarrow_{[v_2,\pi_2,\delta_2]} \cdots G_{n-1} \rightsquigarrow_{[v_n,\pi_n,\delta_n]} G_n$$

the step $G_{n-1} \rightsquigarrow_{[v_n,\pi_n,\delta_n]} G_n$ is called *LSE* iff the following three conditions are satisfied:

(Left-Test) For all $i \in \{0,\ldots,n-1\}$ the subterms of $\lambda_{i,n}(G_i)$ which lie strictly left of v_{i+1} are in normal form.
(Sub-Test) For all $i \in \{0,\ldots,n-1\}$ the proper subterms of $\lambda_{i,n}(G_i/v_{i+1})$ are in normal form.
(Epsilon-Test) For all $i \in \{0,\ldots,n-1\}$ the term $\lambda_{i,n}(G_i/v_{i+1})$ is not reducible at occurrence ϵ with a rule whose number is strictly less than π_{i+1}.

A narrowing derivation is *LSE* iff any single narrowing step is LSE.

Proposition 12. *Consider a system of equations G and a normalized substitution μ. If*

$$H_0 \stackrel{\text{def}}{=} \mu(G_0) \rightarrow_{[v_1,\pi_1]} H_1 \rightarrow_{[v_2,\pi_2]} \cdots \rightarrow_{[v_n,\pi_n]} H_n$$

is a left reduction, then the corresponding narrowing derivation

$$G_0 \rightsquigarrow_{[v_1,\pi_1,\delta_1]} G_1 \rightsquigarrow_{[v_2,\pi_2,\delta_2]} \cdots G_{n-1} \rightsquigarrow_{[v_n,\pi_n,\delta_n]} G_n$$

is a LSE narrowing derivation.

Proof. By Proposition 3 there exists a substitution λ such that $H_i = (\lambda \circ \lambda_{i,n})(G_i)$, for $i = 0,\ldots,n$. We have to show that none of the reducibility tests fails.

Suppose that the step $G_{m-1} \rightsquigarrow_{[v_m,\pi_m,\delta_m]} G_m$ is not LSE, for some $m \in \{1,\ldots,n\}$. Then there exists $i \in \{0,\ldots,m-1\}$ such that either

1. $\lambda_{i,m}(G_i)$ is reducible at an occurrence v strictly left of v_{i+1} or
2. $\lambda_{i,m}(G_i)$ is reducible at an occurrence v strictly below v_{i+1} or
3. $\lambda_{i,m}(G_i)$ is reducible at occurrence v_{i+1} with a rule whose number is strictly less than π_{i+1}.

Since $H_i = (\lambda \circ \lambda_{m,n} \circ \lambda_{i,m})(G_i)$ and $\rightarrow$ is stable under substitutions this implies that one of the properties (1) to (3) must hold with H_i in place of $\lambda_{i,m}(G_i)$. But this means that $H_i \rightarrow_{[v_{i+1},\pi_{i+1}]} H_{i+1}$ is not a left reduction step in contradiction to our assumption. □

As an immediate consequence, we get by Proposition 4 and Proposition 10 the following theorem.

Theorem 13. *LSE narrowing is complete.*

Next we consider the converse of Proposition 12.

Proposition 14. *If*

$$G_0 \rightsquigarrow_{[v_1,\pi_1,\delta_1]} G_1 \rightsquigarrow_{[v_2,\pi_2,\delta_2]} \cdots G_{n-1} \rightsquigarrow_{[v_n,\pi_n,\delta_n]} G_n$$

is a LSE narrowing derivation and $H_i \stackrel{\text{def}}{=} \lambda_{i,n}(G_i)$, *for* $i = 0, \ldots, n$, *then the rewriting derivation*

$$H_0 \rightarrow_{[v_1,\pi_1]} H_1 \rightarrow_{[v_2,\pi_2]} \cdots \rightarrow_{[v_n,\pi_n]} H_n$$

is a left reduction.

Proof. Suppose that the derivation is not a left reduction. Then there exists $i \in \{0, \ldots, n-1\}$ and a rewriting step $\lambda_{i,n}(G_i) \rightarrow_{[v,\pi]} \lambda_{i+1,n}(G_{i+1})$ such that either

1. v lies strictly left of v_{i+1} or
2. v lies strictly below v_{i+1} or
3. the rule number π is stricly less than π_{i+1}

But this implies that the narrowing step $G_{n-1} \rightsquigarrow_{[v_n,\pi_n,\delta_n]} G_n$ is not LSE in contradiction to our assumption. □

This proposition has a number of important consequences. First of all, we can easily prove the following minimality property of LSE narrowing which first appeared in [Wer91].

Theorem 15. *Consider two LSE narrowing derivations*

$$G = G_0 \rightsquigarrow_{[v_1,\pi_1,\delta_1]} G_1 \rightsquigarrow_{[v_2,\pi_2,\delta_2]} \cdots G_{n-1} \rightsquigarrow_{[v_n,\pi_n,\delta_n]} G_n,$$
$$G = G'_0 \rightsquigarrow_{[v'_1,\pi'_1,\delta'_1]} G'_1 \rightsquigarrow_{[v'_2,\pi'_2,\delta'_2]} \cdots G'_{m-1} \rightsquigarrow_{[v'_m,\pi'_m,\delta'_m]} G'_m.$$

If the narrowing substitutions $\sigma \stackrel{\text{def}}{=} \delta_n \circ \ldots \circ \delta_1$ *and* $\sigma' \stackrel{\text{def}}{=} \delta'_m \circ \ldots \circ \delta'_1$, *where* $n \leq m$, *coincide on* $Var(G)$ *up to variable renaming, that is if there exist substitutions* λ *and* λ' *such that* $\sigma = \lambda' \circ \sigma' \,[Var(G)]$ *and* $\sigma' = \lambda \circ \sigma \,[Var(G)]$, *then*

- $\pi_i = \pi'_i$ *and* $v_i = v'_i$ *for* $1 \leq i \leq n$,
- *the narrowing derivation* $G'_n \rightsquigarrow_{[v'_{n+1},\pi'_{n+1},\delta'_{n+1}]} G'_{n+1} \cdots \rightsquigarrow_{[v'_m,\pi'_m,\delta'_m]} G'_m$ *is a left reduction (up to variable renaming).*

Proof. By Proposition 14 the rewriting derivations

$$\sigma(G) = \lambda_{0,n}(G_0) \rightarrow_{[v_1,\pi_1]} \lambda_{1,n}(G_1) \rightarrow_{[v_2,\pi_2]} \cdots \rightarrow_{[v_n,\pi_n]} \lambda_{n,n}(G_n)$$
$$\sigma'(G) = \lambda'_{0,m}(G'_0) \rightarrow_{[v'_1,\pi'_1]} \lambda'_{1,m}(G'_1) \rightarrow_{[v'_2,\pi'_2]} \cdots \rightarrow_{[v'_m,\pi'_m]} \lambda'_{m,m}(G'_m)$$

are both left reductions. Since σ and σ' coincide on $Var(G)$ up to variable renaming, the systems $\sigma(G)$ and $\sigma'(G)$ are identical up to variable renaming. By the unicity of left reductions, this implies $\pi_i = \pi'_i$ and $v_i = v'_i$ for $1 \leq i \leq n$.

Since most general unifiers are unique up to variable renaming we can conclude that G_i and G'_i are identical up to variable renaming, for $i = 1, \ldots, n$. Since $\lambda_{i,n}(G_i)$ and $\lambda'_{i,m}(G'_i)$ are identical up to variable renaming, for $1 \leq i \leq n$, and since $\lambda_{n,n}(G_n) = G_n$, this implies that G'_n and $\lambda'_{n,m}(G'_n)$ are identical up to variable renaming. Therefore the narrowing derivation starting from G'_n and the left reduction starting from $\lambda'_{n,m}(G'_n)$ are the same up to variable renaming. □

If we assume that narrowing derivations starting from the same goal and using the same rules at the same occurrences produce the same narrowing substitution (in any practical implementation, this will be the case), we get the following corollary.

Corollary 16. *If LSE narrowing enumerates two solutions σ and σ' which coincide up to variable renaming, then $\sigma = \sigma'$ holds and the two derivations coincide.*

Proof. We use the same notation as in Theorem 15. Then $G_n = G'_m = true \wedge \ldots \wedge true$ implies $n = m$. □

Another important property of LSE narrowing is that it generates only normalized substitutions.

Proposition 17. *For any LSE narrowing derivation*

$$G = G_0 \rightsquigarrow_{[v_1,\pi_1,\delta_1]} G_1 \rightsquigarrow_{[v_2,\pi_2,\delta_2]} G_2 \quad \cdots \quad \rightsquigarrow_{[v_n,\pi_n,\delta_n]} G_n$$

the narrowing substitution $\delta_n \circ \ldots \circ \delta_1|_{Var(G)}$ is normalized.

Proof. Let x be a variable of G such that $\sigma(x)$ is reducible. Suppose x is instantiated for the first time in the i-th narrowing step. Then there must be an occurrence of x in G_{i-1} which lies below the narrowing occurrence v_i. More formally, there exists an occurrence $v \neq \epsilon$ such that $G_{i-1}/v_i.v = x$. Then $\sigma(G_{i-1}/v_i.v) = \sigma(x)$ is reducible in contradiction to the Sub-Test. □

Corollary 18. *LSE narrowing enumerates only normalized substitutions.*

Note that the last two corollaries no longer hold if we replace the last narrowing step, which uses the ϵ-rule, by a simple unification of the left and the right hand side of G_n. Using the ϵ-rule requires not only that the left and the right hand side are unifiable but also that none of the tests fails.

Finally, let us mention how LSE narrowing is related to left-to-right basic narrowing.

Proposition 19. *Any LSE narrowing derivation is left-to-right basic.*

Proof. We show by induction on n: if $G_0 \rightsquigarrow^* G_n$ is a LSE derivation with narrowing substitution $\lambda_{0,n}$, $L_0 \subseteq FuOcc(G_0)$ and $L_i \stackrel{\text{def}}{=} LRB_i(L_{i-1}, v_i, \pi_i)$ for $i = 1, \ldots, n$, such that $\lambda_{0,n}(G_0)/v$ is normalized[6] for $v \notin L_0$, then $v_i \in L_{i-1}$ for all $i = 1, \ldots, n$. Since $\lambda_{0,n}|_{Var(G_0)}$ is normalized by Proposition 17, this proves the proposition.

The statement trivially holds for $n = 0$. Let $n > 0$. By Proposition 3 and assumption we have $v_1 \in L_0$. In order to conclude $v_i \in L_{i-1}$ for all $i = 2, \ldots n$ by induction hypothesis, we have to show that $\lambda_{1,n}(G_1)/v$ is normalized for $v \notin L_1$. Let $v \in Occ(\lambda_{1,n}(G_1)) \setminus L_1$. If $v \lhd v_1$ then $\lambda_{1,n}(G_1)/v = \lambda_{0,n}(G_0)/v$ is normalized by the Left-Test. If $v \rhd v_1$ then $\lambda_{1,n}(G_1)/v = \lambda_{0,n}(G_0)/v$ is normalized by assumption. Finally, consider the case $v_1 \preceq v$. Since $v \notin L_1$, $\lambda_{1,n}(G_1)/v$ is a subterm of $\lambda_{0,n}(G_0)/v_1$ and therefore normalized. Note that $\omega \in L_1$ for all proper prefixes of v_1, since $\omega \in L_0$ by assumption.

[6] In [Rét87], such a set of occurrences L_0 is called *sufficiently large* on $\lambda_{0,n}(G_0)$.

From Theorem 13 we get immediately the following corollaries.

Corollary 20 (Hullot 80). *Basic narrowing is complete*

Corollary 21 (Herold 86). *Left-to-right basic narrowing is complete.*

6 Normalizing LSE Narrowing

One of the most important optimizations of naive narrowing is normalizing narrowing: after every narrowing step the goal is normalized with respect to the given canonical term rewriting system. This allows us to take advantage of the special properties of rewriting steps compared to narrowing steps. Rewriting steps are special narrowings steps which leave invariant the solution space of the current system of equations and thus do not contribute to the construction of a solution. Naive narrowing does not distinguish rewriting and narrowing steps. Every rewriting step leads to a new path in the search space ("don't know indeterminism"), whereas in a canonical term rewriting system the rewriting steps may be executed in an arbitrary ordering ("don't care indeterminism").

6.1 Normalizing Narrowing

Definition 22 (Normalizing Narrowing). Let G be a normalized system of equations. A *normalizing narrowing step*

$$G \rightsquigarrow^{\downarrow}_{[l\rightarrow r,u,\delta]} G'{\downarrow}$$

is given by a narrowing step $G \rightsquigarrow_{[l\rightarrow r,u,\delta]} G'$ followed by a normalization $G' \xrightarrow{*}_R G'{\downarrow}$ with $G'{\downarrow}$ normalized.

It is not possible to associate with each rewriting derivation a corresponding *normalizing* narrowing derivation where the same rules are applied at the same occurrences. However, for any rewriting derivation $\sigma(G) \xrightarrow{*} \sigma(G){\downarrow}$, where σ is normalized and $\sigma(G){\downarrow}$ is in normal form, there exists another rewriting derivation $\sigma(G) \xrightarrow{*} \sigma(G){\downarrow}$ which has a corresponding normalizing narrowing derivation. Moreover, we can assume that the rewriting steps on $\sigma(G)$ corresponding to narrowing steps on G are left reduction steps. This will be used in the proof of the completeness and minimality of normalizing LSE narrowing.

Theorem 23. *Let G be a normalized system of equations and let μ be a normalized substitution. Then there exists a normalization of $H \stackrel{\text{def}}{=} \mu(G)$*

$$\begin{array}{l} H = H'_0 \rightarrow_{[l_1\rightarrow r_1,v_1]} H_1 \rightarrow_{[l_{11}\rightarrow r_{11},v_{11}]} \cdots \rightarrow_{[l_{1k_1}\rightarrow r_{1k_1},v_{1k_1}]} H'_1 \\ \vdots \\ H'_{n-1} \rightarrow_{[l_n\rightarrow r_n,v_n]} H_n \rightarrow_{[l_{n1}\rightarrow r_{n1},v_{n1}]} \cdots \rightarrow_{[l_{nk_n}\rightarrow r_{nk_n},v_{nk_n}]} H{\downarrow}, \end{array}$$

with left reduction steps $H'_{i-1} \to_{[l_i \to r_i, v_1]} H_i$, $i = 1, \ldots, n$, such that there exists a normalizing narrowing derivation

$$\begin{array}{l} G = G_0{\downarrow} \rightsquigarrow_{[l_1 \to r_1, v_1, \delta_1]} G_1 \to_{[l_{11} \to r_{11}, v_{11}]} \cdots \to_{[l_{1k_1} \to r_{1k_1}, v_{1k_1}]} G_1{\downarrow} \\ \vdots \\ G_{n-1}{\downarrow} \rightsquigarrow_{[l_n \to r_n, v_n, \delta_n]} G_n \to_{[l_{n1} \to r_{n1}, v_{n1}]} \cdots \to_{[l_{nk_n} \to r_{nk_n}, v_{nk_n}]} G_n{\downarrow} \end{array}$$

which uses the same rules at the same occurrences. Moreover, there exists a normalized substitution λ such that

- $\lambda \circ \delta_n \circ \ldots \delta_1 = \mu \ [Var(G)]$
- $H_i = (\lambda \circ \delta_n \circ \ldots \circ \delta_{i+1})(G_i), i = 1, \ldots, n$
- $H'_i = (\lambda \circ \delta_n \circ \ldots \circ \delta_{i+1})(G_i{\downarrow}), i = 0, \ldots, n-1$
- $\lambda(G_n{\downarrow}) = H{\downarrow}$.

Corollary 24 (Réty et al. 85). *Normalizing narrowing is complete.*

6.2 Normalizing LSE Narrowing

Our aim is now to extend the idea of LSE narrowing to the case of normalizing narrowing. We can use essentially the same definition as above. Again, the tests have to be applied to the goals where a narrowing step has taken place, the goals $G_i{\downarrow}$ for $i = 0, \ldots, n-1$.

Definition 25 (Normalizing LSE narrowing). In a normalizing narrowing derivation

$$\begin{array}{l} G_0{\downarrow} \rightsquigarrow_{[v_1, \pi_1, \delta_1]} G_1 \xrightarrow{*} G_1{\downarrow} \\ \vdots \\ G_{n-1}{\downarrow} \rightsquigarrow_{[v_n, \pi_n, \delta_n]} G_n \xrightarrow{*} G_n{\downarrow} \end{array}$$

the step $G_{n-1}{\downarrow} \rightsquigarrow G_n \xrightarrow{*} G_n{\downarrow}$ is called a *LSE step* iff the following three conditions are satisfied

(Left-Test) For all $i \in \{0, \ldots, n-1\}$ the subterms of $\lambda_{i,n}(G_i{\downarrow})$ which lie strictly left of v_{i+1} are in normal form.

(Sub-Test) For all $i \in \{0, \ldots, n-1\}$ the proper subterms of $\lambda_{i,n}(G_i{\downarrow}/v_{i+1})$ are in normal form.

(Epsilon-Test) For all $i \in \{0, \ldots, n-1\}$ the term $\lambda_{i,n}(G_i{\downarrow}/v_{i+1})$ is not reducible at occurrence ϵ with a rule whose number is strictly less than π_{i+1}.

A normalizing narrowing derivation is called a *normalizing LSE narrowing derivation* iff all steps are LSE steps.

Theorem 26. *Normalizing LSE narrowing is complete.*

Theorem 27. *Consider two normalizing LSE narrowing derivations*

$$G_0 \rightsquigarrow^{\downarrow}_{[v_1,\pi_1,\delta_1]} G_1\!\downarrow \rightsquigarrow^{\downarrow}_{[v_2,\pi_2,\delta_2]} \cdots \rightsquigarrow^{\downarrow}_{[v_n,\pi_n,\delta_n]} G_n\!\downarrow$$

$$G'_0 \rightsquigarrow^{\downarrow}_{[v'_1,\pi'_1,\delta'_1]} G'_1\!\downarrow \rightsquigarrow^{\downarrow}_{[v'_2,\pi'_2,\delta'_2]} \cdots \rightsquigarrow^{\downarrow}_{[v'_m,\pi'_m,\delta'_m]} G'_m\!\downarrow.$$

Let $\lambda_{i,n} \stackrel{\text{def}}{=} \delta_n \circ \ldots \circ \delta_{i+1}$, *for* $i = 0, \ldots, n$ *and* $\lambda'_{j,m} \stackrel{\text{def}}{=} \delta'_m \circ \ldots \circ \delta'_{j+1}$ *for* $j = 0, \ldots, m$, *where* $\lambda_{n,n} = \lambda'_{m,m} = id$. *Suppose that* G_0 *and* G'_0 *respectively* $\lambda_{0,n}$ *and* $\lambda'_{0,m}$ *are identical up to variable renaming, that is there exist substitutions* $\tau_0, \tau'_0, \rho, \rho'$ *such that*

- $\tau_0(G_0) = G'_0$, $\tau'_0(G'_0) = G_0$ *and*
- $\lambda_{0,n} = \rho' \circ \lambda'_{0,m} \circ \tau_0\ [Var(G_0)]$, $\lambda'_{0,m} = \rho \circ \lambda_{0,n} \circ \tau'_0\ [Var(G'_0)]$

Then the two derivations are identical up to variable renaming, that is

- $n = m$
- $v_i = v'_i$, *for* $i = 1, \ldots, n$
- $\pi_i = \pi'_i$, *for* $i = 1, \ldots, n$
- *there exist substitutions* τ_i, τ'_i *such that*
  - $\tau_i(G_i\!\downarrow) = G'_i\!\downarrow$, $\tau'_i(G'_i\!\downarrow) = G_i\!\downarrow$,
  - $\lambda_{i,n} = \rho' \circ \lambda'_{i,m} \circ \tau_i\ [Var(G_i)]$, $\lambda'_{i,m} = \rho \circ \lambda_{i,n} \circ \tau'_i\ [Var(G'_i)]$,

  for $i = 1, \ldots, n$.

Corollary 28. *If normalizing LSE narrowing enumerates two solutions* σ *and* σ' *which coincide up to variable renaming, then* $\sigma = \sigma'$ *holds and the two derivations coincide.*

Theorem 29. *For any normalizing LSE narrowing derivation*

$$G_0\!\downarrow \rightsquigarrow_{[v_1,\pi_1,\delta_1]} G_1 \xrightarrow{*} G_1\!\downarrow \rightsquigarrow_{[v_2,\pi_2,\delta_2]} \cdots \rightsquigarrow_{[v_m,\pi_m,\delta_m]} G_m \xrightarrow{*} G_m\!\downarrow$$

the narrowing substitution $\delta_m \circ \ldots \circ \delta_1|_{Var(G_0\downarrow)}$ *is normalized.*

6.3 LSE Normal and Left-to-Right Basic Normalizing Narrowing

Finally, we want to investigate the relationship of normalizing LSE narrowing to normalizing left-to-right basic narrowing as studied in [Rét87, Rét88]. It is well-known that a naive combination of (left-to-right) basic narrowing and normalizing narrowing is not complete.

For rewriting derivations the computation of the sets of basic occurrences is more complicated than for narrowing derivations. We need the notion of *weakly basic rewriting derivation* [Rét87].

Definition 30. Let $t \rightarrow_{[v,l\rightarrow r]} t'$ be a rewriting step. We say that the occurrence ω in t is an *antecedent* of the occurrence ω' in t' iff

- $\omega = \omega'$ and ω is not comparable to v or

– there exists an occurrence ρ' of a variable x in r such that $\omega' = v.\rho'.o$ and $\omega = v.\rho.o$ where ρ is an occurrence of the (same) variable x in l.

Definition 31. Given a rewriting derivation

$$G_1 \rightarrow_{[v_1,l_1\rightarrow r_1]} G_2 \rightarrow_{[v_2,l_2\rightarrow r_2]} \cdots \rightarrow_{[v_{n-1},l_{n-1}\rightarrow r_{n-1}]} G_n$$

and a set $WB_1 \subseteq Occ(G_1)$ of occurrences in G_1 the corresponding sets of *weakly basic occurrences* are inductively defined by

$$\begin{aligned} WB_{i+1} \stackrel{\text{def}}{=} {} & (WB_i \setminus \{v \in WB_i \mid v \succeq v_i\}) \cup \{v_i.o \mid o \in FuOcc(r_i)\} \cup \{v \in Occ(G_{i+1}) \mid \\ & v = v_i.o, o \notin FuOcc(r_i) \text{ and all antecedents of } v \text{ in } G_i \text{ are in } WB_i\}, \end{aligned}$$

for $i = 1, \ldots, n-1$. The rewriting derivation is *weakly based on* WB_1 iff $v_i \in WB_i$, for all $i = 1, \ldots, n-1$. Instead of WB_n we will also write $WB(WB_1, G_1 \xrightarrow{*} G_n)$.

Definition 32 (Normalizing left-to-right basic narrowing). Let $G_0\!\downarrow$ be a normalized system of equations. A derivation

$$\begin{array}{lll} (G_0\!\downarrow, U_0\!\downarrow) & \rightsquigarrow_{[v_1,\pi_1,\delta_1]} (G_1, U_1) & \xrightarrow{*} (G_1\!\downarrow, U_1\!\downarrow) \\ \vdots & & \\ (G_{n-1}\!\downarrow, U_{n-1}\!\downarrow) & \rightsquigarrow_{[v_n,\pi_n,\delta_n]} (G_n, U_n) & \xrightarrow{*} (G_n\!\downarrow, U_n\!\downarrow), \end{array}$$

with $U_0\!\downarrow \stackrel{\text{def}}{=} FuOcc(G_0\!\downarrow)$, $U_i \stackrel{\text{def}}{=} LRB(U_{i-1}\!\downarrow, v_i, \pi_i)$ and $U_i\!\downarrow \stackrel{\text{def}}{=} WB(U_i, G_i \xrightarrow{*} G_i\!\downarrow)$ is called a *normalizing left-to-right basic narrowing derivation* if $v_i \in U_{i-1}\!\downarrow$, $G_i \xrightarrow{*} G_i\!\downarrow$ is weakly based on U_i, and $G_i\!\downarrow$ is normalized for $i = 1, \ldots, n$.

Theorem 33. *Any normalizing LSE narrowing derivation is also a normalizing left-to-right basic narrowing derivation.*

7 Example

Let us finally illustrate the efficiency of our method by a typical example.

Consider the canonical term rewriting system for the integers from [RKKL85] which contains 24 rules and take the goal $x * x + y * y \doteq s^5(0)$. The normalizing left-to-right basic narrowing tree of depth 5 contains 52991 nodes, whereas there are only 693 in the normalizing LSE narrowing tree. To compute this tree we needed 1424 seconds for normalizing left-to-right basic narrowing but only 73 seconds for normalizing LSE narrowing. The computations were done in the Karlsruhe narrowing laboratory KANAL [Kri90], which is implemented in the Prolog dialect KA-Prolog, on a SUN 4/60.

References

[BGM88] P. G. Bosco, E. Giovannetti, and C. Moiso. Narrowing vs. SLD-Resolution. *Theoretical Computer Science*, 59:3 –23, 1988.

[Boc86] A. Bockmayr. Narrowing with inductively defined functions. Technical Report 25/86, Fakultät für Informatik, Univ. Karlsruhe, 1986.

[Boc88] A. Bockmayr. Narrowing with built-in theories. In *First Int. Workshop Algebraic and Logic Programming, Gaußig.* Springer, LNCS 343, 1988.

[DG89] J. Darlington and Y. Guo. Narrowing and unification in functional programming - an evaluation mechanism for absolute set abstraction. In *Proc. Rewriting Techniques and Applications, Chapel Hill.* Springer, LNCS 355, 1989.

[Ech88] R. Echahed. On completeness of narrowing strategies. In *Proc. CAAP88, Nancy.* Springer, LNCS 299, 1988.

[Fay79] M. Fay. First-order unification in an equational theory. In *4th Workshop on Automated Deduction, Austin, Texas*, 1979.

[Fri85] L. Fribourg. Handling function definitions through innermost superposition and rewriting. In *Proc. Rewriting Techniques and Applications, Dijon.* Springer, LNCS 202, 1985.

[Her86] A. Herold. Narrowing techniques applied to idempotent unification. SEKI-Report SR-86-16, Univ. Kaiserslautern, 1986.

[HO80] G. Huet and D. C. Oppen. Equations and rewrite rules, A survey. In R. V. Book, editor, *Formal Language Theory.* Academic Press, 1980.

[Höl89] S. Hölldobler. *Foundations of Equational Logic Programming.* Springer, LNCS 353, 1989.

[Hul80] J. M. Hullot. Canonical forms and unification. In *Proc. 5th Conference on Automated Deduction, Les Arcs.* Springer, LNCS 87, 1980.

[KB91] S. Krischer and A. Bockmayr. Detecting redundant narrowing derivations by the LSE-SL reducibility test. In *Proc. Rewriting Techniques and Applications, Como.* Springer, LNCS 488, 1991.

[Kri90] S. Krischer. Vergleich und Bewertung von Narrowing-Strategien. Diplomarbeit, Fakultät für Informatik, Univ. Karlsruhe, 1990.

[NRS89] W. Nutt, P. Réty, and G. Smolka. Basic narrowing revisited. *Journal of Symbolic Computation*, 7:295–317, 1989.

[Pad88] P. Padawitz. *Computing in Horn Clause Theories*, volume 16 of *EATCS Monograph.* Springer, 1988.

[Rét87] P. Réty. Improving basic narrowing techniques. In *Proc. Rewriting Techniques and Applications, Bordeaux.* Springer, LNCS 256, 1987.

[Rét88] P. Réty. *Méthodes d'Unification par Surréduction.* PhD thesis, Univ. Nancy, 1988.

[RKKL85] P. Réty, C. Kirchner, H. Kirchner, and P. Lescanne. NARROWER: a new algorithm for unification and its application to logic programming. In *Proc. Rewriting Techniques and Applications, Dijon.* Springer, LNCS 202, 1985.

[Wer91] A. Werner. Termersetzung und Narrowing mit geordneten Sorten. Diplomarbeit, Fakultät für Informatik, Univ. Karlsruhe, 1991.

[You88] J. You. Solving equations in an equational language. In *First Int. Workshop Algebraic and Logic Programming, Gaußig.* Springer, LNCS 343, 1988.

[You91] J. You. Unification modulo an equality theory for equational logic programming. *J. Comp. Syst. Sc.*, 42:54–75, 1991.

Set-Of-Support Strategy for Higher-Order Logic

Wenchang Fang and Jung-Hong Kao
Department of Electrical Engineering and Computer Science
Northwestern University
Evanston, IL 60208
fang@eecs.nwu.edu

1 Introduction

Research in automated theorem proving has been concentrated on first-order logic. Higher-order logic (or λ-calculus, Type theory) has been dealt with by only a few people. Though first-order language in which variables range over individual constants can express many statements of mathematics, some of the mathematical concepts can state clearer by higher-order language. The following statements are expressed as second-order language:

1. $\forall x \forall y(x = y) \rightarrow \forall f(f(x) \rightarrow f(y))$
   'If x equals y then y has all properties of x.'
2. $\forall f(Property - of - a - great - general(f) \rightarrow f(Napolean))$
   'Napoleon had all the properties of a great general.'

It is difficult to represent the quantifiers range over the function symbol f in the above sentence in pure first-order logic. One of the approach which is developed by Henschen[6] is to use n-sorted first-order language to represent higher-order concepts.

The $\lambda - calculus$ system was first introduced by Church [1]. Andrews [2] describes a resolution-like refutation system for type theory. He proves that type theory is complete in the sense of using the resolution-based proof procedure. Later he develops the TPS [3], which is an automated theorem proving system for proving theorems of both first and higher-order logic.

One of the problems in higher-order theorem proving is that the predicate variables can be substitute with logical constants. In Huet's [8] [9] higher-order system $\mathcal{L}$, he use a special inference rule, splitting, to resolve this impediment. However, splitting is difficult to control in system $\mathcal{L}$. Fang[5] claims that even the logical constants are embedded in the higher-order language, there are only at some specific situation needed to be splitting. Hence, we may contemplate the strategy that when there is no logical constants replace predicate variables in higher-order theorem proving.

2 Apply the set-of-support strategy to higher-order language

Just like the proof procedure in first-order logic, we need a lifting lemma which can apply to the higher-order situation.

Lemma 1 (Lifting lemma for higher-order logic) If C_1' and C_2' are instances of C_1 and C_2 respectively, where C_1, C_2 are higher-order terms, and if C' is a resolvent of C_1' and C_2', then there is a resolvent C of C_1 and C_2 such that C' is an instance of C.
Proof: Rename the variables in C_1 and C_2 such that the variables are distinguishable. Let the resolution of C_1' and C_2' be of the form:

$$C' = (C_1'\gamma - L_1'\gamma) \cup (C_2'\gamma - (-L_2')\gamma)$$

where γ is a most general unifier of L_1' and L_2'.
Since C_1' and C_2' are instances of C_1 and C_2, let θ be the substitution such that $C_1' = C_1\theta$, $C_2' = C_2\theta$.
Suppose $L_1^1, L_1^2, \ldots, L_1^x$ are the literals in C_1 corresponding to L_1'.
Suppose $L_2^1, L_2^2, \ldots, L_2^y$ are the literals in C_2 corresponding to L_2'.
That is,
$L_1^1\theta = L_1^2\theta = \ldots = L_1^x\theta = L_1'$
$L_2^1\theta = L_2^2\theta = \ldots = L_2^y\theta = L_2'$
Since
$L_1'\gamma = L_2'\gamma$, clearly the set $\{L_1^1, \ldots, L_1^x, L_2^1, \ldots, L_2^y\}$ is unifiable by $\theta\gamma$. Then there must exist a most general unifier, say σ, such that $\sigma\lambda = \theta\gamma$ for some λ. Then C_1 and C_2 have a resolvent using σ as follows:

$$\begin{aligned} C \;=\; & (C_1\sigma - (\{L_1^1, L_1^2, \ldots, L_1^x\})\sigma) \cup \\ & (C_2\sigma - (\{-L_2^1, -L_2^2, \ldots, -L_2^y\}\sigma) \end{aligned}$$

Moreover,

$$
\begin{aligned}
C\sigma\lambda &= [(C_1\sigma - \{L_1^1, L_1^2, \ldots, L_1^x\}\sigma)\cup \\
&\quad (C_2\sigma - \{-L_2^1, -L_2^2, \ldots, -L_2^y\}\sigma)]\lambda \\
&= (C_1\sigma\lambda - \{L_1^1, L_1^2, \ldots, L_1^x\}\sigma\lambda)\cup \\
&\quad (C_2\sigma\lambda - \{-L_2^1, -L_2^2, \ldots, -L_2^y\}\sigma\lambda) \\
&= (C_1\theta\gamma - \{L_1^1, L_1^2, \ldots, L_1^x\}\theta\gamma\cup \\
&\quad (C_2\theta\gamma - \{-L_2^1, -L_2^2, \ldots, -L_2^y\}\theta\gamma) \\
&= (C_1'\gamma - L_1'\gamma)\cup(C_2'\gamma - (-L_2')\gamma) \\
&= C'
\end{aligned}
$$

Notice that there may be many different δ and σ, but at least one $\delta\sigma$ is more general than $\theta\gamma$, C' is an instance of C. Q.E.D

Example 1 This is an example for the higher-order lifting lemma.
Let
$C_1 : hXh = hYh, h(Z(h, w)) = w$
$C_2 : F(u) \neq F(v)$
$C_1' : FG(X(\lambda u \cdot FGu)) = FG(Y(\lambda u \cdot FGu)), FGZ(H, E) = E$
$C_2' : FG(X(\lambda u \cdot FGu)) = FG(Y(\lambda u \cdot FGu))$
So we have
$\theta = \{h/\lambda u \cdot FGu, w/E, u/FGXh, v/FGYh\}$
The resolvent C' of C_1' and C_2' is $FGZ(H, E) = E$
There are two resolvents of C_1 and C_2, which are:
$FGZ(h, E) = E$ if $\sigma = \{h/\lambda uFGu, w/E\}$
$FZ(F, E) = E$ if $\sigma = \{h/F, u/Xh, v/Yh\}$,
and C' is an instance of the first one.

Theorem 1 **Let S be an unsatisfiable set of higher-order clause. If S does not contain any logical constant, then the set-of-support strategy can be used to find a refutation.**
Proof: Since S is unsatisfiable, $S \vdash_{\mathcal{L}} \Box$. Suppose the refutation has the form:
$B_1, B_2, \ldots, B_n, \Box$. Then B_n must be of the form:
$B_n : \phi/\{< e_1, e_1' >, < e_2, e_2' >, < e_3, e_3' >, \ldots\}$
Recall that higher-order unification in general is undecidable. So we just suppose the unifier looks like the following:
$< e_1, e_1' >: \alpha_1, \alpha_2, \ldots$
$< e_2, e_2' >: \beta_1, \beta_2, \ldots$
$< e_3, e_3' >: \gamma_1, \gamma_2, \ldots$
$\vdots$
Since $S \vdash_{\mathcal{L}} \Box$, there must exist a θ such that $\theta = \alpha_i\beta_i\gamma_i \ldots$.
Now we apply the unifier θ to every clause B_i to make B_i ground clauses.
$S' : B_1\theta, B_2\theta, \ldots, B_n\theta$
S' is a set of ground clause. Since the set of support strategy is complete for ground clause. there must be a supported refutation R' of S' for some suitable set T'. Applying the lifting lemma to R' to get a refutation of S supported by the corresponding support set of T.

The following two examples are from Huet [7].

Example 2 Cantor's theorem: N^N is not denumerable.
Some terms and types for this problem are defined as follows:
$\tau(n) = integer$
$\tau(f) = \tau(g) = \tau(S) = integer \rightarrow integer$
where n, f, h are variables.
The negation of the theorem: it is possible to enumerate the functions of integers to integers, i.e.
$\exists h \forall f \exists n h(n) = f$
After reducing to normal form we get clause 3, where N and H are skolem functions.

1. $-f = g, f(n) = g(n)$
2. $-n = S(n)$
3. $H(N(f)) = f$

If f is the same function as g then $f(n)$ and $g(n)$ are equal. The successor of n is never the same as n.

For clarity, we rename the variables and rewrite into the following clauses:

1. $-x = y, x(n) = y(n)$

2. $-m = S(m)$

3. $H(N(f)) = f$

Using clause 3 as the supported clause, we get the following refutation:

(3,1) 4. $x(n) = y(n)/\{< H(N(f)) = f, x = y >\}$

(4,2) 5. $\phi/\{< H(N(f)) = f, x = y >, < x(n) = y(n), m = S(m) >\}$

(un5) 6. $\Box/\{< f, \lambda u \cdot S(H(u, u)) >, < n, N(f) >, < x, H(N(f)) >, < y, f >, < m, H(N(f), n) >\}$

Now we can put back all the variables to see the instances of the original clauses:
Let F be the term $\lambda u \cdot S(H(u, u))$

1. $-H(N(F)) = F, H(N(F), N(F)) = S(H(N(F), N(F))$

2. $-H(N(F), N(F)) = S(H(N(F), N(F)))$

3. $H(N(F)) = F$

(3,1) 4. $H(N(F), N(F)) = S(H(N(F), N(F))$

(4,2) 5. $\Box$

Example 3 The pigeonhole principle: if $n+1$ objects are distributed into n holes, then there is at least one hole which contains more than one object.

The terms and types are as follows:

$\tau(F) = \beta \rightarrow \alpha$
$\tau(G) = \alpha \rightarrow \beta$
$\tau(h) = \alpha \rightarrow \alpha$
$\tau(u) = \tau(v) = \beta$
$\tau(E) = \tau(x) = \tau(y) = \alpha$
$\tau(X) = \tau(Y) = ((\alpha \rightarrow \alpha) \rightarrow \alpha)$
$\tau(Z) = ((\alpha \rightarrow \alpha), \alpha, \alpha)$

After reduction to normal form:

1. $u =_\beta v, Fu \neq_\alpha Fv$
2. $Fu \neq_\alpha E$
3. $Xh \neq_\alpha Yh, h(Z(h,x)) =_\alpha x$
4. $hXh =_\alpha hYh, h(Z(h,x)) =_\alpha x$
5. $x =_\alpha y, Gx \neq_\beta Gy$

There is a one-to-one mapping from the holes to objects, not onto. Every one-to-one mapping is onto. No hole contains two objects.

Clause 5 is the negation of the theorem, so clause 5 has T-support.

(5,3) 6. $-Gx =_\beta Gy, h(Z(h,x')) =_\alpha C$
$x'/\{< x =_\alpha y, Xh =_\alpha Yh >\}$
(6,2) 7. $-Gx =_\beta Gy/C6 + \{< h(Z(h,x)) =_\alpha x,$
$Fu =_\alpha E >\}$
(7,1) 8. $-Fu =_\alpha Fv/C7 + \{< u =_\beta v, Gx =_\beta Gy >\}$
(8,4) 9. $hYh, h(Z(h,x)) =_\alpha x/C8 + \{< Fu =_\alpha Fv,\}$
$hXh =_\alpha hYh >\}$
(9,2) 10. $\phi/\{< x =_\alpha y, Xh =_\alpha Yh >, < h(Z(h,x'))$
$=_\alpha x', Fu =_\alpha E >< u =_\beta v, Gx =_\beta Gy >,$
$< Fu =_\alpha Fv, hXh =_\alpha hYh >\}$
(un10) 11. $\Box/\{< h, \lambda w \cdot FGw >, < u, GXh >,$
$< v, GYh >, < x, Xh >, < y, Yh >,$
$< x', E >\}$

Let H be the term $\lambda x \cdot FGx$. Then the original clauses become more readable:

1. $GXH = GYH, FGXH \neq FGYH$

2. $FGZ(H,E) \neq E$

3. $XH \neq YH, FGXH \neq FGYH$

4. $FGXH = FGYH, FGZ(H,E) = E$

5. $XH = YH, GXH \neq GYH$

(5,3) 6. $FGZ(H,E) = E, GXH \neq GYH$

(6,2) 7. $GXH \neq GYH$

(7,1) 8. $FGXH \neq FGYH$

(8,4) 9. $FGZ(H,E) = E$

(9,2) 10. $\Box$

The following example is from Darlington[4].

Example 4 This is an example to prove that Napolean was a great general. We use the formula described in section 1.

1. $\forall f(Property - of - a - great - general(f) \rightarrow f(Napolean))$

2. $\forall f(Property - of - a - great - general(f) \leftrightarrow \forall x(Great - general(x) \rightarrow f(x)))$

These formulas can be translated to the following clauses:

1. $-P(f), f(N)$

2. $-P(f), -G(x), f(x)$

3. $G(Choice(f)), P(f)$

4. $-f(Choice(f)), P(f)$

5. $-G(N)$

P stands for $Property - of - a - great - general$
N stands for $Napolean$
G stands for $Great - general$

Clause 1 is the skolemization of 1. Clause 2 - clause 4 are from 2. The "Choice" function in higher-order logic corresponds to the skolem function used in first-order logic. Clause 5 is the negation of the conclusion, so clause 5 has T-support.

(5,1) 6. $-P(f)/\{< f(N), G(N) >\}$

(6,3) 7. $G(Choice(f))/\{< f(N), G(N) >\}$

(6,4) 8. $-f(Choice(f))/\{< f(N), G(N) >\}$

(7,8) 9. $\phi/\{< f(N), G(N) >, < f(Choice(f)), G(Choice(f)) >\}$

(un9) 10. $\Box/\{< f, \lambda u \cdot G(u) >\}$

3 Conclusions

We have proven in this paper that set-of-support strategy is complete in higher-order system $\mathcal{L}$. The situation may not apply to other strategies. For example, hyper-resolution which resolve one nuclear clause with several satellite clauses at one time may not be used in system $\mathcal{L}$. Because $\mathcal{L}$ does not contain any inference rules similar to hyper-resolution.

Set-of-support strategy has been worked fine in first-order theorem-proving problem. Though complete in system $\mathcal{L}$, we have not try in too many examples. We have no idea that how the performance is. More experiments need to be done. Of course, other first-order strategy may also work in system $\mathcal{L}$, which need more future research. Finally, find a total new, pure strategy for higher-order logic is one other direction.

References

[1] Church A. A formulation of the simple theory of types. *Journal of symbolic logic*, 5:56–68, 1940.

[2] Peter B. Andrews. Resolution in type theory. *Journal of Symbolic logic*, 36(3):414–432, sept 1971.

[3] Peter B. Andrews, D. Miller, E. Cohen, and F. Pfenning. Automating higher-order logic. In *Contemptary Mathematics 29*, pages 169–192, 1984.

[4] J. L. Darlington. A partial mechanization of second-order logic. In B. Meltser and D Michie, editors, *Machine Intelligence 6*, pages 91–100. American Elsevier Publishing Co., 1971.

[5] Wenchang Fang and Lawrence Henschen. Some heuristics for higher-order logic. In *Golden West International Conference on Intelligent System*, 1992.

[6] Lawrence J. Henschen. *A resolution style proof procedure for higher-order logic*. PhD thesis, University of Illinois at Urbana-Champaign, 1971.

[7] G. P. Huet. *Constrained Resolution: A Complete Method For Higher order Logic*. PhD thesis, Case Western reserve University, 1972.

[8] G. P. Huet. A mechanization of type theory. In *Proceedings of the third IJCAI*, pages 139–146, 1973.

[9] G. P. Huet. A unification algorithm for typed lambda-calculus. *Theoretical Computer Science*, pages 27–57, 1975.

Printing: Weihert-Druck GmbH, Darmstadt
Binding: Buchbinderei Schäffer, Grünstadt

Lecture Notes in Computer Science

For information about Vols. 1–579
please contact your bookseller or Springer-Verlag

Vol. 580: A. Pirotte, C. Delobel, G. Gottlob (Eds.), Advances in Database Technology – EDBT '92. Proceedings. XII, 551 pages. 1992.

Vol. 581: J.-C. Raoult (Ed.), CAAP '92. Proceedings. VIII, 361 pages. 1992.

Vol. 582: B. Krieg-Brückner (Ed.), ESOP '92. Proceedings. VIII, 491 pages. 1992.

Vol. 583: I. Simon (Ed.), LATIN '92. Proceedings. IX, 545 pages. 1992.

Vol. 584: R. E. Zippel (Ed.), Computer Algebra and Parallelism. Proceedings, 1990. IX, 114 pages. 1992.

Vol. 585: F. Pichler, R. Moreno Díaz (Eds.), Computer Aided System Theory – EUROCAST '91. Proceedings. X, 761 pages. 1992.

Vol. 586: A. Cheese, Parallel Execution of Parlog. IX, 184 pages. 1992.

Vol. 587: R. Dale, E. Hovy, D. Rösner, O. Stock (Eds.), Aspects of Automated Natural Language Generation. Proceedings, 1992. VIII, 311 pages. 1992. (Subseries LNAI).

Vol. 588: G. Sandini (Ed.), Computer Vision – ECCV '92. Proceedings. XV, 909 pages. 1992.

Vol. 589: U. Banerjee, D. Gelernter, A. Nicolau, D. Padua (Eds.), Languages and Compilers for Parallel Computing. Proceedings, 1991. IX, 419 pages. 1992.

Vol. 590: B. Fronhöfer, G. Wrightson (Eds.), Parallelization in Inference Systems. Proceedings, 1990. VIII, 372 pages. 1992. (Subseries LNAI).

Vol. 591: H. P. Zima (Ed.), Parallel Computation. Proceedings, 1991. IX, 451 pages. 1992.

Vol. 592: A. Voronkov (Ed.), Logic Programming. Proceedings, 1991. IX, 514 pages. 1992. (Subseries LNAI).

Vol. 593: P. Loucopoulos (Ed.), Advanced Information Systems Engineering. Proceedings. XI, 650 pages. 1992.

Vol. 594: B. Monien, Th. Ottmann (Eds.), Data Structures and Efficient Algorithms. VIII, 389 pages. 1992.

Vol. 595: M. Levene, The Nested Universal Relation Database Model. X, 177 pages. 1992.

Vol. 596: L.-H. Eriksson, L. Hallnäs, P. Schroeder-Heister (Eds.), Extensions of Logic Programming. Proceedings, 1991. VII, 369 pages. 1992. (Subseries LNAI).

Vol. 597: H. W. Guesgen, J. Hertzberg, A Perspective of Constraint-Based Reasoning. VIII, 123 pages. 1992. (Subseries LNAI).

Vol. 598: S. Brookes, M. Main, A. Melton, M. Mislove, D. Schmidt (Eds.), Mathematical Foundations of Programming Semantics. Proceedings, 1991. VIII, 506 pages. 1992.

Vol. 599: Th. Wetter, K.-D. Althoff, J. Boose, B. R. Gaines, M. Linster, F. Schmalhofer (Eds.), Current Developments in Knowledge Acquisition - EKAW '92. Proceedings. XIII, 444 pages. 1992. (Subseries LNAI).

Vol. 600: J. W. de Bakker, C. Huizing, W. P. de Roever, G. Rozenberg (Eds.), Real-Time: Theory in Practice. Proceedings, 1991. VIII, 723 pages. 1992.

Vol. 601: D. Dolev, Z. Galil, M. Rodeh (Eds.), Theory of Computing and Systems. Proceedings, 1992. VIII, 220 pages. 1992.

Vol. 602: I. Tomek (Ed.), Computer Assisted Learning. Proceedings, 1992. X, 615 pages. 1992.

Vol. 603: J. van Katwijk (Ed.), Ada: Moving Towards 2000. Proceedings, 1992. VIII, 324 pages. 1992.

Vol. 604: F. Belli, F.-J. Radermacher (Eds.), Industrial and Engineering Applications of Artificial Intelligence and Expert Systems. Proceedings, 1992. XV, 702 pages. 1992. (Subseries LNAI).

Vol. 605: D. Etiemble, J.-C. Syre (Eds.), PARLE '92. Parallel Architectures and Languages Europe. Proceedings, 1992. XVII, 984 pages. 1992.

Vol. 606: D. E. Knuth, Axioms and Hulls. IX, 109 pages. 1992.

Vol. 607: D. Kapur (Ed.), Automated Deduction – CADE-11. Proceedings, 1992. XV, 793 pages. 1992. (Subseries LNAI).

Vol. 608: C. Frasson, G. Gauthier, G. I. McCalla (Eds.), Intelligent Tutoring Systems. Proceedings, 1992. XIV, 686 pages. 1992.

Vol. 609: G. Rozenberg (Ed.), Advances in Petri Nets 1992. VIII, 472 pages. 1992.

Vol. 610: F. von Martial, Coordinating Plans of Autonomous Agents. XII, 246 pages. 1992. (Subseries LNAI).

Vol. 611: M. P. Papazoglou, J. Zeleznikow (Eds.), The Next Generation of Information Systems: From Data to Knowledge. VIII, 310 pages. 1992. (Subseries LNAI).

Vol. 612: M. Tokoro, O. Nierstrasz, P. Wegner (Eds.), Object-Based Concurrent Computing. Proceedings, 1991. X, 265 pages. 1992.

Vol. 613: J. P. Myers, Jr., M. J. O'Donnell (Eds.), Constructivity in Computer Science. Proceedings, 1991. X, 247 pages. 1992.

Vol. 614: R. G. Herrtwich (Ed.), Network and Operating System Support for Digital Audio and Video. Proceedings, 1991. XII, 403 pages. 1992.

Vol. 615: O. Lehrmann Madsen (Ed.), ECOOP '92. European Conference on Object Oriented Programming. Proceedings. X, 426 pages. 1992.

Vol. 616: K. Jensen (Ed.), Application and Theory of Petri Nets 1992. Proceedings, 1992. VIII, 398 pages. 1992.

Vol. 617: V. Mařík, O. Štěpánková, R. Trappl (Eds.), Advanced Topics in Artificial Intelligence. Proceedings, 1992. IX, 484 pages. 1992. (Subseries LNAI).

Vol. 618: P. M. D. Gray, R. J. Lucas (Eds.), Advanced Database Systems. Proceedings, 1992. X, 260 pages. 1992.

Vol. 619: D. Pearce, H. Wansing (Eds.), Nonclassical Logics and Information Proceedings. Proceedings, 1990. VII, 171 pages. 1992. (Subseries LNAI).

Vol. 620: A. Nerode, M. Taitslin (Eds.), Logical Foundations of Computer Science – Tver '92. Proceedings. IX, 514 pages. 1992.

Vol. 621: O. Nurmi, E. Ukkonen (Eds.), Algorithm Theory – SWAT '92. Proceedings. VIII, 434 pages. 1992.

Vol. 622: F. Schmalhofer, G. Strube, Th. Wetter (Eds.), Contemporary Knowledge Engineering and Cognition. Proceedings, 1991. XII, 258 pages. 1992. (Subseries LNAI).

Vol. 623: W. Kuich (Ed.), Automata, Languages and Programming. Proceedings, 1992. XII, 721 pages. 1992.

Vol. 624: A. Voronkov (Ed.), Logic Programming and Automated Reasoning. Proceedings, 1992. XIV, 509 pages. 1992. (Subseries LNAI).

Vol. 625: W. Vogler, Modular Construction and Partial Order Semantics of Petri Nets. IX, 252 pages. 1992.

Vol. 626: E. Börger, G. Jäger, H. Kleine Büning, M. M . Richter (Eds.), Computer Science Logic. Proceedings, 1991. VIII, 428 pages. 1992.

Vol. 628: G. Vosselman, Relational Matching. IX, 190 pages. 1992.

Vol. 629: I. M. Havel, V. Koubek (Eds.), Mathematical Foundations of Computer Science 1992. Proceedings. IX, 521 pages. 1992.

Vol. 630: W. R. Cleaveland (Ed.), CONCUR '92. Proceedings. X, 580 pages. 1992.

Vol. 631: M. Bruynooghe, M. Wirsing (Eds.), Programming Language Implementation and Logic Programming. Proceedings, 1992. XI, 492 pages. 1992.

Vol. 632: H. Kirchner, G. Levi (Eds.), Algebraic and Logic Programming. Proceedings, 1992. IX, 457 pages. 1992.

Vol. 633: D. Pearce, G. Wagner (Eds.), Logics in AI. Proceedings. VIII, 410 pages. 1992. (Subseries LNAI).

Vol. 634: L. Bougé, M. Cosnard, Y. Robert, D. Trystram (Eds.), Parallel Processing: CONPAR 92 – VAPP V. Proceedings. XVII, 853 pages. 1992.

Vol. 635: J. C. Derniame (Ed.), Software Process Technology. Proceedings, 1992. VIII, 253 pages. 1992.

Vol. 636: G. Comyn, N. E. Fuchs, M. J. Ratcliffe (Eds.), Logic Programming in Action. Proceedings, 1992. X, 324 pages. 1992. (Subseries LNAI).

Vol. 637: Y. Bekkers, J. Cohen (Eds.), Memory Management. Proceedings, 1992. XI, 525 pages. 1992.

Vol. 639: A. U. Frank, I. Campari, U. Formentini (Eds.), Theories and Methods of Spatio-Temporal Reasoning in Geographic Space. Proceedings, 1992. XI, 431 pages. 1992.

Vol. 640: C. Sledge (Ed.), Software Engineering Education. Proceedings, 1992. X, 451 pages. 1992.

Vol. 641: U. Kastens, P. Pfahler (Eds.), Compiler Construction. Proceedings, 1992. VIII, 320 pages. 1992.

Vol. 642: K. P. Jantke (Ed.), Analogical and Inductive Inference. Proceedings, 1992. VIII, 319 pages. 1992. (Subseries LNAI).

Vol. 643: A. Habel, Hyperedge Replacement: Grammars and Languages. X, 214 pages. 1992.

Vol. 644: A. Apostolico, M. Crochemore, Z. Galil, U. Manber (Eds.), Combinatorial Pattern Matching. Proceedings, 1992. X, 287 pages. 1992.

Vol. 645: G. Pernul, A M. Tjoa (Eds.), Entity-Relationship Approach – ER '92. Proceedings, 1992. XI, 439 pages, 1992.

Vol. 646: J. Biskup, R. Hull (Eds.), Database Theory – ICDT '92. Proceedings, 1992. IX, 449 pages. 1992.

Vol. 647: A. Segall, S. Zaks (Eds.), Distributed Algorithms. X, 380 pages. 1992.

Vol. 648: Y. Deswarte, G. Eizenberg, J.-J. Quisquater (Eds.), Computer Security – ESORICS 92. Proceedings. XI, 451 pages. 1992.

Vol. 649: A. Pettorossi (Ed.), Meta-Programming in Logic. Proceedings, 1992. XII, 535 pages. 1992.

Vol. 650: T. Ibaraki, Y. Inagaki, K. Iwama, T. Nishizeki, M. Yamashita (Eds.), Algorithms and Computation. Proceedings, 1992. XI, 510 pages. 1992.

Vol. 651: R. Koymans, Specifying Message Passing and Time-Critical Systems with Temporal Logic. IX, 164 pages. 1992.

Vol. 652: R. Shyamasundar (Ed.), Foundations of Software Technology and Theoretical Computer Science. Proceedings, 1992. XIII, 405 pages. 1992.

Vol. 653: A. Bensoussan, J.-P. Verjus (Eds.), Future Tendencies in Computer Science, Control and Applied Mathematics. Proceedings, 1992. XV, 371 pages. 1992.

Vol. 654: A. Nakamura, M. Nivat, A. Saoudi, P. S. P. Wang, K. Inoue (Eds.), Prallel Image Analysis. Proceedings, 1992. VIII, 312 pages. 1992.

Vol. 655: M. Bidoit, C. Choppy (Eds.), Recent Trends in Data Type Specification. X, 344 pages. 1993.

Vol. 656: M. Rusinowitch, J. L. Rémy (Eds.), Conditional Term Rewriting Systems. Proceedings, 1992. XI, 501 pages. 1993.